Systematics and Conservation of Neotropical Amphibians and Reptiles

Systematics and Conservation of Neotropical Amphibians and Reptiles

Editors

Alessandro Catenazzi
Rudolf von May

MDPI • Basel • Beijing • Wuhan • Barcelona • Belgrade • Manchester • Tokyo • Cluj • Tianjin

Editors
Alessandro Catenazzi
Florida International University
USA

Rudolf von May
California State University
Channel Islands
USA

Editorial Office
MDPI
St. Alban-Anlage 66
4052 Basel, Switzerland

This is a reprint of articles from the Special Issue published online in the open access journal *Diversity* (ISSN 1424-2818) (available at: https://www.mdpi.com/journal/diversity/special_issues/ neotropical_amphibians_reptiles).

For citation purposes, cite each article independently as indicated on the article page online and as indicated below:

LastName, A.A.; LastName, B.B.; LastName, C.C. Article Title. *Journal Name* **Year**, *Volume Number*, Page Range.

ISBN 978-3-0365-0634-0 (Hbk)
ISBN 978-3-0365-0635-7 (PDF)

Cover image courtesy of Luis Coloma.

Contents

About the Editors

Alessandro Catenazzi (Ph.D.) is an Assistant Professor at Florida International University. He is broadly interested in biodiversity conservation and ecology, with active areas of research on chytridiomycosis, frog and lizard systematics and taxonomy, and conservation biology.

Rudolf von May (Ph.D.) is an Assistant Professor at California State University Channel Islands. His interests include ecology, evolution, and conservation of biodiversity, with a focus on amphibians and reptiles in the Andes–Amazon region.

Preface to "Systematics and Conservation of Neotropical Amphibians and Reptiles"

A full-spread photo of the Amazon forest in a pictorial atlas of the world sparked my imagination as a child and I can still see that image nearly sixty years later. I was lucky to visit the Arroyo Cuchuhaqui, east of Alamos, Sonora, Mexico, as a junior college student on a class fieldtrip. An extensive trip through Mexico in the summer of 1974 took me south of the Tropic of Cancer, and I was hooked. I changed my focus from birds to amphibians and reptiles, and studied aspects of their biology as I completed my undergraduate degree in California. For my Ph.D., I was fortunate enough to join a research laboratory that had a Neotropical focus, and I was able to live my dream. My doctoral advisor, Jay M. Savage, was known for his research in Costa Rica, which focused largely on systematics and biogeography of amphibians and reptiles. During my time in Jay's lab, I was one of three ecologists in a systematics lab, and always appreciated the importance of systematics for my ecological investigations. The papers presented in this volume show how well the two fields come together to generate innovative research programs.

When I started in the field of Neotropical herpetology, most of the papers were single authored, and most of the authors were from the north temperate zone. My doctoral research introduced me to field experiments in Costa Rica. While living on site, Craig Guyer and I encountered new records for the field site, and we learned the importance of documenting records and helped build collections. Specimens held in collections were essential for my research. I dissected hundreds of frogs collected as part of another study to determine the body size of sexual maturity for *Oophaga pumilio* before I left to conduct my fieldwork in Costa Rica. My postdoctoral research at the American Museum of Natural History (AMNH) allowed me to use collections to advance my knowledge of feeding in dendrobatid frogs. I also explored tepuis in Venezuela as part of teams from the AMNH in the early 1990s and helped build museum collections. Museums are the libraries of biodiversity, and the papers in this volume show how critical collections have been in testing hypotheses, providing comparative material to diagnose new diversity, and even providing tissues that help resolve systematic relationships.

The book includes papers that describe the state of the field of Neotropical herpetology as we start the second decade of the new century. Conservation case studies, an exceptional monograph on the glass frogs of Ecuador, and systematic studies will be of interest to a variety of scientists. Two attributes of the papers in the book show how far the field has advanced since I started my studies nearly fifty years ago: most of the papers are authored by Latinx scientists, and all the papers represent collaborations among scientists. Taxa emblematic of the Neotropics: harlequin frogs, pitvipers, dendrobatid poison frogs, glass frogs, gymnopthalmid lizards, anoles, snakes, and the amphibian pathogen *Batrachochytrium dendrobatidis*, are included in the book. The papers focus on organisms from Mexico to Paraguay with a heavy focus on South America.

I found every chapter to be informative because of my long-term interest in the amphibians and reptiles of the Neotropics, and am proud of the editors and all the scientists who have done so much to advance the field. The editors brought together a diverse group of scientists working across the Neotropics on key groups to highlight what we know and what research questions await our attention. I would recommend the book to anyone with interests in Neotropical herpetology or any professor looking for a textbook to help anchor a graduate level course in herpetology or conservation.

Maureen A. Donnelly

Department of Biological Sciences and College of Arts, Sciences, and Education, Florida International University, Miami, FL 33199, maureen.a.donnelly@gmail.com

Editorial

Systematics and Conservation of Neotropical Amphibians and Reptiles

Alessandro Catenazzi [1],* and Rudolf von May [2]

[1] Department of Biological Sciences, Florida International University, 11200 SW 8th Street, Miami, FL 33199, USA

[2] Biology Program, California State University Channel Islands, 1 University Dr., Camarillo, CA 93012, USA; rudolf.vonmay@csuci.edu

* Correspondence: alessandro.catenazzi1@fiu.edu or acatenazzi@gmail.com

Received: 14 January 2021; Accepted: 18 January 2021; Published: 25 January 2021

The Neotropics host an exuberant diversity of life forms, including amphibians and reptiles. This diversity is partially unknown. Systematic and taxonomic studies are essential for conservation, because we can only preserve what we know. The first step in documenting biodiversity is to identify species and to name and organize them according to their evolutionary history. The number of species of amphibians and reptiles has increased sharply over the last few decades, but we have much work ahead of us if we want to describe the outstanding biodiversity of the Neotropical herpetofauna. At the same time, traditional and emergent threats are accelerating the erosion of herpetofaunal biodiversity. Traditional threats include habitat loss such as deforestation, wetland drainage, grassland fires, overuse of fertilizers and pesticides, and over-harvesting of wild populations. Emerging threats such as disease and climate change affect species within natural protected areas, where habitat loss and fragmentation may be negligible. Emerging infectious diseases, such as chytridiomycosis, are associated with population decline and the collapse of amphibian communities throughout the Neotropics. Additionally, while climate change threatens many amphibian and reptile species, it remains unclear how species will cope with increasing temperatures, seasonal shifts, and increasing frequency of extreme climatic events. This apparent paradox of species gains in science (i.e., newly named species) amid species loss and population declines is a key element of this special issue.

The endeavor of the special issue was to gather original studies aimed at improving the knowledge of systematics, taxonomy, and conservation of Neotropical amphibians and/or reptiles. We welcomed contributions that examined the evolutionary relationships and geographic distributions of selected Neotropical taxa, helped resolve standing taxonomic issues, and recognized, described, and named new species. We encouraged papers proposing new methods to accelerate taxonomic studies, including those that presented novel molecular techniques. Finally, we welcomed submissions from applied conservation, covering a variety of topics ranging from methods to identify priority areas for conservation and reserve design, to policymaking and assessments of species threat status.

Our issue was very inclusive in terms of welcoming articles from the "core" Neotropical region, but also from adjacent regions. Philip L. Sclater [1] defined the Neotropical region as encompassing central and southern Mexico, Central America, the Caribbean Islands, and South America. Alfred R. Wallace [2] recognized four sub-regions within the Neotropics, based on the distribution and taxonomic relationships of vertebrates. In recent decades, some researchers have redefined the Neotropics on the basis of phylogenetic analyses and geographic range maps [3]. However, not all researchers agree with the updated classification [4]. Some of the most recent analyses recognize the original extent of the Neotropical region, but they exclude the southwestern tip of South America [5–7] (Figure 1). All these classifications can prove useful in explaining patterns of species diversity, diversification, and biotic interchange within the Neotropics.

We are very pleased to present a special issue containing 14 articles that encompass three broad areas: systematics, biogeography, and conservation. The articles cover frogs, salamanders, caecilians, lizards, and snakes from throughout the Neotropics, with specific studies from Mexico, Costa Rica, Panama, Colombia, Ecuador, Peru, Paraguay, and Brazil. Authors in this special issue named a new genus of frogs (*Qosqophryne*), four new species of frogs (one terrestrial-breeding frog and three glassfrogs), and five new lizards (three tropidurid and two gymnophthalmid lizards). A monograph on glassfrogs of Ecuador spans the three thematic areas. Among contributions on biogeography, approaches ranged from species distribution patterns to the use of barcoding at the country level, phylogenomic analyses using ultraconserved elements, and island biology. The remaining studies explored current conservation issues, examining the impact of mining, fungal disease, and the conservation implications of endemism.

The contributions on taxonomy and systematics focus on the Peruvian Andes, a region of exceptional herpetological species richness. The rate of species descriptions for Peru, similar to other mountainous tropical countries, is among the highest for frogs and lizards. The special issue includes a new genus name, for a group of terrestrial-breeding frogs (Strabomantidae). Molecular phylogenetic analyses of three species previously assigned to *Bryophryne*, in addition to bioacoustics and some morphological traits, support the erection of the new genus *Qosqophryne* [8], honoring the region around the city of Cuzco where these frogs are endemic. Santa-Cruz et al. [9] name a related terrestrial-breeding frog in the genus *Noblella* with a distribution extending from the lowlands of southwestern Amazonia to the Andean cloud forests across several protected areas. Aguilar-Puntriano et al. [10] describe three new species of *Liolaemus* (Tropiduridae) from the Pacific coast and the High Andes. Mamani et al. [11] name a new species of *Cercosaura* (Gymnophthalmidae) from a montane forest in central Peru and resolve the taxonomy of *Cercosaura anomala*. Additionally, Mamani et al. hint at the possibility of a new genus for the enigmatic *Cercosaura manicata boliviana*, which is more closely related to the clade containing all known *Potamites* species than to other Cercosaura species [11].

Two other contributions on frog systematics cover wider geographical areas and discuss family-level relationships. Guillory et al. [12] use genomic data from the flanking regions of ultraconserved elements to generate a phylogeny of Neotropical poison frogs (Dendrobatidae). Neotropical poison frogs are famous for their aposematic coloration and associated toxicity, which stimulated curiosity and spurred many investigations on the evolution of skin defenses and coloration [13–15]. Although Guillory et al.'s findings confirm previous phylogenies inferred from combining few mitochondrial and nuclear gene fragments, their approach reveals much potential for scalable genomic techniques that can be applied to help solve conservation problems related to loss of genetic diversity.

Guayasamin et al. [16] offer a superb monograph on all Ecuadorian glassfrogs, covering the taxonomy, morphology, phylogenetic relationships, ecology, and natural history of one of the most charismatic group of frogs. In addition to representing the bulk of this book and special issue, this contribution has all the qualities to become a classic work for people interested in the biology and evolution of glassfrogs. The family Centrolenidae is turning into an excellent model system for studies on parental care, diversification, and evolution [17–20]. Guayasamin et al. [16] provide species accounts for each of the 60 species known from Ecuador, including photographs of living and preserved frogs, drawings, distribution range maps, ecology, and conservation status. Last but not the least, these authors also describe three new species in the genus *Nymphargus*, two of which honor the amphibian biologists Linda Trueb and Luis Coloma [16].

Biogeography is a common thread among five contributions, one on frogs and four on squamates. Ramírez et al. [21] examine the radiation of the highly threatened harlequin frogs (*Atelopus*) into Central America. Their model-based ancestral area estimation supports one or two colonization events from South America. Molecular clock analyses of divergence times suggest that these events occurred prior to 4 million years ago, a slightly older than traditional date for the closure of the Isthmus [21]. In contrast to harlequin frogs and other amphibians, which benefited from early efforts at categorizing their distribution in the context of the Global Amphibian Assessment and IUCN Red List of Threatened Species [22], squamate global threat assessments have experienced a later start, despite studies suggesting a high proportion of threatened species [23]. Therefore, articles in this issue will contribute to ongoing efforts to develop a clear picture of global squamate biodiversity patterns and conservation. As is true for amphibians, it is likely that cryptic species abound among squamates, exposing the benefits of barcoding approaches that can quickly reveal candidate new species and approximate genetic diversity in complexes of cryptic species. Cacciali et al. [24] present the first barcoding analysis of Paraguayan squamates, using sequences of the 16S rRNA mitochondrial gene for 63 native and one introduced species. Although the authors did not sample all species and geographic areas, the study is a first important step in building datasets of molecular genetics that can ameliorate the challenges of taxonomy and conservation. Taxonomic uncertainty, when compounded with low sampling and collecting efforts and rarity, can produce imprecise, if not intriguing, biodiversity patterns. Rabosky et al. [25] discuss one such pattern related to lower species richness of snakes and lizard in southwestern Amazonia compared with northwestern Amazonia. They quantify the reduction at ~25% compared to western equatorial sites and discuss some possible mechanisms for the equatorial to southwestern Amazonia species richness gradient, such as cycles of expansion and contraction of savannah habitats in southwestern Amazonia resulting in the loss of some species [25]. Snake encounters are notoriously serendipitous, and in tropical areas with high species richness, compiling ecological data can take decades. Birskis-Barros et al. [26] contribute natural history information important for conserving pit vipers (Crotalinae), in the Americas. Although most pit vipers have large geographical ranges and narrow habitat breadths, about one tenth of the known species are rare and occur along the Pacific coast of Mexico, in southern Central America, in the Andean region of Ecuador, and in eastern Brazil, driving the inverse correlation between abundance and latitude. Finally, Phillips et al. [27] examine the systematics and ecomorphology of four species of Pacific Island anoles. Anoles are a staple of Caribbean biology and biogeography studies, but much less is known about Pacific Island anoles. The two species from Isla Malpelo and Isla Cocos diverged from mainland ancestors prior to the emergence of their respective islands and, similar to single-island endemic Caribbean anoles, appear to display sexual size dimorphism [27].

Three articles on conservation biology are representative of current challenges in assessing the conservation status of often discreet animals, including possible causes of population declines and local extinctions. The cover of the online special issue illustrates the article discussing conservation implications for enzootic chytridiomycosis, the "covid of frogs," for Costa Rican amphibians [28]. Chytridiomycosis is implicated in the decline of at least 501 species of amphibians worldwide [29]. In this issue, Zumbado-Ulate et al. [28] give an overview of the disease in Costa Rica, where epizootic mass die-offs and declines occurred in the 1980s and 1990s. The chytrid fungus is now common across the country, especially in the Caribbean lowlands and among amphibians with aquatic larvae. Infection loads are generally below theoretical thresholds associated with mortality and highest in direct-developing species [28]. Bornschein et al. [30] discuss the conservation status of the minute but fascinating *Brachycephalus* frogs of the Brazilian Atlantic forest. Several species of *Brachycephalus* suffered population declines or have not been seen in several decades, despite the number of new species continuing to climb. An increasing threat to herpetofauna in many Neotropical countries, the impact of mining and associated habitat loss is discussed for the endemic herpetofauna of Mexico [31]. Mayani-Parás et al. [24] examine the impact of habitat loss and mining activities on potential distributions from ecological niche models of 179 Mexican endemic herpetofaunal species. The daunting conclusion is that the combined effect of habitat loss and mining may exert stronger impacts on extant species distribution than habitat loss alone.

Figure 1. Map of the Neotropical region depicting the transition zones/sub-regions according to the regionalization by Morrone (2014) [6]. Shapefiles downloaded from www.neotropico.com.br/shapefile [32].

We hope this publication will serve multiple functions. It will serve as a field book helping to identify glassfrogs in Ecuador and support species conservation status assessment and design of protected areas. We are optimistic that the book will stimulate more research on the Neotropical herpetofauna and provide strong foundations for proposing or refining hypotheses elucidating its exceptional beauty and diversity.

Funding: This research received no external funding.

Acknowledgments: We thank S. Kupferberg for comments on a draft of the manuscript.

Conflicts of Interest: This research received no external funding.

References

1. Sclater, P.L. On the general geographic distribution of the members of the class Aves. *Zool. J. Linn. Soc.* **1858**, *2*, 130–145. [CrossRef]
2. Wallace, A.R. *The Geographical Distribution of Animals*; Harper and Brothers: New York, NY, USA, 1876; Volumes I & II.
3. Holt, B.; Lessard, J.P.; Borregaard, M.K.; Fritz, S.A.; Araujo, M.B.; Dimitrov, D.; Fabre, P.H.; Graham, C.H.; Graves, G.R.; Jonsson, K.A.; et al. An update of Wallace's zoogeographic regions of the World. *Science* **2013**, *339*, 74–78. [CrossRef]
4. Kreft, H.; Jetz, W. Comment on "An update of Wallace's zoogeographic regions of the World". *Science* **2013**, *341*, 343. [CrossRef] [PubMed]

5. Antonelli, A.; Zizka, A.; Carvalho, F.A.; Scharn, R.; Bacon, C.D.; Silvestro, D.; Condamine, F.L. Amazonia is the primary source of Neotropical biodiversity. *Proc. Natl. Acad. Sci. USA* **2018**, *115*, 6034–6039. [CrossRef] [PubMed]

6. Morrone, J.J. Biogeographical regionalisation of the Neotropical region. *Zootaxa* **2014**, *3782*, 1–110. [CrossRef] [PubMed]

7. Olson, D.M.; Dinerstein, E.; Wikramanayake, E.D.; Burgess, N.D.; Powell, G.V.N.; Underwood, E.C.; D'Amico, J.A.; Itoua, I.; Strand, H.E.; Morrison, J.C.; et al. Terrestrial ecoregions of the worlds: A new map of life on Earth. *Bioscience* **2001**, *51*, 933–938. [CrossRef]

8. Catenazzi, A.; Mamani, L.; Lehr, E.; von May, R. A new genus of terrestrial-breeding frogs (Holoadeninae, Strabomantidae, Terrarana) from southern Peru. *Diversity* **2020**, *12*, 184. [CrossRef]

9. Santa-Cruz, R.; von May, R.; Catenazzi, A.; Whitcher, C.; Tejeda, E.L.; Rabosky, D.L. A new species of terrestrial-breeding frog (Amphibia, Strabomantidae, *Noblella*) from the upper Madre de Dios watershed, Amazonian Andes and lowlands of southern Peru. *Diversity* **2019**, *11*, 145. [CrossRef]

10. Aguilar-Puntriano, C.; Ramírez, C.; Castillo, E.; Mendoza, A.; Vargas, V.J.; Sites, J.W. Three new lizard species of the *Liolaemus montanus* Group from Peru. *Diversity* **2019**, *11*, 161. [CrossRef]

11. Mamani, L.; Chaparro, J.C.; Correa, C.; Alarcón, C.; Salas, C.Y.; Catenazzi, A. A new species of Andean gymnophthalmid lizard (Squamata: Gymnophthalmidae) from the Peruvian Andes, and resolution of some taxonomic problems. *Diversity* **2020**, *12*, 361. [CrossRef]

12. Guillory, W.X.; Muell, M.R.; Summers, K.; Brown, J.L. Phylogenomic reconstruction of the Neotropical poison frogs (Dendrobatidae) and their conservation. *Diversity* **2019**, *11*, 126. [CrossRef]

13. Daly, J.W. Thirty years of discovering arthropod alkaloids in amphibian skin. *J. Nat. Prod.* **1998**, *61*, 162–172. [CrossRef] [PubMed]

14. Santos, J.C.; Coloma, L.A.; Cannatella, D.C. Multiple, recurring origins of aposematism and diet specialization in poison frogs. *Proc. Natl. Acad. Sci. USA* **2003**, *100*, 12792–12797. [CrossRef] [PubMed]

15. Summers, K.; Clough, M.E. The evolution of coloration and toxicity in the poison frog family (Dendrobatidae). *Proc. Natl. Acad. Sci. USA* **2001**, *98*, 6227–6232. [CrossRef]

16. Guayasamin, J.M.; Cisneros-Heredia, D.F.; McDiarmid, R.W.; Peña, P.; Hutter, C.R. Glassfrogs of Ecuador: Diversity, evolution, and conservation. *Diversity* **2020**, *12*, 222. [CrossRef]

17. Castroviejo-Fisher, S.; Guayasamin, J.M.; Gonzalez-Voyer, A.; Vila, C. Neotropical diversification seen through glassfrogs. *J. Biogeogr.* **2014**, *41*, 66–80. [CrossRef]

18. Delia, J.; Bravo-Valencia, L.; Warkentin, K.M. Patterns of parental care in Neotropical glassfrogs: Fieldwork alters hypotheses of sex-role evolution. *J. Evol. Biol.* **2017**, *30*, 898–914. [CrossRef]

19. Delia, J.R.J.; Ramirez-Bautista, A.; Summers, K. Parents adjust care in response to weather conditions and egg dehydration in a Neotropical glassfrog. *Behav. Ecol. Sociobiol.* **2013**, *67*, 557–569. [CrossRef]

20. Hutter, C.R.; Guayasamin, J.M.; Wiens, J.J. Explaining Andean megadiversity: The evolutionary and ecological causes of glassfrog elevational richness patterns. *Ecol. Lett.* **2013**, *16*, 1135–1144. [CrossRef]

21. Ramirez, J.P.; Jaramillo, C.A.; Lindquist, E.D.; Crawford, A.J.; Ibáñez, R. Recent and rapid radiation of the highly endangered Harlequin frogs (*Atelopus*) into Central America inferred from mitochondrial DNA sequences. *Diversity* **2020**, *12*, 360. [CrossRef]

22. Stuart, S.N.; Chanson, J.S.; Cox, N.A.; Young, B.E.; Rodrigues, A.S.L.; Fischman, D.L.; Waller, R.W. Status and trends of amphibian declines and extinctions worldwide. *Science* **2004**, *306*, 1783–1786. [CrossRef] [PubMed]

23. Böhm, M.; Collen, B.; Baillie, J.E.M.; Bowles, P.; Chanson, J.; Cox, N.; Hammerson, G.; Hoffmann, M.; Livingstone, S.R.; Ram, M.; et al. The conservation status of the world's reptiles. *Biol. Conserv.* **2013**, *157*, 372–385. [CrossRef]

24. Cacciali, P.; Buongermini, E.; Kohler, G. Barcoding analysis of Paraguayan Squamata. *Diversity* **2019**, *11*, 152. [CrossRef]

25. Rabosky, D.L.; von May, R.; Grundler, M.C.; Rabosky, A.R.D. The western Amazonian richness gradient for squamate reptiles: Are there really fewer snakes and lizards in southwestern Amazonian lowlands? *Diversity* **2019**, *11*, 199. [CrossRef]

26. Birskis-Barros, I.; Alencar, L.R.V.; Prado, P.I.; Böhm, M.; Martins, M. Ecological and conservation correlates of rarity in New World pitvipers. *Diversity* **2019**, *11*, 147. [CrossRef]

27. Phillips, J.G.; Burton, S.E.; Womack, M.M.; Pulver, E.; Nicholson, K.E. Biogeography, systematics, and ecomorphology of Pacific Island anoles. *Diversity* **2019**, *11*, 141. [CrossRef]

28. Zumbado-Ulate, H.; Nelson, K.N.; Garcia-Rodriguez, A.; Chaves, G.; Arias, E.; Bolanos, F.; Whitfield, S.M.; Searle, C.L. Endemic infection of *Batrachochytrium dendrobatidis* in Costa Rica: Implications for amphibian conservation at regional and species Level. *Diversity* **2019**, *11*, 129. [CrossRef]
29. Scheele, B.; Pasmans, F.; Skerratt, L.F.; Berger, L.; Martel, A.; Beukema, W.; Acevedo, A.A.; Burrowes, P.A.; Carvalho, T.; Catenazzi, A.; et al. Amphibian fungal panzootic causes catastrophic and ongoing loss of biodiversity. *Science* **2019**, *363*, 1459–1463. [CrossRef]
30. Bornschein, M.R.; Pie, M.R.; Teixeira, L. Conservation status of *Brachycephalus* toadlets (Anura: Brachycephalidae) from the Brazilian Atlantic rainforest. *Diversity* **2019**, *11*, 150. [CrossRef]
31. Mayani-Parás, F.; Botello, F.; Castañeda, S.; Sánchez-Cordero, V. Impact of habitat loss and mining on the distribution of endemic species of amphibians and reptiles in Mexico. *Diversity* **2019**, *11*, 210. [CrossRef]
32. Löwenberg-Neto, P. Neotropical region: A shapefile of Morrone's (2014) biogeographical regionalisation. *Zootaxa* **2014**, *3802*, 300. [CrossRef] [PubMed]

Publisher's Note: MDPI stays neutral with regard to jurisdictional claims in published maps and institutional affiliations.

Article

A New Genus of Terrestrial-Breeding Frogs (Holoadeninae, Strabomantidae, Terrarana) from Southern Peru

Alessandro Catenazzi [1,*], Luis Mamani [2,3], Edgar Lehr [4] and Rudolf von May [5]

[1] Department of Biological Sciences, Florida International University, Miami, FL 33199, USA
[2] Museo de Biodiversidad del Perú, Cusco 08000, Peru; luismamanic@gmail.com
[3] Museo de Historia Natural de la Universidad Nacional de San Antonio Abad del Cusco, Cusco 08000, Peru
[4] Department of Biology, Illinois Wesleyan University, Bloomington, IL 61701, USA; elehr@iwu.edu
[5] Biology Program, California State University Channel Islands, Camarillo, CA 93012, USA; rvonmay@gmail.com
* Correspondence: acatenazzi@gmail.com

http://zoobank.org/urn:lsid:zoobank.org:pub:0B8FFBEE-96AA-46E1-BA6F-541DC9FA73BF
Received: 7 April 2020; Accepted: 6 May 2020; Published: 8 May 2020

Abstract: We propose to erect a new genus of terrestrial-breeding frogs of the Terrarana clade to accommodate three species from the Province La Convención, Department of Cusco, Peru previously assigned to *Bryophryne*: *B. flammiventris*, *B. gymnotis*, and *B. mancoinca*. We examined types and specimens of most species, reviewed morphological and bioacoustic characteristics, and performed molecular analyses on the largest phylogeny of *Bryophryne* species to date. We performed phylogenetic analysis of a dataset of concatenated sequences from fragments of the 16S rRNA and 12S rRNA genes, the protein-coding gene cytochrome c oxidase subunit I (COI), the nuclear protein-coding gene recombination-activating protein 1 (RAG1), and the tyrosinase precursor (Tyr). The three species are immediately distinguishable from all other species of *Bryophryne* by the presence of a tympanic membrane and annulus, and by males having median subgular vocal sacs and emitting advertisement calls. Our molecular phylogeny confirms that the three species belong to a new, distinct clade, which we name *Qosqophryne*, and that they are reciprocally monophyletic with species of *Microkayla*. These two genera (*Qosqophryne* and *Microkayla*) are more closely related to species of *Noblella* and *Psychrophrynella* than to species of *Bryophryne*. Although there are no known morphological synapomorphies for either *Microkayla* or *Qosqophryne*, the high endemism of their species, and the disjoint geographic distribution of the two genera, with a gap region of ~310 km by airline where both genera are absent, provide further support for *Qosqophryne* having long diverged from *Microkayla*. The exploration of high elevation moss and leaf litter habitats in the tropical Andes will contribute to increase knowledge of the diversity and phylogenetic relationships within Terrarana.

Keywords: amphibian; Andes; Cusco; high elevation; Neotropical; *Qosqophryne*; tropical mountain; systematic; taxonomy

---

1. Introduction

Terrestrial-breeding frogs of the high Andes display an impressive degree of evolutionary convergence [1–4]. Such convergence is associated with life in the cloud forest and high-Andean grassland. Frogs in many genera of Terrarana have evolved strikingly similar body forms [4,5], typically a small, compact body with very short legs and feet, short arms and hands, loss of toe pads and discs, head wider than long, small eyes directed anterolaterally, and, in many groups, reduction or loss of tympanic structure and function [3]. The high similarity of body forms has delayed obtaining a taxonomic arrangement that reflects the evolutionary history and phylogenetic relationships of most species of small, terrestrial-breeding frogs of the Andes [1,6,7].

Illustrating the complexity within Terrarana of identifying monophyletic groups in presence of ecological convergence, authors originally assigned frogs belonging to different evolutionary lineages to the genus *Phrynopus* [1,8,9]. Indeed, *Phrynopus* might still contain incorrectly classified species of *Pristimantis* that lack vocal sacs, external tympanic apparatus and toe pads [10]. Subsequent molecular analyses revealed a much greater diversity and deeper genetic structure, such that Hedges et al. [1] proposed to split *Phrynopus* into four genera, and to erect the new subfamily Holoadeninae to include the newly described genera *Bryophryne*, *Niceforonia*, and *Psychrophrynella*. Within Holoadeninae, the molecular phylogeny by Hedges et al. [1] recognized *Bryophryne* as a distinct clade on the basis of DNA sequences from a single species, *B. cophites* (formerly *Phrynopus cophites* Lynch, 1975). Hedges et al. [1] used morphological characters to assign to *Bryophryne* a second species, *Phrynopus bustamantei* Chaparro, De la Riva, Padial, Ochoa, and Lehr, 2007. The new genus *Byrophryne*, along with the other genera of Holoadeninae, was recognized using molecular data, despite the lack of morphological synapomorphies [1,2,5,11].

Since Hedges et al. [1] published their molecular phylogeny, researchers have continued discovering terrestrial-breeding frogs: the number of species of *Bryophryne* has increased from two to 14 species [12–17], and the number of species across all Holoadeninae genera from 36 to 151 species [8]. As far as we know, all species of *Bryophryne* have micro-endemic distribution, and are only known to occur at their respective type localities and immediate surroundings [2,12,14–16,18]. The most recent phylogeny included six of the 14 species of *Bryophryne*, and recovered *Bryophryne* as being the sister taxon to the clade containing *Barycholos*, "*Eleutherodactylus bilineatus*", *Euparkerella*, *Holoaden*, and *Noblella* [2]. However, this phylogeny by De la Riva et al. [2] did not include sequences of the three species of *Bryophryne* having an external tympanum and males with subgular vocal sacs, because sequences were unavailable at the time. Additionally, De la Riva et al. [2] erected a new genus, *Microkayla*, to accommodate all species of *Psychrophrynella* from Bolivia (and one species of *Psychrophrynella* from Peru), as well as two new species from Peru. Because of these discoveries, the integration of molecular, acoustic and morphological approaches, and the ongoing revision of existing and new material, we have a better understanding of the diversity in this group of cryptic genera. As part of our ongoing work, we have become aware of (1) uncertainty regarding the evolutionary relationships of *Noblella* and *Psychrophrynella* [2,19,20], (2) an underestimated species richness and endemism in *Noblella* and *Psychrophrynella* [19–22], and (3) three species of *Bryophryne* (*B. flammiventris*, *B. gymnotis*, *B. mancoinca*; Figure 1) having traits not shared with any other species of *Bryophryne*, such as having an external tympanum and males with subgular vocal sacs and emitting advertisement calls. Here we address the latter of these findings, and propose a new genus for the only three species of *Bryophryne* known to produce vocalizations and possessing external tympanic membrane and annulus.

Figure 1. Holotypes of species of *Qosqophryne* gen. n. in dorsolateral and ventral views: (**A,B**) *Q. flammiventris* (MUSM 27613; SVL 19.8 mm): (**C,D**) *Q. gymnotis* (MUSM 25543; SVL 18.4 mm); (**E,F**) *Q. mancoinca* (MUBI 11152; SVL 26.5 mm). Photographs by E. Lehr (**A,B**), A. Catenazzi (**C,D**) and L. Mamani (**E,F**).

2. Materials and Methods

We are familiar with most described species of *Bryophryne*, which we have seen in the field or inspected in collections. We provide a complete list of examined specimens in Appendix A. We used the literature (i.e., original species descriptions) for species whose specimens we could not examine. We have described the advertisement calls of *B. gymnotis* and *B. mancoinca* [14,17], and have heard and provided a short description of the call of *B. flammiventris* [15]. We refer readers to the original publications for details on recording methods.

We combined DNA sequences available from GenBank with sequences from newly collected tissues to generate molecular phylogenies of *Bryophryne* and closely related Holoadeninae taxa (Table 1). We considered sequences for a fragment of the 16S rRNA gene (16S), a fragment of the 12S rRNA gene (12S), the protein-coding gene cytochrome c oxidase subunit I (COI), the nuclear protein-coding gene recombination-activating protein 1 (RAG1), and the tyrosinase precursor (Tyr). All taxa selected for our comparisons belong to the subfamily Holoadeninae [1,23,24].

Table 1. GenBank accession numbers for taxa and genes sampled in this study. Genbank accession codes of the new sequences are highlighted in bold font.

Taxon	16S	12S	COI	RAG1	Tyr	Voucher Nbr	Reference
Barycholos pulcher	EU186709	-	-	-	EU186765	KU 217781	[1]
Barycholos ternetzi	JX267466	-	-	JX267543	JX267680	CFBH 19426	[23]
Bryophryne bakersfield	KT276291	KT276283	-	-	-	MHNC 6007	[12]
Bryophryne bakersfield	MF186344	MF186287	-	KT276278	-	MHNC 6009	[12]
Bryophryne bustamantei	**MT437052**	-	-	**MT431911**	-	MUSM 24537	This study
Bryophryne bustamantei	**CMT437053**	-	-	**MT431912**	-	MUSM 24538	This study
Bryophryne bustamantei	KT276293	KT276286	-	KT276280	KT276296	MHNC 6019	[12]
Bryophryne cf. *zonalis*	**MT437054**	-	**MT435518**	-	-	CORBIDI 17475	This study
Bryophryne cophites	EF493537	-	-	EF493423	EF493508	KU173497	[9]
Bryophryne cophites	KY652641	-	KY672976	KY672961	KY681062	AC 270.07	[22]
Bryophryne hanssaueri	KY652642	-	KY672977	KY681084	KY681063	MUSM 27567	[22]
Bryophryne nubilosus	KY652643	-	KY672978	KY681085	KY681064	MUSM 27882	[22]
Bryophryne phuyuhampatu	MF419259	-	-	-	-	CORBIDI 18224	[16]
Bryophryne phuyuhampatu	MF419259	-	-	-	-	MUBI 14654	[16]
Bryophryne quellokunka	**MT437061**	-	-	-	-	MUSM 27571	This study
Bryophryne quellokunka	**MF186387**	**MF186309**	-	**MF186526**	-	MNCN 43780	[2]
Bryophryne sp.	**MT437062**	-	-	**MT431916**	-	MUSM 27961	This study
Bryophryne sp.	**MT437063**	-	-	**MT431917**	-	AC 41.09	This study
Bryophryne tocra	MF186396	MF186315	-	MF186541	MF186583	MNCN 43786	[2]
Bryophryne wilakunka	MF186349	MF186291	-	-	-	MUBI 5425	[2]
Bryophryne zonalis	**MT437064**	-	-	-	-	MUSM 27939	This study
Eleutherodactylus bilineatus	JX267324	-	-	JX267556	JX267691	MNRJ 46476	[23]
Euparkerella brasiliensis	JX267468	-	-	JX267545	JX267682	-	[23]
Holoaden bradei	EF493366	EF493378	-	EF493449	EU186779	USNM 207945	[9]
Holoaden luederwaldti	EU186710	EU186728	-	EU186747	EU186768	MZUSP 131872	[1]
Holoaden luederwaldti	JX267470	-	-	-	-	CFBH 19552	[23]
Lynchius flavomaculatus	EU186667	EU186667	-	EU186745	EU186766	KU218210	[1]
Lynchius nebulanastes	EU186704	EU186704	-	-	-	KU 181408	[1]
Lynchius oblitus	KX470783	KX470776	-	KX470792	KX470799	MHNC 8614	[25]
Lynchius parkeri	EU186705	EU186705	-	-	-	KU 181307	[1]
Lynchius simmonsi	JF810004	JF809940	-	JF809915	JF809894	QZ 41639	[26]
Microkayla adenopleura	MF186339	-	-	-	-	MNCN 44809	[2]
Microkayla adenopleura	MF186340	MF186283	-	MF186537	MF186565	MNCN 44810	[2]
Microkayla ankohuma	-	MF186288	-	-	-	MNKA 7280	[2]
Microkayla ankohuma	-	MF186289	-	-	-	CBF 5982	[2]
Microkayla boettgeri	MF186352	MF186293	MF186456	-	-	MNCN 43778	[2]
Microkayla boettgeri	MF186353	MF186294	-	-	MF186559	MUBI 5363	[2]
Microkayla boettgeri	MF186354	-	-	-	-	MUBI 5364	[2]
Microkayla cf. *iatamasi*	MF186365	-	-	-	-	MNCN-DNA 20927	[2]
Microkayla chacaltaya	MF186357	-	-	MF186532	-	MNCN 42052	[2]
Microkayla chapi	MF186417	MF186328	-	MF186540	MF186562	MNCN 43762	[2]
Microkayla chilina	MF186411	-	-	-	-	MUBI 5350	[2]
Microkayla chilina	MF186414	MF186327	MF186457	MF186539	MF186561	MNCN 43772	[2]
Microkayla condoriri	MF186358	-	-	-	-	CBF 5988	[2]
Microkayla guillei	AY843720	AY843720	-	-	DQ282995	AMNH A165108	[9]
Microkayla iatamasi	AM039644	AM039712	-	-	-	MTD TD 1231	[9]
Microkayla illampu	MF186373	-	-	-	-	CBF 5999	[2]
Microkayla kallawaya	MF186379	-	-	-	-	MNCN 42509	[2]
Microkayla katantika	MF186380	-	MF186453	-	-	CBF 6012	[2]

Table 1. *Cont.*

Taxon	16S	12S	COI	RAG1	Tyr	Voucher Nbr	Reference
Microkayla kempffi	MF186384	-	-	-	-	MNCN 43646	[2]
Microkayla quimsacruzis	MF186407	-	-	-	-	MNCN 42039	[2]
Microkayla saltator	AM039642	AM039710	-	-	-	MTD TD 1229	[9]
Microkayla sp. Coscapa	MF186399	-	-	-	-	CBF 6564	[2]
Microkayla sp. Khatu River	MF186409	-	-	-	-	MNCN 42034	[2]
Microkayla teqta	MF186400	MF186318	-	-	MF186552	MNCN 45702	[2]
Microkayla utururo	MF186433	-	-	-	-	MNCN 46987	[2]
Microkayla wettsteini	MF186434	MF186338	-	MF186531	MF186551	CBF 6241	[2]
Niceforonia brunnea	EF493357	-	-	-	-	KU 178258	[9]
Niceforonia dolops	EF493394	-	-	-	-	-	[9]
Noblella heyeri	JX267541	JX267463	-	-	-	QCAZ 31471	[23]
Noblella lochites	EU186699	EU186699	-	EU186756	EU186777	KU 177356	[1]
Noblella losamigos	MN366392	-	MN356099	-	-	MVZ 292687	[27]
Noblella losamigos	KY652644	-	-	KY672962	KY681065	MUSA 6973	[22]
Noblella losamigos	MN056358	-	MN356098	-	-	MUBI 17413	[27]
Noblella madreselva	MN064565	-	-	MN355547	-	CORBIDI 15769	[27]
Noblella myrmecoides	JX267542	JX267464	-	-	-	QCAZ 40180	[23]
Noblella myrmecoides	MN056357	-	-	-	-	CORBIDI PV45	[28]
Noblella pygmaea	KY652645	-	KY672979	KY681086	KY681066	MUSM 24536	[22]
Noblella sp.	AM039646	AM039714	-	-	-	MTD 45180	[29]
Noblella sp. R	KY652646	-	KY672980	KY681087	KY681067	MUSM 27582	[22]
Noblella thiuni	MK072732	-	-	-	-	CORBIDI 18723	[28]
Oreobates amarakaeri	JF809996	JF809934	-	JF809913	JF809891	MHNC 6975	[26]
Oreobates ayacucho	JF809970	JF809933	-	JF809912	JF809890	MNCN IDIR5024	[26]
Oreobates cruralis	EU186666	EU186666	-	EU186743	EU186764	KU 215462	[1]
Oreobates gemcare	JF809960	JF809930	-	JF809909	-	MHNC 6687	[26]
Oreobates granulosus	EU368897	JF809929	-	JF809908	JF809887	MHNC 3396	[30]
Phrynopus auriculatus	EF493708	EF493708	-	-	-	KU 291634	[9]
Phrynopus barthlenae	AM039653	AM039721	-	-	-	SMF 81720	[29]
Phrynopus bracki	EF493709	EF493709	-	EF493421	-	USNM 286919	[9]
Phrynopus bufoides	AM039645	AM039713	-	-	-	MHNSM 19860	[29]
Phrynopus heimorum	AM039635	AM039703	MF186462	MF186545	MF186580	MTD 45621	[29]
Phrynopus horstpauli	AM039651	AM039719	-	-	-	MTD 44333	[29]
Phrynopus inti	MF651902	MF651909	-	MF651917	-	MUSM 31968	[3]
Phrynopus kauneorum	AM039655	AM039723	-	-	-	MHNSM 20595	[29]
Phrynopus peruanus	MG896582	MG896605	MG896615	MG896626	MG896631	MUSM 38316	[3]
Phrynopus pesantesi	AM039656	AM039724	-	-	-	MTD 45072	[29]
Phrynopus spI	MG896589	MG896606	-	MG896629	-	MUSM 33261	[3]
Phrynopus tautzorum	AM039652	AM039720	-	-	-	MHNSM 20613	[29]
Phrynopus tribulosus	EU186725	EU186707	-	-	-	KU 291630	[1]
Pristimantis attenboroughi	KY594752	-	KY962779	KY962759	-	MUSM 31186	[10]
Pristimantis pluvialis	KX155577	-	-	KY962769	-	CORBIDI 11862	[31]
Pristimantis reichlei	EF493707	EF493707	-	EF493436	-	MHNSM 9267	[9]
Pristimantis stictogaster	EF493704	EF493704	-	EF493445	-	KU 291659	[9]
Psychrophrynella chirihampatu	KU884559	-	-	-	-	CORBIDI 16495	[19]
Psychrophrynella chirihampatu	KU884560	-	-	-	-	MHNC 14664	[19]
Psychrophrynella glauca	MG837565	-	-	-	-	CORBIDI 18729	[20]
Psychrophrynella sp.	**MT437065**	-	-	-	-	MUSM 27619	This study
Psychrophrynella sp.	**MT437066**	-	-	-	-	MTD 47488	This study

Table 1. *Cont.*

Taxon	16S	12S	COI	RAG1	Tyr	Voucher Nbr	Reference
Psychrophrynella sp. P	KY652660	-	KY672992	KY681089	KY681081	AC116.09	[22]
Psychrophrynella sp. R	KY652661	-	KY672993	KY681090	KY681082	AC148.07	[22]
Psychrophrynella usurpator	KY652662	-	KY672994	KY672975	KY681083	AC186.09	[22]
Qosqophryne flammiventris	MT437055	-	-	-	-	MTD 46890	This study
Qosqophryne flammiventris	MT437056	-	-	MT431913	-	MUSM 27615	This study
Qosqophryne gymnotis	MT437057	-	-	MT431914	-	MUSM 24546	This study
Qosqophryne gymnotis	MT437058	-	-	MT431915	-	MUSM 24543	This study
Qosqophryne mancoinca	MT437059	-	MT435519	-	-	MUBI 16068	This study
Qosqophryne mancoinca	MT437060	-	MT435520	-	-	MUBI 16069	This study

2.1. Laboratory Work

We followed protocols of extraction, amplification, and sequencing of DNA previously used for terrestrial-breeding frogs [1,20,22]. For the focal taxa (the three species members of the new genus), we extracted DNA from tissue samples obtained from six specimens collected in the field (two specimens per species). We also obtained DNA sequences from seven specimens in five other species of *Bryophryne*, and two specimens representing two species in other genera (*Noblella* and *Psychrophrynella*), and the remaining sequences are legacy data from GenBank.

We extracted DNA from liver tissue preserved in 70% ethanol by using a commercial extraction kit (IBI Scientific, Dubuque, IA, USA). We used selected primers (Table 2) to amplify DNA from each gene using the polymerase chain reaction (PCR) [22,32]. We obtained sequence data by running purified PCR products in an ABI 3730 Sequence Analyzer (Applied Biosystems), except sequences of *B. mancoinca* and *B. phuyuhampatu*, which we shipped to MCLAB (San Francisco, CA) for sequencing. We deposited all new sequences in GenBank (Table 1). We provide updated names of 86 terminals included in the analysis for 314 GenBank sequences.

Table 2. Primers used in this study.

Locus	Primer		Sequence (5′-3′)	Reference
16S	16SAR	F	CGCCTGTTTATCAAAAACAT	[33]
	16SBR	R	CCGGTCTGAACTCAGATCACGT	[33]
12S	L25195	F	AAACTGGGATTAGATACCCCACTA	[33]
	H2916	R	GAGGGTGACGGGCGGTGTGT	[33]
COI	dgLCO1490	F	GGTCAACAAATCATAAAGAYATYGG	[34]
	dgHCO2198	R	TAAACTTCAGGGT GACCAAARAAYCA	[34]
RAG1	R182	F	GCCATAACTGCTGGAGCATYAT	[9]
	R270	R	AGYAGATGTTGCCTGGGTCTTC	[9]
Tyr	Tyr1C	F	GGCAGAGGAWCRTGCCAAGATGT	[35]
	Tyr1G	R	TGCTGGGCRTCTCTCCARTCCCA	[35]

2.2. Molecular Phylogenetic Analyses

We inferred the phylogenetic relationships among taxa through analysis of concatenated DNA sequences of the five gene fragments (16S, 12S, COI, RAG1, Tyr). We used *Niceforonia dolops* to root the tree. We aligned sequences with Geneious R6, v. 6.1.8 (Biomatters 2013), using the built-in Geneious Aligner program. We then used PartitionFinder, v. 1.1.1 [36] to select the best partitioning scheme and substitution model for each gene using the Bayesian information criterion (BIC). The best partitioning scheme included the following six subsets (best fitting substitution models are in parentheses): partition subset 1 includes 12S and 16S sequences (GTR + I + G), partition 2 is the first codon position of COI (SYM + G), partition 3 is the second codon position of COI (F81), partition 4 is the third codon position of COI (HKY + G), partition 5 includes the first and second codon positions of RAG together with the first and second codon positions of Tyr (HKY + I + G), and partition 6 includes the third codon position of RAG together with the third codon position of Tyr (K80 + G).

We used MrBayes, v. 3.2.0 [37] to infer a molecular phylogeny for the 106 terminals and 2632 bp concatenated partitioned dataset (16S, 12S, COI, RAG1, Tyr). We performed an MCMC Bayesian analysis that included two simultaneous runs of 10 million generations, sampled once every 1000 generations. Each run had one "cold" chain and three heated chains, and the burn-in was set to discard 25% samples from the cold chain. Upon completion of the MCMC Bayesian analysis, the average standard deviation of split frequencies was 0.003916. We used Tracer version 1.5 [38] to examine the effective sample sizes (ESS), to verify convergence, and to verify that the runs reached stationarity. The observed effective sample sizes were satisfactory for all parameters (ESS > 200). Lastly, we used FigTree v. 1.4.2 [39] to visualize the majority-rule consensus tree and assess node support (based on posterior probability values).

Our research was approved by the Institutional Animal Care and Use Committee of Florida International University (18-009). The Dirección General Forestal y de Fauna Silvestre, Ministerio de Agricultura y Riego issued the permit authorizing this research (collecting permits #292-2014-MINAGRI-DGFFS-DGEFFS, SERNANP-Machu Picchu 054-2012-SERNANP-JEF, Contrato de Acceso Marco a Recursos Genéticos, No 359-2013-MINAGRI-DGFFS-DGEFFS).

The electronic version of this article in portable document format will represent a published work according to the International Commission on Zoological Nomenclature (ICZN), and hence the new names contained in the electronic version are effectively published under that Code from the electronic edition alone. This published work and the nomenclatural acts it contains have been registered in ZooBank, the online registration system for the ICZN. The ZooBank LSIDs (Life Science Identifiers) and the associated information can be viewed through any standard web browser at http://zoobank.org/urn:lsid:zoobank.org:pub:0B8FFBEE-96AA-46E1-BA6F-541DC9FA73BF.

3. Results

We recovered a phylogenetic tree (Figure 2) that was largely congruent with previous analyses [2,24]. However, our tree recovered three species of *Bryophryne* not previously included in phylogenetic analyses (*B. gymnotis, B. flammiventris,* and *B. mancoinca*) as a clade that is sister to the clade containing all species of *Microkayla*. Thus, species of *Microkayla*, instead of other species of *Byrophryne*, share the most common shared ancestor with *B. gymnotis, B. flammiventris,* and *B. mancoinca*. The presence of large, external tympanic membrane and annulus, and males with a median subgular vocal sac and production of vocalizations, immediately distinguishes the newly recognized genus from all other species of *Bryophryne*. At least four species of *Bryophryne* were described as having small, barely visible (under the skin surface) tympanic membranes and annuli (*B. bustamantei, B. quellokunka, B. tocra, B. wilakunka*), but their external appearance does not look that different from the other species of *Bryophryne* known to lack a visible tympanic membrane [2,14,18]. One of these species, *B. bustamantei* was described as producing a short whistle, but there is no recording of the call nor voucher associated with a call [18]. The distribution range of *B. bustamantei* overlaps with that of *B. gymnotis* in the cloud forest near Abra Málaga [14,18,40], and thus it is possible that the call of *B. gymnotis* was erroneously associated with males of *B. bustamantei*. There also seems to be some problems identifying specimens of this species, as shown by our phylogeny where specimens identified as *B. bustamantei* by one of us do not group with sequences from one of the paratypes of *B. bustamantei* (MHNC 6019).

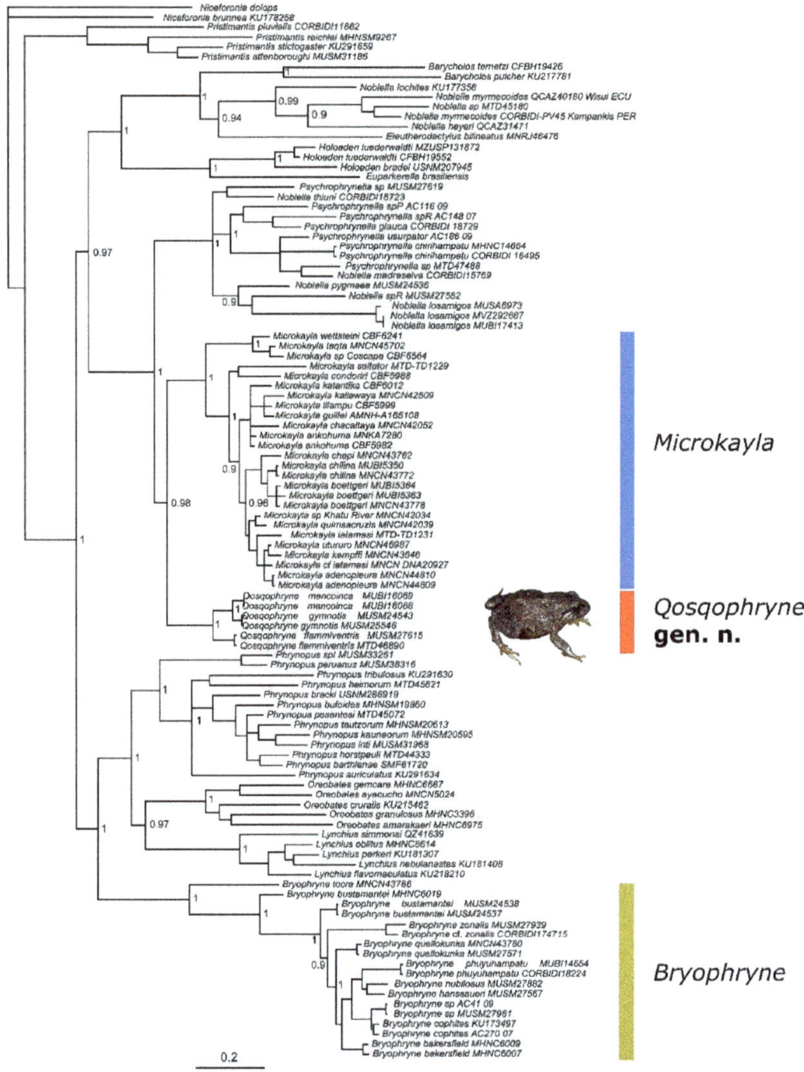

Figure 2. Bayesian maximum clade-credibility tree for 106 species of Holoadeninae (Terrarana) based on a 2646-bp concatenated partitioned dataset (fragments of genes 16S, 12S, COI, RAG1, and Tyr), highlighting the relationships of the three genera *Bryophryne*, *Microkayla* and *Qosqophryne* gen. n. Posterior probabilities are indicated at each node. The frog illustrated here is *Qosqophryne gymnotis*, paratype MUSM 24542 (photograph by A. Catenazzi).

We propose to erect the new genus *Qosqophryne* gen. n. to accommodate *Bryophryne gymnotis*, *B. flammiventris*, and *B. mancoinca*. Several lines of evidence support the idea that *Qosqophryne* is distinct from its sister genus *Microkayla*. The molecular phylogeny indicates there is a degree of divergence comparable to that observed between other genera of strabomantid frogs (Figure 2). Our molecular analyses show strong support for the divergence of *Microkayla* and *Qosqophryne* gen. n. The lack of geographic overlap between the two genera, with a gap region of ~320 km by airline where both genera are absent, further supports this divergence by preventing recent gene flow among species of both genera (Figure 3). Furthermore, several glaciated peaks, including the massive Ausangate mountains and associates peaks of the Cordillera de Vilcanota, are interspersed along this gap region of 320 km.

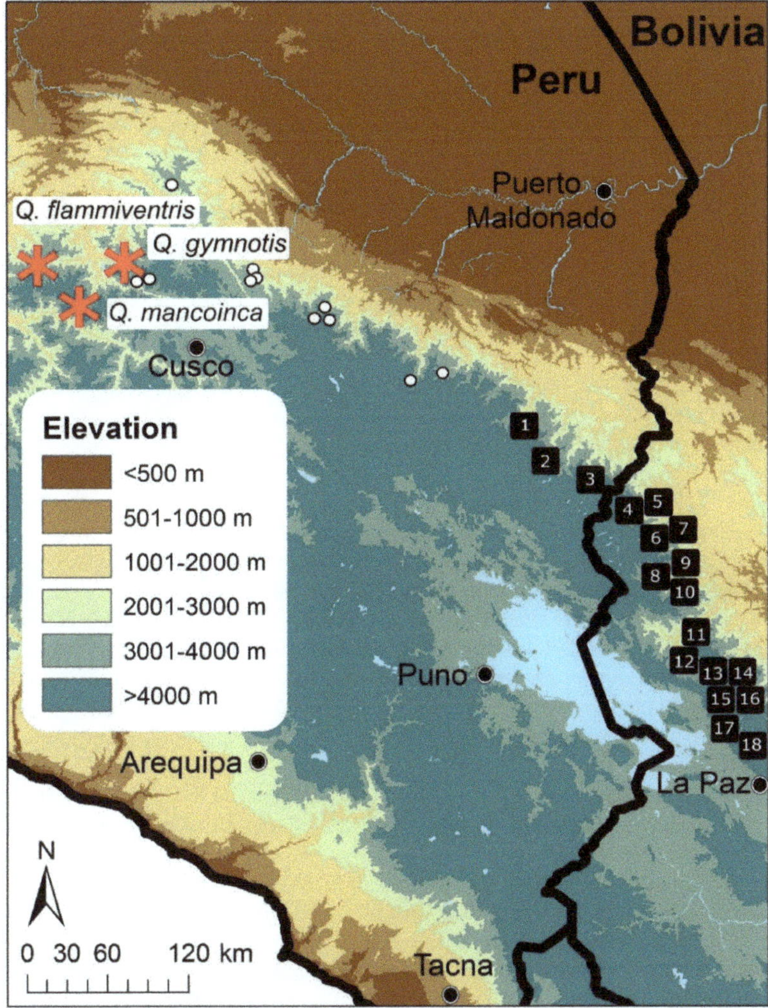

Figure 3. Type localities of frogs in the genera *Bryophryne* (white circles, species details not shown), *Microkayla* (squares) and *Qosqophryne* gen. n. (red asterisks) in southern Peru and northern Bolivia. The known distribution range of these frogs is limited to the type locality and immediate surroundings. For species of *Microkayla*: (1) *M. boettgeri*; (2) *M. chilina*; (3) *M. chapi*; (4) *M. katantika*; (5) *M. chaupi*; (6) *M. melanocheira*; (7) *M. colla*; (8) *M. kallawaya*; (9) *M. guillei*; (10) *M. saltator*; (11) *M. iani*; (12) *M. illampu*; (13) *M. ankohuma*; (14) *M. condoriri*; (15) *M. teqta*; (16) *M. huayna*; (17) *M. chacaltaya*; (18) *M. wettsteini*. The map does not include seven species of *Microkayla* distributed in central and southern Bolivia (type localities outside the limits of this map).

Similarly to recent phylogenies [28,41], we found that *Noblella* is not monophyletic: the species from southern Peru along with species of *Psychrophrynella* form a clade that is sister taxon to *Microkayla* + *Qosqophryne*, whereas the species of *Noblella* from northern Peru and Ecuador are closely related to "*Eleutherodactylus bilineatus*" and *Barycholos* (Figure 2). Because the type species *N. peruviana* occurs in southern Peru, and the most similar species sequenced to date *N. thiuni* is part of the *Noblella/Psychrophrynella* clade [28], our findings support the hypothesis that *Noblella* occurs only in southern Peru and northern Bolivia, and that species from northern Peru and Ecuador belong to a different genus [28,41]. Furthermore, our tree suggests that species of *Noblella* and *Psychrophrynella* belong to the same lineage, as supported by the respective type genera sharing several morphological traits [2,5,20,28,42]. Therefore, the two possibilities are that some species of *Noblella* have been misidentified as *Psychrophrynella* (and vice versa), or that *Psychrophrynella* is a junior synonym of *Noblella*. We will not be able to resolve the taxonomic uncertainty associated with *Noblella* and *Psychrophrynella* until we obtain DNA sequences from the respective type species *N. peruviana* and *P. bagrecito* [2,19,20,28].

Finally, our inferred phylogeny suggests that there are at least seven additional putative new species of *Bryophryne*, *Noblella*, and *Psychrophrynella* (Figure 2), and confirms previous findings of cryptic species diversity particularly in leaf litter, cloud forest frogs in the *Noblella/Psychrophrynella* clade [22]. These putative new species, similarly to most known species of high-elevation Holoadeninae [4], are highly endemic and known from single localities (or, around those localities, from within a narrow elevational range in the same valley, [22]). Of special interest among the putative new species, *Psychrophrynella* MUSM 27619 is the first specimen of the *Noblella/Psychrophrynella* lineage known from the Vilcabamba range.

Taxonomy

Qosqophryne new genus

http://zoobank.org/urn:lsid:zoobank.org:act:7DDB98AD-CCF9-4977-B814-285D25B3D1BF

Type species. *Bryophryne gymnotis* Lehr and Catenazzi, 2009

Included species. *Qosqophryne flammiventris* (Lehr and Catenazzi, 2010), comb. nov.; *Q. mancoinca* (Mamani, Catenazzi, Ttito, Mallqui, Chaparro, 2017), comb. nov.

Diagnosis. (1) Head wider than long, narrower than body, body robust, extremities short; (2) tympanic membrane and annulus present; (3) cranial crests absent; (4) prevomerine teeth and dentigerous process of vomers present (but absent in *Q. flammiventris*); (5) trips of digits narrow, rounded, circumferential grooves absent, terminal phalanges T-shaped to knobbed; (6) Finger I shorter than Finger II, nuptial pads absent; (7) Toe V shorter than Toe III; (8) fingers and toes with lateral fringes (but absent in *Q. flammiventris*); (9) subarticular tubercles small, rounded; (10) dorsolateral folds short, discontinuous or continuous; (11) discoidal fold absent (present in *Q. mancoinca*); (12) trigeminal nerve passing external to *m. adductor mandibulae externus* ('S' condition; Lynch, 1986); (13) snout-vent length from 16.7–19.3 mm in males and 16.0–22.2 mm in females of *Q. gymnotis*, to 19.6–22.9 mm in males and 23.6–26.5 mm in females of *Q. mancoinca*; (14) males with median subgular vocal sac and vocal slits, nuptial pads absent; (15) advertisement call whistle-like, composed of a single, tonal note in *Q. gymnotis*, 2–3 short notes in *Q. mancoinca*, and 3–4 short notes in *Q. flammiventris*.

There are no known morphological synapomorphies for *Qosqophryne*, but the three known species share the following traits (Table 3): (1) males with median subgular vocal sac produce whistle-like tonal calls composed of 1–4 short notes; (2) tongue ovate; (3) skin on venter smooth to weakly areolate (in *Q. flammiventris*); (4) inner tarsal fold absent. Four other genera of Holadeninae occur south of the Apurimac canyon, a proposed biogeographic barrier for high-elevation terrestrial breeding frogs [13–15]. *Bryophryne* differs from *Qosqophryne* in lacking an externally visible tympanum, and having males without vocal sac and not emitting vocalizations [2,12,16]. *Oreobates* have head about the same width as body, smooth venter, subarticular and supernumerary tubercles large, conical or subconical, projecting, and range in snout-vent length from 20–63 mm [1,5]. *Noblella* and *Psychrophrynella* have smooth venter, elongated tongue, two prominent metatarsal tubercles, and in most species facial masks and/or a tarsal fold-like, sigmoid tubercle [2,19,20,28]. *Qosqophryne* is most similar to its sister genus *Microkayla*. Putative synapomorphies of *Microkayla* are a rounded tongue, areolate belly, and absence of prominent metatarsal tubercles [2]. It is presumed that all species of *Microkayla* vocalize, and known calls consist of a simple, short whistle-like tonal note [2,4]. *Qosqophryne* differs from most *Microkayla* in having (except for *Q. flammiventris*) fingers and toes with lateral fringes (absent in *Microkayla*), and having (except *Q. flammiventris*) dentigerous processes of vomers (absent in *Microkayla*). Future examination of osteological characters, for example through computed tomography, might help identify such characters, and resolve the condition of the tympanic apparatus in the three genera *Bryophryne*, *Microkayla* and *Qosqophryne*.

Table 3. Meristic traits (+ = character present, - = character absent) for the three known species of *Qosqophryne* gen. n.

Characters	*Q. gymnotis*	*Q. flammiventris*	*Q. mancoinca*
Skin on dorsum	shagreen	Shagreen with small scattered tubercles	Shagreen with small conical tubercles
Skin on venter	smooth	Weakly areolate	smooth
Dorsolateral folds	Discontinuous, short	Discontinuous, short	Continuous, short
Tympanic membrane	+	+	+
Tympanic annulus	+	+	+
Dentigerous processes of vomers	+	-	+
Vocal sac	+	+	+
Vocal slits	+	+	+
Nuptial pads	-	-	-
Fingers with lateral fringes	+	-	+
Toes with lateral fringes	+	-	+
Inner tarsal fold	-	-	-
Dorsum coloration	Reddish, grayish or purplish brown or dark gray with narrow tan middorsal stripe	Grayish brown	Reddish brown or grayish brown with narrow tan middorsal stripe
Venter coloration	Dark brown, tan, or reddish brown with pale gray flecks	Blackish brown with yellow, orange or pink blotches	Gray or pale bluish gray with reddish-brown reticulation

Etymology. The name refers to the city of Cusco, using the spelling Qosqo which more closely reflects the name in Quechua. *Qosqo* is used in apposition with *phryne*, from the greek for "frog". Thus, the name for the new genus alludes to the geographic distribution of the three known species in the Peruvian Department of Cusco.

Distribution, natural history, and conservation. The three species of *Qosqophryne* occur within a region of ~150 km^2 in the upper montane forests and grasslands of the Cordilleras de Urubamba and Cordillera de Vilcabamba, Provincia La Convención, Department Cusco, Peru. These frogs inhabit cloud forests, elfin forests, montane scrub and humid grasslands (puna) from 3270 to 3800 m a.s.l. Similar to other regions in the high Andes, these habitats and their amphibian communities are threatened by pasture burning, climate change and associated expansion of agricultural activities, deforestation, and the fungal disease chytridiomycosis [43,44]. Although chytridiomycosis has caused the collapse of montane frog communities at several sites in Departamento Cusco [45,46], terrestrial-breeding frogs have generally declined the least, and several species challenged in experimental infection trials appears to resist or tolerate infection [47]. Protection of natural habitats will benefit conservation of these frogs. Two of the three species occur within naturally protected areas: *Q. gymnotis* within the Área de Conservación Privada Abra Málaga, and *Q. mancoinca* within Machu Picchu Historic Sanctuary.

Remarks. The new genus is distinguished from all species of *Bryophryne* by the presence of tympanum and tympanic annulus, and median subgular vocal sacs in males. Furthermore, males of all three species of *Qosqophryne* are known to emit advertisement calls (unknown in all species of *Bryophryne*, except possibly for *B. bustamantei*). We have described the advertisement calls of *Q. gymnotis* and *Q. mancoinca* [14,17]. One of us (LM) has recorded the advertisement call of a male *Q. flammiventris* (MUBI 13365) at the type locality, and this call is composed of 3–4 short notes (~15–35 ms duration) at dominant frequency ~3000 Hz. Females of *Q. gymnotis* attend clutches of 14–16 eggs [39], but unattended clutches of up to 19 eggs have also been found [14].

The new genus *Qosqophryne* is supported by our molecular phylogeny, the most complete to date covering three mitochondrial and two nuclear gene fragments, as well as most described species of *Bryophryne* and *Microkayla*. Despite the absence of known synapomorphies for the sister clades *Microkayla* and *Qosqophryne*, we are confident that our proposed arrangement reflects the evolutionary history of these organisms, and yet still takes into consideration taxonomic stability [48]. There is strong support (bootstrap probabilities) at the node where *Microkayla* and *Qosqophryne* diverge, and the relative branch lengths leading to their respective living species is similar, or in some cases exceed the branch lengths separating other genera within Terrarana (e.g., *Euparkerella* and *Holoaden*, or *Barycholos* and the "northern clade" of *Noblella*).

4. Discussion

Our study integrating molecular, acoustic and morphological information justifies the erection of the new genus of strabomantid frog *Qosqophryne*. The molecular phylogeny we inferred, the most complete phylogeny to date in terms of terminal sampling for genera of Holoadeninae [2,24], provides strong support for this new genus forming a sister clade to *Microkayla*. Furthermore, our phylogeny confirms taxonomic uncertainty regarding the genera *Noblella* and *Psychrophrynella* [2,19,20], suggests the presence of several undescribed species of *Noblella* and *Psychrophrynella*, and generalizes the idea of high species endemism in high elevation Andean strabomantids [2,4,19–22,49].

Morphological synapomorphies for the new genus *Qosqophryne* have not been recognized, and there does not appear to be a unique combination of meristic traits to distinguish all species of *Microkayla* from species of *Qosqophryne*. However, there are some characteristics that help distinguish the two genera. Some of the traits present in *Qosqophryne* but absent in *Microkayla* are fingers and toes with lateral fringes, venter smooth (areolate in *Microkayla*), and presence of dentigerous processes of vomers (but absent in *Q. flammiventris*). The structure of the advertisement call, when known, appears to be similar in both genera, i.e., a whistle-like call, but composed of a single note in *Microkayla* vs. 2–4 notes in *Qosqophryne* (except for *Q. gymnotis*). There is limited information on parental care, but it appears that females attend clutches in *Q. gymnotis* [39], whereas males attend clutches in *M. illimani* and *M. teqta* [50,51]. Similarly to *Qosqophryne*, females attend clutches in *B. cophites* [52], *B. hanssaueri* and *B. nubilosus* (Catenazzi, pers. obs.). However, we lack natural history information from most species of strabomantid frogs, and thus any generalization on parental care is premature.

In support of our proposed new genus, there is a wide gap, both in terms of airline distance and the highly dissected topography, in the distribution range of species of *Microkayla* and *Qosqophryne*. These are all highly endemic, terrestrial-breeding frogs most likely characterized by extreme low vagility, as suggested by their patchy distribution in cloud forests and grasslands. All species of *Microkayla* occur from extreme southern Peru (Department Puno) to the western limits of department Santa Cruz in central Bolivia (Serranía Siberia), whereas the three species of *Qosqophryne* occur in the Vilcabamba mountain range in the Peruvian Department of Cusco. The gap of 320 km by airline between the southernmost locality of *Qosqophryne* (*Q. gymnotis*; −13.07558, −72.38201) and the northernmost locality of *Microkayla* (*M. boettgeri*) overlaps with the distribution range of *Bryophryne*. At the northern limit, *B. abramalagae* and *B. bustamantei* are marginally sympatric with *Q. gymnotis*, whereas at the southern limit, *B. wilakunka* (Ayapata, Puno, −13.85294, −70.31450) occurs ~80 km NW of the type locality of *M. boettgeri* (Phara, Puno, −14.16247, −69.66250). Although many species in these genera of Holoadeninae are likely "micro-endemic", researchers have seldom invested much effort in documenting the distribution ranges of most species, and it is possible that some of these species occur more widely than presently known. Therefore, currently five genera of Holoadeninae occur in the tropical Andes south of the Apurimac canyon in Cusco, Puno and northern Bolivia: *Bryophryne*, *Psychrophrynella* and *Qosqophryne* in the Vilcabamba mountain range; *Bryophryne*, *Noblella* and *Psychrophrynella* in the Vilcanota range; *Bryophryne*, *Microkayala*, *Noblella* and *Psychrophrynella* in the Carabaya range, and *Microkayala* south of the Apolobamba range.

Author Contributions: Conceptualization, A.C., L.M., E.L. and R.v.M.; methodology, A.C., L.M., E.L. and R.v.M.; software, A.C. and R.v.M.; validation, A.C., L.M., E.L. and R.v.M.; formal analysis, A.C. and R.v.M.; investigation, A.C., L.M., E.L. and R.v.M.; resources, A.C. and R.v.M.; data curation, A.C., R.v.M.; writing—original draft preparation, A.C.; writing—review and editing, A.C., L.M., E.L. and R.v.M.; visualization, A.C., L.M., E.L. and R.v.M.; supervision, A.C.; project administration, A.C.; funding acquisition, A.C. All authors have read and agreed to the published version of the manuscript.

Funding: This research was funded by the Amazon Conservation Association, the Rufford Foundation, the Chicago Board of Trade Endangered Species Fund, and the Amphibian Specialist Group.

Acknowledgments: We thank the staff of the Museo de Biodiversidad del Perú (MUBI) for access to the herpetological collection.

Conflicts of Interest: The authors declare no conflict of interest. The funders had no role in the design of the study; in the collection, analyses, or interpretation of data; in the writing of the manuscript, or in the decision to publish the results.

Appendix A. Specimens Examined

Bryophryne abramalagae: PERU: CUSCO: Provincia La Convención: Distrito de Huayopata, Abra de Málaga (13°07'23.8" S, 72°20'51.2" W), 4000 m a.s.l., MUSM 27630–32, MTD 47489–91.

Bryophryne bakersfield: PERU: CUSCO: Provincia La Convención: Distrito de Echarate, Roquerío de Lorohuachana, 3620 m a.s.l. (12°29'43.8" S, 72°04'35.9" W), MHNC 7972.

Bryophryne bustamantei: PERU: CUSCO: Provincia La Convención: Abra de Málaga: MUSM 24537–38.

Bryophryne cophites: PERU: CUSCO: Provincia de Paucartambo: Distrito Kosñipata: S slope Abra Ac[j]anaco, 14 km NNE Paucartambo, 3400 m a.s.l.: KU 138884 (holotype); N slope Abra Ac[j]anaco, 27 km NNE Paucartambo, 3450 m a.s.l.: KU 138885–908, 138911–5 (all paratypes); 2 km NE of Abra Ac[j]anaco, 3280 m a.s.l.: MHNG 2698.24, 5.5 km N of Abra Acanacu [Acjanaco], 3523 m: MUSM 27895, Tres Cruces, 8.5 km N of Abra Ac[j]anaco, 3590 m a.s.l.: MUSM 20855–56, 26283–84, 26264, 26266–67, 26313, 26315, 27896, 30414–17, Pillco Grande, 3865 m a.s.l., near border of Manu NP: CORBIDI 11919.

Bryophryne flammiventris: PERU: CUSCO: Provincia de La Convención, Distrito de Vilcabamba, road between Vilcabamba and Pampaconas, 3800 m a.s.l.: MUSM 27613 (holotype), MUSM 27612, 27614–15, MTD 46890–92 (paratypes).

Bryophryne gymnotis: PERU: CUSCO: Provincia de La Convención, Distrito de Huayopata: 1 km east of San Luis, 3272–3354 m a.s.l.: MUSM 24543 (holotype), MHNG 2710.28, 2710.29, MTD 46860–64, 47288, 47291–92, 47297, MUSM 24541–42, 24544–45, 24546–56, MVZ 258407–10 (paratypes).

Bryophryne hanssaueri: PERU: CUSCO: Provincia de Paucartambo, Distrito de Kosñipata: Acjanaco, Manu National Park, 3266 m a.s.l.: MUSM 27567 (holotype); from near Acjanaco, Manu National Park, 3280–3430 m a.s.l.: MHNG 2698.25, MTD 46865–66, 46887–89, MUSM 24557, 27568–69, 27607–11, MVZ 258411–13 (all paratypes).

Bryophryne mancoinca: PERU: CUSCO: Provincia de La Convención, Hornopampa sector, near Salkantay Mountain, along the road to the Archeological Complex of Choquequirao, 3707 m a.s.l.: MUBI 11152 (holotype), MUBI 11147–11151, 11153, 11154, 11159, 16068, 16069, 16074, 16083 (paratypes).

Bryophryne nubilosus: PERU: CUSCO: Provincia de Paucartambo: Distrito de Kosñipata, 500 m NE of Esperanza, 2712 m a.s.l.: MUSM 26310 (holotype), MUSM 26311; near the type locality, 13°11'33.21" S, 71°35'25.17" W, 3065 m: MTD 47294; near Hito Pillahuata, 2600 m: MUSM 20970; Quebrada Toqoruyoc, 3097 m a.s.l.: MUSM 26312, MTD 47293; Esperanza, 2800 m: MHNSM 26316–17; 13°11'20.2" S, 71°35'07.3" W, 2900 m a.s.l.: MUSM 24539–40.

Bryophryne phuyuhampatu: PERU: CUSCO: Provincia de Paucartambo: Distrito de Paucartambo, Quispillomayo valley, Área de Conservación Privada (ACP) Ukumari Llaqta, 2795–2850 m a.s.l., 13°22'12.14" S; 71°6'49.82" W (WGS84; type locality), CORBIDI 18224–18226, MUBI 14654 and 14655.

Bryophryne quellokunka: PERU: CUSCO: Provincia de Quispicanchis: Distrito de Marcapata: Coline, 3672 m a.s.l.: MUSM 27571, 27573.

Bryophryne zonalis: PERU: CUSCO: Provincia de Quispicanchis, Distrito de Marcapata, Kusillochayoc at 3129 m a.s.l.: MUSM 27570 (holotype), MTD 46867, 46869–70, MUSM 27572, 27574–75, 27861, MVZ 258414 (paratypes); at Puente Coline, 3285 m a.s.l.: MVZ 258415 (paratype).

Microkayla boettgeri: PERU: PUNO: Provincia de Sandia, Distrito de Limbani, Phara, 3466 m a.s.l.: MHNSM 19966 (holotype), MHNSM 19967–76, MTD 46508–9, 46512–19 (paratypes).

Microkayla chapi: PERU: PUNO: Provincia de Sandia, Distrito de Limbani, 3.7 km from Sina, Hirigache River valley, 3466 m a.s.l.: MUBI 5326 (holotype), MUBI 5325, 5327, 5330, 5331, 5328, 5329 (paratypes).

Microkayla chilina: PERU: PUNO: Provincia de Sandia, Distrito de Limbani, 3.7 km from Sina, Hirigache River valley, 3466 m a.s.l.: MUBI 5355 (holotype), MUBI 5350, 5351, 5353, 5354 (paratypes).

Qosqophryne flammiventris: PERU: CUSCO: Provincia de La Convención, Distrito de Vilcabamba, road between Vilcabamba and Pampaconas, 3800 m a.s.l., MUBI 13365.

Qosqophryne gymnotis: PERU: CUSCO: Provincia de La Convención, Distrito de Huayopata: San Luis, MUBI 14315–14319.

References

1. Hedges, S.B.; Duellman, W.E.; Heinicke, M.P. New World direct-developing frogs (Anura: Terrarana): Molecular phylogeny, classification, biogeography, and conservation. *Zootaxa* **2008**, *1737*, 1–182. [CrossRef]
2. De La Riva, I.; Chaparro, J.C.; Castroviejo-Fisher, S.; Padial, J.M. Underestimated anuran radiations in the high Andes: Five new species and a new genus of Holoadeninae, and their phylogenetic relationships (Anura: Craugastoridae). *Zool. J. Linn. Soc.* **2018**, *182*, 129–172. [CrossRef]
3. von May, R.; Lehr, E.; Rabosky, D.L. Evolutionary radiation of earless frogs in the Andes: Molecular phylogenetics and habitat shifts in high-elevation terrestrial breeding frogs. *PeerJ* **2018**, *6*, e4313. [CrossRef] [PubMed]
4. De La Riva, I. Unexpected Beta-Diversity Radiations in Highland Clades of Andean Terraranae. In *Neotropical Diversification: Patterns and Processes*; Rull, V., Carnaval, A., Eds.; Springer: Cham, Switzerland, 2020.
5. Duellman, W.E.; Lehr, E. *Terrestrial-Breeding Frogs (Strabomantidae) in Peru*; Natur und Tier Verlag: Münster, Germany, 2009; p. 382.
6. Heinicke, M.P.; Duellman, W.E.; Trueb, L.; Means, D.B.; MacCulloch, R.D.; Hedges, S.B. A new frog family (Anura: Terrarana) from South America and an expanded direct-developing clade revealed by molecular phylogeny. *Zootaxa* **2009**, *2211*, 1–35. [CrossRef]
7. Heinicke, M.P.; Lemmon, A.R.; Lemmon, E.M.; McGrath, K.; Hedges, S.B. Phylogenomic support for evolutionary relationships of New World direct-developing frogs (Anura: Terraranae). *Mol. Phylogenet. Evol.* **2017**. [CrossRef]
8. Frost, D.R. Amphibian Species of the World: An Online Reference. Version 6.0. Available online: http://research.amnh.org/herpetology/amphibia/index.html (accessed on 23 October 2019).
9. Heinicke, M.P.; Duellman, W.E.; Hedges, S.B. Major Caribbean and Central American frog faunas originated by ancient oceanic dispersal. *Proc. Natl. Acad. Sci. USA* **2007**, *104*, 10092–10097. [CrossRef] [PubMed]
10. Lehr, E.; von May, R. A new species of terrestrial-breeding frog (Amphibia, Craugastoridae, *Pristimantis*) from high elevations of the Pui Pui Protected Forest in central Peru. *ZooKeys* **2017**, *660*, 17–42. [CrossRef]
11. Condori Ccarhuarupay, F.P. *Filogenia Morfológica del Género Bryophryne Hedges, 2008 (Anura: Craugastoridae)*; Universidad Nacional San Antonio Abad del Cusco: Cusco, Peru, 2018.
12. Chaparro, J.C.; Padial, J.M.; Gutierrez, R.C.; De la Riva, I. A new species of Andean frog of the genus *Bryophryne* from southern Peru (Anura: Craugastoridae) and its phylogenetic position, with notes on the diversity of the genus. *Zootaxa* **2015**, *3994*, 94–108. [CrossRef]
13. Lehr, E.; Catenazzi, A. A new species of *Bryophryne* (Anura: Strabomantidae) from southern Peru. *Zootaxa* **2008**, *1784*, 1–10. [CrossRef]
14. Lehr, E.; Catenazzi, A. Three new species of *Bryophryne* (Anura: Strabomantidae) from the Region of Cusco, Peru. *S. Am. J. Herpetol.* **2009**, *4*, 125–138. [CrossRef]
15. Lehr, E.; Catenazzi, A. Two new species of *Bryophryne* (Anura: Strabomantidae) from high elevations in southern Peru (Region of Cusco). *Herpetologica* **2010**, *66*, 308–319. [CrossRef]

16. Catenazzi, A.; Ttito, A.; Diaz, M.I.; Shepack, A. *Bryophryne phuyuhampatu* sp n. a new species of Cusco Andes frog from the cloud forest of the eastern slopes of the Peruvian Andes (Amphibia, Anura, Craugastoridae). *Zookeys* **2017**, *685*, 65–81. [CrossRef] [PubMed]
17. Mamani, L.; Catenazzi, A.; Ttito, A.; Mallqui, S.; Chaparro, J.C. A new species of *Bryophryne* (Anura: Strabomantidae) from the Cordillera de Vilcabamba, southeastern Peruvian Andes. *Phyllomedusa* **2017**, *16*, 129–141. [CrossRef]
18. Chaparro, J.C.; De la Riva, I.; Padial, J.M.; Ochoa, J.A.; Lehr, E. A new species of *Phrynopus* from Departamento Cusco, southern Peru (Anura: Brachycephalidae). *Zootaxa* **2007**, *1618*, 61–68. [CrossRef]
19. Catenazzi, A.; Ttito, A. A new species of *Psychrophrynella* (Amphibia, Anura, Craugastoridae) from the humid montane forests of Cusco, eastern slopes of the Peruvian Andes. *PeerJ* **2016**, *4*, e1807. [CrossRef]
20. Catenazzi, A.; Ttito, A. *Psychrophrynella glauca* sp. n. a new species of terrestrial-breeding frogs (Amphibia, Anura, Strabomantidae) from the montane forests of the Amazonian Andes of Puno, Peru. *PeerJ* **2018**, *6*, e444. [CrossRef]
21. Catenazzi, A.; Uscapi, V.; von May, R. A new species of *Noblella* from the humid montane forests of Cusco, Peru. *Zookeys* **2015**, *516*, 71–84. [CrossRef]
22. von May, R.; Catenazzi, A.; Corl, A.; Santa-Cruz, R.; Carnaval, A.C.; Moritz, C. Divergence of thermal physiological traits in terrestrial breeding frogs along a tropical elevational gradient. *Ecol. Evol.* **2017**, *7*, 3257–3267. [CrossRef]
23. Canedo, C.; Haddad, C.F. Phylogenetic relationships within anuran clade Terrarana, with emphasis on the placement of Brazilian Atlantic rainforest frogs genus *Ischnocnema* (Anura: Brachycephalidae). *Mol. Phylogenet. Evol.* **2012**, *65*, 610–620. [CrossRef]
24. Padial, J.M.; Grant, T.; Frost, D.R. Molecular systematics of terraranas (Anura: Brachycephaloidea) with an assessment of the effects of alignment and optimality criteria. *Zootaxa* **2014**, *3825*, 1–132. [CrossRef]
25. Motta, A.P.; Chaparro, J.C.; Pombal, J.P.; Guayasamin, J.M.; De la Riva, I.; Padial, J.M. Molecular phylogenetics and taxonomy of the Andean genus *Lynchius* Hedges, Duellman, and Heinicke 2008 (Anura: Craugastoridae). *Herpetol. Monogr.* **2016**, *30*, 119–142. [CrossRef]
26. Padial, J.M.; Chaparro, J.C.; Castroviejo-Fisher, S.; Guayasamin, J.M.; Lehr, E.; Delgado, A.J.; Vaira, M.; Teixeira, M., Jr.; Aguay, R.; de la Riva, I. A revision of species diversity in the Neotropical genus *Oreobates* (Anura: Strabomantidae), with the description of three new species from the Amazonian slopes of the Andes. *Am. Mus. Novit.* **2012**, *3752*, 1–55. [CrossRef]
27. Santa-Cruz, R.; von May, R.; Catenazzi, A.; Whitcher, C.; Tejeda, E.L.; Rabosky, D.L. A new species of terrestrial-breeding frog (Amphibia, Strabomantidae, *Noblella*) from the upper Madre de Dios watershed, Amazonian Andes and lowlands of southern Peru. *Diversity (Basel)* **2019**, *11*, 145. [CrossRef]
28. Lehr, E.; Fritzsch, G.; Müller, A. Analysis of Andes frogs (*Phrynopus*, Leptodactylidae, Anura) phylogeny based on 12S and 16S mitochondrial rDNA sequences. *Zool. Scr.* **2005**, *34*, 593–603. [CrossRef]
29. Padial, J.M.; Chaparro, J.C.; De La Riva, I. Systematics of *Oreobates* and the *Eleutherodactylus discoidalis* species group (Amphibia, Anura), based on two mitochondrial DNA genes and external morphology. *Zool. J. Linn. Soc.* **2008**, *152*, 737–773. [CrossRef]
30. Shepack, A.; von May, R.; Ttito, A.; Catenazzi, A. A new species of *Pristimantis* (Amphibia, Anura, Craugastoridae) from the foothills of the Andes in Manu National Park, southeastern Peru. *Zookeys* **2016**, *574*, 143–164.
31. von May, R.; Rabosky, D.L.; Lehr, E. Earless frogs in the Andes: Extraordinary ecological divergence and morphological diversity. *Nat. Hist.* **2018**, *126*, 12–15.
32. Palumbi, S.R.; Martin, A.; Romano, S.; McMillan, W.O.; Stice, L.; Grabawski, G. *The Simple Fool's Guide to PCR (Version 2.0)*; Privately published; Palumbi, S., Ed.; University of Hawaii: Honolulu, HI, USA, 1991.
33. Meyer, C.P. Molecular systematics of cowries (Gastropoda: Cypraeidae) and diversification patterns in the tropics. *Biol. J. Linn. Soc.* **2003**, *79*, 401–459. [CrossRef]
34. Bossuyt, F.; Milinkovitch, M.C. Convergent adaptive radiations in Madagascan and Asian ranid frogs reveal covariation between larval and adult traits. *Proc. Natl. Acad. Sci. USA* **2000**, *97*, 6585–6590. [CrossRef]
35. Lanfear, R.; Calcott, B.; Ho, S.; Guindon, S. PartitionFinder: Combined selection of partitioning schemes and substitution models for phylogenetic analyses. *Mol. Biol. Evol.* **2012**, *29*, 1695–1701. [CrossRef]
36. Ronquist, F.; Huelsenbeck, J. MrBayes 3: Bayesian phylogenetic inference under mixed models. *Bioinformatics* **2003**, *19*, 1572–1574. [CrossRef] [PubMed]

37. Rambaut, A.; Drummond, A. Tracer. Version 1.5. 2007. Available online: http://tree.bio.ed.ac.uk/software/tracer (accessed on 30 October 2019).

38. Rambaut, A. FigTree. Version 1.4.2. 2009. Available online: http://tree.bio.ed.ac.uk/software/figtree (accessed on 30 October 2019).

39. Mamani, L.; Diaz, M.I.; Ttito, J.W.; Condori, F.P.; Ttito, A. Parental care and altitudinal range extension of the endemic frog *Bryophryne gymnotis* (Anura: Craugastoridae) in the Andes of southeastern Peru. *Phyllomedusa* **2017**, *16*, 109–112. [CrossRef]

40. Catenazzi, A.; Ttito, A. *Noblella thiuni* sp. n. a new (singleton) species of minute terrestrial-breeding frog (Amphibia, Anura, Strabomantidae) from the montane forest of the Amazonian Andes of Puno, Peru. *PeerJ* **2019**, *7*, e6780. [CrossRef] [PubMed]

41. Reyes-Puig, J.P.; Reyes-Puig, C.; Ron, S.; Ortega, J.A.; Guayasamin, J.M.; Goodrum, M.; Recalde, F.; Vieira, J.J.; Koch, C.; Yanez-Munoz, M.H. A new species of terrestrial frog of the genus *Noblella* Barbour, 1930 (Amphibia: Strabomantidae) from the Llanganates-Sangay Ecological Corridor, Tungurahua, Ecuador. *PeerJ* **2019**, *7*, e7405. [CrossRef]

42. De la Riva, I.; Chaparro, J.C.; Padial, J.M. The taxonomic status of *Phyllonastes* Heyer and *Phrynopus peruvianus* (Noble) (Lissamphibia, Anura): Resurrection of *Noblella* Barbour. *Zootaxa* **2008**, *1685*, 67–68. [CrossRef]

43. Catenazzi, A.; von May, R. Conservation status of amphibians in Peru. *Herpetol. Monogr.* **2014**, *28*, 1–23. [CrossRef]

44. De La Riva, I.; Reichle, S. Diversity and conservation of the amphibians of Bolivia. *Herpetol. Monogr.* **2014**, *28*, 46–65. [CrossRef]

45. Catenazzi, A.; Lehr, E.; Rodriguez, L.O.; Vredenburg, V.T. *Batrachochytrium dendrobatidis* and the collapse of anuran species richness and abundance in the upper Manu National Park, southeastern Peru. *Conserv. Biol.* **2011**, *25*, 382–391. [CrossRef]

46. Catenazzi, A.; Lehr, E.; Vredenburg, V.T. Thermal physiology, disease and amphibian declines in the eastern slopes of the Andes. *Conserv. Biol.* **2014**, *28*, 509–517. [CrossRef]

47. Catenazzi, A.; Finkle, J.; Foreyt, E.; Wyman, L.; Swei, A.; Vredenburg, V.T. Epizootic to enzootic transition of a fungal disease in tropical Andean frogs: Are surviving species still susceptible? *PLoS ONE* **2017**, *12*, e0186478. [CrossRef]

48. ICZN. *International Code of Zoological Nomenclature*, 4th ed.; International Trust for Zoological Nomenclature: London, UK, 1999.

49. De la Riva, I. Bolivian frogs of the genus *Phrynopus*, with the description of twelve new species (Anura: Brachycephalidae). *Herpetol. Monogr.* **2007**, *21*, 241–277. [CrossRef]

50. Willaert, B.; Reichle, S.; Stegen, G.; Martel, A.; Barrón, S.; Sánchez de Lozada, N.; Greenhawk, N.; Agostini, G.; Muñoz, A. Distribution, ecology, and conservation of the critically endangered frog *Psychrophrynella illimani* (Anura: Craugastoridae) with description of its call. *Salamandra* **2016**, *52*, 317–327.

51. De la Riva, I.; Burrowes, P.A. A new species of *Psychrophrynella* (Anura: Craugastoridae) from the Cordillera Real, Department La Paz, Bolivia. *Zootaxa* **2014**, *3887*, 459–470. [CrossRef] [PubMed]

52. Catenazzi, A. *Phrynopus cophites*. Reproduction. *Herpetol. Rev.* **2006**, *37*, 206.

 *diversity*

Article

A New Species of Terrestrial-Breeding Frog (Amphibia, Strabomantidae, *Noblella*) from the Upper Madre De Dios Watershed, Amazonian Andes and Lowlands of Southern Peru

Roy Santa-Cruz [1],*, Rudolf von May [2],*, Alessandro Catenazzi [3], Courtney Whitcher [2], Evaristo López Tejeda [1] and Daniel L. Rabosky [2]

1   Museo de Historia Natural (MUSA), Universidad Nacional de San Agustín de Arequipa, Área de Herpetología, Av. Alcides Carrión s/n, Arequipa 04001, Peru
2   Museum of Zoology, Department of Ecology and Evolutionary Biology, University of Michigan, 2020 Biological Sciences Building, 1105 North University Ave., Ann Arbor, MI 48109, USA
3   Department of Biological Sciences, Florida International University, Miami, FL 33199, USA
*   Correspondence: chara53@hotmail.com (R.S.-C.); rvonmay@gmail.com (R.v.M.)

http://zoobank.org/urn:lsid:zoobank.org:pub:4A04AFAD-E934-43E5-8F34-8F323A303767

Received: 13 July 2019; Accepted: 18 August 2019; Published: 26 August 2019

Abstract: We describe and name a new species of *Noblella* Barbour, 1930 (Strabomantidae) from southern Peru. Key diagnostic characteristics of the new species include the presence of a short, oblique fold-like tubercle on the ventral part of the tarsal region, two phalanges on finger IV, and an evident tympanum. The elevational distribution of the new species spans 1250 m (240–1490 m) from lowland Amazon rainforest to montane forest on the eastern slopes of the Andes.

Keywords: amphibians; ecomorphology; miniaturization; systematics; taxonomy

---

1. Introduction

The terrestrial-breeding frog genus *Noblella* Barbour, 1930 [1] (Strabomantidae) is distributed in the Andes–Amazon region of Ecuador, Peru, Brazil, and Bolivia and currently includes 14 species: *N. carrascoicola* (De la Riva and Köhler, 1998) [2], *N. coloma* Guayasamin and Terán-Valdez, 2009 [3], *N. duellmani* (Lehr, Aguilar, and Lundberg, 2004) [4], *N. heyeri* (Lynch, 1986) [5], *N. lochites* (Lynch, 1976) [6], *N. lynchi* (Duellman, 1991) [7], *N. madreselva* Catenazzi, Uscapi, and von May, 2015 [8], *N. myrmecoides* (Lynch, 1976) [6], *N. naturetrekii* Reyes-Puig et al. [9], *N. personina* Harvey, Almendáriz, Brito-M., and Batallas-R., 2013 [10], *N. peruviana* (Noble, 1921) [11], *N. pygmaea* Lehr and Catenazzi, 2009 [12], *N. ritarasquinae* (Köhler, 2000) [13], and *N. thiuni* Catenazzi and Ttito, 2019 [14]. Most of these species inhabit montane forests above 1000 m and are morphologically very similar to those in the genus *Psychrophrynella* Hedges, Duellman, and Heinicke 2008 [15]. Recent analyses indicate that there is uncertainty regarding the relationships among species of *Noblella* and *Psychrophrynella* [14,16]. Hedges et al. [15] assigned *N. peruviana* and *P. bagrecito*, respectively, as type species of the two genera. However, the lack of DNA sequences for both *N. peruviana* and *P. bagrecito* has prevented researchers from resolving the phylogenetic relationships between *Noblella* and *Psychrophrynella*. Additionally, recent studies have inferred the non-monophyly of the genus *Noblella* [14] and researchers identified a "northern clade" and a "southern clade" containing species ascribed to *Noblella* [9,14]. This taxonomic issue will only be properly resolved when sequences from *N. peruviana* become available. In the meantime, the description of new species will continue advancing our knowledge of the diversity of these small terrestrial-breeding frogs. Here, we describe and name a new species of *Noblella* on the basis of specimens collected in the lowland Amazon forest and the montane forest of the Amazonian Andes in southern Peru.

2. Materials and Methods

2.1. Fieldwork and Data Collection

We conducted fieldwork at Los Amigos Biological Station (12°34′07″ S, 70°05′57″ W, 250 m a.s.l.), located in the Madre de Dios region, Peru, and at various sites along the Kosñipata Valley, located in the Cusco region, Peru [17]. Specimens were euthanized by immersion in benzocaine hydrochloride solution (250 mg/L), where animals were kept for 10 to 20 min, until movement ceased, or by application of 20% benzocaine paste to the ventral region. After euthanasia, tissue samples (e.g., liver, muscle) were taken from the animals and preserved in 2 mL cryogenic tubes filled with RNAlater or 95% ethanol. Following tissue collection, specimens were fixed in 10% formalin, and permanently stored in 70% ethanol, except for specimens collected in 2018, which were fixed in 95% ethanol and stored in 70% ethanol. Sex and maturity of specimens were determined by observing sexual characters and gonads through dissections. Photographs were taken by R. von May, R. Santa Cruz, C. Whitcher, and A. Catenazzi, and were used for descriptions of coloration in life. We were unable to record calls of the new species.

Use of vertebrate animals was approved by the Animal Care and Use committees of the University of California (ACUC #R278-0412, R278-0413, and R278-0314), the University of Michigan (PRO00008306), Florida International University (IACUC #18-009), and Southern Illinois University (IACUC protocol #16-006).

2.2. Morphological Characters

We followed Duellman and Lehr [18] and Lynch and Duellman [19] for formats of diagnosis and description, except for using the term "dentigerous processes of vomers" instead of "vomerine odontophores" [20]. For taxonomy we follow Padial et al. [21] and Heinicke et al. [22]. We measured the following variables to the nearest 0.1 mm with digital calipers under a stereomicroscope: snout-vent length (SVL), tibia length (TL), foot length (FL, distance from proximal margin of inner metatarsal tubercle to tip of Toe IV), head length (HL, from angle of jaw to tip of snout), head width (HW, at level of angle of jaw), eye diameter (ED), tympanum diameter (TY), interorbital distance (IOD), upper eyelid width (EW), internarial distance (IND), eye–nostril distance (E–N, straight line distance between anterior corner of orbit and posterior margin of external nares), eye to tympanum distance (E-TY), forearm length (ForL), hand length (HaL), finger I length (FIL), finger II length (FIIL), toe I length (TIL), and toe II length (TIIL). Fingers and toes are numbered preaxially to postaxially from I–IV and I–V, respectively. We determined comparative lengths of toes III and V by adpressing both toes against toe IV; lengths of fingers I and II were determined by adpressing the fingers against each other. We compared the new taxon with all described species. Specimens examined are listed in Appendix A; codes of collections are: CORBIDI = Centro de Ornitología y Biodiversidad, Lima, Peru; MUBI = Museo de Biodiversidad del Peru, Cusco, Peru; MUSM = Museo de Historia Natural Universidad Nacional Mayor de San Marcos, Lima, Peru; MUSA = Museo de Historia Natural de la Universidad Nacional de San Agustín, Arequipa, Peru; UMMZ = University of Michigan Museum of Zoology, Ann Arbor, MI, USA.

Additionally, we used the morphological data to examine if body size and body shape vary across elevations. We used principal components analysis (PCA) to examine morphological variation between males and females, and between low (<300 m) and high elevation (1200–1490 m) populations. To remove the possible confounding effect of body size, we performed a body size-correction in which all variables were divided by SVL. We used generalized least-squares (GLS) regression to examine the relationship between SVL and elevation, and fitted a regression line separately for males and females. Subsequently, we calculated PCA using the corrected morphological data. We projected the first two

PC axes on a morphospace using the function princomp in R and displayed differences between male and females taking into account elevation.

2.3. Micro-Computed Tomography

We obtained X-ray micro-computed tomography (µCT) images from two specimens to examine the skull and skeleton of the new species. We scanned two voucher specimens (paratypes) stored in 70% ethanol in the Micro-CT Core facility at the University of Michigan. We placed these specimens inside a glass vial, which in turn was placed in a 34 mm diameter specimen holder, prior to scanning. We used a microCT system, µCT100 Scanco Medical (Bassersdorf, Switzerland), and scan settings were as follows: voxel size 11.4 µm, 55 kVp, 145 µA, 0.5 mm AL filter, 1000 projections around 180°, integration time of 1000 ms, and average data of 3 replicates. We used Scanco's proprietary software to export data to DICOM files. We used the Amira-Avizo software to obtain three-dimensional renderings based on isosurface representations.

The main purpose of the CT-scans was to examine the condition of the tympanic middle ear (inspect if columella was present or not) and the number and shape of phalanges. Both of these characters are key for the diagnosis of species of *Noblella* and *Psychrophrynella*. Thus, images presented here will facilitate osteological comparisons with other species once additional CT-scan data become available.

2.4. Molecular Phylogenetic Analysis

We determined the phylogenetic position of the new species with respect to other closely related taxa through analysis of DNA sequences. Our analysis included sequences obtained from tissue samples collected from type specimens (holotype and paratypes) as well as legacy data from GenBank (Table S1). Sequence data included a fragment of the 16S rRNA gene (16S), a fragment of the 12S rRNA gene (12S), the protein-coding gene cytochrome c oxidase subunit I (COI), the nuclear protein-coding gene recombination-activating protein 1 (RAG1), and the tyrosinase precursor (Tyr). We used *Phrynopus peruanus* to root the tree. We followed previously reported procedures [14,23] for lab work and sequencing, and we deposited new sequences in GenBank (Table S1).

We aligned the sequences with Geneious R6, v. 6.1.8 [24], using the built-in Geneious Aligner program. Subsequently, we used PartitionFinder, v. 1.1.1 [25] to select the appropriate models of nucleotide evolution. We determined the best partitioning scheme and substitution model for each gene using the Bayesian information criterion (BIC). The best partitioning scheme included six subsets (best fitting substitution model in parentheses) as follows. A first partition subset including both 12S and 16S sequences (GTR + I + G). For COI, the best partitioning scheme included three sets of sites (substitution models in parentheses): the first set with first codon position (K80 + G), the second set with second codon position (F81 + I), and the third set with the third codon position (HKY + G). The next subset included the first and second codon positions of RAG together with the first and third codon positions of Tyr (HKY + G). The last subset included the third codon position of RAG together with the second codon position of Tyr (K80 + G).

We used MrBayes, v. 3.2.0 [26] to infer a molecular phylogeny. Our analysis included 36 terminals and a 2571 bp concatenated partitioned dataset (16S, 12S, COI, RAG1, Tyr). We performed an MCMC Bayesian analysis that included two simultaneous runs of 10 million generations, sampled once every 1000 generations. Each run had one "cold" chain and three heated chains, and the burn-in was set to discard 25% samples from the cold chain. The average standard deviation of split frequencies at the end of the runs was 0.001661. Subsequently, we used Tracer version 1.5 [27] to examine the effective sample sizes (ESS), to verify convergence, and to verify that the runs reached stationarity. The observed effective sample sizes were sufficient for all parameters (ESS > 200). Lastly, we used FigTree v. 1.4.2 [28] to visualize the majority-rule consensus tree and assess node support (based on posterior probability values).

2.5. Registration of New Nomenclatural Acts

According to the International Commission on Zoological Nomenclature (ICZN), which produces the International Code of Zoological Nomenclature, the electronic publication of this article in portable document format (PDF) represents a published work. Therefore, the new species name contained in the PDF is effectively published under the International Code of Zoological Nomenclature from the electronic edition alone. This publication and the nomenclatural acts contained in it have been registered in ZooBank, the online official register for the ICZN. The ZooBank Life Science Identifiers (LSIDs) can be accessed and viewed through standard web browsers by appending the LSID to the prefix http://zoobank.org/. The online version of this work is archived and available from the following digital repositories: *Diversity*, CLOCKSS, and e-Helvetica.

3. Results

3.1. Molecular Phylogenetic Analysis

We recovered a phylogenetic tree (Figure 1) that was congruent with previous analyses [16,21], and supported a unique history of divergence of the new species. There was strong support for placing the new species in a clade containing species of *Noblella* distributed in southern Peru as well as several species of *Psychrophrynella*. According to this analysis, the new species is most closely related to *Noblella pygmaea* and to an undescribed species (*Noblella* sp. R in [23]).

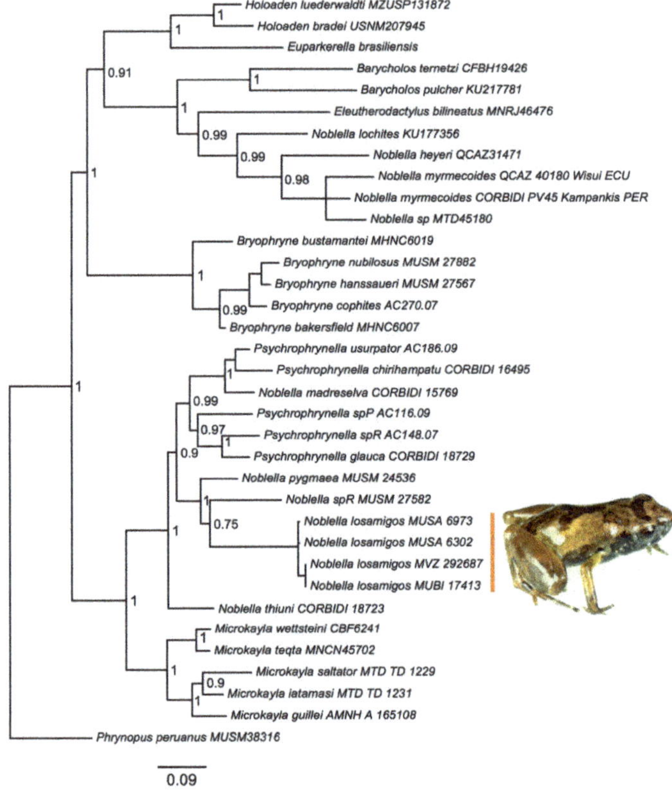

Figure 1. Phylogeny. Bayesian maximum clade-credibility tree for species included in this study based on a 2571 bp concatenated partitioned dataset analyzed in MrBayes (posterior probabilities indicated at each node).

3.2. Taxonomy

***Noblella losamigos* sp. n.** MUSA 6973 (Field number: RvM 3.12)

http://zoobank.org/urn:lsid:zoobank.org:act:EFAADCA4-4649-49AC-869E-4EAF5DB7E905

Phyllonastes myrmecoides Heyer 1977—Rodríguez and Cadle 1990 [29]: p. 413, Table 22.1; Rodríguez 1992 [30]: p. 162, 172, 174, Table I, Table V; Morales and McDiarmid 1996 [31]: p. 511, Table 2; Doan and Arizabal 2002 [32]: p. 114, Appendix 1

Noblella myrmecoides (Lynch 1976)—von May et al. 2009 [33]: p. 18, Table 1; von May et al. 2010a [34]: p. 513, 519, Figure 3, Appendix 1; von May et al. 2010 [35]: p. 10, Figures 193–194 Catenazzi et al. 2013 [17]: p. 274, 280, Table 1; von May et al. 2017 [23]: p. 3261, 3262, Figures 2 and 3, Supplementary Tables S2–S3, Supplementary Figures S2–S4; Whitworth et al. 2016 [36]: Appendices D–E; Villacampa et al. 2017 [37]: p. 4, 58; von May et al. 2018 [38]: Appendix A, Table A1, p. 8, 10–11; von May et al. 2019 [39]: Figures 1 and 2, Figures S1 and S3, Supplementary Information Tables 1 and 2.

Noblella* cf. *myrmecoides (Lynch 1976)—Catenazzi et al. 2013 [17]: p. 274, Table 1.

***Noblella* sp. SP**—Catenazzi and Ttito 2019 [14]: p. 6, 8, Table 1, Figure 3, Appendix 2.

3.2.1. Holotype

MUSA 6973 (Field number: RvM 3.12) (Figure 2), an adult female from 12°34′13.84″ S, 70°04′54.66″ W (Datum WGS 84), Los Amigos Biological Station, 245 m a.s.l., Manu District, Manu Province, Madre de Dios Region, Peru, collected by R. von May and R. Santa Cruz on 16 January 2012.

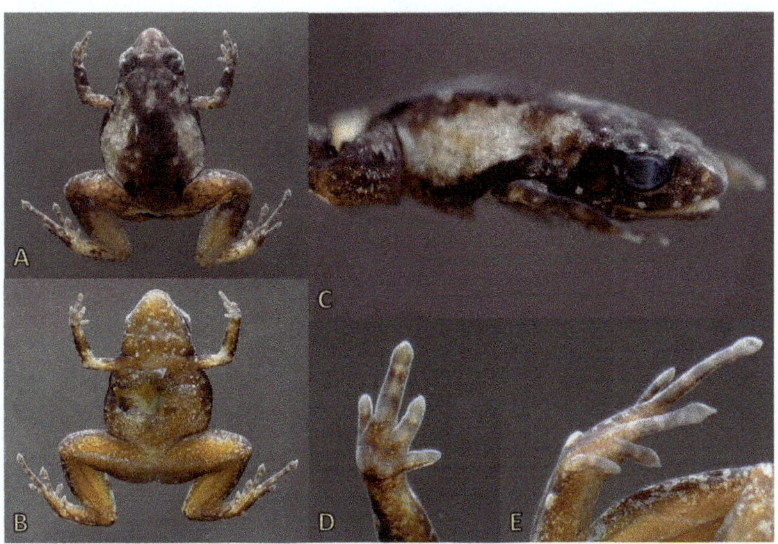

Figure 2. Photographs of preserved specimen of *Noblella losamigos* sp. n.: Adult female holotype MUSA 6973 (SVL 10.7 mm): (**A**) Dorsal view, (**B**) ventral view, (**C**) lateral view, (**D**) right hand, and (**E**) right toe. Photographs by Rudolf von May.

3.2.2. Paratypes

A total of 27 specimens: Three adult males (MUSM 37355, MUSM 37357, UMMZ 246569) and an adult female (MUSM 37356) (Figures 3 and 4); from Los Amigos Biological Station, collected by R. von May and R. Santa Cruz on 26 November 2016; an adult female (UMMZ 246570) from Los Amigos Biological Station, collected by R. von May and R. Santa Cruz on 29 November 2016; three adults (sex unknown) (UMMZ 244945, MUSM 33247, MUSA 6302) from Los Amigos Biological Station, collected by R. von May and R. Santa Cruz on 14–19 May 2014; three specimens male/male/adult

(MUSA 6974, MUSM 24219, MUSM 24251) from Los Amigos Biological Station, collected with the holotype by R. von May and R. Santa Cruz on 16 January 2012. Seven males: MUBI 17412, CORBIDI 17521, from San Pedro, Kosñipata Valley (13.04514S, 71.52922W, 1274 m), collected on 7 April 2018 by A. Catenazzi and M. I. Diaz; CORBIDI 17520, from San Pedro, Kosñipata Valley (13.04833S, 71.53178W, 1341 m), collected on 7 April 2018 by A. Catenazzi and M. I. Diaz; CORBIDI 17524 from San Pedro, Kosñipata Valley (13.03724S, 71.52798W, 1200 m), collected on 26 January 2009 by A. Catenazzi; MVZ:Herp:292685, MVZ:Herp:292686 from San Pedro, Kosñipata Valley (13.0541S, 71.54632W, 1370 m), collected on 24 January 2009 by A. Catenazzi; MVZ:Herp:292687, from San Pedro, Kosñipata Valley (13.05093S, 71.53711W, 1375 m), collected on 28 January 2009 by A. Catenazzi. One juvenile: MUBI 17413, from San Pedro, Kosñipata Valley (13.04514S, 71.52922W, 1274 m), collected on 7 April 2018 by A. Catenazzi and M. I. Diaz. Eight females: MUSM 30429, from San Pedro, Kosñipata Valley (13.04335S, 71.53027W, 1234 m), collected on 26 January 2009 by A. Catenazzi; CORBIDI 17523, from San Pedro, Kosñipata Valley (13.05044S, 71.53373W, 1342 m), collected on 26 January 2009 by A. Catenazzi; MUSM 27578, from San Pedro, Kosñipata Valley (13.05686S, 71.54081W, 1369 m), collected on 2 February 2008 by A. Catenazzi; MUSM 30426, MUSM 30427, MUSM 30428, MVZ:Herp:292684, from San Pedro, Kosñipata Valley (13.05193S, 71.5376W, 1376 m), collected on 24 January 2009 by A. Catenazzi; CORBIDI 17522, from San Pedro, Kosñipata Valley (13.05704S, 71.54778W, 1412 m), collected on 7 February 2008 by A. Catenazzi. Dorsal and ventral views of five paratypes (MUSA 6974, MUSM-37355, UMMZ 246569, MUSM 37356, UMMZ 246570), as well as the holotype, are presented in Figure 5. Three-dimensional reconstructions based on µCT data, from the skull and skeleton of two paratypes (MUSM 37355, MUSM 37356) are presented in Figure 6.

Figure 3. Photographs of live specimen of *Noblella losamigos* sp. n. (**A–C**) Adult male paratype MUSM 37355 (SVL 9.5 mm). Photographs by Rudolf von May.

Figure 4. Photographs of live specimens of *Noblella losamigos* sp. n. (**A,B**) Adult male paratype MUSM 37357 (SVL 10.3 mm). (**C,D**) Adult female paratype MUSM 37356 (SVL 9.7 mm). Photographs by Roy Santa-Cruz.

Figure 5. Photographs of preserved specimens of *Noblella losamigos* sp. n. (**A,B**) Adult female holotype MUSA 6973 (SVL 10.7 mm); (**C,D**) adult male paratype MUSA 6974 (SVL 9.8 mm); (**E,F**) adult male paratype MUSM 37355 (SVL 9.5 mm); (**G,H**) adult male paratype UMMZ 246569 (SVL 9.6 mm); (**I,J**): adult female paratype MUSM 37356 (SVL 9.7 mm); and (**K,L**) adult female paratype UMMZ 246570 (SVL 11.3 mm). Photographs by Rudolf von May, Roy Santa-Cruz, and Courtney Whitcher.

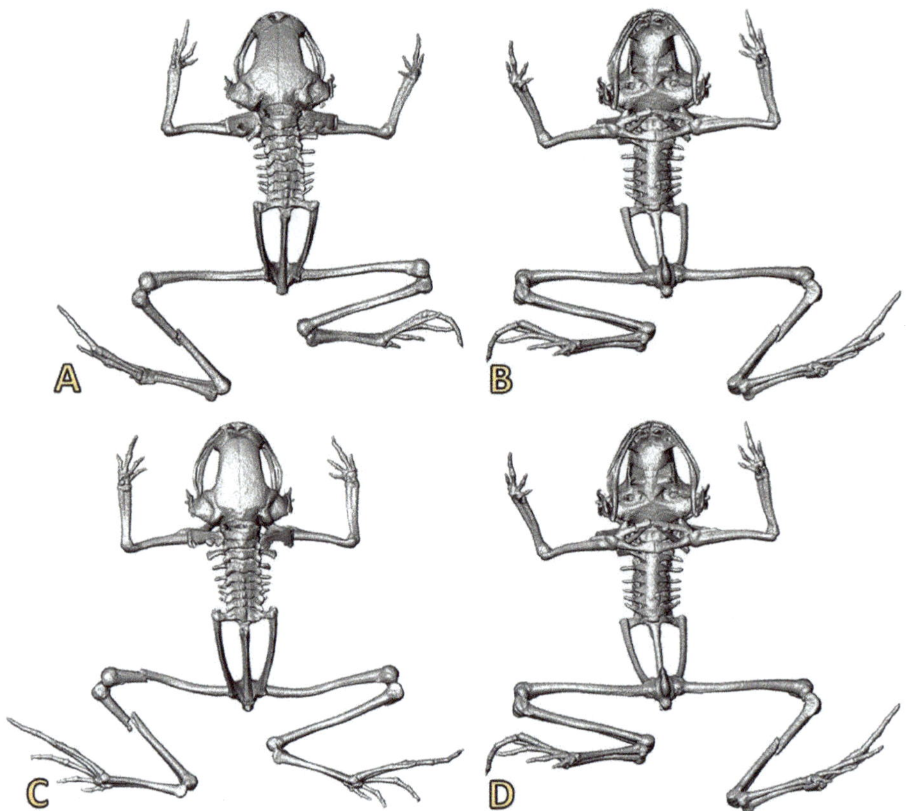

Figure 6. Three-dimensional reconstructions based on μCT data, from the skull and skeleton of two specimens of *Noblella losamigos* sp. n. (**A,B**) Dorsal and ventral views of specimen MUSM 37355 (SVL 9.6 mm). (**C,D**) Dorsal and ventral views of specimen MUSM 37356 (SVL 9.7 mm).

3.2.3. Generic Placement

We assign this species to the genus *Noblella* based on its overall morphological resemblance with other species of *Noblella*, including the presence of T-shaped and pointed terminal phalanges (especially in toes). The genus *Noblella* Barbour, 1930 can be recognized by the following characters [3,7,15,40,41]: tympanic membrane differentiated (except in *N. duellmani* and *N. madreselva*); head narrower than body; cranial crests absent; dentigerous processes of vomers absent; finger I shorter than, or equal in length to, finger II; toe III shorter than toe V; tips of at least toes III–IV pointed; subarticular tubercles not protruding; conspicuous tarsal tubercle; dark inguinal spots present (except in *N. duellmani*); small body size (SVL < 22 mm). Additionally, our phylogenetic analysis indicates that the new species is closely related to other species of *Noblella* distributed in southern Peru (Figure 1), but also to several species of *Psychrophrynella*. However, given that *N. peruviana* and *P. bagrecito* (the type species of *Noblella* and *Psychrophrynella*, respectively) have not been included in phylogenetic analyses, the assignment remains tentative.

3.2.4. Diagnosis

A new species of *Noblella* characterized by (1) skin on dorsum smooth to finely shagreen, skin on belly smooth, discoidal fold absent, dorsolateral folds absent; (2) tympanic annulus visible below skin, with the upper portion (1/4) covered by a supratympanic fold; tympanic membrane evident; columella present (Figure 6); (3) snout short, rounded in dorsal view and bluntly rounded to subtruncate in profile; (4) upper eyelid with minute tubercles, cranial crests absent; (5) dentigerous process of vomers absent; (6) vocal slits present; nuptial pads absent; (7) finger I shorter than finger II; tips of digits rounded, distally ending in papillae; Finger IV having two phalanges (Figures 2 and 7); (8) fingers with narrow lateral fringes; (9) ulnar tubercles absent; (10) short, oblique fold-like tubercle on the ventral part of tarsal region (Figure 2E); (11) no other tubercles on heel and tarsus; (12) inner metatarsal tubercle oval, of higher relief and about one and a half times the size of conical, rounded outer metatarsal tubercle; supernumerary plantar tubercles absent; (13) toes bearing narrow lateral fringes; webbing absent; toe V shorter than toe III; tips of digits weakly acuminate distally and expanded slightly in the digits II, III and IV; tips of the digits II, III, IV and V with discs slightly expanded, elongately acuminate, grooves present distally with papillae; (14) facial mask and lateral band dark brown with cream spots interrupted, extending from tip of snout along the flanks, almost reaching the point of insertion of thighs; (15) dorsum ocher gold to copper brown with or without irregularly-shaped middorsal dark brown marks; some specimens present a clear and slightly evident middorsal line that extends from middle of body to cloaca; dark brown suprainguinal stripes; interorbital bar present or absent; black or gray clear venter, always with irregular white markings; irregular white markings also present on the neck, thighs, and toes; (16) mean SVL 12.18 mm in females (range 9.74–13.60, $n = 9$), 10.08 mm in males (range 9.16–11.40, $n = 12$). Mean values and ranges of other morphological characters are provided in Table 1.

Table 1. Measurements (in mm) of males and females of the type series of *Noblella losamigos* sp. n. See section on morphological characters for definition of each character. Abbreviations: SD = standard deviation, SVL = snout-vent length, TL = tibia length, FL = foot length, HL = head length, HW = head width, ED = eye diameter, TY = tympanum diameter, IOD = interorbital distance, EW = upper eyelid width, IND = internarial distance, E–N = and eye–nostril distance, E-TY = eye to tympanum distance, ForL = forearm length, HaL = hand length, FIL = finger I length, FIIL = finger II length, TIL = toe I length, TIIL = toe II length. Ranges are included in parentheses.

	Females (n = 9)	Range	Males (n = 12)	Range
Character	Mean ± SD	(min–max)	Mean ± SD	(min–max)
SVL	12.18 ± 1.36	(9.74–13.60)	10.08 ± 0.70	(9.16–11.40)
TL	6.16 ± 0.74	(4.95–7.10)	5.34 ± 0.60	(4.50–6.20)
FL	5.44 ± 0.92	(4.52–7.00)	4.56 ± 0.65	(3.68–5.80)
HL	3.56 ± 0.53	(2.55–4.00)	2.97 ± 0.30	(2.44–3.40)
HW	4.09 ± 0.34	(3.55–4.40)	3.62 ± 0.38	(3.19–4.30)
IOD	1.74 ± 0.26	(1.29–2.10)	1.49 ± 0.18	(1.18–1.75)
EW	1.01 ± 0.15	(0.80–1.20)	0.83 ± 0.14	(0.57–1.00)
IND	1.38 ± 0.27	(1.03–1.78)	1.14 ± 0.16	(0.87–1.43)
E–N	0.80 ± 0.07	(0.67–0.93)	0.66 ± 0.10	(0.47–0.80)
S–N	0.34 ± 0.14	(0.05–0.50)	0.24 ± 0.10	(0.05–0.33)
ED	1.55 ± 0.18	(1.26–1.80)	1.33 ± 0.19	(1.02–1.65)
TY	0.56 ± 0.05	(0.45–0.63)	0.52 ± 0.06	(0.42–0.66)
E–TY	0.38 ± 0.12	(0.17–0.50)	0.30 ± 0.10	(0.13–0.41)
ForL	2.95 ± 0.36	(2.35–3.41)	2.60 ± 0.29	(2.16–2.94)
HaL	2.37 ± 0.37	(1.80–2.80)	1.98 ± 0.42	(1.35–2.48)
FIL	0.93 ± 0.13	(0.76–1.16)	0.73 ± 0.07	(0.60–0.80)
FIIL	1.19 ± 0.16	(1.03–1.53)	0.97 ± 0.11	(0.80–1.20)
TIL	1.24 ± 0.20	(0.94–1.43)	0.88 ± 0.09	(0.73–0.98)
TIIL	1.88 ± 0.16	(1.63–2.20)	1.48 ± 0.26	(1.05–1.90)

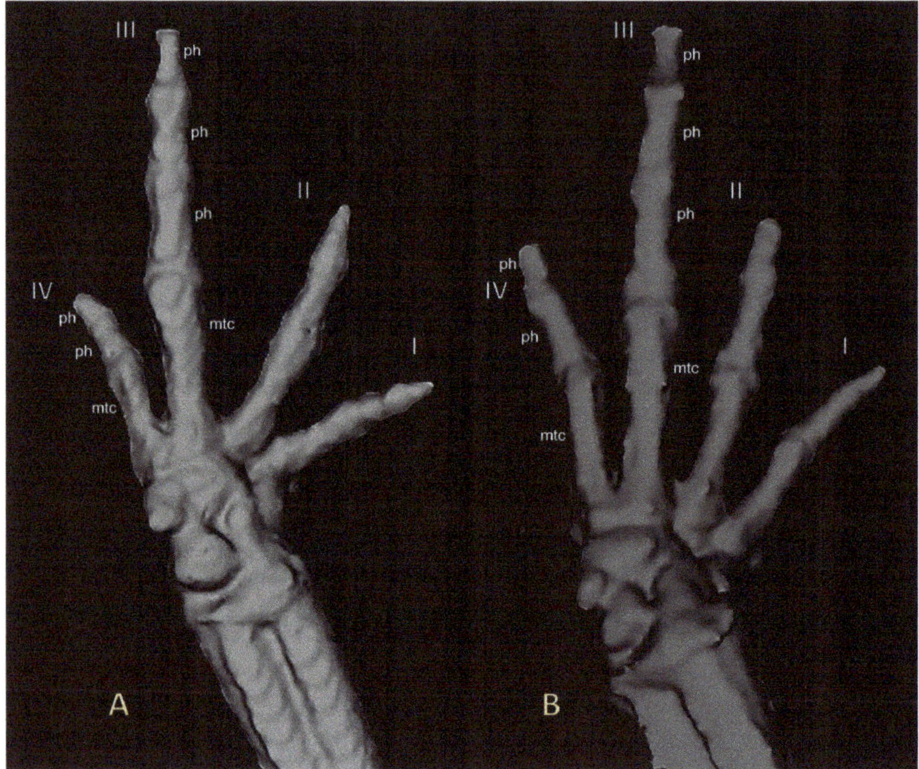

Figure 7. Three-dimensional reconstructions based on µCT data, from (**A**) adult right hand paratype UMMZ 246570 of *Noblella losamigos* sp. n. and (**B**) adult right hand *Noblella myrmecoides* from Loreto UMMZ 246571. Finger phalanges (ph) and metacarpalia (mtc) are noted for fingers III and IV.

3.2.5. Comparisons with Described Species

Noblella losamigos sp. n. has two phalanges in Finger IV like *N. carrascoicola*, *N. lochites*, *N. myrmecoides*, and *N. ritarasquinae*. In contrast, *Noblella losamigos* differs from the following species, which have three phalanges in Finger IV: *N. coloma*, *N. duellmani*, *N. heyeri*, *N. lynchi*, *N. madreselva*, *N. pearsonina*, *N. peruviana*, *N. pygmaea* and *N. thiuni*. Externally, *N. losamigos* has an evident tympanum (absent in *N. carrascoicola*, *N. duellmani* and *N. ritarasquinae*; barely visible below skin in *N. madreselva*, *N. thiuni*, *P. bagrecito*, *P. chirihampatu*, *P. glauca and P. usurpator*). *N. losamigos* sp. n. is similar to *N. myrmecoides* and *N. ritarasquinae* by presence of papillae on the tips of the toes [6,12]. *N. losamigos* sp. n. has a short, oblique fold-like tubercle on the ventral part of tarsal region, similar to *N. ritarasquinae* and *P. glauca* (*N. peruviana* and *N. heyeri* have a prominent tubercle; *N. lochites* and *N. myrmecoides* have a tubercle transversely oriented; *N. coloma*, *N. lynchi*, *N. personina*, *N. thiuni*, *P. chirihampatu*, *P. usurpator* have a elongate tarsal tubercle; *N. carrascoicola* with a poorly marked tubercle; *P. bagrecito* have a smaller and sickle-shaped tubercle; *N. duellmani*, *N. madreselva* and *N. pygmaea* do not have a tubercle or tarsal fold. *N. losamigos* sp. n. is similar to *N. madreselva*, *N. myrmecoides*, *N. thiuni*, *P. chirihampatu*, *P. glauca* in that it has a heel lacking tubercles (*N. pygmaea* present a heel with one minute, round tubercle). The skin on the dorsum of *N. losamigos* is smooth to finely shagreen, similar to *N. carrascoicola*, *N. coloma*, *N. heyeri*, *N. lochites*, *N. myrmecoides*, *N. thiuni*, *P. bagrecito*, *P. chirihampatu*, *P. glauca* and *P. usurpator* (skin on the dorsum with small tubercles or pustules in *N. duellmani*, *N. madreselva*, *N. lynchi*, *N. personina* and *N. pygmaea*). *N. losamigos* presents suprainguinal spots similar to *N. carrascoicola*, *N. coloma*, *N. heyeri*,

N. lynchi, *N. myrmecoides*, *N. ritarasquinae*, *N. thiuni* (diffuse suprainguinal stripes in *N. madreselva* or poorly defined in *N. personina*, longitudinal stripes in *P. chirihampatu and P. glauca*, absent in *N. duellmani*, *N. pygmaea*, and *P. usurpator*). *N. losamigos* has facial mask and lateral band of dark brown with creams spots interrupted and extending from the tip of the snout along the flanks, almost reaching the point of insertion of thighs (*N. duellmani* has a narrow dark brown post orbital stripe; *N. pygmaea* has a broad gray dorsolateral stripe that extends from upper eyelid to insertion of thigh; *N. personina* has a facial mask but lack a lateral dark band extending to the inguinal region [4,9,11]. The new species is larger in SVL (largest known female 13.60 mm, largest known male 11.40 mm) than *N. pygmaea* (largest known female 12.4 mm); it is similar in size to *N. myrmecoides* (largest known female 13.6 mm); and it is smaller than *N. coloma* (largest known female 16.03 mm, largest known male 14.55 mm), *N. duellmani* (largest known female 20.00 mm), *N. heyeri* (largest known female 15.90 mm, largest known male 14.10 mm), *N. lynchi* (largest known female 20.20 mm), *N. lochites* (largest known female 19.4 mm), *N. madreselva* (largest known female 17.6 mm, largest known male 15.6 mm), *N. personina* (largest known female 17.90 mm, largest known male 16.30 mm), *P. bagrecito* (largest known female 18.60 mm, largest known male 16.30 mm), *P. glauca* (largest known female 19.80 mm), *P. chirihampatu* (largest known female 25.80 mm, largest known male 21.70 mm), *P. usurpator* (largest known female 24.1 mm, largest known male 20.3 mm).

3.2.6. Description of Holotype

Adult female (10.7 mm SVL); head narrower than body; head length 29% of SVL; head slightly wider than longer; head width 33% of SVL; snout short, rounded in dorsal view, subtruncate in lateral view (Figure 2); eye large, 46% of head length, its diameter 1.8 times as large as its distance from the nostril; nostrils not protuberant, situated close to snout; canthus rostralis slightly curved in dorsal view, rounded in profile; lores flat; lips rounded; dorsal surface of head and upper eyelids with small tubercles; upper eyelid width 62% of inter-orbital distance; supratympanic fold short; tympanic annulus visible below skin, tympanic membrane evident; postrictal tubercles absent. Choanae round, very small, positioned far anteriorly and laterally, widely separated from each other, slightly concealed by palatal shelf of maxilla; dentigerous process of vomer and vomerine teeth absent; tongue long and narrow; skin on dorsum finely shagreen; discoidal fold absent, dorsolateral folds absent; skin on flanks smooth; skin on ventral surfaces and gular regions smooth to finely areolate; pectoral fold poorly visible, discoidal fold not evident; cloaca protuberant; cloacal region bearing several small tubercles. Outer surface of forearm without tubercles; palmar tubercle flat and oval, approximately twice the size of elongate, thenar tubercle; tarsal tubercle small; supernumerary palmar tubercles present; subarticular tubercles like calluses, flat in ventral and lateral view, largest at the base of fingers; fingers without narrow lateral fringes; Finger IV with two phalanges; when adpressed, Finger 3 > 2 > 4 > 1 (Figure 2); tips of digits rounded, with distal grooves and papillae (Figure 2); forearm lacking tubercles. Hindlimb length moderate, tibia length 51% of SVL; foot length 43% of SVL; upper and posterior surfaces of hindlimbs without tubercles; heel without tubercles; outer surface of tarsus without tubercles; inner metatarsal tubercle, oval, of higher relief and about one time the size of conical, rounded outer metatarsal tubercle; low plantar supernumerary tubercles present; subarticular tubercles not evident in dorsal view; toes bearing narrow lateral fringes, basal webbing absent; tips of digits II, III, IV and V with discs slightly expanded, elongately acuminate, grooves present distally with papillae; digital tip of Toe V smaller than tips of Toes III—IV; when adpressed, relative lengths of toes: 4 > 3 > 5 > 2 > 1 (Figure 2).

Measurements of the holotype (in mm): SVL = 10.72, tibia length (TL) = 5.45, foot length (FL) = 4.60, head length (HL) = 3.12, head width (HW) = 3.55, interorbital distance (IOD) = 1.29, upper eyelid width (EW) = 1.00, internarial distance (IND) = 1.09, eye to nostril distance (E-N) = 0.78, snout to nostril distance (SND) = 0.50, eye diameter (ED) = 1.43, tympanum diameter (TD) = 0.45, eye to tympanum distance (ETD) = 0.19, forearm length (ForL) = 2.75, hand length (HaL) = 1.93, finger I length (FIL) = 0.91, finger II length (FIIL) = 1.30, toe I length (TIL) = 1.15, toe II length (TIIL) = 2.20.

3.2.7. Coloration of Holotype

In alcohol, the dorsal surface of the head (from the interorbital region) and body (to the cloaca) have dark brown marks of irregular shape; the dorsal surface of forelimbs is cream with a dark brown transverse bar in the shape of a wristband. The cloacal and suprainguinal regions present dark brown marks in dorsal view; those of the suprainguinal region are circular. The dorsal surfaces of the hind limbs do not have transverse dark bars, although small irregular dark bars are present in some specimens. The facial mask and lateral band are dark brown with creams spots. The lower part of the flank is dark brown. The iris is dark gray. All ventral surfaces are dark gray with minute cream spots (Figure 2). The coloration of the holotype in life is unknown.

3.2.8. Variation

The description of the coloration in life is based on notes taken in the field and photographs from multiple individuals (Figures 3, 4 and 8–10). The dorsum varies from ocher gold to light brown. Most individuals have an irregularly-shaped dark brown dorsal mark, and an interorbital bar. In some individuals, a light middorsal line extending from the middle of the body to the cloaca replaces the brown dorsal mark. Ventral coloration varies from transparent gray to black. Although all individuals have minute irregular cream flecks on venter, throat, thighs, hands, and feet, the sizes and concentration of these flecks are variable. Some males have black coloration extending from throat to mid venter.

Figure 8. Photographs of live specimens of *Noblella losamigos* sp. n. (**A**) Dorsal view and (**B**) ventral view of MUBI 17413, adult male paratype (SVL 9.6 mm); (**C**) dorsal view and (**D**) ventral view of CORBIDI 17520, adult male paratype (SVL 11.1 mm); (**E**) dorsal view and (**F**) ventral view of CORBIDI 17521, adult male paratype (SVL 9.6 mm). Photographs by A. Catenazzi.

Figure 9. Photographs of live specimens of *Noblella losamigos* sp. n. (**A**) Dorsal view and (**B**) ventral view of MUSM 30427, adult female paratype (SVL 11.6 mm); (**C**) dorsal view and (**D**) ventral view of MVZ:Herp:292684, adult female paratype (SVL 13.2 mm); (**E**) dorsal view and (**F**) ventral view of CORBIDI 17522, adult female paratype (SVL 13.6 mm). Photographs by A. Catenazzi.

Several individuals have evident dark brown circular suprainguinal marks. The forelimb pattern varies from speckled to some individuals having a dark brown transverse bar forming a wristband. A dark facial mask is present in most individuals, though variable in shape and extent. In most individuals, the upper lips have between one and three cream flecks; if present, one of these spots is below the eyes. In some specimens, the facial mask merges with a dark lateral line that extends from the tip of the snout and almost reaches to the point of insertion of the thighs. If present, the lateral line is often broken into blotches by interruptions of the lighter flank color.

Our morphological data indicate that body size and shape of *N. losamigos* sp. n. vary with elevation. Body size in both males and females increases with increasing elevation (Figure 11a; OLS regression model, females $R^2 = 0.81$, df = 7, P < 0.001; OLS regression model, males $R^2 = 0.28$, df = 7, P = 0.045). In addition, the PCA projection of body size-corrected data indicates that body shapes of males and females vary with elevation (Figure 11b). Males and females found in the lowlands occupy a morphological space that is smaller and does not overlap with the highland population, whilst individuals found in the highlands occupy a larger and more variable morphological space.

Figure 10. Photographs of live specimens of *Noblella losamigos* sp. n. (**A**) Dorsal view and (**B**) ventral view of CORBIDI 17524, adult male paratype (SVL 9.8 mm); (**C**) dorsal view and (**D**) ventral view of MUSM 30426, adult female paratype (SVL 13.1 mm); (**E**) dorsal view and (**F**) ventral view of MUBI 17412, juvenile paratype (SVL 7.1 mm).

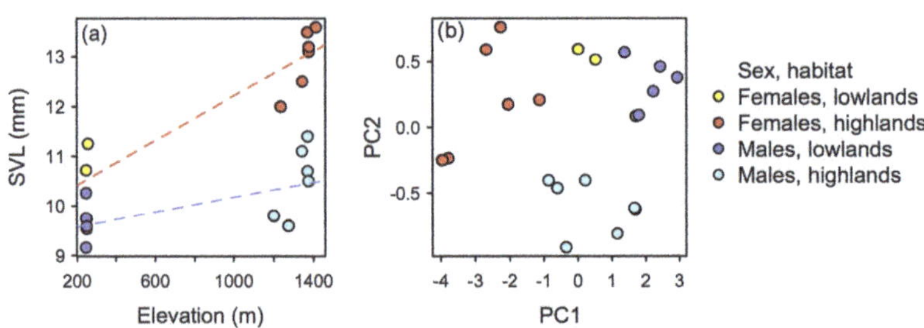

Figure 11. Multi-panel plot displaying variation of body size and body shape in *Noblella losamigos* sp. n. across elevations. (**a**) Body size in females and males tends to increase with increasing elevation. (**b**) Projection of Principal Component Analysis based on body-size-corrected data; body shape differs between females and males and varies with respect to elevation.

3.2.9. Etymology

The specific epithet is a toponym used in apposition and it refers to the type locality. Los Amigos Biological Station is located next to Los Amigos Conservation Concession, on the lower Los Amigos River watershed. Both the station and the conservation concession were established by the Amazon Conservation Association, which is a nonprofit organization that (along with its Peruvian counterpart, Conservación Amazónica—ACCA) promotes scientific research, education, and conservation in the western Amazon.

3.3. Distribution, Natural History, and Threats

Noblella losamigos sp. n. is one of five species in the genus *Noblella* distributed in southern Peru (Figure 12). We found the new species in the leaf litter during surveys conducted from 2003 to 2018 at Los Amigos [33,34,42] and in the Kosñipata Valley [17,43]. The species is known to occur at Cocha Cashu Biological Station and Pakitza in the lowlands of Manu National Park [29,31], and near the Manu Learning Centre in the Andean piedmont [37]. Additionally, the species has been recorded at several lowland sites in the Tambopata Province [32].

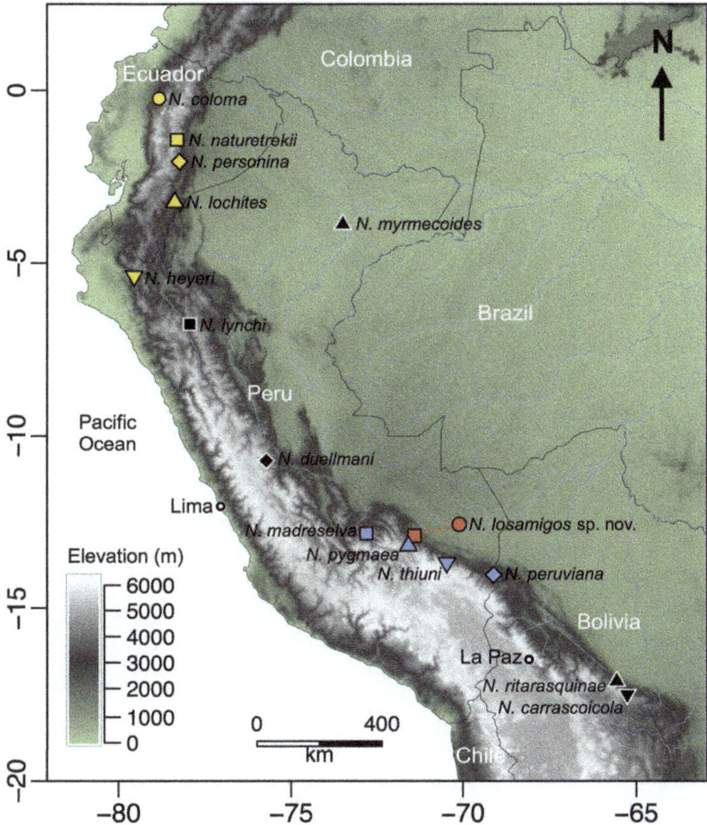

Figure 12. Map of northwestern South America showing the location of the type localities of species in the genus *Noblella*. The red circle indicates the type locality of *Noblella losamigos* sp. n. and the red square indicates the collecting site of paratypes in San Pedro, Kosñipata Valley.

The elevational distribution of *N. losamigos* sp. n. thus spans 1250 m (240–1490 m) from lowland Amazon rainforest to montane forest on the eastern slopes of the Andes. At the type locality in the Amazon lowlands, field notes indicate that the species is more common in the floodplain forest. The species is also present in other forest types including terra firme, bamboo, and palm swamp [29,34]. Sympatric species of leaf-litter frogs include *Adenomera andreae*, *Amazophrynella javierbustamantei*, *Ameerega hahneli*, *Chiasmocleis royi*, *Engystomops freibergi*, *Hamptophryne boliviana*, *Leptodactylus didymus*, and *Pristimantis carvalhoi*. Additionally, several species of gymnophthalmid lizards (Gymnophthalmidae) including *Cercosaura argulus*, *Cercosaura oshaughnessyi*, *Pseudogonatodes guianensis*, and *Ptychoglossus brevifrontalis* are common in the leaf litter in the floodplain forest. At the other localities, including the premontane forest (~450 m; [37]) and montane forest (1200–1485 m) in the Kosñipata Valley, *N. losamigos* sp. n. inhabits the leaf litter of both pristine and secondary forests, including bamboo forest patches. Sympatric leaf litter herptiles in the montane forest include the frogs *Adenomera andreae*, *Ameerega simulans*, *Noblella* sp. R, *Oreobates granulosus*, *Pristimantis danae*, *P. reichlei*, *P. toftae*, *P. salaputium*, *Rhinella leptoscelis*, *R. margaritifera*, and the lizard *Cercosaura argulus*.

The geographic range of *Noblella losamigos* sp. n. overlaps with several natural protected areas including Manu National Park, Amarakaeri Communal Reserve, and Tambopata National Reserve. The species is present in both pristine and secondary forests. The main threats faced by *N. losamigos* sp. n. are habitat loss and modification associated with informal logging and mining activities in the region. According to the IUCN Red List criteria and categories [44], we suggest placing *N. losamigos* sp. n. in the "least concern" category.

4. Discussion

In recent decades, field research conducted in the Andes–Amazon region has uncovered dozens of new species of terrestrial-breeding frogs [12,16,40,45–47], and most of these species have small geographic distributions. Prior to this study, the only species of *Noblella* previously known to occur in lowland rainforest was *N. myrmecoides* and it was assumed that it had a broad geographic distribution across western Amazonia [48]. Our findings indicate that populations of *Noblella* from lowland and montane forest in southern Peru previously ascribed to *N. myrmecoides* represent a new species. Furthermore, our phylogenetic analyses indicate that *N. myrmecoides* belongs to a different clade than the clade including *N. losamigos* sp. n., other species of *Noblella* from southern Peru, and species of *Psychrophrynella*.

Noblella losamigos sp. n. is the only species in the genus *Noblella* that inhabits both lowland and montane rainforest, covering an elevational range of 1250 m, and the only species in the "southern clade" of *Noblella* (likely to represent *Noblella* sensu stricto) that inhabits the lowland Amazon rainforest. Previous research has shown that terrestrial breeding frogs distributed at high-elevations tend to have larger body size and different body shape than species found at lower elevations [49–53]. Our data supported this prediction (Figure 11). It has been unclear as to whether ectotherms follow Bergmann's rule, in which individuals of a species tend to be larger in colder environments and smaller in warmer environments [54,55], though our data suggest that *N. losamigos* sp. n. appears to follow Bergmann's rule, as body size increases with elevation. Our data also suggest that body shape of *N. losamigos* sp. n. varies across elevation, with lowland populations occupying a smaller morphological space than highland populations.

Regardless of the variation in body size and shape exhibited by *N. losamigos* sp. n., all members of the clade containing *Noblella* and *Psychrophrynella* are miniaturized. It is thought that miniaturization may allow species to access microhabitats and food sources that are not available to larger taxa, and may be particularly advantageous to species inhabiting the leaf litter of wet, tropical rainforests [56–59]. However, a smaller body size might lead to higher exposure to water loss resulting from higher surface area-to-volume ratio [59,60]. A further inquiry into the potential effect of ecological factors correlated with elevation (e.g., resource availability, competition) will develop our understanding of the ecomorphology of *Noblella* and *Psychrophrynella*.

Supplementary Materials: The following are available online at http://www.mdpi.com/1424-2818/11/9/145/s1, Table S1: Genbank accession numbers for the taxa sampled in this study.

Author Contributions: Conceptualization, R.S.-C., R.v.M. and A.C.; methodology, R.S.-C., R.v.M. and A.C.; software, R.v.M. and D.L.R.; validation, R.S.-C., R.v.M. and A.C.; formal analysis, R.S.-C. and R.v.M.; investigation, R.S.-C., R.v.M., A.C., C.W., E.L.T. and D.L.R.; resources, R.v.M., A.C., E.L.T. and D.L.R.; data curation, R.S.-C., R.v.M., A.C. and C.W.; writing—original draft preparation, R.S.-C. and R.v.M.; writing—review and editing, R.S.-C., R.v.M., A.C., C.W., E.L.T. and D.L.R.; visualization, R.S.-C., R.v.M. and C.W.; supervision, R.v.M. and D.L.R.; project administration, R.S.-C. and R.v.M.; funding acquisition, R.v.M., A.C. and D.L.R.

Funding: R.S.-C. thanks UNSAINVESTIGA for providing support ("PP-0120-2016" Internship). R.v.M. thanks the National Science Foundation (Postdoctoral Research Fellowship DBI-1103087), the American Philosophical Society, the National Geographic Society Committee for Research and Exploration (Grant # 9191-12), and the Amazon Conservation Association. A.C. was funded with grants from the Mohamed bin Zayed Species Conservation Fund, the Disney Worldwide Conservation Fund, the Foundation Matthey-Dupraz, the Andrew Sabin Family Foundation, and the Amazon Conservation Association. Research was supported in part by a fellowship from the David and Lucile Packard Foundation to D.L.R.

Acknowledgments: We thank Cesar Aguilar (MUSM) and Greg Schneider (UMMZ) for supporting our access to specimens in the MUSM and UMMZ museum collections, respectively. We thank the Amazon Conservation Association and the staffs at Los Amigos Biological Stations and Los Amigos Conservation Concession for facilitating our work at their field stations. We thank Michelle Lynch and Erin Westeen for their help in generating μCT images included in this paper. We thank the Servicio Nacional Forestal y de Fauna Silvestre, Peru (SERFOR) for providing collecting permits (R.D.G. N° 120-2012-AG-DGFFS-DGEFFS, N° 064-2013-AG-DGFFS-DGEFFS, N° 0146-2013-AG-DGFFS-DGEFFS, N° 292-2014-AG-DGFFS-DGEFFS, N° 029-2016-SERFOR-DGGSPFFS, N° 405-2016-SERFOR-DGGSPFFS, and Contrato de Acceso Marco a Recursos Genéticos N° 359-2013-MINAGRI-DGFFS-DGEFFS). R.S.-C. thanks Julieta Cabrera for her help and patience in the coordination and use of research and training funds, and Amaranta Canazas for her support in advancing this manuscript. A.C. also thanks Perú Verde for permission to work at their biological station in San Pedro, and within their protected area (Área de Conservación Privada Bosque Nublado).

Conflicts of Interest: The authors declare no conflict of interest. The funding organizations that provided support for this work had no role in the design of the study, data collection, analyses, or interpretation of data, writing of the manuscript, or the decision to publish the results.

Appendix A

Specimens examined

Noblella duellmani (2 specimens): PERU: Pasco: Santa Bárbara, KU 315004–05.

Noblella heyeri (3 specimens): PERU: Piura: 33 km SW Huancabamba, KU 196529 (holotype), 196530–31 (paratypes).

Noblella lochites (2 specimens): ECUADOR: Morona-Santiago: Río Piuntza, KU 147070 (holotype); ECUADOR: Pastaza: Mera, KU 177356.

Noblella myrmecoides (5 specimens): PERU: Loreto: lower Río Napo region, E bank Río Yanayacu, ca. 90 km N Iquitos, KU 206120; Quebrada Oran, ca. 5 km N Río Amazonas, 85 km NE Iquitos, KU 206121; Quebrada Vásquez, N side of lower Río Tahuayo, KU 220577, 220578, 220579. PERU: Amazonas: Pongo Chinim, Kampankis, CORBIDI 9384.

Noblella pygmaea (15 specimens): PERU: Cusco: Provincia Paucartambo, Kosñipata, MHNG 2725.29–30, MUSM 24535–36, 26306–7, 26318–20, 30423–24, 30453–54, MTD 47286–87.

Noblella thiuni (holotype): PERU: Puno: Provincia Carabaya: Distrito Ollachea, Thiuni, CORBIDI 18723.

Psychrophrynella bagrecito (14 specimens): PERU: Cusco: Quispicanchis: Marcapata, Río Marcapata, below Marcapata, ca. 2740 m, KU 196512 (holotype), KU 196513–18, 196520–21, 196523–25 (all paratypes); La Convención: Hacienda Huyro between Huayopata and Quillabamba, 1830 m, KU 196527–28. (Note: specimens KU 196527–28 from La Convención might not be *P. bagrecito*).

Psychrophrynella chirihampatu (27 specimens): PERU: CUSCO: Provincia Paucartambo, Área de Conservación Privada (ACP) Ukumari Llaqta, Comunidad Campesina de Japu, 2730–3000 m, CORBIDI 16495–16499, CORBIDI 16501–16509, CORBIDI 16696, MHNC 14656, MHNC 14658, MHNC 14661–14662, MHNC 14664, MHNC 14666–14672.

Psychrophrynella glauca (4 specimens): PERU: PUNO: Thiuni, Ollachea, CORBIDI 18729 (holotype), CORBIDI 18730, 16322, 16323 (paratopotypes).

Psychrophrynella usurpator (78 specimens): PERU: Cusco: Provincia Paucartambo, Kosñipata, MUSM 20011, 20873–81, 20896–20913, 20925–33, 20946–47, 20955–57, 21012–18, 26272–73, 26278–79, 26308, 27592, 27906, 27950, 28033–28047, 30303, 30305, 30396–30400, 30405–30409, 30471–30474.

References

1. Barbour, T. A list of Antillean reptiles and amphibians. *Zoologica* **1930**, *11*, 61–116.
2. De la Riva, I.; Köhler, J. A new minute leptodactylid frog, genus *Phyllonastes*, from humid montane forests of Bolivia. *J. Herpetol.* **1998**, *32*, 325–329. [CrossRef]
3. Guayasamin, J.M.; Terán-Valdez, A. A new species of *Noblella* (Amphibia: Strabomantidae) from the western slopes of the Andes of Ecuador. *Zootaxa* **2009**, 47–59. [CrossRef]
4. Lehr, E.; Aguilar, C.; Lundberg, M. A new species of *Phyllonastes* from Peru (Amphibia, Anura, Leptodactylidae). *J. Herpetol.* **2004**, *38*, 214–218. [CrossRef]
5. Lynch, J.D. New species of minute leptodactylid frogs from the Andes of Ecuador and Peru. *J. Herpetol.* **1986**, *20*, 423–431. [CrossRef]
6. Lynch, J.D. Two new species of frogs of the genus *Euparkerella* (Amphibia: Leptodactylidae) from Ecuador and Perú. *Herpetologica* **1976**, *32*, 48–53.
7. Duellman, W.E. A new species of leptodactylid frog, genus *Phyllonastes*, from Peru. *Herpetologica* **1991**, *47*, 9–13.
8. Catenazzi, A.; Uscapi, V.; von May, R. A new species of *Noblella* (Amphibia, Anura, Craugastoridae) from the humid montane forests of Cusco, Peru. *ZooKeys* **2015**, *516*, 71–84. [CrossRef]
9. Reyes-Puig, J.P.; Reyes-Puig, C.; Ron, S.; Ortega, J.A.; Guayasamin, J.M.; Goodrum, M.; Recalde, F.; Vieira, J.J.; Koch, C.; Yánez-Muñoz, M.H. A new species of terrestrial frog of the genus *Noblella* Barbour, 1930 (Amphibia: Strabomantidae) from the Llanganates-Sangay Ecological Corridor, Tungurahua, Ecuador. *PeerJ* **2019**, *7*, e7405. [CrossRef]
10. Harvey, M.B.; Almendáriz, A.; Brito, M.J.; Batallas, R.D. A new species of *Noblella* (Anura: Craugastoridae) from the Amazonian slopes of the Ecuadorian Andes with comments on *Noblella lochites* (Lynch). *Zootaxa* **2013**, *3635*, 1–14. [CrossRef]
11. Noble, G.K. Five new species of Salientia from South America. *Am. Mus. Novit.* **1921**, *29*, 1–7.
12. Lehr, E.; Catenazzi, A. A new species of minute *Noblella* (Anura: Strabomantidae) from southern Peru: The smallest frog of the Andes. *Copeia* **2009**, 148–156. [CrossRef]
13. Köhler, J. A new species of *Phyllonastes* Heyer from the Chapare region of Bolivia, with notes on *Phyllonastes carrascoicola*. *Spixiana* **2000**, *23*, 47–53.
14. Catenazzi, A.; Ttito, A. *Noblella thiuni* sp. n., a new (singleton) species of minute terrestrial-breeding frog (Amphibia, Anura, Strabomantidae) from the montane forest of the Amazonian Andes of Puno, Peru. *PeerJ* **2019**, *7*, e6780. [CrossRef] [PubMed]
15. Hedges, S.B.; Duellman, W.E.; Heinicke, M.P. New World direct-developing frogs (Anura: Terrarana): Molecular phylogeny, classification, biogeography, and conservation. *Zootaxa* **2008**, *1737*, 1–182. [CrossRef]
16. De la Riva, I.; Chaparro, J.C.; Castroviejo-Fisher, S.; Padial, J.M. Underestimated anuran radiations in the high Andes: Five new species and a new genus of Holoadeninae, and their phylogenetic relationships (Anura: Craugastoridae). *Zool. J. Linn. Soc.* **2017**, *182*, 129–172. [CrossRef]
17. Catenazzi, A.; Lehr, E.; von May, R. The amphibians and reptiles of Manu National Park and its buffer zone, Amazon basin and eastern slopes of the Andes, Peru. *Biota Neotropica* **2013**, *13*, 269–283. [CrossRef]
18. Duellman, W.E.; Lehr, E. *Terrestrial-breeding frogs (Strabomantidae) in Peru*; Natur und Tier: Münster, Germany, 2009; p. 382.
19. Lynch, J.D.; Duellman, W.E. Frogs of the genus *Eleutherodactylus* in western Ecuador: Systematics, ecology, and biogeography. *Univ. Kans. Spec. Publ.* **1997**, *23*, 1–236.
20. Duellman, W.E.; Lehr, E.; Venegas, P.J. Two new species of *Eleutherodactylus* (Anura: Leptodactylidae) from the Andes of northern Peru. *Zootaxa* **2006**, *1285*, 51–64.

21. Padial, J.M.; Grant, T.; Frost, D.R. Molecular systematics of terraranas (Anura: Brachycephaloidea) with an assessment of the effects of alignment and optimality criteria. *Zootaxa* **2014**, *3825*, 1–132. [CrossRef]

22. Heinicke, M.P.; Lemmon, A.R.; Lemmon, E.M.; McGrathc, K.; Hedges, S.B. Phylogenomic support for evolutionary relationships of New World direct-developing frogs (Anura: Terraranae). *Mol. Phylogenetics Evol.* **2018**, *118*, 145–155. [CrossRef]

23. von May, R.; Catenazzi, A.; Corl, A.; Santa-Cruz, R.; Carnaval, A.C.; Moritz, C. Divergence of thermal physiological traits in terrestrial breeding frogs along a tropical elevational gradient. *Ecol. Evol.* **2017**, *7*, 3257–3267. [CrossRef]

24. Biomatters. Geneious R6, version 6.1.5. 2013. Available online: http://www.geneious.com/ (accessed on 31 May 2019).

25. Lanfear, R.; Calcott, B.; Ho, S.Y.W.; Guindon, S. PartitionFinder: Combined Selection of Partitioning Schemes and Substitution Models for Phylogenetic Analyses. *Mol. Biol. Evol.* **2012**, *29*, 1695–1701. [CrossRef]

26. Ronquist, F.; Huelsenbeck, J.P. MrBayes 3: Bayesian phylogenetic inference under mixed models. *Bioinformatics* **2003**, *19*, 1572–1574. [CrossRef]

27. Rambaut, A.; Drummond, A.J. Tracer. Version 1.5. 2007. Available online: http://tree.bio.ed.ac.uk/software/tracer/ (accessed on 1 June 2019).

28. Rambaut, A. FigTree, Version 1.4.2. 2009. Available online: http://tree.bio.ed.ac.uk/software/figtree/ (accessed on 1 June 2019).

29. Rodríguez, L.O.; Cadle, J.E. A preliminary overview of the herpetofauna of Cocha Cashu, Manu National Park, Peru. In *Four Neotropical Rainforests*; Gentry, A.H., Ed.; Yale University Press: New Haven, CT, USA, 1990; pp. 410–425.

30. Rodríguez, L.O. Structure et organization du peuplement d'anoures de Cocha Cashu, Parc National Manu, Amazonie Péruvienne. *Rev. De Ecol.* **1992**, *47*, 151–197.

31. Morales, V.R.; McDiarmid, R.W. Annotated checklist of the amphibians and reptiles of Pakitza, Manu National Park Reserve Zone, with comments on the herpetofauna of Madre de Dios, Peru. In *Manu: The Biodiversity of Southeastern Peru*; Wilson, D.E., Sandoval, A., Eds.; Smithsonian Institution: Lima, Peru, 1996; pp. 503–522.

32. Doan, T.M.; Arizabal, W. Microgeographic variation in species composition of the herpetofaunal communities of Tambopata Region, Peru. *Biotropica* **2002**, *34*, 101–117. [CrossRef]

33. von May, R.; Siu-Ting, K.; Jacobs, J.M.; Medina-Müller, M.; Gagliardi, G.; Rodríguez, L.O.; Donnelly, M.A. Species diversity and conservation status of amphibians in Madre de Dios, Peru. *Herpetol. Conserv. Biol.* **2009**, *4*, 14–29.

34. von May, R.; Jacobs, J.M.; Santa-Cruz, R.; Valdivia, J.; Huamán, J.; Donnelly, M.A. Amphibian community structure as a function of forest type in Amazonian Peru. *J. Trop. Ecol.* **2010**, *26*, 509–519. [CrossRef]

35. von May, R.; Jacobs, J.M.; Jennings, R.D.; Catenazzi, A.; Rodríguez, L.O. *Anfibios de Los Amigos, Manu y Tambopata, Perú*; Rapid Color Guide # 236; The Field Museum: Chicago, IL, USA, 2010; p. 12.

36. Whitworth, A.; Downie, R.; von May, R.; Villacampa, J.; McLeod, R. How much potential biodiversity and conservation value can a regenerating rainforest provide? A "best-case scenario" approach from the Peruvian Amazon. *Trop. Conserv. Sci.* **2016**, *9*, 224–245. [CrossRef]

37. Villacampa, J.; Serrano-Rojas, S.; Whitworth, A. *Amphibians of the Manu Learning Centre and Other Areas of the Manu Region*; The Crees Foundation: Cusco, Peru, 2017; p. 282.

38. von May, R.; Catenazzi, A.; Santa-Cruz, R.; Kosch, T.A.; Vredenburg, V.T. Microhabitat temperatures and prevalence of the pathogenic fungus *Batrachochytrium dendrobatidis* in lowland Amazonian frogs. *Trop. Conserv. Sci.* **2018**, *11*, 1–13. [CrossRef]

39. von May, R.; Catenazzi, A.; Santa-Cruz, R.; Gutiérrez, A.; Moritz, C.; Rabosky, D.L. Thermal physiological traits in tropical lowland amphibians: Vulnerability to climate warming and cooling. *PLoS ONE* **2019**, *14*, e0219759. [CrossRef]

40. De la Riva, I.; Chaparro, J.C.; Padial, J.M. A new, long-standing misidentified species of *Psychrophrynella* Hedges, Duellman and Heinicke from Departamento Cusco, Peru (Anura: Strabomantidae). *Zootaxa* **2008**, *1823*, 42–50. [CrossRef]

41. De la Riva, I.; Chaparro, J.C.; Padial, J.M. The taxonomic status of *Phyllonastes* Heyer and *Phrynopus peruvianus* (Noble) (Lissamphibia, Anura): Resurrection of *Noblella* Barbour. *Zootaxa* **2008**, *1685*, 67–68. [CrossRef]

42. von May, R.; Donnelly, M.A. Do trails affect relative abundance estimates of rainforest frogs and lizards? *Austral Ecol.* **2009**, *34*, 613–620. [CrossRef]

43. Catenazzi, A.; Lehr, E.; Vredenburg, V.T. Thermal physiology, disease and amphibian declines in the eastern slopes of the Andes. *Conserv. Biol.* **2014**, *28*, 509–517. [CrossRef]
44. IUCN. Guidelines for Using the IUCN Red List Categories and Criteria—Version 10.1. Prepared by the Standards and Petitions Subcommittee. 2013. Available online: http://www.iucnredlist.org/documents/RedListGuidelines.pdf (accessed on 5 April 2015).
45. Padial, J.M.; De la Riva, I. Integrative taxonomy reveals cryptic Amazonian species of *Pristimantis* (Anura: Strabomantidae). *Zool. J. Linn. Soc.* **2009**, *155*, 97–122. [CrossRef]
46. Lehr, E.; von May, R. New species of *Pristimantis* (Anura: Strabomantidae) from the Amazonian lowlands of southern Peru. *J. Herpetol.* **2009**, *43*, 485–494. [CrossRef]
47. Shepack, A.; von May, R.; Ttito, A.; Catenazzi, A. A new species of *Pristimantis* (Amphibia, Anura, Craugastoridae) from the foothills of the Andes in Manu National Park, southeastern Peru. *ZooKeys* **2016**, *594*, 143–164.
48. Jungfer, K.-H.; Hoogmoed, M.; Angulo, A.; Reynolds, R.; Icochea, J.; Azevedo-Ramos, C. Noblella Myrmecoides. The IUCN Red List of Threatened Species 2010: e.T57235A11606716. 2010. Available online: http://dx.doi.org/10.2305/IUCN.UK.2010-2.RLTS.T57235A11606716.en (accessed on 29 May 2019).
49. Hedges, S.B. Distribution patterns of amphibians in the West Indies. In *Regional Patterns of Amphibian Distribution: A Global Perspective*; Duellman, W.E., Ed.; Johns Hopkins University Press: Baltimore, MD, USA, 1999; pp. 211–254.
50. González-Voyer, A.; Padial, J.M.; Castroviejo-Fisher, S.; De la Riva, I.; Vilà, C. Correlates of species richness in the largest Neotropical amphibian radiation. *J. Evol. Biol.* **2011**, *24*, 931–942. [CrossRef]
51. Narins, P.M.; Meenderink, S.W.F. Climate change and frog calls: Long-term correlations along a tropical altitudinal gradient. *Proc. R. Soc. B* **2014**, *281*, 20140401. [CrossRef]
52. Lehr, E.; von May, R. A new species of terrestrial-breeding frog (Amphibia, Craugastoridae, *Pristimantis*) from high elevations of the Pui Pui Protected Forest in central Peru. *ZooKeys* **2017**, *660*, 17–42. [CrossRef]
53. von May, R.; Lehr, E.; Rabosky, D.L. Evolutionary radiation of earless frogs in the Andes: Molecular phylogenetics and habitat shifts in high-elevation terrestrial breeding frogs. *PeerJ* **2018**, *6*, e4313. [CrossRef]
54. Bergmann, C. Über die Verhältnisse der Wärmeökonomie der Thiere zu ihrer Größe. *Göttingen Stud.* **1847**, *1*, 595–708.
55. Mayr, E. Geographic character gradients and climatic adaptation. *Evolution* **1956**, *10*, 105–108. [CrossRef]
56. Lehr, E.; Coloma, L.A. A minute new Ecuadorian Andean frog (Anura: Strabomantidae, *Pristimantis*). *Herpetologica* **2008**, *64*, 354–367. [CrossRef]
57. Kraus, F. At the lower size limit for tetrapods, two new species of the miniaturized frog genus *Paedo-phryne* (Anura, Microhylidae). *ZooKeys* **2011**, *154*, 71–88. [CrossRef]
58. Rittmeyer, E.N.; Allison, A.; Gründler, M.C.; Thompson, D.K.; Austin, C.C. Ecological guild evolution and the discovery of the world's smallest vertebrate. *PLoS ONE* **2012**, *7*, e29797. [CrossRef]
59. Scherz, M.D.; Hutter, C.R.; Rakotoarison, A.; Riemann, J.C.; Rödel, M.-O.; Ndriantsoa, S.H.; Glos, J.; Roberts, S.H.; Crottini, A.; Vences, M.; et al. Morphological and ecological convergence at the lower size limit for vertebrates highlighted by five new miniaturised microhylid frog species from three different Madagascan genera. *PLoS ONE* **2019**, *14*, e0213314. [CrossRef]
60. Tracy, C.R.; Christian, K.A.; Tracy, C.R. Not just small, wet, and cold: Effects of body size and skin resistance on thermoregulation and arboreality of frogs. *Ecology* **2010**, *91*, 1477–1484. [CrossRef]

 *diversity*

Article

Three New Lizard Species of the *Liolaemus montanus* Group from Perú

César Aguilar-Puntriano [1,*], César Ramírez [1], Ernesto Castillo [1], Alejandro Mendoza [1],
Victor J. Vargas [2,3] and Jack W. Sites Jr. [4,5]

[1] Departamento de Herpetología, Museo de Historia Natural de San Marcos (MUSM), Av. Arenales 1256,
Lima 11, Peru; ramirezperaltac@gmail.com (C.R.); ernesto.cas.95@gmail.com (E.C.);
jagopaul@hotmail.com (A.M.)

[2] Asociación Pro Fauna Silvestre Ayacucho, Jr. José Santos Flores, Mz. A1 Lt.9. Urb. Las Américas San Juan
Bautista, Huamanga, Ayacucho 05001, Peru; vjvargasg@gmail.com

[3] Servicio Nacional Forestal y de Fauna Silvestre (SERFOR), Av. Javier Prado Oeste N° 2442,
Urbanización Orrantía, Magdalena del Mar, Lima 17, Peru

[4] Emeritus Professor, Department of Biology, Emeritus Curator (Herpetology), M.L. Bean Life Science
Museum, Brigham Young University, Provo, UT 84602, USA; jwsites6098@gmail.com

[5] Department of Biology, Austin Peay State University, Clarksville, TN 37043, USA

* Correspondence: caguilarp@unmsm.edu.pe

Received: 1 August 2019; Accepted: 6 September 2019; Published: 11 September 2019

Abstract: Three new species of *Liolaemus* belonging to the *L. montanus* group are described from Perú.
Two new species are restricted to the Ica and Moquegua departments on the Pacific coast, and one
new species is only known from an isolated highland in Ayacucho department. These three new
species differ from closely related species in their coloration patterns and head shape. We comment
on the conservation issues of the new species and other Peruvian species of the *L. montanus* group.

Keywords: Andes; conservation; Pacific coast; Perú; species description; threats

---

1. Introduction

Liolaemus species of the *L. montanus* group comprise about 60 species and several candidate
species [1–3]. The geographic distribution of this group ranges from central Perú and Bolivia to
northern Argentina and Chile, and from sea level to more than five thousand m.a.s.l. Thirteen species in
the northern range of this group are known in Perú: *L. annectens*, *L. aymararum*, *L. etheridgei*, *L. evaristoi*,
L. insolitus, *L. melanogaster*, *L. ortizi*, *L. poconchilensis*, *L. polystictus*, *L. robustus*, *L. signifer*, *L. thomasi*, and
L. williamsi ([2,4] this paper). Recently, two candidate species ("Nazca" and "Abra Toccto") have been
proposed based on an integrative approach in the northern range of the *L. montanus* group, but species
description are still lacking [2].

Moreover, a more comprehensive phylogeny of the *Liolaemus montanus* group identified another
well supported candidate species (*Liolaemus* "Moquegua") based on two mitochondrial and five
nuclear markers. This candidate species forms a clade with *Liolaemus* "Nazca", *L. insolitus* and
L. poconchilensis [3]. *Liolaemus poconchilensis* is one of these rare "toad-like" or "phrynosauroid" head
lizards (eye diameter longer than snout-eye distance, poorly differentiated scale heads, short snout,
short and triangular lower jaw), with eyes surrounded by enlarged ciliary scales or "combs" [5]
Liolaemus "Moquegua" is similar in head shape to *L. poconchilensis*, but it lacks the "combs" surrounding
the eyes and other traits we describe here. In this paper, we also describe two new species previously
identified as *Liolaemus* "Abra Toccto" and *Liolaemus* "Nazca", and comment on conservation issues
threating the new species and other Peruvian species of the *L. montanus* group.

2. Material and Methods

2.1. Sampling

Lizards were collected by hand, photographed and sacrificed with an injection of pentobarbital. Liver tissue was collected for DNA samples, and whole specimens were fixed in 10% formaldehyde and transferred to 70% ethanol for permanent storage in the Bean Life Science Museum at Brigham Young University (BYU) and the Museo de Historia Natural de San Marcos (MUSM). Tissue samples were collected in duplicate, stored in 96% ethanol and deposited in both museum collections.

2.2. Species Descriptions

Species descriptions follow the format of [6] We examined 47 specimens of the new species and 259 specimens of Peruvian species of the *Liolaemus montanus* group. We followed [7] for terminology of scale descriptions and [8] for neck fold terminology. Color descriptions are based on photographs of live animals taken in the field, and specimens examined are provided in Appendix A. Measurements were taken using a digital caliper to the nearest 0.1 mm. Bilateral scale counts and mensurable data were taken from the right side of lizards. Scale state characters were taken using a stereoscope (10×–40×).

3. Results

Below we describe three new species previously recognized as candidate species [2,3]. However, as candidate species without a formal description they probably will not get legal protection. Legal protection is afforded for taxa with scientific names in a formal species description. Once candidate species have names, they will be available and can be incorporated into local protected and international conservation lists. For this reason, we describe three new species of *Liolaemus* lizards and we hope in the near future they will be evaluated and be part of a conservation program if needed.

Species Description

Liolaemus nazca

2017. *Liolaemus* "Nazca" Aguilar et al.

Holotype. MUSM 31523: adult male collected in Marcona District, Nazca Province, Department of Ica, Peru, 15.120 S, 75.338 W, 466 m, on 17 January 2013 by César Aguilar, César Ramírez and Alejandro Mendoza.

Paratypes. MUSM 16100, 31520, 31526, 31541: four adult males, same data as holotype. MUSM 31,521, 31,525, 16,101: three adult females, same data as holotype. MUSM 31524, 31527, 31522: three juveniles, same data as holotype.

Referred specimens. BYU 50471–50472: two males, same data as holotype. BYU 50506–50508, BYU 50510: four females, same data as holotype.

Diagnosis. *Liolaemus nazca* belongs to the *L. montanus* group because it lacks a patch of enlarged scales on the posterior thighs. *Liolaemus nazca* forms a clade with other Pacific coast species, *L. insolitus*, *L. poconchilensis* and *L. chiribaya* sp. nov. It differs from *L. poconchilensis* in being larger (with a maximum SVL of 59.8 mm; 55.9 mm in *L. poconchilensis*) and lacking enlarged serrate ciliary scales. *L. nazca* differs from *L. insolitus* and *L. chiribaya* sp. nov. by having slightly keeled dorsal scales on the body, which become more conspicuous towards the vertebral line. *Liolaemus nazca* also differs from *L. poconchilensis* and *L. chiribaya* sp. nov. by lacking a "phrynosauroid" or "toad-like" head. *Liolaemus nazca* presents an intense orange or yellow ventral region with dark spots, in contrast to *L. insolitus* and *L. poconchilensis*, both of which have clearer ventral regions, and *L. chiribaya* sp. nov. which has two orange lateral stripes on venter.

Liolaemus nazca is distinguished from other Peruvian species of the *L. montanus* group by its bright green and turquoise scales on body flanks surrounded by yellow and black scales. *Liolaemus nazca* also differs from *L. aymararum, L. evaristoi, L. melanogaster, L. polystictus, L. robustus, L. thomasi* and *L. williamsi* in having a smaller SVL (65.9 mm versus 70.1–103.0 mm). *Liolaemus nazca* has fewer scales around midbody (53–62) than *L. signifer* (67–110), and the number of vertebral scales (between the occiput and anterior level of hind limbs) in *L. nazca* is smaller (53–57 scales) than *L. evaristoi* and *L. signifer* (60–129 scales), and greater than those of *L. aymararum, L. ortizii* and *L. thomasi* (30–53). *Liolaemus nazca* differs from *L. etheridgei, L. ortizii* and *L. thomasi*, all of which have noticeably keeled scales. *Liolaemus nazca* females differs from *L. melanogaster, L. polystictus* and *L. thomasi* females by having vestigial precloacal pores. *Liolaemus nazca* males have fewer precloacal pores (3–6) than males of *L. annectens* (6–7) and *L. etheridgei* (6–9).

Description of the holotype. Adult male, SVL 64.5 mm, head length 16.3 mm, head width 13.2 mm, head height 9.5 mm, groin armpit distance 26.4 mm (39.8% SVL), foot length 16.3 mm (25.3% SVL), tail length 77.6 mm (120.3% SVL). 20 dorsal head scales (between the anterior edge of the auditory meatus to the anterior edge of the rostral); dorsal head scales smooth, occipitoparietal scales irregular and convex, frontonasal and parietal area with convex scales; scale organs more abundant in the prefrontal, internasal, lorilabial and loreal regions; supralabial region without scales organs; three organs in the left and one in the right post-rostral. Nasal scale separated from rostral, and separated from the first supralabial by one scale, right nasal bordered by eight scales; cantal separated from nasal by two scales. Six supralabials. Ten lorilabial scales, four in contact with subocular. Six infralabials. Oval auditory meatus (height 2.4 mm, width 1.0 mm) with two small scales on anterior margin. Seven smooth and convex temporal scales. Distance between orbit and auditory meatus 6.5 mm. Rostral almost three times as wide as tall (width 2.8 mm; height 1.1 mm). Mental subpentagonal, almost twice as wide as tall (width 2.7 mm; height 1.5 mm). Hourglass-shaped interparietal, with elongated posterior apex, bordered by eight scales, parietals of similar size as interparietal. Frontal quadrangular. Complete supraorbital semicircles on both sides. Semicircles formed by 13 scales. Four enlarged supraoculars. Six superciliares overlapping on both sides. Ten upper and nine lower ciliary scales. Subocular elongated, larger than eye diameter, separated from supralabials by a single row of lorilabials. Supralabials of similar size. Eight lorilabials, with double and triple rows of scale organs. Eighth, seventh, sixth and fifth lorilabials in contact with subocular. Preocular separated from the lorilabial row by two scales. Postocular as large as preocular. Mental in contact with five scales, three infralabials (on each side) and two enlarged chin scales. Chin scales forming a longitudinal row of four enlarged scales separated one from the other by seven small scales. Gular scales rounded, flat and imbricated. 24 gulars between auditory meatus. Longitudinal neck fold without keeled scales, almost half in size of dorsal scales. Antehumeral pocket and folds well developed. 36 scales between auditory meatus and shoulder (counting along the post-auricular and longitudinal neck fold), 21 scales between the auditory meatus and the neck fold. Gular fold absent. Dorsal scales imbricated, slightly keeled, more conspicuous towards vertebral line. 53 dorsal scales between the occiput and groin level. 54 scales around midbody. Dorsal scales smooth towards flanks and belly. Ventral scales slightly wider than dorsal. 65 ventral scales between mental scale and cloaca; four precloacal pores. Supracarpals smooth and laminar with oval margins. Subdigital lamellae of fingers with three keels, formula I:8; II:12; III:17; IV:17; V:10 (right hand). Supradigital lamellae smooth and imbricated. Infracarpals and infratarsals keeled and imbricated. Supratarsals smooth and angular, but slightly keeled on fourth finger. Subdigital lamellae toe formula I:8; II:13; III:17; IV:21; V:13 (right foot).

Color pattern in life. (Figure 1) Dorsal color light brown with two paravertebral series of eight dark brown spots, more or less symmetrical, between occiput and pelvis, dark brown spots bordered by lighter scales. Lateral region of the body from cheek (postocular region) to the post-cloacal zone (tail base) with patches of emerald green scales on a bright yellow background, interrupted by dark brown transverse spots. Dorsal head brown with dark black spots. Area surrounding loreal, subocular, mental and ocular scales with bright yellow background color; five dark stripes on lateral head, one projects from eye to postocular and temporal region, one to mouth corner, two through subocular and labial region, and one through preocular and nasal. Dorsal limbs light brown with dark spots not reaching phalanges. Tail with dark brown subtriangular spots, which merge towards tip of tail. Ventral head, body and tail bright orange, with dark spots on head, and some on neck and belly.

Figure 1. Holotype (male, SVL = 64.5 mm, MUSM 31523) of *Liolaemus nazca* in dorsal (**A**) and ventral (**B**) views.

Color pattern in preservative. Dorsal background from neck to tip of tail is brown. On dorsal head, dark spots become more evident. Ventral region presents a whitish background coloration; dark spots accentuate and become more conspicuous.

Variation. (Figures 2 and 3). Variation in selected characters is summarized in Table 1. Sexual dichromatism present. Males have two paravertebral series of 6–8 dark brown spots on dorsum surrounded or not by white scales; body flanks with emerald green spots surrounded by yellow and dark brown spots; emerald green spots are present from lower temporal region of head to first third of tail; in young specimens emerald green spots are smaller and do not reach first third of tail. Males with orange or yellow on ventral surface of body, and limbs, with small dark spots in gular region, sometimes forming reticulations. Adult females have throat, belly and base of tail orange, yellowish or whitish, with or without dark spots or reticulations on belly.

Distribution and natural history (Figure 4). *Liolaemus nazca* is only known in Nazca province, Ica department, at elevations of 450–700 m. It was found on the ground or in shallow holes on the ground, mainly in hills ("Lomas") with low shrub vegetation and sandy soil, less frequently in areas with *Tillandsia* sp. In the summer, some individuals were active as early as 7:27 and as late as 16:22; on winter no individuals were observed. Body temperature of ten specimens ranged 22.7–34.0 °C (substrate temperature: 22.8–44.1 °C; air temperature: 21.4–31.0 °C). It was found together with *Ctenoblepharys adspersa*, *Microlophus* sp. and *Phyllodactylus gerrophygus*. This species is viviparous, one female had two embryos on each side at an advanced stage of development. On the Pacific coast, *L. nazca* is the northernmost species of *Liolaemus*.

Table 1. Variation in selected characters among type specimens of *Liolaemus nazca*. All specimens from Museo de Historia Natural Universidad Mayor de San Marcos.

MUSEUM NUMBER	31523	31520	31541	31526	31525	31521	16100	16101	31524	31527	31522
Sex	Male	Male	Male	Male	Female	Female	Male	Female			
Reproductive stage	Adult	Adult	Adult	Adult	Adult	Adult	Subadult	Subadult	Juvenile	Juvenile	Juvenile
SVL	64.5	60.9	63.4	54.4	61.4	56.4	54.2	49.4	28.9	28.1	27.3
Groin armpit distance	26.4	26.0	26.0	23	29.0	26.2	21.0	20.5	11.1	11.3	11.3
Head length	16.3	17.4	16.5	13.6	13.7	13.4	13.8	12.5	8.0	7.8	7.7
Head width	13.2	13.7	14.3	10.5	11.5	11.2	11.3	10.4	6.3	6.0	5.9
Forelimb length	18.5	20.3	17.9	19.6	19.5	19.0	19.5	17.3	9.9	10.8	10.2
Hindlimb length	29.7	31.2	28.5	28.2	27.4	27.4	28.3	26.4	15.6	17.1	13.8
Snout length	6.72	7.2	6.0	5.6	6.2	5.3	5.4	5.2	3.0	3.2	3.0
Scales around midbody	54	58	59.0	57	56	55	59	56	55	57	57
Dorsal scale number	53	56	54	53	53	56	54	50	57	54	55
Head scale number	20	19	19	20	21	21	18	16	18	15	17
Scales around interparietal	8	6	6	5	6	5	7	6	7	7	6
Ventral scale number	65	70	71	70	73	71	74	75	61	75	69
Precloacal pores	4	3	3	6	1 *	3 *	4	6	0	0	0
Supralabial scales	6	7	8	9	8	7	7	9	9	7	8
Gular scales	24	24	21	25	25	23	25	25	21	21	20

* Vestigial precloacal pores.

Etymology

The specific epithet *nazca* is a noun in apposition and is given in honor to the Nazca culture (100–800 A.D.). Among the famous Nazca lines, there is a lizard geoglyph.

Liolaemus chiribaya

2018. *Liolaemus* "Moquegua" Aguilar-Puntriano et al.

Holotype. MUSM 31547: adult male collected near "Cerros Los Calatos", 16.91892S, 70.89596W, Torata District, Mariscal Nieto Province, Moquegua Department, Perú, 2615 m, December 19, 2014 by César Aguilar, Jessie Montalvo and Maribel Angeles.

Paratypes. MUSM 31553: adult male collected in Jaguay Chico, Torata District, Mariscal Nieto Province, Moquegua Department, Peru 16.94567 S, 70.88486 W, 2928 m, December 19, 2014; MUSM 31549–31550: one female and juvenile collected in the same location and date as previous specimen, 2942 and 2913 m respectively; MUSM 31548: a female collected near the Asirune Archaeological Zone, near Jaguay Chico, Torata District, Mariscal Nieto Province, Moquegua Department, Peru, 16.95213 S, 70.87854 W, 2990 m; all above paratypes collected by César Aguilar, Jessie Montalvo and Maribel Angeles; MUSM 31386–31388, 31390–31391: five males and MUSM 31389: one female collected in "Cerro Los Calatos", Torata District, Mariscal Nieto Province, Moquegua Department, Perú, 2794–2988 m, 27–29 December 2012 by Juana Suárez.

Referred specimens. BYU 51568, BYU 51570: two males, collected in Jaguay Chico, Torata District, Mariscal Nieto Province, Moquegua Department, Perú; MUSM 31546: male collected near the Asirune Archaeological Zone, near Jaguay Chico, Torata District, Mariscal Nieto Province, Moquegua Department, Perú; BYU 51564, BYU 51566–51567: two females and one juvenile with same data as MUSM 31546.

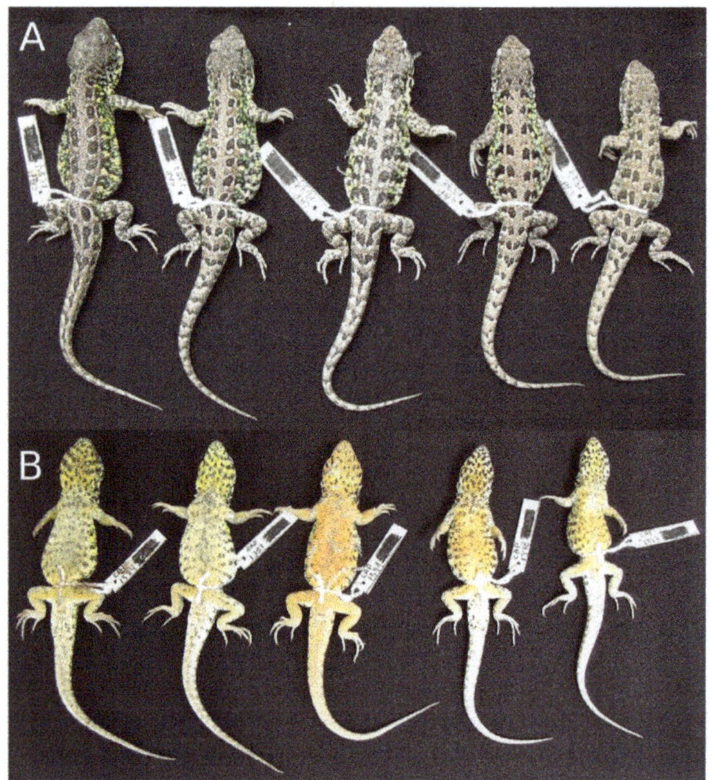

Figure 2. Dorsal (**A**) and ventral (**B**) variation in males of *Liolaemus nazca*. From left to right: MUSM 31520 (SVL = 60.9 mm), BYU 50472 (SVL = 64 mm), MUSM 31523 (holotype, SVL = 64.5 mm), BYU 50471(SVL = 60.0 mm), MUSM 31526 (SVL = 53.1 mm).

Diagnosis. *Liolaemus chiribaya* is identified as a member of the *L. montanus* group by the absence of a patch of enlarged scales on posterior thighs. *Liolaemus chiribaya* forms a clade with *L. insolitus*, *L. poconchilensis* and *L. nazca* sp. nov. It differs from *L. poconchilensis* by having a fourth finger extending beyond the armpit when a hindlimb is brought forward (fourth finger does not exceed past the armpit in *L. poconchilensis*); male *L. chiribaya* further differs from *L. poconchilensis* by the presence of dorsal turquoise spots (absent in males of *L. poconchilensis*), and differs from *L. nazca* by having smooth dorsal body scales (dorsal body scales slightly keeled in *L. nazca*). In addition, *L. chiribaya* lacks an orange or yellow venter with dark spots. *Liolaemus chiribaya* differs from *L. insolitus* by having a greater number of scales around midbody (58–69 vs. 45–53). *Liolaemus chiribaya* differs from *L. nazca*, *L. insolitus* and other Peruvian species of the *L. montanus* group (except *L. poconchilensis*) by having a "phrynosauroid" head. *Liolaemus chiribaya* also differs from other Peruvian species of the *L. montanus* group by having dorsal turquoise scales and a maximum 68.8 mm SVL, being a smaller species than *L. aymararum*, *L. evaristoi*, *L. melanogaster*, *L. polystictus*, *L. robustus*, *L. thomasi* and *L. williamsi* (SVL 70.1–103.0 mm). *L. chiribaya* has fewer scales around midbody (54–66) than *L. signifer* (67–110), fewer maximum number of dorsal scales (between occiput and anterior level of hindlimb; 64) than *L. evaristoi* (75) and *L. signifer* (129), and more than *L. aymararum*, *L. ortizii* and *L. thomasi* (all ≤ 53). It also differs from *L. etheridgei*, *L. ortizii* and *L. thomasi* by lacking strongly keeled scales. Females of *L. chiribaya* also have vestigial precloacal pores, which are absent in females of *L. melanogaster*, *L. polystictus* and *L. thomasi*.

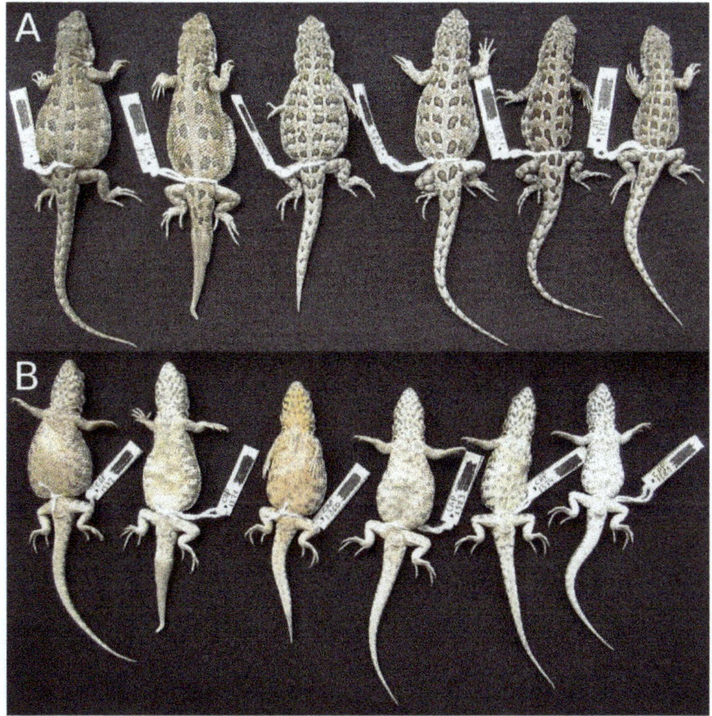

Figure 3. Dorsal (**A**) and ventral (**B**) variation in females of *Liolaemus nazca*. From left to right: MUSM 31525 (SVL = 61.4 mm), BYU 50506 (SVL = 66.1 mm), BYU 50507 (SVL = 54.9 mm), MUSM 31521 (SVL = 56.4 mm), BYU 50508 (SVL = 52.6 mm), BYU 50510 (SVL = 47.2 mm).

Figure 4. Habitat of *Liolaemus nazca*.

Description of the holotype. Adult male, SVL 52.6 mm, head length 14.2 mm, head width 11.7 mm, head height 7.6 mm, groin armpit distance 20.3 mm (38.6% SVL), foot length 14.3 mm (27.2 % SVL), tail length 54.2 mm (103% SVL). 21 dorsal scales on head; smooth dorsal head scales, occipital, parietal and frontonasal area with convex scales, parietal scales polygonal; scale organs numerous in prefrontal, lorilabial and loreal scales; supralabial scales with few scales organs; five organs in the left postrostral and four in the right. Nasal scale separated from rostral, separated by a scale from first supralabial, right nasal limited by eight scales; canthal separated from nasal by two scales. Seven supralabials. Eight lorilabial scales, three in contact with the subocular. Seven infralabials. Oval auditory meatus (height 1.8 mm; width 1.2 mm), with two small protruding scales on anterior margin. Eight smooth and convex temporal scales. Distance between orbit and auditory meatus 4.1 mm. Rostral three times wider than high (width 2.4 mm; height 0.8 mm). Mental subpentagonal, almost twice as wide as high (width 2.4 mm; height 1.3 mm). Interparietal pentagonal, with elongated posterior apex, bordered by five scales, parietals larger than interparietal. Frontal scales polygonal. Complete supraorbital semicircles on both sides. Semicircles formed by 13 scales. Three enlarged supraoculars. Eight superciliares. 14 upper and 12 lower ciliaries. Subocular divided into three, longer than eye diameter, posterior subocular larger and separated from supralabials by a single row of lorilabials. Supralabials of similar size. Lorilabial eighth and seventh in contact with subocular. Preocular of medium size, separated from lorilabial row by a scale. Postoculars of similar size to preocular. Mental scale in contact with four scales: two infralabials (on each side) and two enlarged chin scales. Chin scales forming a longitudinal row of four enlarged scales separated from each other by 12 scales. Gular scales rounded, flat and overlapped with very few scale organs. 25 gular scales between auditory meatus. Longitudinal neck fold without keeled scales, almost half in size of dorsal scales. Antehumeral pocket and well developed antehumeral neck folds. 30 scales between auditory meatus and shoulder (counting along post-auricular and longitudinal neck fold), 25 scales between the auditory meatus and the neck fold. Gular fold absent. Dorsal scales juxtaposed or poorly imbricated. 60 dorsal scales between occiput and groin level. 64 scales around midbody. Dorsal scales smooth on flanks and belly. Ventral scales slightly wider than dorsal scales. 74 ventral scales between mental and cloaca; four precloacal pores present. Smooth laminar supracarpals, with oval or rounded margins. Subdigital lamellae of fingers with three keels, and with formula I:7; II:11; III:15; IV:17; V:8 (right hand). Smooth and imbricated supradigital lamellae. Smooth infratarsals and keeled infracarpals, both strongly imbricated. Supratarsal smooth, oval or rounded. Subdigital lamellae of toes with formula I:9; II:11; III:16; IV:18; V:13 (right foot).

Color pattern in life (Figure 5). Dorsal paravertebral orange and turquoise scales from neck region to tail base. Dorsal head with pale orange and brown scales on occipital and frontal region. Dark brown scales on supraocular, parietal region and snout. Dorsal limbs pale orange, with some orange scales and dark brown scales forming zigzag spots up to base of digits (hand) or on digits (foot). First third of tail with turquoise and orange scales, with dark scales forming transverse bands or spots. Ventral scales on head, limbs and tail white smoke, belly with two orange lateral stripes separated by a central area of pale yellowish scales. Ventral tail white smoke.

Color pattern in preservative. Dorsal light gray scales on vertebral region and dark brown scales on paravertebral region, from neck to tail base. Dorsal head with light gray scales on occipital, parietal, frontal and prefrontal regions. Dark olive scales on supraocular, and some on parietal and temporal region, creamy scales on snout. Dorsal limbs pale gray, with gray scales forming zigzag marks up to base of digits (hand) or on digits (foot). Tail pale gray with dark scales forming transverse bands or spots. Ventral scales white smoke on head, limbs, tail and belly.

Variation. (Figures 5 and 6). Variation in selected characters is summarized in Table 2. Sexual dichromatism present. Females and juveniles on dorsal body have 6 to 8 triangular or quadrangular marks, sometimes bordered by lighter scales; without turquoise blue scales; males with or without triangular or quadrangular marks, but if present marks are usually covered by turquoise blue scales; venter whitish without orange lateral stripes.

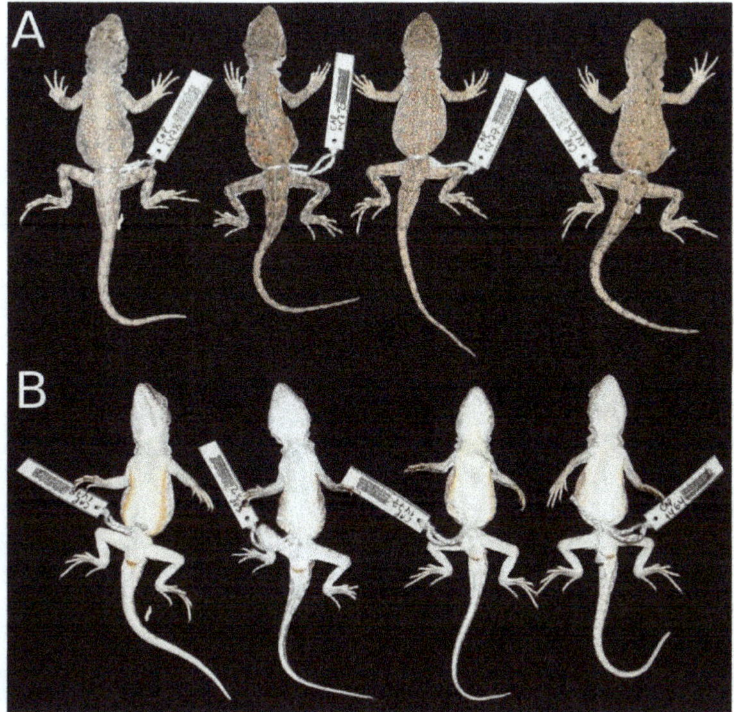

Figure 5. Dorsal (**A**) and ventral (**B**) variation in males of *Liolaemus chiribaya*. From left to right: MUSM 31547 (Holotype, SVL = 52.6 mm), BYU 51568 (SVL unknown), BYU 51570 (SVL unknown), MUSM 31553 (SVL = 52.7 mm).

Table 2. Variation in selected characters among type specimens of *Liolaemus chiribaya* sp. nov. All specimens from Museo de Historia Natural Universidad Mayor de San Marcos.

MUSEUM NUMBER	31547	31386	31387	31391	31390	31388	31548	31549	31389	31553	31550
Sex	Male	Male	Male	Male	Male	Male	Female	Female	Female	Male	
Reproductive stage	Adult	Adult	Adult	Adult	Adult	Adult	Adult	Adult	Adult	Subadult	Juvenile
SVL	52.6	66.8	68.8	61.8	61	64.9	49.6	57.0	66.2	46.8	46.4
Groin armpit distance	20.3	33.0	34.9	31.5	30.6	32.4	20.3	25.0	35.9	21.0	21.0
Head length	14.2	16.0	16.3	15.5	14.6	15.6	13.0	13.5	14.5	12.3	12.4
Head width	11.7	14.0	13.3	13.5	12.7	13.9	11.4	11.7	13.0	10.5	10.4
Forelimb length	23.6	24.4	25.8	24.5	23.2	24.0	20.4	22.5	22.5	19.1	19.6
Hindlimb length	33.8	34.0	33.7	34.6	34.8	34.3	30.9	32.4	32.5	30.1	30.1
Snout length	5	7	6.8	6.8	6.2	6.1	5.1	5.6	5.9	5.1	4.9
Scales around midbody	64	65	66	63	60	65	57	55	60	63	58
Dorsal scale number	60	55	58	59	52	63	56	53	61	57	61
Head scale number	21	20	20	23	20	24	18	18	20	19	21
Scales around interparietal	5	6	7	6	6	6	6	6	7	7	5
Ventral scale number	74	76	72	73	73	73	77	67	73	69	74
Precloacal pores	4	5	5	3	3	3	4 *	2 *	4 *	5	2 *
Supralabial scales	7	9	9	8	9	10	9	7	9	9	7
Gular scales	25	22	21	23	22	19	20	21	20	24	20

* Vestigial precloacal pores.

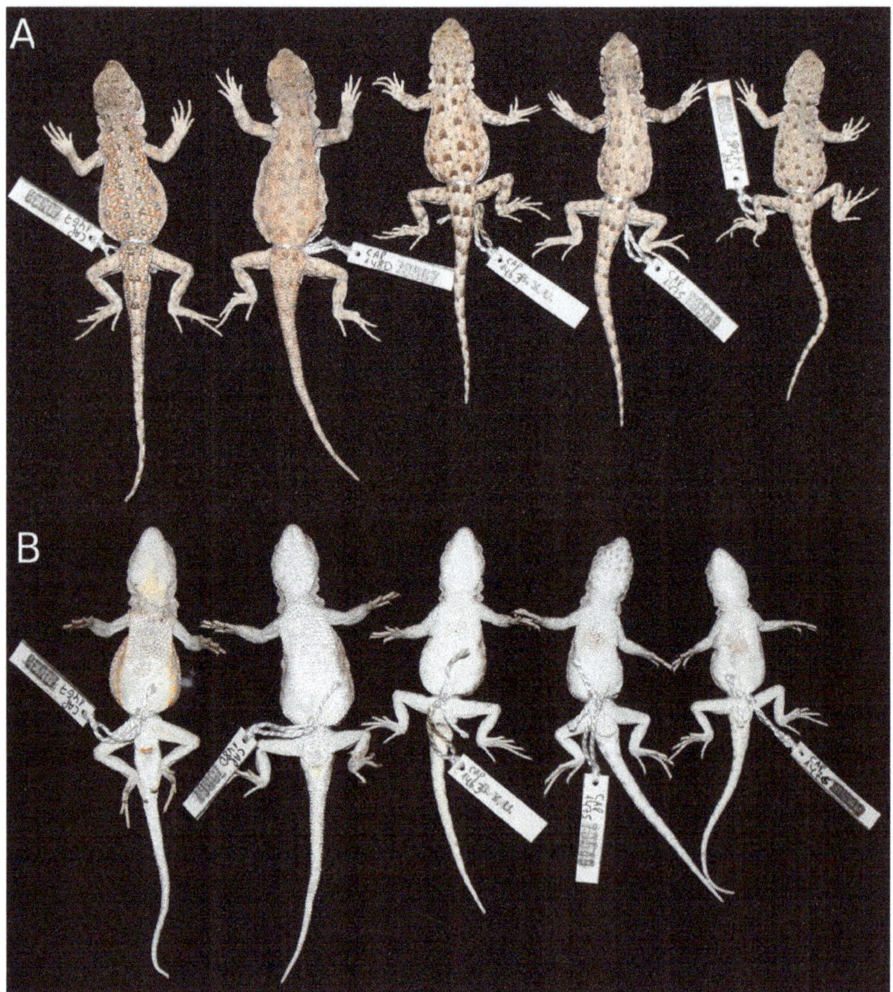

Figure 6. Dorsal (**A**) and ventral (**B**) variation in *Liolaemus chiribaya*. From left to right: MUSM 31546 (male, SVL = 52.5 mm), BYU 51564 (female, SVL unknown), MUSM 31548 (juvenile, SVL = 49.6 mm), BYU 51566 (juvenile, SVL unknown), BYU 51567 (juvenile, SVL unknown).

Distribution and natural history (Figure 7). *Liolaemus chiribaya* is only known in the District of Torata, Province of Mariscal Nieto, Department of Moquegua, at elevations of 2615–3005 m; they were found under rocks or on ground in desert areas with cacti and low shrubs; they were active between 10:00 and 14:00. Other lizard species present in the area were *Phyllodactylus gerrophygus* and *L. tacnae*.

Etymology. The specific epithet *chiribaya* is a noun used in apposition and honors the Chiribaya culture (900–1350 A.D.). Chiribayans were settled in the basin of the Ilo River, and expanded north to the Tambo valley (Arequipa) and the south to the Azapa valley (Chile), including the high altitude regions, up to nearly 3000 m of elevation.

Liolaemus victormoralesii

2017. *Liolaemus* "Abra Toccto" Aguilar et al.
2018. *Liolaemus* "Abra Toccto" Aguilar-Puntriano et al.

Holotype. MUSM 31461: adult male collected at Abra Toccto, Huamanga Province, Ayacucho Department, Perú, 13.346 S, 74.184 W, elevation 4222 m, on 01 June 2012 by César Aguilar, Víctor J. Vargas, Frank Huari and Elver Coronado.

Paratypes. MUSM 31371–31372, 31460: three adult males collected at Abra Toccto, Huamanga Province, Ayacucho Department, Perú, 13.298 S, 74.091 W, elevation 4193–4215 m on 3 December 2012 by Alfredo Guzmán and Víctor J. Vargas; MUSM 31460, 31468: two adult females, same data as holotype; MUSM 31463: juvenile collected at Abra Toccto, Huamanga Province, Ayacucho Department, Perú, 13.35 S, 74.187 W, elevation 4182m, on 04 June 2012 by César Aguilar, Víctor J. Vargas, Frank Huari and Elver Coronado. MUSM 25700, adult male collected at Chiara, Huamanga Province, Ayacucho Department, Perú, 13.341 S, 74.216 W, elevation 4145m, on 30 November 2006 by Margarita Medina.

Figure 7. Habitat of *Liolaemus chiribaya*.

Referred specimens. BYU 50431, BYU 50427, MUSM 31462: three males, same data as holotype; BYU 50433, BYU 50428: two females, same data as holotype.

Diagnosis. *Liolaemus victormoralesii* is identified as a member of the *L. montanus* group by the absence of a patch of enlarged scales on posterior thighs. *Liolaemus victormoralesii* forms a clade with *L. evaristoi*, *L. melanogaster*, *L. polystictus*, *L. robustus* and *L. williamsi*. It differs from closely related *L. evaristoi* by lacking blue scales on the dorsum and flanks, having a larger size (maximum SVL 88.9 mm in *L. victormoralesii* and 70.1 mm in *L. evaristoi*) and by lacking vestigial precloacal pores in females. *Liolaemus victormoralesii* differs from *L. melanogaster* by lacking black belly scales (gray scales in adult *L. victormoralesii*). Adult females of *L. victormoralesii* differ from *L. polystictus* and *L. williamsi* females by having a darker dorsal background coloration and few large contrasting marks dorsally (*L. polystictus* and *L. williamsi* have a lighter dorsal background coloration and large number of small contrasting marks dorsally). *Liolaemus victormoralesii* further differs from *L. williamsi* by having a larger size (maximum SVL 74.9 mm in *L. williamsi*) and by lacking vestigial precloacal pores in females. *Liolaemus victormoralesii* differs from *L. robustus* by lacking dorsal yellow greenish scales. Adult males of *Liolaemus victormoralesii* differs from *L. etheridgei* by lacking light blue dorsolateral scales, from *L. annectens* by having a darker dorsum, and from *L. signifer* by lacking bright yellow and fewer maximum number of dorsal scales (57 vs. 129). *Liolaemus victormoralesii* differs from *L. ortizi* and *L. thomasi* by lacking strongly keeled and by having smaller dorsal scales. *Liolaemus victormoralesii* differs from *L. aymararum* by having more scales around midbody (51–64 vs. 48–52) and smaller dorsal scales. It differs from *L. nazca* sp. nov. by lacking emerald green spots surrounded by black and yellow scales laterally on body. It differs from *L. chiribaya* and *L. poconchilensis* by lacking a "phrynosauroid" head, and from *L. poconchilensis* by lacking well-developed ciliary scales (serrate "combs") surrounding the eyes. It differs from *L. insolitus* by lacking dorsal and lateral blue spots. *L. victormoralesii* further differs from *L. insolitus* and *L. poconchilensis* by lacking dorsal smooth scales. *Liolaemus victormoralesii* also differs from *L. chiribaya* sp. nov., *L. etheridgei*, *L. nazca* sp. nov., *L. ortizii* and *L. poconchilensis* in having a larger SVL (88.9 mm versus 56–77 mm). *Liolaemus victormoralesii* females also differs from *L. annectens*, *L. aymararum*, *L. chiribaya*, *L. etheridgei*, *L. insolitus* and *L. nazca* sp. nov by lacking vestigial precloacal pores.

Description of the holotype. Adul male, SVL 83.8 mm, head length 19.2 mm, head width 18.2 mm, head height 10.8 mm, groin armpit distance 29 mm (34.6% SVL), foot length 19.5 mm (23.4 % SVL), tail length 107 mm (128.5% SVL). 19 dorsal scales on head; smooth dorsal head scales, occipital, parietal and frontonasal area with slightly convex scales, parietal scales polygonal, similar in size to interparietal; numerous scale organs in prefrontal and lorilabial, and few in loreal and supralabial scales; six and four scale organs on left and right postrostral respectively. Nasal scale separated from rostral, separated by a scale from first supralabial, right nasal limited by six scales; canthal separated from nasal by one scale. Nine supralabials. Seven lorilabials, six in contact with subocular. Seven infralabials. Oval auditory meatus (height 3.5 mm; width 2.7 mm). Eight smooth temporal scales. Distance between orbit and auditory meatus 9.4 mm. Rostral two times wider than tall (width 3.2 mm; height 1.5 mm). Mental trapezoidal, almost twice as wide as high (width 3.8 mm; height 2.1 mm). Interparietal hexagonal, with elongated posterior apex bordered by six scales, interparietal similar in size as parietals. Frontal divided into two scales. Complete supraorbital semicircles on both sides. Semicircles formed by 16 scales. Seven enlarged supraoculars. Seven superciliares. 14 upper and 12 lower ciliaries. Subocular not divided, longer than eye diameter. Supralabials of similar size. Some lorilabiales with rows of scale organs. Five lorilabials in contact with subocular. Preocular separated from lorilabial row by two scales. Postoculars similar in size as preocular. Mental scale in contact with four scales: one infralabial (on each side) and two enlarged chin scales (one on each side). Six chinshields (three on each side), second pair separated by four scales. Gular scales rounded, flat and without scale organs. 23 gular scales between auditory meatus. Longitudinal neck fold without keeled, but granular scales. Without antehumeral pocket, well developed antehumeral neck folds. 44 scales between auditory meatus and shoulder (counting along post-auricular and longitudinal neck fold). Gular fold absent. Dorsal scales rhomboid, slightly imbricate or juxtaposed, and slightly keeled. 57 scales between occiput and groin level. 58 scales around midbody. Scales of flanks rhomboidal and smooth. Ventrals slightly larger than dorsals, flat, and imbricated. 74 ventral scales between mental and cloaca; five precloacal pores present. Smooth laminar supracarpals, with oval or angular margins. Subdigital lamellae of fingers with three keels, and with formula I:8; II:13; III:16; IV:17; V:12 (right hand). Smooth and imbricated supradigital lamellae. Infratarsals and infracarpals keeled, both strongly imbricated. Supratarsals keeled, triangular or angular in shape. Subdigital lamellae of toes with formula I:8; II:14; III:18; IV:20; V:14 (right foot).

Color pattern in life (Figure 8). Dorsal head between snout and anterior level of eyes brown with black on scale margins, and between anterior level of eyes and occipital region mainly black. Anterior third of dorsal body with a central band of reddish cream scales and posterior two thirds mostly black with reddish cream scale tips; dorsal flanks mostly black with reddish cream scale tips. Lateral body mostly reddish cream but some with scales half black. Dorsal region of tail, anterior and posterior limbs, similar to dorsal body but also with pale blue scales. No vertebral line or scapular spots. Ventral body and tail gray with pale yellow scales, throat region with gray longitudinal stripes, ventral thighs orange.

Color in preservative. Head darker than body, dorsal head mostly dark beige but some scales completely black, parietal zone dark beige. Trunk with scales having black anteriorly and beige posteriorly, scales close to neck with one third black and two thirds beige, scales close to middle and posterior body two thirds black and one third beige. Fore and hind limbs same color as trunk, but with lighter scales close to hands and foot. Without vertebral line, scapular or paravertebral spots nor dorsolateral stripes. Lateral body with scales having one half black anteriorly and other half beige.

Dorsal tail similar to trunk, but with lighter scales close to scale tips. Ventrally, from mental scale to tail tip, mostly light gray but some scales cream.

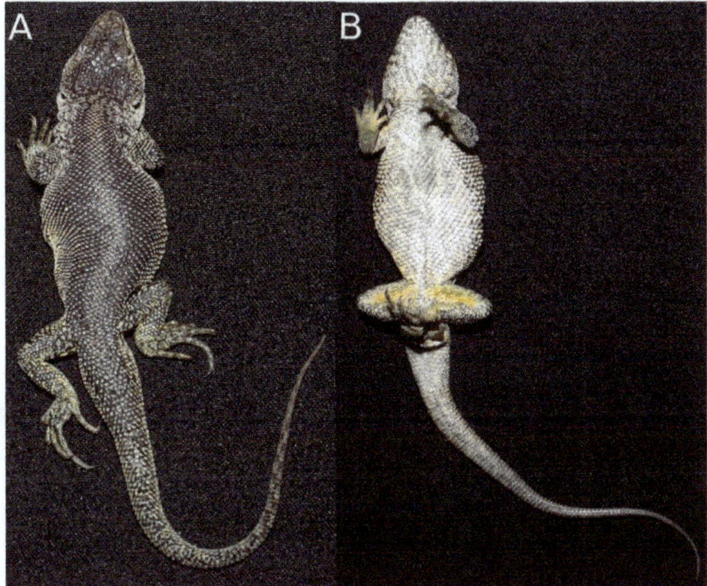

Figure 8. Holotype (male, SVL = 83.3 mm, MUSM 31461) of *Liolaemus victormoralesii* in dorsal (**A**) and ventral (**B**) views.

Variation (Figures 9 and 10). Variation in selected characters is summarized in Table 3. Sexual dichromatism present in ventral coloration, with males having yellow scales on thighs. Juveniles exhibit gray dorsal background coloration. Juveniles with black spots on paravertebral region, sometimes enclosing pale orange scales. Females and juveniles ventrally from mental region to tail tip white or white with gray or black scales, some specimens with reddish or orange scales on body venter.

Distribution and natural history (Figure 11). *Liolaemus victormoralesii* is only known in Huamanga Province, Department of Ayacucho, at elevations of 4175-4252 m; *L. victormoralesii* was found under rocks or on the ground in grassland areas. It was found together with *Liolaemus wari* and the snake *Tachymenis peruviana*. Díaz [9] recorded body temperatures of 104 specimens of *Liolaemus victormoralesii* (as *L. aff. melanogaster*; mean ± standard deviation): 21.3 ± 6.4 °C (substrate temperature: 14.5 ± 4.1 °C; air temperature: 13.2 ± 3.4 °C). However, lower body temperatures were recorded in the summer: 18.9 ± 6.8 °C (substrate temperature: 14.1 ± 3.1 °C; air temperature: 12.1 ± 2.7 °C); and higher body temperatures were recorded in the fall: 29.9 ± 0.1 °C (substrate temperature: 19.5 ± 3.4 °C; air temperature: 15.7 ± 2.2 °C). *Liolaemus victormoralesii* feeds on Araneae, Acari, Collembola, Scorpiones, Coleoptera, Diptera, Hymenoptera, Orthoptera, Lepidoptera, insect larvae and pupae, and vegetal matter [9]. *Liolaemus victormoralesii* feeds more on vegetal matter during summer and winter, and more on arthropods during spring (n = 56; [9]). This species is viviparous, one female had three embryos on each side at an advanced stage of development.

Figure 9. Dorsal and ventral (**B**) variation in *Liolaemus victormoralesii*. (**A**): BYU 50431 (male, SVL = 87.8 mm), (**B**): MUSM 31462 (juvenile, SVL = 60.9 mm), (**C**): BYU 50427 (juvenile, SVL = 52.2 mm.

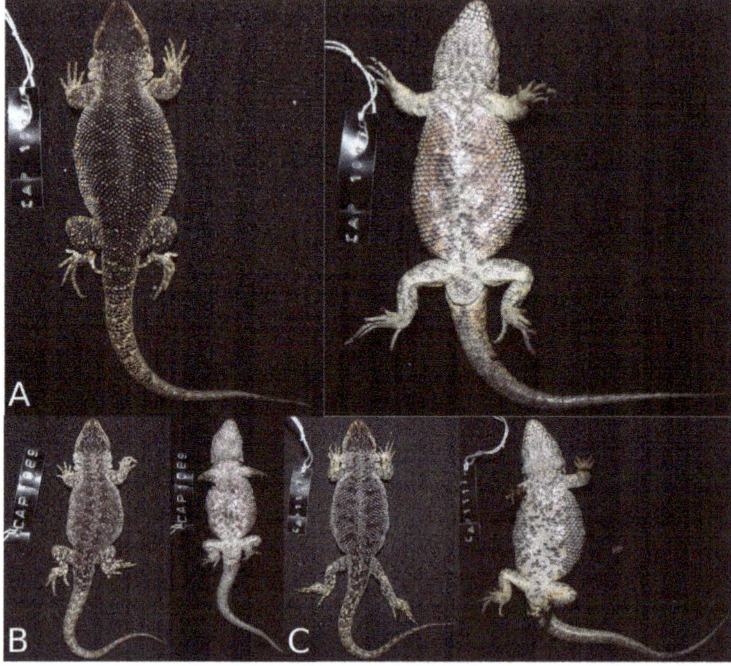

Figure 10. Dorsal and ventral variation in females of *Liolaemus victormoralesii*. (**A**): BYU 50433 (SVL = 79.7 mm), (**B**): BYU 50428 (SVL = 69 mm), (**C**): MUSM 31468 (SVL = 75.4 mm).

Table 3. Variation in selected characters among type specimens of *Liolaemus victormoralesii* sp. nov. All specimens from Museo de Historia Natural Universidad Mayor de San Marcos.

MUSEUM NUMBER	31461	31373	31372	25700	31460	31371	31468	31463
Sex	Male	Male	Male	Male	Female	Female	Female	
Reproductive stage	Adult	Adult	Adult	Adult	Adult	Adult	Adult	Juvenile
SVL	83.8	84.1	81.1	89	79.3	77.7	75.4	63.3
Groin armpit distance	29	32	30.2	33.8	35.6	35.1	34.5	25.8
Head length	19.22	21.2	20.7	20.1	17.7	17.6	16.6	14.4
Head width	18.2	19.1	18.2	18.7	16.4	15.1	15.5	12.8
Forelimb length	26.4	25.3	22.4	34.0	26.8	23.7	25.0	20.1
Hindlimb length	44.1	36.4	34.7	45.0	41.3	33.1	40.1	31.7
Snout length	7.6	8.9	8.2	8.6	7.6	7.8	6.6	5.7
Scales around midbody	58	56	61	52	56	60	51	60
Dorsal scale number	57	50	53	45	53	55	46	56
Head scale number	19	22	21	15	16	21	17	16
Scales around interparietal	7	4	8	7	6	7	7	5
Ventral scale number	74	72	79	66	75	78	70	74
Precloacal pores	7	4	8	7	0	0	0	4
Supralabial scales	11	10	10	10	9	9	10	9
Gular scales	23	26	25	25	24	24	21	27

Figure 11. Habitat of *Liolaemus victormoralesii*.

Etymology. The specific term *victormoralesii* is a noun in apposition and is given to honor our friend and colleague Víctor Morales for his contributions to herpetology. Víctor Morales passed away in December 2015, but his publications and memory live on with us.

4. Discussion

In this paper we describe three new species of *Liolaemus* from Perú (Figure 12), all are assigned to the *L. montanus* group. Two of these new species (*L. nazca* and *L. chiribaya*) inhabit the Pacific Peruvian coast. Although *L. nazca* could be present in San Fernando National Reserve, other populations are outside of any protected area and close to a mining concession in southern Perú. All *L. chiribaya* specimens were found outside any protected area, and some individuals of this species were found in a mining concession in Moquegua department. Other Pacific lowland *Liolaemus* species from the *L. montanus* group face additional threats. *L. insolitus*, for instance, has experienced habitat loss due to increased urbanization of coastal areas, and is categorized as Endangered by the IUCN [10]. Even though *L. poconchilensis* is also categorized as Endangered by the IUCN [11], Peruvian populations of this species lacks any legal protection as they are not included in the local list of protected species [12], and they are not present in any protected area.

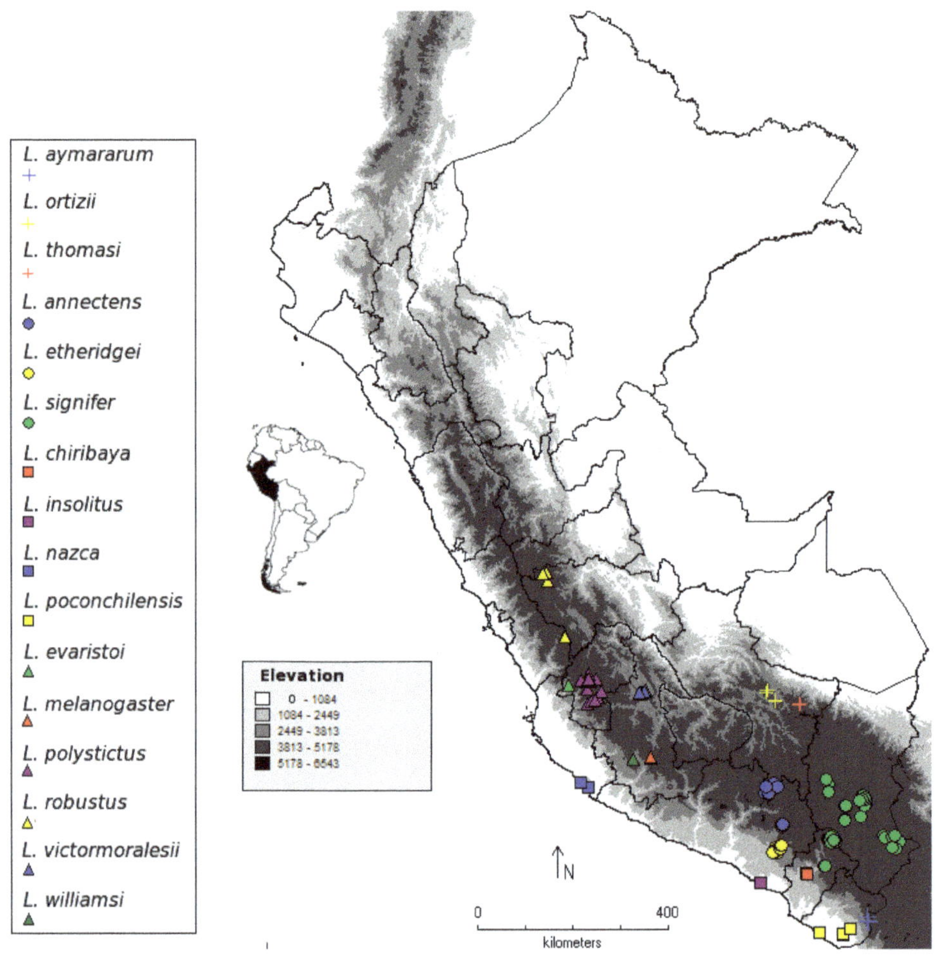

Figure 12. Map showing localities of species of the *Liolaemus montanus* group present in Perú.

Andean *Liolaemus* species of the *L. montanus* group present in Perú face similar threats as their lowland relatives. Some undescribed populations are initially found in mining concessions as a result of consulting activities and as part of environmental impact assessments [13]. For instance, the habitats of likely new, distinct populations related to *L. annectens* in southern Perú might have been destroyed or polluted before these lizards can be described and receive legal protection. Most probably these unnamed lizard populations are not present in any Peruvian protected area either.

Key to the Peruvian species of the *Liolaemus montanus* group.

1a. Dorsal scales larger, maximum of 53 from occiput to anterior level of thighs	2
1b. Dorsal scales smaller, up to 85 from occiput to anterior level of thighs	4
2a. Lateral scales without keels or slightly keeled	*L. aymararum*
2b. Lateral and dorsal scales strongly keeled	3
3a. Adults with dark belly	*L. thomasi*
3b. Not as above	*L. ortizii*
4a. Short snout, short and triangular lower jaw	5
4b. Not as above	6
5a. Ciliary scales well developed, without dorsal green spots	*L. poconchilensis*
5b. Ciliary scales not as above, with dorsal green spots	*L. chiribaya*
6a. Dorsal scales smooth	*L. insolitus*
6b. Dorsal scales keeled or slightly keeled	7
7a. Emerald green spots dorsolaterally, orange or yellow venter	*L. nazca*
7b. Not as above	8
8a. Blue spots dorsally and laterally on body, limbs and tail	9
8b. Not as above	10
9a. No blue spots on head	*L. evaristoi*
9b. Blue spots on head present	*L. etheridgei*
10a. Adult males with orange or bright yellow dorsal body scales (Arequipa, Moquegua and Puno regions)	11
10b. Not as above	12
11a. Small dorsal scales, up to 129 from occiput to anterior level of thighs	*L. signifer*
11b. Fewer scales not as above, maximum of 72 scales from occiput to anterior level of thighs	*L. annectens* (type locality)
12a. Adults with dorsal yellow-greenish scales on body, tail or limbs, and whitish venter	*L. robustus* (type locality)
12b. Not as above	13
13a. Only black scales on belly	*L. melanogaster*
13b. Not as above	14
14a. Females with precloacal pores present	*L. williamsi*
14b. Females without precloacal pores	15
15a. Females with large dorsal marks in a dark background coloration	*L. victormoralesii*
15b. Females with small dorsal marks in a light background coloration	*L. polystictus*

Author Contributions: Conceptualization, C.A.-P., C.R. and J.W.S.J.; Methodology, C.A.-P. and C.R.; Software, C.A.-P.; Validation, C.A.-P., C.R., E.C., A.M., V.V. and J.W.S.J.; Formal Analysis, C.A.-P. and C.R.; Investigation, C.A.-P., C.R., E.C., A.M., V.J.V. and J.W.S.J.; Resources, C.A.-P. and J.W.S.J.; Data Curation, C.A.-P. and C.R.; Writing—Original Draft Preparation, C.A.-P.; Writing—Review & Editing, C.A.-P., C.R., E.C., A.M., V.J.V. and J.W.S.J.; Visualization, C.A.-P.; Supervision, C.A.-P.; Project Administration, C.A.-P. and J.W.S.J.; Funding Acquisition, C.A.-P. and J.W.S.J.

Funding: This research was funded by WAITT FOUNDATION-NATIONAL GEOGRAPHIC SOCIETY, grant number W195-11; NSF-EMERGING FRONTIERS AWARD, grant number EF 1241885; and NSF-DOCTORAL DISSERTATION IMPROVEMENT GRANT, grant number 1501187.

Acknowledgments: We thank A. Resetar (FMNH), J. Losos, J. Rosado (MCZ), and F. Glaw (ZSM) for loans and accessions of specimens under their care. F. Huari, E. Coronado, M. Angeles and J. Montalvo for field assistant. Comments of editors and two anonymous reviewers improved our manuscript considerably. Fieldwork was supported by the BYU Bean Life Science Museum. Permits (RD No. 1280-2012-AG-DGFFS-DGEFFS, RD No. 008-2014-MINAGRI-DGFFS-DGEFFS) were issued by the Ministerio de Agricultura, Lima, Perú. The work was approved by the BYU Institutional Animal Care and Use Committee protocol number 12001 and in accordance with US law.

Conflicts of Interest: The authors declare no conflict of interest.

Appendix A

Specimens used in this study. FMNH= Field Museum of Natural History; ZSM = Zoologische Staatssammlung München; MCZ = Museum of Comparative Zoology.

Liolaemus aymararum (7): Tacna Department: Tacna Province: Palca: MUSM 21336-40; 21529, 21533.

Liolaemus ortizii (14): Cusco Department (Holotype: ZSM 647/1979; Paratype: MCZ 11061); Cusco Department: Paucartambo Province: Abra Huancarani: MUSM 31509-12; BYU 50476-79; Cusco Department: Calca Province: Chaupimayo: BYU 50473, 50475; MUSM 31513-14

Liolaemus thomasi (7): Cusco Department: Quispicanchis Province: Hualla Hualla: MUSM 31515, 31517, 31519; BYU 50466-68, 50470.

Liolaemus insolitus (4): Arequipa Department: Islay Province: Mollendo: MUSM 31489-90; BYU 50500, 50462.

Liolaemus poconchilensis (5): Tacna Department: Tacna Province: MUSM 31542-45; BYU 50176.

Liolaemus evaristoi (4): Huancavelica Department: Castrovirreyna Province: Sinto: MUSM 31454-55, BYU 50434, 50630

Liolaemus melanogaster (10): Ayacucho Department: Lucanas Province: MUSM 31472-76, BYU 50151-55.

Liolaemus polystictus (95): Huancavelica Department: Huancavelica Province: Santa Ines: (Holotype: MCZ 45845; Paratypes: MCZ R-161157, 43782, 45844, 45847, 45849, FMNH 81453-61); Huancavelica Department: Huaytará Province: Santa Ines: (MUSM 31446, 31449-52, 31456; BYU 50346, 50435, 50441, 50443, 50565); Huancavelica Department: Huancavelica Province: Huando: MUSM 29594; Huancavelica Department: Huaytará Province: Pilpichaca (MUSM 28632-38, 28641-56, 28658-77, 28679-701; BYU 50439-40).

Liolaemus robustus (65): Junín Department: (Holotype: FMNH 34242; Paratypes: FMNH 34242/1-15, 34242/19-21, 34242-23, MCZ 45811-12, R-161155-56, 157226); Junín Department: Junín Province: MUSM 13427, 13467-68, 13476-77, 31504, 30822; BYU 50485; Junín Department: Jauja Province: MUSM 13470-71, 13478; Pasco Department: Cerro de Pasco Province: MUSM 18056-58, 18197, 18401, 18403; Lima Department: Yauyos Province: (Paratype: MCZ 45830; MUSM 23633-36, 31505-08; BYU 50480-84); Huancavelica Department: Huancavelica Province: (MUSM 29591-93, 29595-96, 31439,31458, BYU 50438, 50442).

Liolaemus victormoralesii (10): FMNH 81435-42, 81446-47. Without locality data.

Liolaemus williamsi (23): Ayacucho Department: Lucanas Province: Pampas Galeras: (Paratypes: MCZ R-100435, R-145335-37, R-145340, R-157223); Ayacucho Department: Lucanas Province: Pampas Galeras: MUSM 1968, 1972, 1974, 21880-82; Ayacucho Department: Lucanas Province: Chaviña: MUSM 29690, Ayacucho Department: Lucanas Province: Lucanas: (MUSM 31483-87; BYU 50143-44, 50463-65).

Liolaemus annectens (4): Arequipa Department: Caylloma Province: MUSM 31499-500, 31503; BYU 50491.

Liolaemus etheridgei (9): Arequipa Department: Arequipa Province: Pocsi: MUSM 31491-93, 31496-97, BYU 50494-95; Arequipa Department: Arequipa Province: Chiguata: MUSM 31494-95.

Liolaemus signifier (2): Puno Department: Lampa Province: Lampa: Muruhuanca: MUSM 31433; Puno Department: Amantaní Island: MUSM 31434.

References

1. Lobo, F.; Espinoza, R.E.; Quinteros, S. A critical review and systematic discussion of recent classification proposals for liolaemid lizards. *Zootaxa* **2010**, *2549*, 1–30. [CrossRef]
2. Aguilar, C.; Wood, P.L., Jr.; Belk, M.C.; Duff, M.H.; Sites, J.W., Jr. Different roads lead to Rome: Integrative taxonomic approaches lead to the discovery of two new lizard lineages in the *Liolaemus montanus* group (Squamata: Liolaemidae). *Biol. J. Linn. Soc.* **2016**, *120*, 448–467.
3. Aguilar-Puntriano, C.; Avila, L.J.; De la Riva, I.; Johnson, L.; Morando, M.; Troncoso-Palacios, J.; Wood, P.L., Jr.; Sites, J.W., Jr. The shadow of the past: Convergence of young and old South American desert lizards as measured by head shape traits. *Ecol. Evol.* **2018**, *8*, 11399–11409. [CrossRef] [PubMed]

Diversity **2019**, *11*, 161

4.	Gutiérrez, R.C.; Chaparro, J.C.; Vásquez, M.Y.; Quiroz, A.J.; Aguilar-Kirigin, Á.; Abdala, C.S. Descripción y relaciones filogenéticas de una nueva especie de *Liolaemus* (Iguania: Liolaemidae) y notas sobre el grupo de *L. montanus* de Perú. *Cuad. Herpetol.* **2018**, *32*, 1–19.

5.	Valladares, P.J. Nueva especie de lagarto del género *Liolaemus* (Reptilia: Liolaemidae) del norte de Chile previamente confundido con *Liolaemus* (Phrynosaura) reichei. *Cuad. Herpetol.* **2004**, *18*, 41–51.

6.	Quinteros, A.S.; Abdala, C.S. A new species of *Liolaemus* of the *Liolaemus montanus* section (Iguania: Liolaemidae) from Northwestern Argentina. *Zootaxa* **2011**, *2789*, 35–48. [CrossRef]

7.	Smith, H. *Handbook of Lizards: Lizards of the United States and of Canada*; Cornell University Press: New York, NY, USA, 2018; p. 557.

8.	Frost, D.R. Phylogenetic analysis and taxonomy of the *Tropidurus* group of lizards (Iguania: Tropiduridae). *Am. Mus. Novit.* **1992**, *3033*, 1–66.

9.	Díaz Vargas, V. Ecologiía termal y nicho troófico de *Liolaemus wari* y *Liolaemus* aff. *melanogaster* (Sauria: Liolaemidae) en Abra Toccto. Licence of Biologist Thesis, Universidad Nacional de San Cristoóbal de Huamanga, Ayacucho, Peru, 2018.

10.	Aguilar, C.; Quiroz Rodriguez, A.; Perez, J. *Liolaemus insolitus.* The IUCN Red List of Threatened Species, E.T.48442648A48442655. 2017. Available online: http://dx.doi.org/10.2305/IUCN.UK.2017-2.RLTS.T48442648A48442655.en (accessed on 31 July 2019).

11.	Ruiz de Gamboa, M.; Nunez, H.; Valladares, P.; Lobos, G.; Mella, J. *Liolaemus poconchilensis.* The IUCN Red List of Threatened Species, E.T.56086313A56086315. 2017. Available online: http://dx.doi.org/10.2305/IUCN.UK.2017-2.RLTS.T56086313A56086315.en (accessed on 31 July 2019).

12.	Ministerio de Agricultura. Decreto Supremo No 004-2014-MINAGRI que aprueba la actualización de la lista de clasificación y categorización de las especies amenazadas de fauna silvestre legalmente protegidas. *El Peruano* **2014**, *8563*, 520497–520504.

13.	Aguilar, C.; Gamarra, R.; Ramirez, C.; Suarez, J.; Torres, C.; Siu-Ting, K. Anfibios andinos y estudios de impacto ambiental en concesiones mineras de Perú. *Alytes* **2012**, *29*, 88–102.

 *diversity*

Article

A New Species of Andean Gymnophthalmid Lizard (Squamata: Gymnophthalmidae) from the Peruvian Andes, and Resolution of Some Taxonomic Problems

Luis Mamani [1,2,3,*], Juan C. Chaparro [2,3], Claudio Correa [1], Consuelo Alarcón [2,3,4], Cinthya Y. Salas [5] and Alessandro Catenazzi [6]

1   Departamento de Zoología, Facultad de Ciencias Naturales y Oceanográficas, Universidad de Concepción, Barrio Universitario S/N, Concepción P.O. Box 160-C, Chile; ccorreaq@udec.cl
2   Colección de Herpetología, Museo de Historia Natural de la Universidad Nacional de San Antonio Abad del Cusco (MHNC), Cusco 08000, Peru; jchaparroauza@yahoo.com (J.C.C.); consue.alarcon@gmail.com (C.A.)
3   Museo de Biodiversidad del Perú (MUBI), Cusco 08000, Peru
4   Department of Biology, John Carroll University, Cleveland, OH 44118, USA
5   Área de Herpetología, Museo de Historia Natural de la Universidad Nacional de San Agustín de Arequipa (MUSA), Arequipa 04001, Peru; cinthyasalas84@hotmail.com
6   Department of Biological Sciences, Florida International University, Miami, FL 33199, USA; acatenazzi@gmail.com
*   Correspondence: luismamanic@gmail.com

http://zoobank.org/urn:lsid:zoobank.org:pub:FF0EC17F-965E-410D-AF0A-356B19BD4431
http://zoobank.org/urn:lsid:zoobank.org:act:6B3FCF87-82E4-4E2B-8C3B-99FD93E051CD
Received: 23 April 2020; Accepted: 18 September 2020; Published: 21 September 2020

Abstract: The family Gymnophthalmidae is one of the most speciose lineages of lizards in the Neotropical region. Despite recent phylogenetic studies, the species diversity of this family is unknown and thus, its phylogenetic relationships remain unclear and its taxonomy unstable. We analyzed four mitochondrial (12S, 16S, Cytb, ND4) and one nuclear (c-mos) DNA sequences of *Pholidobolus anomalus*, *Cercosaura manicata boliviana* and *Cercosaura* sp., using the maximum likelihood method to give insights into the phylogenetic relationships of these taxa within Cercosaurinae. Our results suggest that *Pholidolus anomalus* is nested within the clade of *Cercosaura* spp., that material we collected near Oxapampa belongs to a new species of *Cercosaura*, and that lizards identified as *Cercosaura manicata boliviana* belong to a separate lineage, possibly a new genus. We assign *Pholidobolus anomalus* to *Cercosaura*, redescribe the species, and designate a neotype to replace the lost holotype. In addition, we describe the new species of *Cercosaura*, and comment about the taxonomic status of "*Cercosaura manicata boliviana*" incertae sedis.

Keywords: Cercosaurinae; *Cercosaura manicata boliviana*; Cusco; diversity; Machupicchu; Oxapampa; *Pholidobolus anomalus*; Peru

1. Introduction

The eastern slopes of the Peruvian Andes are one of the regions with the greatest diversity of flora and fauna [1]. During the last few years, many species of plants and animals from the Peruvian Andes have been discovered (e.g., [2,3]). In particular, researchers have discovered many species of lizards of the family Gymnophthalmidae in poorly explored regions [4–9]. Recent phylogenetic inferences using molecular sequences have uncovered the phylogenetic relationships of many gymnophthalmid taxa [10–13]. However, there are still species whose phylogenetic position remains unclear, such as some populations currently assigned to the genus *Cercosaura* Wagler, 1830 and *Pholidobolus anomalus* Müller, 1923. Furthermore, the existence of poorly explored areas in the Peruvian Andes suggests that knowledge of the species richness of this group is still incomplete.

The genus *Cercosaura* is composed of 16 species that are widely distributed from the Andes to the Amazon [13–17]. The taxonomy of this lineage has long been confusing, but recent phylogenetic studies based on morphological and molecular data [11,15,18–20] have improved our knowledge of the composition and phylogenetic relationships of *Cercosaura*. On the basis of a morphological phylogenetic hypothesis, Doan [18] redefined the genus *Cercosaura* and synonymized it with the genera *Pantodactylus* and *Prionodactylus*. This arrangement was subsequently corroborated by molecular studies [19,20]. Later, Doan and Lamar [17] and Echevarría et al. [16] assigned two more gymnophthalmid taxa to the genus *Cercosaura*, increasing its richness to 14 species. Finally, Sturaro et al. [15] used integrative taxonomy to describe a new species of *Cercosaura*, and to resurrect *C. olivacea*. Despite these advances, the position of some species and populations within *Cercosaura*, such as *C. manicata boliviana*, remained unresolved.

Pholidobolus anomalus was described by Müller [21] from a single male specimen collected in the Department of Cusco, southeastern Peru. The holotype was deposited in the herpetological collection of the Zoologische Staatssammlung München, Germany (ZSM). Bombings during the Second World War damaged the ZSM collection, causing the loss and destruction of many type specimens, including the holotype of *Pholidobolus anomalus* [22]. Müller [21] considered the presence of a pair of small prefrontal scales in an almost rudimentary state as an outstanding character for this species. Due to this character, Müller [21] avoided assigning *P. anomalus* to the genus *Placosoma*. Instead, on the basis of similarities in pholidosis, he assigned the species to the genus *Pholidobolus* [21]. In some gymnophthalmid species, the condition of the prefrontal scales is variable at the intraspecific level [23,24]. Since its description in 1923, *Pholidobolus anomalus* has been reported once in the Department of Cusco [25]. Montanucci [25] analyzed two specimens of *P. anomalus* deposited by Thomas H. Fritts in the herpetological collection at the University of Kansas (KU 134857–58), both collected in the montane forests of Machupicchu (Cusco). On the basis of his morphological observations, Montanucci [25] concluded that *P. anomalus* was erroneously assigned to *Pholidobolus*. Generic reallocation of *P. anomalus* has since been hypothesized by several authors [25,26], but to date no taxonomic change has been proposed.

The molecular phylogenies carried out so far for the genus *Pholidobolus* included almost all species of the genus (*Pholidobolus affinis*, *P. condor*, *P. dicrus*, *P. dolichoderes*, *P. hillisi*, *P. macbrydei*, *P. montium*, *P. paramuno*, *P. prefrontalis*, *P. samek*, *P. ulisesi*, and *P. vertebralis*), but not *P. anomalus* because biological material was unavailable [11,12,27,28]. Previous authors highlighted the need to obtain new material of this taxon to reassess its taxonomic status and to examine its phylogenetic relationships [11,12,27,29].

In this study, we investigated the phylogenetic relationships of two specimens of *P. anomalus*, two specimens of *Cercosaura manicata boliviana*, and a specimen of *Cercosaura* that we identified as a new species. We analyzed four mitochondrial genes (12S, 16S, ND4, Cytb), and one nuclear gene (c-mos) using maximum likelihood inference. As a result of these analyses, and after examination of external morphology of the material assigned to *P. anomalus*, we reassign *P. anomalus* to the genus *Cercosaura*, provide a new description, and designate a neotype. Additionally, we describe the new species of the genus *Cercosaura*, and comment on the taxonomic status of *Cercosaura manicata boliviana* from southern Peru.

2. Materials and Methods

2.1. Biological Material and Taxon Sampling

We analyzed specimens of *Cercosaura* and *Pholidobolus anomalus* from central and southern Peru deposited in the Centro de Ornitología y Biodiversidad, Lima (CORBIDI), Museo de Biodiversidad del Perú, Cusco (MUBI), and Museo de Historia Natural de la Universidad Nacional de San Agustín, Arequipa (MUSA) (Table 1, Appendix A). We examined eight specimens of *P. anomalus* from Cusco, two specimens of *Cercosaura manicata boliviana* from Cusco and Puno, and two specimens of *Cercosaura pacha* sp. nov. from Oxapampa (Figure 1).

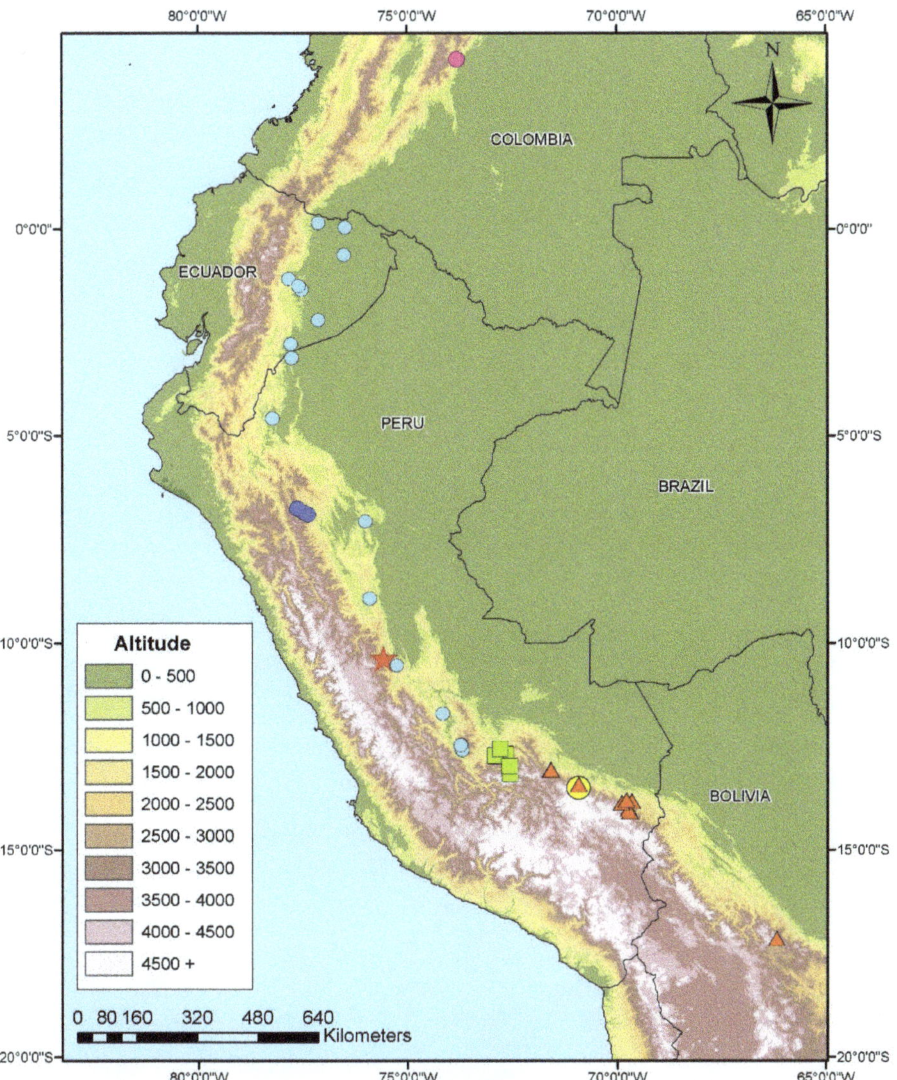

Figure 1. Geographic distribution of species of the genus *Cercosaura* in the Cordillera de los Andes. Orange triangles = "*Cercosaura manicata boliviana*"; light-blue circle = *C. manicata*; purple circle = *C. hypnoides*; blue circle = *C. doanae*; green square = *C. anomala*; red star = *C. pacha* sp. nov.; yellow circle = *Cercosaura* sp. Data taken from the literature [16,17,30] and museum records.

Table 1. Localities (all in Peru), coordinates, voucher numbers, and GenBank accession codes of newly sequenced specimens included in this study. (* described herein)

Species	Locality	Coordinates	Voucher	12S	16S	ND4	Cytb	c-mos
Pholidobolus anomalus	Tucantinas, La Convención, Cusco	12°44′16″ S/72°53′29″ W	MUBI 13328	MT531384	MT524454	MT522845	-	MT512508
Pholidobolus anomalus	Urusayhua, La Convención, Cusco	12°41′32″ S/72°39′18″ W	MUBI 13626	MT531385	MT524455	MT522846	MT512513	MT512509
Cercosaura manicata boliviana	San Pedro, Parque nacional del Manu, Paucartambo, Cusco	13°4′4″ S/71°33′45″ W	CORBIDI 16500	MT531386	MT524452	MT522849	-	MT512512
Cercosaura manicata boliviana	Santo Domingo, Limbani, Sandia, Puno	13°50′1″ S/69°38′29″ W	CORBIDI 18716	MT531387	MT524453	MT522848	MT512515	MT512511
Cercosaura sp. (*)	Lanturachi, Huancabamba, Oxapampa	10°23′01″ S/75°34′49″ W	MUBI 14515	MT531388	MT524456	MT522847	MT512514	MT512510

2.2. DNA Extraction, Amplification, and Sequencing

We obtained DNA sequences from five voucher specimens (Table 1). We extracted DNA from muscle tissues preserved in ethanol 96% using a commercial kit (Catalog #B47282, IBI Scientific). We obtained fragments of the nuclear oocyte maturation factor gene (c-mos), and the four mitochondrial genes: small subunit rRNA (12S), large subunit rRNA (16S), NADH dehydrogenase subunit 4 (ND4), and protein-coding cytochrome b (Cytb). We used standard primer and protocols for the polymerase chain reaction (PCR) (Table 2). We purified PCR products with Exosap-IT (Affymetrix, Santa Clara, CA, USA), and shipped purified products to MCLAB (San Francisco, CA, USA) for sequencing in both directions. We deposited new sequences in GenBank (Table 1). Additionally, we obtained 476 sequences of GenBank (https://www.ncbi.nlm.nih.gov/genbank/) of 124 terminals: 12S (124 sequences), 16S (123 sequences), ND4 (101 sequences), Cytb (19 sequences), and c-mos (109 sequences) (Table S1). We choose outgroups according to Moravec et al. [6].

Table 2. List of primers used in this study.

Gene	Primer	Primers Sequence (5′–3′)	PCR Cycle	Reference
12S	12S1L 12S2H	CAAACTGGGATTAGATACCCCACTAT AGGGTGACGGGCGGTGTGT	94 °C/3 min; 33 × (95 °C/30 s, 57 °C/30 s, 72 °C/90 s); 72 °C/10 min	[31]
16S	16sF.0 16sR.0	CTGTTTACCAAAAACATMRCCTYTAGC TAGATAGAAACCGACCTGGATT	96 °C/3 min; 40 × (95 °C/30 s, 51 °C/60 s, 72 °C/60 s); 72 °C/10 min	[31,32]
ND4	ND412931L ND413824H	CTACCAAAAGCTCATGTAGAAGC CATTACTTTTACTTGGATTTGCACCA	96 °C/3 min; 40 × (95 °C/30 s, 52 °C/60 s, 72°C/60 s); 72 °C/10 min	[33,34]
Cytb	L14841 H15149	AAAAAGCTTCCATCCAACATCTCAGCATGATGAAA AAACTGCAGCCCCTCAGAATGATATTTGTCCTCA	94 °C/5 min; 30 × (94 °C/60 s, 50 °C/60 s, 72 °C/60 s); 72 °C/10 min	[31]
c-mos	G73 G74	GCGGTAAAGCAGGTGAAGAAA TGAGCATCCAAAGTCTCCAATC	96 °C/3 min; 35 × (95 °C/25 s, 52 °C/60 s, 72 °C/120 s), 72 °C/10 min	[35]

2.3. Phylogenetic Reconstruction and Genetic Distances

We aligned the sequences of each fragment independently in MUSCLE [36], implemented in MEGA-X [37]. We concatenated sequences of the five fragments using Mesquite V3.61 [38].

Three phylogenetic analyses were conducted by maximum likelihood (ML) for mitochondrial genes (12S, 16S, Cytb, ND4), a nuclear gene (c-mos), and combined data (mitochondrial + nuclear). We inferred the optimal partition scheme using PartitionFinder 2.1.1 under the Bayesian information criterion (BIC) [39]. The best scheme were nine partitions (12S, 16S, c-mos-pos1 and cmos-pos2, c-mos-pos3, Cytb-pos1, Cytb-pos2, Cytb-pos3 and ND4-pos3, ND4-pos1, and ND4-pos2), and the evolution models were GTR+I+G for 12S, 16S, and ND4-pos1, TIM+G for c-mos-pos1 and c-mos-pos2,

K81UF+G for c-mos-pos3, SYM+I+G for Cytb-pos1, HKY for Cytb-pos2, TRN+G for Cytb-pos3 and ND4-pos3, and TVM+I+G for ND4-pos2. We inferred a phylogenetic tree using IQTREE 2 [40] and branch supports was estimated from 1000 pseudoreplicates using the ultrafast Bootstrap approach [41]. We uses *Alopoglossus viridiceps*, *Bachia flavescens*, *Ecpleopus gaudichaudii*, *Gymnophthalmus leucomystax*, *Rhachisaurus brachylepis* as outgroup taxa [6].

We estimated genetic distances between species (uncorrected p-distances) for 16S, which is the gene most commonly sequenced gene in gymnophthalmid lizards [6], and separately for 12S, ND4, and c-mos genes using Molecular Evolutionary Genetics Analysis (MEGA-X) [36] (Tables 3 and 4). Because the Cytb gene is poorly sampled in gymnophthalmid lizards, and available for only three species of *Cercosaura*, we omitted p-distances for this gene.

Table 3. Uncorrected p-distances for 12S (top) and 16S (bottom) genes between *Cercosaura anomala*, *Cercosaura* sp., and "*C. manicata boliviana*" and other cercosaurine species.

	1	2	3	4	5	6	7	8	9	10	11	12	13	14	15
1. *Cercosaura anomala* MUBI 13626		0.073	0.095	0.06	0.121	0.065	0.069	0.086	0.078	0.073	0.06	0.082	0.073	0.069	0.065
2. *Cercosaura argulus* QCAZ 4888	0.065		0.073	0.073	0.103	0.095	0.091	0.069	0.043	0.039	0.082	0.039	0.034	0.108	0.082
3. *Cercosaura bassleri* CORBIDI 11218	0.078	0.06		0.099	0.112	0.112	0.108	0.06	0.082	0.06	0.103	0.078	0.073	0.121	0.082
4. *Cercosaura doanae* CORBIDI 650	0.067	0.073	0.071		0.116	0.052	0.052	0.095	0.086	0.091	0.069	0.099	0.091	0.052	0.073
5. *Cercosaura eigenmanni* MRT 976979	0.071	0.058	0.048	0.08		0.121	0.116	0.099	0.108	0.108	0.121	0.108	0.116	0.125	0.121
6. *Cercosaura manicata* CORBIDI 8837	0.056	0.065	0.082	0.067	0.089		0.022	0.116	0.091	0.095	0.073	0.095	0.086	0.047	0.078
7. *Cercosaura manicata* QCAZ 5793	0.037	0.048	0.069	0.054	0.071	0.03		0.112	0.086	0.082	0.06	0.082	0.073	0.034	0.073
8. *Cercosaura ocellata* MRT 977406	0.095	0.067	0.065	0.08	0.054	0.095	0.082		0.078	0.073	0.112	0.065	0.078	0.095	0.065
9. *Cercosaura oshaughnessyi* QCAZ 4623	0.076	0.054	0.052	0.078	0.054	0.078	0.06	0.069		0.047	0.078	0.047	0.052	0.095	0.056
10. *Cercosaura parkeri* LG 1560	0.082	0.058	0.063	0.086	0.06	0.073	0.063	0.067	0.058		0.069	0.017	0.013	0.091	0.069
11. *Cercosaura quadrilineata* LG 936	0.086	0.069	0.084	0.089	0.069	0.084	0.069	0.071	0.08	0.076		0.082	0.073	0.078	0.065
12. *Cercosaura schreibersii albostrigatus* LG 1168	0.063	0.054	0.056	0.078	0.048	0.076	0.056	0.071	0.052	0.041	0.069		0.022	0.091	0.078
13. *Cercosaura schreibersii schreibersii* LG 927	0.069	0.06	0.056	0.08	0.054	0.078	0.058	0.076	0.054	0.043	0.08	0.026		0.082	0.073
14. *Cercosaura* sp. MUBI 14515	0.06	0.069	0.067	0.032	0.073	0.058	0.041	0.073	0.067	0.071	0.065	0.065	0.067		0.065
15. *Cercosaura manicata boliviana* CORBIDI 18716	0.056	0.063	0.071	0.069	0.067	0.065	0.045	0.078	0.06	0.071	0.082	0.058	0.065	0.067	

Table 4. Uncorrected p-distances for ND4 (top) and c-mos (bottom) genes between *Cercosaura anomala*, *Cercosaura* sp., and "*C. manicata boliviana*" and other cercosaurine species.

	1	2	3	4	5	6	7	8	9	10	11	12	13	14	15
1. *Cercosaura anomala* MUBI 13626		0.196	0.188	0.187	0.187	0.195	0.185	0.198	0.187	0.176	0.18	0.18	0.179	0.176	0.176
2. *Cercosaura argulus* QCAZ 4888	0.033		0.167	0.174	0.169	0.177	0.193	0.164	0.122	0.153	0.179	0.147	0.159	0.201	0.214
3. *Cercosaura bassleri* CORBIDI 11218	0.033	0.023		0.187	0.167	0.177	0.171	0.179	0.184	0.176	0.187	0.185	0.174	0.196	0.192
4. *Cercosaura doanae* CORBIDI 650	0.016	0.039	0.036		0.164	0.129	0.127	0.185	0.179	0.161	0.176	0.156	0.161	0.093	0.166
5. *Cercosaura eigenmanni* MRT 976979	0.029	0.02	0.01	0.033		0.18	0.184	0.192	0.167	0.166	0.192	0.151	0.166	0.182	0.221
6. *Cercosaura manicata* CORBIDI 8837	0.02	0.042	0.036	0.01	0.036		0.09	0.19	0.192	0.184	0.195	0.164	0.176	0.135	0.187
7. *Cercosaura manicata* QCAZ 5793	0.02	0.042	0.036	0.01	0.036	0		0.184	0.182	0.176	0.169	0.158	0.182	0.127	0.179
8. *Cercosaura ocellata* MRT 977406	0.039	0.029	0.013	0.042	0.016	0.036	0.036		0.177	0.161	0.193	0.184	0.167	0.188	0.2
9. *Cercosaura oshaughnessyi* QCAZ 4623	0.042	0.039	0.036	0.039	0.033	0.036	0.036	0.036		0.135	0.174	0.153	0.159	0.196	0.203
10. *Cercosaura parkeri* LG 1560	0.033	0.016	0.013	0.029	0.01	0.033	0.033	0.02	0.029		0.176	0.113	0.092	0.182	0.18
11. *Cercosaura quadrilineata* LG 936	0.046	0.049	0.046	0.042	0.042	0.046	0.046	0.052	0.049	0.039		0.203	0.198	0.188	0.176
12. *Cercosaura schreibersii albostrigatus* LG 1168	0.033	0.016	0.013	0.029	0.01	0.033	0.033	0.02	0.029	0	0.039		0.118	0.169	0.203
13. *Cercosaura schreibersii schreibersii* LG 927	0.033	0.016	0.013	0.029	0.01	0.033	0.033	0.02	0.029	0	0.039	0		0.182	0.166
14. *Cercosaura* sp. MUBI 14515	0.01	0.039	0.029	0.007	0.026	0.01	0.01	0.036	0.039	0.029	0.042	0.029	0.029		0.179
15. *Cercosaura manicata boliviana* CORBIDI 18716	0.023	0.052	0.042	0.026	0.039	0.029	0.029	0.049	0.052	0.042	0.055	0.042	0.042	0.02	

2.4. Designation of Neotype and Species Descriptions

In the original description of *Pholidobolus anomalus*, Müller [21] was not very precise about the type locality, and mentioned Cusco as the collecting locality. However, the most probable place where the holotype was collected is the Historical Sanctuary of Machupicchu (HSM), in the Cordillera de Vilcabamba. After its discovery by Hiram Bingham in 1911, the HSM became very popular with naturalists. The resulting scientific collections were deposited in different natural history museums [42,43].

We designed a neotype of *Pholidobolus anomalus* because the holotype was lost. This designation is covered by Article 75.3 of the International Code of Zoological Nomenclature (ICZN) [44]. We used the neotype to redescribe the species, and support the generic allocation. We followed Uzzell [45], Kizirian [46], and Doan and Cusi [24] for character definitions and measurements, and Chávez et al. [7] for description format. One of us (LM) observed morphological characters of species and took all measurements using a caliper with a precision of 0.1 mm. We referred to the literature for patterns of scalation and coloration of the following taxa: *Cercosaura anordosquama*, *C. ocellata*, *C. bassleri* and *C. olivacea* [13]; *Cercosaura argulus*, *C. eigenmanni*, and *C. oshaughnessyi* [47]; *C. hypnoides* [17]; *C. quadrilineata*, *C. schreibersii*, and *C. phelpsorum* [18]; *C. doanae* [16]; *C. manicata* [30]; *C. nigroventris* [48]; *C. parkeri* [49]; and *C. steyeri* [50]. We also examined specimens of *Cercosaura* deposited in the CORBIDI, MUBI, and MUSA collections (Appendix A).

This research was approved by the Institutional Animal Care and Use Committee of Southern Illinois University Carbondale (protocol #16-006). The Dirección General Forestal y de Fauna Silvestre, Ministerio de Agricultura y Riego, and Servicio Nacional Forestal y de Fauna Silvestre issued the permit authorizing this research (permits #210-2013-MINAGRI- DGFFS/DGEFFS, 064-2013-AG-DGFFS-DGEFFS, 359-2013-MINAGRI-DGFFS- DGEFFS, 292-2014-MINAGRI-DGFFS-DGEFFS, 024–2017–SERFOR/DGGSPFFS, and 369–2019–MINAGRI-SERFOR-DGGSPFFS).

The electronic version of this article in portable document format will represent a published work according to the International Commission on Zoological Nomenclature, and hence the new names contained in the electronic version are effectively published under that Code from the electronic edition alone. This published work and the nomenclatural acts it contains have been registered in ZooBank, the online registration system for the ICZN. The Life Sciences Identifier (LSID) for this publication is: urn:lsid:zoobank.org:pub:FF0EC17F-965E-410D-AF0A-356B19BD4431.

3. Results

3.1. Phylogenetic Relationships

Our phylogenetic ML (Figure 2) tree inferred using fragments of four mitochondrial and one nuclear gene is congruent with other studies [6,11,12,51,52], except that it recovered the monophyly of *Proctoporus* and that the recently described genus *Wilsonosaura* is nested within the *Proctoporus* lineage (Bootstrap: 76). However, trees using only mitochondrial (Figure S1) and nuclear (Figure S2) markers are not congruent. Our ML analyses using the full dataset recovered the polyphyly of genera *Cercosaura* and *Pholidobolus*. *Pholidobolus anomalus* and *Cercosaura* sp. (MUBI 14515) were nested within *Cercosaura*, whereas *Cercosaura manicata boliviana* was nested with *Potamites* and *Selvasaura*. Likewise, our mixed ML analyses recovered the monophyly of the genera: *Anadia* (98), *Andinosaura* (99), *Cercosaura* (including *C. anomala*, 85), *Dendrosauridion* (99), *Echinosaura* (100), *Gelanesaurus* (100), *Neusticurus* (100), *Macropholidus* (100), *Pholidobolus* (99), *Placosoma* (100), *Potamites* (100), *Proctoporus* (76), *Selvasaura* (100), and *Riama* (100). *Euspondylus* and *Rheosaurus* (both with a single known species) were recovered as independent lineages. Our phylogeny inferred using the full dataset is congruent with other studies, and may better reflect the evolutionary history of the lineages [53].

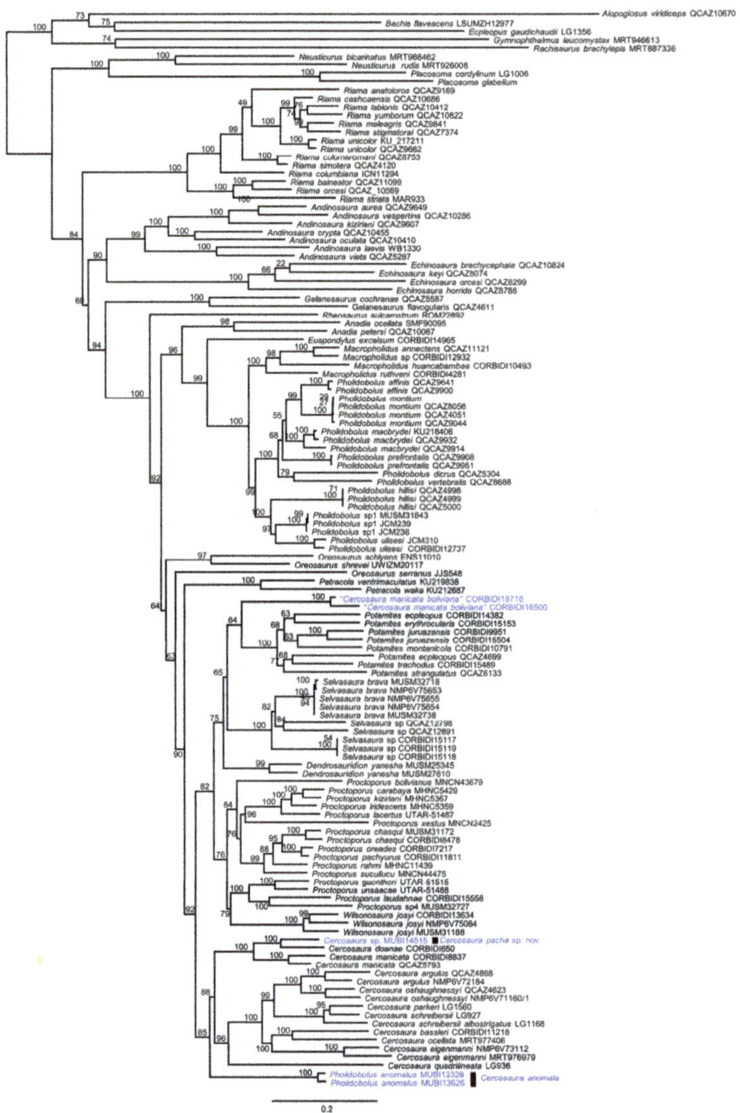

Figure 2. Maximum likelihood tree (log likelihood = −44,674.306, ultrafast boostrap = 1000) showing the phylogenetic relationships of *Cercosaura*, *Pholidobolus anomalus*, and other gymnophthalmid lizards. The numbers next to the nodes are bootstrap values. The analysis was constructed from a concatenated dataset of 2167 bp of four mitochondrial genes (12S, 16S, Cytb, ND4) and a nuclear gene (c-mos). In blue are the samples obtained in this study.

The uncorrected p-distance of the 16S gene between *Pholidobolus anomalus* and other taxa were: *Cercosaura manicata* (3.7–5.6%), *Cercosaura* sp. (6%), and *C. ocellata* (9.5%). *Cercosaura* sp. has a distance of 3.2% with respect to its sister species (*Cercosaura doanae*). Genetic distances among *Cercosaura manicata boliviana* and any other species of *Cercosaura* exceeded 4.5% (Table 3). The uncorrected p-distances between *Cercosaura* sp. and *C. doanae* in 12S (5.2%), ND4 (9.3%), c-mos (0.7%) genes were always greater than the intraspecific distance of *C. manicata* in 12S (2.2%), ND4 (9.0%), and c-mos (0.0%). These uncorrected p-distance between *Cercosaura* sp. and *C. doanae* are outside the intraspecific range, and within the interspecific range for species of gymnophthalmid lizards [11].

Based on our molecular analyses (Figure 2, Tables 3 and 4), we conclude (1) that *P. anomalus* needs to be redescribed and allocated to the genus *Cercosaura*, including designating a neotype; (2) *Cercosaura* sp. (MUBI 14515) is a new species; and (3) *Cercosaura manicata boliviana* is a valid species, the sister lineage of the semiaquatic lizards of the genus *Potamites* [20], but considered here as *incertae sedis*.

3.2. Taxonomy

Cercosaura anomala new. comb. (Müller, 1923)
Pholidobolus anomalus Müller, 1923

Neotype: MUBI 5277 (Figure 3), an adult male from Puente Ruinas, Santuario Histórico de Machupicchu, District of Machupicchu, Province of Urubamba, Department of Cusco, Peru (13°09′42″ S, 72°32′07″ W, at 2060 m), collected by J.C. Chaparro on 20 April 1998.

Figure 3. Neotype of *Cercosaura anomala*, male MUBI 5277 (snout vent length (SVL) = 60.7 mm).

Referred specimens: An adult male (MUBI 641), an adult female (MUBI 640), and a subadult female (MUBI 819) from the same locality as the neotype; two adult females (MUBI 13328, 13626), and a subadult female (MUBI 13529) from Vilcabamba (Figure 4A,B); a subadult male (MUSA 4537) from Maranura; and an adult male (MUBI 16169) from Quellouno (Figure 4C,D). All sites are located in Department of Cusco.

Figure 4. Live specimens of *Cercosaura anomala*. (**A**, **B**) Adult females (Urusayhua, Vilcabamba, (**A**) MUBI 13626, SVL = 68.2 mm, (**B**) MUBI 13328, SVL = 72.1 mm); (**C**, **D**) Adult male (Quellouno, La Convención, MUBI 16169, SVL = 56.2 mm). Photo: (A, B) Luis Mamani, (C, D) Juan C. Chavez-Arribasplata.

Etymology: The specific epithet *"anomalus"* is a nominative adjective in ancient Latin meaning irregular (*anomalus*), which implicitly refers to the presence and irregular shape of the prefrontal scales; then *anomala* (feminine nominative) must follow the genus *Cercosaura* (feminine).

Diagnosis: (1) Body robust, maximum snout vent length (SVL) of females, 72.1 mm; males, 61.7 mm; (2) head flat, elongated, 1.5 times longer than wide; (3) ear opening distinct, slightly recessed; (4) nasals separated by frontonasal; (5) frontonasal undivided; (6) prefrontal, frontal, frontoparietals, parietals and interparietals present, prefrontal scales in contact, occasionally separate; (7) parietal longer than wide; (8) three supraoculars, three postoculars, three infraoculars; (9) three superciliars, complete series; (10) nasal suture present; (11) loreal present; (12) 7–8 supralabials, four supralabials anterior to the posteroventral angle of the subocular, occasionally five, 6–8 infralabials; (13) 5–6 genials in contact; (14) collar fold present; (15) 33–36 transverse rows of dorsal scales, hexagonal, slightly keeled, imbricate; (16) 19–22 transverse rows of ventral scales, quadrangular, smooth, juxtaposed; (17) 35–43 scales around midbody; (18) lateral reduced scales at midbody in 3–4 lines; (19) limbs pentadactyl, all digits clawed; (20) 12–14 subdigital lamellae under finger IV, 17–19 under toe IV; (21) 8–10 femoral pores in males, 4–6 in females; (22) 2 preanal scales, 3–4 postanal scales; (23) tail up to 1.9–2.6 times longer than body; (24) caudals subimbricate, keeled dorsally, smooth ventrally; (25) lower palpebral disc transparent and undivided; (26) in life the dorsum is light-brown with some black spots, flanks dark brown with diffuse ocelli; lips with a cream line that extends to the front of arm insertion; ventral surface cream-reddish with some small scattered black spots, gular region of head cream-reddish with small black spots (Figure 4).

Cercosaura anomala is very similar to *C. hypnoides*, *C. manicata*, and *C. doanae*; all of which present a clear labial bar on both sides of the head and similar dorsal and lateral colorations. However, *C. anomala* differs from all other species of *Cercosaura* by the presence of smooth dorsal scales on the neck. In addition, *C. anomala* can be differentiated from *C. argulus* by the frontonasal scale being undivided (divided in *C. argulus*), 5–6 genial scales in contact (four genials in *C. argulus*); from *C. anodorsquama*, *C. bassleri*, *C. olivacea*, and *C. ocellata* because the keels of dorsal scales do not form lines arranged on the back (the keels of dorsal scales form continuous lines arranged on the back); from *C. hypnoides* by the presence of eight longitudinal ventral scales (six in *C. hypnoides*), absence of dorsolateral stripes (presence of continuous cream dorsolateral stripes in *C. hypnoides*); from *C. doanae* by the presence of dorsal scales with low keels (strongly keeled in *C. doanae*), maximum SVL of 72.1 mm in females (55.6 mm in *C. doanae*); from *C. eigenmanni* by the presence of 35–43 scales around the body (26–32 scales in *C. eigenmanni*), maximum SVL of 72.1 mm in females (47.0 mm in *C. eigenmanni*); from *C. manicata* by the presence of dorsal scales with low keels (strongly keeled in *C. manicata*), dorsal scales with the anterior and posterior edges almost blunt (anterior and posterior edges pointed in *C. manicata*); from *C. nigroventris* by the presence of 5–6 genials in contact (four genials in contact in *C. nigroventris*), maximum SVL of 61.7 mm in males (44.1 mm in *C. nigroventris*); from *C. oshaughnessyi* by the presence of the frontonasal scale not divided (divided in two in *C. oshaughnessyi*), maximum SVL of 72.1 mm in females (51 mm in *C. oshaughnessyi*); from *C. parkeri* by the presence of 35–43 scales around the body (24–30 in *C. parkeri*), 8–10 femoral pores per leg in males (3–5 per leg in *C. parkeri*); from *C. phelpsorum* by the presence of a white line on the upper lip (without a white line in *C. phelpsorum*), maximum SVL of 72.1 mm in females (55 mm in *C. phelpsorum*); from *C. quadrilineata* by the presence of 35–43 scales around the body (26 in *C. quadrilineata*), presence of 8–9 longitudinal ventral scales (four in *C. quadrilineata*); from *C. schreibersii* by the presence of three postoculars (two in *C. schreibersii*), 8–10 femoral pores per leg in males (3–6 in *C. schreibersii*); from *C. steyeri* by the presence of 8–9 longitudinal ventral scales (four in *C. steyeri*), dorsal scales not mucronate (dorsal scales mucronate in *C. steyeri*).

Description of the neotype (MUBI 5277): Adult male, SVL = 60.7 mm, tail length = 126.4 mm; head scales smooth, without striations, or rugosities; rostral scale wider (2.6 mm) than tall (1.5 mm), meeting supralabials on either side at above the height of supralabials, and becoming higher medially, in contact with frontonasal, nasal, and first supralabials; frontonasal pentagonal, wider than longer, widest in the mid, in contact with rostral, nasal, and prefrontals; prefrontals paired, pentagonal in contact with frontonasal, loreal, first superciliar, first supraocular, and frontal; frontal longer than wide, hexagonal, not in contact with superciliars, but in contact with first supraocular, and frontoparietals; frontoparietals polygonal, in contact with the frontal, all supraoculars, parietals, and interparietal; three supraoculars, all in contact with superciliaries, frontal, frontoparietals, parietal, and postocular; interparietal longer than wide, heptagonal, in contact with frontoparietals anteriorly, with parietal laterally, and with postparietals posteriorly; parietals polygonal, anteriorly in contact with frontoparietals, third supraocular, and postocular, laterally in contact with interparietals and supratemporals, and posteriorly with postparietals; three postparietals, smaller than parietals, the mid postparietal is smaller than lateral postparietals. Nasal divided, longer than high, in contact with first, and second supralabials; loreal present, in contact with second and third supralabials, in contact with nasal, first superciliar, and frenocular; four superciliars, first expanded onto surface of head; frenocular triangular in contact with third supralabials, first subocular, and loreal scales; palpebral disc made up of a single transparent scale; three suboculars; three postoculars; temporals smooth, glossy, and polygonal; four anterior supralabials to the posteroventral angle of the third subocular. Mental wider than long, in contact with first infralabial, and postmental; postmental single, polygonal, in contact with first and second infralabial, and first pairs of genials; three pairs of genials, five in contact, anterior pair in contact with second and third infralabials, middle pair in contact with third, fourth and fifth infralabials, posterior pair of genials in contact with fifth infralabial, and pregulars; two enlarged pregulars on left and right side, and 22 small pregulars irregularly distributed among enlarged pregulars; eight rows of gular scales including the collar, and the middle scales enlarged; collar fold distinct; lateral neck scales round, and smooth; dorsal neck scales smooth. Dorsal hexagonal, longer tan wide, juxtaposed, slightly keeled, in 36 transverse dorsal scale rows; 27 longitudinal dorsal scale rows at midbody; continuous lateral scale series, smaller than dorsals; reduced scales at limb insertion regions present; 22 transverse ventral scale rows; eight longitudinal ventral scale rows at midbody; a pair of anterior preanal plate scales; four posterior preanal plate scales; scales on tail rectangular, juxtaposed, and smooth. Limbs pentadactyl; digits clawed; dorsal brachial scales polygonal, imbricate, and smooth; ventral brachial scales small, rounded, and smooth; antebrachial scales polygonal, subequal in size, smooth, and imbricate; ventral antebrachial small, subimbricate, and rounded; dorsal manus scales polygonal, smooth, and subimbricate; palmar scales small, rounded, and domelike; dorsal scale on fingers smooth, quadrangular, imbricate, three on finger I, six on II, eight on III, nine on IV, and five on V; scales on anterodorsal surface of thigh large, polygonal, smooth, and subimbricate; scales on posterior surface of thigh small, rounded, juxtaposed, and keeled; scales on ventral surface of thigh large, roundish, flat, and smooth; nine femoral pores; preanal pores absent; scales on anterior surface of crus polygonal, keeled, and juxtaposed, decreasing in size distally; scales on posterior surface of crus small, roundish, keeled, and subimbricate; scales on dorsal surface of foot polygonal, smooth, and imbricate; scales on ventral surface of foot small, rounded, juxtaposed, and domelike; scales on dorsal surface of toes quadrangular, smooth, overhanging supradigital lamellae, four on toe I, seven on II, 10 on III, 12 on IV, and eight on V; fore and hind limbs overlapping when adpressed against the body.

Coloration in preservative: The dorsal surface of the head, neck, and body is brown with two clear dorsolateral lines on both sides of head that start from the supraoculars and disappear at the middle of body; the lateral surface of the head, neck, and body is dark brown; on both sides of the head and neck there is a cream labial line that extends from the tip of the head to the anterior part of the insertion of brachium; the ventral surface of the head, neck and body is dark gray with irregular cream spots. The dorsal surface of the limbs is brown, and ventral surface of limbs is similar to the ventral surface of the body (Figure 3).

Coloration in life: According to notes and photographs taken by LM of live specimens, the dorsal surface of the head and neck is brown with small black spots; the lateral sides of the head and neck are blackish brown with a cream labial line that extends from the tip of the head to the anterior part of the insertion of brachium; the ventral surface of the head is cream with small brown spots scattered; pregular and gular regions are similar to the ventral surface of head. The dorsal surface of body is brown with scattered black spots; lateral surface of body is blackish brown with black and cream spots that resemble ocelli; the ventral surface of body is reddish cream with scattered black spots. The dorsal surface of the limbs is brown with small black spots, the ventral surface is reddish cream with small black spots. The dorsal surface of the tail is similar to the dorsum, and the ventral surface of the limbs is similar to the ventral surface of the body (Figure 4).

Variation: Morphometric characters and pholidosis are presented in Table 5; *Cercosaura anomala* apparently has sexual dimorphism in size, females (maximum SVL = 72.1 mm, n = 5) are larger than males (maximum SVL = 61.7 mm, n = 4). The condition of the prefrontal scales is variable, all specimens examined have joined prefrontal scales, and only one subadult female (MUBI 819) and the lost specimen (holotype) have small and separate prefrontal scales.

Distribution and natural history: *Cercosaura anomala* inhabits montane forests on the eastern slopes of Cordillera de los Andes, in Department of Cusco, between 1745–2218 m a.s.l. We have observed this lizard on litter and on rocks from 10:00 to 14:00 h on sunny days at five localities in the Cordillera de Vilcabamba: Urusayhua, Tucantinas, Historical Sanctuary of Machupicchu, Maranura, and Quellouno (Figures 1 and 5). Sympatric gymnophthalmid lizards include *Proctoporus machupicchu*, *P. guentheri*, *P. unsaacae*, and *Proctoporus* sp. [8].

Table 5. Morphometric measurements and pholidosis of *Cercosaura anomala* and *C. pacha* sp. nov.

Measurements (mm)	*Cercosaura anomala* (all adults)						*Cercosaura pacha* sp. nov.	
	MUBI 641 Male	MUBI 5277 Male	MUBI 16169 Male	MUBI 13626 Female	MUBI 640 Female	MUBI 13328 Female	MUBI 14515 Female	MUBI 14512 Subadult Female
Snout-vent length	61.7	60.69	56.2	68.22	70.08	72.05	49.7	32.4
Tail length	72.6	126.44	146.2	130.3	107.3	40.87	98.1	41
Head length (chin to eardrum)	13.3	13.22	13.1	14.31	14.18	14.75	10.8	7.8
Head width	10	8.95	8	9.37	9.58	9.09	7.8	5.4
Postoculars	3	3	3	3	3	3	3	3
Superciliars	4/4	4/4	4/4	4/4	4/4	4/4	4/5	3/3
Palpebral disc	entire	entire	entire	entire	entire	entire	divided	1/2
Prefrontal scales	present	present	present	present	present	present	present	present
Nasoloreal suture	present	present	present	present	present	present	present	present
Supralabials	7/8	8/8	7/8	8/8	7/7	7/8	8/8	7/7
Supralabials anterior to the posteroventral angle of the subocular	4/5	4/4	4/4	4/4	4/4	4/4	4/4	4/4
Suboculars	3	3	3	3	3	3	3	3
Infralabials	6/7	7/7	7/7	7/7	8/7	7/7	6/6	6/6
Femoral pores	9/9	9/9	9/10	5/6	4/5	4/4	7/7	6/6
Loreal	present	present	present	present	present	present	present	present
Supraoculars	3	3	3	3	3	3	3	3
Genials in contact	5	5	6	6	6	6	5	4
Gular rows	9	9	8	9	9	9	8	7
Postparietals	5	5	5	5	5	3	3	2
Scales around midbody	38	35	39	39	40	40	38	37
Longitudinal dorsal count	24	27	31	27	24	25	30	29
Longitudinal ventral count	9	8	8	8	8	8	8	8
Transversal dorsal count	34	36	34	35	33	33	35	32
Transversal ventral count	22	21	19	19	21	19	20	19
Lamellae under finger IV	14	13	14	13	14	13	13	12
Lamellae under toe IV	18	18	19	19	19	18	18	18
Anal plate	6	6	5	6	6	5	5	5

Figure 5. Habitat of *Cercosaura anomala* in the montane forest of the Historical Sanctuary of Machupicchu, Department of Cusco.

Cercosaura pacha sp. nov.
Zoobank LSID: urn:lsid:zoobank.org:act:6B3FCF87-82E4-4E2B-8C3B-99FD93E051CD

Holotype: MUBI 14515 (Figure 6), an adult female from Lanturachi, Fundo los Cocos, District of Huancabamba, Province of Oxapampa, Department of Pasco, Peru (10°23′02″ S, 75°34′49″ W, at 1986 m), collected by J.C. Chaparro and C. Alarcón on 21 September 2014.

Figure 6. Holotype of *Cercosaura pacha* sp. nov. female MUBI 14515 (SVL = 49.7 mm).

Paratype: MUBI 14512 (Figure 7), a subadult female from near the type locality (10°23′29″ S, 75°34′12″ W, 1845 m).

Etymology: The specific epithet *"pacha"* is a female noun in Quechua language that means Earth.

Diagnosis: (1) Body robust, SVL 49.7 mm in a single adult female; (2) head flat, elongated, 1.4 times longer than wide; (3) ear opening distinct, slightly recessed; (4) nasals separated by frontonasal; (5) frontonasal undivided; (6) prefrontals, frontal, frontoparietals, parietals and interparietal present; (7) parietals longer than wide; (8) three supraocular; (9) 3–5 superciliar series complete; (10) nasal suture absent; (11) loreal present, in contact with the second supralabial; (12) 7–8 supralabial, four supralabials anterior to the posteroventral angle of the subocular, four infralabials; (13) 4–5 genial, all in contact; (14) collar fold present; (15) 32–35 transverse rows of dorsal, hexagonal, keeled, imbricate; (16) 19–20 transverse ventral rows, quadrangular, smooth, juxtaposed; (17) 37–38 scales around midbody; (18) lateral reduced scales at midbody in three lines; (19) limbs pentadactyl, all digits clawed; (20) 12–13 subdigital lamellae under finger IV, 18 under toe IV; (21) 6–7 femoral pores in females; (22) two preanal scales, three postanal scales; (23) tail up to 2.0 times longer than body; (24) caudals subimbricate, keeled dorsally, smooth ventrally; (25) lower palpebral disc transparent and divided in two; (26) in life the dorsum is brown with two cream dorsolateral stripe that stars over the eyes and join in the middle of the body forming a vertebral dorsal stripe that extends to the tail; lips with a cream line that extend from the third supralabial to the front of back leg; a cream lateral line between arm and leg, below the lateral line; all cream lines are bordered by continuous black spots; the venter is cream-reddish with some small scattered black spots, the gular region of head is cream-reddish with small black spots; tail is orange, with small dark spots ventrally and dorsally, and a cream-orange line laterally that begins at the back of legs and continues to tip of the tail (Figures 6 and 7).

Figure 7. Live specimen of *Cercosaura pacha* sp. nov., subadult female MUBI 14512 (paratype, SVL = 32.4 mm), from Lanturachi, near Oxapampa, Department of Pasco, Peru. Photo: Consuelo Alarcón.

Cercosaura pacha sp. nov. is similar to *C. anomala*, *C. doanae*, *C. hypnoides*, and *C. manicata*. However, *C. pacha* sp. nov. differs from *C. anomala* by having dorsal surface of neck keeled (smooth in *C. anomala*), six genials (4–5); from *C. doanae* by having dorsal scale of neck polygonal, keeled, and the distal edges of scales are blunt (strongly keeled, and the distal edges are pointed in *C. doanae*), dorsolateral stripes forming a vertebral dorsal stripe (not forming a vertebral stripe); from *C. hypnoides* by having loreal scales in contact with supralabials (not in contact with supralabias), eight longitudinal ventral scales (six), dorsal scales of neck polygonals (rounded); from *C. manicata* by having three postoculars (four in *C. manicata*), three suboculars (4–5), eight longitudinal ventral scales (six). Furthermore, *C. pacha* sp. nov. differs from *C. anordosquama*, *C. argulus*, *C. bassleri*, *C. eigenmanni*, *C. nigroventris*, *C. ocellata*, *C. olivacea*, *C. oshaughnessyi*, *C. parkeri*, *C. phelpsorum*, *C. quadrilineata*, *C. schreibersii*, and *C. steyeri* in having a clear labial bar that extends from the third supralabial to the point of insertion of the posterior limbs, and cream dorsolateral stripes that extends over the eyes and join in the middle of the body forming a single vertebral dorsal stripe that reaches the tail. Additionally, *C. pacha* sp. nov. can be distinguished from *C. argulus* and *C. oshaughnessyi* by having an undivided frontonasal (divided in *C. argulus* and *C. oshaughnessyi*); from *C. anordosquama*, *C. bassleri*, *C. ocellata* and *C. olivacea* by having the keels of the dorsal scales not organized in longitudinal rows, and eight longitudinal ventral rows (organized in longitudinal rows, and six in *C. anordosquama*, *C. bassleri*, *C. ocellata* and *C. olivacea*); from *C. eigenmanni* by having 37–38 scales around midbody (26–32 in *C. eigenmanni*); from *C. nigroventris* by having 37–38 scales around the midbody (40–44 in *C. nigroventris*), dorsal scales strongly keeled (weakly keeled in *C. nigroventris*); from *C. parkeri* by having 37–38 scales around the mid-body (24–30 in *C. parkeri*); from *C. phelpsorum* by having dorsal scales strongly keeled (weakly keeled in *C. phelpsorum*); from *C. quadrilineata* by having eight longitudinal ventral scales (four in *C. quadrilineata*); from *C. schreibersii* by having eight longitudinal ventral scales (six in *C. schreibersii*); from *C. steyeri* by having eight longitudinal rows of ventral scales (four in *C. steyeri*) and 37–38 scales around midbody (17).

Description of the holotype (MUBI 14515): Adult female, SVL = 49.7 mm, tail length = 98.1 mm; head scales with some rugosities; rostral scale wider (2.3 mm) than tall (1.2 mm), meeting supralabials on either side at above the height of supralabials, and becoming higher medially, in contact with frontonasal,

nasal, and first supralabials; frontonasal polygonal with blunt edges, wider than longer, widest at the back, in contact with rostral, nasals, loreals, and prefrontals; prefrontals paired, polygonal, in contact with frontonasal, loreal, first superciliar, first supraocular, and frontal; frontal longer than wide, polygonal, in contact with first and second supraocular, and frontoparietals; frontoparietals polygonal, in contact with the frontal, second and third supraoculars, parietals, and interparietal; three supraoculars, all in contact with superciliaries, frontal, frontoparietals, parietal, and postocular; interparietal longer than wide, heptagonal, in contact with frontoparietals anteriorly, with parietal laterally, and with postparietals posteriorly; parietals polygonal, anteriorly in contact with frontoparietals, third supraocular, and postocular, laterally in contact with interparietals and temporals, and posteriorly postparietals; three postparietals, smaller than parietals, the mid postparietal is smaller than laterals postparietals. Nasal undivided, longer than high, in contact with first and second supralabials; loreal present, in contact with second supralabials, nasal, first superciliar, and frenocular; four superciliars, the first expanded onto surface of head; frenocular trapezoidal, in contact with second and third supralabials, infraocular, subocular, and loreal scales; palpebral disc divided in two semitransparent scales; three suboculars; three postoculars; temporals with keeled and smooth scales (the big scales smooth and the small scales keeled), and polygonal; four supralabials anterior to the posteroventral angle of the subocular. Mental wider (2.3 mm) than long (1.5 mm), in contact with first infralabial and postmental posteriorly; postmental single, polygonal, in contact with first and second infralabial, and first pairs of genials; five genials, all in contact, on the left side two and on the right three, in contact with second, third, and fourth infralabials; 35 pregulars irregularly distributed, and small in the mid; seven rows of gular scales including the collar, the middle scales enlarged; collar fold distinct, formed by large scales; lateral neck scales round, upper scales keeled, and lower scales smooth; dorsal neck scales polygonal and keeled. Dorsal hexagonal, longer tan wide, juxtaposed, strongly keeled, in 35 transverse rows; 30 longitudinal dorsal scale rows at midbody; lateral scale series slightly smaller than dorsal; reduced scales at limb insertion regions; 20 transverse ventral scale rows; eight longitudinal ventral scale rows at midbody; four anterior preanal plate scales (the lateral scales are smaller), three posterior preanal plate scales; scales on tail rectangular, juxtaposed, and smooth. Limbs pentadactyl; digits clawed; dorsal brachial scales polygonal, imbricate, and slightly keeled; ventral brachial scales small, rounded, and smooth; dorsal antebrachial scales polygonal, subequal in size, smooth, and imbricate; ventral antebrachial small, subimbricate, and rounded; dorsal manus scales polygonal, smooth, and subimbricate; palmar scales small, rounded, and domelike; dorsal scale on fingers smooth, quadrangular, imbricate, four on finger I, six on II, eight on III, nine on IV, and five on V; scales on anterodorsal surface of thigh large, polygonal, smooth, and sub-imbricate; scales on dorsal surface of thigh large, keeled; scales on posterior surface of thigh small, rounded, juxtaposed, and keeled; scales on ventral surface of thigh large, roundish, flat, and smooth; seven femoral pores; preanal pores absent; scales on anterior surface of crus small, polygonal, keeled, juxtaposed, and decreasing in size distally; scales on posterior surface of crus small, roundish, keeled, and sub-imbricate; scales on ventral surface of crus large, roundish, flat, and smooth; scales on dorsal of foot roundish, smooth, and imbricate; scales on ventral of foot small, rounded, juxtaposed, and domelike; scales on dorsal surface of toes quadrangular, smooth, overhanging supradigital lamellae, three on toe I, six on II, 10 on III, 12 on IV, and eight on V; fore and hind limbs overlapping when adpressed against the body.

Coloration: In preservative, the dorsal surface of the head, neck, and back is dark-brown, the dorsolateral lines are cream grayish and join at midbody to form a vertebral stripe that extends to the tail; the dorsal surface of the tail is dark brown with a dorsal stripe in the anterior part of the tail, and pale orange with some gray spots in the distal part; the lateral sides of the head and neck are blackish brown with a cream labial line that extends from the third supralabial to the anterior part of the insertion of posterior limbs; the ventral surface of the head is gray with small, irregular, brown spots; gular and ventral surfaces of the body are dark gray with cream spots around some scales; the ventral surface of the limbs and tail, are cream with some irregular, dark gray spots (Figure 6). In life, the dorsal surface of the body is brown with scattered black spots; the lateral surface of the body is blackish brown with black and cream spots that resemble ocelli; the ventral surface of the body is reddish cream with scattered black spots. The dorsal surface of the limbs is brown with small black spots, the ventral surface is reddish cream with small black spots. The dorsal and ventral surfaces of the tail are orange, and the ventral surfaces of the limbs are similar to the ventral surface of the body (Figures 6 and 7).

Variation: Table 5 summarizes morphometric characters and pholidosis.

Distribution and natural history: *Cercosaura pacha* sp. nov. inhabits montane forests on the eastern slopes of Cordillera de los Andes, Department of Pasco, central Peru, between 1845–1986 m a.s.l (Figure 1). We captured two specimens using pitfall traps set up for 10 days at the type locality (Figure 8).

Figure 8. Type locality of *Cercosaura pacha* sp. nov, montane forest of Lanturachi, near Oxapampa, Department of Pasco, Peru. Photo: Consuelo Alarcón.

Taxonomic status of *Cercosaura manicata boliviana*

Cercosaura manicata boliviana is considered a subspecies of *Cercosaura manicata*, which is distributed from southeastern Peru to central Bolivia [30]. Werner [54] described *C. manicata boliviana* as *Prionodactylus bolivianus* based on a specimen (Muséum national d'Histoire naturelle, MNHN 00.4) collected in the montane forests of Bolivia (Chacó). Subsequently, *P. bolivianus* and *P. ockendeni* were considered synonyms of *Prionodactylus manicatus*, but both with subspecies status [30]. Finally, according to a phylogenetic study based on morphological data, *P. manicatus bolivianus* was transferred to the genus *Cercosaura* [18]. Echevarría et al. [16] and Uzzell [30] observed clear differences between both subspecies; however, only Echevarría et al. [16] considered this taxon as a putative separate species, and highlighted the need for genetic evidence.

The genus *Prionodactylus* was erected by O'Shaughnessy [55], and the type species was *Prionodactylus manicatus*. However, *P. manicatus* was transferred to the genus *Cercosaura*, and *Prionodactylus* was invalidated and considered as a synonym of *Cercosaura* [18].

According to the molecular evidence obtained in this study using two specimens from southern Peru (Figure 9), *Cercosaura manicata boliviana* is the sister lineage of the genus *Potamites*; therefore, it should be excluded from *Cercosaura*. However, the taxonomic assignation of *Cercosaura manicata boliviana* remains uncertain, because the genus *Prionodactylus* (original genus) is no longer valid. We could assign this species to *Potamites*, but external morphological characters and ecological traits do not support this taxonomic change. *Potamites* is a genus of lizards strongly associated with aquatic ecosystems [20], whereas the individuals of *"Cercosaura manicata boliviana"* have semi-arboreal habits. Thus, we propose to maintain the name *"Cercosaura manicata boliviana"* incertae sedis until a dedicated study can ascertain its phylogenetic relationships.

Figure 9. Live specimen of *"Cercosaura manicata boliviana"*; (**A**) adult male CORBIDI 16500 (SVL = 42.5 mm), from near San Pedro, Kosñipata Valley, Department of Cusco, Peru; (**B**) CORBIDI 18716 (SVL = 44.1 mm), from Santo Domingo, District of Limbani, Province of Sandia, Department of Puno, Peru. Photo: Alessandro Catenazzi.

In conclusion, the molecular, ecological, and morphological evidence support the hypothesis that *"Cercosaura manicata boliviana* is a separate species and a new lineage, which is sister to lizards of the genus *Potamites*. Future studies should ascertain the relationship of this *incertae sedis* with *Potamites*, and determine whether *P. bolivianus* and *P. ockendeni* are conspecifics or separate species.

4. Discussion

Our molecular and morphological evidence solves the taxonomy of *Cercosaura anomala*, reveals *"Cercosaura manicata boliviana"* as *incertae sedis*, and supports the description of a new species of *Cercosaura* from the Andes of Peru. Despite a complex taxonomic history, genetic data have supported recent changes in the systematics and taxonomy of cercosaurine lizards, increasing our understanding of their evolutionary history (e.g., [6,10,12,19,27,53]). However, genetic studies are still incomplete, and many genera and species are pending review and broader sampling of genetic sequences [6,10,12].

The ML topology obtained in this study using concatenated sequences of mitochondrial and nuclear genes recovered the monophyly of *Proctoporus* and included the genus *Wilsonosaura* within *Proctoporus*. This topology contrasts with previous studies that did not support the monophyly of *Proctoporus*, suggesting additional studies are needed to solve the taxonomy and phylogenetic position of *Proctoporus* [6,11,12,51,52]. Moreover, our study considered 129 terminals and addressed the taxonomy and phylogenetic relationships of the three species of *Cercosaura* (*Cercosaura anomala*, *"Cercosaura manicata boliviana"*, and *Cercosaura pacha* sp. nov.).

We designated a neotype for *Cercosaura anomala*, a designation carried out in accordance with article 75.3 of the ICZN, based on a specimen collected in Puente Ruinas, inside the Historical Sanctuary of Machupicchu, Department of Cusco, Peru. The designation of HSM as the type locality of *C. anomala*, and associated genetic data we provided in this work, are important because they will facilitate future taxonomic, phylogenetic, ecological, and evolutionary studies. Moreover, with the generic allocation of *C. anomala* and the description of a new species, we increase the diversity of the genus *Cercosaura*, which now contains 18 species.

In the original description of *Cercosaura anomala*, Müller [21] observed the small size of the prefrontal scales, and the separation between them, stating that these could be rudimentary. Among the material examined in this study, all specimens have large and attached prefrontals, except a subadult female (MUBI 819) with separate prefrontal scales. Variation in the form of prefrontals, and other characters, occurs in different species of gymnophthalmid lizards such as *Pholidobolus vertebralis* [24], *Proctoporus spinalis* [23], *P. machupicchu* [56], and *P. laudhanae* [57]. The high cryptic diversity, and the variation observed in the characters used in taxonomy of these lizards warn us that generic assignments and the description of new species should be undertaken with caution, and if possible, supported by genetic evidence [10,12,20,29].

Despite the similarity of coloration patterns of *Cercosaura anomala* with species of *Pholidobolus*, and *"C. manicata boliviana"* with *C. manicata*, both species were not nested in their designated genera. This result shows that external morphological characters in gymnophthalmid lizards can converge in coloration, and pholidosis [10,12,20,52]. Examples of convergence are the body shape of six divergent lineages of semi-aquatic lizards of the genera *Centrosaura*, *Echinosaura*, *Gelanesaurus*, *Neusticurus*, *Potamites*, and *Rheosaurus*, which share similar body shape ("cocodrile like morphology"), presence of irregulars scales on the back, and a laterally flattened tail that aids in water locomotion [31,58]; body elongations in *Anotosaura*, *Bachia*, *Calyptommatus*, *Heterodactylus*, *Nothobachia*, and *Scriptosaura*, [32,59]; legs reduction in *Bachia*, *Colobosaura*, and *Scriptosaura* [9,32,60]; and external ear loss in *Antonosaura*, *Bachia*, *Heterodactylus*, *Nothobachia*, *Rachisaurus brachylepis*, and *Scriptosaura catimbau* [32,60]. In light of high frequency of evolutionary convergence, it is expected that lizards of the genera *Cercosaura*, *Pholidobolus*, *Macropholidus*, and *"Cerosaura manicata boliviana"* share similarities in their coloration patterns. Future evolutionary studies will further elucidate evolutionary convergence in these lizards.

Supplementary Materials: The following are available online at http://www.mdpi.com/1424-2818/12/9/361/s1: Table S1: Vouchers and accession number codes of taxa included in this study available from GenBank. Figures S1 and S2: Mitochondrial and nuclear maximum likelihood trees respectively showing the phylogenetic relationships of *Cercosaura*, and other gymnophthalmid lizards.

Author Contributions: Conceptualization: L.M., J.C.C., C.C. and A.C.; methodology, L.M., C.C. and A.C.; software, L.M., C.C., C.A., and A.C.; validation, L.M., J.C.C., C.C., C.A., C.Y.S. and A.C.; formal analysis, L.M.

and A.C.; investigation, L.M., J.C.C., C.C., C.A., C.Y.S., and A.C.; resources, A.C.; data curation, L.M., J.C.C., C.C., C.A., C.Y.S., and A.C.; writing—original draft preparation, L.M., C.C. and A.C.; writing—review and editing, L.M., J.C.C., C.C., C.A., C.Y.S. and A.C.; visualization, L.M., C.C. and A.C.; supervision, C.C. and A.C.; project administration, L.M. and A.C.; funding acquisition, L.M. and A.C. All authors have read and agreed to the published version of the manuscript.

Funding: The Asociación para la Conservación de la Cuenca Amazónica (ACCA) provided logistical support at Urusayhua and Tucantinas in Departament of Cusco, Peru.

Acknowledgments: We thank Evaristo López (MUSA), Pablo Venegas (CORBIDI), and MUBI staff for allowing access to their respective herpetological collections during this study; Juan Carlos Chávez-Arribasplata (CORBIDI) for allowing us to use his photographs (Figure 3C,D); Tiffany M. Doan and an anonymous reviewer for their comments and suggestions that improved this manuscript.

Conflicts of Interest: The authors declare no conflict of interest.

Appendix A. Specimens Examined

Cercosaura anomala—Peru: Department of Cusco: Province of Urubamba, District of Machupicchu, sector Puente Ruinas (MUBI 640, 641, 817, 5277); Province of La Convención, District of Santa Ana, sector Urusayhua (MUBI 13626), sector Tucantinas (MUBI 13328, 13529); District of Maranura (MUSA 4537); District of Quellouno (MUBI 16169).

Cercosaura manicata—Peru: Department of Cusco, Province of La Convención, District of Kimbiri, sector Pomoreni (MUBI 6789) and Pichari (MUBI 15734, 15735, 15736).

Cercosaura pacha sp. nov.—Peru: Department of Pasco, Province of Oxapampa, District of Huancabamba, sector Lanturachi (MUBI 14512, 14515).

Cercosaura sp.—Peru: Department of Cusco, Province of Quispicanchi, District of Camanti, sector Sirigua (MUBI 5881).

"Cercosaura manicata boliviana"—Peru: Department of Puno: Province of Paucartambo, District of Kosñipata, Parque Nacional del Manu, Trocha Unión (MUBI 5045), sector San Pedro (CORBIDI 16500); Departament of Puno, Province of Carabaya, sector Gallucunka (MUBI 4657), sector Ollachea (MUBI 11575), Province of Sandia, District of Limbani, Santo Domingo (CORBIDI 18716).

References

1. Myers, N.; Mittermeier, R.A.; Mittermeier, C.G.; Da Fonseca, G.A.; Kent, J. Biodiversity hotspots for conservation priorities. *Nature* **2000**, *403*, 853–858. [CrossRef] [PubMed]
2. Timms, J.; Chaparro, J.C.; Venegas, P.J.; Salazar-Valenzuela, D.; Scrocchi, G.; Cuevas, J.; Leynaud, G.; Carrasco, P. A new species of pitviper of the genus *Bothrops* (Serpentes: Viperidae: Crotalinae) from the Central Andes of South America. *Zootaxa* **2019**, *4656*, 99–120. [CrossRef] [PubMed]
3. Tejedor, A.; Calatayud, G. Eleven new scaly tree ferns (Cyathea: Cyatheaceae) from Peru. *Am. Fern J.* **2017**, *107*, 156–191. [CrossRef]
4. Rodríguez, L.O.; Mamani, L. A new species of *Petracola* (Squamata: Gymnophthalmidae) from Río Abiseo National Park, San Martín, Peru. *Amphib. Reptil. Conserv.* **2020**, *14*, 140–146.
5. Lehr, E.; Moravec, J.; Lundberg, M.; Koehler, G.; Catenazzi, A.; Smid, J. A new genus and species of arboreal lizard (Gymnophthalmidae: Cercosaurinae) from the eastern Andes of Peru. *Salamandra* **2019**, *55*, 1–13.
6. Moravec, J.; Šmíd, J.; Štundl, J.; Lehr, E. Systematics of Neotropical microteiid lizards (Gymnophthalmidae, Cercosaurinae), with the description of a new genus and species from the Andean montane forests. *ZooKeys* **2018**, *774*, 105–139. [CrossRef]
7. Chavez, G.; Catenazzi, A.; Venegas, P.J. A new species of arboreal microteiid lizard of the genus *Euspondylus* (Gymnophtalmidae: Cercosaurinae) from the Andean slopes of central Peru with comments on Peruvian *Euspondylus*. *Zootaxa* **2017**, *4350*, 301–316. [CrossRef]
8. Mamani, L.; Goicoechea, N.; Chaparro, J.C. A new species of Andean lizard *Proctoporus* (Squamata: Gymnophthalmidae) from montane forest of the Historic Sanctuary of Machu Picchu, Peru. *Amphib. Reptil. Conserv.* **2015**, *9*, 1–11.

9. Goicoechea, N.; Padial, J.M.; Chaparro, J.C.; Castroviejo-Fisher, S.; De la Riva, I. A taxonomic revision of *Proctoporus bolivianus* Werner (Squamata: Gymnophthalmidae) with the description of three new species and resurrection of *Proctoporus lacertus* Stejneger. *Am. Mus. Novit.* **2013**, *3786*, 1–32. [CrossRef]

10. Goicoechea, N.; Padial, J.M.; Chaparro, J.C.; Castroviejo-Fisher, S.; De la Riva, I. Molecular phylogenetics, species diversity, and biogeography of the Andean lizards of the genus *Proctoporus* (Squamata: Gymnophthalmidae). *Mol. Phylogenet. Evol.* **2012**, *65*, 953–964. [CrossRef]

11. Torres-Carvajal, O.; Lobos, S.E.; Venegas, P.J. Phylogeny of Neotropical *Cercosaura* (Squamata: Gymnophthalmidae) lizards. *Mol. Phylogenet. Evol.* **2015**, *93*, 281–288. [CrossRef] [PubMed]

12. Torres-Carvajal, O.; Lobos, S.E.; Venegas, P.J.; Chávez, G.; Aguirre-Peñafiel, V.; Zurita, D.; Echevarría, L.Y. Phylogeny and biogeography of the most diverse clade of South American gymnophthalmid lizards (Squamata, Gymnophthalmidae, Cercosaurinae). *Mol. Phylogenet. Evol.* **2016**, *99*, 63–75. [CrossRef] [PubMed]

13. Sturaro, M.J.; Rodrigues, M.T.; Colli, G.R.; Knowles, L.L.; Avila-Pires, T.C. Integrative taxonomy of the lizards *Cercosaura ocellata* species complex (Reptilia: Gymnophthalmidae). *Zool. Anz.* **2018**, *275*, 37–65. [CrossRef]

14. Uetz, P.; Freed, P.; Hošek, J. (Eds.) The Reptile Database. Available online: http://www.reptile-database.org (accessed on 14 April 2020).

15. Sturaro, M.J.; Avila-Pires, T.C.S.; Rodrigues, M.T. Molecular phylogenetic diversity in the widespread lizard *Cercosaura ocellata* (Reptilia: Gymnophthalmidae) in South America. *Syst. Biodivers.* **2017**, *15*, 532–540. [CrossRef]

16. Echevarría, L.Y.; Barboza, A.C.; Venegas, P.J. A new species of montane gymnophthalmid lizard, genus *Cercosaura* (Squamata: Gymnophthalmidae), from the Amazon slope of northern Peru. *Amphib. Reptil. Conserv.* **2015**, *9*, 34–44.

17. Doan, T.M.; Lamar, W.W. A new montane species of *Cercosaura* (Squamata: Gymnophthalmidae) from Colombia, with notes on the distribution of the genus. *Zootaxa* **2012**, *3565*, 44–54. [CrossRef]

18. Doan, T.M. A new phylogenetic classification for the Gymnophthalmid genera *Cercosaura*, *Pantodactylus* and *Prionodactylus* (Reptilia: Squamata). *Zool. J. Linnean Soc.* **2003**, *137*, 101–115. [CrossRef]

19. Castoe, T.A.; Doan, T.M.; Parkinson, C.L. Data partitions and complex models in Bayesian analysis: The phylogeny of gymnophthalmid lizards. *Syst. Biol.* **2004**, *53*, 448–469. [CrossRef]

20. Doan, T.M.; Castoe, T.A. Phylogenetic taxonomy of the *Cercosaurini* (Squamata: Gymnophthalmidae), with new genera for species of *Neusticurus* and *Proctoporus*. *Zool. J. Linnean Soc.* **2005**, *143*, 405–416. [CrossRef]

21. Müller, L. Neue oder seltene Reptilien und Batrachier der Zoologischen Sammlung des bayrischen Staates. *Zool. Anz.* **1923**, *57*, 49–61.

22. Franzen, M.; Glaw, F. Type catalogue of reptiles in the Zoologische Staatssammlung München. *Spixiana* **2007**, *30*, 201–274.

23. Köhler, G.; Lehr, E. Comments on *Euspondylus* and *Proctoporus* (Squamata: Gymnophthalmidae) from Peru, with the description of three new species and a key to the Peruvian species. *Herpetologica* **2004**, *60*, 501–518. [CrossRef]

24. Doan, T.M.; Cusi, J.C. Geographic distribution of *Cercosaura vertebralis* O'Shaughnessy, 1879 (Reptilia: Squamata: Gymnophthalmidae) and the status of *Cercosaura ampuedai* (Lancini, 1968). *Check List* **2014**, *10*, 1195–1200. [CrossRef]

25. Montanucci, R.R. Systematics and evolution of the Andean lizard genus *Pholidobolus* (Sauria: Teiidae). *Univ. Kans. Mus. Nat. Hist. Misc. Publ.* **1973**, *59*, 1–52.

26. Reeder, T.W. A new species of *Pholidobolus* (Squamata: Gymnophthalmidae) from the Huancabamba depression of northern Peru. *Herpetologica* **1996**, *52*, 282–289.

27. Torres-Carvajal, O.; Mafla-Endara, P. Evolutionary history of Andean *Pholidobolus* and *Macropholidus* (Squamata: Gymnophthalmidae) lizards. *Mol. Phylogenet. Evol.* **2013**, *68*, 212–217. [CrossRef]

28. Hurtado-Gómez, J.P.; Arredondo, J.C.; Sales-Nunes, P.M.; Daza, J.M. A new species of *Pholidobolus* (Squamata: Gymnophthalmidae) from the paramo ecosystem in the northern Andes of Colombia. *S. Am. J. Herpetol.* **2018**, *13*, 271–286. [CrossRef]

29. Parra, V.; Nunes, P.M.S.; Torres-Carvajal, O. Systematics of *Pholidobolus* lizards (Squamata, Gymnophthalmidae) from southern Ecuador, with descriptions of four new species. *Zookeys* **2020**, *954*, 109–156. [CrossRef]

30. Uzzell, T.M. A revision of lizard of the genus *Prionodactylus*, with a new genus for *P. leucotictus* and notes on the genus *Euspondylus* (Sauria, Teiidae). *Postilla* **1973**, *150*, 1–67. [CrossRef]

31. Kocher, T.D.; Thomas, W.K.; Meyer, A.; Edwards, S.V.; Pääbo, S.; Villablanca, F.X.; Wilson, A.C. Dynamics of mitochondrial DNA evolution in animals: Amplification and sequencing with conserved primers. *Proc. Natl. Acad. Sci. USA* **1989**, *86*, 6196–6200. [CrossRef]

32. Pellegrino, K.C.M.; Rodrigues, M.T.; Yonenaga-Yassuda, Y.; Sites, J.W. A molecular perspective on the evolution of microteiid lizard (Squamata, Gymnophthalmidae), and a new classification for the family. *Biol. J. Linnean Soc.* **2001**, *74*, 315–338. [CrossRef]

33. Blair, C.; Méndez de La Cruz, F.R.; Ngo, A.; Lindell, J.; Lathrop, A.; Murphy, R.W. Molecular phylogenetics and taxonomy of leaf-toed geckos (Phyllodactylidae: Phyllodactylus) inhabiting the peninsula of Baja California. *Zootaxa* **2009**, *2027*, 28–42. [CrossRef]

34. Arevalo, E.; Davis, S.K.; Sites, J.W., Jr. Mitochondrial DNA sequence divergence and phylogenetic relationships among eight chromosome races of the *Sceloporus grammicus* complex (Phrynosomatidae) in central Mexico. *Syst. Biol.* **1994**, *43*, 387–418. [CrossRef]

35. Saint, K.M.; Austin, C.C.; Donnellan, S.C.; Hutchinson, M.N. C-mos, a nuclear marker useful for squamate phylogenetic analysis. *Mol. Phylogenet. Evol.* **1998**, *10*, 259–263. [CrossRef] [PubMed]

36. Edgar, R.C. MUSCLE: Multiple sequence alignment with high accuracy and high throughput. *Nucl. Acids Res.* **2004**, *32*, 1792–1797. [CrossRef] [PubMed]

37. Kumar, S.; Stecher, G.; Li, M.; Knyaz, C.; Tamura, K. MEGA X: Molecular evolutionary genetics analysis across computing platforms. *Mol. Biol. Evol.* **2018**, *35*, 1547–1549. [CrossRef]

38. Maddison, W.P.; Maddison, D.R. Mesquite: A Modular System for Evolutionary Analysis. Version 3.61. 2019. Available online: http://www.mesquiteproject.org (accessed on 21 March 2020).

39. Lanfear, R.; Calcott, B.; Ho, S.Y.; Guindon, S. PartitionFinder: Combined selection of partitioning schemes and substitution models for phylogenetic analyses. *Mol. Biol. Evol.* **2012**, *29*, 1695–1701. [CrossRef]

40. Minh, B.Q.; Schmidt, H.A.; Chernomor, O.; Schrempf, D.; Woodhams, M.D.; Von Haeseler, A.; Lanfear, R. IQ-TREE 2: New models and efficient methods for phylogenetic inference in the genomic era. *Mol. Biol. Evol.* **2020**, *37*, 1530–1534. [CrossRef]

41. Hoang, D.T.; Chernomor, O.; Von Haeseler, A.; Minh, B.Q.; Vinh, L.S. UFBoot2: Improving the ultrafast bootstrap approximation. *Mol. Biol. Evol.* **2018**, *35*, 518–522. [CrossRef]

42. Barbour, T. Reptiles collected by the Yale Peruvian expedition of 1912. *Proc. Acad. Nat. Sci. Phila.* **1913**, *65*, 505–507.

43. Amaral, A. New genera and species of snakes. *Proc. N. Engl. Zool. Club* **1923**, *8*, 85–105. [CrossRef]

44. ICZN. *The International Trust for Zoological Nomenclature*; Natural History Museum: London, UK, 1999; p. 306.

45. Uzzell, T.M. Teiid lizards of the genus *Proctoporus* from Bolivia and Peru. *Postilla* **1970**, *142*, 1–39.

46. Kizirian, D.A. A review of Ecuadorian *Proctoporus* (Squamata: Gymnophthalmidae) with descriptions of nine new species. *Herpetol. Monogr.* **1996**, *10*, 85–155. [CrossRef]

47. Avila-Pires, T.C.S. *Lizard of Brazilian Amazonia (Reptilia: Squamata)*; Zoologische Verhandelingen: Leiden, The Netherlands, 1995; p. 706.

48. Gorzula, S.; Senaris, J.C. *Contribution to the Herpetofauna of the Venezuelan Guayana*; Scientia Guaianae: Caracas, Venezuela, 1999; p. 269.

49. Soares-Barreto, D.; Martin Valdão, R.; Nogueira, C.; Potter de Castro, C.; Ferrerira, V.L.; Strüssman, C. New locality records, geographical distribution, and morphological variation in *Cercosaura parkeri* (Ruibal, 1952) (Squamata: Gymnophthalmidae) from western Brazil. *Check List* **2012**, *8*, 1365–1369. [CrossRef]

50. Tedesco, M.E. Una nueva especie de *Pantodactylus* (Squamata, Gymnophthalmidae) de la provincia de Corrientes, República de Argentina. *Facena* **1998**, *14*, 53–62.

51. Sánchez-Pacheco, S.J.; Torres-Carvajal, O.; Aguirre-Peñafiel, V.; Nunes, P.M.S.; Verrastro, L.; Rivas, G.A.; Rodrigues, M.T.; Grant, T.; Murphy, R.W. Phylogeny of *Riama* (Squamata: Gymnophthalmidae), impact of phenotypic evidence on molecular datasets, and the origin of the Sierra Nevada de Santa Marta endemic fauna. *Cladistics* **2018**, *34*, 260–291. [CrossRef]

52. Vásquez-Restrepo, J.D.; Ibáñez, R.; Sánchez-Pacheco, S.J.; Daza, J.M. Phylogeny, taxonomy and distribution of the Neotropical lizard genus *Echinosaura* (Squamata: Gymnophthalmidae), with the recognition of two new genera in Cercosaurinae. *Zool. J. Linnean Soc.* **2019**, *189*, 287–314. [CrossRef]

53. Fisher-Reid, M.C.; Wiens, J.J. What are the consequences of combining nuclear and mitochondrial data for phylogenetic analysis? Lessons from *Plethodon* salamanders and 13 other vertebrate clades. *BMC Evol. Biol.* **2011**, *11*, 300. [CrossRef]
54. Werner, F. Beschreibung neuer Reptilien und Batrachier. *Zool. Anz.* **1899**, *22*, 479–484.
55. O'Shaughnessy, A.W.E. An Account of the Collection of Lizards made by Mr. Buckley in Ecuador, and now in the British Museum, with descriptions of the new Species. *Proc. Zool. Soc. Lond.* **1881**, *49*, 227–245. [CrossRef]
56. Díaz, M.I.; Ttito, A.; Mamani, L. Una nueva localidad y reexaminación de la serie tipo de la lagartija Andina de Machu Picchu *Proctoporus machupicchu* Mamani, Goicoechea y Chaparro, 2015 (Squamata: Gymnophthalmidae). *Rev. Peru. Biol.* **2019**, *26*, 503–508. [CrossRef]
57. Chávez, G.; Chávez-Arribasplata, J.C. Distribution and natural history notes on the Peruvian lizard *Proctoporus laudahnae* (Squamata: Gymnophthalmidae). *Phyllomedusa J. Herpetol.* **2016**, *15*, 147–154. [CrossRef]
58. Marques-Souza, S.; Prates, I.; Fouquet, A.; Camacho, A.; Kok, P.J.; Nunes, P.M.; Vechio, F.D.; Recorder, R.S.; Mejia, N.; Junior, M.T.; et al. Reconquering the water: Evolution and systematics of South and Central American aquatic lizards (Gymnophthalmidae). *Zool. Scr.* **2018**, *47*, 255–265. [CrossRef]
59. Grizante, M.B.; Brandt, R.; Kohlsdorf, T. Evolution of body elongation in gymnophthalmid lizards: Relationships with climate. *PLoS ONE* **2012**, *7*, e49772. [CrossRef] [PubMed]
60. Rodrigues, M.T.; Santos, E.M.D. A new genus and species of eyelid-less and limb reduced gymnophthalmid lizard from northeastern Brazil (Squamata, Gymnophthalmidae). *Zootaxa* **2008**, *1873*, 50–60. [CrossRef]

 *diversity*

Article

Phylogenomic Reconstruction of the Neotropical Poison Frogs (Dendrobatidae) and Their Conservation

Wilson X. Guillory [1],*, Morgan R. Muell [1], Kyle Summers [2] and Jason L. Brown [1]

1   Department of Zoology, Southern Illinois University, Carbondale, IL 62901, USA
2   Department of Biology, East Carolina University, Greenville, NC 27858, USA
*   Correspondence: wilson.guillory@siu.edu; Tel.: +1-479-253-7461

Received: 10 June 2019; Accepted: 26 July 2019; Published: 29 July 2019

Abstract: The evolutionary history of the Dendrobatidae, the charismatic Neotropical poison frog family, remains in flux, even after a half-century of intensive research. Understanding the evolutionary relationships between dendrobatid genera and the larger-order groups within Dendrobatidae is critical for making accurate assessments of all aspects of their biology and evolution. In this study, we provide the first phylogenomic reconstruction of Dendrobatidae with genome-wide nuclear markers known as ultraconserved elements. We performed sequence capture on 61 samples representing 33 species across 13 of the 16 dendrobatid genera, aiming for a broadly representative taxon sample. We compare topologies generated using maximum likelihood and coalescent methods and estimate divergence times using Bayesian methods. We find most of our dendrobatid tree to be consistent with previously published results based on mitochondrial and low-count nuclear data, with notable exceptions regarding the placement of Hyloxalinae and certain genera within Dendrobatinae. We also characterize how the evolutionary history and geographic distributions of the 285 poison frog species impact their conservation status. We hope that our phylogeny will serve as a backbone for future evolutionary studies and that our characterizations of conservation status inform conservation practices while highlighting taxa in need of further study.

Keywords: UCE; phylogenetics; amphibians; Dendrobatidae; Aromobatidae; frogs; systematics

1. Introduction

Neotropical poison frogs, represented by the family Dendrobatidae within Anura, are one of the most charismatic and well-studied groups of amphibians. Popularly known for their powerful skin toxins and extravagant aposematism, dendrobatids have featured in scientific studies for decades in fields as diverse as reproductive behavior [1,2], pharmacology [3–5], color evolution [6–8], and biogeography [9–11], as well as recently fueling important studies in the evolution of monogamy [12] and toxin autoresistance [13,14]. In the context of Anura, Dendrobatidae is moderately diverse, inhabiting a range of habitats and ecological niches throughout Central and South America. Many dendrobatids are unfortunately threatened by a variety of factors including habitat destruction [15,16] and smuggling for the pet trade [17,18], making their conservation an important priority for biologists. Despite heavy popular and scientific interest in dendrobatids, in-depth studies of dendrobatid phylogenetic systematics have become scarce despite the rapid progress of phylogenomics. In this paper, we aim to provide the first evolutionary hypothesis of Dendrobatidae derived from genomic-scale data, to put to rest many of the outstanding questions concerning dendrobatid phylogeny.

The first dendrobatid described was *Rana tinctoria* by Cuvier in 1797 [19], later transferred to *Dendrobates* by Wagler in 1830 [20], where it remains to this day. Since then, described dendrobatid diversity has grown significantly, with roughly 198 species in 16 genera as of 2019 (Table 1). Until the 2006 revision of Dendrobatidae by Grant et al. [21], most dendrobatid species were confined to the genera *Dendrobates*, *Phyllobates*, *Colostethus*, *Epipedobates*, and *Minyobates*. Phylogenetic estimates constructed during this time from molecular data [22–30] generally recovered two main clades of dendrobatids: one composed of mostly aposematic frogs in *Phyllobates* and *Dendrobates*, and the other composed of more cryptic frogs in *Colostethus* and *Epipedobates*. A third group, which would later be established as the subfamily Hyloxalinae, was generally placed as sister to the *Dendrobates* clade ([22,23,27], though see [28] for an exception). During this time, the systematics and taxonomy of Dendrobatidae and its sister family Aromobatidae (then regarded as part of Dendrobatidae; this is still the taxonomy used by AmphibiaWeb) were confused and inconsistent. Phylogenies produced during this period were mostly constructed from alignments of a few mitochondrial loci, making them vulnerable to incomplete lineage sorting [31,32].

Table 1. Dendrobatid genera and relevant information. The authority and type species for each genus is given, as well as the number of described species and a very basic description of each genus' geographic range. Species counts and authorities retrieved from Amphibian Species of the World [33].

Subfamily	Genus	Authority	Type Species	No. Species	Range
Dendrobatinae	*Adelphobates*	Grant et al. 2006 [21]	*castaneoticus*	3	Amazonia
	Andinobates	Twomey et al. 2011 [34]	*bombetes*	15	N Amazonia
	Dendrobates	Wagler 1830 [20]	*tinctorius*	5	C America, N Amazonia, Hawaii (introduced)
	Excidobates	Twomey and Brown 2008 [35]	*mysteriosus*	3	NW Peru
	Minyobates	Myers 1987 [36]	*steyermarki*	1	Venezuela
	Oophaga	Bauer 1994 [37]	*pumilio*	12	C America, W Andean versant
	Phyllobates	Bibron 1840 [38]	*bicolor*	5	C America, Colombia
	Ranitomeya	Bauer 1986 [39]	*reticulata*	16	Amazonia
Colostethinae	*Ameerega*	Bauer 1986 [39]	*trivittata*	30	Amazonia, Bolivia
	Colostethus	Cope 1866 [40]	*latinasus*	15	Panama, NW S America
	Epipedobates	Myers 1987 [36]	*tricolor*	8	W Andean versant
	Leucostethus	Grant et al. 2017 [41]	*argyrogaster*	6	W Amazonia
	Silverstoneia	Grant et al. 2006 [21]	*nubicola*	8	C America, Colombia
Hyloxalinae	*Ectopoglossus*	Grant et al. 2017 [41]	*saxatilis*	7	C America, W Andean versant
	Hyloxalus	Jiménez de la Espada 1870 [42]	*fuliginosus*	60	Panama, W Andean versant, NW Amazonia
	Paruwrobates	Bauer 1994 [37]	*andinus*	3	W Andean versant

In 2006, Grant et al. comprehensively revised Dendrobatidae [21], splitting many of the previously paraphyletic genera into a multitude of new, monophyletic ones: *Ranitomeya*, *Adelphobates*, and *Oophaga* from *Dendrobates*; *Ameerega* from *Epipedobates*; and *Silverstoneia* and *Hyloxalus* from *Colostethus*. Most species within *Minyobates* were absorbed into *Ranitomeya*, leaving *M. steyermarki* as the sole member of the now-monotypic genus. The dendrobatid tree was becoming clearer now, with the genera previously in *Dendrobates*, along with *Phyllobates* and *Minyobates*, forming the subfamily Dendrobatinae, *Epipedobates*, *Ameerega*, *Colostethus*, and *Silverstoneia* forming Colostethinae, and *Hyloxalus* forming its own subfamily Hyloxalinae, which Grant et al. recovered as sister to Dendrobatinae rather than Colostethinae, conflicting with most previous phylogenies [21]. After this seminal study, dendrobatid taxonomy continued to fragment, with Twomey and Brown erecting the genus *Excidobates* in 2008 [35], Brown et al. (2011) splitting *Andinobates* from *Ranitomeya* [43], and Grant et al. (2017) establishing *Leucostethus*, a sister genus to *Ameerega*, as well as *Paruwrobates* and *Ectopoglossus*, both members of Hyloxalinae, in another broad systematic review [41].

Since Grant et al.'s 2006 revision, relatively few large-scale phylogenetic studies of dendrobatids have been undertaken [41,44,45]. Santos et al. published a time-calibrated phylogeny of Dendrobatidae in 2009 constructed from ~2400 bp of mitochondrial data [44], and Pyron and Wiens (2011) published an Amphibia supertree constructed via maximum likelihood containing many representatives of Dendrobatidae [45]. Most recently, Grant et al. (2017) published the most comprehensive dendrobatid tree to date [41], constructed using parsimony, and containing representatives of all genera. They provided evidence for the paraphyly of *Colostethus*, as *C. ruthveni* is nested within Dendrobatinae. The latter two studies were based on approximately a dozen mitochondrial and nuclear loci, with the addition of morphological data in the case of Grant et al. (2017). All of these studies recover Hyloxalinae as the sister group to Dendrobatinae, not Colostethinae, consistent with many pre-2006 studies [22–27,29]. Previous estimates differ with respect to the sister genus of *Dendrobates*, which is either *Oophaga* [44] or *Adelphobates* [41,45]. Finally, the problematic taxon *Minyobates steyermarki* is recovered in various places throughout the dendrobatine phylogeny, either as sister to *Adelphobates* [45] or to all other dendrobatines aside from *Phyllobates* and *C. ruthveni* [41].

Many dendrobatid frogs are of conservation concern. The International Union for Conservation of Nature Red List of Threatened Species (also known as the IUCN Red List), is one of the world's most comprehensive inventories of the global conservation status of biological species and has evaluated the status of many dendrobatids. It uses a set of criteria to evaluate the extinction risk of thousands of species globally and is recognized as an authority in the status of biological diversity. Here we present a novel approach for visualizing the relationships between IUCN Red List status, phylogenetic relationships, and spatial distributions. This is a tractable approach for visualizing complex patterns and large quantities of data in relatively simple infographics. These infographics are aimed at summarizing broad patterns, facilitating additional assessment, and complementing more detailed quantitative analyses.

In this study, we attempt to resolve the remaining uncertainties in dendrobatid phylogeny, specifically with regards to relationships between dendrobatid genera and subfamilies. Our primary advance for dendrobatid phylogenetics is the usage of genome-scale molecular markers known as ultraconserved elements (UCEs) [46,47], which provide an order of magnitude more molecular data to work with than previous studies and span the genome across chromosomes [47,48]. UCEs consist of an "ultraconserved" core region with identity or near-identity across the taxon set in question, along with increasingly divergent flanking regions with phylogenetic signal that evidence suggests is greater than in traditional protein-coding loci [49]. UCEs have become popular phylogenomic markers largely thanks to the ease with which thousands of UCE loci can be sequenced from even old museum specimens [50,51]. In recent years, UCEs have been used in many phylogenomic studies of vertebrates, and have been instrumental in resolving difficult phylogenetic problems at both deep and shallow timescales [52–58]. The use of UCEs in phylogenomic studies is appealing due to UCE loci having little overlap with paralogs [59], being found in genomic regions with few transposons [60], and having low saturation rates that decrease the possibility of homoplasy [58]. A study by Gilbert et al. (2015) showed that UCEs contain considerably more net phylogenetic informativeness than traditional protein-coding nuclear loci [49]. All of these factors led us to use UCE sequence capture as our method of choice for generating a phylogenomic dataset of the dendrobatid poison frogs. Here we provide the first dendrobatid phylogeny constructed from genome-scale data, which we hope will anchor future evolutionary studies of this fascinating amphibian group.

2. Materials and Methods

2.1. Data Collection

We gathered 63 dendrobatoid tissue samples from a combination of museum collections, our own field work, and the collections of collaborators (Table S1). Our sample represents 36 species in 13 dendrobatid genera (the newly erected genera *Paruwrobates*, *Ectopoglossus*, and *Leucostethus* were not included) and includes the aromobatid *Allobates femoralis* as an outgroup taxon. In many cases, we include multiple representatives of a given species to account for geographic variation.

For each sample, we performed sequence capture of UCEs in the manner of Faircloth et al. [46]. We extracted genomic DNA from each tissue sample with the Qiagen DNeasy Blood and Tissue Kit (Valencia, CA, USA) and performed quality and yield assessment with a Qubit 3 fluorometer (ThermoFisher Scientific). Extracted DNA was sent to RAPiD Genomics (Gainesville, FL, USA), who performed Illumina sequencing of UCEs, enriching the samples with the Tetrapods-UCE-5Kv1 probe set, which contains 5472 probes that target 5060 UCE loci. Raw reads for each sample are available at the NCBI Sequence Read Archive under project number PRJNA547821.

2.2. Bioinformatics Pipeline

We used the software package PHYLUCE v1.5.0 [61] and associated tools to trim, assemble, and align our sequenced reads. We performed quality trimming on raw reads using Illumiprocessor v2.0.6 [62], a Python wrapper for Trimmomatic v0.36 [63]. We then assembled the trimmed reads with Trinity v1.6 [64] as implemented in PHYLUCE. After assembly, we created two separate taxon sets for later analyses: one containing all samples ($n = 63$, "large dataset"), the other with one sample per species ($n = 37$, "small dataset"). The purpose of the small dataset was to increase computational efficiency for divergence time estimation. After mapping assembled contigs to UCE loci using PHYLUCE, we retained 2733 loci for the large dataset and 2639 for the small one. We performed individual alignments on each locus using MUSCLE v3.8.31 [65] as implemented in PHYLUCE. We filtered for matrix incompleteness by only retaining loci present in 60% or more of taxa and performed additional filtering by calculating the number of parsimony-informative sites (PIS) with PHYLOCH v1.5-5 [66], implemented in a custom R script, and retaining only loci with $10 < \text{PIS} < 120$. Our upper limit on PIS was to filter out outlier loci, while our lower limit was to filter out relatively uninformative loci. After both filtering steps, for the large dataset we retained 1719 of the original 2733 loci, and for the small dataset we retained 1706 of the original 2639 loci. This study was conducted in accordance with the Institutional Animal Care and Use Committee of Southern Illinois University (Protocol number: 18-009, Animal Assurance number: D16-00044).

2.3. Phylogenetic Analyses

For both large and small datasets, we performed both maximum likelihood (ML) and multispecies coalescent-consistent phylogenetic analyses. We used IQ-TREE v1.5.5 [67] to perform our ML analyses, using a general time-reversible (GTR) model and assessing support with 10,000 ultrafast bootstrap replicates [68]. ML analyses were performed on an unpartitioned concatenated matrix (large matrix: 787,199 characters; small matrix: 786,510 characters) to increase computational efficiency. As UCEs are usually not protein-coding, it is unclear which partition schemes should be used for them, or whether they should be used at all [55].

We also inferred the dendrobatid species tree using ASTRAL-III v5.6.1 [69], a summary method consistent with the multispecies coalescent. ASTRAL-III accounts for incomplete lineage sorting by summarizing gene trees constructed separately for each locus, rather than effectively assuming the whole set of loci acts as a single gene, as in a concatenated ML analysis. We made individual gene trees for each UCE locus in IQ-TREE with a GTR model and 1000 ultrafast bootstrap replicates. We contracted near-zero branch lengths to polytomies in the gene trees with IQ-TREE's *-czb* option, an approach recommended by Persons et al. (2016) [70] to reduce downstream bias. We used the gene trees as input for ASTRAL-III. For the large dataset, we assigned each sample to one of 38 putative species in a mapping file used as input with ASTRAL's *-a* option. For the small dataset, we omitted the mapping file so that ASTRAL-III correctly assumed that each sample corresponded to its own species.

2.4. Divergence Time Estimation

We performed divergence time estimation of our dendrobatid phylogeny using BEAST 2 v2.5.0 [71]. Because Bayesian methods are computationally intensive, we reduced our small dataset, consisting of 1706 loci for 37 samples, to four subsets of 50 random loci each for analysis in BEAST, amounting to a total matrix size of 92,742 characters. We analyzed each subset twice in order to ensure that convergence occurred. We concatenated the loci in each subset and did not partition the alignment in order to avoid the intractably long periods of time a partitioned Bayesian analysis can take to converge [72].

We used the utility BEAUti to specify our BEAST settings. We used an HKY model with 4 gamma rate categories, with base frequencies set to "empirical," avoiding the GTR model because it can lead to overparameterization and subsequently low ESS values [73]. We used a relaxed log-normal clock model with a clock rate prior of 1e-10, the same order of magnitude as an estimate of avian UCE substitution rates from Winker et al. [74]. To further reduce computational demands, we fixed the analysis to our small ASTRAL topology by setting the *subtreeSlide, narrowExchange, wideExchange,* and *wilsonBalding* operators to zero, an approach used by Hsiang et al. [75]. We used a Yule tree prior with other priors set to their default values.

For our divergence time calibration, we used an indirect calibration derived from the timetree provided by Santos et al. (2009) [44]. Divergence time estimation in dendrobatids is a difficult problem because they lack a fossil record. Santos et al. calibrated their dendrobatid tree by nesting it within a tree for the whole of Amphibia, which has a voluminous fossil record, and using a combination of paleogeographic, fossil, and molecular clock evidence to date that amphibian tree. For our study, we use Santos et al.'s estimation of the divergence between Dendrobatidae and Aromobatidae (i.e., the node separating *Allobates femoralis* from the rest of our samples). The average of Santos et al.'s three estimations of this node's age was 38.534 Ma with σ = 5.151 Ma. Our calibration assigned a normal distribution with these values to this node.

We ran each analysis with an MCMC chain length of 100,000,000 generations, with a log sampling frequency of 100,000 generations and a tree sampling frequency of 10,000 generations. We assessed convergence between runs of each subset and ESS values for each run with Tracer v1.7.1 [76]. We found that all parameters had ESS values over the popular threshold of 200, and that convergence was reached for each subset. We used LogCombiner v2.5.0 [71] to combine the posterior distributions of trees from each of the eight runs, accounting for a burn-in of 10%. We then used TreeAnnotator v2.5.0 [71] to summarize the combined tree files, targeting the maximum clade credibility tree with mean node heights.

2.5. Visualizing Evolutionary History, Conservation Risk, and Spatial Distributions

To better understand how the evolutionary history and the geographic distribution of poison frogs impacts their conservation status and extinction risk, we downloaded IUCN Red List classifications for all surveyed species of Dendrobatidae and its sister family Aromobatidae (*n* = 285) [77]. For each species, we recorded its population status, Red List status, and its countries of occurrence. The relationships between these factors were visualized in two circular plots created in the R package *circlize* [78]. The phylogenetic results of this study and those of Grant et al. (2017) [41] characterized the genus-level relationships among all input species.

3. Results

We obtained two maximum likelihood trees from IQ-TREE and two species trees from ASTRAL-III. A summarized genus-level phylogeny with divergence times is shown in Figure 1. The large IQ-TREE phylogeny (Figure S1) and small IQ-TREE phylogeny (Figure S2a) had identical topologies in terms of relationships between species. The large ASTRAL-III phylogeny (where 63 samples were coalesced into 38 species; Figure S2b) was nearly identical to the small ASTRAL-III phylogeny (Figure S2c) and the small IQ-TREE phylogeny (Figure S2a), with the exceptions of the additional unidentified *Allobates sp.* sample and the rearrangement of several *Ameerega* species (Figure S2b, gray labels). Topologies for the small IQ-TREE and small ASTRAL-III analyses were also identical (Figure S2), pointing to an overall concordance between methods. In terms of species relationships, the large ASTRAL-III topology differed from the large IQ-TREE topology only in the rearrangement of the three *Ameerega* species mentioned above. Topologies and support values were extremely similar across trees. Each genus was always monophyletic with high support, as well as the three dendrobatid subfamilies. We found Hyloxalinae sister to Colostethinae rather than Dendrobatinae. In all trees, *Oophaga* was recovered as sister to *Dendrobates*, and *Minyobates* was recovered as sister to *Adelphobates* (Figure 1). Regarding other generic relationships, we found that the clade containing *Oophaga* and *Dendrobates* is sister to the clade containing *Adelphobates* and *Minyobates*. *Ranitomeya* and *Andinobates* are recovered as sister genera, with *Excidobates* sister to this clade. *Phyllobates* is recovered as the sister genus to all other dendrobatines. In Colostethinae, we recovered *Epipedobates* and *Silverstoneia* as sister genera, with this clade itself sister to the clade containing *Ameerega* and *Colostethus*.

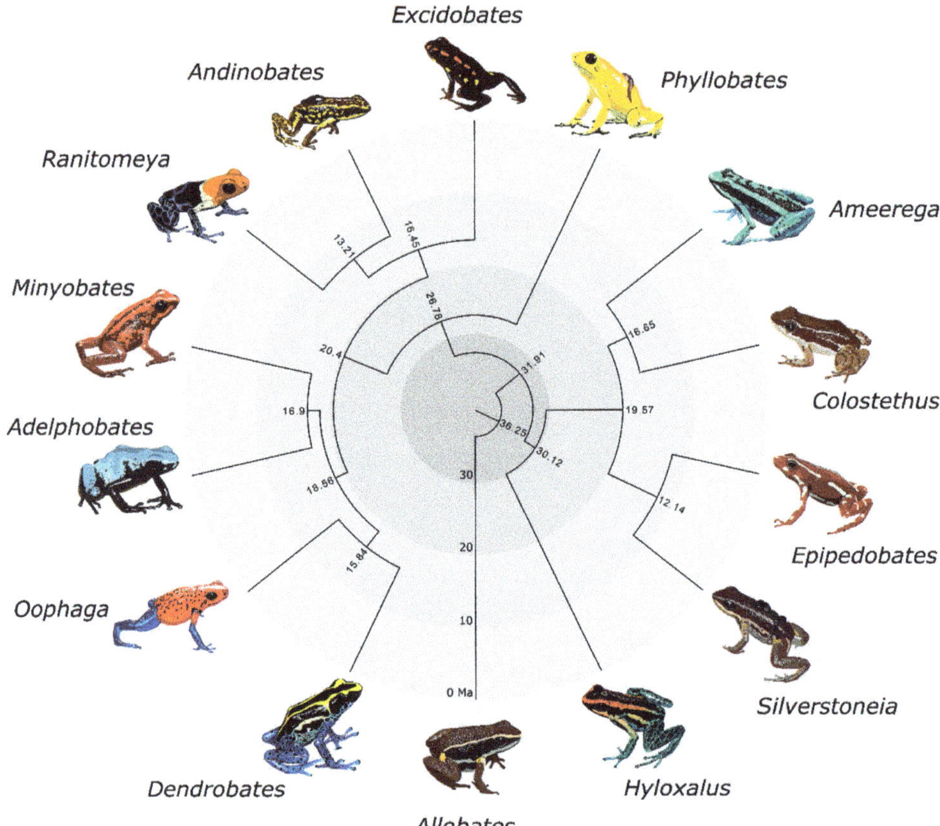

Figure 1. Time-calibrated genus-level phylogeny of Dendrobatidae produced using BEAST. Node labels indicate divergence times (mya). This figure is reduced to one tip per genus from the species-level chronogram in Figure S3. Art by WXG.

Node age estimates and associated error bars (representing uncertainty in node age estimates) are summarized in Table S2 and visualized in a time-calibrated phylogeny in Figure S3. Uncertainty in node age generally increases with deeper time. Our analyses indicate the subfamilies Colostethinae and Hyloxalinae diverged around 30 ± 10 mya. Dendrobatinae diverged from the common ancestor of Hyloxalinae and Colostethinae around 32 ± 10 mya. Dendrobatidae diverged from its sister family Aromobatidae 36 ± 10 mya.

4. Discussion

Establishing a robust understanding of phylogenetic relationships in Dendrobatidae is crucial for addressing questions about dendrobatid evolution and directing conservation efforts towards areas of diversity in maximum need. While a handful of studies have generated phylogenies of the family in recent years, none has used genomic data in their analyses. We used maximum likelihood and coalescent methods in conjunction with a large matrix of genomic markers to construct the first phylogenomic tree of Dendrobatidae. The usage of ML and coalescent methods in conjunction with genome-scale UCE data is intended to bring the status of dendrobatid phylogenetics more in line with current studies in herpetological phylogenetics, which frequently make use of these techniques [53–57]. Concerns with parsimony, which was used to construct the most recent large-scale phylogenetic analysis of dendrobatids [41], as a statistically-consistent phylogenetic method [79–81], and the presence of potential incomplete lineage sorting among large numbers of genes [32,58], also compelled us to use these techniques. Additionally, we estimated divergence times using a Bayesian method, which is currently the most widely-used and accepted type for divergence time estimation [82]. We found that most relationships among dendrobatid genera are largely congruous with the results of past studies, with some exceptions (see below). Our estimated divergence times are very similar to those estimated by Santos et al. (2009), which is to be expected since we used secondary calibrations taken from their study (Figure S3) [44]. However, our divergence time estimation involves different methods (BEAST 2 [71] rather than MULTIDIVTIME [83]) and considerably more genetic data (92,742 characters vs 2380 characters in Santos et al. [44]). Additionally, we recognize that since we used a secondary calibration taken from Santos et al.'s study due to the lack of poison frog fossils, our divergence time estimates may be biased towards younger node dates [84].

Much of our phylogeny is consistent with past phylogenies from mitochondrial and nuclear datasets, but with some key differences. Our analyses place Hyloxalinae sister to Colostethinae rather than Dendrobatinae [22,23], contrary to more recent studies on the family [21,41,44,45]. We also find support for placing the genus *Dendrobates* sister to *Oophaga* [21–23,35,44], in contrast to previous placements of *Dendrobates* as sister to *Adelphobates* [29,41,45], or even the rest of Dendrobatinae [43]. Lastly, we find strong support for placing *Minyobates* sister to *Adelphobates*. This problematic taxon has previously been placed anywhere from sister to *Excidobates* [29], to the rest of Dendrobatinae (excluding *Phyllobates*) [21,41]. Our conclusion for the placement of *Minyobates* corroborates placements recovered by more recent analyses that utilized maximum likelihood and Bayesian methods [35,44,45] rather than parsimony [21,41].

Maximizing efforts to conserve poison frogs (and other species) requires identifying both vulnerable lineages and geographical areas. A crucial step in this process is clarifying the evolutionary relationships of the taxa of interest, followed by the collection of basic population, distributional, and life history data for each taxon. Like many tropical amphibians, poison frogs face threats including habitat destruction [15,16] and smuggling for the pet trade [17,18,43]. Despite being one of the better-studied groups of frogs, a surprising number of poison frog species evaluated by the IUCN were classified as "data deficient" (37.5%, 107 of 285 species), hampering basic aspects of their conservation. Many of these data-deficient taxa belong to understudied genera with mostly cryptic coloration. In particular, the four genera *Hyloxalus*, *Colostethus*, *Allobates*, and *Anomaloglossus* contain a majority (70.1%, 75 of 107) of the "data deficient" taxa (Figure 2). In addition, though comprised of only a few species, little is known of the genera *Paruwrobates* and *Ectopoglossus*, where 3 (of 3) and 7 (of 8) of the contained species are classified as "data deficient", respectively (Table S3). Further, in *Ectopoglossus*, the only species not classified as "data deficient" is classified as "endangered," increasing the urgency for collecting basic life-history data in this group.

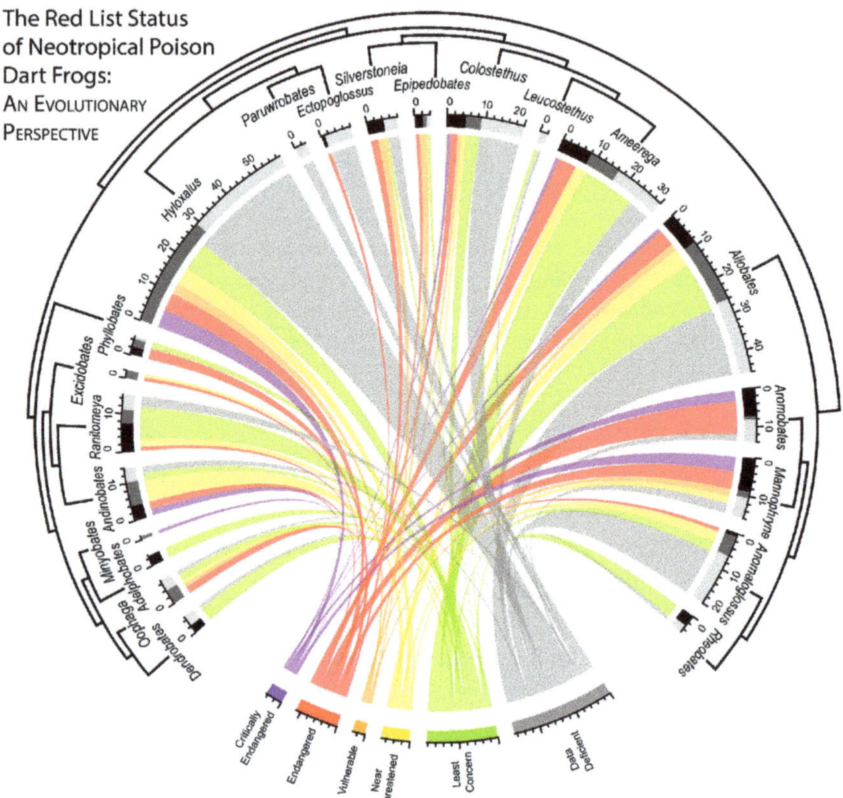

Figure 2. An evolutionary perspective of the Red List status of Aromobatidae and Dendrobatidae. Each species is characterized by a ribbon that is connected to its current Red List status (bottom). The numerical values below each genus depict the number of species with the associated Red List status. Bars on the outer ring depict the current population status of the corresponding genus, either: decreasing, stable or unknown (black, dark grey, or light grey respectively). A tree representing the evolutionary relationships of the genera surrounds the main diagram. Relationships for taxa not included in our study (Dendrobatidae: *Paruwrobates*, *Ectopoglossus*, *Leucostethus*; Aromobatidae: *Aromobates*, *Anomaloglossus*, *Mannophryne*, *Rheobates*) are reproduced from Grant et al. 2017 [41].

Roughly a quarter (22.1%, 63 of 285) of poison frog species were classified as "critically endangered" or "endangered" (18 and 45 species, respectively). Many of these at-risk taxa are concentrated in a few genera, most notably the clade that contains the two Aromobatid genera endemic to the northern Andes, *Aromobates* and *Mannophryne* (containing six "critically endangered" and 14 "endangered" species). The genera *Allobates*, *Hyloxalus*, and *Ameerega* contain a majority of the remaining at-risk species, though the proportions of at-risk species are similar to those of other genera. Additional unique evolutionary lineages of concern, though represented by only a few species, are the genera *Phyllobates*, *Excidobates* and *Minyobates*, where most surveyed taxa are at-risk (Table S3).

Furthermore, some geographic zones possess much higher at-risk diversity than others ("critically endangered" and "endangered" in Figure 3). The northern Andean countries possess both the highest species diversity and the highest diversity of at-risk species, especially Colombia, Peru, and Venezuela. (Figure 3). However, only Venezuela's proportions of 'at-risk' species are much greater than the country average, with 25.5% "endangered" and 16.4% "critically endangered" (average of 11.3% and 5.1%, respectively; Table S4). In contrast, countries with mostly lower elevation species (e.g., Brazil or Bolivia) seem to be well below the average of at-risk species (Table S4).

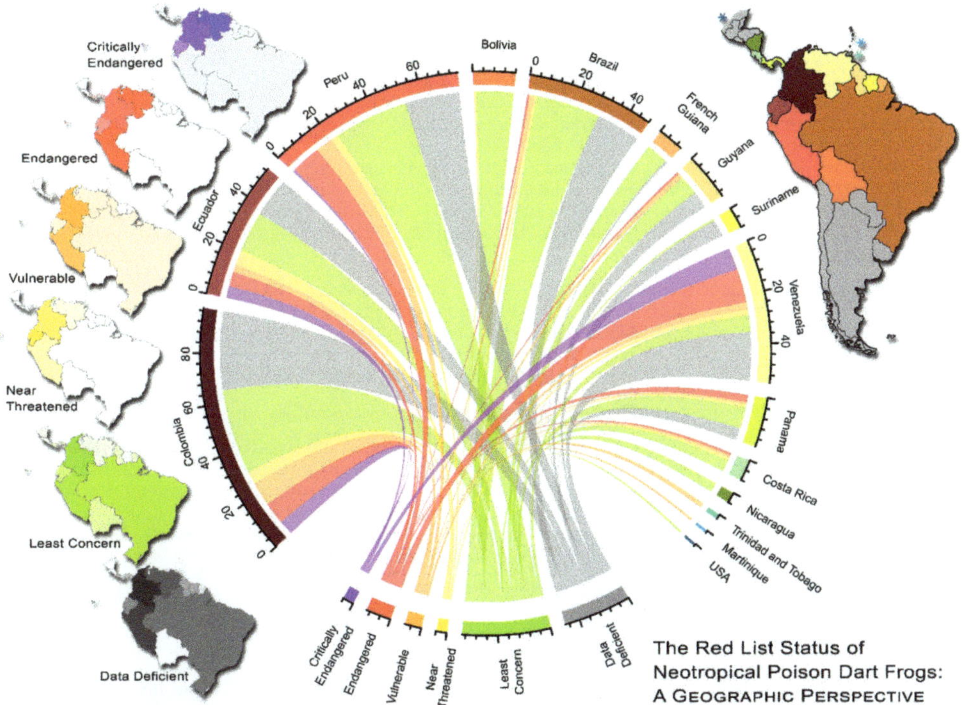

Figure 3. A geographic perspective of the Red List status of Aromobatidae and Dendrobatidae. The species composition of each country is characterized by ribbons connected to the current Red List status for each species (bottom). The numerical values below each country name depict the number of species with the associated Red List status. Bars on the outer ring depict the corresponding color of that country on the main map (top right). Maps to the left display Red List groups, where the intensity of color depicts larger number of species. If a species exists in more than one country, it was represented in each country of occurrence in the plot.

The circular infographics presented here are intended to represent a tractable way to visualize relationships between IUCN Red List categorizations, phylogenetic relationships, and geographic distributions of large numbers of related taxa (here two sister families). It is important to acknowledge that IUCN assessments are updated, on average, every decade. Thus, assessments and population trends represent a coarse temporal grain. Further, the spatial categorization by countries is overly simplistic and does not accurately reflect most species' actual ranges, as environment transcends political boundaries. However, given that environmental policy often occurs at the country level, this remains a practical spatial scope for summarizing assessment data. Lastly, our visualizations are not intended to replace more detailed quantitative assessments e.g., [85,86], but provide a novel perspective of the widely available IUCN data.

Here, we present the first broad-scale phylogenomic reconstruction of Dendrobatidae, furthering the continual study of poison frog systematics. In the future, we hope to improve taxon sampling, as here we were unable to acquire genetic samples for the newly erected genera *Leucostethus*, *Paruwrobates* and *Ectopoglossus* [41], and so their placement in the dendrobatid phylogeny is still predicated on Grant et al.'s (2017) analysis [41]. We were also unable to corroborate the paraphyly of *Colostethus* on account of missing genetic data for *C. ruthveni*. A future phylogenomic reconstruction of this group would benefit from inclusion of these taxa to ensure representation of all groups within Dendrobatidae and its sister family Aromobatidae. Lastly, we hope we have inspired researchers, field biologists, and conservation biologists to help address the highlighted conservation issues in these wonderful amphibians.

Supplementary Materials: Raw reads have been deposited in the NCBI Sequence Read Archive under project number PRJNA547821. The following are available online at http://www.mdpi.com/1424-2818/11/8/126/s1, Figure S1: Species-level phylogeny of all 61 samples included in the study, constructed using IQ-TREE (maximum likelihood), Figure S2: Comparison of (a) maximum likelihood tree made using IQ-TREE with the restricted dataset, (b) species tree made using ASTRAL-III with the comprehensive dataset, and (c) species made using ASTRAL-III with the restricted dataset, Figure S3: Species-level chronogram calibrated with BEAST 2 showing node numbers and uncertainty in divergence time estimation, Table S1: List of dendrobatoid samples included in our phylogenomic analyses and associated locality data, Table S2: Summary of divergence time estimation with BEAST 2, Table S3: IUCN Red List categories of dendrobatoids by genus, Table S4: IUCN Red List categories of dendrobatoids by country.

Author Contributions: Conceptualization: W.X.G. and J.L.B. Methodology: W.X.G. and J.L.B. Validation: W.X.G., M.R.M., and J.L.B. Formal Analysis: W.X.G. and M.R.M. Investigation: J.L.B. Resources: J.L.B., K.S. Data Curation: W.X.G. and M.R.M. Writing—Original Draft Preparation: W.X.G., M.R.M., and J.L.B. Writing – Review and Editing: W.X.G., M.R.M., J.L.B., and K.S. Author Contribution List-Maker: M.R.M. Food Stylist: W.X.G. Visualization: W.X.G. and J.L.B. Supervision: J.L.B. Project Administration: J.L.B. Funding Acquisition: J.L.B.

Funding: This research was funded by start-up funding to J.L.B. from Southern Illinois University Carbondale.

Acknowledgments: We are grateful for the continued support of Southern Illinois University, East Carolina University, Servicio Nacional Forestal y de Fauna Silvestre (Peru), and Centro de Ornitología y Biodiversidad (CORBIDI). We thank Ivan Prates and Miguel T. Rodrigues for generously providing several tissue samples. We are thankful to Peter Larson, Ryan Campbell and Anne Yoder for providing inspiration for Figures 2 and 3. M.R.M. and W.X.G. are grateful for support from the Students United in Preserving, Exploring, and Researching Biodiversity (SUPERB) fellowship, funded by the US National Science Foundation (NSF).

Conflicts of Interest: The authors declare no conflict of interest.

References

1. Wells, K.D. Behavioral ecology and social organization of a dendrobatid frog (*Colostethus inguinalis*). *Behav. Ecol. Sociobiol.* **1980**, *6*, 199–209. [CrossRef]

2. Summers, K.; Symula, R.; Clough, M.; Cronin, T. Visual mate choice in poison frogs. *Proc. R. Soc. B Biol. Sci.* **1999**, *266*, 2141–2145. [CrossRef]

3. Daly, J.W.; Myers, C.W. Toxicity of Panamanian poison frogs (*Dendrobates*): Some biological and chemical aspects. *Science* **1967**, *156*, 970–973. [CrossRef]

4. Daly, J.W.; McNeal, E.T.; Overman, L.E.; Ellison, D.H. A new class of cardiotonic agents: Structure-activity correlations for natural and synthetic analogs of the alkaloid pumiliotoxin B (8-hydroxy-8-methyl-6-alkylidene-1-azabicyclo [4.3.0] nonanes). *J. Med. Chem.* **1985**, *28*, 482–486. [CrossRef]

5. Spande, T.F.; Garraffo, H.M.; Edwards, M.W.; Yeh, H.J.C.; Pannell, L.; Daly, J.W. Epibatidine: A novel (chloropyridyl) azabicycloheptane with potent analgesic activity from an Ecuadoran poison frog. *J. Am. Chem. Soc.* **1992**, *114*, 3475–3478. [CrossRef]

6. Maan, M.E.; Cummings, M.E. Sexual dimorphism and directional sexual selection on aposematic signals in a poison frog. *Proc. Natl. Acad. Sci. USA* **2009**, *106*, 19072–19077. [CrossRef]

7. Wang, I.J.; Shaffer, H.B. Rapid color evolution in an aposematic species: A phylogenetic analysis of color variation in the strikingly polymorphic strawberry poison-dart frog. *Evolution* **2008**, *62*, 2742–2759. [CrossRef]

8. Santos, J.C.; Cannatella, D.C. Phenotypic integration emerges from aposematism and scale in poison frogs. *Proc. Natl. Acad. Sci. USA* **2011**, *108*, 6175–6180. [CrossRef]

9. Brown, J.L.; Twomey, E. Complicated histories: Three new species of poison frogs of the genus *Ameerega* (Anura: Dendrobatidae) from north-central Peru. *Zootaxa* **2009**, *2049*, 38. [CrossRef]

10. Noonan, B.P.; Gaucher, P. Refugial isolation and secondary contact in the dyeing poison frog *Dendrobates tinctorius*. *Mol. Ecol.* **2006**, *15*, 4425–4435. [CrossRef]

11. Gehara, M.; Summers, K.; Brown, J.L. Population expansion, isolation and selection: Novel insights on the evolution of color diversity in the strawberry poison frog. *Evol. Ecol.* **2013**, *27*, 797–824. [CrossRef]

12. Young, R.L.; Ferkin, M.H.; Ockendon-Powell, N.F.; Orr, V.N.; Phelps, S.M.; Pogány, Á.; Richards-Zawacki, C.L.; Summers, K.; Székely, T.; Trainor, B.C.; et al. Conserved transcriptomic profiles underpin monogamy across vertebrates. *Proc. Natl. Acad. Sci. USA* **2019**, *116*, 1331–1336. [CrossRef]

13. Tarvin, R.D.; Santos, J.C.; O'Connell, L.A.; Zakon, H.H.; Cannatella, D.C. Convergent substitutions in a sodium channel suggest multiple origins of toxin resistance in poison frogs. *Mol. Biol. Evol.* **2016**, *33*, 1068–1081. [CrossRef]

14. Tarvin, R.D.; Borghese, C.M.; Sachs, W.; Santos, J.C.; Lu, Y.; O'Connell, L.A.; Cannatella, D.C.; Harris, R.A.; Zakon, H.H. Interacting amino acid replacements allow poison frogs to evolve epibatidine resistance. *Science* **2017**, *357*, 1261–1266. [CrossRef]

15. Whitfield, S.M.; Bell, K.E.; Philippi, T.; Sasa, M.; Bolaños, F.; Chaves, G.; Savage, J.M.; Donnelly, M.A. Amphibian and reptile declines over 35 years at La Selva, Costa Rica. *Proc. Natl. Acad. Sci. USA* **2007**, *104*, 8352–8356. [CrossRef]

16. Angulo, A.; von May, R.; Icochea, J. A reassessment of the extinction risk of the critically endangered Oxapampa poison frog *Ameerega planipaleae* (Dendrobatidae). *Oryx* **2018**, *53*, 1–4. [CrossRef]

17. Auliya, M.; García-Moreno, J.; Schmidt, B.R.; Schmeller, D.S.; Hoogmoed, M.S.; Fisher, M.C.; Pasmans, F.; Henle, K.; Bickford, D.; Martel, A. The global amphibian trade flows through Europe: The need for enforcing and improving legislation. *Biodivers. Conserv.* **2016**, *25*, 2581–2595. [CrossRef]

18. Gorzula, S. The trade in dendrobatid frogs from 1987 to 1993. *Herpetol. Rev.* **1996**, *27*, 116–123.

19. Cuvier, G.L.C.F.D. An VI. In *Tableau Élémentaire de l'Histoire Naturelle des Animaux*; Baudoin: Paris, France, 1797.

20. Wagler, J.G. *Natürliches System der Amphibien, mit Vorangehender Classification der Säugthiere und Vogel: Ein Neitrag zur Vergleichenden Zoologie*; J.G. Cottasche Buchhandlung Nachfolger: Stuttgart, Germany, 1830.

21. Grant, T.; Frost, D.R.; Caldwell, J.P.; Gagliardo, R.; Haddad, C.F.B.; Kok, P.J.R.; Means, D.B.; Noonan, B.P.; Schargel, W.E.; Wheeler, W.C. Phylogenetic systematics of dart-poison frogs and their relatives (Amphibia: Athesphatanura: Dendrobatidae). *Bull. Am. Mus. Nat. Hist.* **2006**, *299*, 1–262. [CrossRef]

22. Vences, M.; Kosuch, J.; Lötters, S.; Widmer, A.; Jungfer, K.-H.; Köhler, J.; Veith, M. Phylogeny and classification of poison frogs (Amphibia: Dendrobatidae), Based on mitochondrial 16S and 12S ribosomal RNA gene sequences. *Mol. Phylogenet. Evol.* **2000**, *15*, 34–40. [CrossRef]

23. Clough, M.; Summers, K. Phylogenetic systematics and biogeography of the poison frogs: Evidence from mitochondrial DNA sequences. *Biol. J. Linn. Soc.* **2000**, *70*, 515–540. [CrossRef]

24. Widmer, A.; Lötters, S.; Jungfer, K.H. A molecular phylogenetic analysis of the neotropical dart-poison frog genus *Phyllobates* (Amphibia: Dendrobatidae). *Naturwissenschaften* **2000**, *87*, 559–562. [CrossRef]

25. Summers, K.; Clough, M.E. The evolution of coloration and toxicity in the poison frog family (Dendrobatidae). *Proc. Natl. Acad. Sci. USA* **2001**, *98*, 6227–6232. [CrossRef]

26. La Marca, E.; Vences, M.; Lötters, S. Rediscovery and mitochondrial relationships of the dendrobatid frog *Colostethus humilis* suggest parallel colonization of the Venezuelan Andes by poison frogs. *Stud. Neotrop. Fauna Environ.* **2002**, *37*, 233–240. [CrossRef]

27. Santos, J.C.; Coloma, L.A.; Cannatella, D.C. Multiple, recurring origins of aposematism and diet specialization in poison frogs. *Proc. Natl. Acad. Sci. USA* **2003**, *100*, 12792–12797. [CrossRef]

28. Vences, M. Convergent evolution of aposematic coloration in Neotropical poison frogs: A molecular phylogenetic perspective. *Org. Divers. Evol.* **2003**, *3*, 215–226. [CrossRef]

29. Roberts, J.L.; Brown, J.L.; von May, R.; Arizabal, W.; Presar, A.; Symula, R.; Schulte, R.; Summers, K. Phylogenetic relationships among poison frogs of the genus Dendrobates (Dendrobatidae): A molecular perspective from increased taxon sampling. 2006, 16, 377–385. 16.

30. Frost, D.R.; Grant, T.; Faivovich, J.; Bain, R.H.; Haas, A.; Haddad, C.F.B.; De Sá, R.O.; Channing, A.; Wilkinson, M.; Donnellan, S.C.; et al. The amphibian tree of life. *Bull. Am. Mus. Nat. Hist.* **2006**, *297*, 1–291. [CrossRef]

31. Maddison, W.P. Gene trees in species trees. *Syst. Biol.* **1997**, *46*, 523–536. [CrossRef]

32. Degnan, J.H.; Rosenberg, N.A. Gene tree discordance, phylogenetic inference and the multispecies coalescent. *Trends Ecol. Evol.* **2009**, *24*, 332–340. [CrossRef]

33. Frost, D.R. Amphibian Species of the World 6.0. Available online: http://research.amnh.org/vz/herpetology/amphibia/ (accessed on 19 July 2019).

34. Twomey, E.; Brown, J.L.; Amézquita, A.; Mejía-Vargas, D. A taxonomic revision of the Neotropical poison frog genus Ranitomeya (Amphibia: Dendrobatidae). *Zootaxa* **2011**, *3083*, 20–39.

35. Twomey, E.; Brown, J.L. Spotted poison frogs: Rediscovery of a lost species and a new genus (Anura: Dendrobatidae) from northwestern Peru. *Herpetologica* **2008**, *64*, 121–137. [CrossRef]

36. Myers, C.W. New generic names from some neotropical poison frogs (Dendrobatidae). *Papéis Avulsos Zool. Mus. Zool. Univ. São Paulo* **1987**, *36*, 301–306.

37. Bauer, L. New names in the family Dendrobatidae (Anura, Amphibia). *RIPA* **1994**, *fall*, 1–6.

38. Bibron, G. pl. 29bis. Type species: Phyllobates bicolor Bibron, 1840, by monotypy. In *Atlas de Zoología de Historia Fisica, Politica y Natural de la Isla de Cuba, Segunda Part. Historia Natural*; Arthur Bertrand: Paris, France, 1840; p. 8.

39. Bauer, L. A new genus and a new specific name in the dart poison frog family (Dendrobatidae, Anura, Amphibia). *RIPA* **1986**, *1986*, 1–12.

40. Cope, E.D. Fourth contribution to the herpetology of tropical America. *Proc. Acad. Nat. Sci. Phila.* **1866**, 123–132.

41. Grant, T.; Rada, M.; Anganoy-Criollo, M.; Batista, A.; Dias, P.H.; Jeckel, A.M.; Machado, D.J.; Rueda-Almonacid, J.V. Phylogenetic systematics of dart-poison frogs and their relatives revisited (Anura: Dendrobatoidea). *South. Am. J. Herpetol.* **2017**, *12*, S1–S90. [CrossRef]

42. Jiménez de la Espada, M. Fauna neotropicalis species quaedam nondum cognitae. *J. Sci. Math. Phys. E Nat.* **1870**, *3*, 57–65.

43. Brown, J.L.; Twomey, E.; Amézquita, A.; de Souza, M.B.; Caldwell, J.P.; Lötters, S.; von May, R.; Melo-Sampaio, P.R.; Mejía-Vargas, D.; Perez-Peña, P.; et al. A taxonomic revision of the Neotropical poison frog genus *Ranitomeya* (Amphibia: Dendrobatidae). *Zootaxa* **2011**, *3083*, 1–120. [CrossRef]

44. Santos, J.C.; Coloma, L.A.; Summers, K.; Caldwell, J.P.; Ree, R.; Cannatella, D.C. Amazonian amphibian diversity Is primarily derived from late Miocene Andean lineages. *PLoS Biol.* **2009**, *7*, e1000056. [CrossRef]

45. Pyron, A.R.; Wiens, J.J. A large-scale phylogeny of Amphibia including over 2800 species, and a revised classification of extant frogs, salamanders, and caecilians. *Mol. Phylogenet. Evol.* **2011**, *61*, 543–583. [CrossRef]

46. Faircloth, B.C.; McCormack, J.E.; Crawford, N.G.; Harvey, M.G.; Brumfield, R.T.; Glenn, T.C. Ultraconserved elements anchor thousands of genetic markers spanning multiple evolutionary timescales. *Syst. Biol.* **2012**, *61*, 717–726. [CrossRef]

47. Bejerano, G.; Pheasant, M.; Makunin, I.; Stephen, S.; Kent, W.J.; Mattick, J.S.; Haussler, D. Ultraconserved elements in the human genome. *Science* **2004**, *304*, 1321–1325. [CrossRef]

48. Siepel, A.; Bejerano, G.; Pedersen, J.S.; Hinrichs, A.S.; Hou, M.; Rosenbloom, K.; Clawson, H.; Spieth, J.; Hillier, L.W.; Richards, S.; et al. Evolutionarily conserved elements in vertebrate, insect, worm, and yeast genomes. *Genome Res.* **2005**, *15*, 1034–1050. [CrossRef]

49. Gilbert, P.S.; Chang, J.; Pan, C.; Sobel, E.; Sinsheimer, J.S.; Faircloth, B.; Alfaro, M.E. Genome-wide ultraconserved elements exhibit higher phylogenetic informativeness than traditional gene markers in percomorph fishes. *Mol. Phylogenet. Evol.* **2015**, *92*, 140–146. [CrossRef]

50. Blaimer, B.B.; Lloyd, M.W.; Guillory, W.X.; Brady, S.G. Sequence capture and phylogenetic utility of genomic ultraconserved elements obtained from pinned insect specimens. *PLoS ONE* **2016**, *11*, e0161531. [CrossRef]

51. McCormack, J.E.; Tsai, W.L.E.; Faircloth, B.C. Sequence capture of ultraconserved elements from bird museum specimens. *Mol. Ecol. Resour.* **2016**, *16*, 1189–1203. [CrossRef]

52. Smith, B.T.; Harvey, M.G.; Faircloth, B.C.; Glenn, T.C.; Brumfield, R.T. Target capture and massively parallel sequencing of ultraconserved elements for comparative studies at shallow evolutionary time scales. *Syst. Biol.* **2014**, *63*, 83–95. [CrossRef]

53. Crawford, N.G.; Faircloth, B.C.; McCormack, J.E.; Brumfield, R.T.; Winker, K.; Glenn, T.C. More than 1000 ultraconserved elements provide evidence that turtles are the sister group of archosaurs. *Biol. Lett.* **2012**, *8*, 783–786. [CrossRef]

54. Crawford, N.G.; Parham, J.F.; Sellas, A.B.; Faircloth, B.C.; Glenn, T.C.; Papenfuss, T.J.; Henderson, J.B.; Hansen, M.H.; Simison, W.B. A phylogenomic analysis of turtles. *Mol. Phylogenet. Evol.* **2015**, *83*, 250–257. [CrossRef]

55. Streicher, J.W.; Wiens, J.J. Phylogenomic analyses of more than 4000 nuclear loci resolve the origin of snakes among lizard families. *Biol. Lett.* **2017**, *13*, 20170393. [CrossRef]

56. Streicher, J.W.; Wiens, J.J. Phylogenomic analyses reveal novel relationships among snake families. *Mol. Phylogenet. Evol.* **2016**, *100*, 160–169. [CrossRef]

57. Streicher, J.W.; Miller, E.C.; Guerrero, P.C.; Correa, C.; Ortiz, J.C.; Crawford, A.J.; Pie, M.R.; Wiens, J.J. Evaluating methods for phylogenomic analyses, and a new phylogeny for a major frog clade (Hyloidea) based on 2214 loci. *Mol. Phylogenet. Evol.* **2018**, *119*, 128–143. [CrossRef]

58. McCormack, J.E.; Faircloth, B.C.; Crawford, N.G.; Gowaty, P.A.; Brumfield, R.T.; Glenn, T.C. Ultraconserved elements are novel phylogenomic markers that resolve placental mammal phylogeny when combined with species-tree analysis. *Genome Res.* **2012**, *22*, 746–754. [CrossRef]

59. Derti, A.; Roth, F.P.; Church, G.M.; Wu, C. Mammalian ultraconserved elements are strongly depleted among segmental duplications and copy number variants. *Nat. Genet.* **2006**, *38*, 1216–1220. [CrossRef]

60. Simons, C.; Pheasant, M.; Makunin, I.V.; Mattick, J.S. Transposon-free regions in mammalian genomes. *Genome Res.* **2006**, *16*, 164–172. [CrossRef]

61. Faircloth, B.C. PHYLUCE is a software package for the analysis of conserved genomic loci. *Bioinformatics* **2016**, *32*, 786–788. [CrossRef]

62. Faircloth, B.C. Illumiprocessor: A trimmomatic wrapper for parallel adapter and quality trimming. 2013. [CrossRef]

63. Bolger, A.M.; Lohse, M.; Usadel, B. Trimmomatic: A flexible trimmer for Illumina sequence data. *Bioinformatics* **2014**, *30*, 2114–2120. [CrossRef]

64. Grabherr, M.G.; Haas, B.J.; Yassour, M.; Levin, J.Z.; Thompson, D.A.; Amit, I.; Adiconis, X.; Fan, L.; Raychowdhury, R.; Zeng, Q.; et al. Full-length transcriptome assembly from RNA-Seq data without a reference genome. *Nat. Biotechnol.* **2011**, *29*, 644–652. [CrossRef]

65. Edgar, R.C. MUSCLE: Multiple sequence alignment with high accuracy and high throughput. *Nucleic Acids Res.* **2004**, *32*, 1792–1797. [CrossRef]

66. Heibl, C. PHYLOCH: R Language Tree Plotting Tools and Interfaces to Diverse Phylogenetic Software Packages. 2008. Available online: http://www.christophheibl.de/Rpackages.html (accessed on 9 June 2019).

67. Nguyen, L.-T.; Schmidt, H.A.; von Haeseler, A.; Minh, B.Q. IQ-TREE: A fast and effective stochastic algorithm for estimating maximum-likelihood phylogenies. *Mol. Biol. Evol.* **2015**, *32*, 268–274. [CrossRef]

68. Minh, B.Q.; Nguyen, M.A.T.; von Haeseler, A. Ultrafast approximation for phylogenetic bootstrap. *Mol. Biol. Evol.* **2013**, *30*, 1188–1195. [CrossRef]

69. Zhang, C.; Rabiee, M.; Sayyari, E.; Mirarab, S. ASTRAL-III: Polynomial time species tree reconstruction from partially resolved gene trees. *BMC Bioinform.* **2018**, *19*, 153. [CrossRef]

70. Persons, N.W.; Hosner, P.A.; Meiklejohn, K.A.; Braun, E.L.; Kimball, R.T. Sorting out relationships among the grouse and ptarmigan using intron, mitochondrial, and ultra-conserved element sequences. *Mol. Phylogenet. Evol.* **2016**, *98*, 123–132. [CrossRef]

71. Bouckaert, R.; Heled, J.; Kühnert, D.; Vaughan, T.; Wu, C.-H.; Xie, D.; Suchard, M.A.; Rambaut, A.; Drummond, A.J. BEAST 2: A Software platform for bayesian evolutionary analysis. *PLoS Comput. Biol.* **2014**, *10*, e1003537. [CrossRef]

72. Ješovnik, A.; Sosa-Calvo, J.; Lloyd, M.W.; Branstetter, M.G.; Fernández, F.; Schultz, T.R. Phylogenomic species delimitation and host-symbiont coevolution in the fungus-farming ant genus *Sericomyrmex* Mayr (Hymenoptera: Formicidae): Ultraconserved elements (UCEs) resolve a recent radiation. *Syst. Entomol.* **2017**, *42*, 523–542. [CrossRef]

73. Drummond, A.J.; Bouckaert, R. *Bayesian Evolutionary Analysis with BEAST 2*; Cambridge University Press: Cambridge, UK, 2015.

74. Winker, K.; Glenn, T.C.; Faircloth, B.C. Ultraconserved elements (UCEs) illuminate the population genomics of a recent, high-latitude avian speciation event. *PeerJ* **2018**, *6*, e5735. [CrossRef]

75. Hsiang, A.Y.; Field, D.J.; Webster, T.H.; Behlke, A.D.; Davis, M.B.; Racicot, R.A.; Gauthier, J.A. The origin of snakes: Revealing the ecology, behavior, and evolutionary history of early snakes using genomics, phenomics, and the fossil record. *BMC Evol. Biol.* **2015**, *15*, 87. [CrossRef]

76. Rambaut, A.; Drummond, A.J.; Xie, D.; Baele, G.; Suchard, M.A. Posterior summarization in bayesian phylogenetics using tracer 1.7. *Syst. Biol.* **2018**, *67*, 901–904. [CrossRef]

77. IUCN 2019. The IUCN Red List of Threatened Species. Version 2019-1. Available online: http://www.iucnredlist.org (accessed on 21 March 2019).

78. Gu, Z.; Gu, L.; Eils, R.; Schlesner, M.; Brors, B. *Circlize* Implements and enhances circular visualization in R. *Bioinformatics* **2014**, *30*, 2811–2812. [CrossRef]
79. Swofford, D.L.; Waddell, P.J.; Huelsenbeck, J.P.; Foster, P.G.; Lewis, P.O.; Rogers, J.S. Bias in phylogenetic estimation and its relevance to the choice between parsimony and likelihood methods. *Syst. Biol.* **2001**, *50*, 525–539. [CrossRef]
80. Felsenstein, J. Parsimony in systematics: Biological and statistical issues. *Annu. Rev. Ecol. Syst.* **1983**, *14*, 313–333. [CrossRef]
81. Felsenstein, J. Cases in which parsimony or compatibility methods will be positively misleading. *Syst. Biol.* **1978**, *27*, 401–410. [CrossRef]
82. Dos Reis, M.; Donoghue, P.C.J.; Yang, Z. Bayesian molecular clock dating of species divergences in the genomics era. *Nat. Rev. Genet.* **2016**, *17*, 71–80. [CrossRef]
83. Thorne, J.L.; Kishino, H. Divergence time and evolutionary rate estimation with multilocus data. *Syst. Biol.* **2002**, *51*, 689–702. [CrossRef]
84. Schenk, J.J. Consequences of secondary calibrations on divergence time estimates. *PLoS ONE* **2016**, *11*, e0148228. [CrossRef]
85. Rosauer, D.; Laffan, S.W.; Crisp, M.D.; Donnellan, S.C.; Cook, L.G. Phylogenetic endemism: A new approach for identifying geographical concentrations of evolutionary history. *Mol. Ecol.* **2009**, *18*, 4061–4072. [CrossRef]
86. González-del-Pliego, P.; Freckleton, R.P.; Edwards, D.P.; Koo, M.S.; Scheffers, B.R.; Pyron, R.A.; Jetz, W. Phylogenetic and trait-based prediction of extinction risk for data-deficient amphibians. *Curr. Biol.* **2019**, *29*, 1557–1563. [CrossRef]

Article

Glassfrogs of Ecuador: Diversity, Evolution, and Conservation

Juan M. Guayasamin [1,2,*], Diego F. Cisneros-Heredia [3,4,5], Roy W. McDiarmid [6], Paula Peña [7,8] and Carl R. Hutter [9,10]

1   Instituto BIÓSFERA-USFQ, Colegio de Ciencias Biológicas y Ambientales COCIBA, Laboratorio de Biología Evolutiva, Campus Cumbayá, Universidad San Francisco de Quito USFQ, Quito 170901, Ecuador
2   Department of Biology, University of North Carolina, Chapel Hill, NC 27599, USA
3   Museo de Zoología & Laboratorio de Zoología Terrestre, Instituto de Diversidad Biológica Tropical iBIOTROP, Colegio de Ciencias Biológicas y Ambientales COCIBA, Universidad San Francisco de Quito USFQ, Quito 170901, Ecuador; dcisneros@usfq.edu.ec
4   Department of Geography, King's College, London WC2R 2LS, UK
5   Instituto Nacional de Biodiversidad INABIO, Quito 170135, Ecuador
6   USGS Patuxent Wildlife Research Center, National Museum of Natural History, Smithsonian Institution, P.O. Box 37012, Washington, DC 20013, USA; mcdiarmr@si.edu
7   Ingeniería en Biodiversidad y Recursos Genéticos, Centro de Investigación de la Biodiversidad y Cambio Climático–BioCamb, Av. Machala y Sabanilla, Universidad Tecnológica Indoamérica, Quito 170103, Ecuador; paula.p.loyola@gmail.com
8   Universidad Internacional Menéndez Pelayo, Isaac Peral, 23-28040 Madrid, Spain
9   Biodiversity Institute and Department of Ecology and Evolutionary Biology, University of Kansas, Lawrence, KS 66045–7561, USA; hutter@lsu.edu or carl.hutter@ku.edu
10   Museum of Natural Sciences and Department of Biological Sciences, Louisiana State University, Baton Rouge, LA 70803, USA
*   Correspondence: jmguayasamin@usfq.edu.ec

Received: 30 December 2019; Accepted: 6 May 2020; Published: 2 June 2020

Abstract: Glassfrogs (family: Centrolenidae) represent a fantastic radiation (~150 described species) of Neotropical anurans that originated in South America and dispersed into Central America. In this study, we review the systematics of Ecuadorian glassfrogs, providing species accounts of all 60 species, including three new species described herein. For all Ecuadorian species, we provide new information on the evolution, morphology, biology, conservation, and distribution. We present a new molecular phylogeny for Centrolenidae and address cryptic diversity within the family. We mploy a candidate species system and designate 24 putative new species that require further study to determine their species status. We find that, in some cases, currently recognized species lack justification; specifically, we place *Centrolene gemmata* and *Centrolene scirtetes* under the synonymy of *Centrolene lynchi*; *C. guanacarum* and *C. bacata* under the synonymy of *Centrolene sanchezi*; *Cochranella phryxa* under the synonymy of *Cochranella resplendens*; and *Hyalinobatrachium ruedai* under the synonymy of *Hyalinobatrachium munozorum*. We also find that diversification patterns are mostly congruent with allopatric speciation, facilitated by barriers to gene flow (e.g., valleys, mountains, linearity of the Andes), and that niche conservatism is a dominant feature in the family. Conservation threats are diverse, but habitat destruction and climate change are of particular concern. The most imperiled glassfrogs in Ecuador are *Centrolene buckleyi*, *C. charapita*, *C. geckoidea*, *C. medemi*, *C. pipilata*, *Cochranella mache*, *Nymphargus balionotus*, *N. manduriacu*, *N. megacheirus*, and *N. sucre*, all of which are considered Critically Endangered. Lastly, we identify priority areas for glassfrog conservation in Ecuador.

Keywords: anura; biogeography; centrolenidae; systematics; taxonomy

1. Introduction

The Neotropical family Centrolenidae Taylor 1951 is a monophyletic taxon that contains about 150 species classified into 12 genera [1] (Figure 1). Historical biogeographic evidence in a phylogenetic context suggests that the group originated in South America and subsequently dispersed to Central America [2,3]. Centrolenids share a unique morphology and behavior that makes them readily identifiable; some of the most evident traits include a green dorsum in most species (Figure 2), completely or partially translucent venter (Figure 3), out-of-water deposition of eggs along streams (Figure 4), and forward-directed eyes (Figure 2A). Other interesting features that have evolved within Centrolenidae include parental care [4] (Figure 5), fighting behavior [5] (Figure 6), and humeral spines in males of some species [6] (Figure 7).

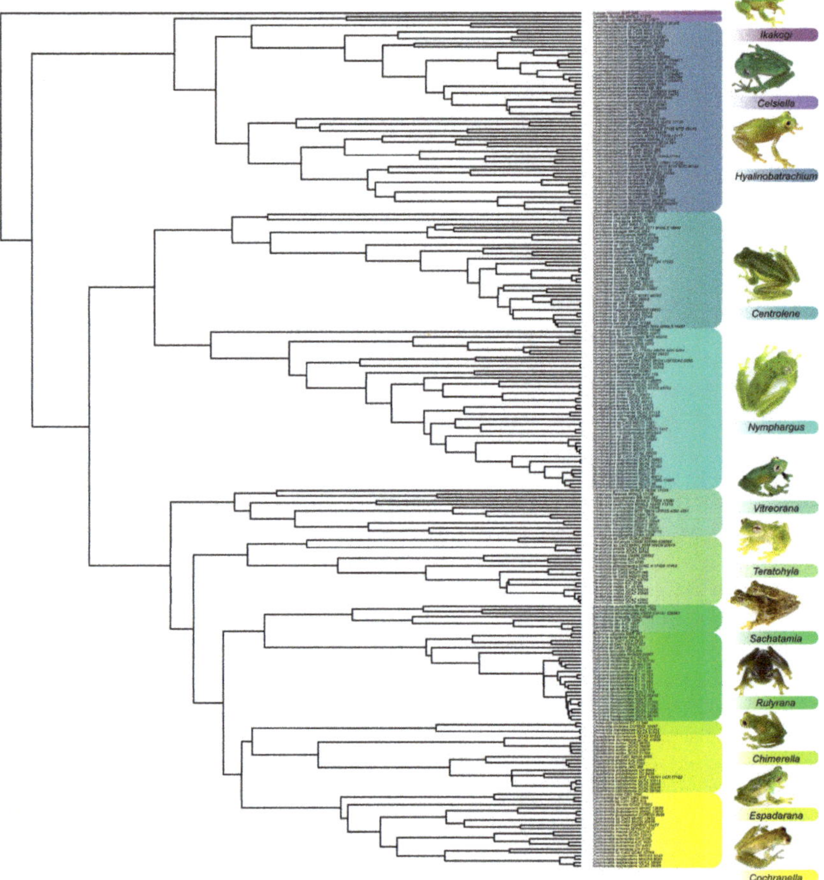

Figure 1. Relationships among glassfrog genera as inferred herein, using maximum likelihood criterium. Taxonomy sensu Guayasamin et al. [1]. Taxon sampling includes 113 named glassfrog species, 24 putative new species, and 49 outgroup taxa (not shown). The dataset contains complete or partial sequences of 10 genes representing 6513 bp of data (mitochondrial: 12S rRNA, 16S rRNA, ND1; nuclear: BDNF, C-MYC exon 2, CXCR4, NCX1, POMC, RAG1, SLC8A3). Sequences were generated in previous studies, as well as this one (see Table S2). Relationships within each genus are shown in additional figures and follow the same methodology.

Figure 2. Dorsal color patterns in glassfrogs. (**A**) Uniform; *Sachatamia ilex*, QCAZ 47193. (**B**) With small and well-defined yellow spots; *Nymphargus humboldti* sp. nov., QCAZ 41084. (**C**) With small and diffuse yellow spots; *Hyalinobatrachium fleischmanni*, QCAZ 45386. (**D**) With large yellow spots; *H. aureoguttatum*, QCAZ 45365. (**E**) With small ocelli, *N. cochranae*, QCAZ 31113. (**F**) With ocelli and dark flecks; *N. anomalus*, QCAZ 47507. (**G**) With irregular light-green marks and small well-defined black spots, *H. iaspidiense*, QCAZ 38438. (**H**) With green reticulation; *H.* cf. *valerioi*, ZSFQ 0544 (photo by Jose Vieira). All photographs by Luis A. Coloma, except when noted.

The monophyly of Centrolenidae is well supported by morphological [7–9] and molecular [2,8,10–14] studies. Osteological characters shared by all glassfrogs (Figure 8) include a dilated medial process on Metacarpal III [7], T-shaped terminal phalanges [15], intercalary element between distal and penultimate phalanges [15], and complete or partial fusion of tibiale and fibulare [15,16]. The diversity of glassfrogs is growing and in constant revision (e.g., [1,6,17–21]). Centrolenid species richness is concentrated in mountain chains, where humidity is high, and streams provide suitable reproductive habitats. Therefore, it is no surprise that the Andes, by far, is the center of diversity and endemism for the group and that Colombia and Ecuador maintain the highest species richness.

In 1973, John D. Lynch and William E. Duellman [22] provided the first review of the glassfrogs of Ecuador, and they reported the presence of 20 species and suggested the occurrence of an additional one. Only 47 years later, the species richness of this family has tripled, reaching the incredible number of 60 species in Ecuador. Since Lynch and Duellman's pioneering work, our understanding of centrolenid biology [23–25], morphology [6,9,17], systematics [1,6,17,26,27], biogeography [3,28], and evolution [2,19,25] has improved substantially. Still, there are novel challenges and conspicuous gaps that need to be filled.

Herein, we provide a new review of Ecuadorian glassfrogs, bringing an update of what is known about the group, highlighting issues pending resolution, and providing a framework that we hope will facilitate further research, particularly with species identification, discovery, and conservation. We employ a candidate species system (e.g., [29]) to identify putative undescribed species to be investigated and to streamline the species discovery process. These tasks are particularly urgent in the current global amphibian extinction phenomenon [30–33].

Figure 3. Ventral transparency in glassfrogs. (**A**) Complete transparency of parietal peritoneum and pericardium; *H. aureoguttatum*. (**B**) Partial transparency: parietal peritoneum is transparent only posteriorly; *N. posadae*. (**C**) Venter opaque (no ventral transparency): ventral parietal peritoneum and the urinary bladder peritoneum are opaque (white); *N. gradisonae*. Photos by Martín Bustamante.

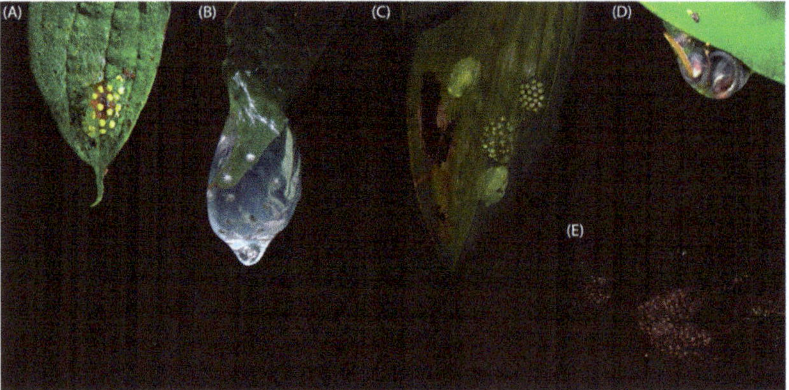

Figure 4. Egg deposition sites in glassfrogs. (**A**) On upper side of leaves (*C. sanchezi*). (**B**) On tip of leave (*N. wileyi*). (**C**) On underside of leaves (*H. cappellei*, photo by C. Barrio-Amorós). (**D**) On the margin of underside of leaves (*Teratohyla spinosa*, photo by R. Puschendorf). (**E**) On rocks (e.g., *Sachatamia albomaculata*, photo by R. Puschendorf). Figure modified from Guayasamin et al. [1].

Figure 5. Parental care in glassfrogs. Adult male of *Hyalinobatrachium aureoguttatum*, with egg clutch. Photo by Luis A. Coloma.

Figure 6. Combat behavior in glassfrogs. Note diversity of positions. Illustrated species: *Nymphargus grandisonae*. Photos by Carl R. Hutter.

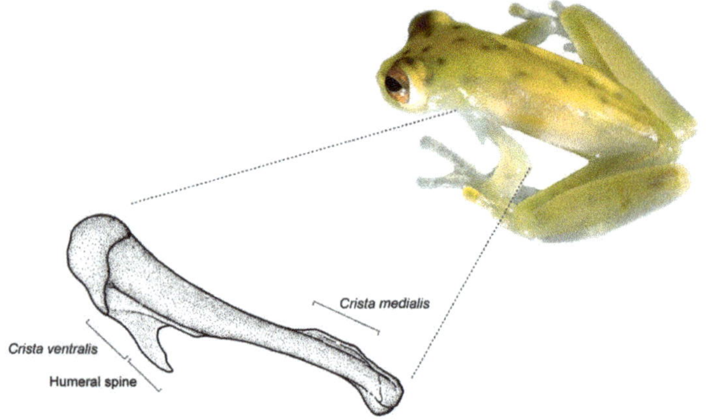

Figure 7. Humeral spine in *Cochranella litoralis*. Note that the humeral spine is formed by the prolongation of the *crista ventralis*. Photo by Luis A. Coloma; drawing by Linda Trueb.

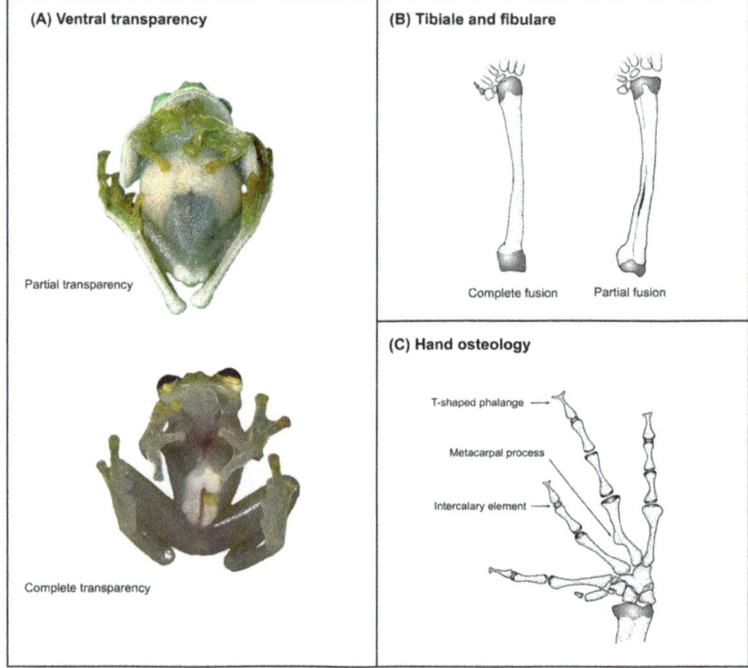

Figure 8. Synapomorphies of Centrolenidae. (**A**) Partial (*Nymphargus posadae*, QCAZ 25090) and complete ventral transparency (*Hyalinobatrachium aureoguttatum*, QCAZ 32070). (**B**) Partial (*N. wileyi*, QCAZ 26029) and complete fusion between tibiale and fibulare (*H. munozorum*, KU 155497). (**C**) Presence of medial process on Metacarpal III and intercalary element (*Teratohyla spinosa*, KU 32935). The presence of T or Y-shaped terminal phalanges is a synapomorphy of Allocentroleniae (Allophrynidae + Centrolenidae). Photos in (**A**) by M. Bustamante. Modified from Guayasamin et al. [1].

2. Methods

2.1. Taxonomy and Species Concept

Throughout this work, we use the name Centrolenidae as originally intended by Taylor [15]. The taxonomic arrangement below the family level (i.e., subfamily and genus) follows the proposal by Guayasamin et al. [1]. We use quotations to denote species with uncertain phylogenetic position to differentiate them from monophyletic clades (e.g., *"Centrolene"*, *"Cochranella"*), as proposed by Guayasamin et al. [1]. An updated taxonomy for all the species in Centrolenidae is provided in Table S1.

For recognizing species, we adhered to the evolutionary species concept first proposed by Simpson [34,35], modified by Wiley [36] and de Queiroz [37,38]. This concept incorporates important theoretical factors such as lineage independence, identity, and evolutionary tendencies, and provides a flexible framework when reproductive isolation is difficult to test (e.g., allopatric populations). Evidence supporting the validity of a species can come from different sources (e.g., morphology, DNA, behavior, ecology), and no trait alone can be considered a biological property that a species must have in order to be recognized [37]. In other words, under the evolutionary species concept, the only necessary property for an entity to be a recognized as a species is that it corresponds to a temporal segment of a metapopulation lineage evolving separately from other lineages [37,38]. Evidence for independent evolution is gathered from different data sources, where integrative taxonomy plays a fundamental role when assessing what represents (or not) a distinctive species (e.g., [39,40]).

2.2. Characters and Terminology

For general terminology and descriptions of morphological characters we follow the proposals by Lynch and Duellman [22] and Cisneros-Heredia and McDiarmid [17]. We illustrate some of the most relevant traits for glassfrog identification, including dorsal color pattern (Figure 2), skin texture (Figure 9), and snout shape (Figure 10). To facilitate comparison with previous literature dealing with anurans, fingers are numbered preaxially to postaxially from I–IV. However, we stress that from an evolutionary perspective, anuran fingers should be numbered from II–V, to reflect the loss of Digit I in anurans [41–43]. Webbing formulae follow the method of Savage and Heyer [44], as modified by Guayasamin et al. [20] (Figure 11). Larval characters follow the terminology recommended by McDiarmid and Altig [45]. The morphology of nuptial excrescences and prepollical spines (Figure 12) follows the types proposed by Flores [46], with the additions and modifications detailed in Cisneros-Heredia and McDiarmid [17] and Guayasamin et al. [1]. Other key traits in centrolenid taxonomy are humeral spines (Figure 7, Figure 13, and Figure 14), ventral transparency (Figure 8), peritoneum and pericardium (with or without iridophores; Figure 13), enlarged subcloacal warts on thighs below vent (Figure 15), and tubercles on the external edge of arm, hand, and foot (Figure 16).

Figure 9. Skin texture in glassfrogs. (**A**) Smooth; *Sachatamia ilex*, QCAZ 47193. (**B**) Shagreen; *Nymphargus humboldti* sp. nov., ZSFQ 0833. (**C**) Pustular; *Centrolene heloderma*, QCAZ 40200. All photographs by Luis A. Coloma, except (**B**) by Jose Vieira/Tropical Herping.

When discussing parental care, we adopt the terminology by Delia et al. [25]. Egg brooding refers to a specific form of ventral contact where the parent positions its body over the egg clutch; this behavior reduces embryonic mortality by protecting embryos from dehydration and, possibly, by preventing fungal development and predation [25]. Parental care is divided into the following behaviors [25]: (i) Short-term maternal care, where brooding is provided for a few hours just after oviposition; (ii) prolonged male care, where parental care is provided for several weeks; and (iii) prolonged female care, where parental care is provided for several weeks (only observed in *Ikakogi tayrona* [47]).

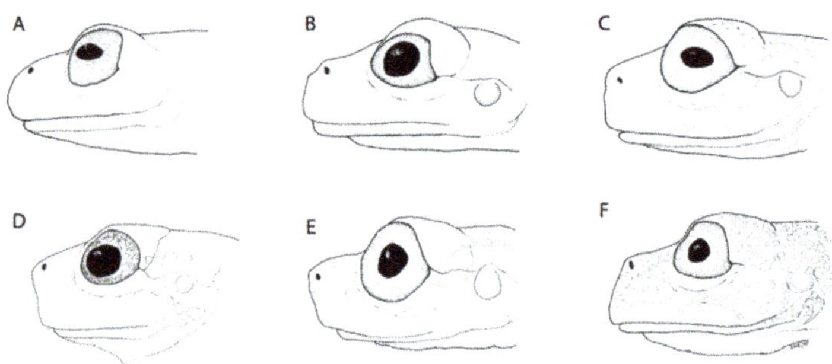

Figure 10. Snout shape in Glassfrogs. (**A**) Round, *Hyalinobatrachium munozorum*, KU 118054. (**B**) Bluntly round, *Centrolene ballux*, KU 164725. (**C**) Truncate, *Nymphargus megacheirus*, KU 143269. (**D**) Slightly protruding, *"Centrolene" medemi*, KU 164493. (**E**) Protruding, *"Cochranella" balionota*, KU 164708. (**F**) Sloping, *Nymphargus grandisonae*, KU 164688. Drawings by Juan M. Guayasamin.

Morphometrics—morphological variables were measured with digital callipers to the nearest 0.1 mm, as follows: (1) Snout–vent length (SVL) = distance from tip of snout to posterior margin of vent; (2) tibia length (TL) = length of flexed leg from knee to heel; (3) foot length (FL) = distance from proximal margin of outer metatarsal tubercle to tip of Toe IV; (4) head length (HL) = distance from tip of snout to posterior angle of jaw articulation; (5) head width (HW) = width of head measured at level of jaw articulations; (6) interorbital distance (IOD) = distance between upper eyelids, representing the width of the underlying frontoparietals; (7) upper eyelid width (UE) = greatest transverse width of upper eyelid; (8) internarial distance (IN) = distance between nostrils; (9) eye–nostril distance (EN) = distance from posterior margin of nostril to anterior margin of eye; (10) snout–eye distance (SE) = distance from tip of snout to anterior margin of eye; (11) horizontal eye diameter (ED) = distance between anterior and posterior borders of eye; (12) tympanum diameter (TD) = distance between anterior and posterior margins of tympanic annulus; (13) eye–tympanum distance (ET) = distance from posterior border of eye to anterior margin of tympanic annulus; (14) radio–ulna length (RUL) = length of flexed forearm from elbow to proximal border of palmar tubercle; (15) hand length (HDL) = distance from the proximal margin of palmar tubercle to tip of Finger III; (16) Finger-I length (F1L) = distance from outer margin of palmar tubercle to tip of Finger I; (17) Finger-II length (F2L) = distance from outer margin of palmar tubercle to tip of Finger II; (18) disc of Finger III (3DW) = greatest width of disc of Finger III; and (19) Finger-III width (F3W) = width of Finger III measured at the level of distal subarticular tubercle, including lateral fringes and excluding webbing. For comparing different body sizes (SVL) among glassfrogs, we considered the average size of males and categorized them according to the following criteria: Minute (SVL < 22 mm), small (SVL 22–25 mm), medium (SVL 25–30 mm), large (SVL 30–50 mm), and giant (SVL > 50). Eye diameter was divided into small (eye diameter < 10% SVL), moderate size (eye diameter 10%–15% SVL), and large (eye diameter > 15% SVL). Tympanum was considered to be very small (tympanum < 20% of eye diameter), small (tympanum 20%–30% of eye diameter), moderate (tympanum 31%–40% of eye diameter), large (tympanum diameter 41%–50% of eye diameter), and very large (tympanum diameter > 50% of eye diameter.

A - Webbing Formula

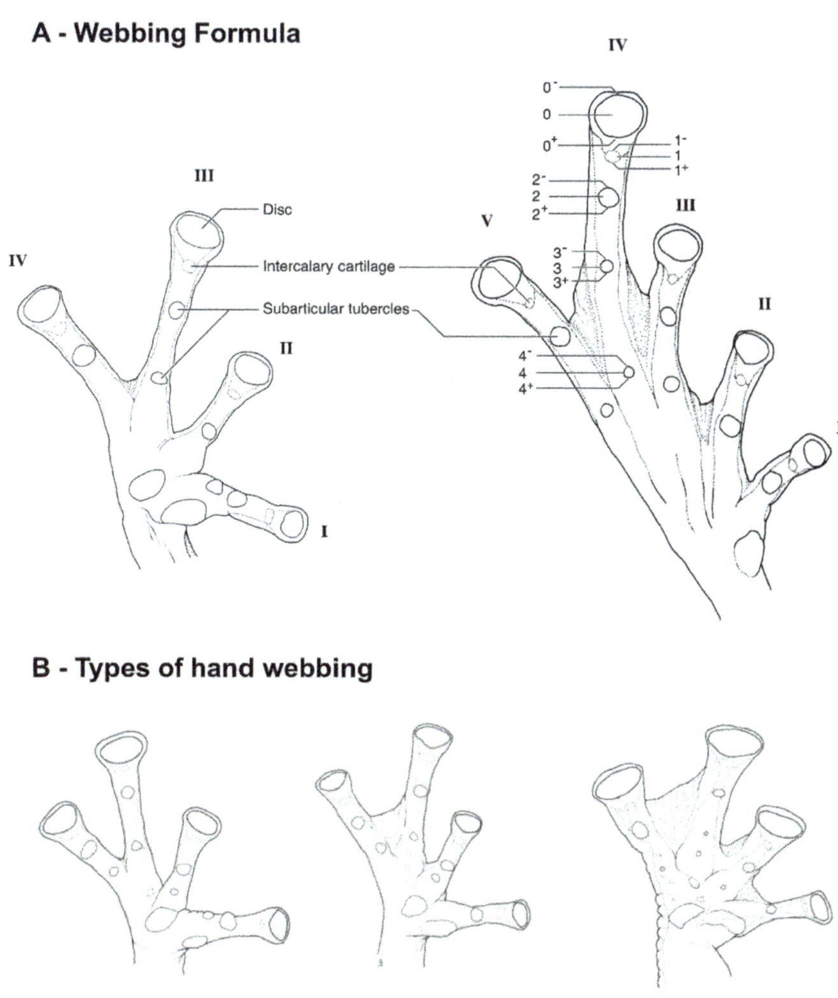

B - Types of hand webbing

Figure 11. Webbing in glassfrogs. (**A**) Terminology used for webbing formula in hands and feet (see Guayasamin et al. [20]; modified from Savage and Heyer [44]). Roman numerals (I, II, III, IV, V) represent fingers or toes. Arabic numerals represent the number of phalanges completely or partially free of webbing. We use 0^- to indicate that webbing reaches the distal margin of the disc; **0** indicates that webbing reaches the middle of the disc; 0^+ indicates that webbing reaches the proximal margin of the disc; 1^- indicates that webbing reaches the distal margin of the intercalary cartilage; **1** indicates that the webbing the middle of the intercalary cartilage; 1^+ indicates that the webbing the proximal margin of the intercalary cartilage; 2^- indicates that webbing reaches the distal margin of the distal subarticular tubercle; **2** indicates that webbing reaches the middle of the distal subarticular tubercle; 2^- indicates that webbing reaches the proximal margin of the distal subarticular tubercle. For example, webbing formula in the illustrated foot is absent between Toes I and II (lateral fringes are not considered as webbing); **II** $1^{2/3}$**—**3^+ **III** $1^{2/3}$**—**3^- **IV** 3^-**—**2 **V**. Figure modified from Guayasamin et al. (2006). (**B**) Simplified type of hand webbing. Absent: *Nymphargus cochranae*, QCAZ 31113. Moderate: *Chimerella mariaelenae*, QCAZ 22363. Extensive: *Centrolene geckoidea*, KU 164490. Drawings by Juan M. Guayasamin.

A - Exposure of Prepollex

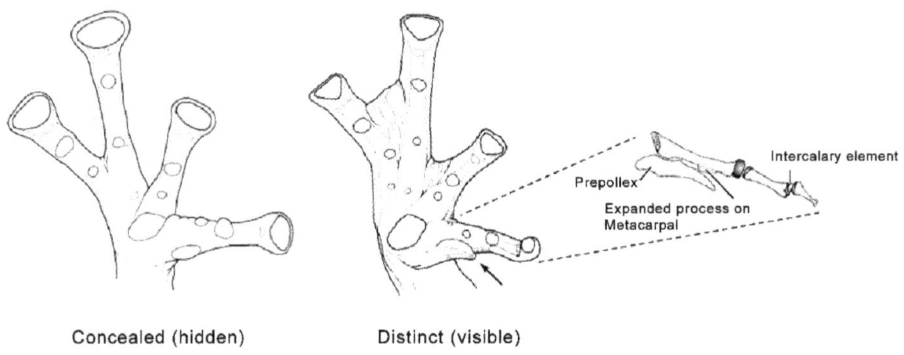

B - Nuptial pad morphology

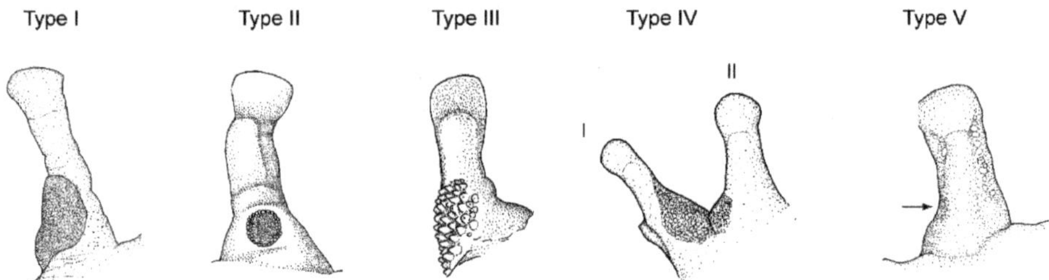

Figure 12. Prepollex and nuptial pad morphology in glassfrogs. (**A**) Condition of the prepollex. Concealed, *Nymphargus cochranae*, QCAZ 31113. Distinct, *Teratohyla spinosa*, KU 164668. (**B**) Nuptial pad morphology. **Type I:** Large to medium-size nuptial excrescence present on the dorsal, lateral, and/or ventral sides of the thumb; *Cochranella posadae*, QCAZ 26023. **Type II:** Small circular or squarish nuptial excrescence present on the dorsal or dorsolateral face of the thumb; *Centrolene lynchi*, MCZ 97846 (figure modified from Flores 1985). **Type III:** Medium-size spinous nuptial excrescence extending from the lateral side of the thumb to its dorsomedial surface; *Nymphargus armatus*, UVC 9400 (modified from Lynch and Ruiz-Carranza 1996). **Type IV:** Large nuptial excrescence formed by a granular pad that extends from the side of the thumb to its dorsomedial surface, and on the proximal dorsolateral surface of finger II; *Cochranella litoralis*, ICN 13821. **Type V:** Medium-size diffuse nuptial excrescence formed by glandular clusters and individual glands; a pad as such is absent; *Hyalinobatrachium aureoguttatum*, QCAZ 27429. **Type VI** (not illustrated): Nuptial excrescences formed by a combination of clustered and individual glands that sparse along the flanks of the body (see text).

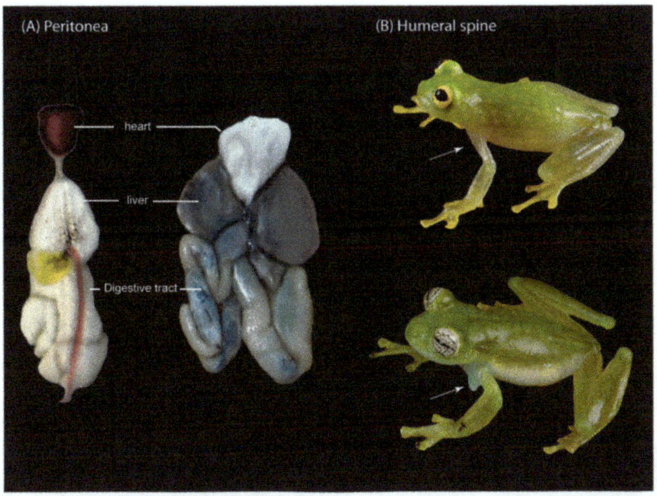

Figure 13. Peritonea and humeral spines in Centrolenidae. (**A,Left**): Pericardium lacking iridophores, hepatic, and visceral peritonea with iridophores (*Hyalinobatrachium aureoguttatum*). (**A,Right**): Pericardium with iridophores, hepatic, and visceral peritonea lacking iridophores (*Centrolene buckleyi*). (**B,Top**): Absence of humeral spine (*H. fleischmanni*). (**B,Bottom**): Presence of humeral spine in males (*Espadarana callistomma*). Figure modified from Guayasamin et al. [1].

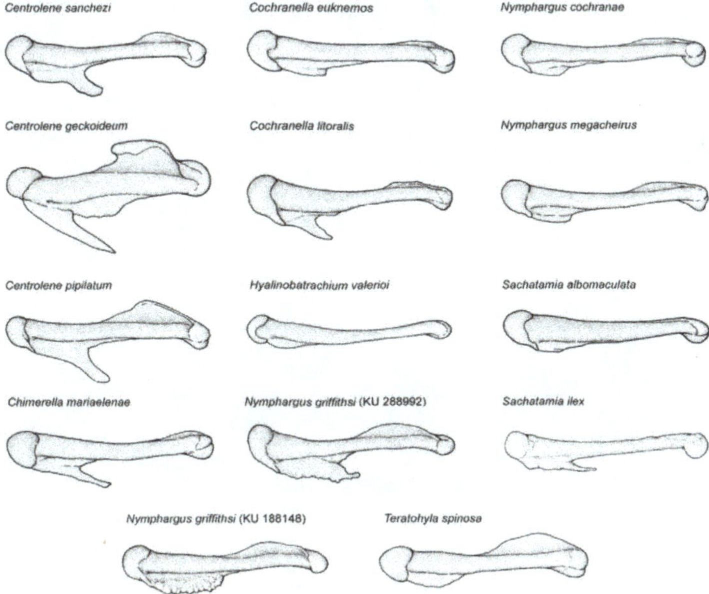

Figure 14. Variation of humeral spines, *crista medialis*, and *crista ventralis*, in adult males of Centrolenidae (modified from Guayasamin et al. [1]). Illustrated specimens are: *Centrolene sanchezi*, KU 170116; *C. geckoidea*, ICN 5598; *C. pipilata*, KU 143286; *Cochranella euknemos*, KU 77534; *Cochranella litoralis*, QCAZ 27693; *Teratohyla spinosa*, KU 32935; *Nymphargus griffithsi*, KU 288992, 188148; *N. cochranae*, KU 123218; *N. megacheirus*, KU 143271; *Sachatamia albomaculata*, KU 65185; *S. ilex*, LACM 72910; *Chimerella mariaelenae*, QCAZ 21252; *H. valerioi*, KU 178091.

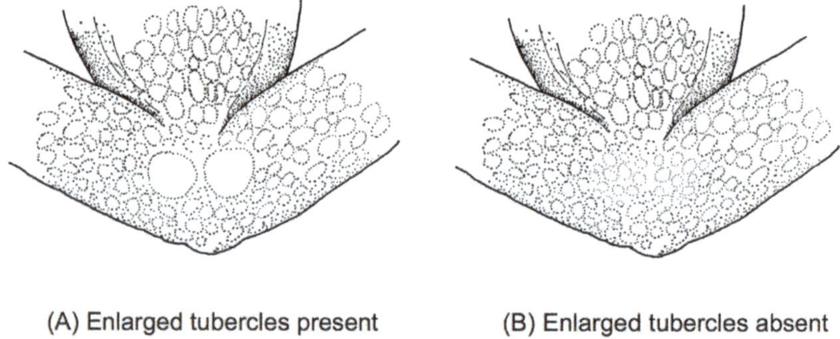

Figure 15. Ventral surfaces of tights in glassfrogs. (**A**) With enlarged subcloacal warts. (**B**) Lacking enlarged warts. Illustrations by Juan M. Guayasamin.

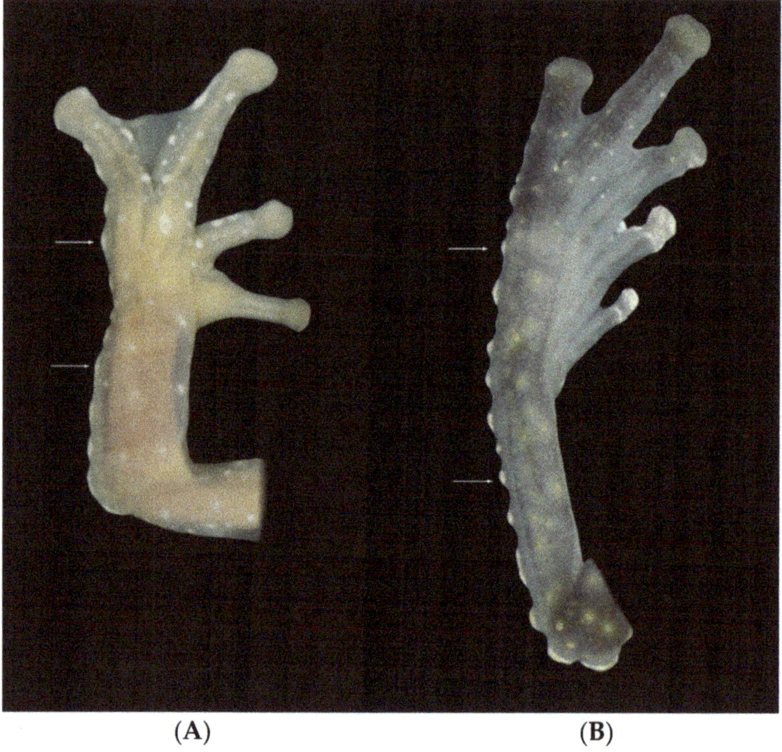

Figure 16. Tubercles on limbs. (**A**) Tubercles on ventrolateral edge of Finger IV and arm, *Cochranella resplendens*, KU 118053. (**B**) Tubercles on ventrolateral edge of Toe V and tarsus, *C. mache*, QCAZ 22412. Photos by Martín Bustamante.

Vocalizations—Calls were recorded in the field by different researchers (see Acknowledgments) using an Olympus Linear PCM Recorder LS-10 tape recorder and a Sennheiser K6-ME67 directional microphone. Recordings obtained by us are stored at the Laboratorio de Biología Evolutiva of the Universidad San Francisco de Quito (LBE). Sounds were recorded in PCM format at a sampling frequency of 44.1 kHz. Audiospectograms and oscillograms were generated in the R package

SeeWave [48]. Frequency information was obtained through fast Fourier transformation, with a 50% window overlap (Hanning window size of 1024 and a frequency window of 43.1 Hz). The following call variables were measured: Call duration, duration between calls, number of notes per call, note duration, duration between notes, dominant frequency, first harmonic, second harmonic [49–51] (Table 1). A call is defined as the sound produced in a single exhalation of air. Calls and notes were divided into two categories, tonal and pulsed, based on distinct morphology. Tonal calls/notes have non-peaked amplitude sustained throughout the duration of the note. Pulsed calls/notes are characterized by having at least one clear amplitude peak. Pulses were defined as a visible increase and decrease of amplitude peaks on the oscillogram within a single note, and notes were defined by a single and complete amplitude rise from and return to the base frequency of the background noise. For a review on the use of bioacoustics in taxonomy and specific variable definitions, see Köhler et al. [52].

Table 1. Definitions of call variables [51].

Call Variables	Definition
Call type	Calls tonal or pulsed; pulsed defined as having amplitude modulation (Dautel et al. [50])
Call/note interval	Measured time between a call/note
Call duration	Call measured from the start of the amplitude rise away from the background noise to return to the background noise
Call rate	(Total number of calls–1)/time measured from start of the first call to the beginning of the last call
Call rise time	Time from start of call to the point at maximum amplitude
Number of pulses	Total number of pulses per call
Pulse/note rate	(Total number of pulses/notes–1)/time from beginning of first pulse/note to the beginning of the final pulse
Pulse length	Time from start to end of one pulse; measured for pulses at the beginning, middle, and end of call
Pulse/note rise time	Time from start of a pulse/note to the point of maximum amplitude
Pulse/note shape	(Pulse rise time/pulse length); unitless variable that describes the overall shape of the amplitude envelope of the pulse. Right or left skewed pulses will have a rise time near the start or end of the call, respectively. This parameter allows comparison of pulses/notes with differing lengths.
Pulse/note amplitude to call peak amplitude ratio	Maximum amplitude of pulses/notes compared to the peak amplitude of the call. Measured between the start, middle, and end of call. Describes amplitude distribution and modulation throughout call.
Pulse/note amplitude change throughout call	Ratio of pulse/notes amplitude compared between the middle and beginning of call, the end and beginning of call, and the end and middle of call
Lower frequency of the fundamental frequency	Lower limit of the fundamental frequency
Upper frequency of the fundamental frequency	Upper limit of the fundamental frequency
Dominant frequency	Frequency of call, which contains the greatest concentration of energy
Frequency modulation	Dominant frequency of the last 0.020 s of call subtracted from the dominant frequency of the first 0.020 of call
Harmonic frequencies	Measured harmonic frequencies
Power, percent of max	Measure for harmonic frequencies, percent of power in harmonic compared to max power of call

Specimens examined—We examined ethanol-preserved specimens from the following herpetological collections: Centro de Biodiversidad y Genética, Cochabamba, Bolivia (CBG); Círculo Herpetológico de Panamá (CHP); Centro Jambatu de Investigación y Conservación de Anfibios, San Rafael, Ecuador (CJ); División de Herpetología, Museo Ecuatoriano de Ciencias Naturales, Quito, Ecuador (DHMECN); Field Museum, Division of Amphibians and Reptiles, Chicago, USA (FMNH); Instituto de Ciencias

Naturales, Universidad Nacional de Colombia, Bogotá, Colombia (ICN); University of Kansas, Museum of Natural History, Division of Herpetology, Lawrence, Kansas, USA (KU); Natural History Museum of Los Angeles County, Section of Herpetology, Los Angeles, California, USA (LACM); Museum of Comparative Zoology, Harvard University, Cambridge, Massachusetts, USA (MCZ); Museo de Historia Natural La Salle, Caracas, Venezuela (MHNLS); Museo de Zoología, Universidad Tecnológica Indoamérica, Quito, Ecuador (MZUTI); Museo de Zoología, Pontificia Universidad Católica del Ecuador, Quito, Ecuador (QCAZ); Colección de Herpetología, Escuela de Biología, Universidad de Costa Rica, San José, Costa Rica (UCR); Museo de Vertebrados, Universidad del Valle, Cali, Colombia (UVC); and Museo de Zoología, Universidad San Francisco de Quito, Ecuador (ZSFQ). When specimens were not available for direct comparison, we relied on descriptions in the literature. Sexual maturity of specimens was determined by the presence of vocal slits and nuptial pads in males and by the presence of eggs or convoluted oviducts in females.

2.3. Evolutionary Relationships

Taxon and gene sampling—we combined genetic sequences available from previous work (mostly [2,3,19]) and new sequences generated during this study (Table S2). Genomic extraction, amplification, and sequencing are as described in Guayasamin et al. [2] and Castroviejo-Fisher et al. [3]. The final dataset contains complete or partial sequences of 10 genes representing 6513 bp of data (mitochondrial: 12S rRNA, 16S rRNA, ND1; nuclear: BDNF, C-MYC exon 2, CXCR4, NCX1, POMC, RAG1, SLC8A3). We sampled 49 outgroup taxa from a large range of families within Hyloidea (Table S2), and include all three species from Allophrynidae, the sister group to Centrolenidae [2,12]. Finally, ingroup taxon sampling includes 251 terminals, representing 113 named species, and 24 putative new species. The percentage of named species sampled (i.e., matrix completeness) for each marker is as follows: 12S = 90%; 16S = 94%; BDNF = 55%; C-MYC exon 2 61%; CXCR4 = 61%; NCX1 = 59%; ND1 = 81%; POMC = 61%; RAG1 = 62%; and SLC8A3 = 60% (complete marker statistics can be found in Table S3). For most species of Ecuadorian glassfrogs, we obtained a total of ~2733 bp from the following mitochondrial markers: 12S rRNA (~907 bp), 16S rRNA (~864 bp), and *ND1* (~960 bp). See Table S2 for genes sequenced for each species and GenBank Accession Numbers.

Candidate species—to assist in identifying putative new species for future study, we employed the candidate species designation system of Vieites et al. [29]. Prior to this study, divergent centrolenid lineages have been designated in inconsistent ways (using sp. or aff. or museum numbers), which can result in confusion in identifying and studying tentative new species. Given the increasing number of putative new species in Centrolenidae, establishing candidate species aims to better organize and maintain the growing discovery of undescribed lineages. Importantly, designating a divergent lineage as a putative new species does not necessarily mean it is a new species; rather, the system is meant to identify these lineages for future study, incorporating multiple lines of evidence (i.e., morphology, bioacoustics, biogeography, or nuclear genes) in order to determine their species status.

Candidate species are typically identified through evidence from genetic distances or phylogenetic lineage divergence through widely used genetic markers that provide a basis for comparison (i.e., 12S, 16S, CO1; [29,53]). In this study, we used the mitochondrial genetic markers 12S, 16S, and ND1 and integrated evidence from genetic distances and phylogenetic relationships to identify candidate species. We first used a threshold of 3% in identifying divergent lineages, which is a threshold that most named centrolenid species exceed (this study). Putative new species were then numbered using a scheme of "sp_CaXX", with numbering beginning at 01 for each genus. Finally, we note that Vieites et al. [29] also uses "unconfirmed" and "confirmed" designations, whereas "confirmed" candidate species have additional evidence (morphology or calls) for their status as species and are simply awaiting detailed analyses and subsequent description. We did not use these designations, as most putative new centrolenid species are supported only through mitochondrial genetic divergence.

Phylogenetics—the analyzed dataset contains complete or partial sequences of 10 genes representing 6513 bp of data (mitochondrial: 12S rRNA, 16S rRNA, ND1; nuclear: BDNF, C-MYC exon 2, CXCR4,

NCX1, POMC, RAG1, SLC8A3). See Tables S2 and S3 for details. Analyses were conducted using maximum likelihood (ML) and Bayesian (BA) criteria. We did not perform searches under the parsimony criterion because of its limitations under certain conditions (long branch attraction; [54,55]) and lack of theoretical support for transition/transversion bias and variation in substitution rates among different nucleotide sites [56]. The 12S and 16S rRNA sequence data were aligned using MAFFT 7.4 [57], using the Q-INS-i algorithm that takes RNA secondary structure into consideration. Protein coding genes were also aligned in MAFFT using the AUTO function and manually inspected for accuracy and open reading frames.

Maximum likelihood was run in the IQ-TREE v1.5.5 software [58]. The data were automatically partitioned, and the best model was implemented using ModelFinder within IQ-TREE [59], which groups partitions with the same model and similar rates and simultaneously searches model and tree space. Node support was assessed via 100 ultra-fast bootstrap replicates [60].

For Bayesian phylogenetic analyses and divergence dating, we used BEAST v2.4.5 [61]. We used a single secondary calibration point using ages estimated from Hutter et al. [62], which estimates divergence dates for Hyloidea using 18 genetic markers and 8 fossil/geographic calibrations. We used median age of Centrolenidae with a normal distribution to capture the 95% confidence interval from this study (Mean = 33.4 Myr; Sigma = 6). We used a single uncorrelated lognormal relaxed clock prior linked across all partitions. We partitioned data by codon position for protein-coding genes and used the Bayesian bModelTest package within BEAST to select the best model for each partition. We estimated the relaxed clock rate using an initial value of 1e-9 and a broad prior (in our case, a gamma distribution with a shape parameter of 0.001 and scale parameter of 1000). We used a Yule speciation process for the tree prior. We ran Markov chain Monte Carlo (MCMC) searches for a total of 100 million generations, sampling every 10,000 generations. Stationarity was assessed by examining the standard deviation of the split frequencies and by plotting the -ln*L* per generation, using Tracer v1.5 [63]; trees generated before stationarity were discarded as "burn-in", which was 20% of trees.

Species conservation status—global and local (Ecuador) conservation status of glassfrogs follow the categorization and criteria established by the International Union for Conservation of Nature (IUCN) [64] (Figure 17), including rate of decline, population size, area of geographic distribution, and degree of distribution fragmentation. The following categories were used: (i) *Not Evaluated:* The species has not yet been evaluated against the criteria, (ii) *Data Deficient:* There is inadequate information to make a direct or indirect assessment of its risk of extinction based on its distribution and/or population status, (iii) *Least Concern:* The species is widespread and abundant and not under immediate risk of extinction, (iv) *Near Threatened:* For species that are not currently threatened, but are close to qualifying for or are likely to qualify for a threatened category in the near future, (v) *Vulnerable:* A taxon is Vulnerable when it is considered to be facing a high risk of extinction in the wild (Criteria A to E for Vulnerable), (vi) *Endangered:* When the species is considered to be facing a very high risk of extinction in the wild (Criteria A to E for Endangered), (vii) *Critically Endangered:* When the species is considered to be facing an extremely high risk of extinction in the wild (Criteria A to E for Critically Endangered), (viii) *Extinct in the Wild:* When the species is known only to survive in cultivation, in captivity, or as a naturalized population, outside its historical range. A taxon is presumed Extinct in the Wild when exhaustive surveys in known and/or expected habitats, at appropriate times, throughout its historic range have failed to record an individual, (ix) *Extinct:* When there is no reasonable doubt that the last individual of the species has died. A taxon is presumed Extinct when exhaustive surveys in known and/or expected habitats, at appropriate times (diurnal, seasonal, annual), throughout its historic range have failed to record an individual.

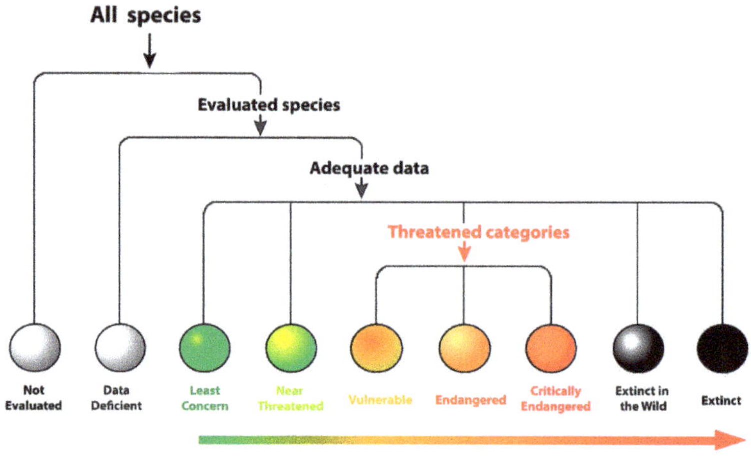

Figure 17. International Union for Conservation of Nature (IUCN) categories of species conservation status.

2.4. Biogeographic Regions of Ecuador

In this study, we applied biogeographic regions of continental Ecuador as a simplification of vegetation types [65,66] (Figure 18). This system has the virtue of combining the characteristics of the vegetation and the historic isolation between the western and eastern slopes of the Andes and the eastern and western lowlands. Mean annual precipitation and mean annual temperature for each of the recognized regions are shown in Table 2. We also summarized the land cover of Ecuador, illustrating which ecosystems have suffered intensive deforestation and which are still preserved (Table 3, Figure 18).

Dry Shrub—characterized by a combination of warm and extremely dry conditions. Annual precipitation can be as low as 60 mm in the westernmost locality (Salinas, Santa Elena Province). This region covers an area of 8033 km^2 and is restricted to the coastal margin of central Ecuador (Figure 18). In some areas, grasses introduced for raising livestock have replaced native plants. In the drier habitats, xerophytic plants are dominant, especially cacti and other thorny plants [67].

Western Deciduous Forest—this forest occurs at an elevation of 50–300 m in central and northern Ecuador (100–400 m in southern Ecuador) and covers 25,673 km^2 (10.3% of Ecuadorian territory, Figure 18). Conditions are drier and the terrain has lower tree densities than in evergreen forests. The trees are generally shorter than 20 m with an understory that can be dense with abundant herbaceous plants. Some tree species lose their leaves during the dry season [67]. More than half of the land cover of this habitat type has been converted for agriculture and grazing cattle.

Chocoan Tropical Rainforest—this rainforest is the second largest biogeographic region in Ecuador, with 31,732 km^2 at elevations ranging from sea level to 300 m (Figure 18). It has a closed canopy with trees that can reach 30 m in height and with an understory dominated by ferns and Araceae [67]. Tree diversity is high, with more than 100 species/ha with diameter at breast height >10 cm, but lower than in the Amazonian Tropical Rainforest [68]. Habitat destruction rate in this region is the highest in Ecuador and only 18.3% of its natural vegetation remains.

Western Foothill Forest—this forest covers 15,305 km^2 on the western Andean slopes with an elevational range of 300–1300 m (400–1000 m in southern Ecuador). Plant endemism is high, especially between latitudes 0° and 3° S [67]. This forest is structurally similar to its counterpart from the eastern Andean slope, although the amphibian communities are highly differentiated.

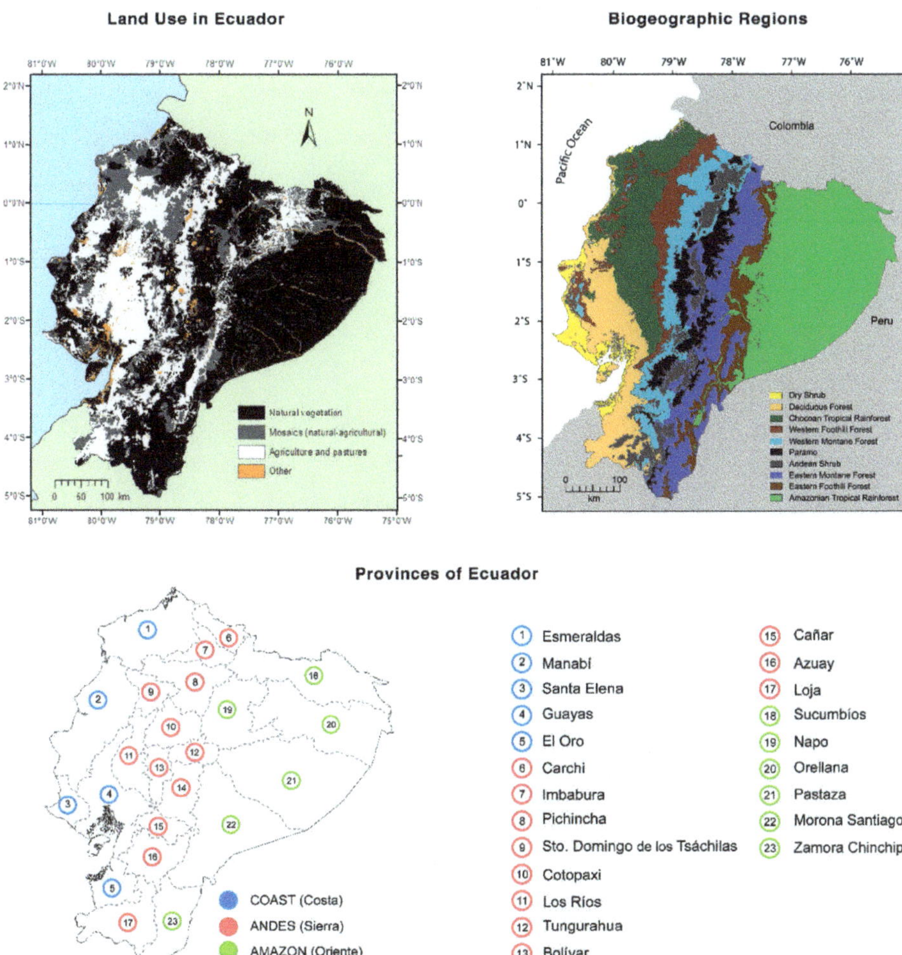

Figure 18. Land use, biogeographic regions, and provinces of Ecuador. Land cover: Modified from Ron et al. [66]; "Mosaics" are mixtures of natural vegetation and either agricultural land or pastures. "Other" includes urban areas, shrimp farms, lakes, rivers, glaciers, and sand banks. Biogeographic regions: Shown as a simplification of vegetation types (Sierra et al. [65], as modified by Ron et al. [66]). Provinces: Provinces are divided into three broad geographic regions: Coast, Andes, and Amazon.

Table 2. Glassfrog diversity and climate in the Ecuadorian biogeographic regions, as defined in Figure 14 [65,66]. Note that any given species can occur in more than one biogeographic region.

Biogeographic Region	Area of the Region (km²)	Mean Annual Temperature (°C)	Mean Annual Precipitation (mm)	No. of Glassfrog Species	Density (Species/1000 km²)
Dry Shrub	8033	24.8 ± 0.66	500 ± 393	0	0
Deciduous Forest	25,673	24.3 ± 1.32	843 ± 316	3	0.12
Chocoan Tropical Rainforest	31,737	25.1 ± 0.59	2086 ± 665	12	0.38
Western Foothill Forest	15,305	22.4 ± 1.71	2,218 ± 907	15	0.98
Western Montane Forest	21,576	15.0 ± 4.04	1187 ± 610	14	0.65
Páramo	15,976	7.3 ± 2.30	803 ± 277	1	0.06
Andean Shrub	11,266	15.8 ± 2.70	817 ± 215	1	0.09
Eastern Montane Forest	31,555	15.8 ± 4.13	1691 ± 799	20	0.63
Eastern Foothill Forest	13,133	21.7 ± 0.91	2923 ± 1023	16	1.21
Amazonian Tropical Rainforest	73,909	24.9 ± 0.81	3349 ± 555	10	0.14
Total: Continental	248,163			60	0.24

Table 3. Land cover (as percentage) in Ecuadorian biogeographic regions [66] (Figure 15). Mosaics: Mixtures of natural vegetation and either agricultural land or pastures. Other: Includes urban areas, shrimp farms, lakes, rivers, glaciers, and sand banks.

Biogeographic Region	Natural Vegetation	Mosaics (Natural-Agricultural)	Agriculture and Pastures	Other
Dry Shrub	53.1	9.9	13.8	23.2
Deciduous Forest	26.7	17.4	51.6	4.3
Chocoan Tropical Rainforest	18.3	23.8	60.0	1.9
Western Foothill Forest	40.3	20.6	38.2	0.9
Western Montane Forest	35.8	28.5	34.6	1.1
Páramo	78.9	7.0	6.7	7.4
Andean Shrub	29.1	7.2	53.7	10.0
Eastern Montane Forest	69.4	16.0	14.2	3.4
Eastern Foothill Forest	59.3	15.6	24.4	0.7
Amazonian Tropical Rainforest	76.2	10	9.5	4.3

Western Montane Forest—this evergreen forest covers 21,576 km² at an elevational range between 1300–3400 m (1000–3000 m in southern Ecuador; Figure 18). The canopy is generally below 25 m with a high abundance of epiphytic plants (especially mosses, ferns, orchids, and bromeliads). At intermediate elevations, especially during the afternoon, the forests become misty and receive horizontal precipitation from low, overhanging clouds. Western Montane Forest is restricted to narrow stretches between the basin of the Mira River (close to the Colombian border) and the basins of the Chanchán and Chimbo rivers (2° S). It is replaced by drier habitats (principally Andean Shrub) south of 4° S, close to the border with Peru. Only 35% of its natural vegetation remains unaltered (Table 3).

Páramo—this is the vegetation type that reaches the highest elevation and covers 15,976 ha (6.1% of the territory; Figure 18). Depending on the region, its lower limit lies between 3000 and 3600 m. Short herbaceous plants generally forming tight clumps dominate the vegetation. The plants are adapted to cold temperatures and to low availability of water. Open grassy areas predominate but are mixed with small patches of forest or shrubs [69]. At higher elevations, the vegetation is restricted to sparse clumps on otherwise bare land. Because of the occurrence of frequent freezes, agriculture is limited, and this has ameliorated anthropogenic habitat degradation. In this region, only 21.1% of the natural vegetation has been cleared or severely fragmented, the lowest proportion for any region (Table 3). However, the Páramo is the region with the highest proportion of endangered amphibians.

Andean Shrub—this biogeographic region lies between 1400 and 3000 m and has an area of 11,266 km^2; it is found in the inter-Andean basins between the Cordillera Occidental and Cordillera Oriental (Figure 18). As a result of rain shadow effects from both mountain chains, the Andean Shrub has a relatively low precipitation (Table 2). Although originally dominated by shrubs, most of the vegetation has been replaced by crops, pastures, or forests of exotic trees (*Eucalyptus* and *Pinus*; [69]). In dry valleys (e.g., Chota, Guayllabamba, and Patate) the native vegetation is spiny. Andean Shrub is almost unrepresented in the Ecuadorian National System of Protected Areas. Habitat degradation is severe; more than half the land cover is devoted to agriculture or to raising cattle (Table 3).

Eastern Montane Forest—this forest covers 31,555 km^2 between 1300 m and 3600 m on the eastern Andean slopes (Figure 18). The vegetation is structurally similar to that from the Western Montane Forest. Above 2900 m, the soil of the forest is covered by moss and the trees are irregularly shaped [69].

Eastern Foothill Forest—this evergreen forest covers 13,133 km^2 between elevations of 600 m and 1300 m (Figure 18), and is a mixture of tree species from the Andes and the lowlands of the Amazon Basin [68]. The canopy reaches up to 30 m in height and encloses a dense sub-canopy and understory. Tree diversity is lower (130 species/ha, >10 cm DBH) [68] than in the Amazonian Tropical Forest. Average annual precipitation is the second highest of all regions (2833 mm).

Amazonian Tropical Rainforest—the Amazonian Tropical Rainforest is the most extensive biogeographic region in Ecuador with a total area of 73,909 km^2 (29.8% of the Ecuadorian continental territory; Figure 18). It is restricted to elevations below 600 m and has the highest average annual precipitation of any region (3349 mm). The dominant forest type, *Terra Firme*, is characterized by well-drained soils. The canopy is 10–30 m high, punctuated with emergent trees up to 40 m (and rarely 50 m); small gaps created by fallen trees [68,70] are common. Tree diversity is high with 200–300 species of trees/ha (>10 cm DBH) [68,70]. Other vegetation types in this region include *Várzea* forest (flooded with white water), *Igapó* forest (flooded with black water), riparian woodland forest, river island scrub, and *Mauritia flexuosa* palm swamps [68,71]. At the local scale (≤100 km^2), amphibian diversity reaches its global peak in the Amazonian Tropical Rainforest of Ecuador [72,73].

2.5. Potential Distribution

Estimating the distribution of a species is challenging, since biotic, abiotic, and historic factors come into play. As an approximation to species distributions and being aware of the associated caveats [74,75], we modelled the ecological niche of all species for which we had at least 10 independent localities (>1 km apart). Given that our ecological models did not include variables such as biotic interactions, random extinction, vagility, influence of diseases or introduced species, we used our results only as proxy, and are fully aware that predicted areas might be, in most cases, overestimated; nevertheless, models are useful for conservations assessments. Locality data were obtained from the following museums: AMNH, BMNH, FHGO, KU, QCAZ, MECN, MZUTI, ZSFQ, and USNM. All specimens were directly examined by Juan M. Guayasamin or Diego F. Cisneros-Heredia.

We used all 19 climatic variables from the WorldClim database (www.worldclim.org). These bioclimatic variables are derived from monthly temperature and rainfall values [76]. A correlation analysis was used to determine independent variables for the distribution of each species. The potential distributions of centrolenid species were estimated using default settings in Maxent v. 3.3.3k [77,78]. We used 70% of the presence records as training data, and the other 30% was used to evaluate the model. Maxent was run 10 times to obtain the ecological model for each species. In order to evaluate models, we employed the area under the curve (AUC), a value that is an indicator of model performance [77,79]. The AUC value was calculated in Maxent and ranges between 0.5 (random classification) and 1 (perfect fit). We discarded models that had AUC values below 0.7 [80]. Maxent continuous models were reclassified in order to obtain a binary map of potential presence or absence, using the maximum training sensitivity plus specificity threshold (MTS + S), a method that has been shown to produce highly accurate predictions [81,82]. The potential distribution models were edited using ArcMap v.10 to obtain more realistic estimates of species distributions by removing areas predicted as present that are located in inaccessible biogeographic regions (e.g., species restricted to the Amazon basin predicted as present in the Chocó and vice versa; species from the Pacific slope of the Andes predicted as present in the Amazonian slope and vice versa).

Impacts of human activities to species—in order to build a *layer of human impact (LHI)* for Ecuador, we used shape files containing information on land use, roads, human settlements, human population density, mining, and oil exploration and concession shape files (ESRI, 2003). The original shape files were converted to raster files in ArcMap v. 10; we then calculated Euclidean distances in each raster file to compute buffer areas around specific features (i.e., roads, oil fields), giving values according to the intensity of human impact and distance from the specific human activity (Table 4). In the case of population density, the intensity value was equal to the logarithmic scale of the population density value for each grid cell. We obtained raster layers with buffer zones and intensity values for each of the impacts, which were added using the map calculator tool (ArcMap v. 10) to obtain a single map summarizing all considered threats.

Table 4. Intensity values and distance of influence of human activities.

Human Impact	Class	Intensity Values (0–100)	Distance of Influence (m)
Roads	1st Order	50	0–1000
	2nd Order	25	
	3rd Order	17	
	Railroad	17	
	Trails	10	
Population density	—	Logarithmic scale	0–7000
Agriculture/Animal husbandry	Monoculture	60	0–2000
	Mixed crops	50	0–1000
	Mosaic (crops plus forest remnants)	40	0–1000
Oil	Active oil fields	70	0–3000
	Oil concession	Present: 20 Future: 20	0–3000 (future)
	Oil exploration	Present: 0 Future: 20	3000 (future)
	Oil exploration- without operator	Present: 0 Future: 20	3000 (future)
Mining Industry	Active mines	Present: 30 Future: 30	10,000
	Mine concession	Present: 0 Future: 30	10,000 (future)

2.6. Registration of New Nomenclatural Acts

According to the International Commission on Zoological Nomenclature (ICZN), which produces the International Code of Zoological Nomenclature, the electronic publication of this article in portable document format (PDF) represents a published work. Therefore, the new species name contained in the PDF is effectively published under the International Code of Zoological Nomenclature from the electronic edition alone. This publication and the nomenclatural acts contained in it have been registered in ZooBank, the online official register for the ICZN. The ZooBank Life Science Identifiers (LSIDs) can be accessed and viewed through standard web browsers by appending the LSID to the prefix http://zoobank.org/. The online version of this work is archived and available from the following digital repositories: Diversity, CLOCKSS, and e-Helvetica.

3. Results

3.1. Phylogenetic Systematics

The relationships among glassfrog genera are shown in Figure 19. We also included a tree for each genus that occurs in Ecuador. These trees are the source of evidence for generic placement and comments on the evolutionary relationships among species discussed below and are summarized in Table S1. Each genus was well supported and congruent with the trees estimated by Guayasamin et al. [1,2], Castroviejo-Fisher et al. [3], and Twomey et al. [19], with only a few differences resulting from our improved taxon sampling. There was significant support (ML: Greater than 95% bootstrap [BS]; BA: Greater than 0.95 posterior probability [PP]) for all genera.

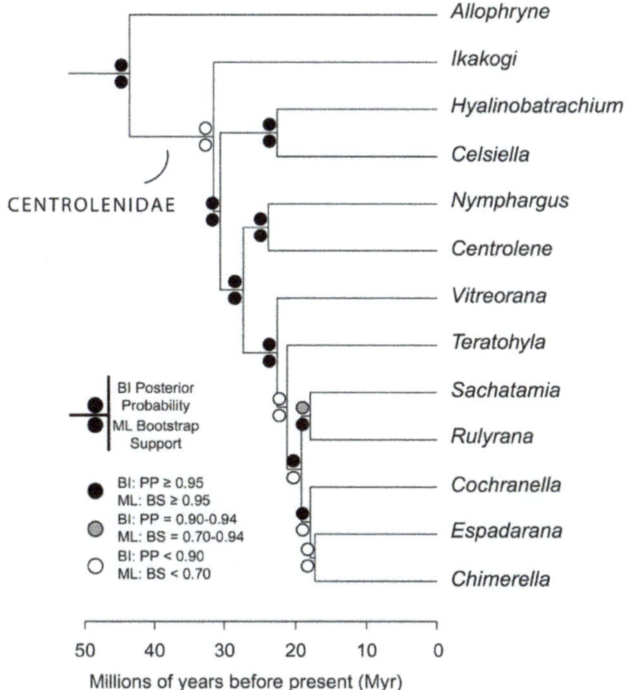

Figure 19. Evolutionary relationships among glassfrog genera (family: Centrolenidae) under maximum likelihood and Bayesian criteria.

3.2. Potential Distribution

The area of the potential distribution of each glassfrog species is summarized in Table S5. AUCs associated to each model are reported in Table S4. The species with the largest area of potential distribution was *Cochranella resplendens* (77,792 km^2), whereas the species with the most restricted predicted distribution was *Centrolene heloderma* (1067 km^2). The areas with the highest impacts were the interandean valleys and the Pacific lowlands; these were also areas with the highest population density. In terms of distribution and range extensions, agriculture and ranching were the activities that had the highest impact on habitats. Other activities with high impact values are related to mining and the oil industry.

4. Species Accounts

Genus *Centrolene* Jiménez de la Espada, 1872 [83]

Etymology: The name *Centrolene* is derived from the Greek words *kéntron* (point or spur) and *ōlénē* (elbow, forearm), alluding to the humeral spine found in males of this genus [84]. Although the name *Centrolene* has been used as neuter in gender for almost 150 years, now, because of an intricate game of words and nomenclatural regulations, it is considered to be feminine [85], creating nomenclatural instability that, in our opinion, was unnecessary. *Centrolene* is the type genus for the family Centrolenidae.

Centrolene ballux (Duellman and Burrowes 1989 [86]; Figures 20–23)

Centrolenella ballux Duellman and Burrowes, 1989 [86]. Holotype: KU 164725.
Type locality: "14 km (by road) west of Chiriboga (00°18′ S, 78°49′ W), 1960 m, Provincia de
    Pichincha, Ecuador" (now in Provincia de Santo Domingo de los Tsáchilas).
Centrolene ballux—Ruiz-Carranza and Lynch, 1991 [6].
"*Centrolene*" *ballux*—Guayasamin, Castroviejo-Fisher, Trueb, Ayarzagüena, Rada, and Vilà, 2009 [1].

Common names: English: Gold-dust Glassfrog. Spanish: Rana de Cristal Polvo de Oro.
Etymology: The specific name *ballux* is Latin; it means "gold dust" and is used in reference to the minute gold flecks on the dorsum [86].
Identification: On the Pacific versant of the Ecuadorian Andes, only *Nymphargus buenaventura* and some populations of *N. griffithsi* are similar to *Centrolene ballux* in having a green dorsum with small light spots (Figure 20), but they differ by possessing basal webbing between outer fingers (moderate webbing in *C. ballux*), concealed prepollex (distinct in *C. ballux*), and lacking humeral spines (present in males of *C. ballux*). The Colombian "*Centrolene*" *robledoi* resembles *C. ballux*; both lack vomerine teeth, have a small series of white spots on flanks, and have a similar webbing and snout shape. However, *C. robledoi* has small dark spots on its dorsum (absent in *C. ballux*), concealed prepollex (distinct in *C. ballux*), and is slightly larger than *C. ballux* (males, SVL 19.9–24.4 mm in *C. robledoi*; 19.2–22.2 mm in *C. ballux*). *Centrolene peristicta* and *C. lynchi* are sympatric with *C. ballux* in several localities, but the two species can be distinguished by having dorsal dark and light minute spots (only light spots present in *C. ballux*; Figure 20).
Diagnosis: (1) Vomers lacking teeth; (2) snout bluntly rounded to truncated in dorsal and lateral profiles (Figure 21); (3) tympanum oriented almost vertically, ED/TD = 31%–34%; tympanic annulus visible except for dorsal border covered by supratympanic fold; tympanic membrane partially pigmented, differentiated from surrounding skin; (4) dorsal surfaces shagreen with small warts; (5) pair of enlarged subcloacal tubercles (Figure 15); (6) anterior two-thirds of ventral parietal peritoneum white, posterior third transparent (condition P3); silvery white pericardium; no iridophores in peritonea covering intestines, stomach, testes, kidneys, gall bladder, and urinary bladder (condition V1); (7) liver tetralobed, lacking iridophores (condition H0); (8) humeral spines present in adult males; (9) no webbing between Fingers I and II, webbing between Fingers II and III basal or absent; webbing formula for outer fingers: III (2$^{1/4}$–2$^{3/4}$)—(2–2$^+$) IV (Figure 21C); (10) webbing formula on foot: I 1—(2–2$^+$) II 1—(2–2$^+$) III (1–1$^+$)—(2–2$^{1/3}$) IV 2$^{1/2}$—(1–1$^{1/3}$) V; (11) ulnar fold low, white; inner tarsal fold

low; outer tarsal fold absent, but small white tubercles evident along ventrolateral margin of tarsus; (12) distinct prepollex, clearly separated from Finger I; in males, nuptial pad Type I; (13) Finger II slightly longer than Finger I (Finger I 91.4%–98.0% of Finger II); (14) disc of Finger III of moderate size, about 47.4%–59.3% of eye diameter; (15) in life, dorsum green with small white warts; upper lip white; bones green; (16) in preservative, dorsal surfaces lavender with small white and/or unpigmented spots, particularly evident on limbs; (17) iris whitish cream with dark-grey thin reticulation and pale yellow hue around pupil; (18) melanophores mostly absent from fingers and toes, except for a few on Toes IV and V and on base of Finger IV; (19) males call from the upper side of leaves; call emitted sporadically and consisting of a single short note (328.5–420.4 ms) with 7–9 pulses; mean dominant frequency at peak amplitude 4833 Hz ± 14 (range 444–464); notes are frequency modulated (Márquez et al. 1996); (20) fighting behavior unknown; (21) eggs deposited on the upper sides of leaves; short-term maternal care unknown; long-term parental care absent; (22) tadpoles undescribed; (23) minute body size; SVL 19.2–22.2 mm ($\overline{X}$ = 20.6 ± 0.911, n = 25) in males; SVL 21.0–23.3 mm (n = 3) in females.

Figure 20. *Centrolene ballux* in life from Reserva Las Gralarias, Pichincha province, Ecuador. (**A,B**) Adult males. (**C**) Adult male in dorsal view, QCAZ 40199. (**D**) Egg clutch on upper side of leaf. (**E**) Metamorph. (**F**) Adult female in ventral view, QCAZ 40196. Photos by Carl R. Hutter (**B,D,E**), Luis A. Coloma (**C,F**), and Alejandro Arteaga/Tropical Herping (**A**).

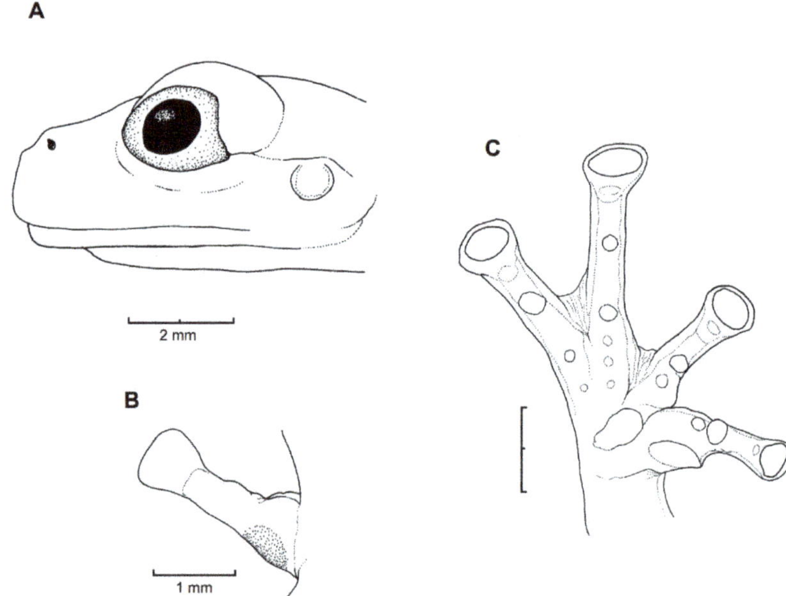

Figure 21. *Centrolene ballux.* (**A**) Head in lateral view, KU 164725. (**B**) Finger I and nuptial pad in dorsal view, KU 200275, adult male. (**C**) Hand in ventral view, KU 164725. Illustrations by Juan M. Guayasamin.

Color in life (Figure 20): Dorsum lime green with small greenish–white to yellowish–white warts; fingers and toes yellowish green; anterior two-thirds of ventral parietal peritoneum white; visceral peritoneum lacking iridophores; pericardium silvery white; bones green; upper lip yellowish white to white; iris cream white with fine black reticulations.

Color in ethanol: Dorsal surfaces of head, body, and limbs lavender with small white and/or unpigmented warts; hands and Toes I–III mostly unpigmented; margin of upper lip white; ulnar fold white; small tubercles on outer ventrolateral margin of tarsus white; small white tubercles posterior to cloacal opening; ventral surfaces cream. Anterior two-thirds of the ventral parietal peritoneum white, posterior third transparent; silvery white pericardium; no iridophores in peritonea covering intestines, stomach, testes, kidneys, gall bladder, and urinary bladder.

Biology and ecology: Individuals are active at night on the upper surfaces of leaves of bushes and trees 50–800 cm above and along small streams, or on ferns over roadside ditches (Arteaga et al. 2013). The reproductive activity of *Centrolene ballux* occurs in the rainy season (December–April [87]). One male (IND-AN 1725) was on a leaf 5 cm below another leaf on which there was a clutch of 18 eggs [86]. Another clutch found at La Planada had 13 eggs. Near Chiriboga, a gravid female (KU 164729) with green eggs and five adult males were found along a stream on 8 May 1975. At Reserva Las Gralarias, on March 2009, a clutch with 21 embryos was found on the upper side of a leaf, which was hidden by another leaf; an adult male was observed nearby (ca. 100 cm from the clutch). *Centrolene ballux* is relatively abundant at Reserva Las Gralarias, where it maintains reproductive populations along several streams (Ballux Creek, Five-frog Creek, Heloderma Creek, Chalguayacu River, Kathy's Creek; [88]). It is unknown if females provide short-term parental care to egg clutches; males do not exhibit parental care [25].

Call (Figure 22): The advertisement call of *Centrolene ballux* was described from Las Palmeras, Ecuador, by Márquez et al. [89]. The call was emitted sporadically and consisted of a single short note (328.5–420.4 ms, *n* = 2) with 7–9 pulses, the first 4–5 being repeated at short, regular intervals, and the last three pulses being emitted at longer intervals (mean pulses per second = 21.0 ± 0.1, range 21.3–21.4). The dominant frequency was high (mean dominant frequency at peak amplitude 4833 Hz ± 14). The notes were frequency modulated (dominant frequency at the beginning of note is lower than at its end). Recordings obtained at Reserva Las Gralarias, Ecuador, match this description.

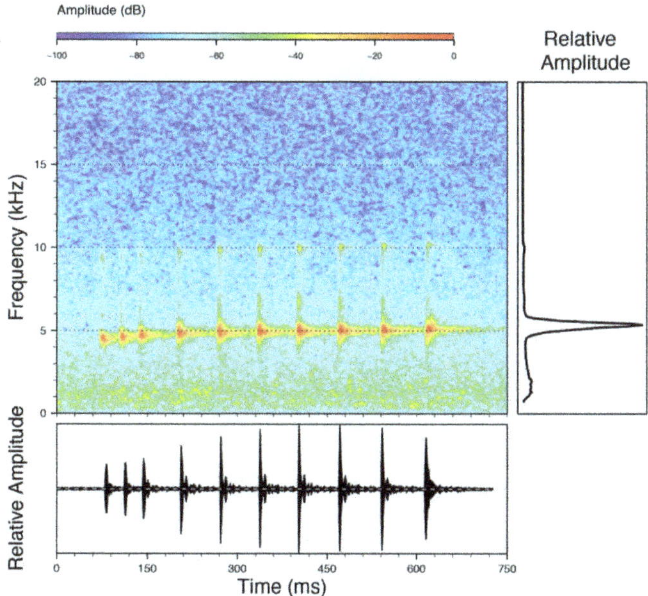

Figure 22. Call of *Centrolene ballux* from Reserva Las Gralarias, Pichincha province, Ecuador.

Tadpole: Not described.

Distribution (Figure 23): *Centrolene ballux* is known from six localities in the cloud forests of Carchi, Imbabura, Pichincha, and Santo Domingo de los Tsáchilas provinces (Ecuador) and one in the department of Nariño (Colombia) on the Pacific slopes of the Cordillera Occidental of the Andes at elevation between 1780 and 2340 m [86,87,89] (Specimens Examined). The species inhabits the Western Montane Forest ecoregion. In Ecuador, it has a potential distribution of 4752 km^2.

Conservation status: Globally, *Centrolene ballux* is currently listed as *Endangered* by the IUCN [90]. In Ecuador, the most recent records of the species are from Reserva Las Gralarias (March 2018) and Reserva Río Manduriacu (2008) [21]. In Colombia, the last published record is from Reserva La Planada on April–June 1986 [86]. Recent surveys at Quebrada Zapadores and Las Palmeras (= Río Guajalito) have been unsuccessful in finding this species [91] (DFCH unpubl. data). During the rainy seasons (December–April) of 2011–2018, reproductive populations were observed in four streams at Reserva Las Gralarias. The species has a very restricted distribution, with only five known localities in Ecuador, in an area where forest fragmentation is common. At Reserva Las Gralarias, the amphibian chytrid fungus *Batrachochytrium dendrobatidis* infects *C. ballux* with a relatively high prevalence, but no recent declines have been recorded [92].

Evolutionary relationships (Figure 24): *Centrolene ballux* and *C. buckleyi* are sister species.

Specimens examined: *Centrolene ballux:* <u>Ecuador</u>: *Provincia de Carchi:* ca. 5 km W La Gruel (0.916667 N, 78.13333 W; 2340 m), KU 202798. *Provincia de Pichincha:* Reserva Las Gralarias (0.00806 S, 78.72433 W, 1852 m), QCAZ 40195–99; Quebrada Zapadores, 5 km ESE of Chiriboga on Chiriboga–Quito road (0.245278 S, 78.7261 W, 2010 m), KU 164733. *Provincia de Santo Domingo de los Tsáchilas:* 14 km we'st of Chiriboga on Chiriboga–Santo Domingo road (0.265278 S, 78.8478 W, 1960 m), KU 164725 (holotype), 164726–32 (paratypes).

Localities from the Literature: *Provincia de Pichincha:* Las Palmeras (=Bosque Protector Río Guajalito) (0.283 S, 78.75 W; 1800 m) [89].

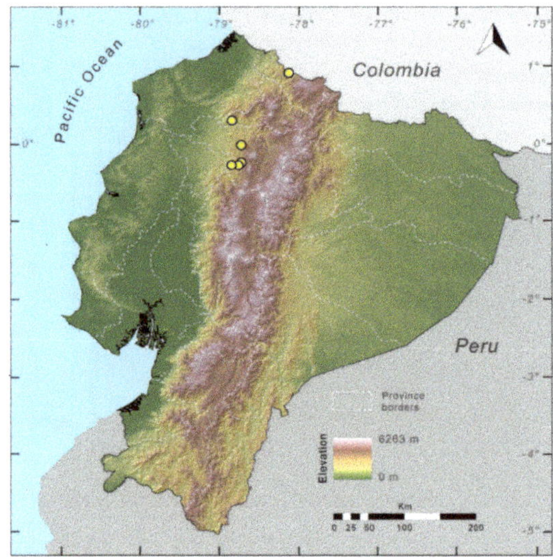

Figure 23. Distribution of *Centrolene ballux* in Ecuador (yellow dots).

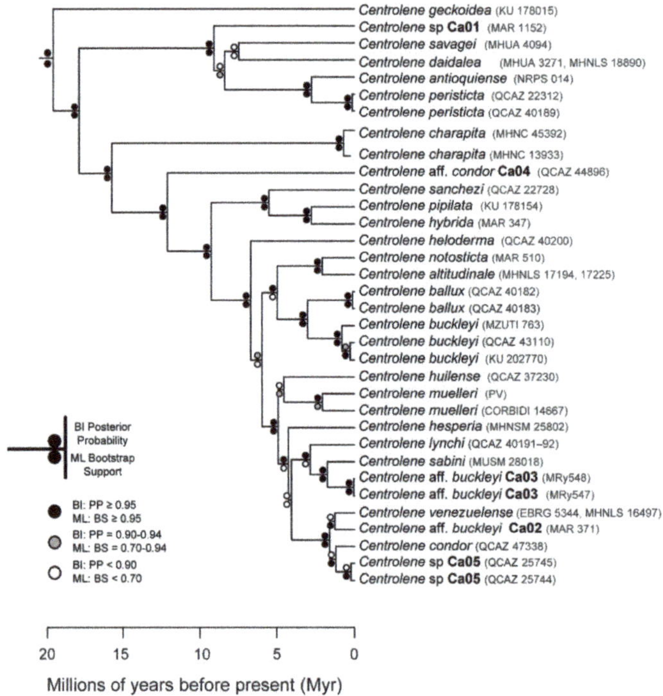

Figure 24. Evolutionary relationships among species in the genus *Centrolene*, inferred using maximum likelihood and Bayesian criteria.

Centrolene buckleyi (Boulenger 1882 [93]; Figures 25–28)

Hylella buckleyi Boulenger, 1882 [93]. Syntypes: BM 78.1.25.16 (Intac), 80.12.5.201 (Pallatanga).
    See comments below on type material. Neotype: KU 202770.
Neotype locality: Isla Wolf of Laguna Cuicocha (0°18′07″ N, 78°22′00″ W; 3070 m), Provincia
    Imbabura, Ecuador, collected on 28 February 1984 by W. E. Duellman.
Centrolenella buckleyi—Noble, 1920 [94].
Cochranella buckleyi—Taylor, 1951 [15]. Rivero, 1961.
Centrolenella buckleyi buckleyi—Rivero, 1968 [95].
Centrolenella johnelsi Cochran and Goin, 1970 [96]. Holotype: MLS 432. Type locality: "San Pedro,
    N of Medellín, Antioquia, Colombia". Synonymy by Ruiz-Carranza and Lynch, 1991 [6].
Centrolene buckleyi—Ruiz-Carranza and Lynch, 1991 [6].

Common names: English: Buckley's Glassfrog. Spanish: Rana de Cristal de Buckley.

Etymology: The specific name *buckleyi* was used to recognize Mr. Buckley, who collected the type series of the species.

Identification: *Centrolene buckleyi* is one of the few species of glassfrogs found in the highlands of Ecuador (2050–3070 m). It is easily recognized by the presence of a large humeral spine (in adult males), white upper lip, inclined snout in lateral profile, and reduced webbing between fingers (Figures 25 and 26). In life, dorsal surfaces are uniform green, but some individuals have small whitish warts. *Centrolene heloderma* and *C. condor* can be confused with *C. buckleyi*; however, *C. heloderma* has a pustular dorsal skin (shagreen in *C. buckleyi*) and *C. condor* has small light and dark spots on the dorsum (dark spots absent in *C. buckleyi*). Similar species from other countries include *Centrolene hesperia*, *C. lemniscata* (Peru), *C. altitudinalis*, and *C. venezuelensis* (Venezuela). We note that, as currently defined, *C. buckleyi* is a species complex that requires further taxonomic studies.

Figure 25. *Centrolene buckleyi* in life. (**Left**): Adult male from locality near Oyacachi, Napo province, 3012 m, MZUTI 763, photo by Eduardo Toral. (**Right**): Adult male from Guarumales, Zamora Chinchipe province, 2070 m, CJ-11364, photo by Diego Acosta-López.

Comments on type material: The description of *Centrolene buckleyi* was based on two specimens housed at The Natural History Museum, London (formerly British Museum of Natural History). As noted by Goin [97] and Lynch and Duellman [22], one syntype (BMNH 80.12.5.201; from Pallatanga, Provincia de Chimborazo, Ecuador) is now almost completely macerated in ethanol and almost no bones remain, and the other (BMNH 78.1.25.16; from Intac, Imbabura province, Ecuador) is missing and recent searches by Jeff Streicher and DFCH were unsuccessful. Both localities are on the western versant of the Andes of Ecuador, on the slopes of the Cordillera Occidental. Because *Centrolene buckleyi* represents a species complex [20,98] (Figure 24), we here designate a neotype for *Centrolene buckleyi* (KU 202770, adult female) collected from Laguna de Cuicocha (near the syntype locality of Intac = Intag), thereby allowing clear comparisons among populations that might represent independent lineages. The morphological characteristics of the neotype fully correspond to those mentioned by Boulenger [93] in the original description of the species. Additionally, mitochondrial sequences of the neotype are included in the phylogeny shown in Figure 24. The designation of the neotype is justified on the Article 75.3 of the International Code of Zoological Nomenclature [99].

Diagnosis: (1) Vomers lacking teeth; (2) snout round in dorsal aspect, slightly sloping to sloping in lateral profile (Figure 26); (3) tympanum partially or completely hidden under skin, when visible oriented almost vertically, its diameter is 29.0%–38.6% of eye diameter; supratympanic fold moderately heavy; tympanic membrane slightly thinner than skin around tympanum; (4) dorsal skin finely shagreen, males with spicules; (5) pair of round subcloacal warts (Figure 26); (6) anterior half to two-thirds of venter covered by white parietal peritoneum, posterior portion translucent (condition P2–P3); silvery white pericardium; translucent peritoneum covering intestines, stomach, testes, kidneys, gall bladder, and urinary bladder (condition V1); (7) liver with four or five lobes, lacking iridophores (condition H0); (8) in males, humeral spines present; (9) webbing absent between Fingers I, II, and III; reduced webbing between outer fingers: III $(2^{1/4}$–$3^-)$—$(2^+$–$2^{1/2})$ IV (Figure 26); (10) webbing formula on foot: I $(1^{1/2}$–$2^-)$—$(2$–$2^{1/4})$ II $(1^-$–$1^+)$—$(2^+$–$2^{1/2})$ III $(1^+$–$1^{2/3})$—$(2^{1/3}$–$3)$ IV $(2^{2/3}$–$3)$—$(1^{2/3}$–$2^-)$ V; (11) ulnar fold low, ventrolateral margin of arm white; inner tarsal fold evident; outer tarsal fold absent, external ventrolateral margin of tarsus white; (12) concealed prepollex; in males, nuptial pad Type I; (13) Finger II slightly longer than Finger I (Finger I 86.5%–98.7% of Finger II); (14) disc of Finger III of moderate size, about 54.2%–64.5% of eye diameter; (15) in life, dorsum uniform green that may have scattered whitish warts; upper lip white, usually continuing as a white line across the flanks; bones green; (16) in preservative, pericardium and anterior half to two-thirds of ventral parietal peritoneum white, visceral peritoneum translucent, peritoneum around kidneys translucent; (17) iris grey-white with thin black reticulation and a horizontal brown stripe; (18) melanophores mostly absent from fingers and toes, except for a few on Toes IV and V and on base of outer fingers; (19) males call from the upper sides of leaves; two call descriptions available (see Call section); (20) males fight upside down, grasping one another venter to venter; (21) eggs deposited on the upper sides of leaves; females provide short-term parental care; prolonged parental care is absent; (22) tadpoles unknown; egg clutches deposited within male's territory, but parental care has not been reported; (23) medium body size, SVL in males 26.5–30.9 mm ($\overline{X}$ = 29.0; n = 25), in females 29.3–34.4 mm ($\overline{X}$ = 31.2; n = 9).

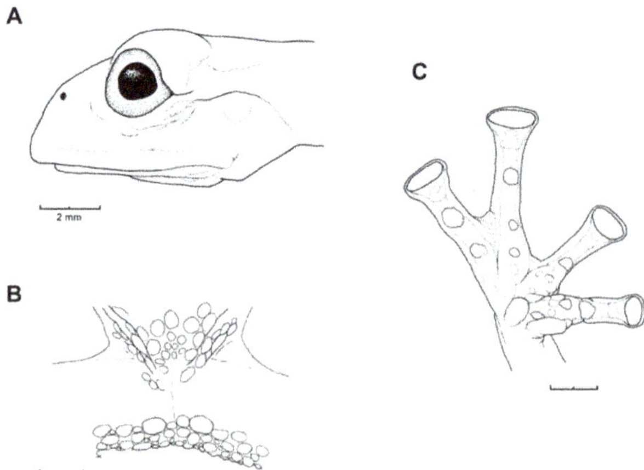

Figure 26. *Centrolene buckleyi.* (**A**) Head in lateral view, adult female, KU 178035. (**B**) Venter and thighs in ventral view, adult female, KU 178040. (**C**) Hand in ventral view, adult female, KU 178055. Illustrations by Juan M. Guayasamin.

Description of Neotype: Adult female, SVL 32.6 mm (KU 202770; Figure 27). Head wider than long; head length 28.5% of SVL; snout rounded in dorsal profile, sloping in lateral profile; canthus indistinct; loreal region slightly concave; upper lip white, slightly flared; nostril protuberant, closer to eye than to tip of snout, directed dorsolaterally; internarial area barely depressed. Eye small (diameter 14% of SVL), directed anterolaterally; transverse diameter of disc of Finger III 50.9% of eye diameter. Supratympanic fold conspicuous; tympanum not visible externally. Vomers lacking teeth, choanae large, round; tongue ovoid, ventral posterior half not attached to floor of mouth and posterior margin notched; vocal slits extending posterolaterally from the posterolateral base of tongue to angle of jaws.

Humeral spine absent; ulnar fold evident, white; relative length of fingers: III > IV > II > I; webbing between Fingers I, II, and III absent, basal webbing for outer fingers: III $2^{2/3}$—2^+ IV; discs and disc pads expanded, elliptical; subarticular tubercles large, round, simple; numerous fleshy supernumerary tubercles present; palmar tubercle elliptical, simple. Length of tibia 51.5% of SVL; inner tarsal fold evident; outer tarsal fold absent, ventrolateral margin of tarsi white; feet about three-fourths webbed; webbing formula on foot I $1^{1/2}$—2 II 1—2^+ III $1^{1/2}$—3 IV 3—$1^{2/3}$ V; discs on toes elliptical; disc on Toe IV narrower that disc on Finger IV; inner metatarsal tubercle large, ovoid; outer metatarsal tubercle not evident; subarticular tubercles small, round; numerous fleshy supernumerary tubercles evident.

Skin on dorsal surfaces of head, body, and lateral surface of head and flanks shagreen, lacking spicules; throat smooth; venter and lower flanks areolate; cloacal opening directed posteriorly at upper level of thighs; small, white tubercles located immediately posterior to cloacae. In ventral view, pair of enlarged subcloacal tubercles not evident.

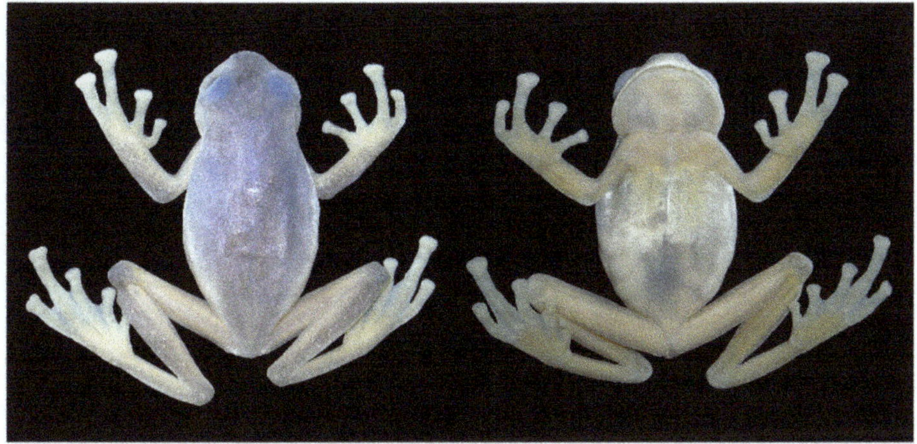

Figure 27. *Centrolene buckleyi*, adult female, neotype, KU 220770. Ecuador, Imbabura province, Isla Grande, Lago Cuicocha, 3070 m. Photos by Juan M. Guayasamin.

Color in life (Figure 25): Dorsal surfaces bright to dark green, sharply demarcated laterally from lower white flanks; throat and most of venter pale green; parietal peritoneum yellowish white; edge of upper lip white; ventrolateral borders of arms and tarsus white; small, white warts posterior to cloacal opening; bones green; grey–white iris with thin black reticulation and a horizontal brown stripe. Some individuals have whitish warts on dorsum (QCAZ 22388, 26031–32) or small white spots on legs and forearms (KU 202770)

Color in ethanol (Figure 27): Dorsum of head and body uniform lavender or with minute whitish or unpigmented spots; limbs cream with slight lavender tone and with or without small white spots; conspicuous white border on the upper lip; dorsally, all fingers, Toes I–III, and most of Toe IV unpigmented; ventrolateral borders of arms and tarsus white; cloacal region mostly unpigmented, except for several white warts posterior to cloacal opening; males with cream nuptial pad on Finger I.

Three specimens (QCAZ 26031–32, KU 178042) were dissected to observe coloration of internal organs: Anterior half to two-thirds of ventral parietal peritoneum white; pericardium white; hepatic peritoneum transparent; peritonea covering viscera and kidneys translucent to opaque.

Biology and ecology: During the day, *Centrolene buckleyi* has been found in terrestrial and arboreal bromeliads near and away from streams in secondary forest and pastures (J. D. Lynch, W. E. Duellman field notes). During the night, *Centrolene buckleyi* were active on terrestrial bromeliads and vegetation 30–160 cm above streams, lakes, and marshes in primary and secondary forests ([20,22], this work). Lynch and Duellman [22] suggested that *C. buckleyi* might breed in the same situations as other centrolenids—rapid, mountain streams—but also in bromeliads or in ciénegas; they based this statement on the observation of an egg clutch (KU 170221) on the inner leaf of bromeliad in an area where *C. buckleyi* was abundant. Males fight dangling upside down while holding onto vegetation with their hind limbs, grasping one another venter to venter (Bolívar et al. 1999). A male (QCAZ 22388) was found on the upper side of a leaf close to another leaf with a clutch of eggs, approximately 160 cm above a stream [20]. Females provide short-term parental care to egg clutches; males do not exhibit parental care [25].

Call (Figure 28): Two descriptions of the call of *Centrolene buckleyi* are available in the literature [20,100]. The differences between these calls, combined with the non-monophyly of *C. buckleyi* (Figure 24), clearly show that *Centrolene buckleyi*, as currently recognized, represents a species complex [20,98]. Below we describe a call from the highlands of Ecuador (MZUTI 763; see Examined Specimens); although the call presented herein is not from the neotype locality, genetically, it clusters together with samples from the type locality; thus, apparently corresponding to the same species. Calls are produced in series, every 45–67 s (mean = 53 ± 9.9; n = 10). Each call has a single, highly pulsed note, with a duration of 0.248–0.288 s (mean = 0.264 ± 13.7; n = 10; Figure 23). Calls have a slight frequency modulation, with an initial dominant frequency at 2816–2941 Hz (mean = 2893 ± 38; n = 10) and a final dominant frequency at 3222–3255 Hz (mean = 3229 ± 12; n = 10). First harmonics are at 5385–6393 Hz.

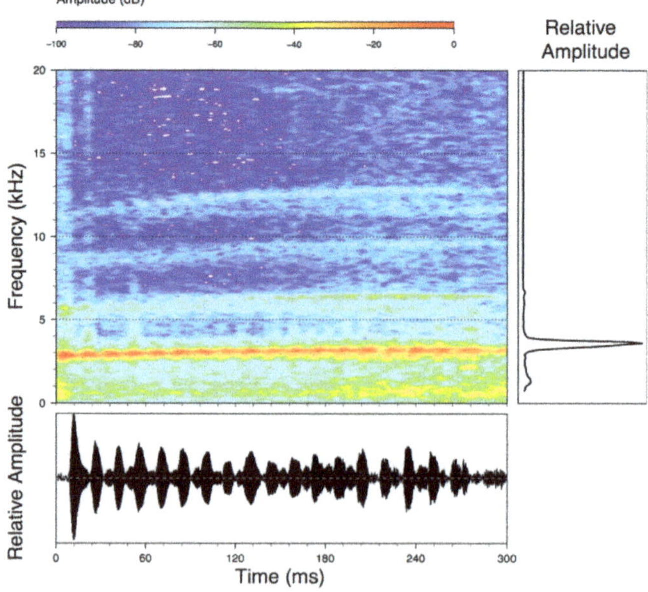

Figure 28. Call of *Centrolene buckleyi*, MZUTI 763. Recorded at páramo wetland nearby Oyacachi, 3012 m, Napo province, Ecuador.

Tadpole: Not described.

Distribution (Figure 29): The *Centrolene buckleyi* species complex occurs through the Andes of Colombia and Ecuador to Huacambamba in the Departamento de Piura in northern Peru at elevations between 2450 and 3300 m ([17,101,102], this work). Records below 2450 m are considered dubious. In Ecuador, *Centrolene buckleyi* is found along the Cordillera Oriental and Cordillera Occidental of the Andes, and inhabits the Western Montane Forest, Andean Shrub, Páramo, and Eastern Montane Forest ecoregions, with a potential distribution of 44,586 km^2.

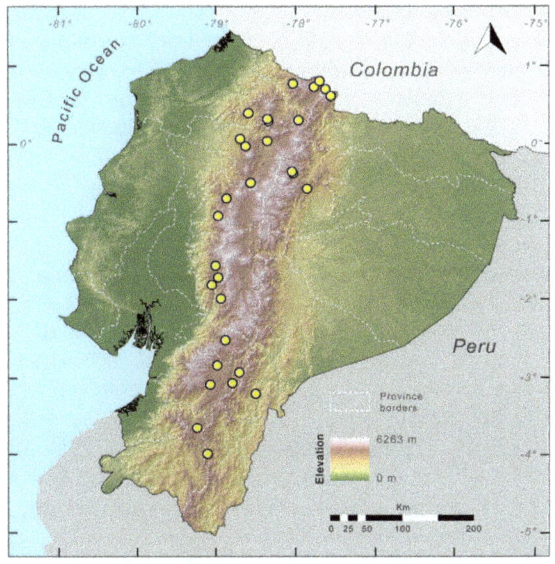

Figure 29. Distribution of the species complex currently recognized as *Centrolene buckleyi* in Ecuador (yellow dots).

Conservation status: *Centrolene buckleyi* is listed as *Vulnerable* at a global level by the IUCN [103]. In Ecuador, *Centrolene buckleyi* was abundant along the high Andes, but recent fieldwork demonstrates population crashes at historical localities (e.g., Pilaló, Cuicocha, Cashca Totoras, Quito, San Pablo de Atenas, Papallacta) ([91], JMG pers. obs.). The last records of the species in Ecuador are from Bosque Protector Cashca-Totoras in 2002 [91], Morán on February 2017 (Diego Batallas, pers. com., Mario Yánez-Muñoz, pers. com.); Yanayacu Biological Station in 2001–2003 [20], Sigchos (0°42′ S, 78°53′ W; 2080 m) on March 2008, Zamorahuaico, near Loja (3°59′ S, 79°12′ W; 2100 m) on August 2008, and nearby Oyacachi (3012 m) on May 2012. The habitat (Páramo, Andean shrub, and montane forest) occupied by the species is heavily impacted by human activities, mainly agriculture and pasture lands. Considering that that populations declines are widespread in the species complex, we suggest the category of *Critically Endangered* in Ecuador (IUCN criteria A2c,e).

Evolutionary relationships (Figure 24): As defined herein, *Centrolene buckleyi* is sister to *Centrolene ballux*. See Taxonomic Remarks.

Taxonomic Remarks: *Centrolene buckleyi* is a species complex that requires formal subdivision [20,98]. Moreover, morphologically, *Centrolene buckleyi* is almost identical to four Andean species, namely *C. altitudinalis*, *C. venezuelensis* (Venezuela), *Centrolene hesperia*, and *C. lemniscata* (Peru). Myers and Donnelly [104] elevated the Venezuelan populations of *Centrolene buckleyi* (*Centrolenella buckleyi venezuelensis* Rivero, 1968 [105]) to the species status, without morphological justification; Señaris and Ayarzagüena [106] provided morphological and acoustic data supporting the recognition of *Centrolene venezuelensis*. The original descriptions of *C. hesperia* and *C. lemniscata* did not include a comparison between these species and *C. buckleyi;* therefore, a reevaluation of their species status is necessary. Genetic [20,98] and morphological [105,106] differences support the validity of *Centrolene altitudinalis;* for example, the lower two-thirds of the tympanic annulus is visible in *C. altitudinalis,* whereas the tympanum is completely or mostly concealed in *C. buckleyi, C. lemniscata,* and *C. hesperia.* Also, we call attention to the conspicuous geographical barriers that have the potential of limiting gene flow among these species. The Peruvian *C. hesperia and C. lemniscata* are isolated from other species by the Huacambamba Depression, and isolated from each other by interandean valleys. Likewise, the Venezuelan *C. venezuelensis* and *C. altitudinalis* are isolated from other species by the Táchira Depression. The distribution of *C. buckleyi* is restricted, mostly, to the Andes of Ecuador and Colombia, but these populations, based on genetic and call data, represent several independent evolutionary lineages (Figure 24). Thus, clearly, further work is necessary within this species complex.

Measurements of neotype (in mm): The morphometric data for the neotype (female, KU 202770) are as follows: SVL = 32.6; tibia length = 16.8; foot length = 15.1; head length = 9.3; head width = 10.3; interorbital distance = 3.3; upper eyelid width = 2.6; internarial distance = 2.8; eye-nostril distance = 2.2; snout-eye distance = 4.3; eye diameter = 4.6; tympanum diameter = —; eye-tympanum distance = —; radio-ulna length = 6.72; hand length = 10.4; Finger I length = 6.3; Finger II length = 6.7; disc of Finger III = 2.3; and Finger III width = 1.2.

Specimens examined: *Centrolene buckleyi:* Ecuador: *Provincia del Azuay:* 10 km N Girón (3.0833 S, 79.0833 W; 2750 m), KU 202777; 11.5 km SE Gualaceo (2.9333 S, 78.71667; 2940 m); Sigsig (3.05 S, 78.8 W; 2450), QCAZ 1245. *Provincia de Bolívar:* Bosque Protector Cashca-Totoras (1.71 S, 78.98 W; 2800–3159 m), QCAZ 740, 21231; Guanujo (1.5667 S, 79.01667 W. 2600 m), KU 182214; San Pablo de Atenas (1.8 S, 79.0667 W), QCAZ 372, 2415–16; Santiago (1.7167 S, 78.9833 W), QCAZ 1531. *Provincia de Carchi:* near Tulcán (0.8 N, 77.7167 W; 2770 m), KU 118005–08; 14 km SW Tulcán (0.72889 N, 77.7958, 3340 m), KU 164516; 5.7 km NW El Carmelo (0.6908 N, 77.63389 W, 2910 m), KU 178053–67; Santa Bárbara (0.61667 N, 77.5833; 2650 m), KU 190017–19; El Goatal, Morán, 0.5 km from Escuela de Morán via La Cortadera, QCAZ 43110; Morán (0.7686 N, 78.056 W, 2785 m), DBR 187, 193. *Provincia de Cañar:* 4 km W Ingapirca (2.51667 S, 78.9 W; 3000 m), KU 178077. *Provincia de Cotopaxi:* near Pilaló (0.933 S, 78.9833 W; 2410 m), KU 178034–50, 202780–83. *Provincia de Imbabura:* La Delicia (0.0667 N, 78.7 W; 2710 m), KU 178079–81, 180311; Lago Cuicocha (0.3 N, 78.3667 W; 3010 m), KU 138822, 178030–33; S shore of Lago Cuicocha (0.29194 N, 78.35389; 3070 m), KU 202703–74; Isla Wolf in Lago Cuicocha (0.3019 N, 78.366 W; 3070 m), KU 202770–72; Mariano Acosta (0.3 N, 77.9833 W; 3000 m), QCAZ 12172. *Provincia de Loja:* 13 km E Loja, Abra de Zamora (3.9744 S, 79.1114 W; 2800 m), KU 164511–15, 166321; 2 km SSW Saraguro (3.6397 S, 79.24 W; 2560 m), KU 178068–76; 3.7 km S Saraguro (3.6469 S, 79.245 W; 2800 m), KU 202778–79. *Provincia de Morona Santiago:* 25.5 km WSW Plan de Milagro (3.21667 S, 78.5 W; 2600 m), KU 202775. *Provincia de Napo:* 9.2 km ESE Papallacta (0.3761 S, 78.0683 W; 2750 m), KU 178052; 11 km ESE Papallacta (0.3869 S, 78.0569; 2660 m), KU 164507–09; 11.2 km ESE Papallacta (0.388 S, 78.055; 2660 m), KU 178051; 12 km ESE Papallacta (0.39194 S, 78.04944 W; 2630 m), KU 155481–92, 164505–06; 31 km N Jondachi (0.5711 S, 77.869 W; 2190 m), QCAZ 2740; Oyacachi–El Chaco road (0.219 S, 78.044 N; 3012 m), MZUTI 763. *Provincia de Pichincha:* 9.5 km NW Nono (0.026389 S, 78.6403 W; 2530 m), KU 164510; Hacienda El Beaterio, KU 178078; near Machachi (0.5 S, 78.5667 W; 2950 m), KU 148429–30. *Provincia de Zamora Chinchipe:* 13.5 km E Loja (3.973 S, 79.107 W; 2800 m), KU 142648; Guarumales (3.9405 S, 78.987 W; 2070 m), CJ-11364.

Localities from the literature: *Centrolene buckleyi:* Ecuador: *Provincia del Azuay:* Sinicay, 2560 m, AMNH 17464. *Provincia de Chimborazo:* Pallatanga (1.9833 S, 78.95 W; 1520 m), BMNH 80.12.5.201. *Provincia de Imbabura:* Intac (0.4 N, 78.6 W), BMNH 78.1.25.16. *Provincia de Zamora Chinchipe:* Sabanilla (4.033 S, 79.0 W), AMNH 13530 [22,93].

Centrolene charapita Twomey, Delia, and Castroviejo-Fisher 2014 [19] (Figures 30 and 31).

Centrolene charapita Twomey, Delia, and Castroviejo-Fisher, 2014 [19]. Holotype: MHNCP 13933. Type locality: "near the village of La Oliva, past the village of Muyo (a larger village roughly 49 km N from Bagua), Amazonas, Peru (5°18′3.86″ S, 78°23′44.57″ W, 682 m)".

Common names: English: Charapita Glassfrog. Spanish: Rana de Cristal Charapita.

Etymology: The specific name *charapita* derives from a type of yellow chili pepper (ají charapita), which resembles the dorsal ocelli of the species [19].

Identification: *Centrolene charapita* is the only glassfrog with yellow dorsal ocelli, scalloped and enameled ulnar and tarsal folds (Figure 30), and relatively large size (adult male SVL = 34.7–37.0).

Figure 30. *Centrolene charapita* in life. Type series from near La Olivia, 682 m, Departamento Amazonas, Peru. Photos by Santiago Castroviejo-Fisher and Evan Twomey.

Diagnosis: *Centrolene charapita* is recognized by (modified from Twomey et al. [19]): (1) Vomers with 4–10 vomerine teeth; (2) in Peruvian population, snout truncated in dorsal view and truncated to slightly rounded in profile; in Ecuadorian population, snout round in dorsal view and sloping in profile; (3) tympanum small, partially hidden under skin; supratympanic fold present; (4) dorsal skin smooth with microspicules and low enameled warts on ocelli; venter and thighs coarsely areolate, other ventral surfaces smooth; (5) cloacal ornamentation conspicuous, formed by enameled folded skin (flaps) surrounded by lower warts; pair of enlarged subcloacal warts; (6) anterior third-to-half of parietal peritoneum white, posterior portion transparent (state P2); pericardium, hepatic, renal, and gonadal peritonea transparent, visceral peritoneum white; (7) liver trilobed, lacking iridophores (state H0); (8) humeral spines absent; (9) no webbing between Fingers I and II; webbing formula for outer fingers: II 2⁻—3^{1/3} III 2—(1⁺–1^{1/3}) IV; (10) webbing formula on foot: I (1^{1/3}–1^{1/2})—(2⁻–2) II 1—(2⁻–2^{1/3}) III (1–1⁺)—(2–2⁺) IV (2–2^{1/3})—1 V; (11) scalloped enameled ulnar and tarsal folds, extending from fringe on postaxial edge of Finger IV to elbow, and from fringe on postaxial edge of Toe V to ankle, respectively; (12) concealed prepollex; in males, nuptial pad Type I, not pigmented;

(13) when appressed, Finger I about equal length or slightly longer than Finger II; (14) eye diameter about 2.5 times the width of disc on Finger III; (15) in life, dorsal and dorsolateral surfaces covered with yellow or pale green spots of different size (ocelli); upper lip and tip of finger and toes I, II, and III white (enameled); ventral surfaces not pigmented; bones green; (16) in preservative, dorsal surfaces and dorsolateral surfaces with cream spots of different size (ocelli), bearing enameled warts and microspicules set in a lavender-greyish reticulum; upper lip and tip of finger and toes I, II, and III white (enameled); ventral surfaces not pigmented; (17) iris in life off-white to light grey with yellow tones and black dots and reticulation, with black ring delimiting iris; in preservative, iris silvery white with black dots and reticulation, with black ring delimiting iris; (18) in life and preservative, tip of fingers and toes white (enameled), Finger IV and Toes IV and V pigmented, Finger II and Toe III not pigmented, but for a small group of melanophores towards the tip, Fingers and Toes I and II not pigmented; (19) males call from upper side of leaves; call not described; (20) fighting behavior unknown; (21) eggs deposition site unknown; parental care unknown; (22) tadpoles undescribed; (23) large size in adult males; SVL 34.7–37.0 mm (n = 5) in males; females unknown.

Color in life (Figure 30): Modified from Twomey et al. [19]. Dorsal and dorsolateral surfaces green, with yellow or pale green spots of different size (ocelli). Upper lip and discs of fingers and toes I, II, and III white (enameled). Ventral surfaces, flanks, upper arms, Fingers I and II, and Toes I and II not pigmented. Iris delimited by a black ring, background coloration off white to light grey with yellow shades and black dots and reticulation. Anterior third-to-half of parietal peritoneum white, posterior portion transparent. Visceral peritoneum white. Hepatic peritoneum transparent.

Color in ethanol: Obtained from Twomey et al. [19]. Dorsal and dorsolateral surfaces with relatively large cream spots bearing enameled warts and microspicules set in a lavender–greyish reticulum, which is formed by minute melanophores that are darker around the ocelli. Upper lip and tips of fingers and toes I, II, and III white (enameled). Ventral surfaces, flanks, upper arms, Fingers I and II, and Toes I and II not pigmented.

Biology and ecology: Very little is known about *Centrolene charapita*. Twomey et al. [19] report that, at the type locality, individuals were found perched on riverine vegetation during the night, and that sympatric amphibians included *Ameerega trivittata*, *Bolitoglossa altamazonica*, *Cochranella erminea*, *Hyloscirtus* sp., *Rulyrana mcdiarmidi*, *Pristimantis* aff. *acuminatus*, *Pristimantis* sp., and *Rhaebo glaberrimus*. Parental care is unknown.

Call: Unknown.

Tadpole: Not described.

Distribution (Figure 31): *Centrolene charapita* is known from the type locality, a stream (5°18′3.86″ S, 78°23′44.57″ W, 682 m) near La Oliva, Peru [19], and Reserva Natural Maycu (4.2° S, 78.6° W, 940–1219 m) in southeast Ecuador.

Figure 31. Distribution of *Centrolene charapita* in Ecuador (yellow dots).

Conservation status: The conservation status of *Centrolene charapita* has not been evaluated by the IUCN. In Ecuador, populations are within a private reserve (Reserva Natural Maycu). The whole distribution range of the species is within a mining concession. Based on IUCN criteriaB2, Ba, Bb(iii), we suggest that *C. charapita* should be considered as *Critically Endangered* in Ecuador.

Evolutionary relationships (Figure 24): Our phylogeny places *Centrolene charapita* as the sister species to a clade formed by several *Centrolene* species. A previous analysis had inferred *C. charapita* as sister to *C. geckoidea* [19].

Specimens examined: *Centrolene charapita:* <u>Ecuador</u>: *Provincia de Zamora Chinchipe:* Reserva Natural Maycu (4.207° S, 78.630° W, QCAZ 66783; 4.222° S, 78.645° W, QCAZ 66786–87; 4.229° S, 78.616° W, QCAZ 66785; 4.226° S, 78.620° W, QCAZ 66784).

Centrolene condor Cisneros-Heredia and Morales-Mite 2008 [107] (Figures 32 and 33).

Centrolene condor Cisneros-Heredia and Morales-Mite, 2008 [107]. Holotype: QCAZ 37279.
Type locality: "Destacamento Militar Cóndor Mirador, western slope of the Cordillera del
  Cóndor (03°18′25″ S, 78°23′36″ W, between 1750–1850 m elevation), Provincia de Zamora
  Chinchipe, República del Ecuador".

Common names: English: Condor Glassfrog. Spanish: Rana de Cristal del Cóndor.

Etymology: The name of this species is in reference to its type locality, Cordillera del Cóndor, a mountain chain shared by Ecuador and Peru [107].

Identification: *Centrolene condor* can be distinguished from all other glassfrogs by having a green dorsum with many small yellowish–white flecks and dark bluish-black/brown flecks and punctuations (Figure 32), sloping snout in lateral view, light labial stripe, vomerine teeth, small humeral spine in males, enameled ulnar fold, and enameled metatarsal fold followed by a row of distinct enameled tubercles along the outer edge of the tarsus. Among Ecuadorian centrolenids, only *C. pipilata* shares a similar dorsal pattern, but *C. condor* has vomerine teeth (absent in *C. pipilata*) and a small, curved humeral spine (larger and straight in *C. pipilata*).

Figure 32. *Centrolene condor* in life. Ecuador, Zamora Chinchipe province, Paquisha, EPN 11343. Photo by Ana Almendáriz.

Diagnosis: *Centrolene condor* is diagnosed by the following traits (modified from Cisneros-Heredia and Morales-Mite [107]): (1) Vomerine teeth present; (2) snout subacuminate in dorsal view and sloping in profile; nostrils slightly elevated, producing depression in the internarial area; (3) tympanic annulus visible, vertical, with slight dorsolateral orientation; tympanic membrane not differentiated from surrounding skin; (4) dorsal skin shagreen, but with low warts and abundant spicules; (5) ventral skin granular; subcloacal area also granular with several enameled warts; (6) upper two-thirds of parietal peritoneum covered by iridophores (condition P3), pericardium white, all other peritonea not covered by iridophores (condition V1); (7) liver lobed, lacking iridophores (condition H0); (8) small humeral spines present in adult males; (9) webbing absent between Fingers I and II, basal between II and III, moderate between outer fingers: III 2—2⁻ IV; (10) webbing between toes moderate: I 1$^{1/2}$—2^{+} II 1—2^{+} III 1—2 IV 2—1 V; (11) enameled ulnar fold; enameled fringe on edge of toe V that continues into thin enameled metatarsal fold and then into row of enameled tubercles along outer edge of tarsus; (12) nuptial excrescences present, Type I; concealed prepollex; (13) first finger shorter than second; (14) eye diameter larger than width of disc on Finger III; (15) in life, green dorsum with abundant yellowish–white flecks and abundant dark flecks; green bones; (16) in preservative, dorsal surfaces greyish with slight lavender hue and abundant light and dark flecks; (17) in life, iris cream–yellow with fine dark reticulation; in preservative, iris light grey with fine dark reticulation; (18) melanophores absent from fingers and toes, except for few on dorsal surfaces of outer fingers and outer toes; (19) males call from upper side of leaves; call composed by two pulsed notes, with dominant frequency at 2.62–2.97 KHz; (20, 21, 22) parental care, fighting behavior, egg clutches, and tadpoles unknown; and (23) medium body size, SVL in adult males 23.2–27.6 mm ($\overline{X}$ = 25.4 ± 1.826; *n* = 5); females unknown.

Color in life (Figure 32): Green dorsum with abundant yellowish–white flecks and dark punctuations. Enameled spots and yellowish–white flecks on arms and legs. Whitish–yellow labial line present. Enameled arm and leg folds. Green bones [107].

Color in ethanol: Dorsal surfaces greyish with slight lavender tint, with abundant light and dark flecks. Anterior two-thirds of ventral parietal peritoneum white, pericardium white, all other peritonea lack white lining (modified from Cisneros-Heredia and Morales-Mite [107]).

Biology and ecology: *Centrolene condor* is a recently described species and little information is available on its natural history. It is nocturnal and males call from amidst the leaves of riverine vegetation a few centimeters over water in mature elfin forest. The species seems to be fairly common at Loma Tigres Bajo, where about 30 males were heard calling [108]. Tadpoles have a bright red coloration and are benthic [108].

Call: The advertisement call has a duration of 650–700 ms ($n = 3$); it is similar to a trill, composed of two pulsed notes; the first note has a duration of 435–510 ms and is conspicuously longer than the second (101–102 ms); the time between notes is 66–118 ms, and the dominant frequency is 2.62–2.97 KHz [108].

Tadpole: Not described.

Distribution (Figure 33): *Centrolene condor* is endemic to Ecuador, from localities in the Cordillera del Cóndor, Ecuador, at an elevation of 1737–2920 m [107,108]. The species is distributed within the Eastern Montane Forest ecoregion. Additional searches in the Cordillera del Cóndor have not yielded additional localities for the species [108].

Conservation status: *Centrolene condor* is currently listed as *Data Deficient* by the IUCN [109]. Given that the species is known from few localities in an isolated mountain range (Cordillera del Cóndor) with extensive mining activities and associated deforestation and contamination, we suggest considering the species as *Endangered*—EN B1ab(iii), at the global and local levels, in agreement with Almendáriz and Batallas [108].

Evolutionary relationships (Figure 24): Terminals identified as *Centrolene condor* are placed in different positions in the tree. Thus, it is likely that more than one species is found in what is currently recognized as *C. condor*.

Specimens examined: *Centrolene condor:* <u>Ecuador</u>: Provincia Zamora Chinchipe: Destacamento Militar Cóndor Mirador, western slope of the Cordillera del Cóndor (03°18′25″ S, 78°23′36″ W, 1750–1850 m), QCAZ 37279.

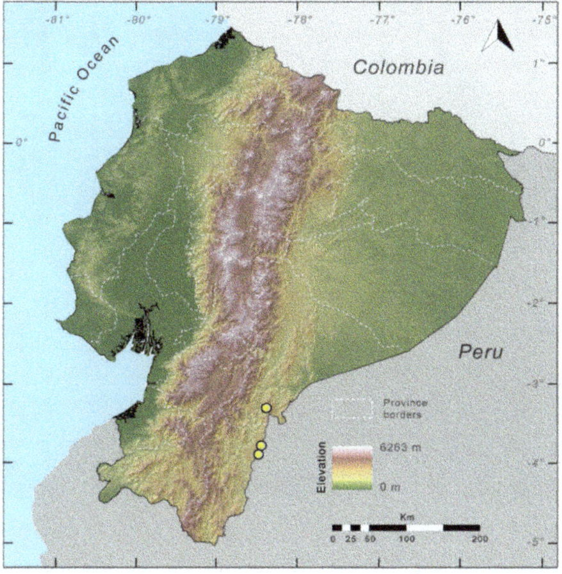

Figure 33. Distribution of *Centrolene condor* in Ecuador (yellow dots).

Centrolene geckoidea Jiménez de la Espada 1872 [83] (Figures 34–36).

Centrolene geckoideum Jiménez de la Espada, 1872 [83]. Holotype: MNCN 1596.
Type locality: "las riberas del río Napo en el Ecuador". See comment in Distribution.
Centrolene geckoidea—Barrio-Amorós, Rojas-Runjaic, and Señaris, 2019 [85].

Common names: English: Gecko Glassfrog. Spanish: Rana de Cristal Geco.

Etymology: The specific epithet *geckoidea* refers to the enormous size of the discs on fingers and toes of this species, which resemble those of gecko lizards (Gekkonidae). For almost 150 years, this species was known as *C. geckoideum*, but its specific ephited was recently modified to *geckoidea* [85].

Identification: *Centrolene geckoidea* is unique among centrolenids in having a giant body size (adult males, SVL 70.2–80.7 mm; females, SVL 61.8–72.9 mm) and webbing between the two innermost fingers (Figure 35). Males also have a conspicuous humeral spine (Figure 34), which in some individuals, perforates the skin of the arm. Only *C. paezorum*, endemic of the high Andes of Colombia, could be confused with *C. geckoidea*; *C. paezorum* is known only from a single female and is reported to be smaller (SVL 44.5 mm; see Remarks) and with less hand webbing than *C. geckoidea* [110].

Figure 34. *Centrolene geckoidea* in life. Ecuador, Carchi province, Río La Plata (00°48′ N, 78°02′ W; 2525 m), on the Maldonado–Tulcán road, DHMECN 0900. Photo by Doug Wechsler (25 July 1988).

Diagnosis: (1) Each vomer with four or five teeth; (2) snout truncated in dorsal and lateral profiles (Figure 35); (3) tympanum visible and small, oriented almost vertically, its diameter 40.3%–50.0% of eye diameter; supratympanic fold moderate; tympanic membrane completely pigmented, differentiated from surrounding skin; (4) dorsal surfaces of males and females covered with warts, spicules evident only in males; (5) ventral surfaces of thighs below vent lacking pair of enlarged warts; (6) anterior two-thirds or the entire ventral parietal peritoneum white (conditions P3–P4); silvery white pericardium; no iridophores in peritonea covering intestines, stomach, testes, kidneys, gall bladder, and urinary bladder (condition V1); (7) liver tetralobed, lacking iridophores (condition H0); (8) in males, humeral spines present; (9) webbing present, but reduced between Fingers I, II, and III; extensive webbing between Fingers III and IV; hand webbing formula I 2—(2–2$^+$) II (1–1$^+$)—(2$^{3/4}$–3$^+$) III (1$^{1/3}$–1$^{2/3}$)—1 IV (Figure 35); (10) webbing formula on foot I 0$^+$—(0$^+$–1) II 0$^+$—(0$^+$–1) III 0$^+$—(1–1$^{1/2}$) IV (1–1$^{1/2}$)—0$^+$ V; (11) ulnar fold with small, white tubercles; inner tarsal fold low, short; outer tarsal fold low, with small, white tubercles; (12) concealed prepollex; in males, nuptial pad Type I; (13) Finger I about same length as Finger II (Finger I 91.7%–100.0% of Finger II); (14) disc of Finger III wide, about 115%–130% of eye diameter; (15) in life, dorsum dull green to dark grey; upper lip yellow; bones green; (16) in preservative, dorsal surfaces grey to dark grey; (17) iris greenish gold with fine black reticulation; (18) melanophores covering dorsal surfaces of hands and feet; (19) males call from rocks behind waterfalls or within or near spray zones of fast-flowing streams; call loud, high-pitched, trilled whistle that lacks any consistent pattern of amplitude modulation; weakly to moderately pulsed, duration 155–373 ms; emitted infrequently at intervals of 1.48–5.05 min; dominant frequency modulated, 3468–4187 Hz; (20) fighting behavior unknown; (21) black eggs deposited on rocks in spray zones of streams; males provide prolonged parental care; (22) tadpoles dark, labial tooth row formula (LTRF) 2/3, but see below; (23) giant body size, males SVL 70.2–80.7 mm ($\overline{X}$ = 75.2, *n* = 12); females, SVL 61.8–72.9 mm ($\overline{X}$ = 68.1, *n* = 9).

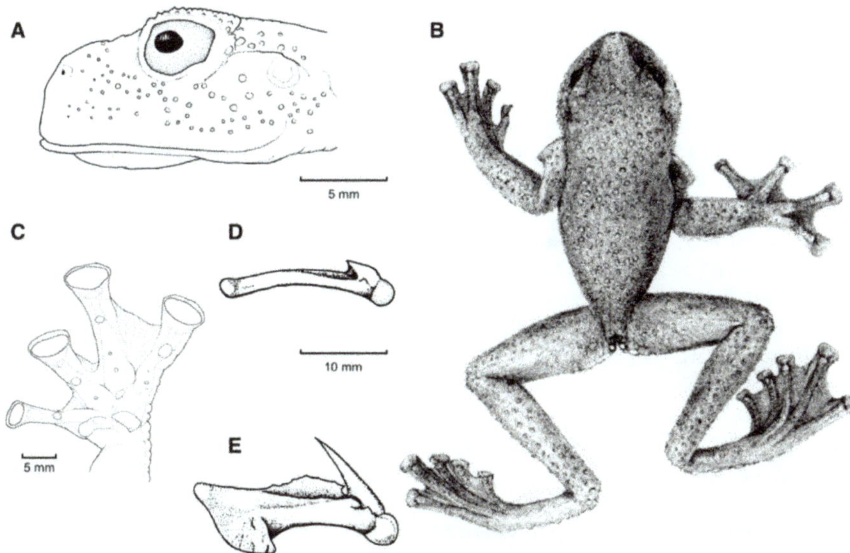

Figure 35. *Centrolene geckoidea.* (**A**) Head in lateral view, male, KU 116447. (**B**) Drawing of holotype, not to scale [83]. (**C**) Hand in ventral view, male, KU 116447; drawing by Juan M. Guayasamin. (**D**) Humeral spine of adult female; modified from Rueda-Almonacid [111]. (**E**) Humeral spine of adult male; modified from Rueda-Almonacid [111].

Color in life (Figure 34): Dorsum dull green, with enameled greenish or bluish warts; throat greenish yellow; margin of upper lip yellow; venter cream or yellow cream; ulnar, tarsal, and cloacal tubercles creamy white; white flecks on flanks; iris pale greenish gold with fine black reticulation; palpebrum clear (W. E. Duellman field notes, 9 April 1975; this work); bones green [111]. Males and females are dark grey to brownish green during the night [112].

Color in ethanol: Dorsal surfaces of head, body, and limbs cream grey to dark grey with tips of spicules being white; cloacal, ulnar, and tarsal tubercles white; ventral surfaces cream. Anterior two-thirds of the ventral parietal peritoneum white, posterior third transparent; pericardium white; no iridophores in peritonea covering liver, kidneys, and digestive tract.

Biology and ecology: Males and females of *Centrolene geckoidea* have been found clinging to vertical or overhanging rock surfaces in the spray zone behind waterfalls [112–114]. Males have remarkable call-site fidelity. Egg clutches are deposited on rocks within the spray zone and contain 89–112 black eggs, which are glued to the rocks by a brittle jelly unknown in other centrolenids [112,113]; an "empty space" is evident in the center of the attached clutch [113]. Although oviposition has not been observed, Grant et al. [112] suggested that the ova may be deposited as the female rotates through a circle with the head at the center, a scenario that would explain the origin of the "empty space" near the center of the egg mass. During the night, males have been observed sitting on or near up to four clutches of eggs while calling. During the day, males also have been observed sitting near or on clutches on the rockface behind a waterfall or hidden in spaces between rocks [112]. Thus, all observations by Grant et al. [112] suggest that males provide prolonged parental care to clutches.

Call: Grant et al. [112] described the call of *Centrolene geckoidea* from El Queremal (3.483 N, 76.7 W), Valle del Cauca, Colombia. The information shown below is based on their study. Males call from behind waterfalls or within or near spray zones of fast-flowing streams. Males have call-site fidelity and have been observed vocalizing at the same site for up to a month. The call is a loud, high-pitched, trilled whistle that lacks any consistent pattern of amplitude modulation. The call is weakly to moderately pulsed and has a duration of 155–373 ms. Calls were emitted infrequently at intervals or 1.48–5.05 min; however, males typically call less frequently (i.e., as little as less than once per hour). The dominant frequency is at 3468–4187 Hz, and it is modulated, beginning at low frequencies (3593–3781 Hz) and rising to a maximum of 3718–4187 Hz, after which it falls, ending at 3468–3718 Hz.

Tadpole: Lynch et al. [113] and Rueda-Almonacid [111] provide descriptions and illustrations of the tadpole of *Centrolene geckoidea*. Lynch et al. [113] briefly described a tadpole (total length of 22.3 mm) as having a typical centrolenid body form; mouth lacking upper tooth rows, but with two evident lower rows (P_1 with a wide medial gap); upper jaw sheath thick with minute serrations; lower jaw sheath thin and poorly keratinized; one row of large subconical marginal papillae borders the oral disc ventrolaterally and posteriorly. Rueda-Almonacid [111] mentioned that tadpoles in Gosner stage 22, one month after hatching, had two incomplete tooth rows on the anterior labium and three tooth rows on the posterior labium. However, three months after hatching, all the tooth rows were lost, and only dermal ridges were evident [111]. The dorsal surfaces of the body and the tail musculature were black, and the venter lacked pigmentation [111].

Distribution (Figure 36): *Centrolene geckoidea* occurs across the three Andean Cordilleras of Colombia (Cordillera Occidental, Central, and Oriental; in the departments of Antioquia, Valle del Cauca, Caldas, Quindio, Risaralda, Tolima, Boyacá, and Caquetá), south to the Pacific Andean slope of Ecuador (Carchi and Pichincha provinces) at elevations of 1750–2525 m [111–116] (Figure 35). In Ecuador, the potential distribution of the species covers an area of 10,579 km^2 within the Western Montane Forest region.

Jiménez de la Espada [83] mentioned that the species was found at *"las riberas del Río Napo en el Ecuador"*. During his trip in South America, Jiménez de la Espada visited several eastern Andean localities (e.g., Basin of Río Quijos, San José de Moti, Cordillera de Guacamayos, Cosanga) before reaching the lowlands near the Río Napo. It has been suggested that the type locality is in error [110], but the holotype of *C. geckoidea* could have been collected in the headwaters of the Napo river, or at one of the eastern Andean localities [17].

Figure 36. Distribution of *Centrolene geckoidea* in Ecuador (yellow dots).

Conservation status: Globally, *Centrolene geckoidea* is listed by the IUCN as a *Critically Endangered* species [117]. In Ecuador, the species has a severely fragmented distribution, and evidence of population declines. Intensive fieldwork in historical localities such as Quebrada Zapadores has failed in finding the species [91] (JMG, pers. obs.; DFCH, pers. obs.). The last reports of *C. geckoidea* in Ecuador are from Río La Plata (00°48′ N, 78°02′ W; 2525 m), on the Maldonado–Tulcán road, on 25 July 1988, and from Bosque Protector Río Guajalito in January to May between 1998 and 1999 [17]. A similar situation exists for populations of *C. geckoidea* found at Valle del Cauca in the Cordillera Occidental (Colombia), as they have not been seen for more than 10 years (W. Bolivar and J.J Sarria-Ospina, pers. com.).

Evolutionary relationships (Figure 24): The molecular tree reported by Twomey et al. [19] places *Centrolene geckoidea* as the sister species of *C. charapita*. However, our inferred tree recovers *C. geckoidea* as sister to all other *Centrolene* species. Since *C. geckoidea* is the type species for the genus *Centrolene*, its phylogenetic placement has fundamental taxonomic implications.

Remarks: The external morphology, osteology, and myology of *Centrolene geckoidea* was studied in detail by Rueda-Almonacid [111]. Given the remarkable morphological similarly between *C. geckoidea* and *C. paezorum*, we consider the possibility that *C. paezorum* represents a junior synonym of *C. geckoidea*. Differences between these species include the absence of vomerine teeth in *C. paezorum*, smaller size, and reduced webbing [110]. Also, the two species occupy different elevations; *C. geckoidea* is found at elevations between 1750–2525 m, whereas *C. paezorum* occurs at 3030 m. Additional specimens from the type locality of *C. paezorum* (Colombia: Departamento del Cauca: km 55–56 on the Popayán–Inza road, 3030 m) are necessary to clarify its status. Until recently, the holotype of *Centrolene geckoidea* was thought to be lost [118], but in a recent publication, González-Fernández [119] provided dorsal and ventral photographs of the holotype (MNCN 1596).

Specimens examined: *Centrolene geckoidea*: <u>Ecuador</u>: *Provincia de Pichincha:* 1 km SW of San Ignacio (0.4486 S, 78.7478 W, 1920 m), KU 178015–17; Quebrada Zapadores, 5 km ESE of Chiriboga on Chiriboga–Quito road (0.245278 S, 78.7261 W; 2010 m), KU 164492; 9 km SE of Tandayapa (0.0167 S, 78.683 W; 2150 m), KU 164490–91. *Provincia del Carchi:* Río La Plata, on the Maldonado-Tulcán road (0.8 N, 78.033 W; 2525 m), DHMECN 0900.

Centrolene heloderma (Duellman, 1981 [120]; Figures 37–40).

Centrolenella heloderma Duellman, 1981 [120]. Holotype: KU 164715.
Type locality: "Quebrada Zapadores, 5 km east-southeast of Chiriboga, 2010 m, Provincia de Pichincha, Ecuador (00°17′ S, 78°47′ W)".
Centrolene helodermum—Ruiz-Carranza and Lynch, 1991 [6].
Centrolene heloderma—Guayasamin, Castroviejo-Fisher, Trueb, Ayarzagüena, Rada, and Vilà, 2009 [1].

Common names: English: Warty Glassfrog. Spanish: Rana de Cristal Verrugosa.

Etymology: The specific epithet *heloderma* combines the Greek words "helos" (wart) and "derma" (skin) and is used in allusion to the dorsal texture of the species.

Identification: *Centrolene heloderma* differs from all glassfrogs by having a green pustular dorsum and males with humeral spines (Figure 34). *Centrolene heloderma* mostly resembles *Centrolene buckleyi*; both species have humeral spines in males, snout inclined in lateral profile, and a narrow white labial stripe. The most conspicuous difference between these species is that *Centrolene buckleyi* lacks a pustular dorsal skin. Further, *C. buckleyi* differs by having a concealed tympanum (completely visible in *C. heloderma*; Figure 37).

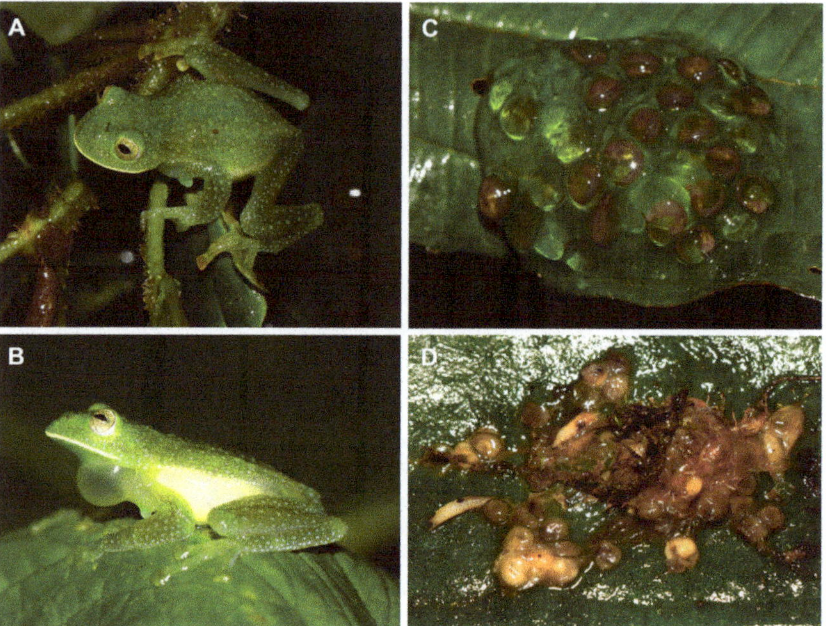

Figure 37. *Centrolene heloderma* from Reserva Las Gralarias, Ecuador. (**A,B**) Adult male. (**C**) Egg clutch. (**D**) Egg clutch parasitized by larvae of Drosophilidae fly. Photos by Jaime Culebras.

Diagnosis: (1) Vomers lacking teeth; (2) snout subacuminate in dorsal profile, inclined anteriorly from nostrils to margin of lip in lateral profile (Figures 37 and 38); (3) tympanum completely visible, oriented almost vertically, relatively large, its diameter 38.5%–44.5% of eye diameter; supratympanic fold moderate; tympanic membrane pigmented, clearly differentiated from surrounding skin; (4) dorsal surfaces of males and females pustular; in males, minute spicules evident only on flanks and tympanic region; (5) pair of enlarged subclocacal warts; (6) anterior three-fourths of ventral parietal peritoneum with white iridophores (condition P3); white pericardium; no iridophores in peritonea covering intestines, stomach, kidneys, testes, gall bladder, and urinary bladder (condition V1); (7) liver tetralobed, lacking iridophores (condition H0); (8) in males, humeral spines present; (9) webbing absent between Fingers I, II, and III; moderate webbing between Fingers III and IV; webbing formula III $(2-2^{1/2})$—$(1^{2/3}-2^+)$ IV (Figure 37); (10) webbing formula on foot: I $(1^{1/4}-1^{1/2})$—2 II $(1-1^+)$—$(2^+-2^{1/4})$ III $(1-1^+)$—$(2-2^+)$ IV $(2^+-2^{1/2})$—$(1-1^{1/4})$ V; (11) external ulnar fold evident, white; inner tarsal fold low, short; outer tarsal fold long, with low white tubercles; (12) concealed prepollex; in males, nuptial pad Type I; (13) Finger II slightly longer than Finger I (Finger I 91.0%–95.5% of Finger II); (14) disc of Finger III of moderate width, about 59.3%–74.2% of eye diameter; (15) in life, dorsum green with green to bluish white warts; upper lip white; bones green; (16) in preservative, dorsal surfaces dull lavender, sometimes with minute white or cream spots; (17) in life, iris yellow to pale golden yellow with fine black reticulations; (18) melanophores mostly absent from fingers and toes, except for some on Finger IV, and Toes IV and V; (19) males call from upper sides of leaves near streams; call short (133–188 ms, mean = 161 ms, SD = 15.4 ms), with two notes per call; notes strongly pulsed; dominant frequency at 4393–4823 (mean = 4682, SD = 104) Hz; (20) fighting behavior unknown; (21) egg placed on the upper surfaces of leaves; parental care unknown; (22) tadpoles unknown; (23) medium body size, male SVL 26.8–31.5 mm ($\overline{X}$ = 29.0, n = 17); in one female, SVL 32.3 mm.

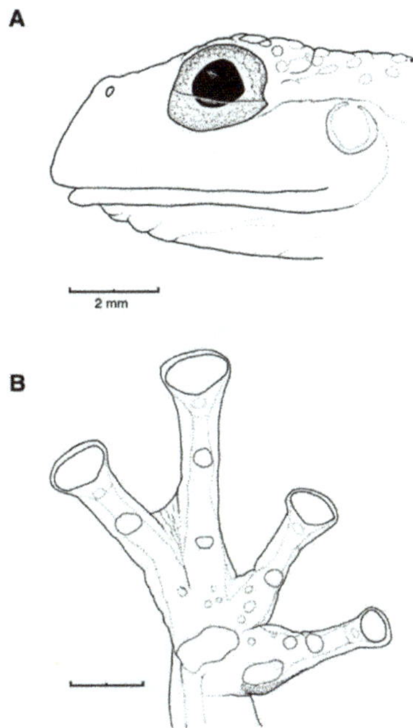

Figure 38. *Centrolene heloderma*, male, KU 164718. (**A**) Head in ventral view. (**B**) Hand in ventral view. Illustrations by Juan M. Guayasamin.

Color in life (Figure 37): Dorsum yellow green to dark green with green to bluish white pustules; margin of lip whitish yellow; ventral parietal peritoneum whitish yellow; throat pale greenish yellow; cloacal, heel, and ulnar tubercles white; heart not visible; bones green; iris yellow to pale golden yellow with fine black reticulations [120].

Color in ethanol: Dorsal surfaces of head, body, forearms, thighs, and shanks dull lavender; other surfaces dull cream; flanks white; margin of upper lip and cloacal tubercles white (Duellman, 1981).

White parietal peritoneum covering the anterior three-fourths of the venter; pericardium white; no iridophores in peritonea covering liver, intestines, stomach, kidneys, and urinary bladder.

Biology and ecology: During the rainy season (December–April), at night, *Centrolene heloderma* has been found active on vegetation near streams in primary and disturbed forest. Occasionally, males have been observed away from streams, probably looking for new territories [87]. Males call from the upper surfaces of leaves near streams. Egg clutches, with 25–29 brownish eggs, are placed on the upper side of leaves near streams [87]. At Reserva Las Gralarias, *C. heloderma* is abundant at Five-frog Creek, but has also been observed at Ballux Creek, Heloderma Creek, and Hercules Creek [88,92]. Parental care is unknown.

Call (Figure 39): The following description is based on the analysis of 58 notes contained within 29 calls from 3 individuals (LBE-C-001, 012, 015); individuals are from Río Alambí (2390 m), and Reserva Las Gralarias (1822–1864 m), in Pichincha province, Ecuador. The typical advertisement call is relatively short (range = 133–188 ms, mean = 161 ms, SD = 15.4 ms), and has two notes per call. Notes are strongly pulsed. The first note is longer than the second note (first note duration: Range = 38–76 ms, mean = 60.2 ms, SD = 9.89 ms; second note duration: Range = 17–44 ms, mean = 27.8 ms, SD = 6.283 ms). The two notes are also separated by an internote duration of 50–99 (mean = 72.9, SD = 9.5) ms. Notes are pulsed and similarly variable in number, with 2–7 (mean = 3.741, SD = 1.906) amplitude peaks throughout the note. The first note tends to have more pulses than the second note. Pulses within a note have a rate of 45.5–117.6 (mean = 83, SD = 15) pulses per second. Notes generally have their peak amplitude in the first 50% of the note (relative peak time: Range = 0.032–0.8697, mean = 0.273, SD = 0.242), but infrequently the first note can have its peak amplitude in the last 50% of the note. Frequencies between both notes are also highly similar. The dominant frequency of a note measured at peak amplitude is 4393–4823 (mean = 4682, SD = 104) Hz and is contained within the fundamental frequency. The fundamental frequency has a lower limit of 4221–4651 (mean = 4513, SD = 138) Hz and a higher limit of 4651–5082 (mean = 4875, SD = 104) Hz.

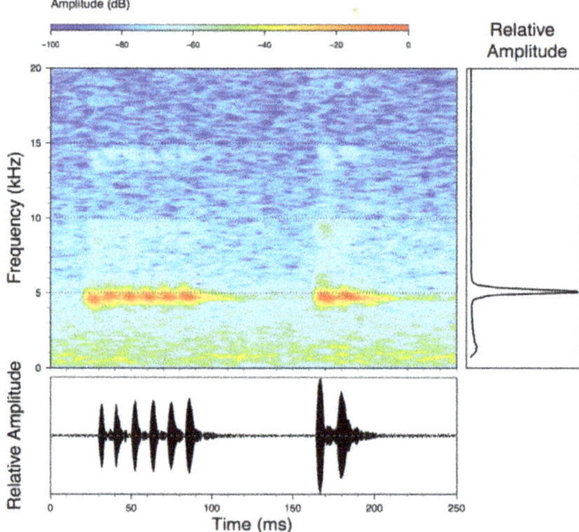

Figure 39. Call of *Centrolene heloderma* (LBE-C-012) recorded from Reserva Las Gralarias, Pichincha province, Ecuador. Note that each call has two distinctive pulsed notes.

Tadpole: Not described.

Distribution (Figure 40): *Centrolene heloderma* occurs on the Pacific slopes of the Cordillera Occidental in Colombia (departments of Antioquia, Cauca, Valle del Cauca, and Risaralda) south to the Tandayapa and Saloya Valleys, in Ecuador (provinces of Imbabura, Pichincha, and Santo Domingo de los Tsáchilas) at elevations of 1850–2575 m ([87,120,121], this work). In Ecuador, *Centrolene heloderma* has been recorded from seven localities at elevations between 1960 and 2575 m ([87,120,121]) in this work and has a potential distribution of 1067 km^2, within the Western Montane Forest ecoregion.

Figure 40. Distribution of *Centrolene heloderma* in Ecuador (yellow dots).

Conservation status: Globally, *Centrolene heloderma* is currently listed as *Vulnerable* by the IUCN [122]. In Ecuador, major threats include habitat loss, introduction of exotic predatory fish (Trout), climate change, and emerging diseases [87,123]. The type locality (Quebrada Zapadores), as well as nearby localities, have been visited numerous times in the last years (2000–2015; wet and dry seasons) with no new records of the species [91] (JMG, pers. obs.; DFCH, pers. obs.); in this area, the last record of the species dates from March 1979, when three individuals where collected at Quebrada Zapadores. In 2006, an individual apparently assignable to *C. heloderma* was photographed in the western versant of the Pichincha volcano (M. H. Yánez-Muñoz, pers. com.). On March 2009, *C. heloderma* was discovered at Reserva Las Gralarias (0.00806 S, 78.72433 W, 1852 m), Pichincha Province, where it maintains three nearby, reproductive populations. During September 2015, a population of *C. heloderma* was found at Cordillera de Toisán. Krynak et al. [121] report on a relatively large population (20+ individuals) at Río Alambi observed on 24 April 2017. The potential distribution of the species in Ecuador is 1067 km^2, 34% of which is affected by human activities. The amphibian chytrid fungus *Batrachochytrium dendrobatidis* has been found infecting *C. heloderma* at Reserva Las Gralarias, but no recent declines have been observed [92]. Given the current information of the species, we suggest that it should be considered as *Endangered* in Ecuador, following IUCN criteria B2, Ba, Bb(iii).

Evolutionary relationships (Figure 24): *Centrolene heloderma* is sister to a clade formed by several *Centrolene* species.

Specimens examined: *Centrolene heloderma:* Ecuador: *Provincia de Imbabura:* Cordillera de Toisán (0.50276° N, 78.5515° W; 2575 m), MZUTI 4234. *Provincia de Pichincha:* Quebrada Zapadores, 5 km ESE of Chiriboga on Chiriboga–Quito road (0.2453° S, 78.726° W, 2010 m), KU 164714–15; 9 km SE Tandayapa (0.01667° S, 78.683° W, 2160 m); 8.6 km SE Tandayapa (0.0333° S, 78.7° W, 2000 m), USNM 211218; 13.1 km SW Nono (0.0025° S, 78.659° W, 2140 m), MCZ 97834, USNM 211216–17; Reserva Las Gralarias (0.00806° S, 78.72433° W, 1852 m), QCAZ 40200. *Provincia de Santo Domingo de los Tsáchilas:* 14 km west of Chiriboga on Chiriboga–Santo Domingo road (0.2653° S, 78.848° W, 1960 m), KU 164716–21.

Centrolene huilensis (Ruiz-Carranza and Lynch, 1995 [26]) (Figure 41).

Centrolene huilense Ruiz-Carranza and Lynch, 1995 [26]. Holotype: ICN 7462.
<u>Type locality</u>: "Colombia, Departamento de Huila, Municipio San José de Isnos, 1 km NW Isnos, vertiente oriental de la Cordillera Central, 1°57′ Latitud N, 76°15′ W Greenwich, 2190 m".
Centrolene huilensis—Barrio-Amorós, Rojas-Runjaic, and Señaris, 2019 [85].

Common names: English: Huila Glassfrog. Spanish: Rana de Cristal de Huila.
Etymology: The specific epithet *huilensis* refers to the type locality of the species, within the Huila Department, Colombia [26].
Identification: *Centrolene huilensis* (Figure 41) can be differentiated from other glassfrogs by having a green dorsum with a combination of dark green to dark lavender and white spots, white tubercles on the ventrolateral edges of the Finger V, forearm, elbow, Toe V, tarsus, and heel, a humeral spine in adult males, and a relatively large body size (SVL = 23.6–26.7 mm in males; SVL = 28.7 mm in 1 female). Species with a similar dorsal color pattern include *Centrolene peristicta*, *C. daidalea*, *C. condor*, *C. lynchi*, *C. muelleri*, *C. pipilata*, *C. savagei*, *C. solitaria*, and *Nymphargus truebae*. Body size of *C. huilensis* is larger than *C. peristicta*, *C. pipilata*, *C. savagei*, and *C. solitaria* (in *C. peristicta*, male SVL = 17.9–22.0 mm, female SVL = 20.8–20.9 mm; in *C. pipilata*, male SVL 19.7–22.6 mm, female SVL 22.6–23.6 mm; in *C. savagei*, male SVL = 23.3–23.9 mm, female SVL = 19.8–22.6 mm; in *C. solitaria*, holotype male SVL = 19.3). Males of *C. huilensis* have humeral spines, which are absent in males of *C. daidalea*, *C. savegei*, *C. solitaria*, and *N. truebae*. Additionally, some of the species are located in different biogeographic regions; *C. peristicta* and *C. lynchi* are found on the Pacific slopes of the Andes of Ecuador and Colombia (*C. peristicta* is also found on the western slopes of the Cordillera Central, Colombia), and *C. muelleri* and *N. truebae* are only known from the Andes of Peru.

Figure 41. *Centrolene huilensis* in life. Ecuador, Napo province, Yanayacu Biological Station, QCAZ 45905. Photos by Santiago R. Ron (BioWebEcuador).

Diagnosis: (1) Vomers lacking teeth; (2) snout short, round in dorsal aspect, slightly sloping in lateral view; (3) tympanum moderate, oriented almost vertically, with slight lateral and posterior inclinations, its diameter 31%–40% of eye diameter; tympanic annulus completely visible, except for upper border; supratympanic fold evident; tympanic membrane translucent and pigmented as surrounding skin; (4) dorsal skin shagreen, males with low warts and spicules uniformly distributed; (5) pair of enlarged subcloacal warts (Figure 15); (6) anterior 40%–60% of the ventral parietal peritoneum white, posterior portion translucent (condition P3); white pericardium; no iridophores in peritonea covering intestines, stomach, kidneys, gall bladder, and urinary bladder (condition V1); (7) lobed liver, lacking iridophores (condition H0); (8) males with conspicuous humeral spines; (9) webbing absent between Fingers I, II, and III; moderate between outer fingers; webbing formula: III $(2^+–2^{1/2})$—$(2–2^+)$ IV; (10) webbing between toes extensive; foot three/fourths webbed: I $(1–1^{2/3})$—$(1^{3/4}–2^+)$ II $(1–1^+)$—$(2^-–2^{1/2})$ III $(1–1^{1/2})$—$(2^+–2^{2/3})$ IV $(2–3^-)$—$(1^-–1^{1/2})$ V; (11) ulnar and tarsal folds with white tubercles; (12) concealed prepollex; nuptial pad Type I; (13) Finger I slightly shorter than Finger II (Finger I length 94%–97% of Finger II); (14) disc of Finger III width 51%–63% of eye diameter; (15) in life, dorsum green with small diffuse dark green to dark lavender spots and smaller white spots; bones green; (16) in preservative, dorsum cream to light lavender with dark lavender spots and smaller white spots; (17) in life, iris cream with a slight yellow hue and thin black reticulations, yellowish–cream circumpupilar ring; (18) melanophores absent from dorsal surfaces of fingers and toes, except for Finger IV, and Toes IV and V; (19) males call from upper side of leaves; call undescribed; (20) fighting behavior unknown; (21) eggs deposition site unknown; parental care unknown; (22) tadpoles unknown; (23) small to medium body size; in Colombian populations, adult males, SVL 23.6–26.7 mm ($\overline{X}$ = 25.1 ± 1.703, n = 7 [26]); in Ecuador, SVL 23.8 in one adult male and 28.7 mm in one adult female.

Color in life (Figure 41): Dorsum green with dark lavender and dark green spots of different sizes, and smaller white spots; upper lip white; region below eye with small white warts; bones green. Upper flanks with same color pattern as dorsum; lower flanks with numerous small white warts. Ulnar and tarsal folds with white tubercles; small white cloacal tubercles. Iris cream with a slight yellow hue and thin black reticulations, yellowish–cream circumpupilar ring.

Color in ethanol: Dorsal surfaces of head, body, and limbs cream to light lavender with dark lavender spots and smaller white spots; margin of upper lip white; region below eye with small white warts; white tubercles just posterior to cloaca. White parietal peritoneum covers anterior 40%–60% of venter; white pericardium; iridophores absent from peritonea covering digestive tract, liver, kidneys, and gall and urinary bladders.

Biology and ecology: In Ecuador, individuals were found active during the night on upper surfaces of leaves along slow-flowing streams. Parental care is unknown.

Call: Not described.

Distribution (Figure 42): *Centrolene huilensis* is known from two localities, one in the Amazonian slope of the Andes of Ecuador (Yanayacu Biological Station, 2000 m, Napo Province), and the type locality in the Cordillera Central of Colombia (near Isnos, Huila Department) at elevations between 2000–2190 m ([26], this work).

Figure 42. Distribution of *Centrolene huilensis* in Ecuador (yellow dot).

Conservation status: Globally, *Centrolene huilensis* is currently listed as *Endangered* by the IUCN [124]. In Ecuador, assessing the conservation status of this species remains challenging because only a single additional locality (Yanayacu Biological Reserve) has been registered since its description in 1995. Yanayacu Biological Reserve is a relatively well-studied site [20,125] and we have been unable to find it during recent surveys (August 2014, June 2016; January 2017); thus, it is possible that *C. huilensis* spends most of its time in the canopy and/or is extremely rare. We suggest the that *C. huilensis* should be placed in the *Data Deficient* category for Ecuador.

Evolutionary relationships (Figure 24): With the current gene and taxon sampling, *Centrolene huilensis* is inferred as sister to *C. muelleri*.

Specimens examined: *Centrolene huilensis:* <u>Ecuador</u>: *Provincia de Napo:* Yanayacu Biological Station (0°41′ S, 77°53′ W; 2100 m), QCAZ 37230, 45905.

Centrolene lynchi (Duellman, 1980 [126]; Figures 43–46)

Centrolenella lynchi Duellman, 1980 [126]. Holotype: KU 164691.
Type locality: "a stream 4 km northeast (by road) of Dos Ríos, Provincia Pichincha, Ecuador,
    1140 m (0°21′ S, 78°54′ W)" (now in Provincia de Santo Domingo de los Tsáchilas).
Centrolene lynch—Ruiz-Carranza and Lynch, 1991 [6].
Centrolenella grandisonae—Lynch and Duellman, 1973 [22].
Centrolenella gemmata Flores, 1985 [46]. Holotype: MCZ 104073. Type locality: "San Francisco de
    las Pampas, 1500 m in elevation, Provincia Cotopaxi, Ecuador (00°25′ S, 78°57′ W, just NW
    of junction of Río Las Juritas and Río Toachi)". **New synonymy.**
Centrolene gemmatum—Ruiz-Carranza and Lynch, 1991 [6]. Guayasamin, Castroviejo-Fisher,
    Trueb, Ayarzagüena, Rada, and Vilà, 2009 [1].
Centrolenella scirtetes In part, Duellman and Burrowes, 1989 [86]. Holotype: KU 202720. Type
    locality: "1.4 km (by road) southwest of Tandayapa (00°07 S, 78°40 W), 1820 m, Provincia
    de Pichincha, Ecuador". **New synonymy.**
Centrolene scirtetes—Ruiz-Carranza and Lynch, 1991 [6]. Guayasamin, Castroviejo-Fisher, Trueb,
    Ayarzagüena, Rada, and Vilà, 2009 [1].

Common names: English: Lynch's Glassfrog. Spanish: Rana de Cristal de Lynch.

Etymology: The specific epithet *lynchi* honors Dr. John D. Lynch, who has made multiple and meaningful contributions to the field of amphibian systematics.

Identification: *Centrolene lynchi* is distinguished from most glassfrogs by having, in life, a green dorsum with minute yellowish–white flecks and diffuse small black spots (Figure 43). Among centrolenids found on the Pacific slopes of the Andes, only *C. peristicta* has a similar dorsal color pattern; however, body size is clearly different between the two species (in males, SVL 23.3–26.5 mm in *C. lynchi*; 17.9–21.2 mm in *C. peristicta*). The Colombian *C. antioquiensis* lacks black spots on the dorsum and has a white gastrointestinal peritoneum. *Centrolene lynchi* is also similar to the Colombian *C. quindianum*, which invariably has iridophores on the digestive tract (iridophores absent in *C. lynchi*). Other species, such as *Nymphargus truebae* (from the Amazonian slopes of the Peruvian Andes) and *N. garciae* (from the Cordillera Central of the Colombian Andes and the Amazonian slope of the Ecuadorian Andes), have a dorsal coloration similar to that found in *C. lynchi*, but lack humeral spines in males and webbing between Fingers III and IV.

Diagnosis: (1) Vomerine teeth absent; (2) snout round or truncated in dorsal aspect and truncated to slightly inclined in lateral profile; (3) tympanic annulus visible, oriented almost vertically, with slight lateral and posterior inclinations, its diameter about 31.3%–39.4% of eye diameter; supratympanic fold evident; tympanic membrane translucent, partially pigmented, differentiated from surrounding skin; (4) dorsal skin shagreen in males and females, males have low, white warts, and spicules and spiculated warts on sides of head; (5) pair of enlarged subcloacal warts (Figure 15); (6) anterior half to two-thirds of ventral parietal peritoneum white, remaining posterior portion transparent (condition P2–P3); white pericardium; no iridophores in peritonea covering intestines, stomach, and kidneys; transparent peritoneum around gall bladder and urinary bladder (condition V1); (7) liver with four clearly defined lobes, lacking iridophores (condition H0); (8) males with conspicuous humeral spines; (9) webbing absent between Fingers I and II, absent or greatly reduced between Fingers II and III, and moderate between outer fingers; hand webbing formula III (2–2$^{1/4}$)—(2$^-$–2$^+$) IV; (10) webbing between toes extensive; foot webbing formula: I (1–1$^+$)—(2–2$^+$) II (1–1$^+$)—(2–2$^+$) III (1–1$^+$)—(2–2$^+$) IV 2$^+$—(1–1$^+$) V; (11) ulnar fold present, white; external tarsal fold absent, internal tarsal fold short, low; (12) concealed prepollex, except in a few individuals; nuptial pad Type II; (13) Finger I slightly shorter than Finger II or about equal its length (Finger I 94.5%–100.0% of Finger II); (14) disc of Finger III width about 48.4%–57.0% of eye diameter; (15) in life, dorsum dull green with minute yellowish–white warts and small diffuse black spots (Figure 43); green bones; (16) in preservative, dorsum lavender with numerous minute white spots and larger dark lavender spots (Figure 44); (17) in life, iris greyish–white, with a yellow hue around the pupil; (18) melanophores mostly absent from fingers and toes, except for a few on base of Fingers III and IV, and along Toes IV and V; (19) males call from upper sides of leaves; call is relatively short and consists of a tonal note followed by one to three peaked notes; notes separated by 9.0–138.0 ms; mean tonal note dominant frequency 5296 Hz (SD = 58, range = 4995–5599 Hz); mean peaked note dominant frequency 5264 Hz (SD = 72, range = 4995–5513); notes lack frequency modulation; (20) males fight while hanging upside down, grasping one another venter to venter; (21) eggs placed on upper surfaces of leaves; females provide short-term parental care; prolonged parental care absent; (22) tadpoles unknown; (23) small body size; male SVL 23.3–26.5 mm ($\overline{X}$ = 24.7, *n* = 22); in two females SVL 24.6–25.0 mm.

Figure 43. *Centrolene lynchi* in life from Reserva Las Gralarias, Pichincha province, Ecuador. (**A**) Adult male, not collected. (**B**) Adult male in ventral view, QCAZ 40192; photo by Luis A. Coloma. (**C**) Fight between males; photo by Henry Imba and Rebecca Abuza. (**D**) Egg clutch.

Color in life (Figure 43): Dorsum yellow green with minute yellowish–white warts (spiculated in males) and small and diffuse black spots; bones green; throat and venter cream white; fingers and toes dull yellow; upper lip white; ulnar and outer tarsal folds with thin white line or low white tubercles; flanks and venter cream white; small white tubercles just posterior to cloaca. White parietal peritoneum covers anterior half to two-thirds of venter; white pericardium; visceral and hepatic peritonea lacking iridophores. Iris greyish white with black reticulation, and a yellow hue surrounding the pupil.

Color in ethanol (Figure 44): Dorsal surfaces of head, body, and limbs lavender with minute white warts and small black spots; margin of upper lip white; region below eye with small white warts; small white tubercles just posterior to cloaca. White parietal peritoneum covering the anterior half to two-thirds of the venter; pericardium white; no iridophores in peritonea covering liver, intestines, stomach, kidneys, gall bladder, and urinary bladder.

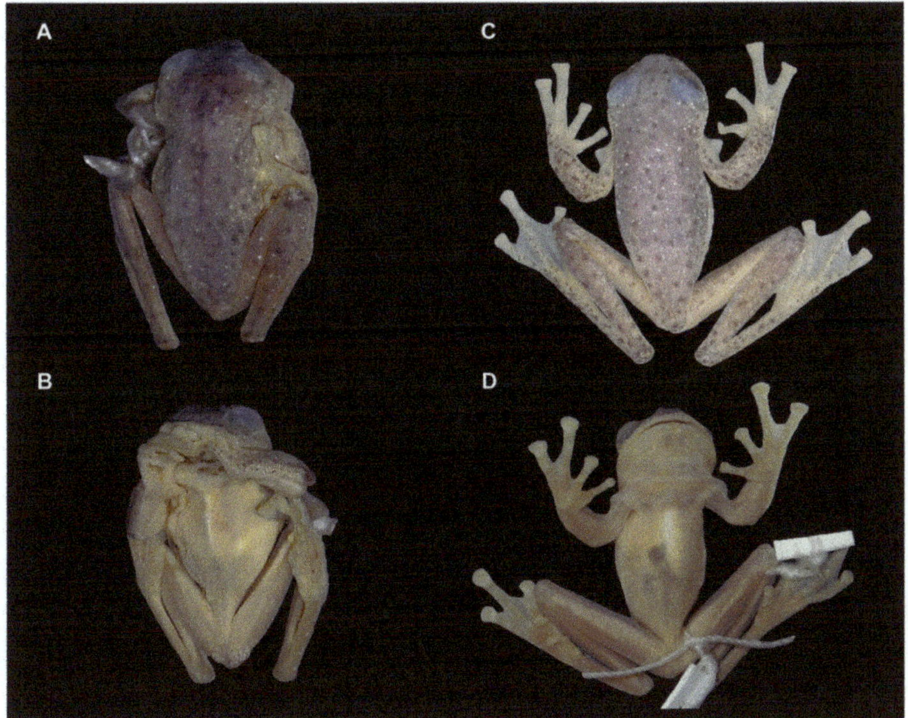

Figure 44. Type material of *Centrolene lynchi and C. gemmata* in preservative. (**A,B**) *C. gemmata*, MCZ 104077, paratype. (**C**) *C. lynchi*, KU 164698, paratype, dorsal view. (**D**) *C. lynchi*, KU 164691, holotype, ventral view.

Variation: The dorsum of most examined specimens except one (KU 164695) have spicules and spiculated warts; it is possible that this variation is a result of the reproductive condition of this particular male individual. Females of *Centrolene lynchi* lack spicules, normal sexual dimorphism in Centrolenidae.

Biology and ecology: The following description is from Dautel et al. [50]. At Reserva Las Gralarias (Lucy's Creek), females place clutches of 21–24 neon green eggs on vegetation in males' territories ($n = 4$, mean = 22.25, SD = 1.08); most egg clutches are placed on top of large leaves or ferns above a fast-flowing stream at a height of 165–600 cm ($n = 14$, mean height = 313.3 cm, SD = 78.7). Dautel et al. [50] suggested that males provide long-term parental care; however, a more detailed study by Delia et al. [25] showed that only females provide parental care (short term) and that prolonged parental care is absent in the species.

Calls (Figure 45): The following description is from Dautel et al. [50]. The typical advertisement call is relatively short and consists of a tonal note followed by one to three peaked notes. The first note is tonal and generally longer than the following notes, which are pulsed and show clear frequency peaks. Notes are separated by 9.0–138.0 ms. The dominant frequencies of tonal and peaked notes are similar (mean tonal note dominant frequency = 5296 Hz, SD = 58, range = 4996–5599 Hz; mean peaked note dominant frequency = 5264 Hz, SD = 73, range = 4996–5513) and show no frequency modulation. The aggressive call in this species is markedly different in structure from the advertisement call. It consists of a single short note containing two pulses ($n = 5$, mean length duration = 150.4 ms, SD = 6.7, range 140.0–156.0 ms), at a dominant frequency significantly lower than the advertisement call (mean frequency = 4892 Hz, SD = 39, range = 4823–4910 Hz); like the advertisement call, the note of the aggressive call is not frequency modulated.

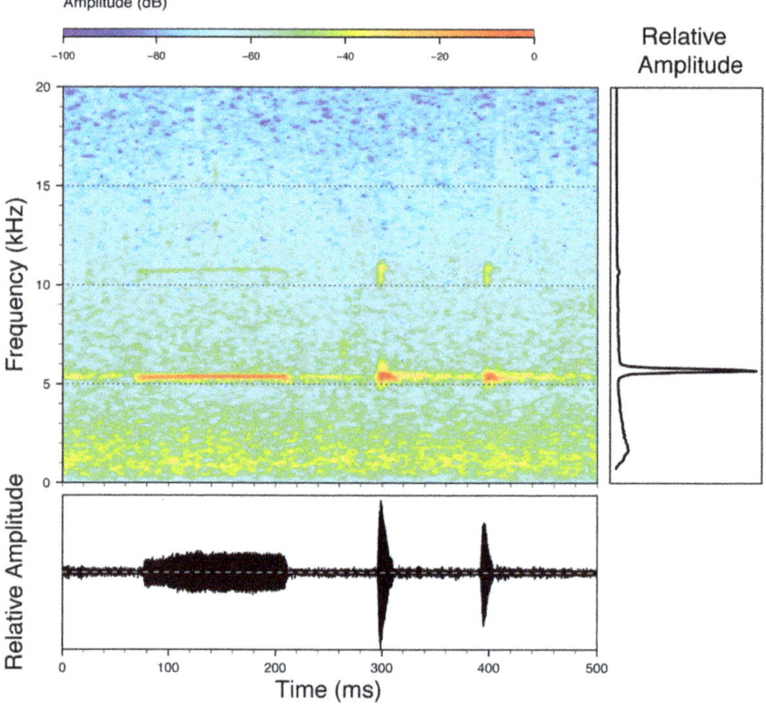

Figure 45. Call of *Centrolene lynchi* (LBE-C-013) recorded at Reserva Las Gralarias, 1822 m, Pichincha province, Ecuador.

Tadpole: Not described.

Distribution (Figure 46): *Centrolene lynchi* is known from the Pacific slope of the Cordillera Occidental of the Andes in Ecuador and southern Colombia. In Colombia, the species has been reported in only one locality (Reserva La Planada, 7 km route of Chucunés, 1780 m; as *C. scirtetes* [86], but see Taxonomic Remarks); specimens cited as *C. lynchi* by Coloma et al. [126] from Risaralda Department, Colombia, actually corresponds to *C. quindianum* (Marco Rada, pers. obs.). In Ecuador, *Centrolene lynchi* is known from seven localities from the Pacific slope of the Cordillera Occidental of the Andes at elevations of 1140–1852 m (see Specimens Examined), with a potential distribution of 1442 km^2. The habitat of the species in Ecuador is within the Western Foothill Forest and the Western Montane Forest regions.

Conservation status: Globally, *Centrolene lynchi* is currently listed as *Endangered* by the IUCN [126]. Arteaga et al. [87] also suggested the *Endangered* category for Ecuadorian populations. Although the distribution of the species is larger than previously thought (see Distribution and Remarks), the species has not been found in some of its historic localities in the last 20 years, suggesting that the species has suffered population declines (i.e., San Francisco de Las Pampas). In recent years, reproductive populations of *C. lynchi* have been found at Reserva Las Gralarias (2009–2019) and at Bosque Protector Río Guajalito (2006) ([87], JMG pers. obs., DFCH pers. obs., Mario Yánez-Muñoz pers. comm.). The amphibian chytrid fungus *Batrachochytrium dendrobatidis* has been found infecting *C. lynchi* at Reserva Las Gralarias, but no recent declines have been observed [92].

Figure 46. Distribution of *Centrolene lynchi* in Ecuador (yellow dots).

Evolutionary relationships (Figure 24): *Centrolene lynchi* is sister to a clade formed by *C. sabini* and an unidentified species of *Centrolene*.

Taxonomic remarks: *Centrolene lynchi* was described by Duellman [127] from a locality 4 km NE (by road) of Dos Ríos, Provincia de Pichincha. After a few years, Flores [46] described *Centrolene gemmata* from San Francisco de Las Pampas, Provincia de Cotopaxi. Flores [46] mentioned that *Centrolene lynchi* differs from *Centrolene gemmata* mainly by: (i) Having a snout truncated in dorsal view and round in profile (snout round in dorsal view and slightly anteroventrally sloping in *C. gemmata*); (ii) having slightly more webbing on the hands (webbing formula of IV 2—2 V in *Centrolene lynchi*; webbing formula of IV (2–2$^{1/3}$)—(2$^-$–2$^+$) V in *C. gemmata*); and (iii) in overall head shape, with a squatter, more semicircular head shape when viewed from above, with little or no post-cephalic constriction, and nostrils only very slightly protuberant, in contrast to the more elongated, circular head shape of *gemmata* when viewed from above, with a marked post-cephalic constriction and very protuberant nostrils. We have examined the complete type series of *Centrolene lynchi* and three paratypes of *C. gemmata* (MCZ A-104397, A-104074, A-104077) and, as noted by Cisneros-Heredia and McDiarmid [17], find that the differences mentioned above are not consistent. Several of the characteristics present in *C. gemmata* fall into the variation observed in *Centrolene lynchi*, and others are the product of preservation artifacts. It is clear that the type material of *C. gemmata* was poorly preserved, causing dehydration of the specimens, which usually produces changes in head shape (i.e., post-cephalic constriction and protuberant nostrils present in the type series of *C. gemmata*). The examined type series of *C. gemmata* has a snout that is truncated to slightly sloping in profile; the exact same variation is present in the type material of *Centrolene lynchi*. Also, the hand webbing formula in the two species are nearly identical [III $(2–2^{1/4})$—$(2^{-}–2^{+})$ IV in *Centrolene lynchi*, and III $(2–2^{1/3})$—$(2^{-}–2^{+})$ IV in *C. gemmata*]. Therefore, none of the characters listed by Flores (1985) to differentiate *C. gemmata* from *Centrolene lynchi* are valid. Based on the evidence mentioned above, we consider these to represent the same species and place *Centrolene gemmata* in the synonymy of *Centrolene lynchi*.

Duellman and Burrowes [86] described *Centrolenella scirtetes* from a locality 1.4 km SW Tandayapa (Ecuador, male holotype KU 202720) and from Reserva La Planada (Colombia, two female paratypes IND-AN 1405 and 1533). For unknown reasons, *C. scirtetes* was never compared with specimens of *Centrolene lynchi*. As pointed out by Cisneros-Heredia and McDiarmid [17], the holotype of *C. scirtetes* is undistinguishable from the holotype of *C. lynchi* (KU 164691). Consequently, herein we formally place *C. scirtetes*, as defined by its holotype, in the synonymy of *Centrolene lynchi*. We have not examined the paratypes of *C. scirtetes*, but material from Colombia identified as *C. scirtetes* (ICN 12172–74) and resembling the description and photograph of the paratypes provided by Duellman and Burrowes (1989) are indistinguishable from individuals of *Nymphargus griffithsi*, in which the dorsum has black flecks and the humeral crista ventralis presents a distal prolongation. Accordingly, we conclude that the Colombian material assigned to *C. scirtetes* is in fact representative of *N. griffithsi*.

Specimens examined: *Centrolene lynchi*: <u>Ecuador</u>: *Provincia de Cotopaxi:* San Francisco de Las Pampas, just NW of junction of río Las Juritas and río Toachi (0.433 S, 78.9667 W; ca. 1500 m), MCZ A-104397, A-104074, A-104077; *Provincia de Pichincha:* 1.4 km SW of Tandayapa (0.033 S, 78.7667 W; 1820 m), KU 202720; Tandapi (0.4164 S, 78.7989 W; 1460 m), KU 118036, 118047–50, 178095–104; 2.1 km E Tandapi (0.4258 S, 78.7853 W; 1500 m), MCZ A-93313–14, 95742; Reserva Las Gralarias (0.01675 S, 78.73165; 1852 m), QCAZ 40191–2, 40194. *Provincia de Santo Domingo de los Tsáchilas:* stream 4 km northeast (by road) of Dos Ríos (0.3028 S, 78.8678 W; 1140 m), KU 164691 (holotype), KU 164692–99 (paratypes); 14.4 km ENE La Palma on the road La Palma-Chiriboga (0.25 S, 78.846 W; 1380 m), MCZ A-91455.

Centrolenella medemi Cochran and Goin, 1970 [96]. Holotype: USNM 152277.

Type locality: "Puerto Asís, upper Río Putumayo, [Comisaría] Putumayo, Colombia"
   (apparently in error; see Distribution).
Centrolene medemi—Ruiz-Carranza and Lynch, 1991 [6].
"Centrolene" medemi—Guayasamin, Castroviejo-Fisher, Trueb, Ayarzagüena, Rada, and Vilà,
   2009 [1].

Common names: English: Medem's Glassfrog. Spanish: Rana de Cristal de Medem.
Etymology: The epithet *medemi* honors Dr. Fred Medem, who discovered the species [96].
Identification: *"Centrolene" medemi* is unique among Ecuadorian glassfrogs by having a dark
olive-green dorsum with large greenish–cream spots (Figure 47), wide disc on Finger III (>80% of eye
diameter), small tympanum (<20% of eye diameter), and fully webbed foot. Adults are robust and
relatively large (SVL 25.5–44.3 mm).

Figure 47. *"Centrolene" medemi* in life. Adult female (KU 164493) from 2 km SSW of junction between
Río Reventador and Baeza-Lumbaqui road, Ecuador. Photo by William E. Duellman.

Diagnosis: (1) Teeth on dentigerous process of the vomer present or absent, each process
bearing zero to four teeth; (2) snout round in dorsal profile, truncated to slightly protruding in lateral
profile (Figure 48); (3) tympanum small, tympanum diameter 17.1%–18.5% of eye diameter, dorsal
third of tympanum covered by supratympanic fold, tympanic membrane pigmented and not clearly
differentiated from surrounding skin; (4) dorsal surfaces of males and females smooth to shagreen;
males with small spicules on dorsum and flanks; (5) pair of slightly enlarged subcloacal warts;
(6) anterior half of the ventral parietal peritoneum white (condition P2); silvery white pericardium; no
iridophores in peritonea covering the intestines, stomach, testes, kidneys, gall bladder, and urinary
bladder (condition V1); (7) liver tetralobed, two large ventral lobes covering two smaller lobes; hepatic
peritoneum lacking iridophores (condition H0); (8) in males, small humeral spines present; (9) hand

webbing: Webbing between Fingers I, II, and III absent or basal, extensive webbing between Fingers III and IV; webbing formula as follows: II (1$^+$–2)—3$^+$ III (1$^{1/4}$–2)—(0–1$^{1/4}$) IV; (10) fully webbed foot: I (0–0$^+$)—(0$^+$–1) II (0–0$^+$)—(0–1) III (0–0$^+$)—(0$^+$–1) IV (1$^-$–1)—0$^+$ V; (11) ulnar and tarsal folds low; (12) concealed prepollex; in males, nuptial pad Type I; (13) Finger I slightly shorter than Finger II (Finger I about 91%–98% length of Finger II); (14) disc of Finger III large, 80%–91% of eye diameter; (15) in life, dorsal surfaces of head, body, and limbs olive green to greyish brown with large (up to 2.8 mm) cream spots (Figure 47); bones bluish green or green; (16) in ethanol, dorsal surfaces of head, body, and limbs pale brown with large cream spots; (17) iris greenish brown with black reticulation; (18) melanophores covering dorsal surfaces of fingers and toes; (19) calling behavior unknown; (20) fighting behavior unknown; (21) eggs deposited on rocks along streams; parental care unknown; (22) tadpoles unknown; (23) medium body size; in adult males, SVL 25.5–30.8 mm; in adult females, SVL 34.7–44.3 mm.

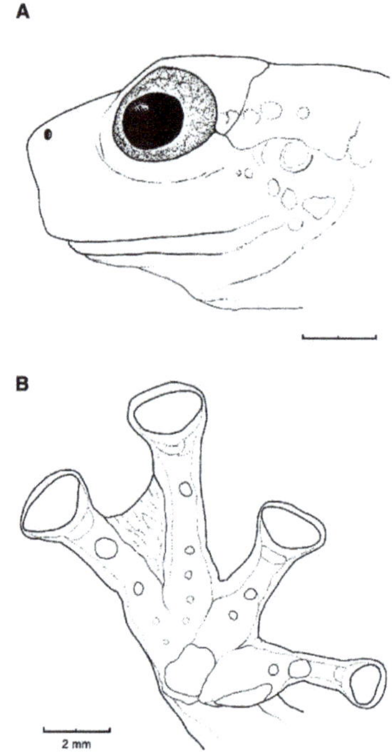

Figure 48. *"Centrolene" medemi.* (**A**) Head in lateral view, KU 164493. (**B**) Hand in ventral view, KU 164494. Illustrations by Juan M. Guayasamin.

Variation: The size of the nuptial pad varies in males and can be restricted to the dorsolateral area of the base of Finger I or extend until reaching the typical Type I morphology.

Color in life (Figure 47): At night, dorsum black with bluish green spots; venter dull blue. By day, dorsum dark olive-brown with pale green spots; flanks cream. Lining of mouth and tongue pale blue. Bones green. Iris dull greyish–bronze with minute black flecks. Webbing pale yellowish tan (W. E. Duellman field notes, 19 March 1975).

Color in ethanol: Dorsal surfaces of head, body, and limbs pale brown with large cream spots. White lining on the anterior half of the ventral parietal peritoneum; silvery white pericardium; no iridophores in peritonea covering the liver, intestines, stomach, testes, kidneys, gall bladder, and urinary bladder.

Biology and ecology: During the night, adults have been found on rocks, in rock crevices, and on rock cliffs along streams; juveniles were found on palm leaves or in rock crevices [128]. The oviductal eggs of one female (KU 164493) are dark brown. It is likely that females of *"Centrolene" medemi* deposit their eggs on rocks, as reported in species with similar microhabitats (e.g., *Centrolene geckoidea*, *"Centrolene" petrophilum* [113,129]). Parental care is unknown.

Call: Not described.

Tadpole: Not described.

Distribution (Figure 49): *"Centrolene" medemi* is known from localities on the eastern and western slopes of the Cordillera Oriental of the Colombian Andes and the Amazonian slopes of the Cordillera Oriental of Ecuador at elevation between 790 and 1800 m [101,128,130]. In Ecuador, *"Centrolene" medemi* has been registered from a single stream nearby Volcán Reventador at 1490 m (Specimens Examined), within the Eastern Montane Forest ecoregion. Although this species was described from a locality in lowlands of Amazonian Colombia (Puerto Asís, upper Río Putumayo at about 280 m [96]), all subsequent records are from Andean localities. Based on this evidence, Ruiz-Carranza et al. [101] rejected the record from Amazonian Colombia.

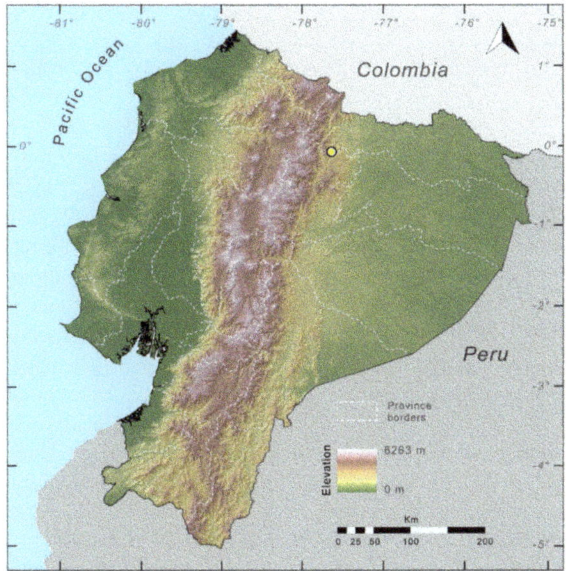

Figure 49. Distribution of *"Centrolene" medemi* in Ecuador (yellow dot).

Conservation status: Globally, *"Centrolene" medemi* is currently listed as *Endangered* by the IUCN [131]. In Ecuador, the only record is from a stream nearby Volcán Reventador on 19 March 1975 [130], and area that has been visited several times during the last 10 years. We suggest that the species should be locally listed as *Critically Endangered*.

Evolutionary relationships (Figure 24): Based on morphological traits, *"Centrolene" medemi* was placed in the *Centrolene geckoidea* species group by Ruiz-Carranza and Lynch [6]. Guayasamin et al. [1] considered the generic placement of *"Centrolene" medemi* as *incertae sedis* within the subfamily Centroleninae. No genetic data are available for this species.

Specimens examined: *"Centrolene" medemi*: <u>Ecuador</u>: *Provincia de Napo*: Stream 2 km SSW of junction between Río Reventador and Baeza-Lumbaqui road (0.1 S; 77.6 W, 1490 m), KU 164493–94.

Centrolene peristicta (Lynch and Duellman, 1973 [22]; Figures 50–53).

Centrolenella peristicta Lynch and Duellman, 1973 [22]. Holotype: KU 118051.
Type locality: "Tandapi, 1460 m, Provincia Pichincha, Ecuador".
Centrolene peristictum—Ruiz-Carranza and Lynch, 1991 [6]. Guayasamin, Castroviejo-Fisher, Trueb, Ayarzagüena, Rada, and Vilà, 2009 [1].
Centrolene peristicta—Barrio-Amorós, Rojas-Runjaic, and Señaris, 2019 [85].

Common names: English: Dappled Glassfrog. Spanish: Rana de Cristal Punteada.

Etymology: The specific epithet is derived from the Greek *peristiktos*, meaning dappled, and refers to the spotted color pattern of the species [22].

Identification: *Centrolene peristicta* is distinguished from most glassfrogs by its minute body size (SVL < 21.3 mm) and having, in life, a yellowish–green dorsum with minute white flecks and small diffuse black spots (Figure 50). Additionally, adult males of *C. peristicta* have a small, curved humeral spine that is morphologically different from the spines present in most glassfrogs (i.e., not curved). Similar species include *C. antioquiensis*, *C. lynchi*, *C. pipilata*, and *Nymphargus truebae*. *Centrolene antioquiensis*, an endemic species from Colombia, is remarkably similar to *C. peristicta* (see Taxonomic Remarks), but *C. antioquiensis* has less hand webbing: hand webbing formulae in *C. antioquiensis*, III $(2^+–2^-)$—$(2–2^-)$ IV; in *C. peristicta*, III $(2^+–2^-)$—$(1^{1/2}–1^+)$ IV. *Centrolene lynchi* is larger than *C. peristicta* (in males, SVL 23.3–26.5 mm in *Centrolene lynchi*) and has a very different call. *Centrolene pipilata* occurs on the Amazonian slopes of the Andes, whereas *C. peristicta* is restricted to the Pacific slopes of the Andes. *Nymphargus truebae*, a species only known from the Andes of southern Peru, has the same color pattern as *C. peristicta*, but *N. truebae* lacks humeral spines and webbing between Fingers III and IV (present in *C. peristicta*).

Diagnosis: (1) Vomers lacking teeth; (2) snout round in dorsal aspect, round or truncated in lateral profile (Figure 51); (3) tympanum large, oriented almost vertically, with slight lateral and posterior inclinations, its diameter 44.0%–51.9% of eye diameter; tympanic annulus visible, supratympanic fold evident; tympanic membrane translucent, partially pigmented, differentiated from surrounding skin; (4) dorsal skin shagreen with small warts corresponding to yellowish–white spots, males and females lack spicules; (5) pair of enlarged subcloacal warts (Figure 15); (6) anterior half of the ventral parietal peritoneum white, posterior half transparent (condition P2); white pericardium; iridophores partially or completely covering stomach and colon; no iridophores in peritonea covering kidneys, gall bladder, and urinary bladder (condition V2); (7) liver with four clearly defined lobes, lacking iridophores (condition H0); (8) males with conspicuous humeral spines; (9) webbing absent between Fingers I and II, reduced between Fingers II and III, and moderate to extensive between outer fingers (Figure 51); webbing formula: II $(1^{3/4}–2^-)$—$(3^+–3^{1/4})$ III $(2^+–2^-)$—$(1^{1/2}–1^+)$ IV; (10) webbing between toes extensive; webbing formula on foot I 1—$(1^{1/2}–2)$ II 1—$(2^-–2^+)$ III $(1^-–1^+)$—$(2^-–2^+)$ IV $(2^-–2^+)$—$(1–1^{1/3})$ V; (11) ulnar fold present, with low white tubercles; outer tarsal fold present, with low white tubercles; internal tarsal fold low and short; (12) prepollex usually exposed; nuptial pad Type I or Type III; (13) Finger I slightly shorter or as long as Finger II (Finger I length 90.9%–102.7% Finger II); (14) disc of Finger III width about 45.6%–67.3% of eye diameter; (15) in life, dorsum green with minute yellowish–white spots and larger black spots (Figure 50); green bones; (16) in preservative, dorsum lavender with numerous minute white spots and larger diffuse dark lavender spots; (17) in life, iris white grey with a yellow hue and thin black reticulation; thin yellow line surrounds the pupil; (18) melanophores mostly absent from fingers and toes, except for a few on dorsal surfaces of Finger IV and proximal portion of Finger III; (19) males call from mostly from the lower sides of leaves; each call has one pulsed note with a duration of 0.036–0.087 s and a dominant frequency of 6471–7278 Hz; (20) fighting behavior unknown; (21) egg clutches usually deposited on the underside of leaves; short-term maternal care absent; prolonged parental care provided by males; (22) tadpoles undescribed; (23) minute body size; in adult males, SVL 17.9–21.2 mm ($\overline{X}$ = 19.9, n = 14); in two adult females SVL 20.8–20.9 mm.

Figure 50. *Centrolene peristicta* in life. (**Top row**): Male guarding two egg clutches; Reserva Las Gralarias. (**Bottom left**): Adult male guarding eggs, from Mindo Biological Reserve, QCAZ 22313; photo by Martín Bustamante. (**Bottom right**): Individual from Reserva Las Gralarias; photo by Marco Rada.

Diagnosis: (1) Vomers lacking teeth; (2) snout round in dorsal aspect, round or truncated in lateral profile (Figure 51); (3) tympanum large, oriented almost vertically, with slight lateral and posterior inclinations, its diameter 44.0%–51.9% of eye diameter; tympanic annulus visible, supratympanic fold evident; tympanic membrane translucent, partially pigmented, differentiated from surrounding skin; (4) dorsal skin shagreen with small warts corresponding to yellowish–white spots, males and females lack spicules; (5) pair of enlarged subcloacal warts (Figure 15); (6) anterior half of the ventral parietal peritoneum white, posterior half transparent (condition P2); white pericardium; iridophores partially or completely covering stomach and colon; no iridophores in peritonea covering kidneys, gall bladder, and urinary bladder (condition V2); (7) liver with four clearly defined lobes, lacking iridophores (condition H0); (8) males with conspicuous humeral spines; (9) webbing absent between Fingers I and II, reduced between Fingers II and III, and moderate to extensive between outer fingers (Figure 51); webbing formula: II $(1^{3/4}$–$2^-)$—$(3^+$–$3^{1/4})$ III $(2^+$–$2^-)$—$(1^{1/2}$–$1^+)$ IV; (10) webbing between toes extensive; webbing formula on foot I 1—$(1^{1/2}$–2) II 1—$(2^-$–$2^+)$ III $(1^-$–$1^+)$—$(2^-$–$2^+)$ IV $(2^-$–$2^+)$—$(1$–$1^{1/3})$ V; (11) ulnar fold present, with low white tubercles; outer tarsal fold present, with low white tubercles; internal tarsal fold low and short; (12) prepollex usually exposed; nuptial pad Type I or Type III; (13) Finger I slightly shorter or as long as Finger II (Finger I length 90.%–102.7% Finger II); (14) disc of Finger III width about 45.6%–67.3% of eye diameter; (15) in life, dorsum green with minute yellowish–white spots and larger black spots (Figure 50); green bones; (16) in preservative, dorsum lavender with numerous minute white spots and larger diffuse dark lavender spots; (17) in life, iris white grey with a yellow hue and thin black reticulation; thin yellow line surrounds the pupil; (18) melanophores mostly absent from fingers and toes, except for a few on dorsal surfaces of Finger IV and proximal portion of Finger III; (19) males call from mostly from the lower sides of leaves; each call has one pulsed note with a duration of 0.036–0.087 s and a dominant frequency of 6471–7278 Hz; (20) fighting behavior unknown; (21) egg clutches usually deposited on the underside of leaves; short-term maternal care absent; prolonged parental care provided by males; (22) tadpoles undescribed; (23) minute body size; in adult males, SVL 17.9–21.2 mm ($\overline{X}$ = 19.9, n = 14); in two adult females SVL 20.8–20.9 mm.

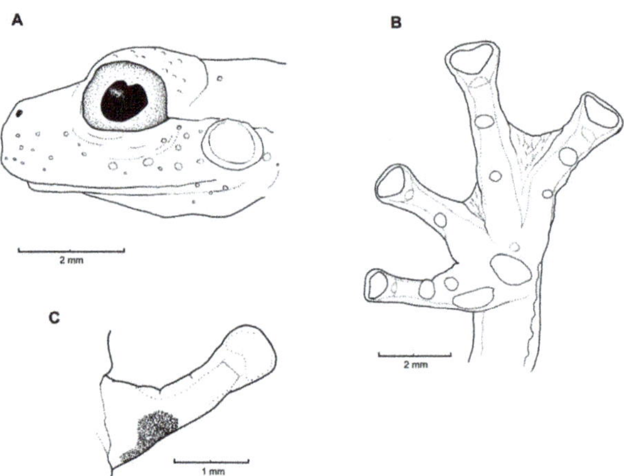

Figure 51. *Centrolene peristicta*, KU 178148. (**A**) Head in lateral view. (**B**) Hand in ventral view. (**C**) Dorsal view of Finger I, showing nuptial pad. Illustrations by Juan M. Guayasamin.

Color in life (Figure 50): Dorsum green with minute yellowish–white flecks and small dark grey spots; upper lip white; region below eye with small white warts; bones green; vocal sac green; upper flanks green with minute white spots; lower flanks cream with minute white spots; venter yellowish cream; fingers and toes dull yellow green; ulnar fold with thin white line or with low white ulnar tubercles; outer tarsal fold with low white tubercles; small white tubercles just posterior to cloaca.

Color in ethanol: Dorsal surfaces of head, body, and limbs lavender with minute white spots and small black spots; margin of upper lip white; region below eye with small white warts; white tubercles just posterior to cloaca. White parietal peritoneum covers anterior half of venter; white pericardium; clear peritoneum on liver and kidneys; digestive tract partially or completely covered with white lining.

Variation: One individual (QCAZ 16313) has considerably less webbing on the hands (III $2^{1/2}$—2^+ IV). Individuals from streams nearby the town of Mindo lack iridophores on the gastrointestinal peritonea.

Biology and ecology: The information shown below is from Salgado and Guayasamin [132]. At Reserva Las Gralarias, during the night, *Centrolene peristicta* is active on vegetation 40–600 cm above permanent streams in primary evergreen lower-montane forests and cloudforests. Its breeding season is in the rainy (December–April), but peaks in February–April. Males vocalize to advertise themselves and defend territories in which females place clutches containing 6–41 eggs. Most of the time, egg clutches are placed on the underside of leaves and ferns 70–500 cm directly above streams, or under dead leaves nearby streams. Males are polygynous and exhibit high site fidelity; some males were observed simultaneously guarding egg clutches at different stages of development. Parental care by males was demonstrated experimentally; unattended clutches have a significantly lower eclosion rate than attended clutches, and prolonged hatching time. Clutch mortality was mainly because of desiccation, predation, and parasitism. Embryos develop for 17–27 days, then hatching as free-swimming larvae. At Reserva Las Gralarias, the species is abundant at Lucy's Creek, but it is also found at Kathy's Creek and Santa Rosa River [88].

Call (Figure 52): The information shown below is from Salgado and Guayasamin [132]. Advertisement call of males of *Centrolene peristicta* consists of one short and pulsed note that resembles the sound of a cricket chirp. The note has a duration of 0.036–0.087 s ($X + SE = 0.062 + 1.07, N = 35$) and each call has one note. The dominant frequency is at 6471–7278 Hz ($X + SE = 6878 + 177$ Hz., $N = 35$). The fundamental frequency is the same as the dominant frequency; there is one (*range* = 3235–6198 Hz) or two harmonics (*range* = 6471–7278 Hz). There is no conspicuous frequency modulation during the call. The repetition rate of the advertisement call varies from two to 20 calls per minute.

Tadpole: Not described.

Distribution (Figure 53): *Centrolene peristicta* is known from the Pacific slope of the Cordillera Occidental of the Andes in Ecuador and Colombia at elevations between 1380 and 1900 m ([22,87,133,134], this work). In Ecuador, *C. peristicta* is known from localities in the provinces of Carchi, Pichincha, and Santo Domingo de los Tsáchilas, at elevations of 1400–1852 m, with a potential distribution of 12,603 km². All localities are within the Western Montane Forest ecoregion.

Conservation status: Globally, *Centrolene peristicta* is currently listed as *Least Concern* by the IUCN [135]. In Ecuador, Arteaga et al. [87] suggested the category of *Near Threatened*. Several recent observations show that the species has a wider distribution than previously thought. Reproductive populations have been observed at Reserva Las Gralarias (December 2009–May 2018; JMG, pers. obs.), Mindo Biology Station (February 2002; December 2014; M. R. Bustamante and L. Bustamante, pers. com.), and at Río Pachijal on August 2001 (I. Tapia, pers. com.). The potential distribution of the species covers an area of 12,603 km², 44.6% of which is affected by human activities. At Reserva Las Gralarias, the species is infected by the amphibian chytrid fungus *Batrachochytrium dendrobatidis* [92], but populations look healthy and there is no evidence of declines. The conservation category of *Least Concern* seems too optimistic, given the fragmented distribution of the species. We suggest that the species should be considered as *Near Threatened*, in agreement with Arteaga et al. (2013).

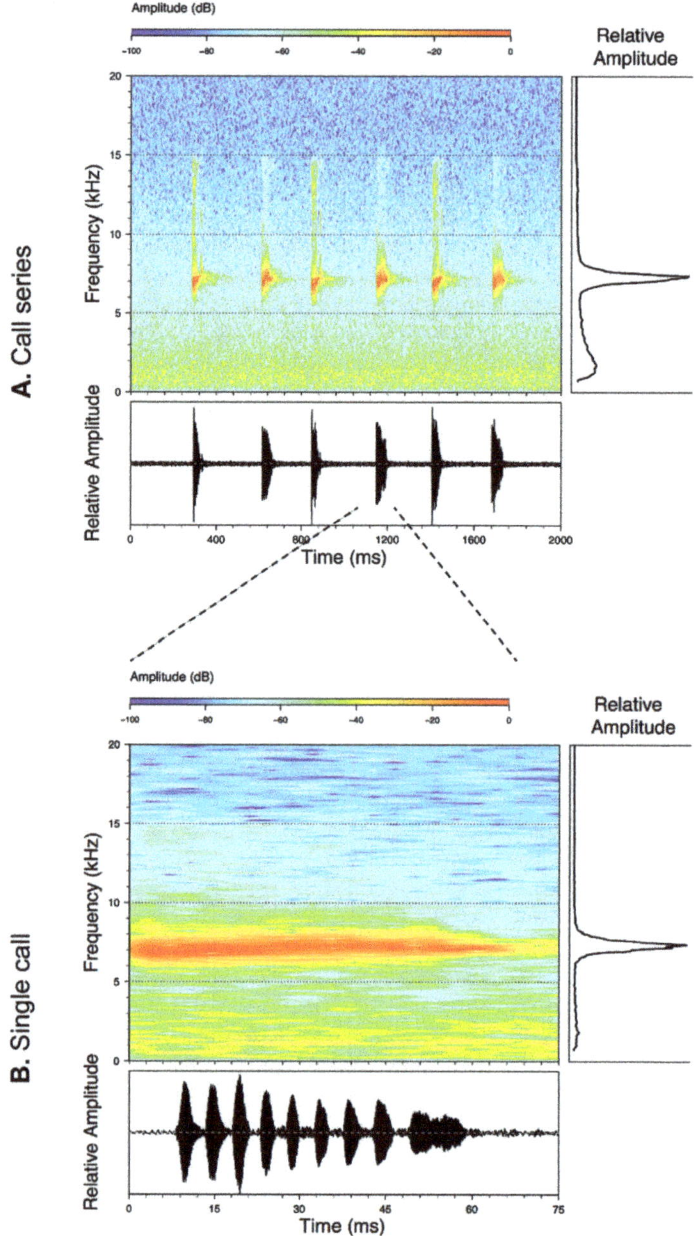

Figure 52. Call of *Centrolene peristicta* (LBE-C-047), recorded at Reserva Las Gralarias, 1800 m, Pichincha province, Ecuador. (**A**) Series of calls. (**B**) Single, pulsed call.

Figure 53. Distribution of *Centrolene peristicta* in Ecuador (yellow dots).

Evolutionary relationships (Figure 24): Molecular evidence places *Centrolene peristicta* and *C. antioquiensis* as sister species.

Taxonomic Remarks: *Centrolene antioquiensis* and *C. peristicta* are almost identical in morphology and could represent a single species. The subtle differences found between these two species might be a consequence of incomplete sampling or intraspecific geographic variation. The analysis of calls from *C. antioquiensis* is critical to assess the validity of the species status of *C. peristicta*.

Specimens examined: *Centrolene peristicta:* Ecuador: *Provincia de Carchi:* Maldonado (0.9 N, 78.1 W, 1410 m), KU 178137–44, 178145–51, 180325; *Provincia de Pichincha:* Bosque Protector Río Guajalito (0.233 S, 78.817 W; 1900 m), QCAZ 6446; stream near Mindo Biology Station (0.07805 S, 78.7319 W; 1600 m), QCAZ 22757–59, 22312–14; Tandapi (0.416389 S, 78.7989 W, 1520 m), KU 118051–52, 121053; 1.6 km W of Tandapi (0.40472 S, 78.8058 W; 1400 m), KU 178152; 5 km W of Tandapi on the Tandapi–Atenas road (0.3954 S, 78.8326 W; 1670 m), QCAZ 15901, 15922; Río Pachijal (0.2 S, 78.75 W; 1740 m), QCAZ 16316; Reserva Las Gralarias (0.00806 S, 78.72443 W; 1852 m), QCAZ 47298. *Provincia de Santo Domingo de los Tsáchilas:* Río Faisanes, ca. 15 km NE of La Palma, on the Quito–Chiriboga–Santo Domingo road (0.3167 S, 78.817 W; 1380 m), USNM 286714. Colombia: *Departamento de Nariño*, Reserva La Planada, ICN 12114 (PR 7871).

Centrolene pipilata (Lynch and Duellman 1973 [22]; Figures 54–56).

Centrolenella pipilata Lynch and Duellman, 1973 [22]. Holotype: KU 143278.
Type locality: "16.5 km NNE of Santa Rosa, 1700 m, on Quito-Lago Agrio road, Provincia de Napo, Ecuador".
Centrolene pipilatum—Ruiz-Carranza and Lynch, 1991 [6].
Centrolene pipilata—Barrio-Amorós, Rojas-Runjaic, and Señaris, 2019 [85].

Common names: English: Peeping Glassfrog. Spanish: Rana de Cristal Piadora.

Etymology: The specific epithet *pipilata* is an adjectival derivative of the Latin verb *pipila*, meaning to peep, and refers to the call of this centrolenid frog [22].

Identification: *Centrolene pipilata* is distinguished from most glassfrogs by having, in life, a green dorsum with yellowish–white flecks and diffuse dark green marks, and a distinct prepollex (Figures 54 and 55). On the Amazonian slopes of Ecuador, the only species that have a similar dorsal color pattern are *C. sanchezi* and *C. huilensis*. *Centrolene sanchezi* has a short, thin humeral spine (large, wide, and usually with the tip projected anteriorly in *C. pipilata*) and lacks a distinct prepollex. *Centrolene huilensis* is conspicuously larger than *C. pipilata* (in *C. huilensis*, male SVL = 23.6–26.7 mm; SVL = 28.7 mm in 1 female; in *C. pipilata*, male SVL 19.7–22.6 mm, female SVL 22.6–23.6 mm). *Nymphargus truebae*, a species only known from the Andes of southern Peru, has the same color pattern as *C. pipilata*, however, males in *N. truebae* lack humeral spines (present in males of *C. pipilata*). The Colombian *Centrolene hybrida* has a white gastrointestinal peritoneum (iridophores absent in *C. pipilata*), and lacks a distinct prepollex.

Figure 54. *Centrolene pipilata.* (**Top row**): Coloration in life, adult male, holotype, KU 143278; photos by William E. Duellman. (**Bottom row**): Coloration in preservative of KU 178155 (**left**) and KU 143287 (**right**); photos by Juan M. Guayasamin.

Diagnosis: (1) Vomers lacking teeth; (2) snout round in dorsal aspect, round or truncated in lateral profile (Figure 55); (3) tympanum moderate, oriented almost vertically, with slight lateral and posterior inclinations, its diameter 31.0%–39.4% of eye diameter; tympanic annulus completely visible; supratympanic fold evident; tympanic membrane translucent, partially pigmented, clearly differentiated from surrounding skin; (4) dorsal skin shagreen, males with spicules coinciding with white spots; (5) pair of enlarged subcloacal warts (Figure 15); (6) anterior 50%–60% of the ventral parietal peritoneum white, posterior portion transparent (condition P2–P3); white pericardium; no iridophores in peritonea covering intestines, stomach, kidneys, gall bladder, and urinary bladder (condition V1); (7) liver with four clearly defined lobes, lacking iridophores (condition H0); (8) males with conspicuous humeral spines; (9) webbing absent between Fingers I, II, and III; moderate between outer fingers (Figure 55); webbing formula: III $(2^+-2^{1/2})$—$(2-2^+)$ IV; (10) webbing between toes extensive; webbing formula on foot: I (1^+-2^-)—$(2-2^+)$ II $(1-1^{1/3})$—$(2-2^{1/3})$ III $(1-1^{1/2})$—2^+ IV $(2-2^{1/2})$—$(1-1^{1/2})$ V; (11) white ulnar fold present; outer tarsal margin with low white tubercles: Internal tarsal fold low; (12) exposed prepollex; nuptial pad Type II; (13) Finger I as long as Finger II or slightly shorter (Finger I 90.0%–100% of Finger II); (14) disc of Finger III width 44.8%–56.5% of eye diameter; (15) in life, dorsum green with small diffuse black spots and yellowish–white flecks (Figure 54); green bones; (16) in preservative, dorsum lavender with numerous minute white spots and larger dark lavender spots (Figure 54); (17) in life, iris greyish–white with thin black reticulations and a yellow hue around pupil; (18) melanophores mostly absent from fingers and toes, except for a few on proximal half of Finger IV, and along Toes IV and V; (19) males call from upper side of leaves; call undescribed; (20) fighting behavior unknown; (21) eggs placed on upper surface of leaves; parental care unknown; (22) tadpoles unknown; (23) minute body size; in adult males, SVL 19.7–22.6 mm ($\overline{X}$ = 21.5, *n* = 9); in two adult females SVL 22.6–23.6 mm.

Color in life (Figure 54): Dorsum green with minute yellowish–white flecks and larger, diffuse black spots; upper lip white; region below eye with small white warts that are spiculated in males; bones green; upper flanks green with minute white spots; lower flanks whitish cream; ulnar fold with thin white line; outer tarsal fold with low white tubercles; small white tubercles just posterior to cloaca; iris greyish–white with thin black reticulations and a yellow hue around pupil.

Color in ethanol: Dorsal surfaces of head, body, and limbs lavender with small white spots and larger dark lavender spots (Figure 54); margin of upper lip white; region below eye with small, spiculated white warts; white tubercles just posterior to cloaca. White parietal peritoneum covers anterior 50%–60% of venter; white pericardium; no iridophores in peritonea covering digestive tract, liver, kidneys, gall bladder, and urinary bladder.

Variation: In preservative, some individuals lack dorsal dark spots.

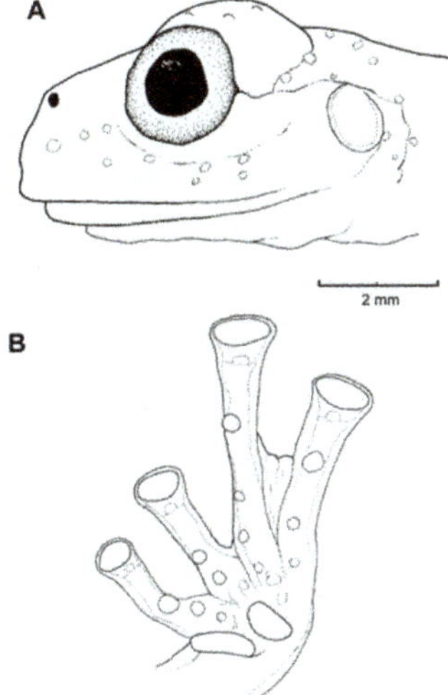

Figure 55. *Centrolene pipilata*, KU 143278, holotype. (**A**) Head in lateral view. (**B**) Hand in ventral view. Illustrations by Juan M. Guayasamin.

Biology and ecology: All individuals have been found at night on vegetation along cascading mountain streams. In captivity, a female deposited a clutch of 18 eggs. Males call from the upper side of leaves. At Río Azuela, *Centrolene pipilata* was found in sympatry with *Nymphargus anomalus*, *N. megacheirus*, *N. siren*, and *Hyalinobatrachium pellucidum*. At 16.5 NNE of Santa Rosa, the species was found with *Espadarana audax*, *N. megacheirus*, and *N. siren* [22]. Eggs have clear jelly and pale green yolks [22]. Parental care is unknown.

Call: Not described.

Tadpole: Not described.

Distribution (Figure 56): *Centrolene pipilata* is endemic to the cloud forest on the Amazonian slope of the Ecuadorian Andes at elevations between 1300–1910 m ([17,22], this work). The species has been recorded from four localities in the provinces of Napo and Sucumbíos. The habitat of the species is within the Eastern Montane Forest region.

Figure 56. Distribution of *Centrolene pipilata* in Ecuador (yellow dots).

Conservation status: Globally, *Centrolene pipilata* is listed by the IUCN as *Endangered* [136] because of its limited distribution and the continuing decline in the extent and quality of suitable habitat. The last confirmed report of *C. pipilata* was at 14.7 km NE of Río Salado on February 1979 [17]. Recent surveys at Río Azuela [91] and the type locality (JMG, pers. obs.) have failed to find the species. In Ecuador, based on the IUCN criteria A2c,e, we suggest that the species should be considered as *Critically Endangered*.

Evolutionary relationships (Figure 24): Given the current taxon and gene sampling, *Centrolene pipilata* and *C. hybrida* are sister species.

Specimens examined: *Centrolene pipilata:* Ecuador: *Provincia de Napo:* 16.5 km NNE Santa Rosa (0.2186 S, 77.732 W, 1700 m), KU 143278–83, 143554; 3.2 km NNE Oritoyacu (0.4597 S, 77.8672 W; 1910 m), KU 178153; Río Azuela (0.11667 S, 77.61667 W; 1740 m), KU 143284–87, 155498, 166331; Río Salado, 1 km upstream from Río Coca (0.19167 S, 77.6997 W; 1420 m), KU 178154–55.

Centrolene sanchezi Ruiz-Carranza and Lynch, 1991 [129] (Figures 57–61).

Centrolene sanchezi Ruiz-Carranza and Lynch, 1991 [129]. Holotype: ICN 24293.

Type locality: "Departamento de Caquetá, municipio de Florencia, vereda Gabinete, 3.1 Km por carretera abajo del Alto Gabinete, vertiente oriental, Cordillera Oriental, 1°4′ latitud N, 75° 4′ W de Greenwich, 2190 m", Colombia.

Centrolene bacatum Wild, 1994 [137]. Holotype: KU 202803. Type locality: "11.2 km WSW Plan de Milagro (03°02′ S, 78°35′ W, 2350 m), Provincia Morona Santiago, Ecuador". **New synonymy.**

Centrolene guanacarum Ruiz-Carranza and Lynch, 1995 [26]. Holotype: ICN 11686. Type locality: "Departamento del Cauca, municipio de Inzá, Km. 84 carretera Popayán a Inzá, Río Guanacas, quebrada afluente, Internado Indígena Río Guanacas, vertiente oriental Cordillera Central, 2°34′ Latitud N, 76°05′ W de Greenwich, 1800–1900 m", Colombia. **New synonymy.**

"Centrolene" guanacarum—Guayasamin, Castroviejo-Fisher, Trueb, Ayarzagüena, Rada, and Vilà, 2009 [1]. **New synonymy.**

Centrolene bacata—Barrio-Amorós, Rojas-Runjaic, and Señaris, 2019 [85].

Common names: English: Sánchez's Glassfrog. Spanish: Rana de cristal de Sánchez.

Etymology: The specific epithet *sanchezi* honors Ricardo Sánchez, who, with John D. Lynch, discovered the species.

Identification: *Centrolene sanchezi* is easily distinguished from other glassfrogs by its minute body size (SVL 18.9–22.3 mm), green dorsum with small white spots, and the presence of white warts in an area that extends from below the eye to the insertion of the arm (Figures 57 and 58A). Additionally, adult males have conspicuous humeral spines. The only other centrolenid from eastern Ecuador that can be confused with *Centrolene sanchezi* is *Centrolene pipilata*, which has a green dorsum with small, diffuse black spots and yellowish–white flecks (Figure 54), and a distinct prepollex. Also, the two species occupy different elevations on the Amazonian slope of the Andes; *C. sanchezi* is found at elevations of 1950–2350 m, whereas *C. pipilata* occurs at 1420–1910 m.

Diagnosis: (1) Vomerine teeth absent; (2) snout rounded in dorsal aspect, bluntly rounded or truncated in lateral profile (Figure 58); (3) tympanum oriented almost vertically, its diameter 31.4%–37.8% of eye diameter; tympanic annulus visible except for dorsal border covered by supratympanic fold; tympanic membrane partially pigmented, but differentiated from surrounding skin; (4) dorsal surfaces shagreen, with small spicules evident in most males; (5) pair of enlarged subcloacal warts (Figure 15); (6) anterior half of ventral parietal peritoneum white, posterior half translucent (condition P2); white pericardium, translucent visceral peritoneum (condition V1); (7) liver tetralobed, lacking iridophores (condition H0); (8) humeral spines present in adult males; (9) no webbing between Fingers I and II; webbing formula on hand: II 2—3$^{1/3}$ III 2$^{1/2}$—2$^{1/4}$ IV (Figure 58); (10) webbing formula on foot: I 1$^{1/2}$—2$^+$ II (1–1$^{1/2}$)—(2–2$^{1/3}$) III 1$^+$—(2$^+$–2$^{1/4}$) IV 2$^{1/2}$—(1–1$^+$) V (Figure 58); (11) ulnar and inner tarsal folds low or absent; outer tarsal fold absent; (12) nuptial pad Type I, concealed prepollex; (13) Fingers I and II about equal in length (FII/FI = 91.3%–104.5%); (14) disc of Finger III of moderate size, about 40.3%–50.6% of eye diameter; (15) in life, dorsum dark green with small white spots (Figure 57); conspicuous white warts below the eye and tympanum; upper lip white; bones green; (16) in preservative, dorsum lavender with small white spots; (17) iris pale bronze with black reticulation; (18) melanophores mostly absent from fingers and toes, except for a few on Toes IV and V and on base of outer fingers; (19) males call from the upper side of leaves; calls are produced in series; each call has one or two notes, with a duration of 6–21 ms (mean = 11, SD = 2.8); dominant frequency at peak amplitude is 5719–6188 Hz (mean = 5996, SD = 131); (20) fighting behavior unknown; (21) eggs deposited on upper side of leaves; short-term maternal care unknown; parental care by males absent; (22) tadpoles unknown; (23) minute body size, SVL 18.9–22.3 mm in males (*n* = 15); 20.9 mm in one female.

Figure 57. *Centrolene sanchezi* in life. (**A,B**) Male and female from Yanayacu Biological Station, Ecuador. (**C**) Holotype of *C. sanchezi*, ICN 24293. (**D**) Paratype of *C. guanacarum*, ICN 11685. Photos by A. Arteaga (**A,B**) and P. Ruiz-Carranza (**C,D**).

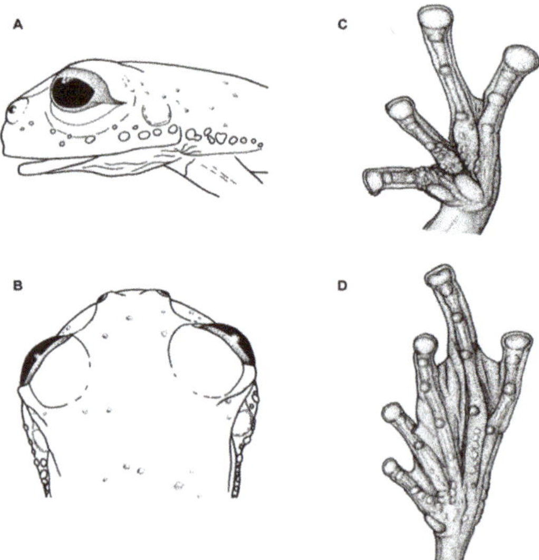

Figure 58. *Centrolene sanchezi*, KU 202803 (holotype of *Centrolene bacata*). (**A**) Head in lateral view. (**B**) Head in dorsal view. (**C**) Hand in ventral view. (**D**) Foot in ventral view. Drawings not to scale. Modified from Wild [137].

Color in life (Figure 57): Based on field notes by W. E. Duellman (4 March 1984), reported in Wild [137], and observation by authors. Dorsum dark green; series of white–cream tubercles under eye;

throat and ventral surfaces of limbs green; digits pale green; cloacal region with white warts; ventral parietal peritoneum white anteriorly and translucent posteriorly; visceral peritoneum translucent; bones green; iris pale bronze with thin black reticulation.

Color in ethanol (Figure 59): Dorsal surfaces lavender with small, unpigmented spots and white warts; limbs cream lavender with numerous small, unpigmented spots and few white warts; white warts on lateral surface of head; upper lip white; tympanum pigmented with purple specks; cloacal region with white or cream warts; iris silvery white with dark purple reticulation. Fingers I and II and Toes I–III dorsally unpigmented; some pigmentation visible on Fingers III and IV, and Toes IV and V. White parietal peritoneum covering anterior half of venter, posterior half translucent; silvery-white pericardium; translucent peritonea covering liver, gastrointestinal track, and renal capsules (dissected male: QCAZ 22386).

Figure 59. Types of *Centrolene sanchezi* and *C. bacata* in preservative. (**A,B**) *Centrolene bacata*, holotype KU 202803. (**C,D**) *C. sanchezi*, paratype ICN 24294. Photos by Juan M. Guayasamin.

Biology and ecology: During the night, *Centrolene sanchezi* has been found on leaves approximately 130–300 cm above streams in primary and secondary forest. Males call from the upper sides of leaves. A male (QCAZ 22728) was found nearby two egg clutches, one on the underside and the other on the upper side of a single leaf; egg clutches have 14–18 pale yellowish–green eggs. During February 2013, males have been heard calling at Yanayacu Biological Station. At this locality, *Centrolene sanchezi* (reported as *C. bacata*) is the most abundant centrolenid; other sympatric species at Yanayacu include *Centrolene* aff. *buckleyi*, *C. huilensis*, *Nymphargus posadae*, *N. siren*, and *N. wileyi* [20,125]. Short-term maternal care unknown; parental care by males absent ([25], as *C. bacata*).

Call (Figure 60): We analyzed 63 notes contained within 11 calls from 1 individual (LBE-C-023). Calls are produced in series, which can be relatively long (range = 618–3085 ms, mean = 1462.6 ms,

SD = 639.5 ms); each can have one or two notes; note repetition rate is 3–9 (mean = 5.7, SD = 1.6) notes per call. Notes sometimes occur in pairs (i.e., much shorter note interval than compared to the rest of the call). Each call is very short, with a duration of 6–21 (mean = 11, SD = 2.8) ms. Notes are strongly pulsed and have one or two (mean = 1.4, SD = 0.5) amplitude peaks throughout the note, where the second amplitude peak is generally weaker than the first. Pulses within a note have a rate of 83–200 (mean = 122, SD = 29) pulses per second. Notes have their peak amplitude in the first 50% of the note (relative peak time: Range = 0.0987–0.5497, mean = 0.216, SD = 0.076). The dominant frequency measured at peak amplitude is 5719–6188 (mean = 5996, SD = 131) Hz and is contained within the fundamental frequency. The fundamental frequency has a lower limit of 4688–6000 (mean = 5714, SD = 222) Hz and a higher limit of 5906–6656 (mean = 6268, SD = 144) Hz.

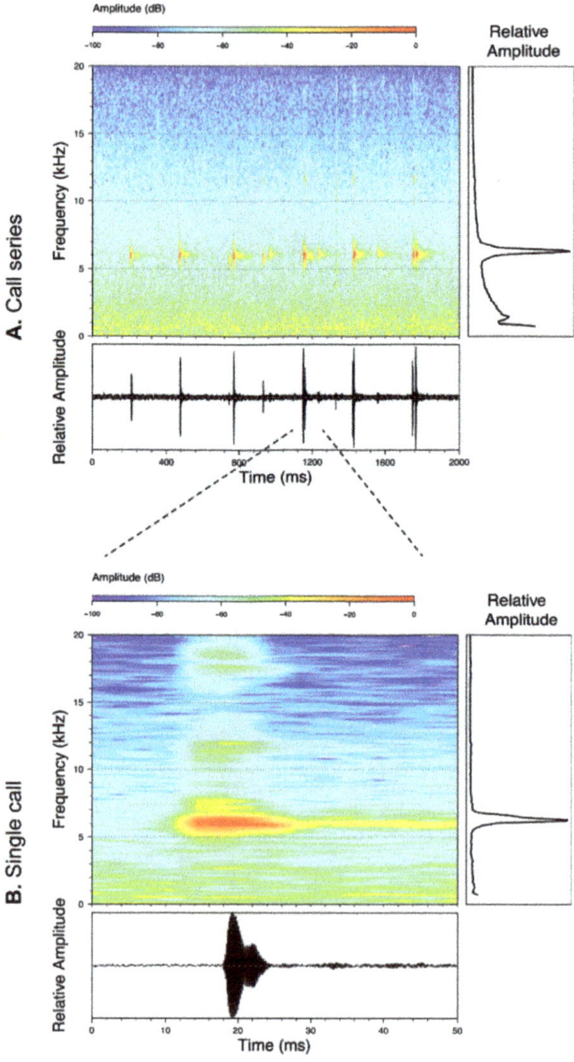

Figure 60. Call of *Centrolene sanchezi* (LBE-C-023), recorded at Reserva Yanayacu, 2150 m, Napo province, Ecuador. (**A**) Series of calls. (**B**) Single, pulsed call.

Tadpole: Not described.

Distribution (Figure 61): *Centrolene sanchezi* is known from a few localities on the Amazonian slopes of the Andes of Ecuador and Colombia and from one locality on the eastern slope of the Colombian Cordillera Central (see Specimens Examined), at elevations between 1800 and 2350 m [20,101,129,137]. In Ecuador, the habitat of the species is within the Eastern Montane Forest region.

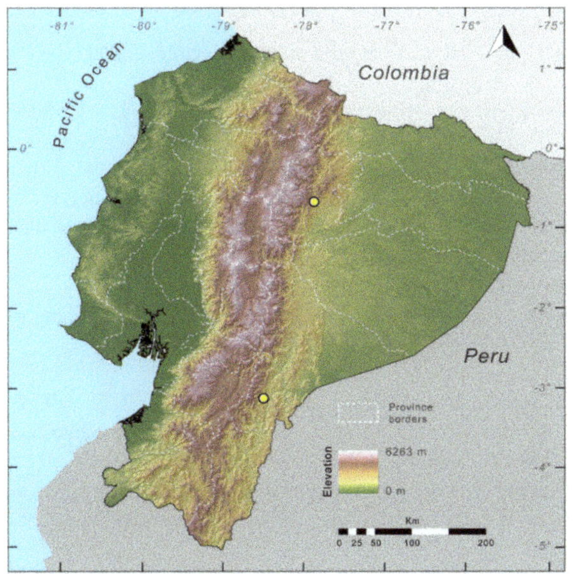

Figure 61. Distribution of *Centrolene sanchezi* in Ecuador (yellow dots).

Conservation status: Globally, *Centrolene sanchezi* is listed as *Data Deficient* by the IUCN [138], but the evaluation does not account for the synonym presented herein. The habitat of the species is fragmented by agriculture and pastureland and threatened by mining (mainly in southern Ecuador). We suggest that the species should be considered as *Endangered*, following IUCN criteria B2, Ba, Bb(iii).

Evolutionary relationships (Figure 24): *Centrolene sanchezi* is the sister species to a clade formed by *C. pipilata* plus *C. hybrida*. These three species are found on the Amazonian slopes of the Andes.

Taxonomic Remarks: Examination of the type material of *Centrolene sanchezi*, *C. bacata*, and *C. guanacarum* reveals no morphological differences among them. Moreover, all specimens share two distinctive traits; a small, laminar humeral spine in males and the presence of a lateral row of enameled warts that extends from below the eye to just posterior to the insertion of the arm (Figure 51). Therefore, we place *Centrolene bacatum* Wild 1994 [137] and *Centrolene guanacarum* Ruiz-Carranza and Lynch 1995 [26] under the synonymy of *Centrolene sanchezi* Ruiz-Carranza and Lynch, 1991 [129].

Specimens examined: *Centrolene sanchezi*: <u>Ecuador</u>: *Provincia de Morona Santiago*: 11.2 km WSW Plan de Milagro (03°07′ S, 78°30′ W; 2350 m), KU 202803 (holotype of *Centrolene bacata*), 202804, 202807–12 (paratypes of *C. bacata*); *Provincia de Napo*: Yanayacu Biological Station (0°41′ S, 77°53′ W; 2100 m), QCAZ 16212, 17807, 22386–87, 22728, 26025–27, 26056, 27438. <u>Colombia</u>: *Departamento de Caquetá*: Municipio de Florencia, Vereda Gabinete, 3.1 Km por carretera abajo del Alto Gabinete, vertiente oriental, Cordillera Oriental, 1°4′ latitud N, 75°4′ W de Greenwich, 2190 m, ICN 24293 (holotype of *C. sanchezi*); *Departamento del Cauca*: Municipio de Inzá, Km. 84 carretera Popayán a Inzá, Río Guanacas, quebrada afluente, Internado Indígena Río Guanacas, vertiente oriental Cordillera Central, 2°34′ Latitud N, 76°05′ W de Greenwich, 1800–1900 m, ICN 11685 (paratype of *C. guanacarum*).

Genus *Chimerella* Guayasamin, Castroviejo-Fisher, Trueb, Ayarzagüena, Rada, & Vilà, 2009 [1].

Etymology: The name *Chimerella* comes from the Greek *Khímaira* and the suffix–*ella* diminutive. In Greek mythology, the Chimera is a creature composed of parts of multiple animals; the name refers to the peculiar combination of morphological characteristics of *Chimerella mariaelenae* [1].

Chimerella mariaelenae (Cisneros-Heredia and McDiarmid, 2006 [139]; Figures 62–65).

Centrolene mariaelenae Cisneros-Heredia and McDiarmid, 2006 [139]. Holotype: DFCH-USFQ
 D125.
Type locality: "small stream, tributary of the Jambue River, ca. 16 km S from Zamora,
 Podocarpus National Park (ca. 04°15′ S, 78°56′ W, 1820 m), on the western slope of
 Contrafuerte de Tzunantza, Cordillera Oriental, eastern slopes of the Andes, Provincia de
 Zamora Chinchipe, Republic of Ecuador".
Cochranella parabambae (in part)—Goin, 1961 [97].
Chimerella mariaelenae—Guayasamin, Castroviejo-Fisher, Trueb, Ayarzagüena, Rada, and Vilà,
 2009 [1].

Common names: English: María Elena's Glassfrog. Spanish: Rana de Cristal de María Elena.

Etymology: The specific epithet *mariaelenae* is a noun in the genitive case and a patronym for María Elena Heredia, Diego F. Cisneros-Heredia's mother [139].

Identification: *Chimerella mariaelenae* (Figures 62 and 63), *Chimerella corleone*, and *Vitreorana gorzulae* are the only known glassfrog that have the following combination of traits: Humeral spine in adult males, transparent ventral parietal peritoneum, and white pericardial, hepatic, and visceral peritonea. *Vitreorana gorzulae*, an endemic to the Guiana Shield, is distinguished by lacking dark spots on the dorsum (present in *C. mariaelenae*). *Chimerella mariaelenae* is easily differentiated from the Peruvian *C. corleone* by having an orange to reddish iris (silvery white in *C. corleone*) and a dorsum with small black dots (dorsum with yellow dots in *C. corleone* [19]). On the Amazonian slopes of Ecuador, only species in the genus *Hyalinobatrachium* and *Teratohyla amelie* could be confused with *C. mariaelenae*. However, these species lack the small dark spots that characterize the dorsum of *C. mariaelenae*. Additionally, adult males of *C. mariaelenae* have small humeral spines, which are absent in all *Hyalinobatrachium* species and *Teratohyla amelie*.

Diagnosis: (1) Vomers lacking teeth; (2) snout truncated in dorsal aspect, truncated to slightly protruding in lateral profile (Figure 63); (3) tympanum oriented posterolaterally with slight dorsal inclination, its diameter 23.0%–27.4% of eye diameter; tympanic membrane translucent, clearly differentiated from surrounding skin; (4) dorsal skin shagreen; males with minute dorsal spicules (only visible under magnification ×250); (5) pair of enlarged subcloacal warts; (6) ventral parietal peritoneum transparent (condition P0); iridophores in pericardium and peritonea covering digestive tract and testes; kidneys and urinary bladder lacking iridophores (condition V5); (7) liver bulbous and covered by iridophores (condition H2); (8) males with small humeral spines; (9) webbing absent or basal between inner fingers, moderate between outer fingers (Figure 63); webbing formula IV $(2^+–2^{1/2})$—$(2^+–2^{1/2})$ V; (10) webbing between toes extensive; webbing formula on foot I $(1^{1/2}–2^-)$—$(2–2^{1/2})$ II 1—2^+ III $(1^+–1^{1/2})$—$(3^-–3^+)$ IV $(2^{1/2}–3^{1/4})$—$(1^+–1^{1/4})$ V; (11) ulnar and tarsal folds absent; (12) concealed prepollex; nuptial pad Type I; (13) Finger I slightly longer than Finger II (Finger II length 95.6%–97.3% of Finger I); (14) disc of Finger III relatively narrow, its width about 22.3%–25.7% of eye diameter; (15) in life, dorsum yellowish green with small grey to black spots (Figure 62); bones light green; (16) in preservative, dorsum pale lavender with small, dark, lavender spots; (17) in life, iris white with abundant dark flecks, pupil surrounded by a rufous to orange ring, outlined by a dark grey to blue ring (Figure 62); (18) dorsal surfaces of fingers and toes lacking melanophores, except for some on base of Toe V; (19) males call from the upper surface of leaves; typical call with two notes; each note extremely short at 4–7 (mean = 6, SD = 0.9) ms; dominant frequency measured at peak amplitude is 6718–8010 (mean = 7510, SD = 408) Hz; (20) fighting behavior unknown; (21) eggs laid on the upper surfaces of leaves; short-term maternal care present; parental care by males absent; (22) tadpoles with elongated body, head with dorsoventral compression; labial tooth row formula 2(1)(2)/3; (23) minute body size; adult males, SVL 17.9–19.7 mm (*n* = 3); one adult female SVL 20.8 mm.

Color in life (Figure 62): Dorsal surfaces yellowish green, with several grey to black spots. No white coloration is evident on upper lip or ulnar and tarsal regions. Iris bicolored, with dark brown circumpupilar area separated from pupil by orange to red–brown ring; whitish external background with abundant dark flecks on uppermost and lowermost portions of iris. Venter completely transparent (clear ventral parietal peritoneum), showing white heart (covered by white pericardium); white digestive tract, liver, testes, and gall bladder. Transparent peritoneum covering urinary bladder. Bones white to light green.

Color in ethanol: Dorsal surfaces of head, body, and limbs pale lavender, with small dark lavender spots. Ventral parietal peritoneum and kidneys lack iridophores. White pericardial, hepatic, and visceral peritonea.

Figure 62. *Chimerella mariaelenae* in life. (**Top row**): Adult male from Bigal River Biological Station, 931 m, Ecuador; photos by Ross Maynard. (**Bottom row**): Amplectant pair, Quebrada Pangayaku, 930 m, Napo province, Ecuador, MZUTI 1680–81; photo by Eduardo Toral.

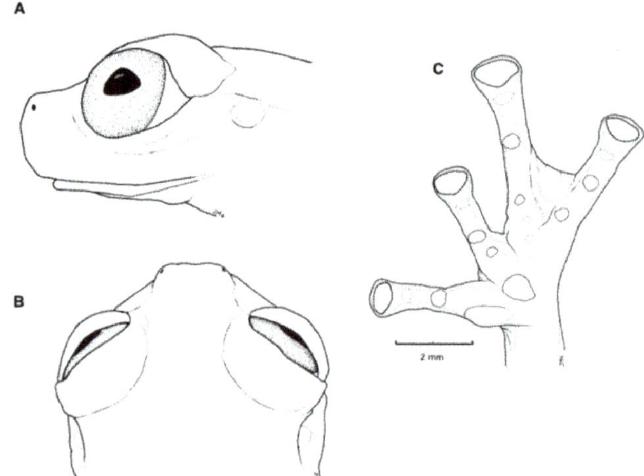

Figure 63. *Chimerella mariaelenae.* (**A**) Head in lateral view, QCAZ 31729. (**B**) Head in dorsal view, QCAZ 21252. (**C**) Hand in ventral view, QCAZ 22363. Illustrations by Juan M. Guayasamin.

Biology and ecology: The biology of *Chimerella mariaelenae* is poorly known. All individuals have been found on the upper surfaces of leaves along small streams and ditches in cloud forest. At the type locality, *C. mariaelenae* is sympatric with *Nymphargus cochranae* and *Hyloscirtus phyllognathus* [139]. Reproductively active individuals have been observed nearby Cascada San Rafael in Quebrada Pangayacu in August 2012. Short-term maternal care present; parental care by males absent [25].

Tadpole: The following information is summarized from Terán-Valdez and Guayasamin [140] and is based on a specimen at Gosner Stage 39 raised in laboratory conditions. Body elongated and oval-depressed, wider (body width = 5.1 mm) than higher (body height = 3.8 mm). Chondrictial elements not visible. Snout rounded in dorsal and lateral views. Lateral line system visible, formed by several stitches parallel or perpendicular to longitudinal axis of body. Short, single, sinistral spiracle located at posterolateral region of body; spiracular aperture with dorsoposterior orientation, with inner wall present as a low ridge. Vent tube short and abdominal, free posteriorly, opening directed posteriorly. Tail long, with subacute tip. Myotomes visible throughout length of tail; straight medial line visible, separating dorsal and ventral myotomes. Dorsal fin originating at about mid-length of tail; height relatively uniform until distal end, where it decreases abruptly. Proximally, ventral fin originating at base of tail muscle, reaching its maximum height after mid-length of tail. Oral disc non-emarginated and surrounded by 49 marginal uniserial papillae. Only ventral and lateral papillae present, lacking dorsal papillae. Upper and lower jaw sheaths nearly straight and fully keratinized, with serrated edge. Labial tooth row formula 2(1)(2)/3; gap in tooth row A-1 could be artificial (because of teeth loss). In life, dorsally, tadpoles at Stage 39 brownish with two areas without pigmentation at anterolateral border of eye. Reddish coloration visible on anterior half of dorsum because of its transparency. Anterior-most part of head is grey. Iridophore aggregations present on dorsum, along vertebral column. Iris bronze. Ontogenetic variation is provided in Terán-Valdez and Guayasamin [140].

Call (Figure 64): We analyzed 26 notes contained within six calls from one individual (LBE-C-021). The typical advertisement call is variable from being short to a moderate length call (range = 231–1761 ms, mean = 679 ms, SD = 623.3 ms). The typical call has two notes, but some call can have up to 10 (mean = 4.3, SD = 3.4) notes. Each note is extremely short at 4–7 (mean = 6, SD = 0.9) ms. Notes have a single amplitude peak. The dominant frequency measured at peak amplitude is 6718–8010 (mean = 7510, SD = 408) Hz and is contained within the fundamental frequency. The fundamental frequency has a lower limit of 6460–7752 (mean = 7222, SD = 387) Hz and a higher limit of 7063–8441 (mean = 7828, SD = 398) Hz.

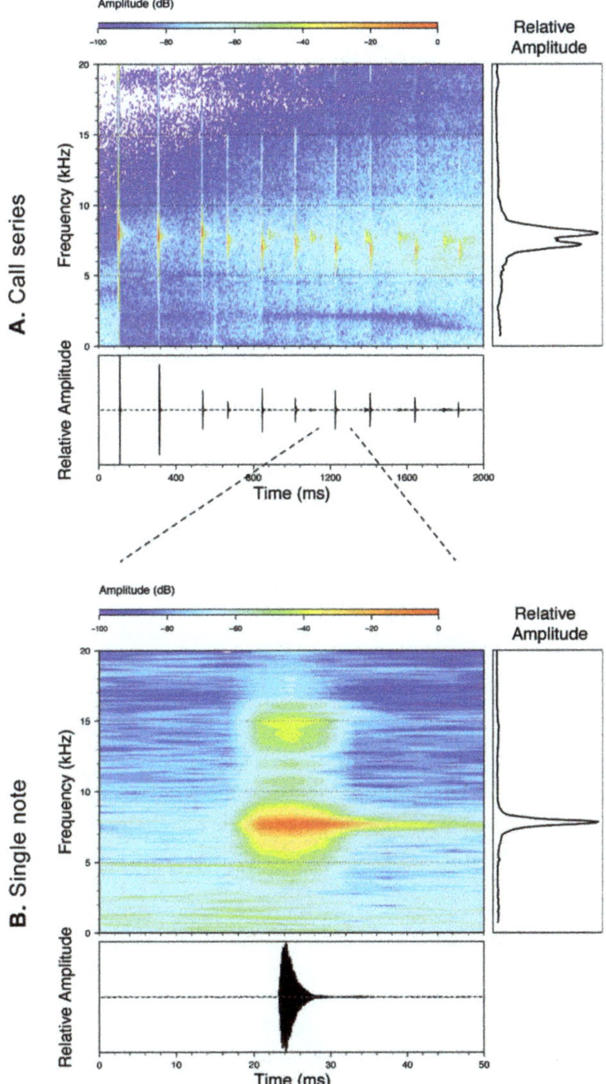

Figure 64. Call of *Chimerella mariaelenae* (LBE-C-021), recorded at Pangayaku creek, 929 m, Napo province, Ecuador. (**A**) Series of calls. (**B**) Single, pulsed note.

Distribution (Figure 65): *Chimerella mariaelenae* occurs on the Amazonian slopes of the Ecuadorian Andes at elevations between 813 and 1820 m ([139,141,142], this work), and was recently reported from Peru [143]. The species is known from a few localities in the provinces of Napo, Morona Santiago, Orellana, Tungurahua, and Zamora Chinchipe (Specimens Examined). The potential distribution of the species is 25,472 km^2 within the Eastern Foothill and Montane Forest regions.

Figure 65. Distribution of *Chimerella mariaelenae* in Ecuador (yellow dots).

Conservation status: Globally, *Chimerella mariaelenae* is currently listed as *Least Concern* by the IUCN [144]. The species is threatened by human activities, although it tolerates some habitat disturbance and has a relatively large distribution, including populations within protected areas (i.e., Bigal River Biological Station, Parque Nacional Podocarpus, Parque Nacional Sangay, Reserva Ecológica Cayambe-Coca, Reserva Narupa). In Ecuador, major threats for the species include deforestation and mining; thus, we suggest that the species should be considered as *Near Threatened* at the local level.

Evolutionary relationships (Figures 19 and 66): In the original description and based on morphological data, *Chimerella mariaelenae* was assigned to the *Centrolene gorzulai* group, which, otherwise, consisted of species from the Guiana Shield. Under this hypothesis, the distribution of *C. mariaelenae* was thought to support a biogeographical connection between the Andes and the Guiana Shield [139]. Subsequently, Cisneros-Heredia and McDiarmid [17], based on differences in the chromatophore organization, considered that the relationships of *C. mariaelenae* were uncertain. Phylogenetic trees based on molecular data show that *Chimerella mariaelenae* and *C. corleone* are sister species [19] (Figure 66); both species are restricted to the Amazonian versant of the Andes, with no close relationship with species from the Guiana Shield. The genus *Chimerella* is herein inferred as sister to *Espadarana* (Figure 19), but with low nodal support.

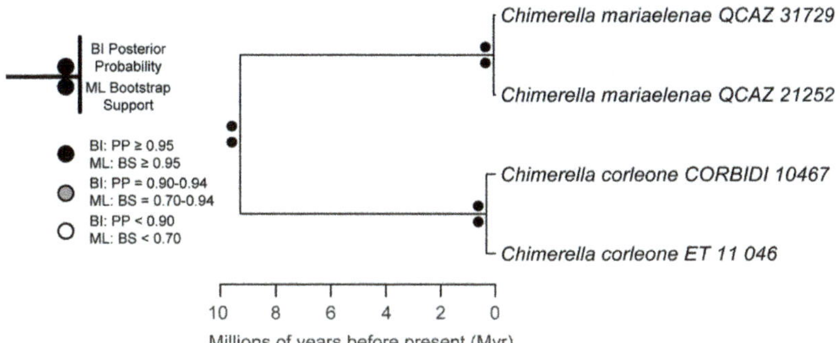

Figure 66. Evolutionary relationships between species in the genus *Chimerella*, inferred using maximum likelihood and Bayesian criteria.

Remarks: One of the specimens reported by Goin [97] as *Centrolenella parabambae* from the Topo River is an adult female *Chimerella mariaelenae*.

Specimens examined: *Chimerella mariaelenae:* <u>Ecuador</u>: *Provincia de Napo:* Río Hollín (ca. 00°58′ S, 77°45′ W, ca. 1400 m), QCAZ 18618–19, 22363. Quebrada Pangayaku (0.78255 S, 77.791 W, 929 m), MZUTI 1680–81; Reserva Narupa (0.684 S, 77.741 W, 1179–1208 m), ZSFQ 0437–38; Reserva Narupa (0.671 S, 77.774 W, 1502 m), ZSFQ 0439–41. *Provincia de Morona Santiago:* 6.7 km W of 9 de Octubre (02°13′30.5″ S, 78°17′25.6″ W, 1715 m), QCAZ 32643. *Provincia de Tungurahua:* near Río Negro (01°24′ S, 78°15′ W, 1423 m), on the Río Negro–Río Verde road, QCAZ 21252, 31729; Río Topo, BM 1912.11.1.69. *Provincia de Zamora Chinchipe:* stream tributary of the Río Jambue (ca. 4°15′ S, 78°56′ W, ca. 1820 m), ca. 16 km S from Zamora, Parque Nacional Podocarpus, DFCH-USFQ D125; gravel road E to Sarsa (3.80783976 S, 78.60593076 W; 1500 m), QCAZ 47053.

Photographic records: *Provincia de Orellana:* Reserva Río Bigal (0.532913° S, 77.423228° W; 981 m). Photo by Morley Read.

Genus *Cochranella* Taylor, 1951 [15].

Etymology: Named in honor of Doris M. Cochran, herpetologist and curator of the Smithsonian Institution, Washington, D.C.

Cochranella granulosa (Taylor, 1949 [145]; Figures 67 and 68).

Centrolenella granulosa Taylor, 1949 [145]. Holotype: FMNH 178269.
Type locality: "Los Diamantes, one mile south of Guápiles, (Cantón de Pococí, Provincia Limón), <u>Costa Rica</u>".
Cochranella granulosa—Taylor, 1951 [15].

Common names: English: Granular Glassfrog. Spanish: Rana de Cristal Granulosa.
Etymology: The specific epithet *granulosa* comes from the Latin word *granum* (grain), in allusion to the granular dorsal skin texture of the species.
Identification: *Cochranella granulosa* can be distinguished from most glassfrogs by having a dark green to bluish–green dorsum with whitish granules and faint to well-defined dark spots (Figure 67), yellowish–green hands and feet, a reduced white ventral parietal peritoneum (which covers less than

40% of the venter), white digestive tract, and by lacking humeral spines. Another distinctive trait is the granular texture of its dorsal skin. Species that might have a sympatric distribution with *C. granulosa* and that present a similar dorsal coloration include *Cochranella litoralis* and some populations of *Espadarana prosoblepon*. However, *C. granulosa* has a granular skin (shagreen in *E. prosoblepon* and *C. litoralis*) and lacks humeral spines (present in males of *E. prosoblepon* and *C. litoralis*). Also, *C. litoralis* has an orange to red iris (yellow to golden in *C. granulosa*) and *E. prosoblepon* has a digestive tract that lacks white iridophores (present in *C. granulosa*).

Figure 67. *Cochranella granulosa* in life. (**Top row**): Reserva Guayacán, Costa Rica; photos by Brian Kubicki. (**Bottom row**): Reserva El Jardín de los Sueños, 461 m, Cotopaxi province, Ecuador, MZUTI 4811; photos by Jaime Culebras (dorsolateral) and Javier Aznar (ventral).

Diagnosis: (1) Vomerine teeth present, each vomer with one to three teeth; (2) snout truncated in dorsal aspect, slightly sloping in lateral profile; (3) tympanum oriented dorsolaterally, with slight posterior inclination, its diameter about 35% of eye diameter; tympanic annulus visible, low supratympanic fold evident, tympanic membrane partially pigmented and clearly differentiated from surrounding skin; (4) dorsum granular, lacking spicules; (5) venter areolate, lacking pair of enlarged subcloacal warts; (6) white lining on the anterior 20%–40% of the ventral parietal peritoneum (condition P1); peritoneum with white iridophores covering digestive tract (condition V2); (7) liver tetralobed; hepatic peritoneum lacking iridophores (condition H0); (8) humeral spines absent; (9) webbing absent between Fingers I–II, vestigial to moderate between Fingers II–III, extensive between III–IV; webbing formula on hand: II $(1^+$–2$)$—$(3$–$3^+)$ III $(1^{1/3}$–$2^{1/2})$—$(1$–$2^+)$ IV; (10) webbing on foot extensive, formula: I 0^+—$1^{1/2}$ II 0^+—$1^{2/3}$ III $(1$–$1^+)$—2^- IV 2—1^- V; (11) ulnar fold low; inner and outer tarsal folds low and thin; (12) concealed prepollex; nuptial pad Type I; (13) Finger I about same length as Finger II; (14) disc of Finger III narrow, about 40% of eye diameter; (15) in life, dorsum dark green to bluish green with fine, yellowish–white granules and faint to clearly marked dark spots (Figure 67); bones green; (16) in preservative, dorsum cream to light lavender with faint to clearly defined dark spots; (17) iris yellow to golden with few, faint brownish streaks; (18) melanophores absent from dorsal surfaces of hands and toes; (19) males call from the upper side of leaves; each call consists of three to five notes, the first note being slightly longer that the others; average duration for a three-note call is 0.8–0.9 s; dominant frequency at 3700 Hz; (20) males fight venter to venter, hanging from the vegetation by the tips of their toes; (21) dark egg are placed on the upper sides of leaves; short-term maternal care present; parental care by males absent; (22) tadpoles with a flattened body and slender tail; tadpole in early stages have a brown pigmentations that is lost in older, red larvae; jaw sheaths are strongly serrated, the upper jaw forms a smooth M-shape; the lower sheath is V-shaped; labial tooth row formula 2/3; (23) medium body size; in males, SVL 22.5–29.0 mm; in females, SVL 29–32 mm.

Color in life (Figure 67): Dorsal surfaces green to bluish green, with yellowish or bluish granules and faint to clearly marked blue to black spots. Anterior 20%–40% of venter white, posterior portion translucent. Iris yellow to golden. Pale or whitish lip in populations from Central America (Savage, 2002; Kubicki, 2007). One individual from Ecuador (MZUTI 4811) lacks dark dorsal spots.

Color in ethanol: Dorsum cream to light lavender to dark lavender with faint to clearly marked dark spots (the Ecuadorian specimen lacks dark spots, MZUTI 4811); iridophores on the anterior 20%–40% of the ventral parietal peritoneum; pericardium and gastrointestinal peritonea with white iridophores; hepatic peritoneum lacking iridophores.

Biology and ecology: The information below is from Kubicki [24] and RWM (pers. obs.) *Cochranella granulosa* is active at night in riparian habitats of streams through pristine tropical forest, but also sometimes along streams in disturbed areas. Males fight venter to venter, hanging from the vegetation by the tips of their toes. Reproduction usually occurs in riparian habitats, but Kubicki [24] also reported reproductive activity in a small forest pond about 10 m away from the nearest stream. Egg clutches consist of 40 to 104 eggs [24,146,147] that are attached on masses on the upper distal half of a leaf overhanging the stream. Following hydration, the eggs form a pendent mass (i.e., drip tip) that hangs off the leaf and thereby ensures a constant flow of water over the developing larvae [4,146]. There is no long-term parental care in this species [148]. Delia et al. [25] experimentally demonstrated that females provide short-term parental care (brooding) to their egg clutches just after oviposition; this behavior, although brief (i.e., few hours), reduces embryo mortality from dehydration.

Call: Males often call in a chorus, with one male initiating calling and then being answered by several males. The call consists of a rapid high-pitched pulsed trill, usually composed by three notes (range three to five notes; [24]). Usually, the first note is longer (average duration = 0.3 s) than the subsequent notes (average duration = 0.15 s). The mean duration for a three-note call is 0.8–0.9 s [24]. The dominant frequency is 3700–4500 Hz [24,149,150]. In Ecuador (Los Laureles, Cotopaxi province) males were actively calling in February 2016 (Jaime Culebras, pers. obs.).

Tadpoles: The tadpole of *Cochranella granulosa* was first described by Starrett [146]. Hoffmann [147] recently provided an additional and detailed description; below, we present a summary based on Hoffmann's description. Tadpoles of *C. granulosa* and *C. euknemos* share a particular slim shape. Recently hatched tadpoles (one to five days old) are markedly slender and dark brown. In the laboratory, older tadpoles are slender with a very long tail; they lack much body pigment and their bright red coloration is from the highly vascularized internal anatomy anterior to the body cavity and visible through the body wall. The oral disc is not emarginate and is bordered laterally and along the posterior margin with 28–41 marginal papillae (mean = 33 ± 4 papillae). Jaw sheaths are strongly serrate; the upper jaw forms a smooth M-shape, and the lower sheath is V-shaped. Generally, all five tooth rows are well developed; LTRF is 2(2)/3; the A2 row has a large central gap with the two halves extending as short wings on each side of the lateral descendent ends of the upper jaw sheath.

Distribution (Figure 68): The distribution of *Cochranella granulosa* includes the humid lowland and premontane slopes from the Atlantic drainage of eastern Honduras to central Panama and on the Pacific versant in humid upland or gallery forests from northern Costa Rica to northern Ecuador, 40–1500 m elevation ([148,151], this work). Records of *Cochranella granulosa* in Ecuador are from El Jardín de los Sueños reserve, Cotopaxi province [151], and Durango, 238 m, Esmeraldas province.

Conservation status: Globally, *Cochranella granulosa* is listed as *Least Concern* by the IUCN [152]. We suggest *Data Deficient* in Ecuador, since it is known from two recent records (Figure 68).

Evolutionary relationships (Figure 69): *Cochranella granulosa* is sister to *C. resplendens*, being an example of allopatric speciation mediated by the uplift of the Andes.

Figure 68. Distribution of *Cochranella granulosa* in Ecuador (yellow dot).

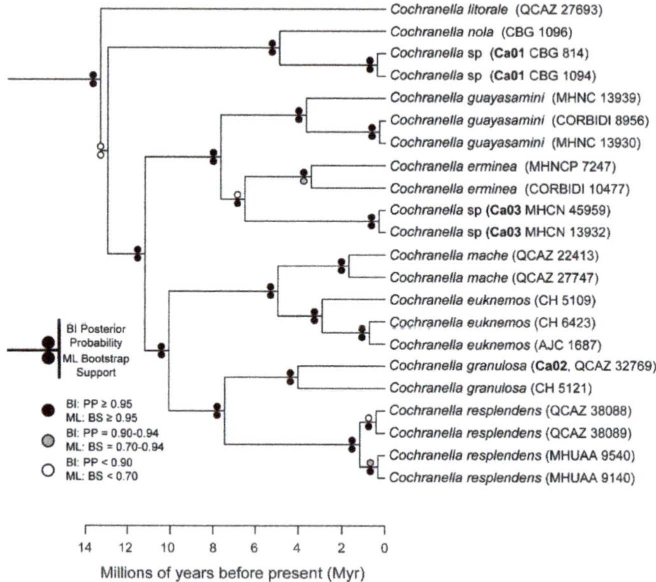

Figure 69. Evolutionary relationships among species in the genus *Cochranella*, inferred using maximum likelihood and Bayesian criteria.

Remarks: Karyotype of *Cochranella granulosa* is 2N = 20 [153].

Specimens examined: *Cochranella granulosa*: Ecuador: *Provincia de Esmeraldas:* 4 km W Durango (1.042° N, 78.1081° W, 253 m), QCAZ 32769. *Provincia de Cotopaxi:* Reserva El Jardín de los Sueños (0.8416° S, 79.2006° W, 461 m), MZUTI 4811.

Cochranella litoralis (Ruiz-Carranza and Lynch, 1996 [154]; Figures 70–73).

Centrolene litoralis Ruiz-Carranza and Lynch, 1996 [154]. Holotype: ICN 13821.
Type locality: "Departamento de Nariño, municipio de Tumaco, La Guayacana, Litoral Pacífico, 1°49.8′ Latitud N, 78°46.2′ W de Greenwich, 100 m, (Colombia)".
Centrolene litorale—Frost, 2004 [155].
Cochranella litoralis—Guayasamin, Castroviejo-Fisher, Trueb, Ayarzagüena, Rada, and Vilà, 2009 [1].

Common names: English: Litoral Glassfrog. Spanish: Rana de Cristal del Litoral.

Etymology: The specific epithet *litoralis* makes reference to the proximity of the type locality to the ocean [154].

Identification: *Cochranella litoralis* is the only species in Centrolenidae that has a bright orange to red iris (Figures 70 and 71) and adult males with humeral spines and a nuptial pad that partially covers the two innermost fingers (Figure 72). Additionally, its dorsum is pale yellow green with small grey to black dots and faint yellow–cream dorsolateral stripes. Males are small (SVL < 20 mm) and have humeral and distinct prepollex. In the Chocoan lowlands, only some populations of *Espadarana prosoblepon* and *Nymphargus griffithsi* have a green dorsum with dark dots and humeral spines. Both *E. prosoblepon* and *N. griffithsi* are considerably larger, have a cream to white iris, and lack prepollical spines. The Peruvian *Cochranella guayasamini* also has a red iris and humeral spines, but lacks dark dorsal spots [19].

Figure 70. *Cochranella litoralis* in life. Adult male, QCAZ 27693, from near Durango, 220 m, Esmeraldas province, Ecuador. Photos by Luis A. Coloma.

Diagnosis: (1) Vomers lacking teeth; (2) snout truncated in dorsal and lateral profiles (Figure 72); (3) tympanum of moderate size, oriented almost vertically, with slight lateral and posterior inclinations, its diameter about 30% of eye diameter; tympanic annulus mostly visible, with supratympanic fold covering its posterodorsal margin; tympanic membrane translucent, partially pigmented, clearly differentiated from surrounding skin; (4) dorsal skin smooth anteriorly and shagreen in sacral region, lacking spicules; (5) lacking pair of enlarged subcloacal warts; (6) anterior half of ventral parietal peritoneum with iridophores, posterior half translucent (condition P2); iridophores in pericardium and gastrointestinal peritoneum; no iridophores on testes, gall bladder, urinary bladder, and kidneys (condition V2); (7) liver with clearly defined lobes covered by transparent peritoneum (condition H0); (8) males with small and pointy humeral spines; (9) webbing absent between Fingers I and II, absent or vestigial between Fingers II and III, moderate between outer fingers; webbing formula IV 2—2⁻ V; (10) webbing between toes moderate; webbing formula on foot I 1—2 II 1—2⁺ III 1—2¹ᐟ² IV 2¹ᐟ²—1 V; (11) ulnar and tarsal folds absent or low and inconspicuous, lacking white coloration; (12) distinct prepollex; nuptial pad Type V; (13) Finger I as long as Finger II; (14) disc of Finger III width about

33% of eye diameter; (15) in life, dorsum green usually with small dark spots and faint yellow–cream dorsolateral stripes (Figures 70 and 71); pale green bones; (16) in preservative, dorsum lavender with dark spots; (17) in life, iris orange to red; (18) dorsal surfaces of fingers and toes lacking melanophores, except for some on base of Toe V; (19) calling behavior unknown; call undescribed; (20) fighting behavior unknown; (21) egg deposition site unknown; parental care unknown; (22) tadpoles unknown; (23) minute body size; in two adult males, SVL 19.4–20.0 mm; females unknown.

Figure 71. *Cochranella litoralis* in life. Male from Tundaloma Lodge (1.182 N, 78.749 W; 74 m), Esmeraldas province, Ecuador (3 January, 2014; not collected). Photo by Lucas Bustamante/Tropical Herping.

Color in life (Figures 70 and 71): *Cochranella litoralis* has a pale yellowish–green dorsum with dark spots that vary in size and conspicuousness; the anterior half of the venter is white, turning transparent posteriorly; the iris is bright orange to red.

Color in ethanol: Dorsal surfaces cream lavender with small dark spots, and two faint white dorsolateral lines. White parietal peritoneum covering anterior half of venter. White pericardium; white peritonea covering stomach and lower colon. Liver, testes, kidneys, and urinary and gall bladders covered by translucent peritonea. Iris silvery white, with black punctuations.

Biology and ecology: A male of *Cochranella litoralis* and an egg clutch presumed to be from the same species were found on riverine vegetation (L. Bustamante, pers. comm.). The egg clutch was placed on the upper side of a leaf and contained 25 embryos (Figure 71). Parental care unknown.

Call: Not described.

Tadpole: Not described.

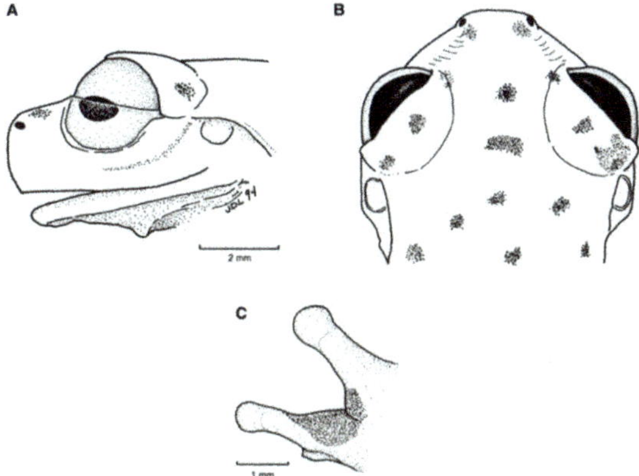

Figure 72. *Cochranella litoralis*, holotype, adult male, ICN 13821. (**A**) Head in lateral view. (**B**) Head in dorsal view. (**C**) Fingers I and II in dorsal view; note that nuptial pad extends on the two fingers; illustrated by Juan M. Guayasamin. (**A**,**B**) Modified from Ruiz-Carranza and Lynch [154].

Distribution (Figure 73): *Cochranella litoralis* has been reported from the type locality in Colombia and four localities in Ecuador at elevations below 260 m ([130,154], this work). In Ecuador, this species has a potential distribution of 5784 km^2 within the Chocoan Tropical Forest region.

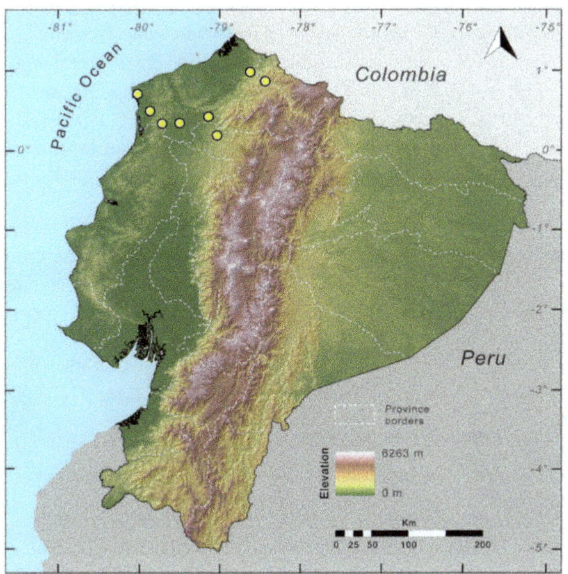

Figure 73. Distribution of *Cochranella litoralis* in Ecuador (yellow dots).

Conservation status: Globally, *Cochranella litoralis* is currently listed as *Vulnerable* by the IUCN [156]. In Ecuador, it has a distribution restricted to the provinces of Esmeraldas and Cotopaxi (Chocó ecoregion), where logging is causing continuous habitat reduction and fragmentation. The most recent reports of this species are from Río Cachabí (September 2005), Tundaloma (January 2014), and Jardín de los Sueños (2018). In Ecuador, because of habitat loss and mining, we suggest that the species should be considered as *Endangered*, following IUCN criteria B2a, B2(iii).

Evolutionary relationships (Figure 69): *Cochranella litoralis* is sister to all other member of the genus *Cochranella*, but with low nodal support. Twomey et al. [19] recovered a sister relationship between *C. litoralis* and *C. nola*.

Specimens examined: *Cochranella litoralis:* <u>Colombia</u>: *Departamento de Nariño*: Municipio de Tumaco, La Guayacana (1.83 N, 78.77 W; 100 m), ICN 13821. <u>Ecuador</u>: *Provincia de Esmeraldas:* stream near Durango (1.047 N, 78.618 W; 220 m), QCAZ 27693; Pichiyacu, Comunidad Chachi, Río Cayapas (0.9397° N, 79.005° W; 260 m), QCAZ 31705; Río Cachabí (1.03 N, 78.77 W, 200 m), 2 km NE Urbina on the San Lorenzo—Lita road, DHMECN 3198; Tundaloma Lodge (1.17868° N, 78.7497° W; 74 m), MZUTI 3481.

Localities from the literature: *Cochranella litoralis:* Ecuador: *Provincia de Esmeraldas:* Tsejpu, Río Zapallo (0.7 N, 78.9 W, 150 m) [156].

Cochranella mache Guayasamin and Bonaccorso, 2004 [157] (Figures 74–76).

Cochranella mache Guayasamin and Bonaccorso, 2004 [157]. Holotype: QCAZ 22412.

<u>Type locality:</u> "Riachuelo La Ducha (0°20'41" N, 79°42'36" W; 510 m), tributary of Río Aguacatal, Reserva Biológica Bilsa, 27.4 km W (airline distance) of the town of Quinindé, Montañas del Mache, Provincia Esmeraldas, Ecuador".

Common names: English: Mache Glassfrog. Spanish: Rana de Cristal de Mache.

Etymology: The specific name *mache* is a noun in apposition and refers to the Montañas de Mache, the type locality of the species [157].

Identification: *Cochranella mache* is differentiated from most glassfrogs by having a snout gradually inclined in profile and dermal folds with white tubercles on the ventrolateral edges of Finger V, forearm, elbow, Toe V, tarsus, and heel (Figures 74 and 75). Species sharing similar characteristics are: *Centrolene daidalea, C. savagei, C. solitaria, Cochranella resplendens,* and *C. euknemos.* Three of these species (*C. daidalea, C. savagei,* and *C. solitaria*) are restricted to the Andes, and *Cochranella resplendens* is found only in the Amazon Basin. The closely related *C. euknemos* lacks the tubercles that are characteristic in *C. mache* [157].

Figure 74. *Cochranella mache* in life. Adult male, QCAZ 27764, from Río La Carolina, 500 m, Esmeraldas province, Ecuador. Photos by Luis A. Coloma.

Diagnosis: (1) Each vomer with three or four teeth on dentigerous process; (2) snout subacuminate in dorsal aspect and gradually inclined in lateral profile (Figure 75); (3) tympanum oriented almost vertically, with slight lateral and posterior inclinations, its diameter about 33%–37% of eye diameter; tympanic annulus visible, low supratympanic fold evident, tympanic membrane translucent and pigmented as surrounding skin; (4) dorsal skin shagreen with warts that usually correspond to light spots; males with numerous minute spicules; (5) ventral skin granular; several round, enameled warts around cloaca;

cloacal fold present; pair of enlarged subcloacal warts; (6) anterior one-third of parietal peritoneum covered by iridophores (condition P1); iridophores over pericardium and visceral peritonea (digestive tract, gonads); renal capsules covered by iridophores in some individuals (condition V2); (7) liver tri- or tetralobed, hepatic peritoneum clear or with small, isolated patches of iridophores (condition H0); (8) humeral spines absent; (9) webbing absent between Fingers I and II, reduced between Fingers II and III, and extensive between outer fingers (Figure 75); webbing formula II (1⁻–1⁺)—(3⁻–3⁺) III (2⁻–2)—(1–1⁺) IV; (10) webbing between toes extensive (Figure 75); webbing formula on foot I (1⁻–1)—(2–2⁻) II 1—2 III (1–1⁻)—2⁻ IV (2⁻–2)—(1–1⁻) V; (11) ventrolateral edges of Finger IV, forearm, elbow, Toe V, tarsus, and heel with dermal enameled folds and tubercles; (12) concealed prepollex; nuptial pad large; (13) Finger I about same length as Finger II (Finger I about 94%–98% of Finger II); (14) disc of Finger III width about 44%–54% of eye diameter; (15) color in life, dark olive-green to lavender–blue or pale blue dorsum with numerous small yellow to orange spots, large dull to bright yellow patch on top of head (Figure 74); (16) color in preservative, dorsal surfaces pale lavender with small white or cream spots; tubercles on dermal folds of limbs, fingers, and toes cream-white; inner fingers and toes white or unpigmented; (17) in life, iris whitish cream to beige with thin brown reticulation; white to golden circumpupilary ring; (18) melanophores covering dorsal surfaces of Fingers III and IV and Toes IV and V; (19) males call from the upper sides of leaves; each call formed by two notes; dominant frequency at 5383–5426 Hz; (20) fighting behavior unknown; (21) brown eggs placed on upper sides of leaves; parental care unknown; (22) tadpoles unknown; (23) small body size; in adult males, SVL 22.0–27.0 mm (*n* = 43); in adult females, SVL 28.0–33.4 mm (*n* = 4).

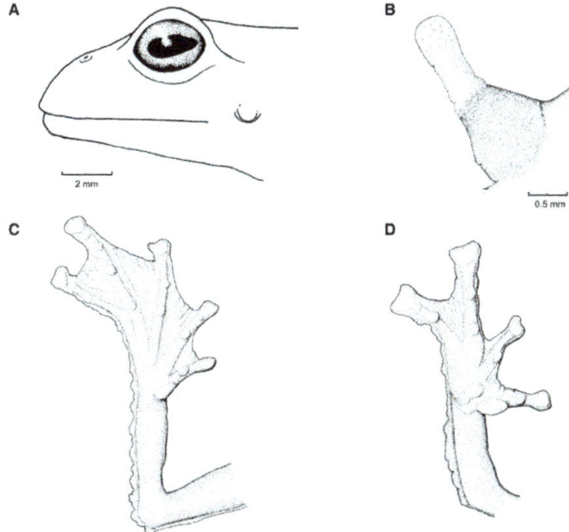

Figure 75. *Cochranella mache*. (**A**) Head in lateral view, paratype, adult male, KU 291176. (**B**) Nuptial excrescences on dorsal surface of Finger I, paratype, adult male, QCAZ 22413. (**C**) Foot in ventral view, KU 291176. (**D**) Hand in ventral view, KU 291176. Hand and foot not drawn to scale. Modified from Guayasamin and Bonaccorso [157].

Color in life (Figure 74): Dorsal surfaces vary from green to lavender blue or pale blue, with small yellow to orange spots. Upper lip with a thin, white margin. White tubercles visible on dermal folds of hind- and forelimbs. Innermost fingers and toes (Fingers I and II, Toes I and II) white in males and partially white in females. Throat and ventral surfaces of limbs blue green. White parietal peritoneum covering anterior half of venter, posterior portion translucent; white pericardial and visceral peritonea. Transparent hepatic peritoneum, but in some individuals, a small patch of white iridophores covers part of liver. White warts surround cloaca. Iris whitish cream to beige with thin brown reticulation; white to golden ring surrounding pupil. Bones green ([157,158], this work).

Color in ethanol: Dorsum of head, body, and limbs lavender with small white or cream spots. Anterior one-third of parietal peritoneum covered by iridophores; iridophores also cover pericardium and visceral peritonea (digestive tract and gonads); hepatic peritoneum lacking iridophores or with small, isolated patches of iridophores; renal capsules covered by iridophores in some specimens ([157,158], this work).

Biology and ecology: The information shown below was obtained mainly from Ortega-Andrade et al. [159] and, to a lesser extent, from Guayasamin and Bonaccorso [157]. *Cochranella mache* seems to be restricted to small streams and rivulets in primary and secondary forests in lowlands and piedmont forest. Abundance of the species is correlated with rainfall, being higher during rainy season. Males call from the upper surface of leaves or branches on trees and bushes; reproduction seems to be restricted to the rainy season. Although individuals have been observed on different strata of the forest up to 6 m high, *C. mache* prefers the midstory vegetation. Amplexus is axillary and a gravid female was observed to have about 30 eggs. The diversity of amphibians and reptiles at the type locality (Reserva Biológica Bilsa) of *C. mache* is described by Ortega-Andrade et al. [160]. Parental care unknown.

Call: The advertisement call of *Cochranella mache* was described by Ortega-Andrade et al. [159]. The recorded male was calling from the upper side of a dead *Heliconia* leaf, about 2.5 m above ground, horizontally separated from the stream by about 3m. A call consists of two notes, separated by an interval of 0.107–0.130 s. Each note has a duration of 0.038 ± 0.008 (0.029–0.049) s. The dominant frequency is at 5410.2 ± 17.9 (5383–5426) Hz. The harmonics are not visible. The call rate is 1.46 calls per minute.

Tadpole: Not described.

Distribution (Figure 76): *Cochranella mache* has been reported from few localities in northwestern Ecuador, provinces of Esmeraldas, Imbabura, and Manabí, at elevations between 38 and 800 m ([157–159], this work), and from the Pacific lowlands of Colombia, Departments of Antioquia and Valle del Cauca, at elevations of 750–1030 m [161]. Three localities lie within protected areas (Reserva Biológica Bilsa, Reserva Biológica Canandé, Reserva Jama-Coaque). In Ecuador, the potential distribution of *C. mache* is 27,433 km^2 (see also Ortega-Andrade et al. [159]) within the Chocoan Tropical Forest and the Western Foothill Forest regions. When deforestation is taken into account, ~70% of the predicted distribution is reduced [159].

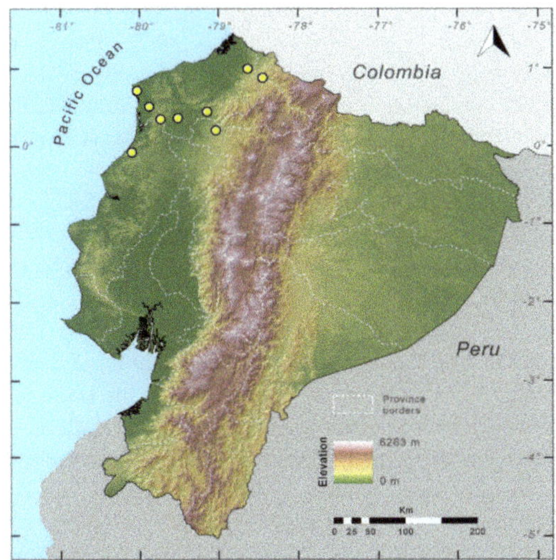

Figure 76. Distribution of *Cochranella mache* in Ecuador (yellow dots).

Conservation status: Listed as *Endangered* by the IUCN at a global level [162]. Other studies suggest that the species should be classified into the category of *Critically Endangered* [159,163]. The distribution of the species is likely to correspond to the remaining lowland forest in northwestern Ecuador and southwestern Colombia, an area under the constant pressure by wood companies. Although the range of *C. mache* is partially within the Mache-Chindul Ecological Reserve and three private reserves (Reserva Biológica Bilsa, Reserva Biológica Canandé, Reserva Jama-Coaque), most conservation measures are ineffective because of institutional and funding restrictions and a lack of law enforcement. Some of the larger fragments are preserved by private organizations but many remain unprotected. Habitat degradation is mainly caused by unsustainable timber extraction, uncontrolled expansion of the agricultural frontier, and replacement by non-native plantations [159,163]. Records from Colombia, however, expand considerably the distribution of *C. mache*. More than half of this distribution area of the species is affected by human activities [159]. Under scenarios of climate changes, the potential distribution of *C. mache* suffers a reduction of 13%–21% of its predicted range [159].

Evolutionary relationships (Figure 69): *Cochranella mache* is the sister species of *C. euknemos*.

Specimens examined: *Cochranella mache:* <u>Ecuador</u>: *Provincia de Esmeraldas:* Montañas de Mache, Reserva Biológica Bilsa, Riachuelo La Ducha (0°20′41″ N, 79°42′36″ W; 510 m), tributary of Río Aguacatal, 27.4 km W (airline distance) of the town of Quinindé, QCAZ 22412 (holotype), QCAZ 22413, KU 291176; Río Balthazar (0.9745 N, 78.61675 W; 645 m), QCAZ 27747, 31327; Monte Saíno, Punta Galeras region (0.700 N, 80.017 W; 100 m), DHMECN 2611; 3 km NW of Quinindé (0.350 N, 79.483 W; 150 m), DFCH-USFQ LQ23; Reserva Biológica Canandé (0.433 N, 79.133 W; 270 m), DHMECN 3560. *Provincia de Imbabura:* Río La Carolina (0.70449 N, 78.20115 W; 500 m), on the Ibarra–San Lorenzo Road, nearby Jijón y Caamaño, QCAZ 27764. <u>Colombia</u>: *Departamento de Antioquia:* Municipio Dabeiba, Río Amparradó, Quebrada Iotó, 805 m, ICN 10689–90, 8665; Municipio Frontino, Vereda Venados, PNN Las Oriquideas, Quebrada La Miguera, 1030 m, ICN 19638–9.

Records from the literature: Ortega-Andrade et al. [159]: Ecuador: Comunidad San Salvador (0.496710 N, 79.85298 W; 38 m); Hacienda Shangrilá (0.18630 N, 79.03019 W; 499 m).

Photographic record: Ryan Lynch: Ecuador: *Provincia de Manabí:* Reserva Jama-Coaque (0.0978 S, 80.147 W).

Cochranella resplendens Lynch and Duellman, 1973 [22] (Figures 77–82).

Cochranella resplendens Lynch and Duellman, 1973 [22]. Holotype: KU 118053.
Type locality: "Santa Cecilia, 340 m, Provincia de Napo (Sucumbíos), Ecuador".
Cochranella resplendens—Ruiz-Carranza and Lynch, 1991a. Guayasamin, Castroviejo-Fisher, Trueb, Ayarzagüena, Rada, and Vilà, 2009 [1].
Cochranella phryxa Aguayo and Harvey, 2006 [164]. Holotype: CBG 778. Type locality: "approximately 20 km west of Población de la Cascada (Territorio Comunitario de Origen y Reserva de la Biósfera Pilón Lajas), La Paz Department, Bolivia (15°22′37″ S, 67°12′4″ W), 1000 m". **New synonymy.**

Common names: English: Resplendent Glassfrog. Spanish: Rana de Cristal Resplandeciente.

Etymology: The specific name *resplendens* is derived from the Latin verb *resplendo*, meaning to glitter, and is used in allusion to the jewel-like appearance of this frog [22].

Identification: *Cochranella resplendens* is distinguished from most glassfrogs by having a snout gradually inclined in profile and ventrolateral edges of Finger IV, forearm, elbow, Toe V, tarsus, and heel with dermal folds and enameled tubercles (Figure 77, Figure 78, Figure 81). Additionally, the species lacks humeral spines and has a white venter except for the posterior one-fourth, which is transparent. The following species could be confused with *Cochranella resplendens*: *Centrolene daidalea*, *Cochranella mache*, *Centrolene savagei*, and *Centrolene solitaria*. However, none of these species is found in the Amazon basin, where *Cochranella resplendens* occurs; *C. mache* is distributed in the Pacific lowlands of Ecuador, whereas *C. daidalea*, *C. savagei*, and *C. solitaria* are endemic to the Colombian Andes.

Figure 77. *Cochranella resplendens* in life. (**Left**): Amplectant pair. (**Right**): Female brooding clutch. Locality: 2 km N from HW 20 on a dirt road out of Guagua Sumaco, Napo province, Ecuador. Photos by Jesse Delia.

Diagnosis: (1) Each vomer with three or four teeth on dentigerous process; (2) snout round in dorsal aspect and gradually inclined in lateral profile (Figure 78); (3) tympanum oriented almost vertically, with slight lateral and posterior inclinations, its diameter about 38% of eye diameter; tympanic annulus partially visible, low supratympanic fold evident, tympanic membrane barely translucent, pigmented as surrounding skin; (4) dorsal skin shagreen with elevated and spiculated warts corresponding to white spots; (5) venter granular; pair of slightly enlarged subcloacal warts; enlarged cloacal fold present; (6) white lining (iridophores) on the anterior three-fourths of the ventral parietal peritoneum, posterior one-fourth transparent (condition P3); iridophores in pericardium and peritonea covering intestines and stomach; renal capsules and gall bladder and urinary bladder lacking iridophores (condition V2); (7) liver lobate, covered by transparent peritoneum (condition H0); (8) humeral spines absent; (9) webbing absent between inner fingers, and extensive between outer fingers (Figure 78); webbing formula III $(1^{1/2}$–$2^{-})$—$(1$–$1^{1/2})$ IV; (10) webbing between toes extensive (Figure 81); webbing formula on foot I (1–1$^+$)—(1$^{2/3}$–2) II (0–0$^+$)—(2–2$^+$) III (1–1$^{1/2}$)—(2–2$^+$) IV (2$^-$–2$^+$)—(1$^-$–1$^+$) V; (11) ventrolateral edges of Finger IV, forearm, elbow, Toe V, tarsus, and heel with dermal folds and enameled tubercles (Figures 77, 79 and 81); (12) concealed prepollex; nuptial pad not evident; (13) Finger I about same length as Finger II; (14) disc of Finger III width about 55% of eye diameter; (15) in life, dorsum dark green with numerous small white spots (Figures 77 and 79); bones green; (16) in preservative, dorsum lavender with small white spots (Figure 81); (17) iris whitish cream to beige, with a fine grey to brown reticulation; pale yellow circumpapilar ring; (18) dorsal surfaces of fingers and toes with few small, enameled spots; melanophores only on surfaces of Finger IV and Toe V; (19) calling behavior unknown; call undescribed; (20) fighting behavior unknown; (21) egg clutches placed on upper sides of leaves; short-term maternal care present; parental care provided by males absent; (22) tadpoles with non-emarginate oral apparatus; tooth row formula 2(2)/3; upper jaw slightly curved; (23) medium body size; in two adult males, SVL 26.5–26.6 mm; females unknown.

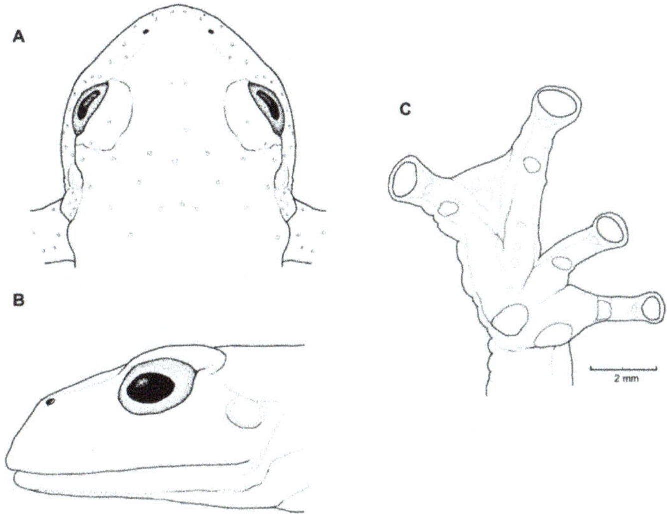

Figure 78. *Cochranella resplendens*, holotype KU 118053. (**A**) Head in dorsal view. (**B**) Head in lateral view. (**C**) Hand in ventral view. Drawings of head not to scale. Illustrations by Juan M. Guayasamin.

Color in life (Figures 77 and 79): Dorsum green with numerous, small white to bluish–white spots. Ventral surfaces mostly white, except posterior portion, which is translucent. Upper lip, tubercles on forearm and foot, and cloacal warts white. Bones green. Iris whitish cream to beige, with fine grey to brown reticulation; pale yellow circumpapilar ring ([22,164], this work).

Color in ethanol (Figure 81): Dorsal surfaces of head, body, and limbs slate grey to lavender with small white spots, often corresponding to small, flat tubercles. Upper lip white; tympanum pigmented as surrounding skin. White tubercles present on Fingers III and IV and Toes II–V. White on dermal folds of forearm and tarsus and on tips of Finger IV and Toe V. White cloacal tubercles. Pericardium and anterior three-fourths of the ventral parietal peritoneum white. White peritonea covering intestine, stomach, and colon. Transparent peritonea covering kidney, gall bladder, and urinary bladder. Hepatic peritoneum clear in holotype and most known specimens, but with a patch of iridophores in two Ecuadorian specimens and in Bolivian specimen ([22,164], this work).

Biology and ecology: *Cochranella resplendens* is a rare species. Since its description in 1973, only eight additional individuals have been reported (Specimens Examined). The scarcity of *Cochranella resplendens* is not related to low sampling effort. At the type locality (Santa Cecilia), fieldwork from 1967 through 1972 resulted in the collection of 7765 specimens of amphibians and reptiles, but no additional specimens of *Cochranella resplendens* [165]. However, canopy surveys may reveal that its rarity is an artifact of microhabitat sampling [166]. Sixty-six species of amphibians, including *Teratohyla midas* and *Hyalinobatrachium munozorum*, are known to occur at Santa Cecilia [165]. At the Bolivian locality (20 km W of Población La Cascada), *Cochranella resplendens* was found sympatric with *Rulyrana spiculata* and *Cochranella* sp. [164]. At the Tiputini Biodiversity Station, *Cochranella resplendens* was found sympatric with *Teratohyla midas*, *Hyalinobatrachium munozorum*, and *Vitreorana ritae* ([134], this work). Females provide short-term parental care; male parental care is absent [25].

Figure 79. Life cycle in *Cochranella resplendens*. (**A**) Juvenile, QCAZ 38088. (**B**) Ontogenetic variation of tadpoles. (**C**) Egg clutch. Photos by Luis A. Coloma. Figure modified from Terán-Valdez et al. [167].

Tadpole (Figures 79 and 80): At Río Napinaza, a clutch with 74 embryos was found on the upper side of a leaf [167]. According to Terán-Valdez et al. [167], the tadpole of *Cochranella resplendens* has the following traits (based on tadpole in Gosner Stage 36): Body elongated, oval-depressed (sensu [168]), wider than high; snout rounded in dorsal and lateral views. Eyes located on dorsal surface of head; in early stages, eyes C-shaped; after Stage 35, eyes become round. Short, single, sinistral spiracle, at the posterolateral region of the body; vent tube short and abdominal, free posteriorly, opening directed posteriorly; myotomes visible throughout length of tail; dorsal fin originating at about mid-length of tail; ventral fin originating almost at base of tail muscle and reaching its maximum height posterior to mid-length of tail. Oral disc non-emarginate; marginal papillae uniserial, distributed around oral disc, but larger on the lower labium; upper labium with papillae only on lateral extremes. Upper jaw sheath completely keratinized with serrated edge; slightly curved. Lower jaw sheath keratinized, U-shaped, and with serrated edge. Labial tooth row formula 2(2)/3; A-2 with medial gap. For ontogenetic variation, see Terán-Valdez et al. [167]. Dorsally, the tadpole is mostly red, with some aggregations of iridophores that form an interorbital line and a middorsal band from behind the eyes to nearly the end of the body (based on a tadpole in Gosner Stage 36). Dorsally, the anterior-most part of the body is pale yellow. Laterally, the tadpole is translucent. The iris is mostly black with a yellow ring around the pupil [167].

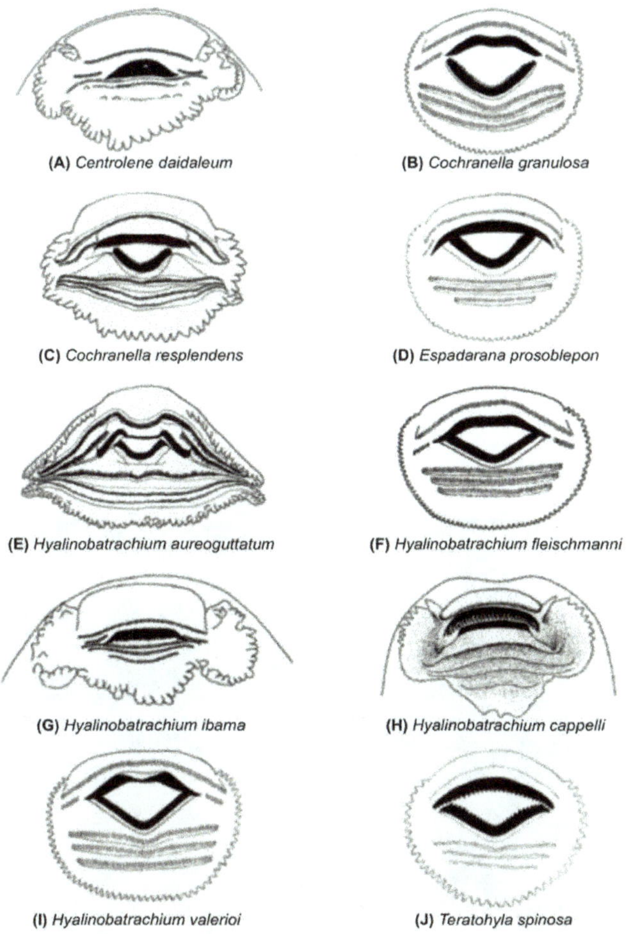

Figure 80. Oral apparatus of different species of Centrolenidae (from Terán-Valdez et al. [167]). *Cochranella granulosa* and *Hyalinobatrachium ibama* are in Gosner Stage 24, *Cochranella resplendens* in Stage 36, *H. cappellei* in Stage 25, *H. aureoguttatum* in Stage 35. Gosner stages of the remaining species are not provided in the original descriptions. Figures modified from: (**A,G**) Rada et al. [169]; (**B,D,F,I,J**) Starrett [146]; (**C,E**) Terán-Valdez et al. [167], (**H**) Noonan and Bonett [170].

Taxonomic Remarks: Aguayo and Harvey [164] described *Cochranella phryxa* from the Amazonian slopes of the Bolivian Andes. These authors mentioned that *C. phryxa* can be distinguished from *Cochranella resplendens* (characteristics in parenthesis) by having a hidden tympanum (tympanum visible), a first finger longer than the second (second finger longer than the first), and a straight cloacal fold (U-shaped cloacal fold). After examining the holotype and six additional specimens (QCAZ 38099, MHNSM 19507, DFCH-USFQ D103–4; FHGO 1305, 1324) of *Cochranella resplendens*, we find that the tympanic membrane in *Cochranella resplendens* is only slightly differentiated from the skin surrounding the tympanum. Therefore, one could interpret the tympanum in *Cochranella resplendens* as hidden, which is the character state that Aguayo and Harvey [164] described for *C. phryxa*. The relative length of the two innermost fingers has been used for many authors when comparing centrolenid species. Although we agree that the character is useful when the differences are obvious (e.g., comparing species of *Hyalinobatrachium*), it is problematic when differences are minor, as in *C. resplendens*. We have tried to minimize the error in the assessment of this character by measuring the fingers with a digital caliper. Consistently, we found that these fingers have almost the same length in *Cochranella resplendens* (Finger I length 96%–102% of Finger II). The last character used by Aguayo and Harvey [164] is the shape of the cloacal fold, which they described as straight in *C. phryxa* and U-shaped in *Cochranella resplendens*. Differences in shape of the cloacal ornaments may be either artificial, associated with how the specimens are preserved (i.e., posterior extremities fixed with an anterior, posterior, or parallel orientation in relation to the sagittal axis of body), or due to intraspecific variation. The cloacal ornamentation (i.e., white warts and folds) is the same in the two species (compare Lynch and Duellman [22]: Figure 2c; Aguayo and Harvey [164]: Figure 4). *Cochranella mache*, a species with U-shaped cloacal folds, shows the same intraspecific variation observed between *C. resplendens* and *C. phryxa*. The only difference that we find between the holotype of *Cochranella resplendens* and the Bolivian specimen is the presence of a small patch of iridophores on the hepatic peritoneum of the latter. However, it is possible that this patch is either an artifact of preservation (i.e., iridophores from the ventral parietal peritoneum can be attached to the liver) or intraspecific variation (i.e., the presence/absence of patches of iridophores in the peritoneum has been observed in other centrolenids (e.g., *Cochranella mache* [17,158]). For the reasons discussed above, we formally place *Cochranella phryxa* Aguayo and Harvey, 2006 [164], in the synonymy of *Centrolenella resplendens* Lynch and Duellman, 1973 [22].

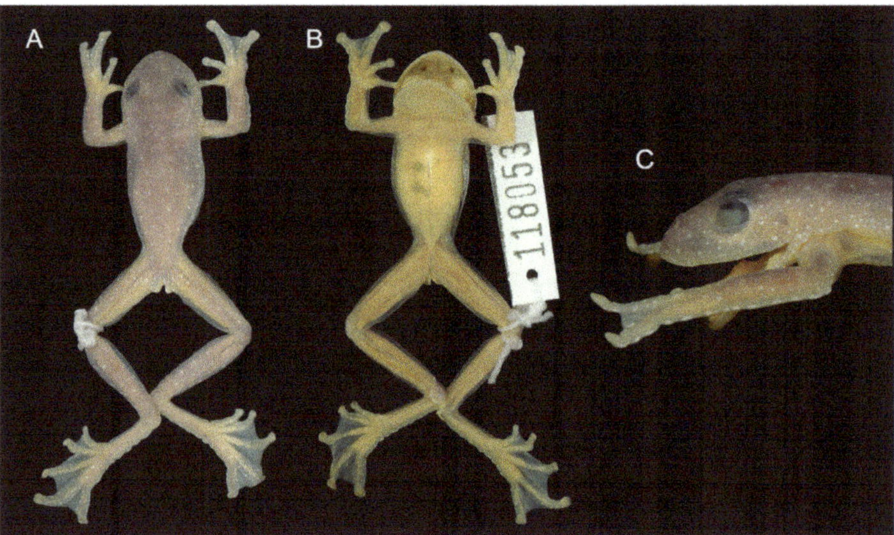

Figure 81. *Cochranella resplendens* in preservative. Holotype (KU 118053) from Santa Cecilia, Sucumbíos province, Ecuador. (**A**) Dorsal view. (**B**) Ventral view. (**C**) Lateral view. Photos by Martín Bustamante.

Distribution (Figure 82): *Cochranella resplendens* in known from the Amazon basin of Colombia, Ecuador, Peru, and Bolivia, at elevations between 190 and 1100 m ([22,134,164,167,171], this work).

A disjunct population of the species was recently found on the eastern slope of the central Andes in the Departamento Antioquia, Colombia, at much higher elevations (1309–1699 m) [172]. In Ecuador, this species is known from localities in the provinces of Morona Santiago, Napo, Orellana, Pastaza, and Sucumbíos at elevations between 250–1100 m. In Ecuador, the potential distribution of *C. resplendens* is 77,793 km^2 within the Amazonian Tropical Rainforest and Eastern Foothill Forest ecoregions.

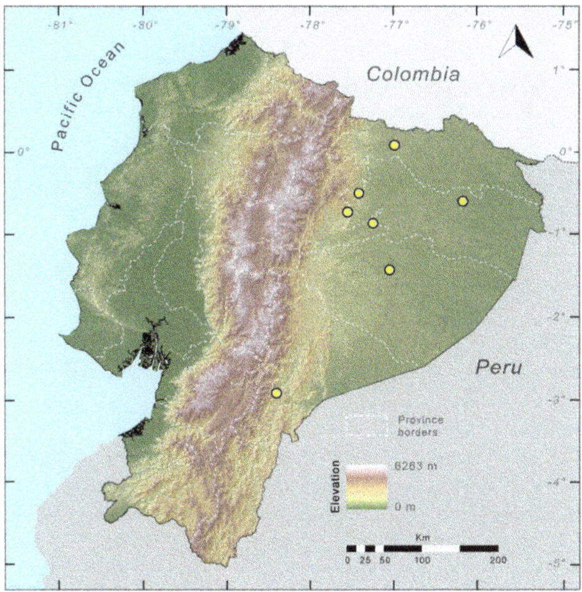

Figure 82. Distribution of *Cochranella resplendens* in Ecuador (yellow dots).

Conservation status: Globally, *Cochranella resplendens* is currently listed as *Least Concern* by the IUCN [173]. Although it is evident that *C. resplendens* is rare in collections (see above), we presume that this is a consequence of inadequate sampling [166]. The potential distribution of this species in Ecuador is 77,793 km^2, 14% of which is affected by human activities. Given the broad distribution of *C. resplendens* and the lack of immediate threats, the category of *Least Concern* is justified.

Evolutionary relationships (Figure 69): With the current genetic and taxon sampling, the sister species of *C. resplendens* is *C. granulosa*. These two species are an example of vicariant speciation, mediated by the uplift of the Andes.

Specimens examined: *Cochranella resplendens:* <u>Colombia</u>: *Departamento de Putumayo:* Santa María de Sucumbíos (ca. 00°16′ N, 76°55′ W), AMNH 88083 (Lynch and Duellman, 1973). <u>Ecuador</u>: *Provincia de Sucumbíos:* Santa Cecilia (00°03′ N, 76°58′ W; 340 m), KU 118053 (holotype). *Provincia de Morona Santiago:* Río Napinaza (2.927° N, 78.407° W; 1100 m), QCAZ 38088. *Provincia de Napo:* Reserva Yachana (00°52′21.71″ S, 77°14′13.43″ W; 300–350 m), QCAZ 38099; *Provincia de Orellana:* San José Viejo de Sumaco (0.5333 S, 77.4167 W; ca. 810 m), USNM 288460; Tiputini Biodiversity Station (00°37′ S, 76°10′ W; 190–270 m), DFCH-USFQ D103–04; *Provincia de Pastaza:* Pozo Garza, Oryx (01°26′ S, 77°03′ W; 300 m), FHGO 1305, 1324. <u>Peru</u>: *Departamento de San Martín:* Cainarachi Valley (6°43′10 S, 76°29′13 W; 550 m), Km 33, Carretera Tarapoto-Yurimaguas, MHNSM 19507. <u>Bolivia</u>: *Departamento de La Paz:* approximately 20 km west of Población de la Cascada (15°22′37″ S, 67°12′4″ W; 1000 m), CBG 778 [164].

Photographic records: Ecuador: *Provincia de Napo:* ca. 2 km north from HW 20 on a dirt road out of Guagua Sumaco (0.688° S, 77.6° W; ca. 900 m); photographs by Jesse Delia (Figure 77).

Genus *Espadarana* Guayasamin, Castroviejo-Fisher, Trueb, Ayarzagüena, Rada, & Vilà 2009 [1].

Etymology: The name *Espadarana* honors Marcos Jiménez de la Espada, a Spanish zoologist who was part of the Comisión Científica del Pacífico that explored America between 1862 and 1865. Jiménez de la Espada described the first centrolenid frog, *Centrolene geckoideum* in 1872. In Spanish, the word *Espada* means sword, in allusion to the humeral spines present in males of species in this genus. *Espadarana* is a combination of the words *Espada* and *rana* (frog) and is feminine in gender [1].

Espadarana audax (Lynch and Duellman, 1973 [22]; Figures 83–87).

Centrolenella audax Lynch and Duellman, 1973 [22]. Holotype: KU 146624.
Type locality: "Salto de Agua, 2.5 km NNE of Río Reventador on Quito–Lago Agrio road,
    1660 m, Provincia de Napo, Ecuador".
Centrolene audax—Ruiz-Carranza and Lynch, 1991 [6].
"Centrolene" audax—Guayasamin, Castroviejo-Fisher, Trueb, Ayarzagüena, Rada, Vilà, 2009 [1].
Centrolene fernandoi—Duellman and Schulte, 1993 [174]. Synonymy by Cisneros-Heredia and
    Guayasamin, 2014 [175].
Espadarana audax—Twomey, Delia, and Castroviejo-Fisher, 2014 [19].

Common names: English: Daring Glassfrog. Spanish: Rana de Cristal Audaz.
Etymology: The specific name *audax* is Latin, meaning daring, and is used in allusion to the precipitous regions inhabited by this species [22].
Identification: *Espadarana audax* is unique among Ecuadorian glassfrogs by having small yellow spots on the dorsum, short and distally curved humeral spines in males, and extensive webbing only between Fingers III and IV (Figures 83–85). In Ecuador, species with a similar dorsal coloration include *Nymphargus buenaventura*, *N. siren*, *N. humboldti* sp. nov., *Rulyrana flavopunctata*, and *Teratohyla midas*; however, males of these species lack humeral spines. In addition, *Teratohyla midas* has visceral peritonea covered by iridophores; *N. buenaventura*, *N. siren*, and *N. humboldti* sp. nov. have considerably less hand webbing; and *R. flavopunctata* has more webbing between Fingers III and IV. The Colombian *Centrolene notosticta* has a similar dorsal color pattern but lacks vomerine teeth (present in *E. audax*) and has less webbing between the outer fingers. *Espadarana audax* is most similar to *E. durrellorum* (see Taxonomic Remarks), but the latter differs by lacking yellow dorsal spots. Also, the two species are found at slightly different, although overlapping, elevations; *E. audax* inhabits the Amazonian slopes of the Andes (800–1900 m), whereas *E. durrellorum* is found mostly in the Amazonian lowlands and foothills (220–1150 m).

Figure 83. *Espadarana audax* in life. Adult male, QCAZ 37871, from Gral. Leonidas Plaza Gutiérrez (Limón), Quebrada del Río Napinaza, Morona Santiago, Ecuador. Photos by Luis A. Coloma.

Diagnosis: (1) Dentigerous process of the vomer bearing two to four teeth; (2) snout rounded in dorsal profile, rounded to truncated in lateral profile; (3) tympanum of moderate size, tympanum diameter 23%–29% of eye diameter, low supratympanic fold evident, tympanic membrane partially pigmented, clearly differentiated from surrounding skin; (4) dorsal surfaces of males and females shagreen, with minute spicules in males; (5) pair of enlarged subcloacal warts, other cloacal ornamentation absent; (6) anterior two-thirds of the ventral parietal peritoneum covered with white iridophores, posterior third transparent (condition P2); white pericardium; no iridophores in peritonea covering intestines, stomach, testes, kidneys, gall bladder, and urinary bladder (condition V1); (7) liver tetralobed, hepatic peritoneum lacking iridophores (condition H0); (8) in adult males, humeral spines present; (9) webbing absent between Fingers I, II, and III; webbing formula for outer fingers III $(2^--2^{1/3})$—(2^--2^+) IV; (10) extensively webbed foot: I $(1–1^+)$—$(2–2^+)$ II (1^--1^+)—$(2–2^{1/4})$ III $(1–1^{1/4})$—$(2^+-2^{1/3})$ IV $(2^+-2^{1/3})$—$(1–1^+)$ V; (11) ulnar fold present, inner tarsal fold present, outer tarsal fold absent; (12) concealed prepollex; in males, nuptial pad Type I; (13) Finger I about same length as Finger II (Finger II 95.6%–102% length of Finger I); (14) disc of Finger III moderate, its width 43%–56% of eye diameter; (15) in life, dorsal surfaces of head, body, and limbs green with small yellow spots; bones green; (16) in ethanol, dorsal surfaces of head, body, and limbs lavender with small white spots; (17) in life, iris white with yellowish hue and thin black reticulations; (18) melanophores on outer fingers and outer toes; (19) males call from upper sides of leaves; calls are produced in series, with each series having four or five calls; each call is composed by a single, pulsed note; dominant frequency is at 5426–6718 (mean = 6146, SD = 368) Hz; (20) fighting behavior unknown; (21) eggs deposited in moss on branches; short-term maternal care present; parental care by males absent; (22) tadpoles unknown; (23) small body size; in adult males, SVL 21.6–25.5 mm ($\overline{X}$ = 23.5 ± 0.722, n = 41); in adult females, SVL 24.5–28.8 ($\overline{X}$ = 27.0 ± 1.468, n = 7).

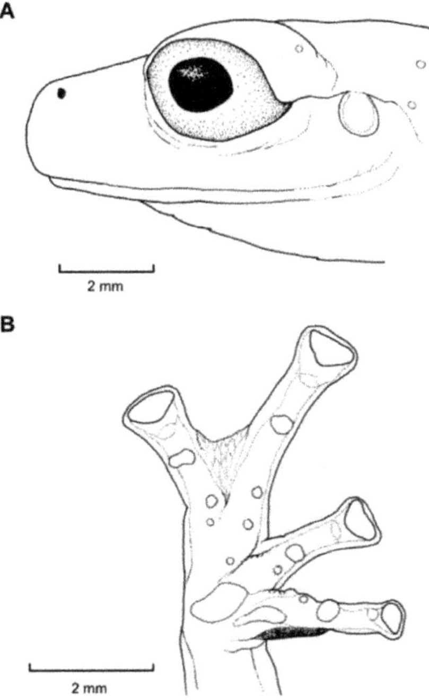

Figure 84. *Espadarana audax*, adult males. (**A**) Head in lateral view, KU 164503. (**B**) Hand in ventral view, KU 178018. Illustrations by Juan M. Guayasamin.

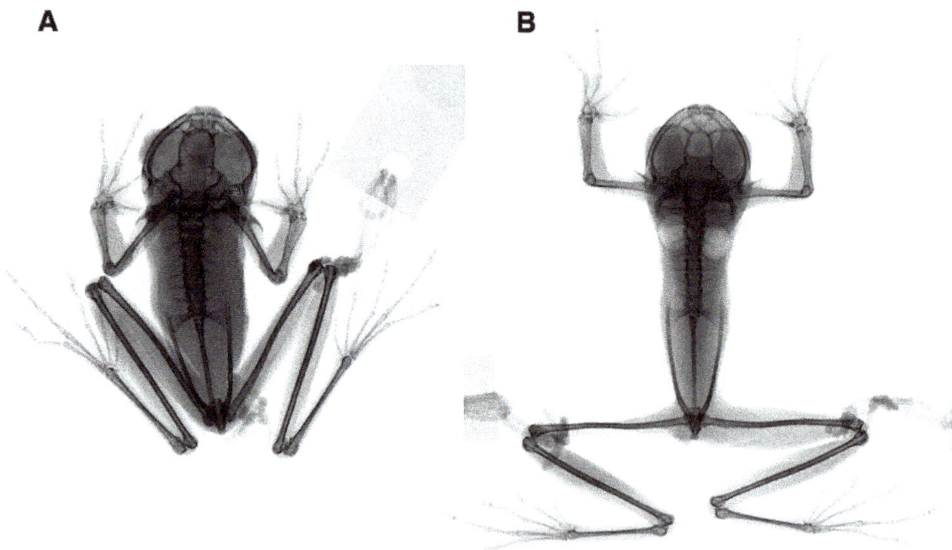

Figure 85. Humeral spines of *Espadarana audax* and *E. durrellorum*. (**A**) *Espadarana audax*, KU 164500. (**B**) *Espadarana durrellorum*, QCAZ 47909. Photos by Juan M. Guayasamin.

Variation: In several males, the dorsal spicules are minute and only visible under magnification. A population from Leonidas Plaza Gutiérrez (Limón) has a dorsum with very few and small dorsal yellow spots.

Color in life (Figure 83): Dorsum green, with small yellow spots; fingers and toes pale yellow; parietal peritoneum white; heart not visible; visceral peritoneum lacking iridophores. Ventral surfaces of limbs unpigmented; bones green; iris pale bronze to mustard with thin black reticulation ([22], this work).

Color in ethanol: Dorsal surfaces of head, body, and limbs lavender with small white dots; Fingers I–III and Toes I–III cream (without melanophores); upper lip white. White lining on the anterior two-thirds of ventral parietal peritoneum; white pericardium; no iridophores in peritonea covering liver, intestines, stomach, testes, kidneys, gall bladder, and urinary bladder. Some preserved individuals (KU 164497–98) have lost their dorsal white spots; however, close examination revealed the presence of collapsed chromatophores (apparently iridophores).

Biology and ecology: By day, one individual (KU 146624) was found in a bromeliad. During the night, adults are found on the upper sides of leaves along streams, a site from which males call (W. E. Duellman field notes, 19 March 1975; this work). On 18 July 1981, 12 individuals were found in Río Salado; the males were calling and one female (KU 178023) had dark brown oviductal eggs visible through the body wall. Females provide short-term parental care; male parental care is absent [25].

Call (Figure 86): We analyzed 14 notes contained within four calls from two individuals (MZUTI 1492–93). Calls are emitted in series; a typical call series has four or five calls (mean = 4.3, SD = 0.6) and is relatively long (range = 2398–3285 ms, mean = 2711 ms, SD = 497.8 ms). Each call has a single note with a duration of 26–53 (mean = 38, SD = 8.2) ms. Notes have clearly defined pulses with 5–11 (mean = 7.4, SD = 1.7) amplitude peaks throughout the note. Pulses within a note have a rate of 171–208 (mean = 189, SD = 10) pulses per second. Notes have their peak amplitude in the first 50% of the note (relative peak time: Range = 0.0144–0.0722, mean = 0.043, SD = 0.02), whereby the first note usually has the highest amplitude. The dominant frequency of a note measured at peak amplitude is 5426–6718 (mean = 6146, SD = 368) Hz and is contained within the fundamental frequency. The fundamental frequency has a lower limit of 5254–6288 (mean = 5925, SD = 374) Hz and a higher limit of 6115–7063 (mean = 6608, SD = 297) Hz.

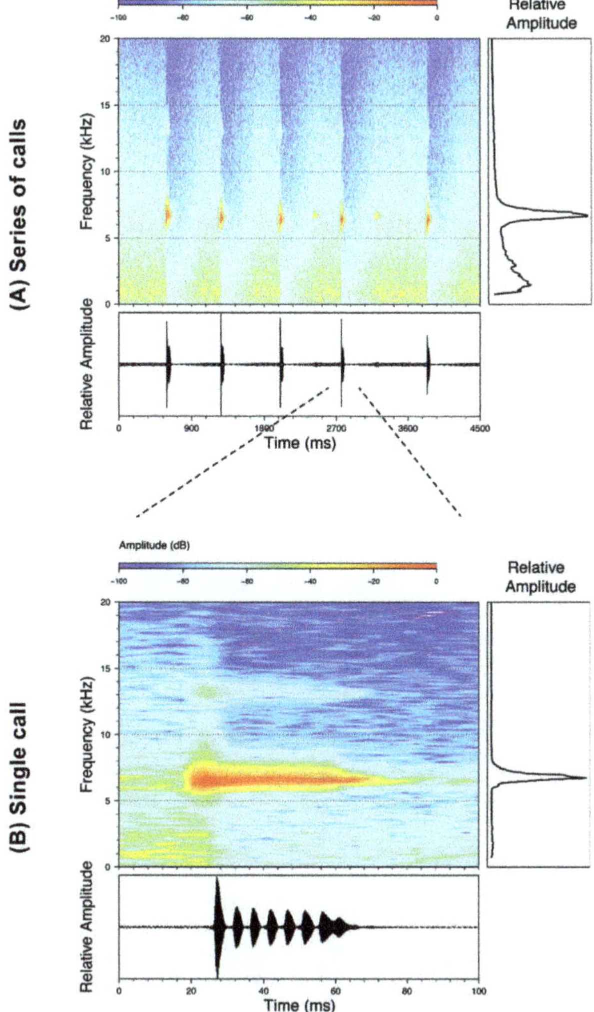

Figure 86. Calls of *Espadarana audax*, MZUTI 1492, recorded at Río Hollín, 1038 m, Napo province, Ecuador. (**A**) Call series. (**B**) Single call.

Tadpole: Not described.

Distribution (Figure 87): *Espadarana audax* occurs through the Eastern Montane region of southeastern Colombia, Ecuador, and northern Peru at elevations between 800 and 1900 m. This species has been recorded from the provinces of Napo, Morona Santiago, Sucumbíos, and Zamora Chinchipe in Ecuador, the departments of Huila (eastern slope of Cordillera Central), Putumayo, Cauca, Caquetá, and Boyacá (eastern slope of Cordillera Oriental) in Colombia, and the Department of San Martín in Peru ([19,22,174–176], this work). In Ecuador, the species is found within the Eastern Foothill Forest and Eastern Montane Forest ecoregions. The potential distribution of *E. audax* is 6738 km^2, 9% of which is affected by human activities.

Figure 87. Distribution of *Espadarana audax* in Ecuador (yellow dots).

Conservation status: Globally, *Espadarana audax* is currently listed as *Least Concern* by the IUCN [177]. In Ecuador, *E. audax* has not been found in some of its historical localities (e.g., Río Azuela, Río Salado ([91], JMG, pers. obs.)), but has been discovered recently at several new localities (Miazi Alto, General Leonidas Plaza Gutiérrez, Zamora); see Specimens Examined. The distribution of the species is fragmented because of agriculture, pasture lands, and mining. We suggest that, in Ecuador, the species should be considered as *Near Threatened*.

Evolutionary relationships (Figure 88): Given the current gene and taxon sampling, *Espadarana audax* and *E. durrellorum* are sister species.

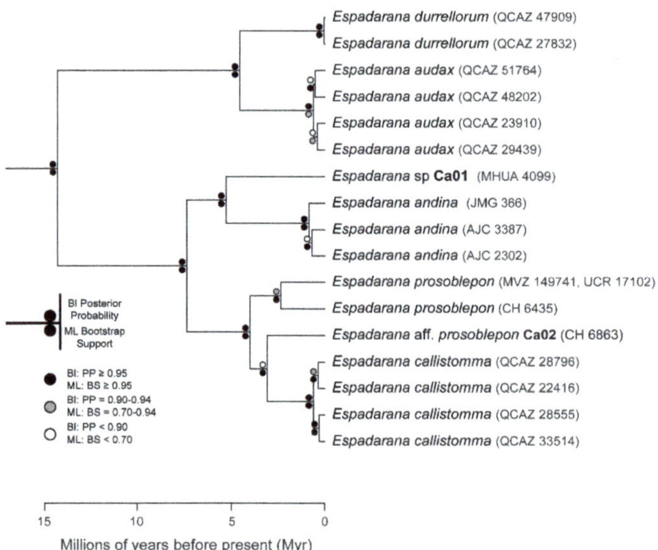

Figure 88. Evolutionary relationships among species in the genus *Cochranella*, inferred using maximum likelihood and Bayesian criteria.

Taxonomic Remarks: The morphological differences that separate *Espadarana durrellorum* and *E. audax* are subtle, but given the available genetic and morphological data, we consider them as valid species. *"Centrolene" fernandoi* was recently placed in the synonymy of *E. audax* [175].

Specimens examined: *Espadarana audax:* <u>Ecuador</u>: *Provincia Napo:* 2.5 km NNE of Río Reventador on Quito–Lago Agrio road (0.133 S, 77.633 W, 1600 m), KU 146624 (holotype); 16.5 km NNE Santa Rosa (0.2186 S; 77.7319 W, 1700 m), KU 143290, 143292; 2 km SSW of junction between Río Reventador and Baeza-Lumbaqui road (0.1 S, 77.6 W, 1700 m), KU 164497–504; 7 km SW Río Azuela on Quito–Lago Agrio road (0.1667 S, 77.667 W), KU 155502–03; Río Salado (0.19167 S, 77.6997 W, 1420 m), KU 178018–27; 43 km NE Santa Rosa (0.1186 S, 77.6003 W, 1490 m), KU 190015; 8.9 km NE Santa Rosa on Quito–Lago Agrio road (0.1667 S, 77.667 W), KU 190016; Río Hollín (0.72 S, 77.639 W; 1100 m), QCAZ 6898; Reserva Narupa (0.6848 S, 77.742 W, 1164 m), ZSFQ 382–86. *Provincia Sucumbíos:* Río Azuela (0.11667 S, 77.6167 W, 1740 m), KU 164496. *Provincia Morona Santiago:* Quebrada del Río Napinaza, 6.6 km N in the road Limón–Macas (2.92665 S, 78.40701 W; 1100 m), QCAZ 37871, 29439; La Y (3.43236 S, 78.60449 W, 835 m), QCAZ 23910. *Provincia Zamora Chinchipe:* Alto Miazi (4.25026 S, 78.61746 W; 1250 m), QCAZ 41653; Cordillera del Cóndor, Centro Shuar El Tink, ca. 1050 m, QCAZ 48202. <u>Colombia</u>: *Departamento Huila:* Parque Arqueológico San Agustín, 3 km SW of San Agustín, 1750 m. <u>Peru</u>: *Departamento San Martín:* W slope of Abra Tangarana, 7 km (by road) NE of San Juan de Pacaysapa (6.2 S, 76.733 W; 1080 m), KU 211770–75.

Espadarana callistomma (Guayasamin and Trueb 2007 [9]; Figures 89–92).

Centrolene callistommum—Guayasamin and Trueb, 2007 [9]. Holotype: QCAZ 25832.
<u>Type locality</u>: "stream affluent of Río Bogotá (1°05'13.8" N, 78°41'25.8" W, 83 m), nearby San Francisco de Bogotá, Provincia de Esmeraldas, Ecuador."
Espadarana callistomma—Guayasamin, Castroviejo-Fisher, Trueb, Ayarzagüena, Rada, and Vilà, 2009 [1].

Common names: English: Beautiful eyes Glassfrog. Spanish: Rana de cristal de ojos bellos.

Etymology: The specific name *callistomma* is derived from the Greek *kallistos–*, meaning "most beautiful" and *omma*, meaning "eye", as a reference to the fantastic iris pattern in this species [9].

Identification: *Espadarana callistomma* is easily distinguished from most glassfrogs by its large size (in males, SVL 26.7–29.6 mm; in females, SVL 29.5–31.8 mm), uniform green dorsal coloration, and silvery white iris with black reticulations (Figure 89). *Espadarana callistomma* closely resembles *Sachatamia ilex*; however, males of *E. callistomma* have a conspicuous humeral spine, whereas males of *S. ilex* have a small humeral spine that is embedded in the arm musculature (Figure 201). Some populations of *Espadarana prosoblepon* have the same color pattern as *E. callistomma*, but they have a significantly smaller body size (in males, SVL 23.2–27.5 mm; in females, SVL 25.3–27.8 mm) and a two-note call (three to four notes in *E. callistomma*). See Taxonomic Remarks.

Figure 89. *Espadarana callistomma* in life from Durango, Esmeraldas province, Ecuador. Adult male, QCAZ 32055. Photo by Martín Bustamante.

Diagnosis: (1) Vomerine teeth present, each vomer with two to seven teeth; (2) snout truncated in dorsal and lateral views (Figure 90); (3) tympanum small (tympanum diameter 20%–31% eye diameter), oriented vertically, with lateral inclination; tympanic annulus visible except for upper border, which is covered by supratympanic fold; tympanic membrane partially pigmented, differentiated from surrounding skin; (4) dorsal surfaces of males and females shagreen, with minute spicules evident in males (visible under magnification); (5) pair of enlarged subcloacal warts (Figure 90); (6) anterior two-thirds of ventral parietal peritoneum covered with white iridophores, posterior third transparent (condition P3); white pericardium; translucent peritoneum covering intestines, stomach, gall bladder, testes, and urinary bladder; kidneys covered with translucent peritoneum (condition V1); (7) liver tetralobed, covered by transparent peritoneum (condition H0); (8) humeral spines present in adult males; (9) no webbing between Fingers I and II, webbing between Fingers II and III reduced, webbing between Fingers III and IV extensive (Figure 90), webbing formula: II ($1^{2/3}$–2)—(3^+–$3^{1/4}$) III ($1^{1/2}$–$1^{2/3}$)—(1–$1^{1/2}$) IV; (10) webbing on foot extensive (Figure 90), webbing formula: I (0^+–1)—(2^-–2^+) II (0^+–1)—(2^-–2^+) III (0^+–1)—2^- IV (2^-–$2^{1/3}$)—(1^-–1^+) V; (11) ulnar and inner tarsal folds low; outer tarsal fold absent; (12) concealed prepollex; in males, nuptial pad Type I; (13) Finger II slightly longer than Finger III (Finger III 93.3%–100% length of Finger II); (14) disc of Finger III of moderate size, about 29%–34% eye diameter; (15) in life, dorsum olive green (Figure 89); bones green; (16) in preservative, dorsum dark lavender; (17) iris silvery white with thin black reticulations; (18) melanophores covering dorsal surfaces of Fingers III and IV and Toes IV and V; (19) males call from upper side of leaves; each call consists of three to four pulsed notes, dominant frequency at 5343–5812 Hz; (20) fighting behavior unknown; (21) eggs deposited on upper side of leaves; parental care unknown; (22) tadpoles unknown; (23) in males, SVL 26.7–29.6 mm ($\overline{X}$ = 27.8 ± 0.762; *n* = 19); in females, SVL 29.5–31.8 mm ($\overline{X}$ = 30.5 ± 0.796; *n* = 13).

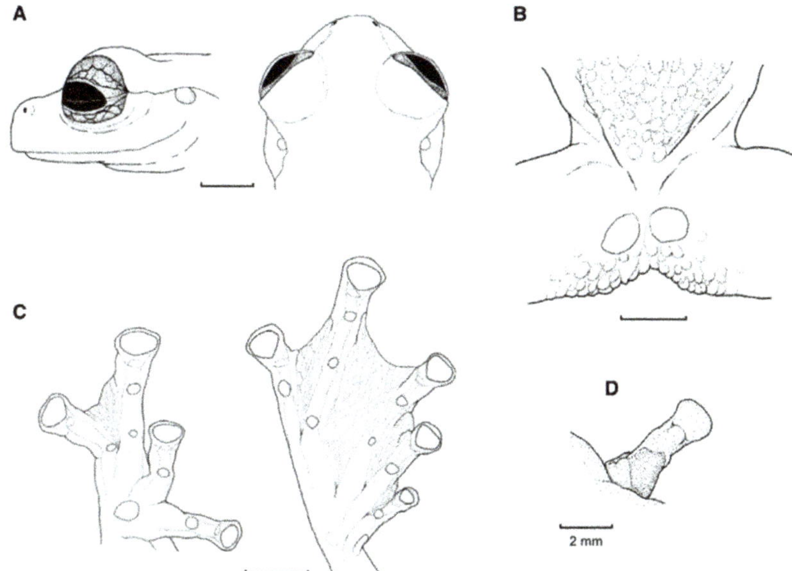

Figure 90. *Espadarana callistomma*. (**A**) Head in lateral and dorsal views, paratype, adult male, QCAZ 28803. (**B**) Ventral surfaces of thighs showing enlarged subcloacal tubercles, holotype, adult male, QCAZ 25832. (**C**) Hand and foot in ventral view, paratype, adult male, QCAZ 28803. (**D**) Nuptial pad and thumb in dorsal view, holotype, adult male, QCAZ 25832. Modified from Guayasamin and Trueb [9].

Osteology: A detailed osteological description of *Espadarana callistomma* was provided by Guayasamin and Trueb [9]; they mentioned that the frontoparietals were exceedingly slender along the orbital margin of the braincase. After reexamining the cleared-and-stained material, it seems that the frontoparietals extend medially and their medial margins partially delimit the frontal fontanelle (area shown as cartilage in [9]: Figure 8A).

Color in life (Figure 89): Dorsum uniform olive green, without spots; upper lip with whitish–cream coloration; iris silvery white with marked black reticulations; flanks white; parietal peritoneum white, covering anterior two-thirds of abdomen (heart not visible); bones green [9].

Color in preservative: Dorsal surfaces of head, body, and limbs uniform lavender; upper lip cream; iris white with marked dark lavender reticulations; nuptial pad on Finger II cream; dorsally, Fingers II and III and Toes I–III unpigmented; venter cream [9].

Biology and ecology: The natural history of *Espadarana callistomma* is poorly known. The species is active during the night and has been found on leaves along streams. Males call from the upper side of leaves, and females deposit pigmented eggs on the upper side of leaves. It is unknown if males or females provide parental care [9].

Call (Figure 91): Description based on a call (LBE-C-022) of a male *E. callistomma* recorded by JMG on 17 May 2010 at Reserva Otokiki, Ecuador. Each call consists of three to four pulsed notes ($\overline{X}$ = 3.7 ± 0.483, *n* = 10) and lasts 0.28–0.44 s ($\overline{X}$= 0.38 ± 0.066, *n* = 10). Notes are produced at a rate of 9.1–11.0 notes/s ($\overline{X}$= 9.76 ± 0.596, *n* = 10); each note has a duration of 0.014–0.04 s ($\overline{X}$ = 0.03 ± 0.007, *n* = 10). The call is not frequency modulated; dominant frequency at 5343–5812 Hz (*n* = 10).

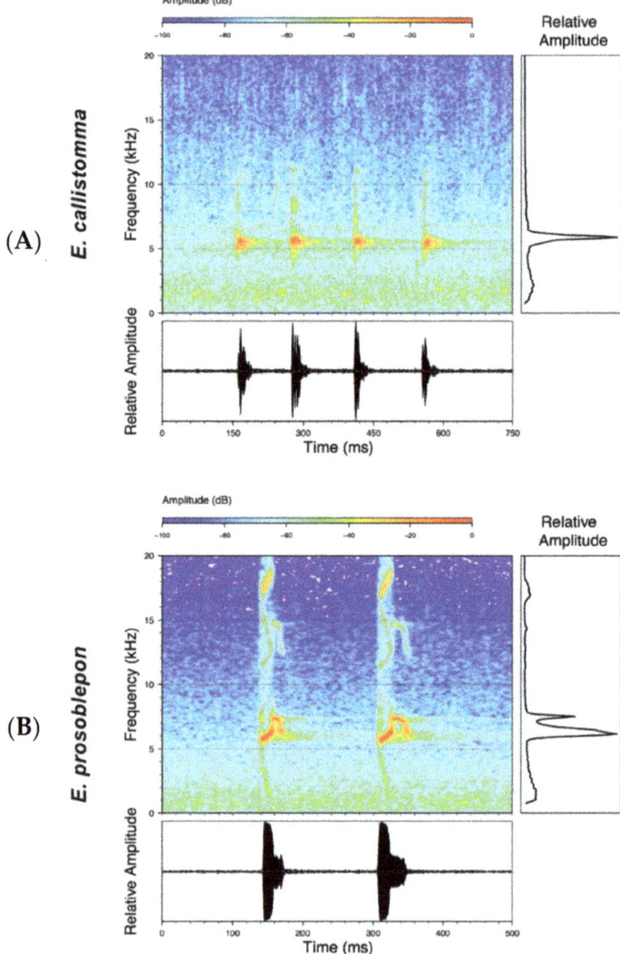

Figure 91. Calls of *Espadarana*. (**A**) Call of *Espadarana callistomma* from Reserva Otokiki, 706 m, Esmeraldas province, Ecuador, LBE-C-022. (**B**) Call of *E. prosoblepon* from Cordillera de Chontilla, 908 m, Pichincha province, Ecuador, MZUTI 601.

Tadpole: Not described.

Distribution (Figure 92): *Espadarana callistomma* is known from four localities in the lowland rainforest of northwestern Ecuador (provinces of Esmeraldas and Carchi) at elevations below 500 m ([9], this work) and the Pacific lowlands of Colombia [178]. The habitat of the species is within the Chocoan Tropical Forest and the Western Foothill Forest regions (see Biogeographic Regions).

Figure 92. Distribution of *Espadarana callistomma* in Ecuador (yellow dots).

Conservation status: Globally, *Espadarana callistomma* is currently listed as *Least Concern* by the IUCN [179]. At present, *E. callistomma* is known from a few localities in the Pacific lowlands of Ecuador and Colombia [9,178]. The most likely scenario, however, is that *E. callistomma* is restricted to portions of the Chocó ecoregion, an area with high rates of deforestation and mining; we suggest that the species is considered as *Endangered* in Ecuador, following IUCN criteria B1, B2a, B2b(iii).

Evolutionary relationships (Figure 88): *Espadarana prosoblepon* is paraphyletic with respect to *E. callistomma*, although *E. callistomma* is monophyletic. The most likely explanations to the observed pattern are: (i) Divergence between the two species is recent and reciprocal monophyly was not yet achieved, (ii) *E. callistomma* is a synonym *of E. prosoblepon*, or (iii) *E. prosoblepon* is a species complex that requires further subdivision. Given the morphological and acoustic data at hand, we favor the first and third hypotheses.

Taxonomic Remarks: The specific status of *Espadarana callistomma* is uncertain because of the conflict among different sets of data. Genetically, *Espadarana callistomma* and *E. prosoblepon* are not reciprocally monophyletic (Figure 88). On the other hand, body size of males and females of *E. callistomma* is significantly larger than those of *E. prosoblepon* (*t-test*, $P < 0.001$), and the dorsal and iris color patterns of *E. callistomma* are different from most *E. prosoblepon* (Figure 89 and Figure 95). In terms of vocalization, *E. callistomma* has a call with more notes (three to four notes per call) than *E. prosoblepon* (two notes per call; Figure 91). Given that we find biologically meaningful differences among the species (i.e., body size, vocalizations), we recognize both species as valid. The lack of molecular divergence could be the product of a recent divergence, but this is a hypothesis that remains untested. Another pending taxonomic problem is that the type of *Hylella parabambae* matches the spotless coloration of *E. callistomma* and some populations of *E. prosoblepon*.

Specimens examined: *Espadarana callistomma:* <u>Ecuador</u>: *Provincia de Esmeraldas:* stream affluent of Río Bogotá (1°05′13.8″ N, 78°41′25.8″ W, 83 m), nearby San Francisco de Bogotá, QCAZ 25832 (holotype), 27776–8, 28555–56, 28557 (C&S), 28558; stream affluent of Río Bogotá (1°05′9.06″ N, 78°41′8.7″ W, 77 m), 2 km E of San Francisco de Bogotá on the San Francisco–Durango road, QCAZ 28803; 2.1 km E of Durango (1.02477 N, 78.61746 W; 284 m), QCAZ 32169. *Provincia de Carchi:* Río La Carolina (0°42′16.16″ N, 78°12′4.14″ W, 500 m), on the Ibarra–Lita road, nearby Jijón y Caamaño, QCAZ 27744–45, 27768.

Espadarana durrellorum (Cisneros-Heredia, 2007 [180]; Figures 85, 93 and 94).

Centrolene durrellorum Cisneros-Heredia, 2007 [180]. Holotype: DFCH-USFQ D131.
Type locality: "small rivulet tributary of the Jambue River, ca. 6 km S from Zamora (ca. 04°03′ S,
   78°56′ W, 1150 m), on the western slope of Contrafuerte de Tzunantza, Cordillera Oriental,
   eastern slopes of the Andes, Provincia de Zamora Chinchipe, República del Ecuador".
"Centrolene" durrellorum—Guayasamin, Castroviejo-Fisher, Trueb, Ayarzagüena, Rada, and
   Vilà, 2009 [1].
Espadarana durrellorum—Twomey, Delia, and Castroviejo-Fisher, 2014 [19].

Common names: English: Durrell's Glassfrog. Spanish: Rana de Cristal de Durrell.
Etymology: The specific name *durrellorum* honours Gerald Durrell and Lee Durrell (Durrell
Wildlife Conservation Trust) for their contributions to the conservation of biodiversity [180].
Identification: *Espadarana durrellorum* can be distinguished from all other glassfrogs by having
a green dorsal coloration (Figure 93), humeral spines short and straight in adult males (Figure 85),
moderate body size (25.7–26.1 mm adult males), and lacking iridophores on the visceral peritonea.
Among Ecuadorian centrolenids, *E. durrellorum* is most similar to *E. audax*. The main trait to differentiate
between the two species is the absence of yellow dorsal spots in *E. durrellorum* (present in *E. audax*).
E. audax inhabits higher elevations (800 to 1900 m) than *E. durrellorum* (220–1150 m).

Figure 93. *Espadarana durrellorum* in life. Ecuador, Napo province, Comunidad Ñukanchi Allpa, 403 m,
QCAZ 47909. Photos by Luis A. Coloma.

Diagnosis: (1) Vomerine teeth present; (2) snout rounded in dorsal view, bluntly truncated in
lateral view; (3) tympanic annulus evident, oriented dorsolaterally; very weak supratympanic fold
above tympanum; (4) dorsal skin shagreen, with minute spicules uniformly distributed; (5) ventral
skin granular; cloacal area granular, with two large, rounded, flat subcloacal tubercles, and two
folds on the sides of the cloacal opening; distinct cloacal sheath; (6) upper two-thirds of the parietal
peritoneum covered by iridophores (condition P3), pericardium white, all other peritonea lacking
iridophores (condition V1); (7) liver lobed, lacking iridophores (condition H0); (8) in adult males,
humeral spine short and straight, with sharp point; (9) webbing absent between Fingers I and II, basal
or absent between II and III, moderate between outer fingers: In males, III 2—2⁻ IV; in one female III
1²ᐟ³—(1¹ᐟ²–2⁻) IV; (10) webbing formula on feet: In males, I 2—2⁻ II 1—2 III 1⁺—2⁻ IV 2¹ᐟ³—1⁺ V; in one
female, I 1—1³ᐟ⁴ II 1—2⁻ III 1⁺—2 IV 2—1 V; (11) no dermal folds on hands, forearms, feet, or tarsus;
(12) nuptial excrescences present, Type I; concealed prepollex; (13) first finger slightly longer than
second; (14) eye diameter larger than width of disc on finger III; (15) in life, dorsal surfaces uniform
green; (16) in preservative, dorsal surfaces uniform pale lavender; (17) in life, iris golden with thin
dark reticulations; in preservative, lavender with darker reticulation; (18) melanophores widespread
on outer fingers and outer toes; (19) males call from upper surfaces of leaves; call undescribed; (20, 21,

22) fighting behavior, egg clutches, tadpoles, and parental care unknown; (23) small body size; SVL 24.8–26.1 mm in adult males (*n* = 3); 27.0 in one female.

Color in life (Figure 93): Uniform green dorsum without flecks or spots. Iris golden with dark reticulations. Green bones.

Color in ethanol: All dorsal surfaces pale lavender (no light or dark spots). Ventral surfaces cream. Parietal peritoneum covered by iridophores to the level of the lower stomach; pericardium white, all other peritonea lacking iridophores.

Biology and ecology: *Espadarana durrellorum* is a recently described species and little information is available on its natural history. It is nocturnal, and males call from the tops of leaves of riverine vegetation in mature Foothill and Lowland Evergreen forests. Parental care unknown.

Call: Not described.

Tadpole: Not described.

Distribution (Figure 94): *Espadarana durrellorum* is known from few localities on the foothills of Cordillera Oriental and the Amazonian lowlands of Ecuador, at elevations of 300–1267 m [176,180].

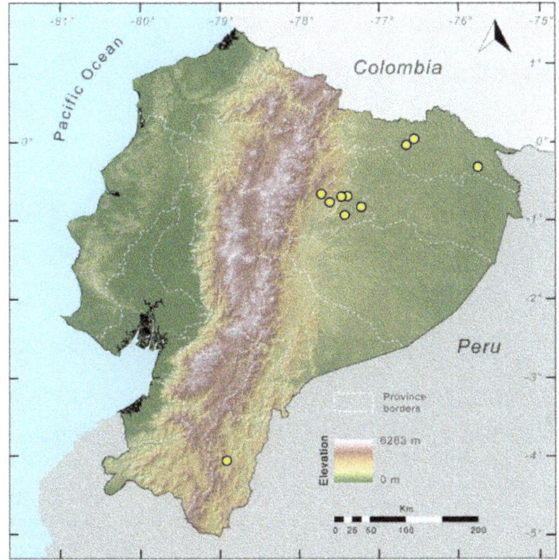

Figure 94. Distribution of *Espadarana durrellorum* in Ecuador (yellow dots).

Conservation status: Globally, *E. durrellorum* is *Least Concern* by the IUCN [181]; the species is known from few confirmed records, and a re-evaluation of its conservation status may be required.

Evolutionary relationships (Figure 88): *Espadarana durrellorum* and *E. audax* are sister species.

Taxonomic Remarks: See Taxonomic Remarks under *Espadarana audax*.

Specimens examined: *Espadarana durrellorum:* <u>Ecuador</u>: *Provincia de Napo:* ca. 45 km E of Narupa, on the Hollín–Loreto road (ca. 800 m), DFCH-USFQ D291; Parroquia Chontapunta, comunidad Ñukanchi Allpa, cabecera del río Canoayacu (0.99965° S, 77.39619° W; 403 m), QCAZ 47909; Reserva Yachana (00°52′21.7″ S, 77°14′13.4″ W; 300–350 m), DHMECN 03492, 06790–91; Pungarayacu (00°42′27.8″ N, 77°44′26.2″ W; 1267 m), DHMECN 03476. *Provincia de Sucumbíos:* Zábalo (0.3181333° S, 75.76625° W; 220 m), QCAZ 27832; Shuara (00°00′26″ N, 76°33′55.1″ W; 300 m), DHMECN 06794–95; Plataforma Shushufindi (00°05′14.8″ S, 76°40′03.8″ W; 330 m), DHMECN 06793. *Provincia de Zamora Chinchipe:* tributary of the Jambue River, ca. 6 km S from Zamora (ca. 04°03′ S, 78°56′ W; 1150 m), on the western slope of Contrafuerte de Tzunantza, Cordillera Oriental, eastern slopes of the Andes, DFCH-USFQ D131.

Espadarana prosoblepon (Boettger, 1892 [182]; Figures 91 and 95–97).

Hyla prosoblepon Boettger, 1892 [182]. Syntypes: SMF 3756 (formerly 1400.1a, according to
    Boettger, 1892 [182]), and ZMB 28019, according to Duellman, 1977 [118]; SMF 3756
    designated lectotype by Mertens, 1967.
Type locality: "Plantage Cairo (La Junta) bei Limon, (Cantón de Siquirres, Provincia de Limón)
    atlantische Seite von Costa Rica". Savage, 1974 [183], commented on the type locality.
Hylella puncticrus Boulenger, 1896 [184]. Syntypes: BM 96.10.8.70–71. Type locality: "La Palma"
    San José Province, Costa Rica. Placed in synonymy by Günther, 1901 [185].
Hyla parabambae Boulenger, 1898. Holotype: BM 98.4.28.163. Type locality: "Paramba" Imbabura
    Province, Ecuador. Noted elsewhere in the original publication as "Paramba, a farm on the
    W. bank of the River Mira, at 3500 feet altitude; it is still in the forest region, but the open
    country commences two or three miles higher up the Mira". Placed in synonymy by Lynch
    and Duellman, 1973 [22].
Centrolene prosoblepon—Noble, 1924 [186].
Cochranella parabambae—Taylor, 1951 [15].
Centrolenella parabambae—Goin, 1964 [187].
Centrolenella prosoblepon—Goin, 1964 [187].
Hyla ocellifera Boulenger, 1899. Holotype: BM 98.5.19.3. Type locality: "Paramba, (Provincia
    Imbabura,) N. W. Ecuador". Placed in synonymy by Cisneros-Heredia and McDiarmid,
    2007 [17].
Espadarana prosoblepon—Guayasamin, Castroviejo-Fisher, Trueb, Ayarzagüena, Rada, and Vilà,
    2009 [1].

Common names: English: Variable Glassfrog. Spanish: Rana de Cristal Variable.

Etymology: The specific name *prosoblepon* is apparently derived from the Greek words *proso*
(forward, onward, in front) and *blepo* (look or see), probably referring to the forward orientation of the
eyes in centrolenid frogs [84].

Identification: *Espadarana prosoblepon* is distinguished from most glassfrogs from the Pacific
versant of the Andes by having a large, flat, and projected humeral spine, and green dorsum with
small black spots (Figure 95); however, *E. prosoblepon* has a highly variable dorsal coloration, including
individuals that have both black and yellows spots, only yellow spots, only black spots, or absence
of spots all together. In Ecuador, the only species that have a green dorsum with dark spots are
Cochranella litoralis, *Vitreorana ritae*, *Nymphargus cochranae*, and *N. megacheirus*. In three of these species
(*V. ritae*, *N. cochranae*, *N. megacheirus*), males lack humeral spines; additionally, they occur on the
Amazonian lowlands or Amazonian slope of the Andes and are never in sympatry with *E. prosoblepon*.
Cochranella litoralis has a bright orange iris and is smaller than *E. prosoblepon* (in males, SVL < 22.1 mm in
C. litoralis; 23.2–27.5 mm in *E. prosoblepon*). Unspotted *E. prosoblepon* can be confused with *E. callistomma*
and *C. buckleyi*; *E. callistomma* is distinguished by having a larger body size (in males, SVL 26.7–29.6
mm; in females, SVL 29.5–31.8 mm), contrasting black-and-white iris, and a call with three to four notes
(call with two notes in *E. prosoblepon*; see Taxonomic remarks). *Centrolene buckleyi* has an inclined snout
in lateral view (Figure 26) and is only found at elevations above 2000 m. Individual of *E. prosoblepon*
with uniform dorsum and white iris with black reticulations resemble *Sachatamia ilex*; however, males
of *E. prosoblepon* have a conspicuous humeral spine, whereas males of *S. ilex* have a small humeral
spine that is embedded in the arm musculature (Figure 201). The closely related *E. andina* is similar
to *E. prosoblepon* by having dark spots on the dorsum but has a pointy humeral spine that is almost
parallel to the humerus, whereas in *E. prosoblepon*, the spine is broad, laminar, and projected at an angle
of about 45° from the humerus.

Figure 95. *Espadarana prosoblepon* in life. Note variation of dorsal patterns. Frogs from Mindo and surroundings, Pichincha province, Ecuador; not collected. Photos by Alejandro Arteaga/Tropical Herping.

Diagnosis: (1) Each vomer with two to seven teeth on dentigerous process; (2) snout usually truncated in dorsal and lateral profiles (Figure 96); (3) tympanum small, oriented almost vertically, with slight lateral and posterior inclinations, its diameter 20.5%–30.8% of eye diameter; tympanic annulus mostly visible, with supratympanic fold covering its posterodorsal margin; tympanic membrane translucent, partially pigmented, clearly differentiated from surrounding skin; (4) dorsal skin shagreen; males and females may possess minute spicules on flanks and tympanic region; (5) pair of enlarged subcloacal warts; (6) white iridophores on the anterior 50%–70% of the ventral parietal peritoneum, posterior portion transparent (condition P2–P3); white pericardium; translucent peritonea covering intestines, stomach, kidneys, gall bladder, and urinary bladder (condition V1); (7) liver with four clearly defined lobes covered by transparent peritoneum (condition H0); (8) males with conspicuous humeral spines; (9) webbing absent between Fingers I and II, absent or basal between Fingers II and III; moderate between outer fingers (Figure 96); webbing formula III ($1^{2/3}$–2)—(1–2) IV; (10) webbing between toes moderate; webbing formula on foot I (0^{+}–1)—(2^{-}–2^{+}) II (0^{+}–1)—(2^{-}–2^{+}) III (0^{+}–1)—(2^{-}–2^{+}) IV (2^{-}–$2^{1/3}$)—(1–1^{+}) V; (11) ulnar and tarsal folds absent or low and inconspicuous, lacking white coloration; (12) concealed prepollex; nuptial pad Type I; (13) Finger I as long as Finger II or slightly longer (Finger II length 93.8%–101% Finger I); (14) disc of Finger III moderate, width 28.6%–49.0% of eye diameter; (15) in life, dorsum green usually with small dark spots (Figure 95; but see *Color in life* section); bones green; (16) in preservative, dorsum lavender with or without cream and/or dark spots; (17) in life, iris varies from grey with fine dark reticulations to white with thick black reticulations; (18) melanophores covering dorsal surfaces of Fingers III and IV and Toes IV and V; (19) males call from upper side of leaves or small branches; (20) males fight upside down, grasping one another venter to venter; (21) black eggs are deposited on the upper surface of leaves, moss-covered rocks, or branches; short-term maternal care present; males do not provide parental care; (22) tadpoles with non-emarginated oral apparatus; tooth row formula 2/3; upper jaw curved; (23) small to medium body size; in adult males, SVL 23.2–27.5 mm ($\overline{X}$ = 24.8 ± 0.982, *n* = 53); in adult females, SVL 25.3–27.8 mm ($\overline{X}$ = 26.6 ± 0.859, *n* = 15).

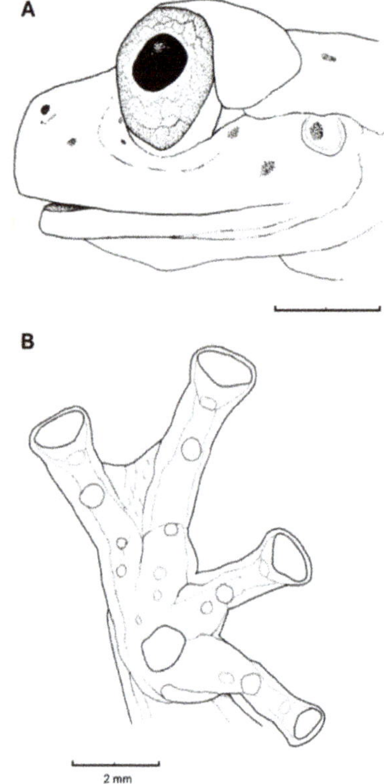

Figure 96. *Espadarana prosoblepon*, KU 132462. (**A**) Head in lateral view. (**B**) Hand in ventral view. Illustrations by Juan M. Guayasamin

Color in life (Figure 95): In Costa Rica, dorsal surfaces green dorsum with dark spots [148]. About one-half of Honduran specimens lack small dark dorsal spots [84]. In Ecuador, several color patterns have been reported: (1) Uniform green dorsum; (2) dorsum green with dark spots; (2) dorsum green with yellow spots; (3) dorsum green with dark and yellow spots, separated from each other; and (4) dorsum green with dark spots surrounding yellow spots, forming false ocelli. Iris varies from greyish white to pale bronze with fine dark reticulation, to silvery white with thick black reticulations. Green bones; whitish cream upper lip; white venter that is transparent posteriorly; ventrolateral margins of forearm and tarsus lacking white coloration.

Color in ethanol: As described above, but green turns into lavender and yellow spots turn white. Some individuals (e.g., KU 132463–64) present black and smaller white spots on dorsum. White parietal peritoneum covering about anterior 60% of venter. Translucent peritonea cover digestive tract, liver, kidneys, and urinary and gall bladders; white pericardium.

Biology and ecology: The following information is based on studies by Jacobson [188], Hayes [189], Savage [148], Hoffmann [147], and Arteaga et al. [87]. At night, during the breeding season, *Espadarana prosoblepon* is found on vegetation along margins of streams in primary to slightly degraded evergreen forests and, sporadically, even in pastures. Reproduction peaks in the rainy season, when males become more vocal and often engage in aggressive territorial combats. Males initiate amplexus by jumping on the back of approaching gravid females. Females lay 20–52 black eggs on a variety of substrates and directly above the water. Females provide short-term parental care; male parental care is absent [25]. After hatching, tadpoles usually hide under leaf litter or sand at the bottom of streams. Males are territorial, and the advertisement call is used to space them along the stream. If an intruder moves into an occupied territory, the resident male produces a rapid series of calls; if the intruder approaches too closely, a fight ensues with both males dangling upside down while holding onto vegetation with their legs and grappling with their arms and humeral spines. The combat concludes when the loser drops from the fight site or signals submission by flattening his body against the leaf surface. *Espadarana prosoblepon* feeds on a variety of small arthropods including beetles, moths, spiders, and mites. As a defense mechanism, this glassfrog produces pungent odors when grabbed. *Espadarana prosoblepon* is the only glassfrog species for which survival information is available; McCaffery and Lips [190] report that mean annual survival probability is 0.46 (0.41–0.52, 95% CI). These authors also provide abundance data on several populations from Panama.

Call (Figure 91): The following description is based on a call of a male *Espadarana prosoblepon* (MZUTI 601) recorded by Italo Tapia during the night of 4 July 2012 at Río Sune Chico (908 m, Cordillera Chontilla, Pachijal-Mashpi) Pichincha province, Ecuador. Air temperature at the moment of the recording was 18.8 °C. Calls are produced every 4–7 min. Each call consists of two pulsed notes and lasts 0.207–0.223 s ($\overline{X}$ = 0.212 ± 0.007, n = 4). Each note has a duration of 0.032–0.054 notes/s ($\overline{X}$ = 0.040 ± 0.008, n = 8); time between the two notes of each call is 0.130–0.137 s ($\overline{X}$ = 0.134 ± 0.004, n = 4). Calls are frequency modulated; the dominant frequency at the beginning of each note is 5712–5825 Hz ($\overline{X}$ = 5759 ± 56.1, n = 4), whereas at the end of the note is at 6196–6394 Hz ($\overline{X}$ = 6252 ± 95.7, n = 4). Harmonics are also frequency modulated, showing complex spectral patterns.

The call of *E. prosoblepon* from Costa Rica shows some differences from the Ecuadorian call described above. The following information is from Jacobson [188] and describes a call from Monteverde, Costa Rica. Males of *Espadarana prosoblepon* call at irregular intervals (17 calls/h) from the upper surfaces of leaves during the night. The advertisement call consists of two to five short "beeps". When encountering another frog within 15 cm, the resident males often give a series of rapid short beeps. Males of *Espadarana prosoblepon* call vigorously while in amplexus and also immediately after egg deposition. Kubicki [24] described the call of *E. prosoblepon* as a rapid three-note series, but sometimes males emit four to five notes; each note has an average duration of 0.03 s; the dominant frequency is 5.5–5.8 kHz. Typical call series have a total duration lasting 0.5 s (average three-note series) or 0.75 s (average four-note series). Because males call from exposed sites on the upper surfaces of leaves and twigs, these periodic calling bouts, often synchronized among a few, nearby males, may make it more difficult for predatory bats (i.e., *Trachops cirrhosus*) to locate and eat calling male *E. prosoblepon* (RWM, pers. obs.).

Tadpole: Females lay egg clutches on the upper side of leaves, moss-covered rocks, or branches; clutch size varies from 20 to 52 eggs that have a dark brown to black coloration; females remain with the eggs for some time (up to 131 min), but once they leave neither parent returns to the clutch [87,147,148,188,189]. Embryos are black but quickly turn yellow. By the time the tadpoles hatch and fall into the stream, they are red colored [148]. At Stage 25 [191], larvae have a sinistral spiracle, a relatively simple oral disc with medium-sized jaw sheaths, 2/3 tooth rows, and a single row of marginal papillae with a wide dorsal gap. The second anterior (upper) tooth row (A-2) also has a wide gap [148]. Drawings of the mouth and body can be found in Starrett [146], Savage [148]), and McCranie and Wilson [84].

Distribution (Figure 97): *Espadarana prosoblepon* is found from eastern Honduras, Nicaragua, Costa Rica, and Panama, south along the Pacific slopes of the Andes of Colombia and Ecuador from sea level to about 1600 m elevation. In Colombia, it occurs on the northern and eastern flanks of the Eastern Cordillera south to Caldas, and in the Magdalena Valley [192]. In Ecuador, *Espadarana prosoblepon* has been reported from several localities below 1620 m [87] (Specimens examined). The habitat of the species in Ecuador is mainly within the Chocoan Tropical Forest and the Western Foothill Forest regions, but it is also found in Deciduous Forest and the lower limit of the Western Montane Forest region. In Ecuador, it has a potential distribution of 41,612 km^2.

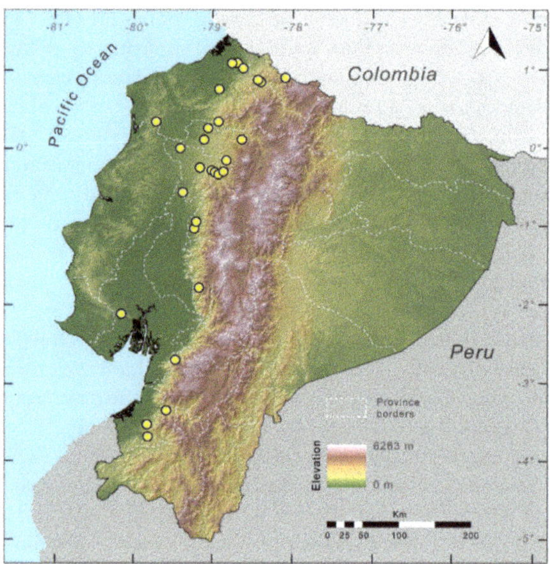

Figure 97. Distribution of *Espadarana prosoblepon* in Ecuador (yellow dots).

Conservation status: Globally, *Espadarana prosoblepon* is listed as *Least Concern* [192] in view of its large distribution and presumed large populations. In Ecuador, this species is common and is found even in habitats with moderate alterations; the category of *Least Concern* is justified.

Evolutionary relationships (Figure 88): *Espadarana prosoblepon* is paraphyletic, containing populations currently assigned to *E. callistomma*, a pattern that can be explained, mainly, by recent species divergence or the two species actually representing a single evolutionary lineage. See Taxonomic Remarks below.

Internal morphology: A detailed osteological and myological description of *Espadarana prosoblepon* was provided by Eaton [193]. The hyoid apparatus illustrated by Eaton lacks the anterolateral processes that are evident in cleared-and-stained material from Ecuador (KU 178163) and Costa Rica (KU 65178). We assume that the normal condition is the presence of these processes.

Taxonomic remarks: While the general morphology of *Espadarana prosoblepon* is mostly conservative, its color pattern is extremely variable, and because of an insufficient understanding of the species' variation, different names have been assigned to distinctive morphs. For example, populations whose individuals have dorsal coloration consisting of black spots surrounding yellow spots and forming ocelli were called *ocellifera*, while others with dorsum lacking spots have been variously named *parambae* and *callistomma*. Several authors, including Lynch and Duellman [22], Savage [42], and Cisneros-Heredia and McDiarmid [17], have described some of the interspecific variation characteristic of *E. prosoblepon*, but much remains to be done. For example, most descriptions and color photographs of *E. prosoblepon* depict the iris of this species as grey with fine dark reticulations, but Kubicki [24] reported individuals with a white iris with black reticulations. Unfortunately, most centrolenid descriptions have paid little attention to iris patterns; in fact, with the exception of the descriptions of *Sachatamia ilex*, *Hyalinobatrachium ignioculus*, and *H. eccentricum* (the last two taxa now under the synonymy of *H. cappellei*), iris coloration, if reported at all, usually is based on examination of photographs; thus, the inter and intraspecific variation in iris color is poorly known. Genetic data suggest that *E. prosoblepon* and *E. callistomma* might represent one evolving lineage (Figure 88). Morphological differences between *E. callistomma* and *E. prosoblepon* are in body size (*callistomma* being larger), dorsal coloration (without spots in *callistomma*; usually with spots in *prosoblepon*), and iris coloration (white with contrasting black reticulation in *E. callistomma*; less contrasting in *E. prosoblepon*). In terms of vocalization, *E. callistomma* has a call with more notes (three to four notes per call) than *E. prosoblepon* (two notes per call; Figure 91). At the moment, we consider that the morphological and acoustic differences justify recognition of *E. callistomma* as a valid species.

Specimens examined: *Espadarana prosoblepon*: Ecuador: *Provincia de Azuay:* 12.9 km W Luz María (2.6889 S, 79.474 W, 740 m), KU 217502, QCAZ 12603–04; *Provincia de Bolívar:* Balzapamba (1.7667 S, 79.1833 W, 800 m), KU 132555–56, 132462–65; *Provincia de Carchi:* near Maldonado (0.9 N, 78.1 W, 1410 m), KU 178156–57; *Provincia de Cotopaxi:* near La Mana (0.933 S, 79.2167 W, 300 m), QCAZ 8641; near Sigchos (0.7 S, 78.883 W), USNM 288441; km 8 of the Pucayacu-Sigchos road (1.0097 S, 79.23769 W), QCAZ 40674. *Provincia de Esmeraldas:* Reserva Biológica Bilsa (0.3447 S, 79.71 W, 500 m), KU 291165–75, QCAZ 22414–17; Viruela, QCAZ 10267; Reserva Biológica Bilsa (0.3447 S, 79.71 W, 500 m), KU 291165–75; Reserva Ecológica Cotacachi-Cayapas, Charco Vicente (0.792 N, 79.1978 W; 60 m), QCAZ 11364–65; 5 km W of Durango (1.0858 N, 78.74 W), QCAZ 13206, 13212, 13242. *Provincia de El Oro:* near Valle Hermoso (3.50194 S, 79.81722 W; 379 m), QCAZ 37249; Río Chillayacu (3.32834 S, 79.58102 W; 395 m), 16.8 km W of Piñas (3.667 S, 79.667 W, 600 m), USNM 286738–39. *Provincia del Guayas:* Chongon-Colonche hills near Guayaquil, (ca. -2.1 S, -80.15 W), USNM 288438; *Provincia de Imbabura:* near Lita (0.833 N, 78.4667 W, 520 m), KU 133482–83; 6 km E of Lita (0.79472 N, 78.4286 W), QCAZ 4318–19; Zona de amortiguamiento de Reserva Cotacachi Cayapas, near Rio Aguas Verdes (0.331010° N, 78.93152° W; 670 m), QCAZ 46009; km 5 in the Lita-Ibarra road (0.84773 N, 78.42175 W), QCAZ 39919. *Provincia de Los Ríos:* Estación Biológica Río Palenque, 56 km N of Quevedo (0.55 S, 79.3667 W, 220 m), KU 164616–22. *Provincia de Santo Domingo de los Tsáchilas:* Río Orito, on the Toachi-Chiriboga road (0.30561 S, 78.882 W, 1315 m), QCAZ 15356; Río Faisanes, on the Toachi-Chiriboga road (0.2608 S, 78.845 W; 1400 m), QCAZ 15357, 15360, 15362; La Florida (0.28361 S, 79.0189 W), QCAZ 20726–28, 20730–32; Otongachi, near La Unión del Toachi (0.3167 S, 78.95 W; 900 m), QCAZ 25094; 5 km NE of La Florida (0.25694 S, 79.0539 W), QCAZ 7184–893; 4 km NE of the Dos Ríos–Chiriboga road (0.305139 S, 78.884333 W; 1270 m), QCAZ 31982; Santo Domingo de los Colorados (0.25 S, 79.15 W, 660 m), KU 121054–55; 4 km NE of Dos Ríos (0.30278 S, 78.8678 W, 1140 m), KU 164623–34; 2 km E and 1 km S of Santo Domingo de los Colorados (0.24512 S, 79.15509 W, 600 m), KU 178158–66; La Palma (0.3167 S, 78.9167 W, 920 m), KU 178167. *Provincia de Pichincha:* Reserva Maquipucuna (0.12429 N, 78.62936 W; 1343 m), QCAZ 42179; 6 km NW (by air) of Pedro Vicente Maldonado (0.10421 N, 79.10279 W; 544 m), QCAZ 35430; La Concordia, Bosque Protector La Perla (0.01 N, 79.4 W, 190 m), QCAZ 12602, Reserva Mashpi, 18 km N of San Miguel de Los Bancos (0.15 S, 78.883 W, 1100 m), DFCH-USFQ 293–95.

Genus *Hyalinobatrachium* Ruiz-Carranza and Lynch, 1991 [6].

Etymology: The generic name *Hyalinobatrachium* comes from the Greek words *hyalos* (glass) and *batrachion* (frog), alluding to the translucent and delicate appearance of the species in this genus [6].

Hyalinobatrachium adespinosai Guayasamin, Vieira, Glor, Hutter 2019 [194] (Figures 98–101).

Hyalinobatrachium adespinosai Guayasamin, Vieira, Glor, Hutter, 2019 [194]. Holotype: ZSFQ
    1648.
Type locality: "San Jacinto River (1.3447 S, 78.1814 W; 1795 m asl), Tungurahua province,
    Ecuador".

Common names: English: Adela's Glassfrog. Spanish: Rana de Cristal de Adela.

Etymology: The specific epithet *adespinosai* honors Adela Espinosa, an Ecuadorian conservationist and board member of the Jocotoco Foundation (http://www.jocotoco.org/wb#/EN/LaFundacion). Adela has concentrated her work to the conservation of species and ecosystems. *Hyalinobatrachium adespinosai* is found only within the limits of a natural reserve owned by Adela and her husband, Antonio Páez [194].

Identification: Among glassfrogs, *H. adespinosai* (Figure 98) is diagnosable mainly by having a transparent pericardium. However, it is morphologically cryptic with four closely related taxa (*H. anachoretus, H. pellucidum, H. esmeralda)*; these species display a similar size and color pattern (pale green dorsum with yellow dots and a transparent venter and pericardium; red heart visible ventrally). However, calls have diverged noticeably (Figure 99); the major difference is the structure of the call, with two species (*H. adespinosai* and *H. anachoretus*) having pulsed calls and the others having tonal vocalizations. The call of *H. adespinosai* sp. nov. is further differentiated from that of *H. anachoretus* by being longer, having more pulses per note, and being produced at a higher rate [194]. Genetic distances for the 16S marker between *H. adespinosai* and closely related species (*H. pellucidum, H. esmeralda, H. yaku)* are >2.5%. However, *H. adespinosai* and *H. anachoretus* only have a 1% genetic distance, although they are separated by a major biogeographic barrier (i.e., Marañon river valley) [194].

Diagnosis: The following characters are found in *Hyalinobatrachium adespinosai:* (1) Dentigerous process of the vomer lacking teeth; (2) snout truncated in dorsal and lateral profiles; (3) tympanum barely visible, hidden under skin, with coloration similar to that of surrounding skin; (4) dorsal skin shagreen, lacking tubercles; (5) ventral skin areolate; cloacal ornamentation absent, paired round subcloacal warts absent; (6) parietal peritoneum transparent; pericardium with thin layer of iridophores (in life, a red heart is mostly visible ventrally); liver, viscera, and testes covered by iridophores; (7) liver white, bulbous; (8) humeral spines absent; (9) hand webbing formula: I (2–3)—(2–2$^+$) II (1–1$^+$)—3$^{1/3}$ III (2–2$^+$)—(2$^-$–2) IV; (10) foot webbing moderate; webbing formula: I 1—(1$^{2/3}$–2$^-$) II (1–1$^-$)—(2$^-$–2$^{1/3}$) III (1–1$^+$)—(2$^+$–2$^{1/3}$) IV 2$^+$—(1$^+$–1$^{1/3}$) V; (11) fingers and toes with thin lateral fringes; ulnar and tarsal folds present, but difficult to distinguish, with thin layer of iridophores that extends to ventrolateral edge of Finger IV and Toe V; (12) nuptial excrescence present as a small pad on Finger I (Type V), concealed prepollex; (13) when appressed, Finger I longer than II; (14) diameter of eye about two times wider than disc on Finger III; (15) coloration in life: Dorsal surfaces pale yellowish green with small pale yellow spots and minute grey to black melanophores; bones white; (16) coloration in preservative: Dorsal surfaces pale cream with minute melanophores; (17) iris coloration in life: White with pale yellow hue and minute lavender spots; (18) melanophores absent from most fingers and toes, but present on Finger IV and Toes IV and V; (19) males call from underside of leaves; advertisement call consisting of single note, distinctly pulsed (9–13 pulses per call), with duration of 0.382–0.430 s, and dominant frequency at 4645–5001 Hz; (20) fighting behavior unknown; (21) males attend egg clutches located on the underside of leaves overhanging streams; clutch size of 22 embryos (*n* = 1); (22) tadpoles unknown; (23) SVL in adult males 20.5–22.2 mm (*n* = 3), females unknown.

Figure 98. *Hyalinobatrachium adespinosai* in life, holotype, ZSFQ 1648. Photos by Jose Vieira/Tropical Herping.

Color in life (Figure 98): Dorsal surfaces apple green to yellowish green with diffuse yellow spots and minute grey to black melanophores. Melanophores absent from fingers and toes, except Finger IV and Toes IV and V. Ventrally, parietal peritoneum and pericardium transparent, with a red heart always visible, even when a very thin layer of iridophores is present on the pericardium of some individuals. Visceral peritoneum of gall bladder and urinary bladder transparent; hepatic and visceral peritonea white; ventral vein red. Iris pale yellowish white, with numerous minute lavender spots. Bones white [194].

Color in ethanol: Dorsal surfaces cream dotted with minute dark lavender melanophores; venter uniform cream; visceral peritoneum lacking iridophores; pericardium with a very thin layer of iridophores. Iris silvery white with minute lavender melanophores [194].

Biology and Ecology: All individuals were found on the underside of leaves of riverine vegetation along the San Jacinto River. The section of river was fast-flowing and had visible rapids. Although the population is locally abundant, individuals are difficult to observe because they are usually found at the canopy level (4–16 m above ground level). Males were calling in the months of July and August. One male (ZSFQ 1648) was apparently guarding an egg clutch containing 22 embryos; both the adult male and the egg clutch were on the same leaf most of the time, but the male also moved to nearby leaves [194].

Call (Figure 99): The description is based on recording from nine individuals. The call of *Hyalinobatrachium adespinosai* has a striking resemblance to the chirp of a cricket. Each call is composed by a single and high-pitched pulsed note and has a duration of 0.38–0.44 s ($\overline{X}$ = 0.38 ± 0.017). Time between calls varied from 2.0–11.0 s ($\overline{X}$ = 4.58 ± 2.3). The fundamental frequency, the same as the dominant frequency, is at 4645–5203 Hz ($\overline{X}$ = 4855 ± 152). There is no frequency modulation. The first harmonic is at 9336–9754 Hz and the second harmonic is at 14,159–14,444 Hz [194].

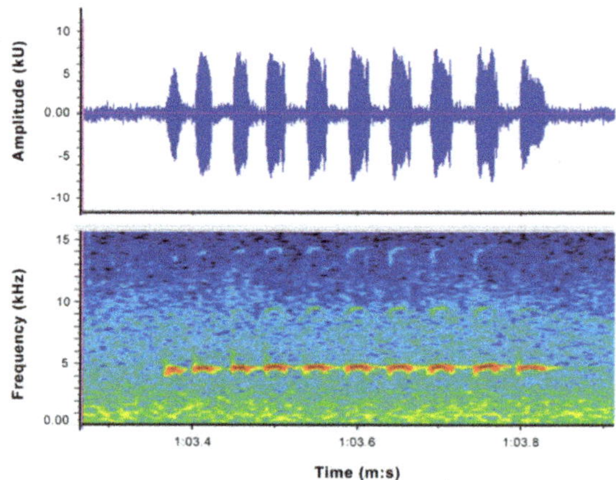

Figure 99. Call of *Hyalinobatrachium adespinosai*, holotype, ZSFQ 1648, recorded in field conditions at the type locality (San Jacinto River, 1795 m asl), Tungurahua province, Ecuador). Air temperature = 18 °C. Obtained from Guayasamin et al. [194].

Tadpole: Not described.

Distribution. *Hyalinobatrachium adespinosai* is only known from the type locality: San Jacinto River (1.3447 S, 78.1814 W; 1795 m asl), Tungurahua Province, Ecuador (Figure 100) [194].

Figure 100. Distribution of *Hyalinobatrachium adespinosai* in Ecuador (yellow spot).

Evolutionary relationships. *Hyalinobatrachium adespinosai* is sister to the Peruvian *H. anachoretus* [194]. Since the species was recently described, it is not included in the *Hyalinobatrachium* tree shown in Figure 101. The most closely related species to *H. adespinosai* share several morphological traits, including a red heart exposed ventrally (*H. adespinosai* + *H. anachoretus* + *H. pellucidum* + *H. yaku*) [194].

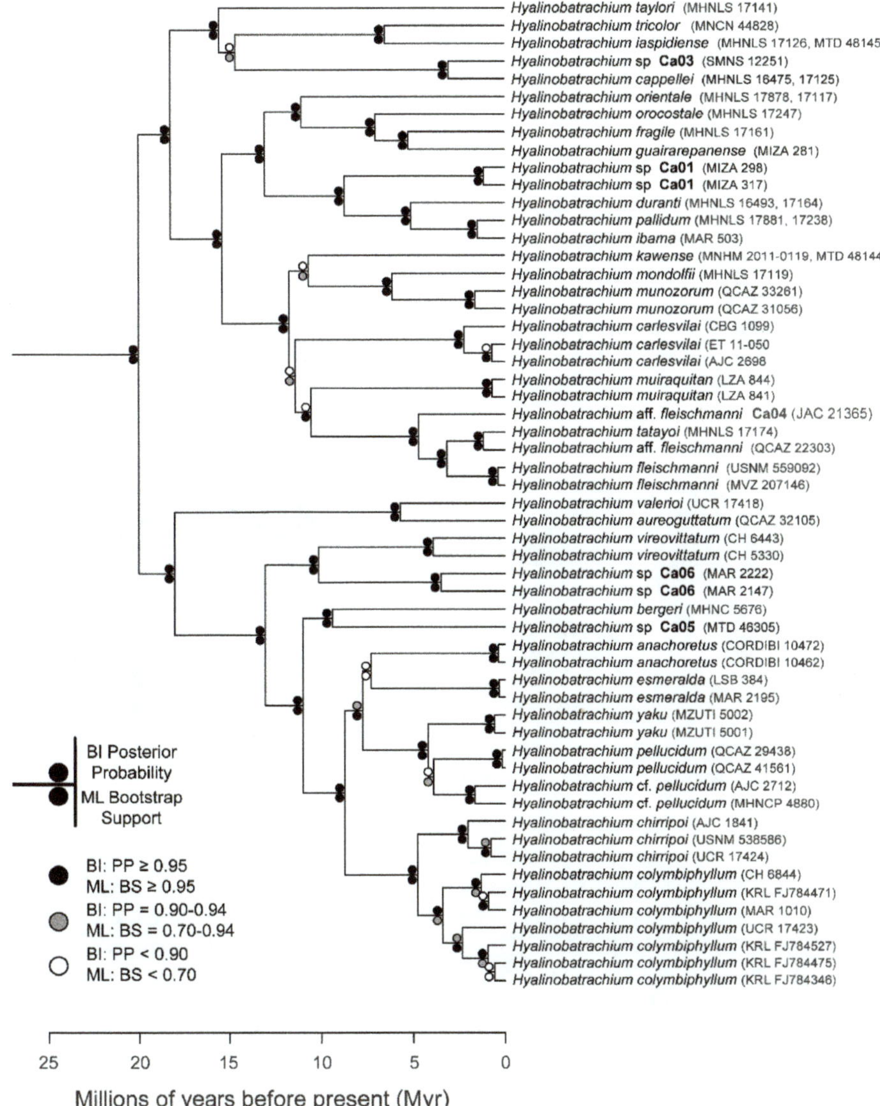

Figure 101. Evolutionary relationships among species in the genus *Hyalinobatrachium*, inferred using maximum likelihood and Bayesian criteria.

Conservation status: *Hyalinobatrachium adespinosai* has not been evaluated by the IUCN. However, Guayasamin et al. [194] suggested that it should be considered as *Data Deficient*.

Specimens examined. *Hyalinobatrachium adespinosai*: <u>Ecuador</u>: *Provincia de Tungurahua:* San Jacinto River (1.3447 S, 78.1814 W; 1795 m asl), ZSFQ 1647–48, 1650–52 (type series).

Hyalinobatrachium aureoguttatum (Barrera-Rodríguez and Ruiz-Carranza, 1989 [195];
    Figures 101–105).
Centrolenella aureoguttata Barrera-Rodriguez and Ruiz-Carranza, 1989 [195]. Holotype: ICN
    17506.
Type locality: "Colombia, Departamento del Chocó, vertiente occidental de la cordillera
    Occidental, Municipio El Carmen de Atrato, Km 23 carretera El Carmen–Quibdó, 5°47′
    latitud N y 76°20′ W, 1030 m.s.n.m.".
Hyalinobatrachium aureoguttatum—Ruiz-Carranza and Lynch, 1991 [6].

Common names: English: Sun Glassfrog. Spanish: Rana de Cristal del Sol.

Etymology: The specific name *aureoguttatum* is derived from the Latin words *aureus* (gold) and
guttatus (dappled, speckled, spotted), referring to the dorsal pattern of the species [195].

Identification: *Hyalinobatrachium aureoguttatum* is easily distinguished from most glassfrogs by
having a completely transparent ventral parietal peritoneum, a white bulbous liver, white visceral
peritoneum, and by the absence of humeral spines. It differs from all other *Hyalinobatrachium* by having,
in life, large yellow spots on the dorsum (Figure 102). The most similar and related species is *H. valerioi*,
from which is differentiated mainly by having large dorsal yellow spots produced by xanthophores
and iridophores (absent in *H. valerioi*; see Figure 121).

Figure 102. *Hyalinobatrachium aureoguttatum* in life. (**Left and center**) Adult male, QCAZ 45365; photos
by L. A. Coloma. (**Right**) Adult male, QCAZ 32068; photo by Martín Bustamante.

Diagnosis: (1) Vomers lacking teeth; (2) snout truncated in dorsal and lateral profiles (Figure 103);
(3) tympanum of moderate size (tympanum diameter 30%–34% of eye diameter), with dorsolateral
orientation and posterior inclination, supratympanic fold low, tympanic membrane clearly differentiated
from surrounding skin; (4) dorsal surfaces smooth to shagreen, lacking spicules; (5) venter smooth;
lacking pair of enlarged subcloacal warts; (6) ventral parietal peritoneum transparent (condition P0);
pericardium polymorphic, with or lacking iridophores; white peritonea covering intestines, stomach,
and testes; transparent peritoneum covering kidneys, and urinary and gall bladders (Figure 102,
Condition V5–V6); (7) liver bulbous, covered by white peritoneum (condition H2); (8) humeral
spines absent; (9) basal webbing between Fingers I and II, extensive webbing between outer finger:
I (2–2$^{1/4}$)—(2–2$^{1/4}$) II (1$^-$–1$^+$)—(2–3$^+$) III (1–2$^+$)—(1–1$^+$) IV; (10) foot webbing extensive: I (0$^+$–1)—(1–1$^{2/3}$)
II (0$^+$–1)—(1–1$^{1/2}$) III (0$^-$–1)—(1$^{1/2}$–2) IV (1$^{1/2}$–2$^+$)—(1$^-$–1) V; (11) ulnar and tarsal folds absent;
(12) concealed prepollex; in males, nuptial pad Type III; (13) Finger I longer than Finger II (Finger II
about 87%–95% length of Finger I); (14) disc of Finger III of moderate size, 35%–56% of eye diameter;
(15) in life, dorsum greenish yellow with two to seven large yellow spots, and with or without brown
flecks; ventral parietal peritoneum transparent, pericardium polymorphic (white or transparent),
gastrointestinal peritoneum white; bones white; (16) in ethanol, dorsum cream with large white spots,
and with or without brown flecks; (17) in life, iris white to yellow, with minute dark lavender flecks
around the pupil or forming an horizontal stripe; (18) fingers and toes lacking melanophores, except for

few melanophores on proximal portions of Toes IV and V; (19) males usually call from the underside of leaves; call undescribed; (20) fighting behavior unknown; (21) eggs deposited on the underside of leaves; prolonged parental care provided by males; maternal care absent; (22) oral apparatus of tadpoles with an emarginate disc; M-shaped upper jaw sheath; tooth row formula 2(2)/3; (23) minute body size; in males, SVL 20.4–24.0 mm ($\overline{X}$ = 21.8 ± 0.631, n = 36); in females, SVL 22.9–23.9 mm ($\overline{X}$ = 23.3, n = 3).

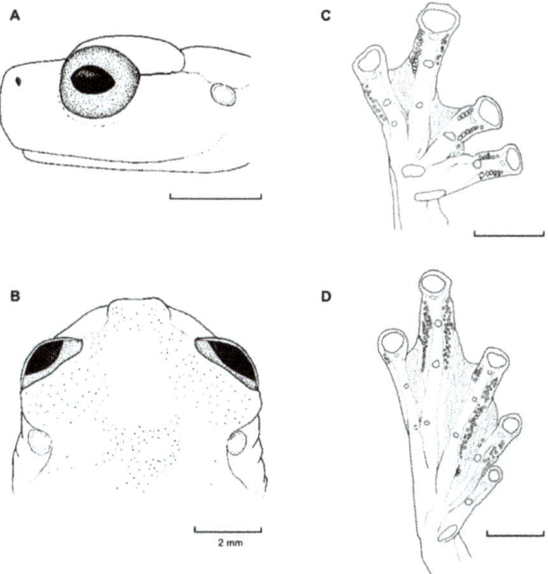

Figure 103. *Hyalinobatrachium aureoguttatum*, adult male, QCAZ 27429. (**A**) Head in lateral view. (**B**) Head in dorsal view. (**C**) Hand in ventral view. (**D**) Foot in ventral view. Illustrations by Juan M. Guayasamin

Color in life (Figure 102): Dorsum greenish yellow with two to seven large yellow spots (diameter 0.5–2.1 mm), and with or without black flecks; upper lip unpigmented; ventral parietal peritoneum transparent; pericardium polymorphic, transparent (red heart visible ventrally) or white (heart not visible ventrally); transparent peritonea covering the kidneys, and urinary and gall bladders; white peritonea covering the liver, intestines, stomach, and testes; bones white; iris white to yellow, with minute dark lavender flecks around the pupil or forming a horizontal stripe.

Color in ethanol: Dorsum cream with large white spots and with or without dark flecks; iris white with dark lavender pigment; white peritonea covering the liver, intestines, stomach, and testes; pericardium polymorphic (white, partially white, or transparent).

Biology and Ecology: According to Barrera-Rodríguez and Ruiz-Carranza [195], most individuals of *Hyalinobatrachium aureoguttatum* were found on the underside of leaves (*Heliconia* spp.) 100–700 cm above small streams during the night. Amplectant pairs and egg clutches were found during July and August 1987. Amplexus is axillary, and eggs are deposited on the underside of leaves. In Durango (Provincia de Esmeraldas, Ecuador), individuals were reproductively active (amplectant pairs and egg clutches) 24–26 May 2006. At Estero Piedras, the species was reproducing in August 2007; a clutch with 36 eggs was found on the underside of a leaf on a bush along a stream [167]. Valencia-Aguilar et al. [196] studied populations from the Pacific lowlands of Colombia. They found that males exhibit high fidelity to their territory; each male repeatedly uses the same leaf (*Heliconia* sp., *Anthurium* sp., *Philodendron* sp., *Cyclanthus* sp., *Calathea* sp., *Musa* sp.) for perching, calling, mating, and clutch attendance. Territoriality seems to be low, since several males where found in close proximity and fights or aggressive behaviors were not observed in intrusion events by co-specific males.

Female place green eggs on the underside of leaves; clutches 25–49 eggs [167,195–197]. Males provide parental care during the day and night until hatching (mean = 17.1 days ± 1.8). Males are polygynous and simultaneously attend up to five clutches [196].

Call: Most males call from the undersides of leaves, but some have been observed calling from leaf tops. The specific spectral and temporal characteristics of the call are unknown.

Tadpole: A description of the tadpole can be found in Ibáñez et al. [197] and Terán-Valdez et al. [167] (Figures 80 and 104).

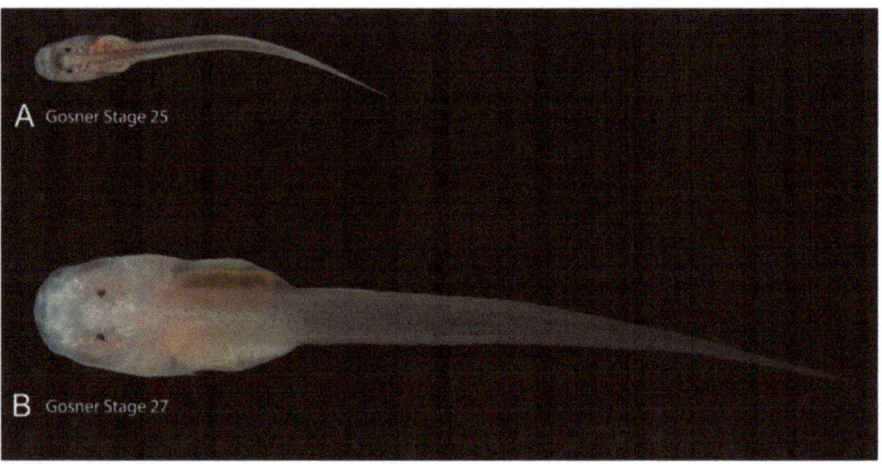

Figure 104. *Hyalinobatrachium aureoguttatum*, ontogenetic variation of tadpoles. (**A**) Gosner Stage 25, QCAZ 37752. (**B**) Gosner Stage 27, QCAZ 32072. Photos by L. A. Coloma. Figure modified from Terán-Valdez et al. [167].

Distribution (Figure 105): *Hyalinobatrachium aureoguttatum* is known from extreme southwestern Panama and the Pacific lowlands and western slopes of the Cordillera Occidental of Colombia and Ecuador at elevations below 1340 m [101,167,195,197–199]. In Ecuador, this species has been reported from the provinces of Esmeraldas and Imbabura at elevations below 600 m (Specimens Examined). In Ecuador, the species has a potential distribution of 4,481 km^2 within the Chocoan Tropical Forest and Western Foothill Forest ecoregions.

Conservation status: Listed globally as *Near Threatened* by the IUCN [200]. Given that Ecuadorian populations of the species are fragmented because of agriculture, pasture lands, and mining, we suggest placing the species in the *Endangered* category based on IUCN criteria B1, B2a, B2b(iii). The species is partially protected within the Mache-Chindul reserve.

Evolutionary relationships (Figure 101): *Hyalinobatrachium aureoguttatum* and *H. valerioi* are sister species.

Remarks: Some populations from Colombia usually have small (but visible) brown flecks on the dorsal surfaces of the head and body; Ecuadorian populations lack these flecks (Figure 102). Also, all Colombian specimens examined by us have a white pericardium, a character that is polymorphic in the Ecuadorian populations (white, partially white, or transparent pericardium). It is possible that these differences correspond to independent evolutionary lineages; however, we prefer to maintain these populations as one species until more data (e.g., acoustic, molecular) are available.

Figure 105. Distribution of *Hyalinobatrachium aureoguttatum* in Ecuador (yellow dots).

Specimens examined: *Hyalinobatrachium aureoguttatum:* <u>Ecuador</u>: *Provincia de Esmeraldas:* stream affluent of the Río Durango (1.05 N, 78.6167 W; 100–150 m), QCAZ 27429, 6302, 6303, 6441–42, 28802; 2 km E of San Francisco (1.0872 N, 78.6905 W; 60–80 m), on the San Francisco-Durango road, QCAZ 32101–02, 32105, 32129, 32132–33; Río Quingue, nearby Caimito (0.72096 S, 80.09117 W, 47 m), QCAZ 37306. *Provincia de Imbabura:* 6 km SE of Lita (0.79 N, 78.43 W; 600 m), QCAZ 4323. <u>Colombia</u>: *Departamento del Chocó:* Municipio El Carmen de Atrato, km 23 on road El Carmen–Quibdó (5°47′ N; 76°20′ W, 1030 m), ICN 17507, 17509–10, 17512; km 44 on road El Carmen–Quibdó, 630 m, ICN 17252–54, 17515–16; km 53 on road El Carmen–Quibdó, 420 m, ICN 17248, 17257, 17260, 17262, 17266–67, 17520–21, 17525, 17527–28, 17531–34, 17536–37.

Hyalinobatrachium chirripoi (Taylor, 1958 [201]; Figures 106–108).

Cochranella chirripoi Taylor, 1958 [201]. Holotype: KU 36865.
Type locality: "Cocales Creek, Suretka, (Cantón de Talamanca,) Limón Province", Costa Rica.
<u>Savage [183]</u> commented on the type locality.
Centrolenella chirripoi—Savage, 1967 [202].
Hyalinobatrachium chirripoi—Ruiz-Carranza and Lynch 1991 [6].
Hyalinobatrachium cardiacalyptum McCranie and Wilson, 1997 [203]. Holotype: USNM 342161.
Type locality: "Caño El Cajón (14°21′ N, 85°29′ W), at its junction with the Río Patuca, Departamento de Olancho, Honduras, elevation 200–225 m". Placed in synonymy by Cisneros-Heredia and McDiarmid, 2007 [17].

Common names: English: Chirripó Glassfrog. Spanish: Rana de Cristal Chirripó.

Etymology: The specific name *chirripoi* is named for the Chirripó Indians, local inhabitants of the area where the species was first found [201].

Identification: *Hyalinobatrachium chirripoi* is easily distinguished from most glassfrogs by having, in life, a lime green dorsum with small yellow spots, a completely transparent ventral parietal peritoneum (red heart visible ventrally; Figure 106), and by lacking humeral spines. Additionally, it differs from most *Hyalinobatrachium* by having more webbing between Fingers II and III.

Figure 106. *Hyalinobatrachium chirripoi* in life. Reserva Canandé, 372 m, Esmeraldas province, Ecuador. Photos by Jaime Culebras.

Diagnosis: (1) Vomers lacking teeth; (2) snout truncated in dorsal view, truncated to slightly protruding in lateral profile; (3) tympanum almost completely concealed, small when visible (tympanum diameter 2.3%–2.8% of SVL); tympanic membrane not differentiated from surrounding skin; (4) dorsal surfaces shagreen, lacking spicules; (5) venter areolate; lacking pair of enlarged subcloacal warts; (6) ventral parietal peritoneum transparent (condition P0); pericardium transparent; white peritonea covering the intestines and stomach; transparent peritoneum covering urinary and gall bladders; kidneys covered by a mostly translucent peritoneum, except for some iridophores on their ventral portions (condition V6); (7) liver bulbous, covered by white peritoneum (condition H2); (8) humeral spines absent; (9) basal to moderate webbing between Fingers I and II, extensive webbing between outer finger: I (2^+–$2^{1/3}$)—(2–2^-) II 1—($2^{2/3}$–3^-) III (2^-–2)—($1^{1/3}$–$1^{1/2}$) IV; (10) foot webbing extensive: I (0^+–1)—($1^{1/2}$–$1^{2/3}$) II 1—($1^{1/2}$–$1^{2/3}$) III (0^+–1)—2^- IV (2^-–2)—(1–1^+) V; (11) ulnar and tarsal folds present, with minute iridophores; (12) concealed prepollex; in males, nuptial pad Type V; (13) Finger I longer than Finger II (Finger II about 85%–90% length of Finger I); (14) disc of Finger III of moderate size, 38%–47% of eye diameter; (15) in life, dorsum line green with small yellow spots, venter transparent (red heart visible; Figure 106); bones white; (16) in ethanol, dorsum cream with minute lavender spots and iridophores; (17) in life, iris yellowish white, with minute dark lavender flecks; (18) fingers and toes lacking melanophores, except for few on Toes IV and V; (19) males call from the upper and undersides of leaves; call undescribed; (20) fighting behavior unknown; (21) eggs deposited on the underside of leaves; maternal care absent; prolonged parental care provided by males; (22) elongate tadpole, with emarginate oral disc; tooth row formula 2/3; (23) small body size; in Ecuadorian males, SVL 25.0–25.5 mm (*n* = 2); Savage [148] reported that adult males have a SVL = 24–26 mm and that females are probably larger.

Color in life (Figure 106): Dorsum lime green with several small yellow spots; ventral parietal peritoneum transparent; pericardium transparent (red heart visible ventrally); transparent peritonea covering urinary bladder; white peritonea covering liver, intestines, and stomach; transparent peritoneum covering most of kidneys; bones white; iris yellowish white, with minute dark lavender flecks.

Color in ethanol: Dorsum cream with minute lavender spots and iridophores; ventral parietal pericardium translucent; white peritonea covering liver, intestines, and stomach; pericardium transparent; iris silvery white with dark lavender flecks.

Biology and Ecology: According to Kubicki [24,204], in southeastern Costa Rica, *Hyalinobatrachium chirripoi* deposits 65–80 greenish–white eggs in a single layer on the underside of leaves overhanging streams. Males were most often seen calling from below palm fronds or other smooth leaves between 1–4 m above the water. Males were also seen calling from the upper sides of vegetation, but much less frequently. One male was seen guarding eggs during the night, and another male was observed guarding eggs during daylight hours [24]. At Reserva Itapoa, Ecuador, in June 2014, a male was observed calling and guarding an egg clutch on the underside of a leaf. Maternal care is absent; prolonged parental care is provided by males [25].

Call (Figure 107): Most males call from the undersides of leaves, but some have been observed calling from the upper sides [24,204]. The call is a high-pitched insect-like buzz, very similar to that of *H. colymbiphyllum* [204]. We analyzed two notes from one individual (LBE-019). The typical advertisement call is composed by a single note. Note duration is 230–270 (mean = 250, SD = 28.3) ms. Notes are strongly pulsed and have 14–16 (mean = 15, SD = 1.4) amplitude peaks throughout the note, with pulses becoming difficult to distinguish near the end of the note. Pulses within a note have a rate of 59–61 (mean = 60, SD = 1) pulses per second. Notes have their peak amplitude in the first 50% of the note (relative peak time: Range = 0.0623–0.0638, mean = 0.063, SD = 0.001), where the peak amplitude occurs in the first several pulses. The dominant frequency of a note measured at peak amplitude is 4565 (mean = 4565, SD = 0) Hz and is contained within the fundamental frequency. The fundamental frequency has a lower limit of 1895–3273 (mean = 2584, SD = 974) Hz and a higher limit of 4651–5082 (mean = 4866, SD = 305) Hz.

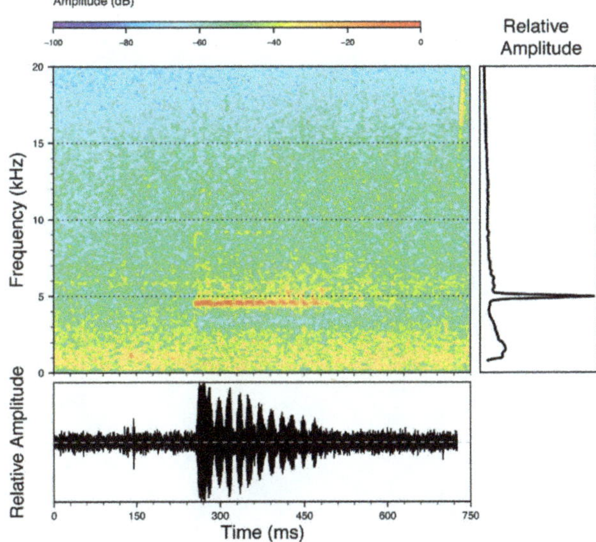

Figure 107. Call of *Hyalinobatrachium chirripoi* recorded at Reserva Itapoa, 321 m, Esmeraldas province, Ecuador, LBE-C-019.

Egg masses and tadpoles: Greenish–white eggs (53–80) are deposited in a single layer on the underside of leaves; as development continues the embryos become pale tannish red [24,147,204]. Below, we present a summary of the tadpole description by Hoffmann [147]. Hatchlings (Gosner stages 25 and 26) have an unusual pattern on their dorsum: Two rows of brown-pigmented, spider-like dots of melanophores stretch longitudinally over the entire body and join in a semicircle at the tip of snout; another line of pigmented spots extends dorsally over the tail musculature. Because of the general lack of pigment and high transparency, the body coloration is partially determined by the red color of the gills and heart. As development continues (Gosner stages 25–41), the tadpoles become very elongated and are among the most elongated centrolenid larvae known (ratio of body length to body width of 2.41 ± 0.22); they have a completely transparent skin and the dorsolateral lines are mostly lost; the body has a pale rose shine and the tail is cream or, sometimes, yellowish. The oral disc is emarginate and, except for a moderate dorsal gap, is bordered by a single row of about 45 marginal papillae; the ventral ones being flatter than the lateral ones. The LTRF is 2(2)/3; the tooth rows are about of equal width (P-3 is slightly narrower) and extend across most of the oral disc. The A-2 gap is about as wide as the upper jaw, which is broadly arched; the lower jaw is V-shaped; the edges of both jaw sheaths are edged by many narrow serrations [147].

Distribution (Figure 108): *Hyalinobatrachium chirripoi* is known from lowland forest localities in Central America in southeastern Costa Rica and eastern Panama, and south into South America in western Colombia and Ecuador at elevations below 600 m ([24,148,204], this work). In Ecuador, this species has been observed in few localities within the Esmeraldas province (Río Quingue, Río Bogotá, Reserva Itapoa, Tesoro Escondido, Reserva Canandé), in the Chocó ecoregion, at elevations below 320 m.

Conservation status: Listed globally as *Least Concern* by the IUCN [205]. In Ecuador, the species has been recently found in five localities in the Chocó Ecoregion, an area under constant deforestation pressure. Although the *Least Concern* category seems accurate at the global level, in Ecuador, the species is threatened by habitat destruction because of agriculture, pasture lands, and mining. Thus, we suggest placing the species in the *Endangered* category at the local level, based on IUCN criteria B1, B2a, B2b(iii).

Figure 108. Distribution of *Hyalinobatrachium chirripoi* in Ecuador (yellow dots).

Evolutionary relationships (Figure 101): Molecular data support a sister relationship between *Hyalinobatrachium chirripoi* and *H. colymbiphyllum*.

Specimens examined: *Hyalinobatrachium chirripoi*: <u>Ecuador</u>: Provincia de Esmeraldas: Río Quingue (0.722° N, 80.08° W; 47 m), QCAZ 37309, 48271; Río Bogotá, DFCH-USFQ C1903; Reserva Itapoa (0.513° N, 79.134° W; 320 m), MZUTI 3609–10; Tesoro Escondido (0.542° N, 79.145° W; 225 m), MZUTI 3625; Reserva Canandé (0.526° N, 79.209° W; 310 m), MZUTI 4745. <u>Costa Rica</u>: Limón Province: Suretka, along Cocales Creek, KU 36862–64, 36866–70.

Hyalinobatrachium fleischmanni (Boettger, 1893 [206]; Figures 109–111).

Hylella fleischmanni Boettger, 1893 [206]. Lectotype: SMF 3760, designated by Mertens (1967) [207].
Type locality: "San José, Costa Rica".
Hylella chrysops Cope, 1894 [208]. Neotype: SMF 3760 (= holotype of *Hylella fleischmanni*), designated by Starrett and Savage (1973) [209]. Placed in synonymy by Boulenger (1895) [210].
Centrolenella fleischmanni—Noble (1924) [186].
Centrolenella viridissima Taylor, 1942 [211]. Holotype: EHT-HMS 27725 (now FMNH 100093). Type locality: "Agua de Obispo, Guerrero", Mexico. Placed in synonymy by Starrett and Savage (1973) [209].
Cochranella fleischmanni—Taylor (1951) [15].
Cochranella decorata Taylor, 1958. Holotype: KU 36896. Type locality: "Hda. La Florencia, about 3 miles west of Turrialba, Cartago Province, Costa Rica". Placed in synonymy by Starrett and Savage (1973) [209].
Cochranella millepunctata Taylor, 1958 [201]. Holotype: KU 36887. Type locality: "La Palma, San José Province, Costa Rica". Placed in synonymy by Starrett and Savage (1973) [209].
Centrolenella fleischmanni—Goin (1964) [187].
Hyalinobatrachium fleischmanni—Ruiz-Carranza and Lynch (1991) [6].

Common names: English: Fleischmann's Glassfrog (Liner, 1994). Spanish: Rana de Cristal de Fleischmann.

Etymology: The specific name *fleischmanni* is a patronym for Carl Fleischmann, a German collector–naturalist who donated the specimens used by Boettger to describe the species [84].

Identification: *Hyalinobatrachium fleischmanni* is easily recognizable by having, in life, a lime green dorsum with small yellowish spots, a completely transparent ventral parietal peritoneum, white heart and digestive tract (pericardium and gastrointestinal peritoneum covered by iridophores), lacking humeral spines, and having reduced webbing between inner fingers (Figure 110). In the Ecuadorian Pacific lowlands, similar species include *H. aureoguttatum*, which has large yellow spots on the dorsum; *H. valerioi*, distinguished by having a green dorsal reticulum; and *H. chirripoi*, which, in life, has a transparent pericardium (red heart visible through ventral skin). We note that *H. fleischmanni* represents a species complex that requires taxonomic revision [212].

Figure 109. *Hyalinobatrachium fleischmanni* in life. Ecuador, near Durango, 77 m, QCAZ 32107. Photos by Martín Bustamante.

Diagnosis: (1) Vomerine teeth absent; (2) snout subacuminate in dorsal aspect and rounded in profile; (3) tympanum not visible; supratympanic fold absent; (4) dorsal skin shagreen; (5) venter smooth; lacking pair of enlarged subcloacal warts; (6) ventral parietal peritoneum transparent (condition P0); white peritoneum covering heart, intestines and stomach; transparent peritoneum on urinary bladder (condition V5); (7) liver bulbous, hepatic peritoneum white (condition H2); (8) humeral spines absent; (9) webbing reduced between Fingers I, II and III, moderate between outer fingers (Figure 110); webbing formula I (2–2$^+$)—2 II (1–1$^+$)—(3) III (2)—(1$^+$–2) IV (1–2) IV; (10) webbing between toes moderate; webbing formula on feet I (1–1$^{1/2}$)—(2) II (1)—(2–2$^+$) III (1)—(2$^+$–2$^-$) IV (2–2$^+$)—(1–1$^+$) V; (11) ulnar and tarsal fold absent; (12) concealed prepollex; nuptial pad Type IV; (13) Finger II shorter than Finger I; (14) eye diameter larger than width of disc on Finger III (disc of Finger III width 33%–37% of eye diameter); (15) in life, lime green dorsum with yellow spots (Figure 109); venter transparent, pericardium usually white (but see Color in life); bones white; (16) in preservative, dorsum cream; (17) in life, iris yellowish to greyish white with some dark punctuations; (18) fingers and toes lacking melanophores; (19) males call from the underside of leaves; the call consists of a single "wheet" note, with duration of 150–300 ms, and a dominant frequency of 3800–4500 Hz at the beginning and 4800–5300 HZ at the end; (20) fighting behavior varies from vent to vent to amplexus-like; (21) egg clutches usually laid on the underside of leaves; maternal care absent; males provide prolonged parental care; (22) oral apparatus complete; oral disc with single row of marginal papillae laterally and ventrally, wide dorsal gap; jaw sheaths normal; 2(2)/3 labial tooth rows, A-2 with wide gap, tooth rows situated nearly lateral to mouth and upper jaw sheath; (23) minute body size; snout–vent length in adult males 19.3–26.8 mm (n = 13), and in adult females 22.4–31.1 mm (n = 7).

Color in life (Figure 109): Dorsum lime green with pale yellow or greenish spots. Venter transparent, pericardium usually white, visceral and hepatic peritonea white. Twomey et al. [19] report that within a single population of *H. fleischmanni* near San Gabriel Mixtepex (Oaxaca, Mexico), adults exhibited variation in the condition of the pericardium (from white to transparent).

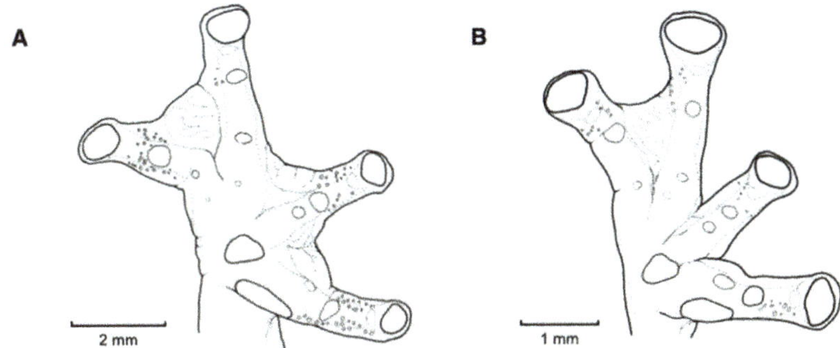

Figure 110. *Hyalinobatrachium fleischmanni*, variation in hand webbing, ventral view. (**A**) KU 116447. (**B**) QCAZ 22301. Illustrations by Juan M. Guayasamin.

Color in ethanol: Cream dorsum with dark melanophores in the places where green coloration was in life. Venter translucent. Parietal peritoneum completely transparent, heart and all viscera covered by white lining.

Biology and ecology: The natural history of *Hyalinobatrachium fleischmanni* has been reviewed by several studies, including Greer and Wells [213], Villa [214], Jacobson [188], Hayes [189], Savage [148], Kubicki [24], and Delia et al. [215]. We summarize the essential information here but refer readers to those papers for more data. Individuals of *Hyalinobatrachium fleischmanni* have been found on vegetation up to 10 m above streams at night. Males are territorial and call from the lower (mainly) and upper surfaces of leaves. Females usually deposit eggs clutches on the lower surfaces of leaves, although some variation has been reported [215]; maternal care is absent [25]. Physical combat occurs when a male intrudes into an occupied territory. Initially the owner of the territory will call vigorously, but if that strategy fails, male–male combat starts usually with both males adopting an amplexus-like position. Both males may give quick calls. Although venter to venter combat was thought to be absent in *Hyalinobatrachium*, *H. fleischmanni* has been observed to adopt this combat position, with both males dangling upside down while holding vegetation with their hind limbs [215]. Males attend egg clutches at night, and may continue to advertise, sometimes ending with multiple clutches from different females. During the day, the male parent retreats to nearby vegetation. Only males are involved in parental care of the eggs. Delia et al. [25] observed that males of this species are attentive to individual embryo needs. Eggs are pale greenish white, and clutches are deposited on the underside of leaves. Each egg clutch has one layer of 10–50 eggs [189,216].

Call: The call consists of a single "wheet" note that rises at the end, with duration of 150–300 ms. The dominant frequency is 3800–4500 Hz at the beginning and 4800–5300 HZ at the end. Calls are emitted 4–19 per minute [148,188,209,213].

Tadpole: A description of the tadpole can be found in Starrett [146], Savage [148], and Hoffmann [147].

Distribution (Figure 111): *Hyalinobatrachium fleischmanni* is known from southern Mexico, across Mesoamerica, south to central Ecuador at elevations below 800 m [24,148,212,217,218]. In Ecuador, the species is known from localities in the Pacific lowlands (mostly below 300 m) in the provinces of Esmeraldas, Pichincha, Santo Domingo de los Tsáchilas, Los Ríos, and Guayas. In Ecuador, the potential distribution of the species is 39,738 km^2. The distribution of *H. fleischmanni* is an overestimation because, as currently recognized, it is a species complex [212].

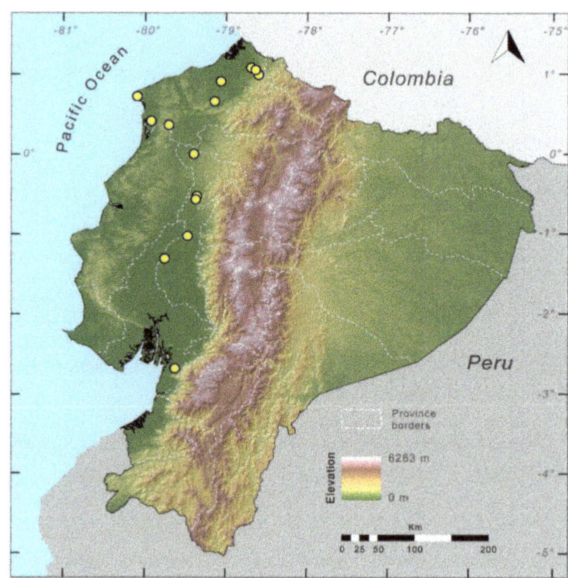

Figure 111. Distribution of *Hyalinobatrachium fleischmanni* in Ecuador (yellow dots).

Conservation status: Globally, *Hyalinobatrachium fleischmanni* is considered as *Least Concern* by the IUCN because of its wide distribution and tolerance of habitat modifications [219]. However, recent studies suggest that this taxon represents a species complex that requires subdivision [212]. In Ecuador, about 55% of the potential distribution of the species is affected by human activities. Even though habitat destruction is considerable, *H. fleischmanni* is known to tolerate a degree of disturbance [24]. In Ecuador, we suggest maintaining the category of *Least Concern* until new taxonomic studies are available.

Evolutionary relationships (Figure 101): Our tree shows *Hyalinobatrachium fleischmanni* as the sister species of *H. muiraquitan*, which is endemic to the Brazilian Amazon basin. Although the Venezuelan *H. tatayoi* falls within the genetic variation of *H. fleischmanni*, recent studies suggest that *H. fleischmanni* represents a species complex that requires taxonomic revision [212].

Specimens examined: *Hyalinobatrachium fleischmanni:* <u>Costa Rica</u>: Alajuela, USNM 219249–61. San José, USNM 219263–80. Guanacaste, USNM 219282–303. <u>Ecuador</u>: *Provincia de Los Ríos:* Quevedo, USNM 60520; Río Palenque, USNM 286639–40; Patricia Pilar, USNM 286645; Hacienda Cerro Chico, USNM 286646. *Provincia de Esmeraldas:* 4 km W Durango, QCAZ 23549 km^2 in the San Francisco–Durango road, QCAZ 32073; Río Quingue, nearby Caimito (0.72096 S, 80.09117 W, 47 m), QCAZ 37308; Río Onzole, QCAZ 10433; Tesoro Escondido (0.542 N, 79.145 S, 225 m), MZUTI 3621–3625. *Provincia de Santo Domingo de los Tsáchilas:* Bosque Protector La Perla, QCAZ 12606. <u>Honduras</u>: Olancho: USNM 342162–342213. Nicaragua: Matagalpa: USNM 220013–18. Nueva Segovia: USNM 220019–36. <u>México</u>: Chiapas: USNM 115499.

Localities from the literatura: <u>Ecuador</u>: *Provincia de Guayas:* Cerro de Hayas (2.7299° S, 79.6297° W, 127 m), MZUA.AN.1693–1694; *Provincia de Los Ríos:* Macul (1.2279° S, 79.7531° W, 84 m), MZUA.AN.660–661 [218].

Hyalinobatrachium iaspidiense (Ayarzagüena, 1992 [220]; Figures 112 and 113).

Centrolenella iaspidiensis Ayarzagüena, 1992 [220]. Holotype: EBD 28803.
Type locality: "Quebrada Jaspe. San Ignacio de Yuruaní. Edo. Bolívar. Venezuela."
Centrolene iaspidiensis—Duellman, 1993 [221]. Unintended combination.
Hyalinobatrachium iaspidiense—Myers and Donnelly, 1997 [104].
Hyalinobatrachium nouraguensis—Lescure and Marty, 2001 [222]. Holotype: MNHNP 1999.8604.
  Type locality: "Saut Arataye (environs du camp de base), Réserve des Nouragues (bassin
  de l'Approuague), Guyane française". Placed in synonymy by Yánez-Muñoz, Pérez-Peña,
  and Cisneros-Heredia, 2009 [223].

Common names: English: Jaspe Glassfrog. Spanish: Rana de Cristal de Jaspe.

Etymology: The specific name *iaspidiense* is derived from the Greek word *iaspis*, meaning jasper, in reference to the type locality of the species, Quebrada de Jaspe.

Identification: *Hyalinobatrachium iaspidiense* is one of the most easily recognizable centrolenid. In life, it has a dorsum with large, lime green, irregular blotches and small, black spots (Figure 112). The only species with a similar dorsal coloration is *H. mesai*, a species known only from the southern slope of Sarisariñama-tepui, Venezuela [224]. It has been suggested, however, that *H. mesai* and *H. iaspidiense* represent the same species [223].

Figure 112. *Hyalinobatrachium iaspidiense* in life. (**Left**): Adult female (QCAZ 38438) from Ecuador, Yachana Reserve, 300–350 m; photo by Susan North. (**Right**): Adult male (SMNS 12247) from Guyana, Mabura Hill Forest Reserve, 60 m; photo by Raffael Ernst.

Diagnosis: (1) Vomerine teeth absent; (2) snout truncated in dorsal aspect and slightly protruding in profile; (3) tympanum not visible; supratympanic fold absent; (4) dorsal skin shagreen; (5) venter smooth; pair of enlarged subcloacal warts, but low and difficult to see; (6) ventral parietal peritoneum transparent (condition P0); pericardium transparent; white peritoneum covering intestines and stomach; transparent peritoneum on urinary bladder (condition V6); (7) liver bulbous, hepatic peritoneum white (condition H2); (8) humeral spines absent; (9) webbing absent between Fingers I, II, and III, moderate between outer fingers; webbing formula III $(2-2^+)$—(1^+-2^-) IV; (10) webbing between toes moderate; webbing formula on foot I $(1-1^+)$—$(2^+-2^{1/3})$ II $(1^+-1^{1/3})$—$(2^+-2^{1/4})$ III $(1-1^+)$—$(2-2^{3/4})$ IV $(2^+-2^{1/4})$—$(1^+-1^{1/4})$ V; (11) ulnar fold present, enameled; external tarsal fold present, enameled; internal tarsal fold short and low; (12) concealed prepollex; nuptial pad Type IV; (13) Finger II shorter than Finger I (Finger II about 93% of Finger I); (14) disc of Finger III width 39%–57% of eye diameter; (15) in life, dorsum translucent with a yellowish–green background coloration, large lime green blotches, and small black spots (Figure 112); bones white; (16) in preservative, dorsum cream with large, irregular white blotches and small, black spots; (17) in life, iris yellowish to greyish white; (18) fingers and toes lacking melanophores, except few present on Finger IV and Toes IV and V; (19) males call from the underside of leaves; the call consists of a single pulsed note, with a dominant frequency at 4440–4710 Hz; (20) fighting behavior unknown; (21) egg deposition site unknown; parental care unknown; (22) tadpole unknown; (23) minute body size; in Venezuelan specimens: SVL in females 22.7 mm ($n = 1$), in males 19.8–21.8 mm ($n = 8$) [106]; in Ecuador, male SVL = 19.9 mm ($n = 1$), and female SVL = 21.9 mm ($n = 1$).

Color in life (Figure 112): Dorsum translucent, with a yellowish–green background coloration, large lime-green blotches, and small black spots. Venter transparent, pericardium transparent (red heart visible), visceral and hepatic peritonea white [223,225].

Color in ethanol: Dorsum cream lavender, with large irregular white marks and small black spots. Ulnar and tarsal folds enameled; some of the warts below cloaca enameled. Venter transparent, pericardium transparent, visceral and hepatic peritonea white; peritonea around urinary bladder transparent; gall bladder white (description based on two adults, MECN 4033, QCAZ 38438).

Variation: Ulnar and tarsal folds less evident and with few iridophores (QCAZ 38438).

Biology and ecology: Found on vegetation 4–7 m above streams [106]. In Ecuador (Yachana Reserve), an adult female was found on a long blade of tall grassy shrub overhanging water, located in the middle of a stream (4 m wide). Stream habitat is variable, but the female was found in a shallow area (~20 cm deep) where the stream becomes a riffle and is more fast-flowing. Stream bottom here is composed of about 80% pebbles and small rocks and 20% sand near edges. The forest adjacent to the stream is near pristine and about 25–30 m high [225]. Parental care unknown.

Call: The call consists of a single note, with 16 pulses and a duration of 62.7–75.1 ms. Dominant frequency at 4440–4710 Hz. The total frequency range for the call is between 3710 and 5850 Hz, with a secondary frequency at 8600–9200 Hz. The call was emitted frequently (7–10 calls per minute) [106].

Tadpole: Not described.

Distribution (Figure 113): *Hyalinobatrachium iaspidiense* is known from the Guiana region of Venezuela, central Guyana, Surinam (Sipaliwini District), and French Guiana, as well as from the Amazon Basin of Ecuador, Peru, and Brazil at elevations below 1000 m [106,222,225–229]. In Ecuador, the species is known from few localities in the Amazonian raiforest.

Figure 113. Distribution of *Hyalinobatrachium iaspidiense* in Ecuador (yellow dots).

Conservation status: *Hyalinobatrachium iaspidiense* is considered as *Data Deficient* by the IUCN [230]. However, given the large distribution of the species and the absence of immediate threats, we suggest that it should be considered as *Least Concern*, both globally and in Ecuador.

Evolutionary relationships (Figure 101): *H. iaspidiense* is the sister species of *H. tricolor*.

Specimens examined: *H. iaspidiense:* <u>Ecuador</u>: *Provincia Napo:* Yachana Reserve (0°52′21.71″ S, 77°14′13.43″ W; 300–350 m), QCAZ 38438; Estación Biológica Jatun Sacha (1.066 S, 77.617 W, 405 m), QCAZ 53023. *Provincia Orellana:* Km 66 on the Pompella Sur–Iro road (0.8022 S, 76.398 W, 280 m), QCAZ 54947. *Provincia Sucumbíos:* Totoa Nai'qui (0.03442° S, 76.75278° W, ca. 280 m), MECN 4033.

Hyalinobatrachium munozorum (Lynch and Duellman, 1973 [22]; Figures 114–116).

Centrolenella munozorum Lynch and Duellman, 1973 [22]. Holotype: KU 118054.
<u>Type locality:</u> "Santa Cecilia, 340 m, Provincia Napo (Sucumbíos), Ecuador".
Hyalinobatrachium munozorum—Ruiz-Carranza and Lynch, 1991 [6].
Hyalinobatrachium ruedai Ruiz-Carranza and Lynch, 1998 [27]. Holotype: ICN 40409. Type
    locality: "(Colombia,) Departamento de Caquetá, municipio de Miraflores, Parque Nacional
    Natural de Chiribiquete, campamento base, 530 m". **New synonymy.**

Common names: English: Muñoz's Glassfrog. Spanish: Rana de Cristal de Muñoz.

Etymology: The specific name *munozorum* is a patronym for Ildefonso Muñoz B. and Blanca Muñoz, hosts of the extensive herpetological work carried out in Santa Cecilia by W. E. Duellman and his students [22].

Identification: *Hyalinobatrachium munozorum* can be distinguished from most glassfrogs by having a green dorsum with pale yellow spots, a completely transparent ventral parietal peritoneum, white liver and digestive tract, and by lacking humeral spines (Figure 114). Similar species with a completely transparent venter that inhabit the Amazon basin include *H. bergeri*, *H. carlesvilai*, *H. iaspidiense*, *H. mondolfii*, *Chimerella mariaelenae*, and *Teratohyla amelie*. *Hyalinobatrachium munozorum* differs from *C. mariaelenae* by having, in life, a green dorsum with pale yellow spots (dorsum green with dark grey spots in *C. mariaelenae*) and by lacking humeral spines (small humeral spines present in adult males of *C. mariaelenae*). *Teratohyla amelie* has a uniform green dorsum, lacking the pale-yellow spots visible in *H. munozorum*. *Hyalinobatrachium iaspidiense*, in life, has a dorsum with large, green, irregular marks and small, black spots. Morphological differentiation among *H. munozorum*, *H. mondolfii*, and *H. carlesvilai* is minimal, although molecular divergence is considerable [22,231,232]; the pupil of *H. bergeri* and *H. carlesvilai* is surrounded by a dark grey ring (absent in *H. munozorum*).

Figure 114. *Hyalinobatrachium munozorum* in life. Adult male from stream near Tena, 708 m, MZUTI 1616. Photos by Eduardo Toral.

Diagnosis: (1) Vomers lacking teeth; (2) snout varies from truncated to rounded in dorsal aspect and round to truncated in lateral view; (3) tympanum usually obscured by skin, small when evident, its diameter about 25% of eye diameter; supratympanic fold low; (4) dorsal skin shagreen; males lacking spicules; (5) ventral skin texture areolate; lacking pair of enlarged subcloacal warts; (6) ventral parietal peritoneum transparent, white pericardium, white peritonea covering intestines and stomach, transparent urinary bladder; (7) liver bulbous, covered by white peritoneum; (8) humeral spines absent; (9) webbing between inner fingers absent; webbing formula for outer fingers: III $(1^{1/2}-2^-)$—$(1-2^-)$ IV; (10) feet about three-fourths webbed; webbing formula: I (0^+-1^-)—$(1^{1/3}-2^-)$ II (0^+-1)—$(1^{1/2}-2)$ III $(0^+-1^{1/3})$—$(2^--2^{1/3})$ IV $(2^--2^{1/3})$—(0^+-1) V; (11) ulnar and tarsal folds variable, from inconspicuous to pronounced; (12) concealed prepollex; in males, nuptial pad not evident; (13) Finger I slightly longer than Finger II (Finger II length 93.7%–97.6% of Finger I); (14) disc of Finger III of moderate size, its width 37.9%–56.2% of eye diameter; (15) in life, dorsum green with diffuse yellow spots, venter transparent; bones white; (16) in preservative, dorsum cream lavender with small unpigmented spots; (17) in life, iris pale to bright gold with dark punctuations; (18) dorsal surfaces of fingers and toes lacking melanophores; (19) call composed by a single note with a duration of 145–178 (mean = 170, SD = 9.9) ms; each call is mostly tonal, with a dominant frequency at 5719–5906 (mean = 5812, SD = 64) Hz; (20) fighting behavior unknown; (21) egg clutches laid on the underside of leaves; parental care unknown; (22) tadpoles unknown; (23) minute body size; in males, SVL 19.9–21.9 mm ($\overline{X}$ = 20.6 ± 0.691; n = 8); in females, SVL 20.9–23.6 mm (n = 3).

Color in life (Figure 114): Dorsal surfaces of head and body green with small, pale yellow spots and minute dark brown flecks. Parietal peritoneum transparent. White peritonea covering heart, liver, digestive tract, and testes; transparent urinary bladder. Iris pale gold with minute dark flecks. See Taxonomic Remarks.

Color in ethanol: Dorsal surfaces of head, body, and limbs cream with numerous minute lavender flecks. White visceral and hepatic peritonea. Layer of iridophores covering heart.

Variation: Specimens in the type series have a round snout in lateral view, and low ulnar and tarsal folds that are inconspicuous and unpigmented. Additional specimens from the type locality (Santa Cecilia) show the following variation: Snout truncated in lateral view, ulnar and tarsal folds conspicuous and white (KU 155493–96); conspicuous tarsal fold and low ulnar fold (KU 152488, 152489, 175215). Specimens identified as *Hyalinobatrachium ruedai* (a synonym of *H. munozorum*; see below) by Ruiz-Carranza and Lynch [27] and Cisneros-Heredia and McDiarmid [233] show the following variation: Snout truncated to rounded in dorsal and lateral views, ulnar and tarsal folds vary from inconspicuous to low and white. A female (KU 154749) has a truncated snout and conspicuous white ulnar and tarsal folds. The pericardium of all specimens from Ecuador and Colombia (including the type series) is white. In some individuals (KU 123225, 152488–89, 155494, 155496, 175504), the layer of iridophores is thinner. Cisneros-Heredia and McDiarmid [17] reported that one specimen showed a bicolored iris in life, with a light grey circumpupilary zone. These authors also reported variation in the iris coloration in preservative, from lavender–cream background with dense dark punctuations and lavender mid-line to uniform white.

Biology and ecology: The type series was found on leaves of bushes and trees at night: One over a pond, one away from water in primary forest, one on a palm frond 2 m above a stream, and one on an herbaceous leaf more than 2 m above a stream. The specimen from Lago Agrio was obtained from the foliage of a large tree that was felled during the clearing of primary forest [22]. A specimen from Tena was collected at night on the leaf of a bush at a rivulet in secondary forest [17]. Males called from riverine vegetation at a locality nearby Tena on August 2012. Parental care is unknown.

Call (Figure 115): We analyzed 18 notes from 1 individual (LBE-C-020) recorded at Río Bigal, Orellana province, Ecuador. The call is composed by a single note. Note duration is 145–178 (mean = 170, SD = 9.9) ms. Notes are mostly tonal, but they exhibit one to seven (mean = 4.5, SD = 1.6) low-amplitude peaks. Notes usually have their peak amplitude in the first 50% of the note (relative peak time: Range = 0.1699–0.661, mean = 0.448, SD = 0.123). The dominant frequency of a note measured at peak amplitude is 5719–5906 (mean = 5812, SD = 64) Hz and is contained within the fundamental frequency.

The fundamental frequency has a lower limit of 5625–5812 (mean = 5708, SD = 63) Hz and a higher limit of 5812–6000 (mean = 5906, SD = 64) Hz.

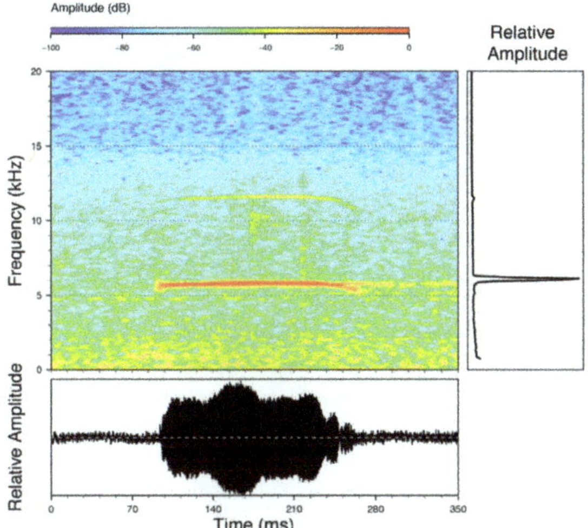

Figure 115. Vocalization of *Hyalinobatrachium munozorum* recorded at Río Bigal, 930 m, Orellana province, Ecuador, by Morley Read (LBE-C-020).

Tadpole: Not described.

Distribution (Figure 116): *Hyalinobatrachium munozorum* is known from localities in the Amazonian lowlands and Andean foothills of Colombia, Ecuador, Peru, and Bolivia at elevations below 980 m ([17,22,27,134,232], this work). In Ecuador, this species has been recorded from localities below 300 m; although, there is one juvenile (identification unconfirmed), found at 920 m (Specimens examined). In Ecuador, the potential distribution of the species is 101,986 km^2.

Figure 116. Distribution of *Hyalinobatrachium munozorum* in Ecuador (yellow dots).

Conservation status: Globally, *Hyalinobatrachium munozorum* is listed as *Least Concern* by the IUCN [234]. Although *H. munozorum* is rare in collections, we presume that this is a consequence of inadequate sampling at the canopy level [166]. The distribution of the species is large and lacks immediate threats; thus, the *Least Concern* conservation status is justified.

Evolutionary relationships (Figure 101): *H. mondolfii* is the closest relative of *H. munozorum*.

Taxonomic Remarks: Lynch and Duellman [22] used the terms "heart visible" and "heart not visible" to illustrate the condition of the ventral parietal peritoneum; species that have a transparent (=clear) peritoneum were considered as having a "heart visible", whereas species with a white parietal peritoneum had a "heart not visible". Ruiz-Carranza and Lynch [27] discussed the character states of the pericardium; they applied the term "heart visible" for species that have a transparent parietal peritoneum as well as a transparent pericardium (red heart visible in life). In the same work, Ruiz-Carranza and Lynch [27] described *H. ruedai*, which was diagnosed by having a white pericardium; *H. ruedai* was compared with congenerics with a white heart, a list that did not include *H. munozorum*. We assume that Ruiz-Carranza and Lynch [27] interpreted the condition of "heart visible" for *H. munozorum* as equivalent to a visible red heart. As mentioned above, the type series of *H. munozorum* and *H. ruedai* have a white heart. The only other characteristic that could separate *H. ruedai* from *H. munozorum* is the hand webbing; however, the range observed in *H. ruedai* falls within the variation of *H. munozorum*. Therefore, herein we place *Hyalinobatrachium ruedai* Ruiz-Carranza and Lynch, 1998 [27], in the synonymy of *Centrolenella munozorum* Lynch and Duellman, 1973 [22].

Specimens examined: *H. munozorum*: Ecuador: *Provincia Sucumbíos:* Santa Cecilia (00°03′ N, 76°58′ W; 340 m), KU 118054 (holotype), 105251, 123225, 150620 (paratypes), 152488–89, 155493–96, 175504. *Provincia Orellana:* Río Yasuní (00°51′ S, 76°23′ W; ca. 250 m), KU 175215; Tiputini Biodiversity Station (00°37′ S, 76°10′ W; 190–270 m), DFCH-D105. *Provincia Zamora Chinchipe:* Shaime (4°20′ S, 78°40′ W; ca. 920 m), QCAZ 31056 (identification not certain). *Provincia Napo:* Tena (00°59′ S, 77°49′ W; 500 m), DFCH-USFQ 0735. *Provincia Pastaza:* Río Manderoyacu, EPN 6427; km 6 vía San Ramón-El Triunfo (1.355S, 77.86456), QCAZ 33261. Colombia: *Departamento Caquetá:* Municipio de Miraflores, Parque Nacional Natural Chiribiquete, 530 m, ICN 40409–11 (type series of *H. ruedai*), IND-AN 5448-52. Peru: *Departamento de Huanuco:* Finca Panguana, Río Llullapichis, 4–5 km upstream from Río Pachitea (9°36′ S, 74°55′ W; 200 m), KU 154749, 172167–69; *Departamento de Cuzco:* 40 km E Quince Mil on road to Puerto Maldonado, above Río Marcapata (13°09′ S, 71°25′ W), KU 197028–29.

Hyalinobatrachium pellucidum (Lynch and Duellman, 1973 [22]; Figures 117–120).

Centrolenella pellucida Lynch and Duellman, 1973 [22]. Holotype: KU 143298.
Type locality: "Río Azuela, 1740 m, Quito–Lago Agrio road, Provincia Napo (Sucumbíos), Ecuador."
Hyalinobatrachium pellucidum—Ruiz-Carranza and Lynch, 1991 [6].
Hyalinobatrachium lemur Duellman and Schulte, 1993. Holotype: KU 211768. Type locality: "west slope of Abra Tangarana, 7 km (by road) northeast of San Juan de Pacaysapa (06°12′ S, 76°44′ W, 1080 m), Provincia Lamas, Departamento San Martín, Perú". Synonymy by Castroviejo-Fisher, Padial, Chaparro, Aguayo, and De la Riva, 2009 [231].

Common names: English: Andean Glassfrog. Spanish: Rana de Cristal Andina.

Etymology: The specific name *pellucidum* is derived from the Latin word *pellucidus* (clear, transparent) and refers to the transparent parietal peritoneum of this centrolenid frog [22].

Identification: *Hyalinobatrachium pellucidum* is distinguishable from most glassfrogs by having a completely transparent ventral parietal peritoneum, a white liver, and transparent pericardium (in life, red heart visible ventrally; Figure 117). On the Amazonian slopes of the Andes and Amazon basin, the only other species with a visible red heart in life is *H. iaspidiense*, which differs from *H. pellucidum* by having large lime green blotches and small black spots on the dorsum (both absent in *H. pellucidum*).

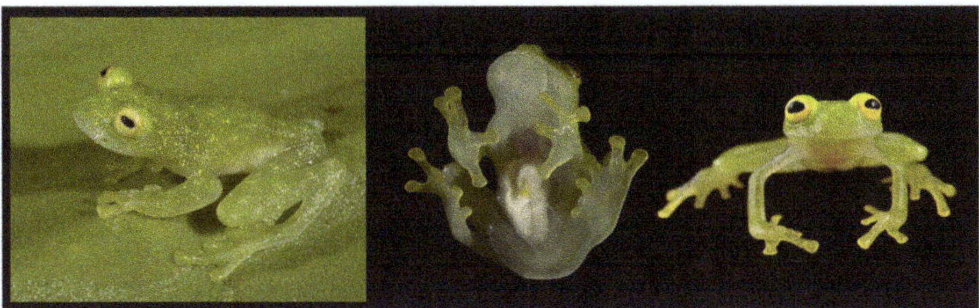

Figure 117. *Hyalinobatrachium pellucidum* in life. (**Left and center**): Male from Río Napinaza, Morona Santiago province, Ecuador, QCAZ 42000. (**Right**): Male from Miazi Alto, Zamora Chinchipe province, 1250 m, QCAZ 41648. Photos by Luis A. Coloma.

Diagnosis: (1) Vomerine teeth absent; (2) snout truncated in dorsal aspect and profile (Figure 118); (3) tympanum partially hidden under skin, its diameter about 30% of eye diameter; supratympanic fold low; (4) dorsal skin shagreen; (5) ventral skin texture areolate; lacking pair of enlarged subcloacal warts; (6) ventral parietal peritoneum transparent (condition P0); pericardium transparent; white peritoneum covering intestines and stomach; transparent peritoneum on urinary bladder (condition V6); (7) liver bulbous, hepatic peritoneum white (condition H2); (8) humeral spines absent; (9) webbing absent between Fingers I and II, extensive between outer fingers (Figure 118); webbing formula II $1^{1/2}$–3^{+} III 2^{+}—(2^{-}–2) IV; (10) webbing between toes extensive; webbing formula on foot I 1—$1^{1/2}$ II 1^{+}—2^{-} III 1^{+}—2^{+} IV 2—1 V; (11) ulnar and tarsal folds present, usually white (but see Variation); (12) concealed prepollex; nuptial pad type unknown; (13) Finger I slightly longer than Finger II (Finger II about 95% of Finger I); (14) disc of Finger III width about 50% of eye diameter; (15) in life, dorsum pale green with diffuse yellow spots; venter transparent, exposing the red heart (Figure 117); color of bones white; (16) in preservative, dorsum creamy white with minute purple flecks visible under magnification; (17) iris pale silvery bronze in life; (18) dorsal surfaces of fingers and toes lacking melanophores; (19) each has a single, tonal note with a duration of 98–140 (mean = 128, SD = 8.7) ms; dominant frequency is at 5599–5857 (mean = 5690, SD = 58) Hz; (20) fighting behavior unknown; (21) egg clutches placed on the underside of leaves; maternal care absent; prolonged parental care provided by males; (22) tadpoles undescribed; (23) minute body size; in males, SVL 20.4–21.4 mm; in one female SVL 21.6 mm.

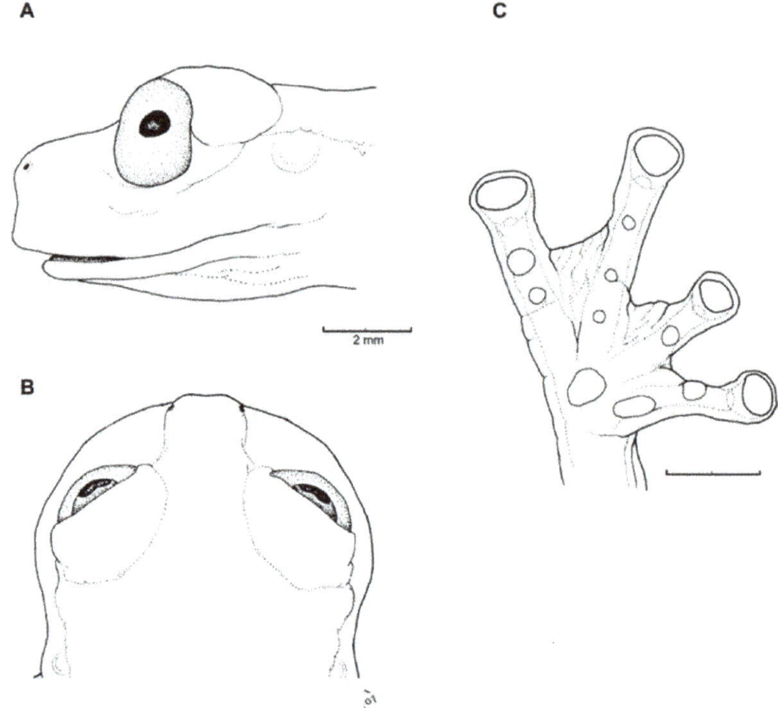

Figure 118. *Hyalinobatrachium pellucidum*, holotype, KU 143298. (**A**) Head in lateral view. (**B**) Head in dorsal view. (**C**) Hand in ventral view. Illustrations by Juan M. Guayasamin.

Color in life (Figure 117): Dorsum pale green with diffuse yellow spots; venter and hidden surfaces of limbs lacking pigment; fingers and toes yellow; parietal peritoneum clear; red heart visible; bones white; iris pale silvery bronze [22].

Color in ethanol: Dorsum creamy white with minute purple flecks and small white granules visible under magnification; digestive tract and liver covered by white peritonea; pericardium lacking iridophores; ulnar and tarsal folds white.

Variation: Castroviejo-Fisher et al. [231] reported the following variation: The holotype of *H. pellucidum* has marked and enameled ulnar, tarsal, and cloacal folds, while specimens previously identified as *H. lemur* (a synonym of *H. pellucidum*) only show weak and non-enameled folds. The variation is interpreted as being the product of preservation artifacts and intraspecific variation. Two additional specimens from the same locality and near the type locality of *H. lemur* (~45 Km straight line) show intermediate states regarding ulnar, tarsal, and cloacal folds. The specimen KU 217297 has an enameled but weak ulnar fold, a weak but non-enameled tarsal fold, and an enameled and weak cloacal fold; KU 217295 has an enameled and marked ulnar fold, an enameled but weak tarsal fold, and an enameled and marked cloacal fold. A specimen (MHNCP 4880) collected in Cusco, Peru, and assigned to *H. pellucidum*, has very weak and barely enameled ulnar, tarsal, and cloacal folds.

Biology and ecology: The holotype of *Hyalinobatrachium pellucidum* was found on the leaf of an herb over a small stream at night. The following species were in the same stream and in other small streams nearby: *Nymphargus megacheirus*, *N. anomalus*, *N. siren*, *Centrolene pipilata*, *Hyloscirtus phyllognathus* [22]. Short-term maternal care is absent; males provide prolonged parental care [25].

Call (Figure 119). We analyzed 36 notes from one individual (USNM 286708), recorded at the type locality of the species, Río Azuela, 1740 m, Napo province, Ecuador, by Roy McDiarmid. The advertisement call is relatively short which has a single note per call (i.e., call = note). Notes are moderate in duration and the note duration is 98–140 (mean = 128, SD = 8.7) ms. Notes are tonal and do not show any clear amplitude peaks. The dominant frequency is 5599–5857 (mean = 5690, SD = 58) Hz and is contained within the fundamental frequency. The fundamental frequency has a lower limit of 5163–5685 (mean = 5381, SD = 678) Hz and a higher limit of 5685–5943 (mean = 5778, SD = 56) Hz.

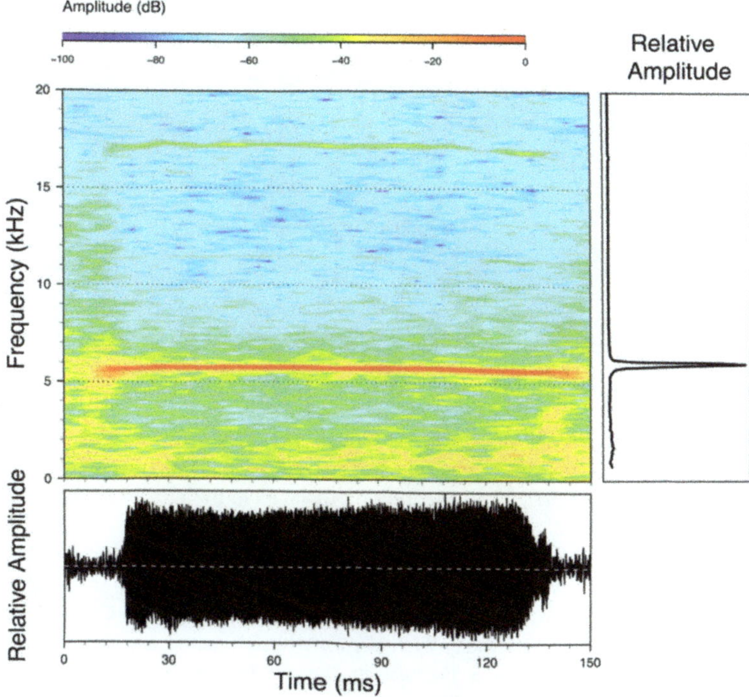

Figure 119. Call of *Hyalinobatrachium pellucidum* recorded at Río Azuela, 1740 m, Napo province, Ecuador, recorded by Roy McDiarmid (USNM 286708).

Tadpole: Not described.

Distribution (Figure 120): *Hyalinobatrachium pellucidum* is known from a few records in the lower montane rainforest on the Amazonian Andean slopes of Ecuador and Peru, at elevations between 1000–1740 m ([22,231], this work). In Ecuador, *Hyalinobatrachium pellucidum* inhabits the Eastern Foothill Forest and Eastern Montane Forest ecoregions, and it is known from localities between 1013 and 1740 m; these localities are Río Azuela, Río Reventador, km 6.6 on the Limón–Macas road, and Cordillera del Cóndor (Specimens Examined).

Figure 120. Distribution of *Hyalinobatrachium pellucidum* in Ecuador (yellow dots).

Conservation status: Globally, *Hyalinobatrachium pellucidum* is listed as *Near Threatened* by the IUCN [235]. In Ecuador, the species is severely fragmented because of agriculture, pasture lands, and mining. Thus, we suggest that, locally, it should be placed in the *Vulnerable* category, following IUCN criteria b1, B2a, B2biii.

Evolutionary relationships (Figure 101): *Hyalinobatrachium pellucidum* is sister to *H. yaku*.

Specimens examined: *Hyalinobatrachium pellucidum:* Ecuador: *Provincia de Sucumbíos:* Río Azuela (0.1167 S, 77.6167 W; 1740 m), Quito–Lago Agrio road; KU 164691 (holotype), USNM 286708–10; Río Reventador, USNM 286711–12. *Provincia de Morona Santiago:* km 6.6 on the Limón-Macas road (ca. 2.92816 S, 78.344 W; 1013 m), QCAZ 29438. *Provincia de Zamora Chinchipe:* Cordillera del Cóndor, Miazi Alto (4.25044 S, 78.61356 W; 1282 m), QCAZ 41560–61.

Hyalinobatrachium esmeralda: Colombia: Departamento de Boyacá, Municipio de Pajarito, Inspección Policía Corinto, finca 'El Descanso', quebrada 'La Limonita', 1600–1650 m, ICN 9592–94, 9596, 9602–03 (type series of *H. esmeralda*).

Hyalinobatrachium valerioi (Dunn, 1931 [236]; Figures 121–123).

Centrolene valerioi Dunn, 1931 [236]. Holotype: MCZ 16003.
Type locality: "La Palma, Costa Rica, 4500 feet".
Cochranella valerioi—Taylor, 1951.
Cochranella reticulata—Taylor, 1958 [201]. Holotype: KU 32922. Type locality: "near bridge across
  Río Reventazón at the Inter-American Institute of Agriculture, Turrialba, Cartago Province,
  Costa Rica". Placed in synonymy by Starrett and Savage, 1973 [209].
Centrolenella valerioi—Starrett and Savage, 1973 [209].
Hyalinobatrachium valerioi—Ruiz-Carranza and Lynch, 1991 [6].

Common names: English: Reticulated glassfrog [237], Valerio's glassfrog. Spanish: Rana de
Cristal Reticulada, Rana de Cristal de Valerio.
   Etymology: The specific name *valerioi* is a patronym for Manuel Valerio, who, with Emmett R.
Dunn, discovered the species in the field.
   Identification: Ecuadorian population tentatively assigned to *Hyalinobatrachium valerioi* can be
distinguished from all other glassfrogs by its green dorsum with numerous large yellowish–green
spots (Figure 121), transparent ventral parietal peritoneum, and most visceral peritonea covered by
iridophores. The pericardium varies from white to mostly transparent (red in life). Among Ecuadorian
centrolenid, only *H. aureoguttatum* is similar, but differs by having large yellow to golden dorsal spots
(Figure 102), which are produced by the interaction of xanthophores (yellow) and iridophores (white);
the spots in *H. valerioi* are structurally different since they lack iridophores.
   Diagnosis: (1) Vomerine teeth absent; (2) snout truncated in dorsal view, truncated to
protruding in lateral view; (3) tympanum with a dorsolateral orientation, of moderate size
(tympanum diameter 30%–43% of eye diameter); tympanic membrane evident; only lower border
of tympanic annulus visible; (4) dorsal skin finely shagreen; (5) ventral skin granular; no cloacal
ornamentation; (6) parietal peritoneum transparent (condition P0), iridophores covering all visceral
peritonea; pericardium variable, from white to mostly translucent (condition V5–V6); (7) liver
bulbous, covered by white peritoneum (condition H2); (8) humeral spine absent; (9) webbing
between Fingers I and II reduced, moderate between Fingers II and III, expanded between outer
fingers, webbing formula: I $(2^+–2^{1/3})$—2 II $(1–1^+)$—$(3^-–3^{1/3})$ III $(1^{1/2}–2^{2/3})$—$(1^+–2^-)$ IV; (10) webbing
on feet: I $(1–1^{1/2})$—$(2–2^+)$ II $(1–1^{2/3})$—$(2–2^{1/4})$ III $(1^-–1^{1/3})$—$(2^+–2^{1/3})$ IV $(2–2^{1/2})$—$(1–1^+)$ V; (11) ulnar
and tarsal fold absent; (12) nuptial excrescences Type V in adult males; concealed prepollex;
(13) first finger slightly longer than second; (14) eye diameter larger than width of disc on Finger
III (disc of Finger III 32%–36% of eye diameter); (15) in life, dorsal surfaces yellowish green with
green reticulum (or green with numerous yellowish–green spots); ventral parietal peritoneum
transparent; bones white; (16) in preservative, dorsal surfaces cream with dark melanophores;
(17) in life, iris golden with dark pigmentation scattered throughout, especially around the pupil;
(18) melanophores absent from fingers and toes, except for few on Toes IV and V; (19) males usually
call from underside of leaves, but sometimes they call from the upper leaf surface; call is a short *seet*,
with a duration of 200–250 milliseconds, and a modulated frequency; (20) amplexus-like fighting
behavior; (21) egg clutches laid on underside of leaves; nocturnal and diurnal parental care by
males; maternal care absent; (22) tadpole with ventral mouth, spiracle posterior and mid-lateral;
oral disc complete, moderate, with single row of marginal papillae laterally and ventrally, large
dorsal gap; tooth row formula 2(2)/3, A-2 with large gap above mouth; (23) minute body size; SVL
in adult males 18.1–24.4 mm (*n* = 23), in adult females 20.2–25.1 mm (*n* = 7).

Figure 121. *Hyalinobatrachium valerioi* in life. (**Top row**): Male from Reserva de Biodiversidad Mashpi, 1080 m, Ecuador; photos by Lucas Bustamante/Tropical Herping. (**Bottom row**): Male guarding eggs, Selva Verde Lodge, Costa Rica; photo by Carlos Martínez. Note differences in coloration.

Color in life (Figure 121): According to Kubicki [24], dorsum yellowish green with green reticulation. Transparent venter; pericardium varies from white to almost transparent; white liver and digestive tract. Iris golden with dark pigmentation scattered throughout, especially surrounding pupil; several individuals with dark pigmentation restricted to lateral regions of pupil. Bones white.

Color in ethanol: Cream dorsum with dark melanophores in the places that where green in life. Venter translucent cream. Parietal peritoneum completely transparent, all visceral peritonea covered by white lining. Pericardium polymorphic and white or translucent.

Biology and ecology: The following information was obtained from Savage [42], McDiarmid and Adler [238], McDiarmid [216], and Hayes [189]. A nocturnal frog that inhabits forested streams. Males are territorial and vigorously call from under leaves where females deposit their egg clutches. Physical combat occurs between males if another male intrudes into a male's territory despite the calls of the owner; both males may squeak during the fight; once one of the males is pinned venter down, he is held for a while, then leaves. Pale greenish–white eggs (25–40 in number) are deposited in a single-layer on the underside of a leaf. Males continue to advertise and may end up with as many as seven clutches from different females. The males attend the eggs both during the day and at night. At night, the male continues to call and, occasionally exhibits hydric brooding behavior, in which he apparently empties his bladder over the eggs. Only males are involved in parental care of the eggs. McDiarmid [4] attributed the higher survivorship to hatching in this species than in the syntopic *Hyalinobatrachium colymbiphyllum* to the diurnal parental care. Vockenhuber et al. [239] experimentally verified the positive effect that parental care has on embryonic survivorship; also, they identified arthropod predation as the main cause for embryonic mortality.

Call (Figure 122): Males usually call from the underside of leaves, but sometimes they vocalize from the upper leaf surface. In Costa Rica, the call is a high-pitched *seet* lasting 200 to 250 milliseconds, with a dominant frequency initially of 7.0 kHz rising to 7.5 kHz and then dropping again to 7.2 kHz, repeated every 7 to 10 s [209]. Kubicki [24] reported call variation among Costa Rican populations. Below, we describe a call of an individual tentatively assigned to *Hyalinobatrachium valerioi* (USNM 201475) recorded near Santo Domingo de los Colorados by RWM; we analyzed 19 notes from one individual (Figure 122). Each note is composed by a single note. The note duration is 46–78 (mean = 72, SD = 7.7) ms. Notes are generally tonal and do not show any clear amplitude peaks throughout the note. The dominant frequency at peak amplitude is 5771–6632 (mean = 6197, SD = 253) Hz and is contained within the fundamental frequency. The fundamental frequency has a lower limit of 5599–5943 (mean = 5771, SD = 91) Hz and a higher limit of 6374–6718 (mean = 6632, SD = 76) Hz.

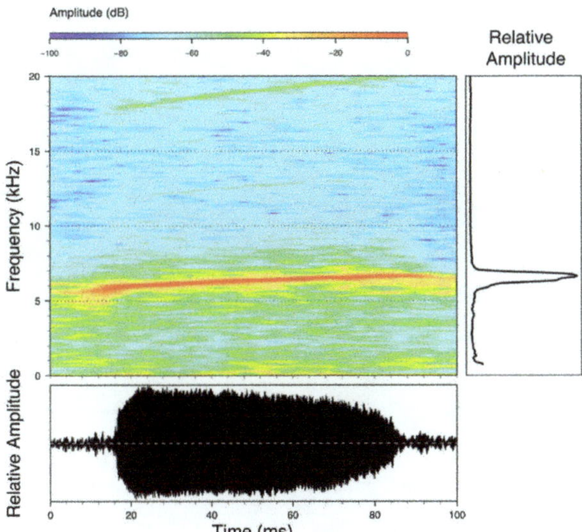

Figure 122. Call of a male (USNM 201475) tentatively assigned to *Hyalinobatrachium valerioi*, recorded near Santo Domingo de los Colorados, Ecuador, recorded by RWM.

Egg masses and tadpoles: The following information was obtained from Savage (2002) and incorporates data from McDiarmid and Adler [238], McDiarmid [216], and Hayes [189]. Green egg masses are deposited in a single-layer jelly mass on the underside of leaves; average clutch size is 35, with 40 eggs being the upper limit. Egg masses are attacked by diurnal wasps, which remove eggs one at a time and carry them away, presumably to a nest, until the egg mass is depleted [4]. Starrett [146] described the tadpole of *H. valerioi* as follows (as *Cochranella reticulata*): Tadpole similar to that of *H. fleischmanni*, but tail longer at hatching and color paler; tail of older tadpole two- and three-fifths times as long as body; dorsal part of body with brown blotches; tail musculature with brown blotches dorsally and laterally; pigment is in depressions between myotomes posteriorly; tail fins clear. Mouth ventral, nearly terminal; papillae surrounding all but anterior portion of the disc. LTRF 2(2)/3; A-1 complete, A-2 rows very short, occurring only lateral to the jaw sheaths; outermost posterior row not well developed at this stage (probably, in larger tadpoles, nearly as long as inner two); anterior jaw sheath is more heavily pigmented than the posterior one that is doubly arched and weakly pigmented medially, often appearing broken; posterior jaw sheath armed with small, equal length serrations. Hoffmann [147] recently provided a very detailed description of the tadpole of the species. He mentioned that tadpoles of *H. valerioi* are distinguished by their accented slenderness and by the noticeably longer tail, about three times longer than the body; additionally, the oral disc of *H. valerioi* is more ventrally positioned than in other glassfrog species [147].

Distribution (Figure 123): As currently recognized, *Hyalinobatrachium valerioi* occurs from Costa Rica to southwestern Ecuador [156]. In Ecuador, it occurs in the northern and southern Pacific lowlands, foothills, and slopes below 1500 m in the provinces of Azuay, Carchi, Esmeraldas, Los Ríos, Pichincha, and Santo Domingo de los Tsáchilas (Specimens examined), within the Western Foothill Forest and Western Montane Forest ecoregions. In Ecuador, it has a potential distribution of 49,705 km^2.

Figure 123. Distribution of populations tentatively assigned to *Hyalinobatrachium valerioi* in Ecuador (yellow dots).

Conservation status: Globally, *Hyalinobatrachium valerioi* is categorized as *Least Concern* by the IUCN [240]. The Ecuadorian populations currently assigned to *H. valerioi* (see Figure 123) might represent an undescribed species. The habitat of the species is severely fragmented, and the species has exhibited populations declines in several historical localities (e.g., La Florida, Río Palenque, Manta Real; Luis A. Coloma, pers. comm.; JMG, pers. obs.). During the last year (2015–2019), we have found reproducing populations at Reserva Los Cedros and Reserva Mashpi. Based on the current available information, we suggest that *H. valerioi* should be considered as *Endangered* in Ecuador (IUCN criteria A2c, A2e).

Evolutionary relationships (Figure 101): *Hyalinobatrachium valerioi* is the sister species to *H. aureoguttatum*.

Taxonomic remarks: The coloration of *Hyalinobatrachium valerioi* is variable across its distribution. Population may differ in having the pericardium completely transparent or covered with iridophores [24]. Also, at least in Costa Rica, there is notorious call variation [24]. At the moment, we are still uncertain about the taxonomic position of the populations from Ecuador; further studies are needed to assess if *H. valerioi*, as currently defined, is composed by several morphologically similar species.

Specimens examined: *Hyalinobatrachium valerioi:* <u>Ecuador</u>: *Provincia de Azuay:* 12.9 km W of Luz María, KU 217503–04. *Provincia de Carchi:* Maldonado, KU 178082–91. *Provincia de Cañar:* Manta Real (03°10′ S 80°26′ W), DHMECN 0134. *Provincia de Imbabura:* Reserva Los Cedros (0.3026° N, 78.781° W; 1360 m), MZUTI 3268. *Provincia de Los Ríos:* Río Palenque (0.55° S, 79.36667° W), USNM 286746–49, KU 147580, 164650. *Provincia de Santo Domingo de los Tsáchilas:* 5 km W of La Florida (0.25694° S, 79.0538° W; 860 m), QCAZ 12600, 27652; 4 km NE of Dos Ríos (0.25694° S, 79.0538° W), KU 164651–56; Río Faisanes (0.26083° S, 78.845° W). *Provincia de Pichincha:* Tandapi (0.41638° S, 78.7988° W), KU 178092; 1 km E of Pedro Vicente Maldonado (0.833° S, 79.0347° W; 670 m), QCAZ 12605; Reserva de Biodiversidad Mashpi (0.164° N, 78.867° W; 1080 m), MUTI 3921.

Localities from the literature: *Hyalinobatrachium valerioi:* <u>Ecuador</u>: *Provincia de Esmeraldas:* Bilsa Biological Station [160].

Hyalinobatrachium yaku Guayasamin, Cisneros-Heredia, Maynard, Lynch, Culebras, Hamilton 2017 [241] (Figures 124–126).

Hyalinobatrachium yaku Guayasamin, Cisneros-Heredia, Maynard, Lynch, Culebras, Hamilton, 2017 [241]. Holotype: MZUTI 5001.

<u>Type locality</u>: "Stream affluent of the Kallana river (1.4696° S, 77.2784° W, 325 m), nearby the Kichwa community of Kallana, province of Pastaza, Ecuador".

Common names: English: Yaku Glassfrog. Spanish: Rana de Cristal Yaku.

Etymology: The epithet *yaku* is the Kichwa word for *water* [241].

Identification: In life, *Hyalinobatrachium yaku* is distinguishable from all other glassfrogs by having middorsal dark green spots on the anterior half of the body and, ventrally, a completely transparent peritoneum and pericaridum (Figure 124).

Diagnosis: (1) Dentigerous process of the vomer lacking teeth; (2) snout truncated in dorsal and lateral views; (3) lower half of tympanic annulus visible; tympanic membrane clearly differentiated and with coloration similar to that of surrounding skin; (4) dorsal skin shagreen; (5) ventral skin areolate; cloacal area glandular, with tubercular and slightly enameled patch on each side of cloaca, lacking paired round subcloacal warts; (6) parietal peritoneum, pericardium, kidneys, and urinary bladder transparent (lacking iridophores); hepatic, gastrointestinal, and testicular peritonea covered by iridophores; (7) liver white, bulbous; (8) humeral spines absent; (9) basal webbing between Fingers I and II, moderate webbing between external fingers; hand webbing formula: I 2—2 II 0^+—3^+ III 2^-—$(1-2^-)$ IV; (10) foot webbing moderate; webbing formula: I $(1-1^+)$—$(2-2^-)$ II (0^+-1)—$(2^+-2^{1/3})$ III 1—$2^{1/3}$ IV $2^{1/3}$—$(1-1^{1/3})$ V; (11) fingers and toes with thin lateral fringes; ulnar and tarsal folds present, but difficult to distinguish, with thin layer of iridophores that extends to ventrolateral edge of Finger IV and Toe V; (12) nuptial excrescence present as a small pad on Finger I (Type V), concealed prepollex; (13) when appressed, Finger I longer than Finger II; (14) diameter of eye about 2.1 times wider than disc on Finger III; (15) coloration in life: Dorsal surfaces apple green to yellowish green with small yellow spots and minute grey to black melanophores; posterior head and anterior half of the body with few small, well-defined dark green spots placed mid-dorsally; bones white; (16) coloration in preservative: Dorsal surfaces pale cream with minute lavender to black melanophores; (17) iris coloration in life: Silver to yellow, with minute dark spots concentrated around pupil; (18) melanophores present on Finger IV and Toes IV–V, absent on other fingers and toes; in life, hands and feet are cream with a light green hue, with tips of fingers and toes being yellowish green; (19) males call from the undersides of leaves; advertisement call consisting of a single tonal note; call duration note 0.3–0.4 s, dominant frequency 5219–5330 Hz, with no frequency modulation; (20, 21) clutch size unknown; maternal care unknown; prolonged parental care provided by males; (22) tadpoles unknown; (23) adults small; SVL in adult males 20.8–22.3 mm (n = 3), in adult female 21.1 mm (n = 1).

Figure 124. *Hyalinobatrachium yaku* in life. (**Top row**): adult male, MZUTI 5001, holotype, in dorsal and ventral view. (**Bottom row**): adult male, paratype, QCAZ 55628. Obtained from Guayasamin et al. [241].

Color in life (Figure 124): In adults, dorsum apple green to yellowish green with small yellow spots and minute dark melanophores; posterior head and anterior half of the body with few small, well-defined dark green spots placed mid-dorsally, the anterior-most spot generally being the largest. Hands and feet cream with a light green hue, with tips of fingers and toes being yellowish green;

melanophores absent from fingers and toes, except Finger IV and Toes IV and V. Ventrally, parietal peritoneum and pericardium transparent (red heart fully visible); visceral peritoneum of gall bladder and urinary bladder transparent; hepatic and visceral peritonea white. Ventral vein red. Iris silver to yellow, with minute dark spots that encircle the pupil, giving the impression of diffuse rings. Bones white [241].

Color in ethanol: Dorsal surfaces cream dotted with minute dark lavender to black melanophores; venter uniform white; peritonea as in life. Iris white with lavender melanophores that become more numerous near the pupil. There are no traces of the characteristic middorsal dark-green spots in preserved specimens [241].

Variation: Juveniles have the same color pattern as adults; the number and extent of the middorsal green dots varies, but they are usually smaller and less pronounced posteriorly [241].

Biology and ecology: Only basic biological information of *H. yaku* is known. The data presented below are from Guayasamin et al. [241]. Males call from the underside of leaves of riverine vegetation. One male (holotype) was on the same leaf as two egg clutches, approximately 3 m above the stream. Observations suggest that prolonged parental care is provided by males. Syntopic species at the type locality are: *Nymphargus mariae, Teratohyla midas, Agalychnis hulli, Callimedusa tomopterna, Boana calcarata, B. geographica, Osteocephalus fuscifacies, Pristimantis enigmaticus*, and *P. peruvianus*. At the locality of Ahuano, the species was found along a small stream, tributary of the Arajuno River; the stream was slow flowing, very narrow (approximately 1 m wide), shallow (approximately 40 cm deep), and covered by secondary forest. At Ahuano, syntopic species were *Teratohyla midas* and *H. ruedai*. Individuals from San José de Payamino were found perched on leaves of small shrubs, ferns, and grasses (30–150 cm above ground) in disturbed secondary forest. Additionally, all individuals recorded at San José de Payamino were found >30 m from any stream. Syntopic species at San José de Payamino are reported by Maynard et al. [242].

Call (Figure 125): The information provided below is taken from Guayasamin et al. [241]. The advertisement call of *Hyalinobatrachium yaku* is a single and high-pitched tonal note, with no frequency nor amplitude modulation. The call lasts 0.27–0.4 s (0.3 ± 0.03) and has an average call rate of 9.0 calls/minute. Time between calls varied from 5.3–8.9 s (7.1 ± 1.1). The dominant frequency ranges from 5219–5330 Hz (5284 ± 35.0).

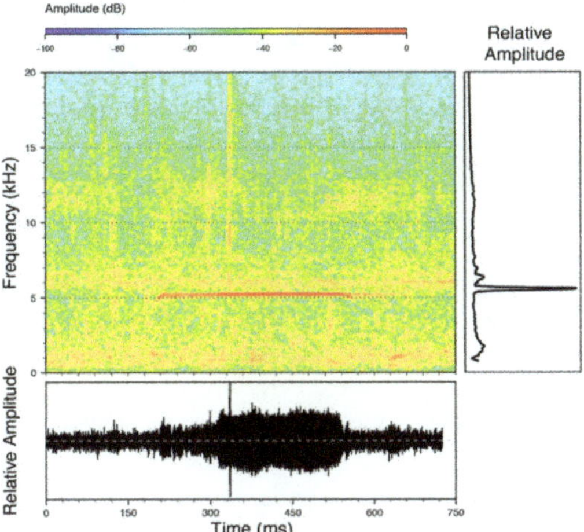

Figure 125. Call of *Hyalinobatrachium yaku* from Kallana, 325 m, Pastaza province, Ecuador; holotype, MZUTI 5001.

Tadpole: Not described.

Distribution (Figure 126): *Hyalinobatrachium yaku* is known from three localities on the Amazonian lowlands of Ecuador at elevations between 300–360 m. It is likely that *H. yaku* has a broader distribution, including areas in Amazonian Colombia and Peru [241].

Figure 126. Distribution of *Hyalinobatrachium yaku* in Ecuador (yellow dots).

Conservation status: According to Guayasamin et al. [241], available information is insufficient to suggest a conservation category, thus *H. yaku* is a *Data Deficient* species.

Evolutionary relationships (Figure 101): *H. yaku* and *H. pellucidum* are sister species.

Specimens examined: *Hyalinobatrachium yaku*: <u>Ecuador</u>: *Provincia Pastaza:* stream affluent of Kallana river (1.4696° S, 77.2784° W, 325 m), MZUTI 5001 (holotype), 5002 (paratypes). *Provincia Orellana:* Timburi-Cocha Research Station (0.4800° S, 77.2829° W, 300 m) near San José de Payamino, QCAZ 55628, 53352 (paratypes). *Provincia Napo:* Ahuano (1.0632° S, 77.5265° W, 360 m), ZSFQ 2322.

Genus *Nymphargus* Cisneros-Heredia and McDiarmid 2007 [17].

Etymology: The name *Nymphargus* is formed from the Greek *nymphae* in allusion to the nymphs, beautiful goddesses in Greek mythology that personify the creative and fostering activities of nature, living in mountains, valley, springs, and rivers; and, *argus* in allusion to the mythological Greek Argus, nephew of the nymph Io, a giant with a hundred eyes, whose eyes became the ocelli in the peacock's tail. The name is masculine and alludes to the ocelli found on the dorsum of some of the species of the genus [17].

Nymphargus anomalus (Lynch and Duellman, 1973 [22]; Figures 127–130).

Centrolenella anomala Lynch and Duellman, 1973 [22]. Holotype: KU 143299.
Type locality: "Río Azuela, 1740 m, Quito–Lago Agrio road, Provincia Napo, Ecuador".
Cochranella anomala—Ruiz-Carranza and Lynch, 1991 [6], 57:21.
Nymphargus anomalus—Cisneros-Heredia and McDiarmid, 2007 [17].

Common names: English: Anomalous Glassfrog. Spanish: Rana de Cristal Anómala.
Etymology: The specific name *anomalous* refers to the unusual dorsal coloration of this species [22].
Identification: *Nymphargus anomalus* is unique by having, in life, a tan dorsum with black ocelli with orangish–tan centers (Figures 127 and 128), instead of the usual green that characterizes most glassfrogs. Among centrolenids, only one species, *N. ignotus*, has a similar dorsal color pattern; however, *N. ignotus* is found on the Pacific slopes of the Andes, whereas *N. anomalus* is restricted to the Amazonian slopes of the Andes. *Nymphargus anomalus* further differs from other species having ocellated patterns (*N. cochranae*, *N. ignotus*, *N. laurae*) by having scattered black and lavender flecks between ocelli (Figure 127).
Diagnosis: (1) Vomers lacking teeth; (2) snout truncated in dorsal and lateral profiles (Figure 129); (3) tympanum relatively small, oriented almost vertically, with evident lateral and posterior inclinations, its diameter about 20% of eye diameter; tympanic annulus mostly visible, with supratympanic fold covering its dorsal margin; tympanic membrane differentiated and translucent, partially pigmented; (4) in males, dorsal skin shagreen with or without minute spiculae; in females, dorsal skin shagreen without spiculae; (5) venter areolate; pair of enlarged subcloacal warts (Figure 15); (6) anterior half of ventral parietal peritoneum white, posterior half transparent (condition P2); white pericardium; translucent peritoneum covering intestines (condition V1); (7) liver with four clearly defined lobes covered by transparent peritoneum (condition H0); (8) humeral spines absent; (9) webbing absent between Fingers I–III, basal or moderate between Fingers III and IV; webbing formula: III $(2^{1/3}$–$3^{+})$—$(2^{+}$–$2^{2/3})$ IV (Figure 129); (10) webbing between toes moderate; webbing formula on foot I 2^{-}—$(2$–$2^{1/4})$ II $(1$–$1^{1/4})$—$(2^{+}$–$2^{1/3})$ III $(1$–$1^{+})$—$(2^{+}$–$2^{1/4})$ IV $(2^{+}$–$2^{2/3})$—$(1^{+}$–$1^{1/2})$ V; (11) ulnar fold low; inner and outer tarsal folds present, but low and difficult to distinguish; (12) concealed prepollex; nuptial pad Type I; (13) Finger I slightly shorter than Finger II (Finger I length 91.5%–99.2% of Finger II); (14) disc width of Finger III about 50% of eye diameter; (15) in life, dorsum pale brown or tan with dark brown flecks and black ocelli with brownish–orange centers (Figure 127); color of bones white; (16) in preservative, dorsum cream to tan with black dots and lavender flecks and black and lavender ocelli enclosing whitish centers; (17) iris pale yellow with thin black reticulation; bright yellow circumpupilar ring; (18) dorsal surfaces of all fingers and toes usually with few melanophores, except Finger I that usually lacks melanophores; (19) males call from the upper surfaces of leaves; calls undescribed; (20) fighting behavior unknown; (21) eggs deposited on mossy branches over streams; parental care unknown; (22) tadpoles unknown; (23) small body size; SVL in males 21.2–24.8 mm ($\overline{X}$ = 23.1 mm; n = 16); SVL in three adult females: 25.7, 27.0, and 27.0 mm.

Figure 127. *Nymphargus anomalus* in life, ZSFQ 2123, from a tributary of Río San Jacinto, 1672 m, Tungurahua province. Photos by Jose Vieira/Tropical Herping.

Figure 128. Ontogenetic color change in *Nymphargus anomalus*, QCAZ 48107. Photos by Luis A. Coloma.

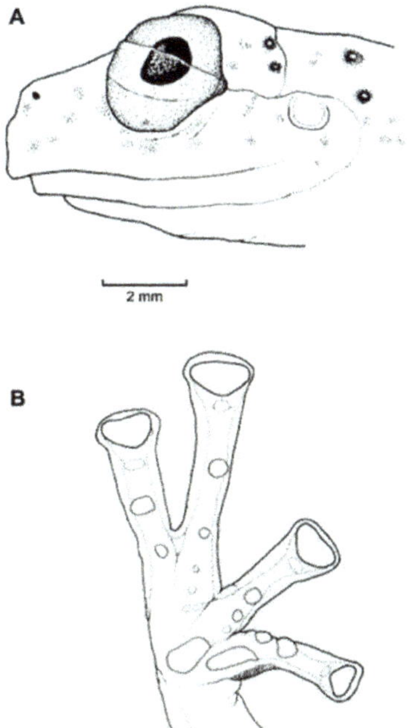

Figure 129. *Nymphargus anomalus,* adult male, KU 143299. (**A**) Head in lateral view. (**B**) Hand in ventral view. Illustrations by Juan M. Guayasamin.

Color in life (Figures 127 and 128): Dorsum pale brown or tan with slight yellow to pink hue; dorsal surfaces of head, body, and limbs with small flecks and black ocelli that enclose orange spots. Anterior part of venter white, posterior half of venter translucent. Iris pale yellow with black reticulations; color of iris becomes whitish towards margins. *Nymphargus anomalus* exhibits ontogenetic variation in its dorsal coloration, being green in postmetamorphs and then gradually acquiring the tan to pale brown coloration that is characteristic of adults. The spotted pattern also changes ontogenetically; postmetamorphs have yellow spots that progressively change towards ocelli composed of black rings with orange centers.

Color in ethanol: Dorsum cream to tan with lavender flecks and black and lavender ocelli enclosing whitish centers. Eyelids bear some iridophore aggregations visible through the skin. White pericardium; hepatic and visceral peritonea are translucent. Peritonea covering urinary and gall bladders, kidneys, and testes without iridophores.

Biology and ecology: The holotype of *Nymphargus anomalus* was found on a mossy limb of a bush about 1.5 m above a cascading rivulet at night. The following species were found in the same stream and in other small streams nearby: *Hyloscirtus phyllognathus, Centrolene pipilata, Nymphargus megacheirus, N. siren,* and *Hyalinobatrachium pellucidum* [22]. At Río Yana Challuwa Yaku, males where found calling on leaves and branches during the night. Two egg clutches were on a mossy branch above the stream. Parental care is unknown.

Call: Not described.

Tadpole: Not described.

Distribution (Figure 130): *Nymphargus anomalus* is known only from four localities (Volcán Sumaco, Río Azuela, Río Yana Challuwa Yaku, nearby Río Jacinto) on the Amazonian slopes of the Ecuadorian Andes, at elevations between 1668–1795 m. The habitat of the species is within the Eastern Montane Forest region ([22], this work).

Conservation status: *Nymphargus anomalus* is listed as *Critically Endangered* by the IUCN [243]. Until recently, the species was known from a single individual collected on 23 October 1971 (Río Azuela). Surveys at the type locality have failed in find this species ([91], RWM and JMG, pers. obs.), but other populations have been discovered (Figure 130; Specimens examined). In Ecuador, because of its restricted distribution and habitat fragmentation, we suggest that it should be considered as *Endangered* (IUCN criteria B1, B2a, B2biii).

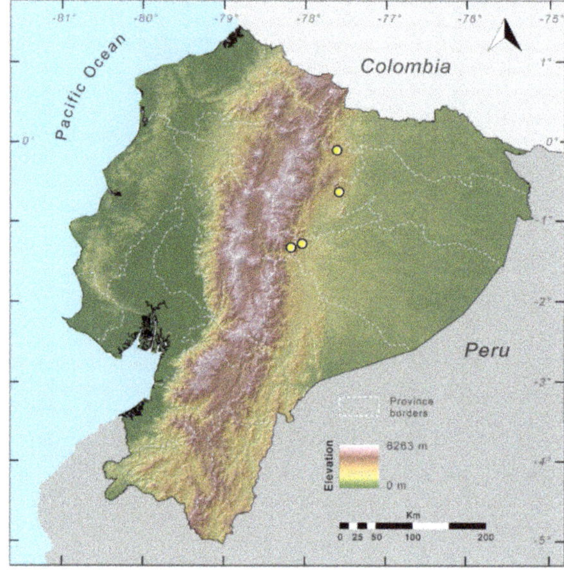

Figure 130. Distribution of *Nymphargus anomalus* in Ecuador (yellow dots).

Evolutionary relationships (Figure 101): Lynch [244] suggested that *N. anomalus* is the sister species of *N. ignotus*. The hypothesis by Lynch is based on the presence of two putative synapomorphies: Pale brown coloration and small ocelli on elevated warts, with the corresponding plesiomorphic conditions being green coloration and no ocelli. Based on mitochondrial genes and a taxon sampling that does not include *N. ignotus*, we found that the sister species of *N. anomalus* is *N. megacheirus*, which has a coloration that matches the plesiomorphic states.

Specimens examined: *Nymphargus anomalus:* <u>Ecuador</u>: *Provincia de Napo:* Río Azuela (0.11667 S, 77.6167 W, 1740 m), KU 143299. Parque Nacional Sumaco, Volcán Sumaco, near the Pavayacu refuge (0.61497 S; 77.59065 W; 1771 m), QCAZ 41312–13. *Provincia de Pastaza*: Reserva Comunitaria Ankaku, zona de amortiguamiento del Parque Nacional Llanganates, cabecera del Río Yana Challuwa Yaku (01.26764 S; 78.04797 W; 1668 m), males: QCAZ 45696–97, 45699, 45701, 45703–11, 45728–29; females: QCAZ 45698, 45700, 45702. *Provincia de Tungurahua:* stream tributary of the San Jacinto River (1.3447 S, 78.1814 W; 1795 m asl), ZSFQ 899.

Nymphargus balionotus (Duellman, 1981 [120]; Figures 131–133).

Centrolenella balionota Duellman, 1981 [120]. Holotype: KU 164702.
Type locality: "3.5 km (by road) northeast of Mindo, 1540 m, Provincia de Pichincha, Ecuador (00°01′ S, 78°44′ W)".
Cochranella balionota—Ruiz-Carranza and Lynch, 1991 [6].
Centrolene balionotum—Cisneros-Heredia and McDiarmid, 2006 [17].
"Cochranella" balionota—Guayasamin, Castroviejo-Fisher, Trueb, Ayarzagüena, Rada, Vilà, 2009 [1].
Nymphargus balionotus—Guayasamin, Cisneros-Heredia, Vieira, Kohn, Gavilanes, Lynch, Hamilton, Maynard, 2019 [21].

Common names: English: Mindo Glassfrog. Spanish: Rana de Cristal de Mindo.
Etymology: The specific epithet *balionotus* combines the Greek words *balios* (spotted, dappled) and *notos* (back), and is used in allusion to the dorsal color pattern of the species [120].
Identification: *Nymphargus balionotus* is unique by having, in life, a pale green dorsum with reddish–brown dorsolateral stripes, small reddish–brown spots, and larger yellow spots (Figure 131). Additionally, males are small (SVL < 23 mm) and possess a blade-like ventral crest on the humerus that can be confused with a humeral spine. See Remarks.

Figure 131. *Nymphargus balionotus* in life, ZSFQ 533. Ecuador, Reserva Río Manduriacu, 1254 m, Tungurahua province. Photos by Jose Vieira/Tropical Herping.

Diagnosis: (1) Vomers lacking teeth; (2) snout truncated in dorsal view, protruding lateral profile (Figure 132); (3) tympanum relatively small, oriented almost vertically, with evident lateral and posterior inclinations, its diameter 23%–30% of eye diameter; tympanic annulus mostly visible, with supratympanic fold covering its dorsal margin; tympanic membrane differentiated and translucent, not pigmented; (4) in males, dorsal skin smooth to shagreen, lacking spicules; females unknown; (5) venter slightly granular; pair of enlarged subcloacal warts; (6) anterior half of ventral parietal peritoneum with iridophores, posterior half transparent (condition P2); pericardium covered by iridophores; peritoneum covering digestive tract lacking iridophores (condition V1); (7) liver with four or five lobes, lacking iridophores (condition H0); (8) males with a bladelike ventral crest on the humerus that resembles a small humeral spine; (9) webbing absent between inner fingers; webbing formulae for outer fingers IV (2–2$^{1/2}$)—(2$^-$–2) V (Figure 132); (10) webbing between toes moderate; webbing formula on foot: I (1$^{1/3}$–1$^{1/2}$)—(2–2$^+$) II (1$^+$–1$^{1/2}$)—(1–1$^{1/2}$) III (1$^{1/3}$–2)—(2$^-$–3) IV (2$^-$–3)—1$^{1/2}$ V; (11) ulnar fold low; inner and outer tarsal folds present, but low and difficult to distinguish; (12) concealed prepollex; nuptial pad in males variable, ranging from Type I to Type III; (13) Finger I slightly longer than Finger II (Finger II length 91.8%–97.4% Finger I); (14) disc of Finger III width about 36%–44% of eye diameter; (15) in life, dorsum pale green with reddish–brown dorsolateral stripes, small reddish–brown spots, and larger yellow spots (Figure 131); color of bones pale green; (16) in preservative, dorsum cream with lavender and white dorsolateral stripes, small dark lavender spots, and larger white spots; (17) iris greyish white with numerous minute brown spots; (18) dorsal surfaces of fingers and toes lacking melanophores; (19) males call from the upper sides of leaves overhanging streams; call undescribed; (20) fighting behavior unknown; (21) egg deposition site unknown; parental care unknown; (22) tadpoles unknown; (23) minute body size; in adult males, SVL 20.1–21.8 mm ($\overline{X}$ = 20.8 ± 0.534, *n* = 11); females unknown.

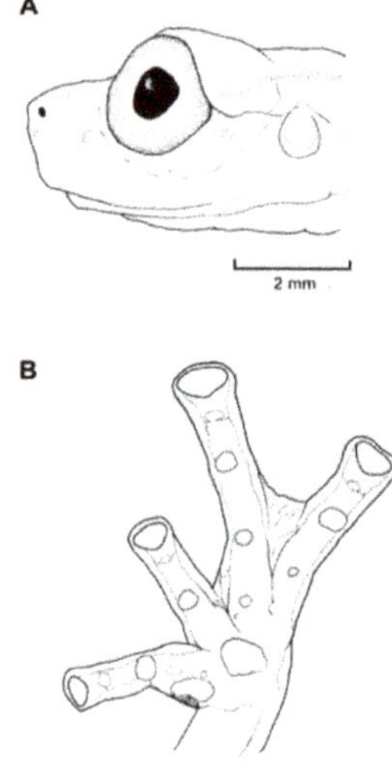

Figure 132. *Nymphargus balionotus*, adult male, KU 164708. (**A**) Head in lateral view. (**B**) Hand in ventral view. Illustrations Juan M. Guayasamin.

Color in life (Figure 131): Pale green dorsum with reddish–brown dorsolateral stripes, small reddish–brown spots, and elevated larger yellow spots. Head with reddish–brown interorbital bar and postorbital stripe that is continuous with dorsolateral stripe. Large, elevated yellow spot on anteromedial edge of each eyelid. Pale green bones ([120], this work).

Color in ethanol: Dorsal surfaces cream with lavender and white dorsolateral stripes, small dark lavender spots, and larger white spots. White lining covers about anterior half of ventral parietal peritoneum. White pericardium; digestive tract, liver and kidneys covered by translucent peritonea.

Biology and ecology: During the night, all individuals at the type locality were calling less than 1 m above the water from the upper surfaces of leaves of herbs and ferns overhanging a trickling stream. Other stream-breeding frogs at the type locality included *Nymphargus grandisonae*, *N. griffithsi*, and *Hyloscirtus alytolylax* [120]. *Nymphargus balionotus* is nocturnal, arboreal, epiphyllous, and usually found on vegetation along clear-water streams in pristine or in moderately disturbed foothill and cloud forests [87]. Parental care unknown.

Call: Not described.

Tadpole: Not described.

Distribution (Figure 133): *Nymphargus balionotus* occurs between 400 and 1540 m along the western slope of the Cordillera Occidental of Colombia, from El Tambito in the Departamento de Cauca, south to Ecuador [21,87,115,116,120,245]. In Ecuador, it is known from localities in the provinces of Carchi, Imbabura, Pichincha, and Cotopaxi at elevations between 1400 and 1540 m (Specimens Examined). It is found within the Western Foothill and Montane Forest ecoregions.

Figure 133. Distribution of *Nymphargus balionotus* in Ecuador (yellow dots).

Conservation status: *Nymphargus balionotus* is listed by the IUCN as *Vulnerable* at a global level [245]. The species was abundant at the type locality (3.5 km NE of Mindo), where 13 individuals were observed in two nights (7–8 April 1975) [120]. A single reproductive population is known in Ecuador, Reserva Río Manduriacu, which is threatened by mining activities [21]. Thus, in Ecuador, we suggest the category of *Critically Endangered*, following IUCN criteria B2a + B2b(iii).

Evolutionary relationships: Lynch and Ruiz-Carranza [246] suggested that *Nymphargus balionotus* is the nearest relative of *Nymphargus armatus* and *N. griffithsi,* based on the following characters: (i) Snout short and truncate, (ii) slender habitus, (iii) similar webbing and dentition, and (iv) a large bladelike ventral crest present on the humerus. Based on mitochondrial data, a recent study found that *N. balionotus* is sister to *N. manduriacu* [21].

Remarks: Ecuadorian individuals of *Nymphargus balionotus* have a small, pointed humeral spine. The Colombian populations have a variable *crista ventralis* that, in some cases, takes the form of a pointed humeral spine. The polymorphism in this characteristic has created confusion in the generic placement of the species [6,21,139]. It also has been suggested that at least two species are currently confused under the name *Nymphargus balionotus* (M. Rada, pers. com.)

Specimens examined: *Nymphargus balionotus:* Ecuador: *Provincia de Carchi:* Cabeceras del Río Baboso (00°53″ N, 78°27″ W; 1400 m), DH-MECN 0865. *Provincia de Pichincha:* 3.5 km NE Mindo (0.0322 S, 78.761 W, 1540 m), KU 164702 (holotype), 164701, 164703–11. *Provincia de Imbabura:* Reserva Río Manduriacu, 1240–1254 m, ZSFQ 0531–33.

Photographic record: Ecuador: *Provincia de Cotopaxi:* Río Lomapi, NE of La Maná, Reserva Ecológica Illinizas (0.8270275 S, 79.0843359 W, ca. 1300 m); photo by Martín Bustamante.

Nymphargus buenaventura Cisneros-Heredia and Yánez-Muñoz, 2007 [115] (Figures 134 and 135).

Cochranella buenaventura Cisneros-Heredia and Yánez-Muñoz, 2007 [115]. Holotype: DHMECN 3563.
Type locality: "Reserva Buenaventura (03°38′ S, 79°45′ W, 1200 m elevation), canton Piñas, Provincia de El Oro, República del Ecuador".
Nymphargus buenaventura—Cisneros-Heredia and McDiarmid, 2007 [17].

Common names: English: Buenaventura Glassfrog. Spanish: Rana de Cristal de Buenaventura.
Etymology: The specific epithet refers to the type locality, Reserva Buenaventura [115].
Identification: *Nymphargus buenaventura* can be distinguished from most centrolenids by its green dorsum with small pale yellow to cream spots, reduced webbing between fingers, absence of humeral spine, and absence of iridophores on the digestive visceral peritonea, but with iridophores covering the renal capsules. The last character is shared with at least three other species, *N. cariticommatus*, *N. griffithsi*, and *N. wileyi*. *Nymphargus cariticommatus* occurs only on the Amazonian Andean slopes of Ecuador, whereas *N. buenaventura* is restricted to the Pacific Andean slopes. *Nymphargus griffithsi* differs by having dark dorsal spots. *Nymphargus wileyi* differs by having a uniform green dorsum and inhabiting the Amazonian slopes of the Ecuadorian Andes. Another species that could be confused with *N. buenaventura* is *N. lasgralarias*, which lacks the yellow spots characteristic of the former.

Figure 134. *Nymphargus buenaventura* in life from Reserva Buenaventura, 1200 m, El Oro province, Ecuador. Photo by Mario Yánez-Muñoz.

Diagnosis: *Nymphargus buenaventura* has the following traits: (1) Vomerine teeth absent; (2) snout truncated in dorsal view and profile; nostrils elevated, slightly concave internarial area; (3) lower half of tympanic annulus evident, oriented dorsolaterally with dorsoventral inclination; supratympanic fold absent or very weak; (4) dorsal skin slightly shagreen with scattered flat tubercles corresponding to light spots; (5) ventral skin areolate; pair of large, flat subcloacal warts; subcloacal skin granular and enameled; (6) upper half of parietal peritoneum covered by iridophores (condition P2), all other peritonea transparent, except for renal capsules and pericardium covered by iridophores (condition V1); (7) liver tetralobed, uncovered by iridophores (condition H0); (8) humeral spine absent; (9) webbing absent between Fingers I, II, and III; basal between Fingers III and IV; webbing formulae: III $2^{2/3}$—$2^{1/2}$ IV; (10) webbing on feet moderate; webbing formulae: I 2—2^+ II 1^+—$2^{1/2}$ III 1—$2^{1/2}$ IV $2^{1/2}$—$1^{2/3}$ V; (11) outer ventral edges of forearms and tarsi with low folds; (12) unpigmented nuptial pad Type I, concealed prepollex; (13) first finger shorter than second, (14) eye diameter larger than width of disc on Finger III; (15) color in life: Dorsum green with scattered pale yellow spots; bones green; (16) color in preservative: Dorsal surfaces pale lavender with scattered cream spots; (17) in life, iris yellowish silver with thin dark maroon reticulations; (18) yellow–green hands with bright yellow discs, melanophores present on outer fingers and outer toes; (19) males call from upper sides of leaves along streams; call undescribed; (20) fighting behavior unknown; (21) egg clutches placed on upper sides of leaves along streams; parental care unknown; (22) tadpoles unknown; (23) small body size; SVL of adult males 20.9–22.4 mm (*n* = 4); SVL in one adult females 23.5 mm.

Color in life (Figure 134): Pale green dorsum with pale yellow to cream spots on all dorsal surfaces, including arms and legs. Hands and feet yellowish green with pale yellow discs. Venter cream. Iris yellowish silver with fine maroon reticulations.

Color in ethanol: Lavender dorsum with cream spots. Anterior two-thirds of parietal peritoneum white, posterior section translucent; pericardium white; most visceral peritonea clear except for white peritoneum covering the renal capsules.

Biology and ecology: Very little information is known on the natural history of *Nympharugus buenaventura*. It has been found on leaves of shrubs and low trees, 1–3 m above streams. A male was found near an egg clutch with 38 embryos [115]. Parental care is unknown.

Call: Not described.

Tadpole: Not described.

Distribution (Figure 135): *Nymphargus buenaventura* is known from three localities in El Oro province (Reserva Buenaventura, Marcabelí, Cascadas de Manuel) and one in Azuay province (Luz María), southwestern Ecuador, within the Deciduous Forest ecoregion. The localities are: Reserva Buenaventura (03°38′43″ S, 79°45′48″ W, 1200 m), Luz María (02°41′02″ S, 79°25′01″ W, 770 m), Marcabelí (03°44′48″ S, 79°53′36″ W, 667 m), and Cascadas de Manuel (03°12′22″ S, 79°43′34″ W, 800 m) [115,247].

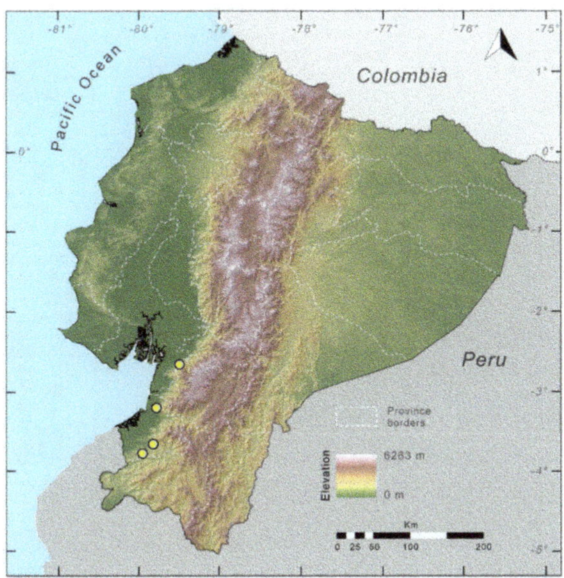

Figure 135. Distribution of *Nymphargus buenaventura* in Ecuador (yellow dots).

Conservation status: *Nymphargus buenaventura* is classified as *Data Deficient* by the IUCN [248]. The species seems to be endemic to southwestern Ecuador, an area with severe deforestation. Thus, we suggest placing the species in the *Endangered* conservation category, following IUCN criteria B1, B2a,

B2biii. The species is partially protected by a private reserve (Reserva Buenaventura, managed by Fundación Jocotoco).

Evolutionary relationships (Figure 136): *Nymphargus buenaventura* has not yet been included in any molecular phylogeny. However, because of its close morphological resemblance to *N. griffithsi*, *N. wileyi*, and *N. cariticommatus*, it is considered as part of the *Nymphargus* clade (sensu Guayasamin et al. [1]).

Specimens examined: *Nymphargus buenaventura:* <u>Ecuador:</u> *Provincia de El Oro:* Reserva Buenaventura (03°38′ S, 79°45′ W, 1200 m), DHMECN 3563 (holotype), 2524, 3561, 3562 (paratypes).

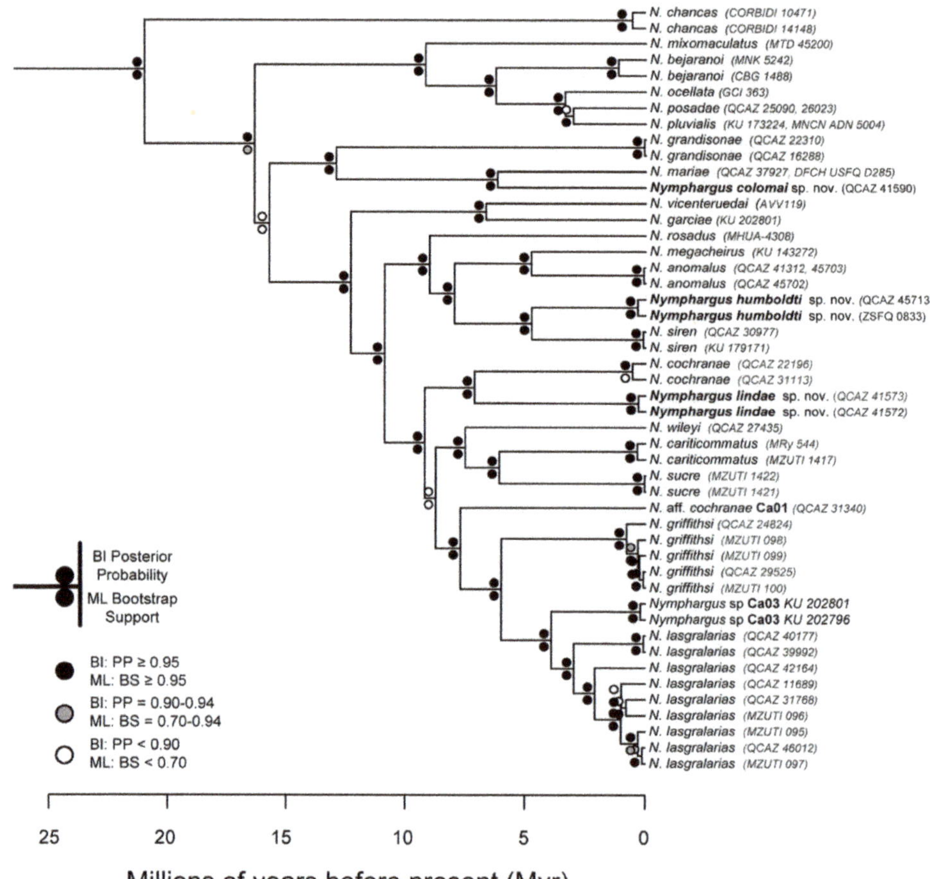

Figure 136. Evolutionary relationships of glassfrogs in the genus *Nymphargus*, inferred using maximum likelihood and Bayesian criteria.

Nymphargus cariticommatus (Wild, 1994 [137]; Figures 137–139).

Cochranella cariticommata Wild, 1994 [137]. Holotype: KU 202806.
Type locality: "11.2 km WSW Plan de Milagro (03°02′ S, 78°35′ W, 2350 m), Provincia Morona-Santiago, Ecuador".
Nymphargus cariticommatus—Cisneros-Heredia and McDiarmid, 2007 [17].

Common names: English: Unadorned Glassfrog. Spanish: Rana de Cristal Escueta.

Etymology: The specific epithet, *cariticommatus*, is an adjective formed from the Latin word *caritus*, meaning without, joined with *Kommos*, a Greek work meaning decoration or embellishment. The name refers to the species' plain appearance and lack of distinctive features [137].

Identification: *Nymphargus cariticommatus* is likely to be confused with *Nymphargus wileyi*, *N. griffithsi*, and *N. buenaventura*. However, in life, *N. cariticommatus* has a green dorsum with small yellow dots (Figure 137; uniform green dorsum in *N. wileyi*), an esophageal peritonea covered by iridophores (translucent in *N. griffithsi* and *N. buenaventura*, and with a thin layer of iridophores in *N. wileyi*; but see Remarks), and slightly less hand webbing on the outer fingers than *N. buenaventura*. Additionally, *N. cariticommatus* is found on the Amazonian versant of the Andes, whereas *N. griffithsi* and *N. buenaventura* inhabit the Pacific versant.

Figure 137. *Nymphargus cariticommatus* in life from Ecuador, Zamora Chinchipe province, Reserva Tapichalaca, 2200 m. Photos by Mario Yánez (**left**) and Marco Reyes (**right**).

Diagnosis: (1) Vomers lacking teeth; (2) snout round in dorsal profile, and truncated in lateral profile (Figure 138); (3) tympanum evident and of moderate size (tympanum diameter 26.7%–27.1% of eye diameter), dorsal border of tympanic annulus covered by supratympanic fold, tympanic membrane differentiated from surrounding skin; (4) dorsal surfaces shagreen, with minute spicules in males; (5) pair of enlarged subcloacal warts; (6) anterior two-thirds of ventral parietal peritoneum white, posterior third transparent (condition P2); silvery white pericardium; translucent peritonea covering the intestines, stomach, testes, gall bladder, and urinary bladder; esophagus white; kidneys white with unpigmented spots (condition V1); (7) liver tetralobed, three ventral lobes partially covering one smaller lobe; hepatic peritoneum transparent (condition H0); (8) humeral spines absent; (9) no webbing between inner fingers; webbing formula as follows: IV $2^{3/4}$—$2^{2/3}$ V in one male, IV 3^{+}—$2^{3/4}$ V in one female; (10) foot webbing moderate: I 2—$2^{1/4}$ II (1–$1^{1/2}$)—($2^{1/3}$–$2^{2/3}$) III ($1^{1/3}$–$1^{1/2}$)—($2^{1/2}$–$2^{2/3}$) IV ($2^{2/3}$–$2^{3/4}$)—2 V (Figure 138); (11) low ulnar fold in males, not visible in females; low inner tarsal fold present in both sexes; (12) concealed prepollex; nuptial pad Type I in males; (13) Finger II slightly longer than Finger I (Finger I about 90%–94% length of Finger II); (14) disc of Finger III of moderate size, 50%–56% of eye diameter; (15) in life, dorsum green with small yellow spots (Figure 137); (16) in ethanol, dorsum lavender with small white spots; (17) iris silvery white with thin black reticulation, yellow ring around pupil; (18) dorsal surfaces of fingers and toes lacking melanophores; (19) males call from the upper side of leaves; call described as soft 'zeet, zeet, zeet', but specific characteristics unknown; (20) fighting behavior unknown; (21) eggs deposited on the upper side of leaves; parental care unknown; (22) tadpoles unknown; (23) small body size; one adult male, SVL 23.3 mm; one adult female, SVL 25.5 mm.

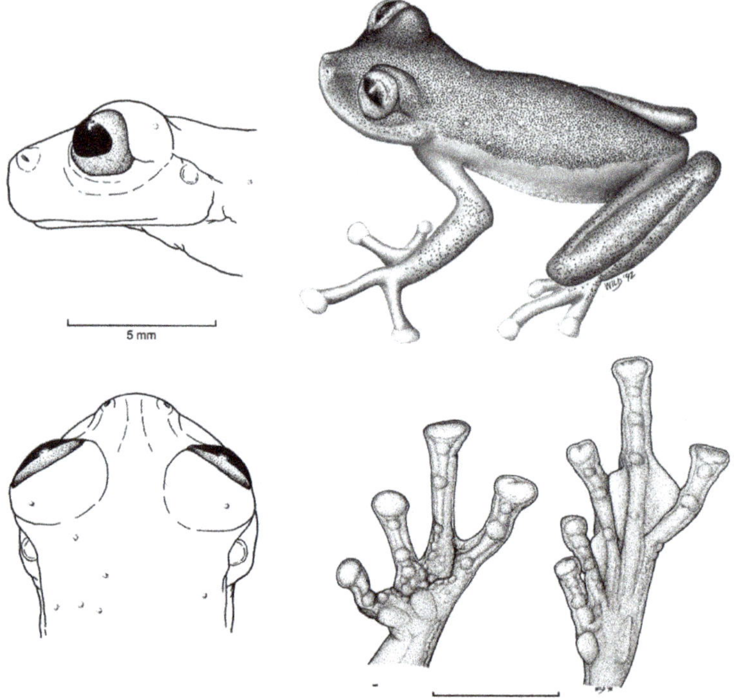

Figure 138. *Nymphargus cariticommatus.* Drawings of the holotype, adult male, KU 202806. Modified from Wild [137].

Color in life (Figure 137): Dorsal surfaces of head, body, and limbs green with small yellow spots. Dorsal surfaces of fingers and toes yellow. Iris silvery white with black reticulations.

Color in ethanol: Dorsum of body and limbs lavender with few, small, scattered white spots (fewer in females). Cheek with pale white flecks; margin of lip unpigmented; tympanum with few lavender flecks in males, lacking in females. Hands and feet lacking pigment [137]. Liver covered by transparent peritoneum; white parietal peritoneum covering anterior two-thirds of venter; silvery white pericardium; transparent peritonea covering the intestines, stomach, testes, gall bladder, and urinary bladder; peritoneum around kidneys with a thin silvery white lining; iris silvery white, with purple reticulations. In the original description, Wild [137] stressed the white esophagus in *N. cariticommatus*, similar to that seen in *Teratohyla midas*. In the preserved type series, only a thin layer of iridophores is evident.

Biology and ecology: *Nymphargus cariticommatus* was found on the upper sides of leaves and ferns within 2 m of cascading streams [137]. Parental care is unknown.

Call: Not described.

Tadpole: Not described.

Distribution (Figure 139): *Nymphargus cariticommatus* is endemic to the Amazonian slopes of the southern Andes of Ecuador at elevations between 2200 and 2700 m ([115,116,137,139], this work).

The species has been recorded from localities in the provinces of Zamora Chinchipe and Morona Santiago. The habitat of the species is within the Eastern Montane Forest ecoregion.

Figure 139. Distribution of *Nymphargus cariticommatus* in Ecuador (yellow dots).

Conservation status: Globally, *Nymphargus cariticommatus* is listed as *Data Deficient* by the IUCN [249]. The last reports of the species come from surveys at the Reserva Biológica Tapichalaca on 2004–2005 ([115], Yánez-Muñoz and Meza-Ramos, pers. com.). In Ecuador, the species is threatened by habitat destruction and fragmentation because of agriculture, pasture lands, and mining. We suggest considering the species as *Endangered* (IUCN criteria B1, B2a, B2biii).

Evolutionary relationships (Figure 136): Based on mitochondrial data, *Nymphargus cariticommatus* is sister to *N. sucre*.

Taxonomic Remarks: As mentioned in the diagnosis, subtle morphological differences separate *Nymphargus cariticommatus* from *N. wileyi*. We have examined all the available material of the two species (see Specimens Examined), including the type series, and observed that none of the specimens of *N. wileyi* have dorsal spots, whereas all individuals of *N. cariticommatus* have them. Also, in the preserved material, the esophageal peritoneum of *N. cariticommatus* is more pronounced and extended than in *N. wileyi*, in which the iridophores are restricted to the anterior portion of the esophageal peritoneum. Considering the data at hand, we maintain the species status of *N. wileyi*.

Specimens examined: *Nymphargus cariticommatus*: <u>Ecuador</u>: *Provincia de Morona Santiago:* 11.2 km WSW Plan de Milagro (03°07′ S, 78°30′ W; 2350 m), KU 202806 (holotype), KU 202805; El Cruzado (3.050 S, 78.517 W; 2194 m), USNM 288435–36. *Provincia de Zamora Chinchipe:* Reserva Biológica Tapichalaca (4.49208 S, 79.128389 W; 2200 m), DH-MECN 1974, 2429; 18.1 km E of Loja on the old road to Zamora, 2700 m, QCAZ 33977; Shucos, on the old road from Loja to Zamora, MRy 544.

Nymphargus cochranae (Goin 1961 [97]; Figures 140–144).

Cochranella cochranae Goin, 1961 [97]. Holotype: BM 1912.11.1.68.
<u>Type locality:</u> "El Topo, Río Pastaza, Eastern Ecuador, 4200 feet."
Centrolenella cochranae—Goin, 1964 [187].
Cochranella cochranae—Ruiz-Carranza and Lynch, 1991 [6].
Nymphargus cochranae—Cisneros-Heredia and McDiarmid, 2007 [17].

Common names: English: Cochran's Glassfrog. Spanish: Rana de Cristal de Cochran.

Etymology: The specific epithet honors Doris M. Cochran (1898–1968), herpetologist and curator of the Smithsonian Institution [97].

Identification: *Nymphargus cochranae* can be distinguished from all other glassfrogs by its green dorsum with small dark blue to black ocelli enclosing orange dots (Figure 140), relatively large size (adult males, SVL 24.0–26.2 mm; adult females, SVL 27.8–30.3 mm), absence of webbing between fingers (Figure 141), and lacking humeral spines. *Nymphargus cochranae* further differs from other species with dorsal ocellated patterns (*N. anomalus*, *N. lindae* sp. nov., *N. ignotus*, *N. laurae*, *N. ocellatus*) by having relatively smaller ocelli (Figure 142). Additionally, *N. ignotus* is found on the Pacific slopes of the Andes, whereas *N. cochranae* is restricted to the Amazonian slopes of the Andes. *N. anomalus* further differs by having scattered black and lavender flecks between ocelli (absent in *N. cochranae*); *N. ocellatus* has a green dorsal coloration with large dark rings that, in life, have greenish white centers; *N. laurae*, known from a single adult male is smaller (SVL = 19.7 mm), has pointed papillae on Toes I–IV (absent in *N. cochranae*), and larger ocelli. *Nymphargus cochranae* is most similar to its sister species, *N. lindae* sp. nov., in female body size (*N. cochranae*, SVL = 27.8–30.3 mm; *N. lindae* sp. nov., SVL 27.2–27.8 mm) and the relative size of ocelli (minute *in N. cochranae* and clearly larger in *N. lindae* sp. nov.), which is the most conspicuous difference between the two species. The only other species similar to *N. cochranae* is *N. megacheirus*, the dorsal dots of which coincide with spicules and could be confused with ocelli. However, *N. megacheirus* is larger (SVL of males = 26.8–31.5 mm; SVL of females = 31.2–32.9 mm) and the males have conspicuous spicules and spiculated warts on the dorsum (minute spicules in males of *N. cochranae*).

Figure 140. *Nymphargus cochranae* in life. Ecuador, Volcán Sumaco, 1500 m, QCAZ 31113. Photos by Luis A. Coloma.

Diagnosis: (1) Vomers usually with teeth, but see Variation; (2) snout truncated in dorsal profile, and truncated to slightly protruding in lateral profile (Figure 141); (3) tympanum relatively small, its diameter 19.5%–27.8% eye diameter, dorsal border of tympanic annulus covered by supratympanic fold, tympanic membrane pigmented as surrounding skin; (4) dorsal skin of males and females shagreen, minute spicules usually present in males; (5) pair of enlarged subcloacal warts; (6) anterior two-thirds of ventral parietal peritoneum white, posterior third transparent (condition P2); white pericardium; translucent peritonea covering intestines, stomach, testes, kidneys, gall bladder, and urinary bladder (condition V1); (7) liver tetralobed, two large ventral lobes partially covering two smaller lobes; hepatic peritoneum transparent (condition H0); (8) humeral spines absent; (9) hand webbing absent; (10) foot about two-thirds webbed: I $(2–2^+)$—$(2^{1/3}–2^{1/2})$ II $(1^{1/2}–2^-)$—$(3^-–3)$ III $(1^{3/4}–2)$—$(3^-–3^+)$ IV $(3^-–3^+)$—$(2–2^+)$ V; (11) ulnar and tarsal folds low; (12) concealed prepollex; nuptial pad Type I in males; (13) Finger I about same length as Finger II (Finger II 96.3%–101.8% of length of Finger I); (14) disc of Finger III small, 19.4%–28.3% of eye diameter; (15) in life, dorsum green with small dark blue to black ocelli enclosing orange dots (Figure 140); bones green; (16) in ethanol, dorsal surfaces lavender with small black ocelli enclosing white dots; (17) iris creamy white with thin dark grey reticulation, pupil surrounded by pale brownish–yellow coloration; (18) dorsal surfaces of hands and feet lacking melanophores, except for proximal portion of Finger IV and Toe V; (19) males call from the upper side of leaves or branches overhanging streams; single, high-pitched note; spectral and temporal characteristics of call unknown; (20) fighting behavior unknown; (21) eggs deposition site unknown; parental care unknown; (22) tadpoles unknown; (23) small to medium body size; SVL of adult males 24.0–26.2 mm ($\overline{X}$ = 25.0 ± 0.812, n = 6); SVL of adult females 27.8–30.3 mm ($\overline{X}$ = 28.9 ± 1.014, n = 7).

Color in life (Figure 140): Green dorsum with small dark blue to black ocelli enclosing bright orange spots. Ventral surfaces mostly white, except for posterior third, which is translucent. Bones green. Iris greyish white with a thin black reticulation; area immediately around pupil pale brownish yellow.

Color in ethanol (Figure 143): Dorsal surfaces lavender with small black ocelli enclosing white spots. Pericardium and anterior two-thirds of ventral parietal peritoneum white (presence of iridophores). Liver, intestines, stomach, testes, kidneys, gall bladder, and urinary bladder without white lining.

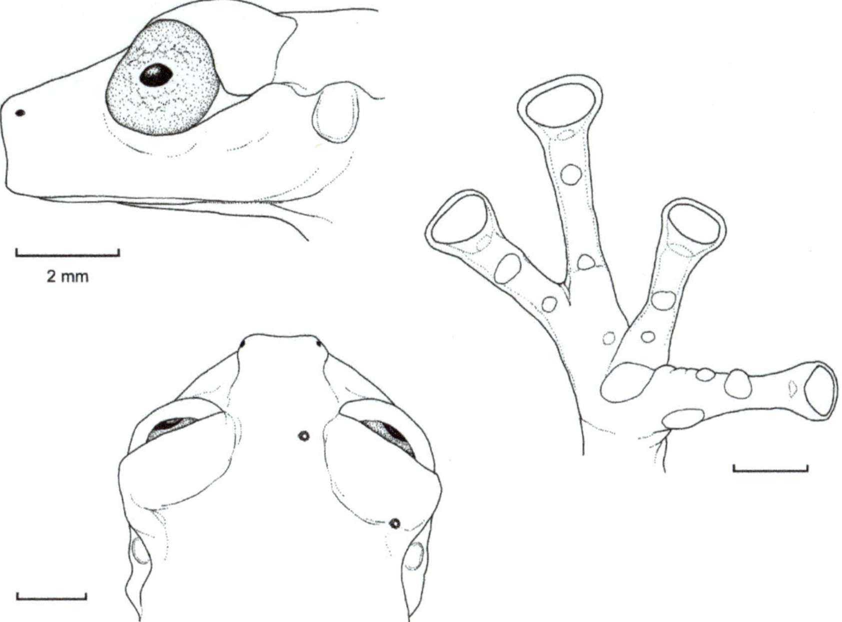

Figure 141. *Nymphargus cochranae.* Head in lateral and ventral views, adult female, QCAZ 22197. Hand in ventral view, adult male, QCAZ 31113. Illustrations by Juan M. Guayasamin.

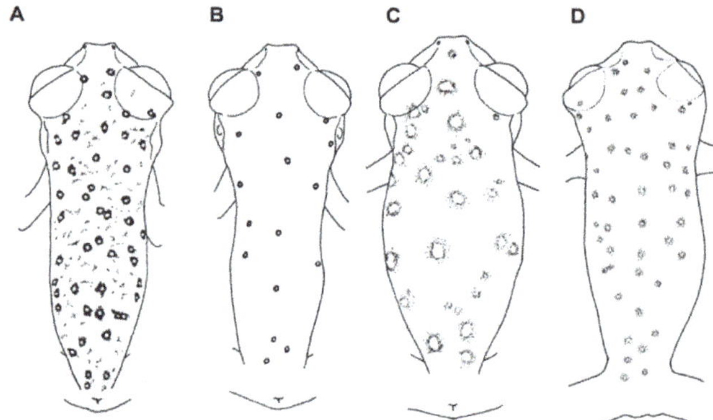

Figure 142. Dorsal patterns of ocellated glassfrogs, excluding *Nymphargus laurae* and *N. lindae* sp. nov. (**A**) *N. anomalus*, KU 143299. (**B**) *N. cochranae*, KU 121035. (**C**) *N. ocellatus*, LSU 25990. (**D**) *N. ignotus*, ICN 14750. Figure modified from Lynch and Duellman [22].

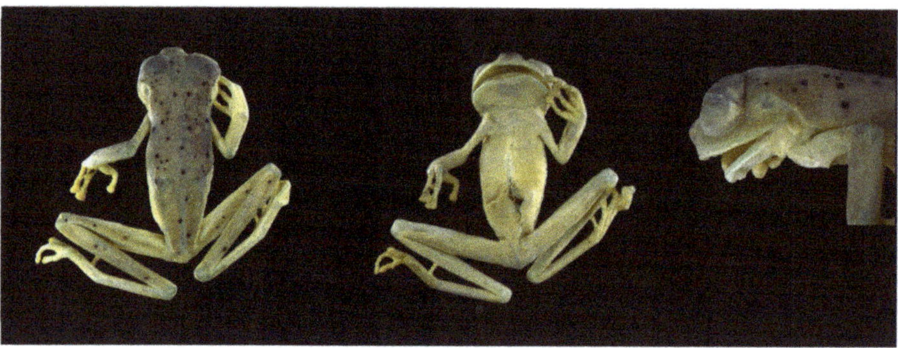

Figure 143. *Nymphargus cochranae*, holotype in preservative, BMNH 1912.11.1.68. Ecuador, El Topo, Río Pastaza. Photos by Martín Bustamante.

Variation: Goin [97] described *Nymphargus cochranae* as lacking teeth on the dentigerous process of the vomer, but also noted the presence of teeth in six other specimens [187]. Most specimens (n = 13) examined by us present teeth (USNM 286634 lacks teeth) and we assume that this is the usual character state in the species.

Biology and ecology: *Nymphargus cochranae* is active during the night. In the reproductive season, males call at night from the upper surfaces of leaves or branches near streams. Parental care is unknown.

Call: Described by Lynch and Duellman [22] as a single, high-pitched note. The specific spectral and temporal characteristics of the call are unknown.

Tadpole: Not described.

Distribution (Figure 144): *Nymphargus cochranae* occurs in Ecuador and Colombia on the Amazonian slopes of the Andes at elevations between 1170 m and 1960 m. The species has been reported from localities in the provinces of Napo, Orellana, Pastaza, Tungurahua, Sucumbíos, and Zamora Chinchipe (Specimens Examined). There is an unconfirmed record from the Ecuadorian Amazonia (Garza 1, at 300 m; QCAZ 1216). The potential distribution of the species is 21,723 km^2, within the Eastern Foothill Forest and Eastern Montane Forest regions. The only record of the species in Colombia comes from the Serranía de Churumbelos [250].

Figure 144. Distribution of *Nymphargus cochranae* in Ecuador (yellow dots).

Conservation status: Globally, *Nymphargus cochranae* is considered as *Vulnerable* by the IUCN [251]. In Ecuador, the species remains abundant at several localities, including the Sumaco and Cayambe Coca National Parks. However, part of the species' range is fragmented; thus, we suggest considering *N. cochranae* as *Near Threatened*.

Evolutionary relationships (Figure 136): *Nymphargus cochranae* is sister to *N. lindae* sp. nov. However, one sequenced individual (QCAZ 31340) is inferred as sister to a clade formed by *N. griffithsi* + *N. lasgralarias*.

Taxonomic Remarks: *Nymphargus cochranae* might be a species complex (see Evolutionary relationships). Also, see taxonomic remarks of *N. laurae* and *N. lindae* sp. nov.

Specimens examined: *Nymphargus cochranae:* Ecuador: *Provincia de Orellana:* km 13 on Loreto–Coca road (0.5836 S, 77.234 W), QCAZ 22196–97; *Provincia Napo:* Cascada San Rafael (0.1127 S, 77.596 W, 1280–1372 m), QCAZ 002, USNM 286632–37, 284304–10; 14.7 km (by road) NE of Río Salado (0.1289 S, 77.608 W, 1310 m), USNM 286638; 14 km by road SW of Reventador (ca. 0.11278 S, 77.596 W; 1400–1500 m), USNM 284304–06; Pacto Sumaco (0.726 S, 77.566 W; 1400 m), QCAZ 31113; Río Salado, 1 km upstream from Rio Coca (0.1916 S, 77.6997 W, 1420 m), KU 164517–18. *Provincia de Pastaza:* Abitagua, 8 km NW Mera (1.41667 S, 78.1667 W, 1300 m), KU 121033–35; *Provincia de Tungurahua:* Río El Topo (1.4166 S, 78.1667 W, 1220 m), BMNH 1912.11.1.68; 11 km E Río Negro (1.433 S, 78.13 W, 1170 m), KU 146605; *Provincia de Sucumbíos:* S slope Cordillera del Due above Río Coca (0.0833 N, 77.66 W, 1150 m), KU 123216–18. *Provincia de Zamora Chinchipe:* Estación Científica San Francisco (3.967 S, 79.066 W, 1960 m), QCAZ 31340–41.

Localities from the literature: *Nymphargus cochranae:* Ecuador: *Provincia de Zamora Chinchipe:* Alto Nangaritza (4.25026 S, 78.61746 W; 1256–1430 m) [252]; Contrafuerte de Tzunantza (4.046 S, 78.922 W) [253]; Alto Machinaza 1 (3.8976 S, 78.482 W) [254], Alto Machinaza 2 (3.7708 S, 78.554 W) [254].

Nymphargus colomai new species Guayasamin and Hutter (Figure 145).

http://zoobank.org/urn:lsid:zoobank.org:act:46978D55-22DF-409E-870C-F3C2203A5E6E

Common names: English: Coloma's Glassfrog. Spanish: Rana de Cristal de Coloma.

Etymology: The specific name is a patronym for Luis A. Coloma in recognition of his pioneer and continual efforts in studying and protecting amphibians, as well as mentoring numerous students, including JMG. Luis Coloma is the Director of the Jambatu Center for Research and Conservation of Amphibians (Centro Jambatu de Investigación y Conservación de Anfibios; see http://www.anfibiosecuador.ec/). The Jambatu Center houses dozens of critically endangered amphibians, in an outstanding effort to conserve frogs and toads. Luis received the *Sabin Award* for Amphibian Conservation in 2007 (http://www.amphibians.org/grants/sabin-award/).

Holotype: QCAZ 41590, adult male, collected from Miazi Alto (4.25044° S, 78.61356° W; 1260 m), Cordillera del Cóndor, Provincia de Zamora Chinchipe, Ecuador, by Juan M. Guayasamin and Elicio Tapia on 8 April 2009.

Paratypes: QCAZ 41591–92, adult males, and QCAZ 41641, juvenile, collected by Juan M. Guayasamin, Elicio Tapia, Silvia Aldás, and Holger Braun on 8–12 April 2009, at the same locality as holotype.

Generic placement: The new species is placed in the clade *Nymphargus* (sensu Guayasamin et al. [1]) based on morphological and molecular data. All species in *Nymphargus* share an absence of webbing between Fingers I–III and an absence or reduced webbing between Fingers III and IV; additionally, males lack humeral spines (except *N. grandisonae*). *Nymphargus colomai* sp. nov. has the aforementioned traits. Molecular analyses of mitochondrial genes unambiguously place the new species in the clade *Nymphargus* (Figure 16).

Identification: *Nymphargus colomai* sp. nov. is unique by having a white iris with a contrasting black horizontal stripe and a dorsum that varies from dull yellowish green to tangerine yellow, brown, or grey olive, always with numerous small yellow to orange spots (Figure 145). Additionally, *N. colomai* sp. nov. lacks humeral spines and hand webbing. Among glassfrogs that inhabit the Amazonian slopes of the Andes, only *N. cariticommatus*, *N. chancas*, *N. humboldti* sp. nov., *N. siren*, and *N. sucre* could be confused with *N. colomai* sp. nov. However, *N. cariticommatus*, *N. chancas*, *N. siren*, and *N. humboldti* sp. nov. have a green dorsum with few and well-defined dorsal spots, whereas *N. colomai* sp. nov. has a non-green dorsum (see above) with numerous and diffuse yellow dorsal spots. Moreover, *N. siren* is smaller than *N. colomai* sp. nov. (*N. siren*, SVL < 23.3 mm; *N. colomai* sp. nov. SVL = 24.7–25.7 mm); and *N. cariticommatus* has white esophagus and renal capsules (cream in *N. colomai* sp. nov.) Both *N. colomai* sp. nov. and the Peruvian *N. chancas* share the clearly defined black horizontal stripe on a silver background, but the two species are easily differentiated by their dorsal color patterns (in *N. colomai* sp. nov.: Dull yellowish green to tangerine yellow, brown, or grey olive, with numerous small yellow to orange spots; in *N. chancas:* Yellowish–green dorsum with small yellow dots; compare Figures 145 and 146) and genetics (Figure 136).

Figure 145. *Nymphargus colomai* sp. nov. in life from stream nearby Miazi Alto, Zamora Chinchipe province, Ecuador. (**Top left**): QCAZ 41590, holotype; other frogs are part of the type series. Photos by Holger Braun and Juan M. Guayasamin.

Figure 146. *Nymphargus chancas* in life from 5.9 km SE from the type locality of Abra Tangarana, San Martín, 6°19′15.65″ S 76°41′44.16″ W, 973 m, CORBIDI 10471. Photos by Jesse Delia (**upper row**) and Evan Twomey (**lower row**).

Diagnosis: (1) Dentigerous process of the vomer lacking teeth or with few teeth (up to 3); (2) snout truncated to slightly protruding in lateral profile; truncated in dorsal view; (3) tympanum oriented almost vertically, with slight lateral and posterior inclinations, its diameter about 23%–27% of eye diameter; upper half of tympanic annulus obscured by supratympanic fold and warts; tympanic membrane pigmented as surrounding skin; (4) dorsal skin shagreen, with numerous spiculated warts and spicules in males; females unknown; (5) venter areolate; pair of enlarged subcloacal warts; (6) white parietal peritoneum covering about anterior half of venter (condition P2); white pericardium; translucent peritonea on kidneys, intestines, stomach, gall and urinary bladders (condition V1); (7) liver lobate, covered by translucent peritoneum (condition H0); (8) humeral spines absent; (9) webbing absent between Fingers I, II, and III, absent or basal between Fingers III and IV; webbing formula III $(2^{3/4}$–3)—$(2^{1/2}$–$2^{2/3})$ IV; (10) feet about two-thirds webbed; webbing formula: I 2^-—$2^{1/3}$ II $(1^+$–$1^{1/3})$—$2^{1/2}$ III $1^{1/2}$—$(2^{2/3}$–$2^{3/4})$ IV 3^-—$(1^{1/2}$ – $2^-)$ V; (11) ulnar and tarsal folds present, low; (12) concealed prepollex; in males, nuptial pad Type I; (13) Finger II slightly longer than Finger I; (14) disc of Finger III width about 46%–50% of eye diameter; (15) in life, dorsum varies from dull yellowish green to tangerine yellow, brown, or grey olive with numerous small yellow to orange spots; bones green; (16) in preservative, dorsum grey to greyish lavender with numerous unpigmented spots; (17) in life, iris silvery white with slight yellow hue and a clearly marked horizontal black stripe; (18) melanophores on dorsal surfaces of Fingers III and IV and Toes IV and V; (19) males call from the upper surfaces of leaves; call unknown; (20) fighting behavior unknown; (21) egg deposition site unknown; parental care unknown; (22) tadpole unknown; (23) small body size; in males, SVL 25.0–25.7 mm (n = 3); females unknown.

Description of holotype: Adult male, SVL 25.0 mm. Head slightly wider than long (head length 98% of head width); snout truncated in dorsal and lateral profiles; canthus rostralis indistinct, slightly concave; loreal region slightly concave; lips slightly flared; nostril protuberant, closer to tip of snout than to eye, directed frontolaterally; internarial area barely depressed. Eyes large, directed anterolaterally at an angle ~50°; transverse diameter of disc of Finger III 48% eye diameter. Supratympanic fold low, obscuring upper portion of tympanic annulus; tympanum oriented mostly vertically, but with slight posterolateral inclination; tympanic membrane translucent, pigmented as surrounding skin. Dentigerous process of vomer low, situated transversely between choanae, with one or two teeth; choanae large, longitudinally rectangular; tongue ovoid, with ventral posterior fifth not attached to floor of mouth and posterior margin slightly notched; vocal slits extending posterolaterally from about the lateral margin of tongue (at about half the length of tongue) to angle of jaws. Humeral spine absent. Low ulnar folds evident on external and internal ventrolateral margins of arm; relative lengths of fingers: III > IV > II > I; webbing absent between Fingers I–III, basal between Fingers III and IV, webbing formula III 3–$2^{2/3}$ IV; discs expanded, nearly elliptical; disc pads nearly triangular shaped; subarticular tubercles small, round, simple; low supernumerary tubercles present; palmar tubercle elliptical, simple; nuptial pad large (Type I), ovoid, granular, extending from ventrolateral base to dorsal surface of Finger I, covering proximal half of Finger I. Length of tibia 57% SVL; low inner tarsal fold evident; outer tarsal fold absent; foot two-thirds webbed; webbing formula of foot: I 2^-—$2^{1/3}$ II $1^{1/4}$—$2^{1/2}$ III $1^{1/2}$—$2^{3/4}$ IV 3^-—2^- V; discs on toes round to elliptical, lacking papillae; disc on Toe IV slightly narrower that disc on Finger III; disc pads triangular; inner metatarsal tubercle large, ovoid; outer metatarsal tubercle small and inconspicuous; subarticular tubercles small, round; supernumerary tubercles low. Skin on dorsal surfaces of head, body, and lateral surface of head and flanks shagreen with numerous spiculated warts and minute spicules; lower flanks show glandular cells; throat smooth; belly and lower flanks areolate; cloacal opening directed posteriorly at upper level of thighs; cloacal warts present, unpigmented. Ventral surface of thighs with pair of enlarged tubercles.

Coloration of holotype in life (Figure 145): The holotype has a grey olive dorsal coloration, with numerous yellow to orange spots. The venter is white anteriorly and translucent posteriorly. The iris is silvery white, with a clearly marked black horizontal stripe. Dorsally, at the posterior end of the head, there is a circular, concave area with dark grey collocation.

Color variation in life (Figure 145): Dorsum can be dull yellowish green, tangerine yellow, brown, or grey olive, always with numerous small and diffuse yellow to orange spots. Anterior half of venter white, posterior portion transparent. Iris might have a slight yellow hue.

Coloration of holotype in ethanol: Dorsal surfaces of head, body, and limbs greyish brown, with small, unpigmented spots. Middorsal dark lavender spot at scapular level. Anterior half of ventral parietal peritoneum white, posterior half cream. White pericardium; translucent peritonea covering digestive tract, liver, kidneys, and gall and urinary bladders.

Color variation in ethanol: Dorsal surfaces of head, body, and limbs vary from greyish brown to lavender with small, unpigmented spots.

Measurements of holotype (mm): *Nymphargus colomai* sp. nov., adult male, QCAZ 41590: SVL = 25.0, head length = 8.1, head width = 8.3, snout length = 3.3, IOD = 2.6, upper eyelid width = 2.5, eye diameter = 3.3, tympanum diameter = 0.9, tympanum–eye distance = 1.7, femur = 13.1, tibia = 14.2, foot length = 11.5, hand length = 5.7, disc of Finger III width = 1.6, disc of Toe IV width = 1.5.

Biology and ecology: *Nymphargus colomai* sp. nov. is active during the night. All individuals were found on vegetation 100–170 cm above fast-flowing streams. Males call from the upper surfaces of leaves. Parental care is unknown.

Call: Not described.

Tadpole: Not described.

Distribution (Figure 147): *Nymphargus colomai* sp. nov. is known only from its type locality at the Cordillera del Cóndor, Zamora Chinchipe province, Ecuador.

Figure 147. Distribution of *Nymphargus colomai* sp. nov. in Ecuador (yellow dot).

Conservation status: Following UICN criteria, we consider *Nymphargus colomai* sp. nov. as *Endangered*, following IUCN criteria B2a, B2(iii). The main threats for this species are habitat destruction (i.e., cattle, agriculture) and contamination associated with mining.

Evolutionary relationships (Figure 136): *N. colomai* sp. nov. is inferred as sister to *N. mariae*.

Taxonomic comments: *Nymphargus colomai* sp. nov. was previously confused with *N. chancas* [252], which we restrict to Peru. Preserved specimens are morphologically similar, but in life, several differences are evident (Figures 145 and 146). The two species are not closely related (Figure 136).

Nymphargus garciae (Ruiz-Carranza and Lynch, 1995 [255]; Figures 148 and 149).

Cochranella garciae Ruiz-Carranza and Lynch, 1995 [255]. Holotype: ICN 11752.
Type locality: "Departamento de Cauca, municipio de Inzá, Km 64–73 carretera Popayán a Inzá, vertiente oriental Cordillera Central, 2°34′ latitud N, 76°4′ W de Greenwich, 2590–2660 m [Colombia]".
Nymphargus garciae—Cisneros-Heredia and McDiarmid, 2007 [17].

Common names: English: García's Glassfrog. Spanish: Rana de Cristal de García.

Etymology: The specific epithet is a patronym for Dr. Evaristo García, author of the pioneer book *Los ofidios venenosos del Cauca*, published in 1896 [255].

Identification: *Nymphargus garciae* can be differentiated from other glassfrogs by its moderate body length (SVL 25.9–28.4 mm in adult females, SVL 25.1–29.9 mm in adult males) ([255], this work), green dorsum usually with dark green to blue spots, and absence of humeral spines. In Ecuador, *N. garciae* could be confused with *Centrolene buckleyi* and *N. megacheirus*. However, *C. buckleyi* has basal webbing between Fingers III and IV (absent in *N. garciae*), lacks dark spots on the dorsum (usually present in *N. garciae*), and males with prominent humeral spines (spines absent in *N. garciae*). *Nymphargus megacheirus* has a similar dorsal coloration and hand-webbing pattern as *N. garciae*, but the two species have allopatric distributions, *N. garciae* inhabiting higher elevations (1900–2700 m) than *N. megacheirus* (1300–1740 m). Additionally, *N. megacheirus* is slightly larger than *N. garciae* (in *N. megacheirus*, SVL 26.8–31.5 mm in adult males, 31.2–32.9 mm in adult females). Two species endemic to Colombia, *N. nephelophilus* and *N. oreonympha*, are extremely similar to *N. garciae*. *Nymphargus oreonympha* is reported to have basal webbing between Fingers III and IV (webbing absent in *N. garciae*), whereas *N. nephelophilus* has a smooth skin (shagreen in *N. garciae*) and lacks ulnar and tarsal folds (present in *N. garciae*). See Remarks.

Figure 148. *Nymphargus garciae* in life from Ecuador, Napo province, trail from Oyacachi to El Chaco, 3012 m, MZUTI 764. Photos by Eduardo Toral.

Diagnosis: (1) Vomerine teeth absent; (2) snout rounded in dorsal aspect, truncated to slightly sloping in lateral profile; (3) tympanum oriented almost vertically, with slight posterior inclination, small in size (tympanum diameter 22.9%–30.6% of eye diameter); tympanic annulus visible except for dorsal border covered by supratympanic fold; tympanic membrane partially pigmented but differentiated from surrounding skin; (4) dorsal surfaces shagreen, with spicules evident in males; (5) pair of enlarged subcloacal warts (Figure 15); (6) anterior half of parietal peritoneum white (condition P02; white pericardium and renal capsules, translucent visceral peritoneum (condition V1); (7) tetralobed liver covered by transparent peritoneum (condition H0); (8) humeral spines absent; (9) no webbing between fingers, although junction of lateral fringes resemble basal webbing; (10) webbing formula on foot: I (2⁻–2⁺)—(2⁺–2²ᐟ³) II (1–1³ᐟ⁴)—(2¹ᐟ²–3⁻) III (1¹ᐟ⁴–1³ᐟ⁴)—(2¹ᐟ³–3⁻) IV (2¹ᐟ³–3⁺)—(1³ᐟ⁴–2⁺) V; (11) ulnar and tarsal folds present, enameled; (12) nuptial pad Type I, concealed prepollex; (13) Finger II slightly longer than Finger I (FI/FII = 0.87–0.95); (14) disc of Finger III relatively large (3DW/ED = 0.56–0.77); (15) in life, dorsum green with or without green to blue spots; white upper lip and ulnar and tarsal folds; bones green; (16) in preservative, dorsum lavender with or without dark lavender spots; (17) iris greyish white with thin black reticulation; (18) fingers lacking melanophores dorsally; melanophores restricted to proximal dorsal areas of Toes IV and V; (19) males call from upper side of leaves; call unknown; (20) fighting behavior unknown; (21) eggs deposition site unknown; parental care unknown; (22) tadpoles unknown; (23) medium body size; in Ecuador, males SVL 24.9–29.9 mm ($\overline{X}$ = 26.7 ± 1.723; n = 7); in Colombia, males SVL 25.1–27.7 mm ($\overline{X}$ = 26.3 ± 0.745; n = 15), females SVL 25.9–28.4 mm (n = 3).

Color in life (Figure 148): Dorsum dark green with cream tubercles and dark green to blue spots; hands and feet yellowish green. Glandular pericloacal region white. Pericardium and anterior half of ventral parietal peritoneum white; translucent visceral peritonea. Based on observations on preserved specimens, we assume that part of the type series of *Nymphargus garciae* (ICN 11743, 11759) was uniform green in life, corresponding to the uniform lavender coloration in ethanol. Some of the photographs of Ecuadorian individuals of *N. garciae* show a uniform green dorsal coloration.

Color in ethanol: Dorsal surfaces lavender with or without small dark spots; spicules resemble minute white spots in males. White upper lip; tympanic membrane pigmented with purple specks. Cloacal region with white warts. Dorsally, fingers and toes mostly unpigmented, but some pigmentation visible on proximal half of Toes IV and V. White parietal peritoneum covering anterior half of venter; silvery–white pericardium; transparent peritoneum covering liver and urinary bladder; gastrointestinal peritoneum translucent; renal capsules white with minute unpigmented spots, as those in *N. wileyi* ([20]: Figure 12). Internal traits based on a dissected male (ICN 11759).

Biology and ecology: Adults of *Nymphargus garciae* were found on vegetation and rocks along streams; one gravid female (ICN 7495) had numerous eggs, which had a dark brown animal and a cream vegetal pole [255]. Parental care is unknown.

Call: Not described.

Tadpole: Not described.

Distribution (Figure 149): *Nymphargus garciae* is known from localities on the eastern slope of the Cordillera Central at elevations of 1900–2700 m in Colombia [101,255], and from the Cordillera Oriental of the Ecuadorian Andes at 2550–3012 m. In Ecuador, this species is found in the Eastern Montane Forest ecoregion.

Figure 149. Distribution of *Nymphargus garciae* in Ecuador (yellow dots).

Conservation status: *Nymphargus garciae* is classified as *Vulnerable* by the IUCN [256]. We consider the current conservation status justified.

Evolutionary relationships (Figure 136): *N. garciae* is the sister species to *N. vicenteruedai*.

Remarks: Given the morphological similarities between *Nymphargus garciae*, *N. nephelophilus*, and *N. oreonympha*, it is possible that they are actually one species with geographic variation.

Specimens examined: *Nymphargus garciae:* Ecuador: *Provincia de Napo*: 11 km ESE Papallacta (0.38694° S, 78.05694° W; 2660 m), KU 164658–62; 60 km E San Miguel de Salcedo (0.96667° S, 78.21667°; 2550 m), KU 202793; trail from Oyacachi to El Chaco (0.21891° S, 78.04442°; 3012 m), MZUTI 764. *Provincia de Sucumbíos*: 18 km E Santa Bárbara (0.575° N, 77.51138° W), KU 202796.

Nymphargus grandisonae (Cochran and Goin, 1970 [96]; Figures 150–153).

Centrolenella grandisonae Cochran and Goin, 1970 [96]. Holotype: BM 1910.7.11.68.
Type locality: "Pueblo Rico, (Departamento) Caldas, southwestern Colombia, 5000 feet."
Centrolene grandisonae—Ruiz-Carranza and Lynch, 1991 [6].
Nymphargus grandisonae—Guayasamin, Castroviejo-Fisher, Trueb, Ayarzagüena, Rada, and Vilà, 2009 [1].

Common names: English: Red-spotted Glassfrog, Measles' Glassfrog, Grandison's Glassfrog. Spanish: Rana de Cristal Sarampiona, Rana de Cristal de Grandison.

Etymology: The specific epithet is a patronym for Alice G. C. Grandison, former curator of the British Museum [96].

Identification: *Nymphargus grandisonae* is unique by having a green dorsum with small bright red spots (Figure 150). Another diagnostic trait is the presence of a white urinary bladder, a characteristic otherwise known only in glassfrogs from the Atlantic forest of Brazil and Argentina.

Figure 150. *Nymphargus grandisonae* in life. (**A**) Adult male, QCAZ 40001. (**B**) Adult female, ventral view, QCAZ 32282. (**C**) Metamorph. (**D**) Egg clutch, QCAZ 40004. (**E**) Fight between males. Photos by: (**A,C,D**) Luis A. Coloma, (**B**) Martín Bustamante, (**E**) Carl R. Hutter.

Diagnosis: (1) Dentigerous process of vomer usually bearing teeth, each process with zero to seven teeth; (2) snout round in dorsal view, truncated in lateral profile (Figure 151); (3) tympanum partially hidden under skin, oriented almost vertically, its diameter 27.0%–34.5% of eye diameter; supratympanic fold moderate; tympanic membrane pigmented, barely differentiated from surrounding skin; (4) dorsal surfaces of males and females finely pustular, minute spicules evident only in males; (5) pair of enlarged subcloacal warts; (6) white iridophores covering all or most of ventral parietal peritoneum (condition P4); white pericardium; translucent peritoneum covering intestines, stomach, testes, and gall bladder; kidneys dorsally and laterally covered with white lining; urinary bladder completely covered with white iridophores (condition V4); (7) liver tetralobed, covered by transparent peritoneum (condition H0); (8) in males, humeral spines present, small; (9) webbing absent between Fingers I, II, and III; moderate webbing between Fingers III and IV; webbing formula III ($2^{1/3}$–$2^{1/2}$)—(2–$2^{1/4}$) IV (Figure 151); (10) webbing formula on foot, I (1–$1^{3/4}$)—(2–$2^{1/4}$) II (1–$1^{1/2}$)—(2–$2^{1/2}$) III (1–$1^{1/2}$)—(2–$2^{1/4}$) IV (2–$2^{1/3}$)—(1–$1^{1/2}$) V; (11) ulnar fold low, white; inner tarsal fold low, short; outer tarsal fold absent or low and inconspicuous; (12) concealed prepollex; in males, nuptial pad Type I; (13) Finger I about same length as Finger II (Finger I 92.0%–102.2% of Finger II); (14) disc of Finger III of moderate width, about 55.2%–64.0% of eye diameter; (15) in life, dorsum green with small red spots; upper lip white; bones green; (16) in preservative, dorsal surfaces lavender with cream white spots; (19) males call from upper sides of leaves near streams; advertisement call formed by a single note, note duration 0.056–0.158 s, 18–22 pulses per note, dominant frequency at 3100–4048 Hz; (20) fighting behavior variable (see Biology and ecology section); (21) females deposit eggs on upper side of leaves near streams; short-term maternal care present; parental care by males absent; (22) tadpoles with M-shaped upper jaw sheath, small gap in row A-2, and non-emarginate oral disc; (23) medium body size; males, SVL 25.1–29.3 mm ($\overline{X}$ = 27.2, *n* = 44); females, SVL 28.9–30.7 mm ($\overline{X}$ = 29.8, *n* = 4).

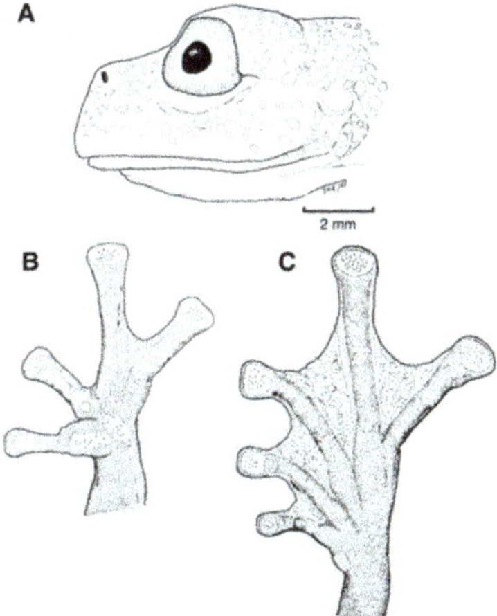

Figure 151. *Nymphargus grandisonae.* (**A**) Head in lateral view, KU 164688, drawing by Juan M. Guayasamin. (**B**) Hand in ventral view, KU 164688. (**C**) Foot in ventral view, KU 118047. (**B,C**) Modified from Lynch and Duellman [22], not to scale.

Color in life (Figure 150): Dorsum green with small red spots; throat pale green; margin of upper lip white; flanks and venter white; iris yellowish grey; bones green; peritonea on liver and digestive tract lacking iridophores.

Color in ethanol: Dorsal surfaces of head and body lavender with creamy–white spots; flanks white. White parietal peritoneum covering all or most of venter; pericardium white; translucent peritoneum covering liver, intestines, and stomach; kidneys dorsally and laterally covered with white iridophores; white urinary bladder.

Biology and ecology: *Nymphargus grandisonae* is a nocturnal and epiphyllous glassfrog that prefers riparian vegetation in evergreen and cloud forests. The species is active on vegetation near streams in primary and secondary forest, and on shrubs along the borders of pastures [51,87]. *Nymphargus grandisonae* is a highly territorial species, with males having site fidelity for prolonged periods of time [51]. Also, males have frequent fights and complex combat behaviors (Figure 6), including the following types: (1) Dangling amplexus-like; (2) dangling venter to venter; (3) amplexus-like; (4) a head to vent wrestle; and (5) reverse dangling amplexus-like. As a consequence of fighting, males usually present injuries produced, likely, by humeral spines (see Hutter et al. [51]; Figure 6). Interspecific combat has also been observed between *N.* aff. *grandisonae* and *Espadarana prosoblepon* [257]. *Nymphargus grandisonae* reproduces in the rainy season (December–April); at this time of the year, males call from the upper sides of leaves. Amplexus occurs within male's territories and egg clutches, containing 30–71 eggs, are deposited on vegetation above streams [51]. At Reserva Las Gralarias, the species has been found at Ballux Creek, Five-frog Creek, Heloderma Creek, Hercules Creek, Chalguayacu River, Kathy's Creek, Lucy's Creek, Waterfall Trail, and Santa Rosa River Trail ([88], this work). Short-term parental care is provided by females; no parental care is provided by males [25].

Call: Hutter et al. [51] presented a detailed description of the advertisement, courtship, territorial, encounter, distress, and release calls of *Nymphargus grandisonae*. The information of these calls is summarized in Tables 5–7 and Figure 152.

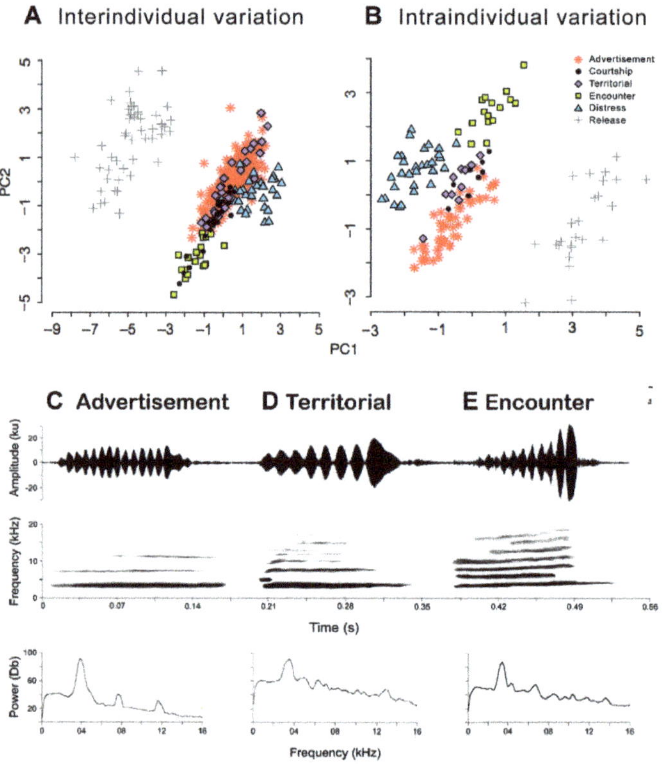

Figure 152. Call types in *Nymphargus grandisonae*. (**A**) Principal component analyses (PCA) of call variables for different types of calls of all individuals. (**B**) PCA plot showing call variation within an individual. (**C–E**) Oscillogram, spectrogram, and power spectrum of different types of calls. Modified from Hutter et al. [51].

Table 5. Call types and associated behaviors described for *Nymphargus grandisonae* [51]. Definitions follow Bogert [258], Duellman and Trueb [259], and McDiarmid and Adler [238].

Call Type	Description
Advertisement	Call intended to advertise a male's location to other nearby males in addition to attracting females
Courtship	Communicative call emitted close-range (0.5 m–1 m) in the presence of a female
Territorial	Modification of the advertisement call; emitted by a male territory holder in response to the presence of another invading male; invading male may or may not offer a response call; emitted long range
Encounter	Further modification of the advertisement call; Produced during close-range antagonist interactions between males; prior to, or during combat; emitted short range
Release	Call given during combat, in which a male is clasping another male; accompanied by vibrations
Distress	Call usually emitted in antagonistic inter-specific interactions; in situations of duress. Here, they are random, loud variable peeps emitted during combat or during the observed "throating grasping while dangling" behavior.

Table 6. Description of combat types observed for *Nymphargus grandisonae* [51].

Combat Type	Description
Venter-to-venter	Two males dangle from the substrate by their hind limbs, positioned venter-to-venter; may include males grasping the other around the head
Head-to-vent wrestle	One male's head is positioned near the vent of the other; one or both males grasp its opponent around the waist with its forelimb
Amplexus-like	One male is position upon the dorsum of another; accompanied by struggling and jumping
Dangling amplexus-like	Two males are dangling facing vent to dorsum, often grasping the head of the other male; results from tumbling off the substrate in an amplexus-like position
Dangling reverse amplexus-like	Two males are dangling from the substrate with the head of one male positioned near the vent of another; results from tumbling off the substrate where one male climbs the other
Limb grasping	One male immobilizes the other by grasping its limbs; occurred while one male remains dangling from the substrate while the other attempts to exit onto the substrate surface

Table 7. Comparisons of call types recorded for *Nymphargus grandisonae*. Data are the mean ± standard deviation, followed by the range (modified from Hutter et al. [51]).

Call Variables	Call Type			
	Advertisement	Courtship	Territorial	Encounter
n-calls (individuals)	417 (22)	27 (3)	30 (3)	19 (2)
Call duration (s)	0.115 ± 0.018	0.120 ± 0.023	0.110 ± 0.015	0.115 ± 0.016
	0.056–0.158	0.082–0.170	0.076–0.141	0.091–0.148
Call rise time (s)	0.061 ± 0.037	0.069 ± 0.023	0.065 ± 0.030	0.066 ± 0.033
	0.003–0.482	0.022–0.113	0.010–0.109	0.004–0.109
Pulse rate (/s)	143.2 ± 18.8	140.2 ± 12.1	137.3 ± 17.1	131.4 ± 12.3
	107.9–320.8	111.8–168.8	105.3–190.9	115.7–169.8
Dominant frequency (Hz)	3588 ± 189.6	3541 ± 206.0	3353 ± 101.0	3441 ± 142.0
	3101–4048.2	3187–3876	3101–3445.3	3187–3618
Frequency modulation (Hz)	142 ± 110.4	469 ± 198.9	213 ± 116.9	639 ± 135.6
	0–517	258–861	0–430	345–861
Lower dominant frequency (Hz)	3159 ± 185	3076 ± 218.7	2959 ± 118.1	2950 ± 117.2
	2694–3645	2615–3407	2700–3109	2700–3109
Higher dominant frequency (Hz)	4039 ± 210.3	4003 ± 197.3	3736 ± 136.4	3832 ± 165.7
	3442–4651	3721–4372	3527–4000	3527–4070
First harmonic (Hz)	7180 ± 409.1	7057 ± 377.5	6524 ± 296.0	6700 ± 234.3
	5857–8613	6503–7752	6029–7149	6202–7020

Tadpole: The tadpole of *Nymphargus grandisonae* was described by Ospina-Sarria et al. [260]. Characteristic traits include M-shaped upper jaw sheath, small gap in row A-2, and a non-emarginated oral disc. In life, tadpoles (Gosner stage 36) exhibit a cream coloration; hyobranchial apparatus and heart are pinkish. Caudal musculature is reddish cream; caudal and ventral fins are translucent, with small darks spots at the distal end of the tail.

Distribution (Figure 153): *Nymphargus grandisonae* occurs on the western slopes of the Cordillera Occidental and Cordillera Central in Colombia, from Departamento de Antioquia, south to the Pacific versant of the Andes of north and central Ecuador [51,87,96,127,261,262]. In Ecuador, *N. grandisonae* has been found in the provinces of Carchi, Cotopaxi, Pichincha, El Oro, and Santo Domingo de los Tsáchilas, at elevations between 1140–2150 m (Specimens Examined). In Ecuador, the potential distribution of the species is 8087 km^2, mostly within the Western Montane Forest ecoregion, but also including the Deciduous and Western Foothill ecoregions.

Figure 153. Distribution of *Nymphargus grandisonae* in Ecuador (yellow dots).

Conservation status: *Nymphargus grandisonae* is listed as a *Least Concern* species by the IUCN [261] and Arteaga et al. [87]. The species tolerates a certain level of forest disturbance. In Ecuador, there are several recent reports of *N. grandisonae* in Chiriboga, Reserva Río Guajalito, Reserva Las Gralarias, Tandayapa, Tandapi, and Reserva Otonga. At Reserva Las Gralarias, this glassfrog is infected by the chytrid fungus *Batrachochytrium dendrobatidis*, but populations have not shown evidence of drastic declines and the species is fairly common in the area [92]. Experimental studies have shown that tadpoles of *N. grandisonae* are susceptible to introduced trout, by having a higher mortality and changing its morphology [123]. We consider that the *Least Concern* conservation status is adequate.

Evolutionary relationships (Figure 136): *Nymphargus grandisonae* is sister to a clade formed by *N. mariae* and *N. colomai* sp. nov.

Taxonomic remarks: The name "*Centrolenella grandisonae*" was misapplied by Lynch and Duellman [22], who confused *N. grandisonae* with *Centrolene lynchi*. Duellman [127] provided a detailed description of these two species. There are reports [257] of *N. grandisonae* at Reserva Buenaventura (El Oro province, Ecuador) that, most likely, represent a new species (Paul Szekely, Diego F. Cisneros-Heredia, pers. comm.)

Specimens examined: *Nymphargus grandisonae:* Ecuador: *Provincia de Carchi:* Quebrada Naranjo, near Maldonado (0.9 N, 78.1 W; 1410 m), KU 178168–69. *Provincia de Cotopaxi:* Reserva Otonga (0.4189 S, 79.004 W; 1800 m), QCAZ 20718, 20725, 11683; 18.2 km on the Quillotuña–Pucayacu road (0.6784 S,

79.01564 W), QCAZ 40388. *Provincia de Pichincha:* 3.5 km NE of Mindo (0.0322 S, 78.761 W, 1340 m), KU 164686–90; Quebrada Zapadores, 5 km ESE of Chiriboga on Chiriboga–Quito road (0.2375 S, 78.735278 W; 2010 m), QCAZ 16288; near Tandapi (0.4164 S, 78.7989 W; 1520 m), KU 180319–22; 5 km Won the Tandapi-Atenas road (0.3954 S, 78.7989 W; ca. 1700 m), QCAZ 14292–93, 17744, 17753–58; 25.7 km ENE La Palma on La Palma-Chiriboga road, 1820 m, MCZ 93587; 2 km E Tandapi, (0.4258 S, 78.7853 W; 1550 m), MCZ 93023–26, 97848–51, USNM 211211; 1 km SW Tandayapa (0.033 S, 78.7667; 1640 m), MCZ 97847, USNM 211214; 2.9 km SW Tandayapa on Tandayapa-Mindo road (0.05 S, 78.783 W; 1820 m), USNM 211212–13; 5.1 km SE Tandayapa on Tandayapa-Nono road (0.033 S, 78.7167 W; 1850 m), USNM 211215; Reserva Las Gralarias, Five-frog Creek (0.03098 S, 78.70853 W; 2150 m), MZUTI 430–31. *Provincia de Santo Domingo de los Tsáchilas:* 4 km NE Dos Ríos (0.30278 S, 78.8678 W; 1140 m), KU 164670–85; 1 km W of Río Faisanes and its intersection with Chiriboga-La Palma road (0.3035 S, 78.869 W), QCAZ 15364.

Localities from the literature: *Nymphargus grandisonae:* <u>Colombia</u>: Serranía de los Paraguas (Boquerón; 04°44.2′ N, 76°18.3′ W, 2000 m) [260]. <u>Ecuador</u>:

Nymphargus griffithsi (Goin, 1961 [97]; Figures 154–157).

Cochranella griffithsi Goin, 1961 [97]. Holotype: BM 1940.2.20.4.
<u>Type locality</u>: "Río Saloya, Ecuador, 4000 feet".
Centrolenella griffithsi—Goin, 1964 [187].
Cochranella griffithsi—Ruiz-Carranza and Lynch, 1991 [6].
Nymphargus griffithsi—Cisneros-Heredia and McDiarmid, 2007 [17].

Common names: English: Griffiths' Glassfrog. Spanish: Rana de Cristal de Griffiths.

Etymology: The specific epithet honors Dr. Ivor Griffiths, of Birkbeck College, London, in recognition of his contributions to the understanding of relationships among frogs [97].

Identification: Among glassfrogs found on the Pacific slope of the Andes, *Nymphargus griffithsi* can be distinguished by having a uniformly green dorsum with small dark flecks (Figure 154), and by lacking webbing between fingers. The species also lacks humeral spines, although enlarged humeral crests are apparent in some specimens (Figure 155). Only two species, *N. buenaventura*, a species restricted to southwestern Ecuador (Reserva Buenaventura, Provincia de El Oro), and *N. lasgralarias*, from Reserva Las Gralarias (Pichincha Province), can be confused with *N. griffithsi*, and no distinctive morphological traits distinguish among these species. However, *N. buenaventura* is slightly smaller and has a green dorsum with yellow spots, whereas populations of *N. griffithsi* lack these yellow spots and have dark flecks on the dorsum (see Taxonomic Remarks). The most reliable traits to differentiate between *N. lasgralarias* and *N. griffithsi* are the calls; *N. griffithsi* produces a single tonal or multi-pulsed (i.e., two or more pulses) call, while the calls of *N. lasgralarias* are always pulsed. Also, *Nymphargus griffithsi* emits its advertisement call as a single note, whereas *N. lasgralarias* emits its calls singly or in a series. In addition, *N. lasgralarias* has shorter call duration than *N. griffithsi* (call duration in *N. lasgralarias* = 0.016–0.044 s; call duration in *N. griffithsi* = 0.103–0.148 s) [88]

Diagnosis: (1) Vomers lacking teeth; (2) snout truncated in dorsal view and truncated to protruding in profile; (3) lower half of the tympanic annulus evident, oriented dorsolaterally with dorsoventral inclination; tympanic membrane pigmented as surrounding skin; supratympanic fold absent or very weak; (4) dorsal skin slightly shagreen; (5) pair of enlarged subcloacal warts; (6) upper half of ventral parietal peritoneum white, posterior half transparent (condition P2); white pericardium, all other peritonea clear (condition V1); (7) liver tetralobed; hepatic peritoneum transparent (condition H0); (8) humeral spines absent, although enlarged humeral crests are apparent in some specimens (Figure 155); (9) hand webbing absent; (10) foot about two-thirds webbed: I (2^--2^+)—$(2^+-2^{1/4})$ II

$(1^+-1^{2/3})$—$(2^{1/2}-3^-)$ III $(1^{1/2}-1^{2/3})$—$(2^{1/3}-3^-)$ IV $(2^{1/3}-3^-)$—$(1^{1/2}-2^-)$ V (Figure 155); (11) ulnar and tarsal folds enameled and low; (12) concealed prepollex; nuptial pad Type I in adult males; (13) Finger I shorter than Finger II; (14) disc of Finger III 20.5%–26.5% of eye diameter; (15) in life, dorsum varies from uniform green to green with small dark flecks; bones green; (16) in ethanol, dorsal surfaces lavender with small dark marks; (17) iris background white to cream beige with brown to black punctuations; (18) fingers and toes lacking melanophores, except some on Toes IV and V; (19) males call from the upper side of leaves overhanging streams; call tonal and composed by single note with a duration of 0.103–0.148 s, and a dominant frequency at 3790–4307 Hz; (20) males show dangling fighting behavior; (21) eggs deposited on tips of leaves overhanging streams; short-term maternal care present; parental care by males absent; (22) tadpoles unknown; (23) small body size; in adult males, SVL 22.5–24.2 mm ($\overline{X} = 23.0 \pm 0.7$, $n = 5$); SVL adult females unknown.

Color in life (Figure 154): Dorsum varies from uniformly green to green with small dark flecks. Ventral surfaces greenish white. Bones green. Iris from white to cream beige, with brown to black punctuations forming thin reticulation; in some individuals, a yellow hue is visible around pupil.

Figure 154. (**A,B**) *Nymphargus griffithisi* in life, MZUTI 099. (**C,D**) *Nymphargus lasgralarias* in life, MZUTI 094. Photos of both species from Reserva Las Gralarias, 2175–2200 m, Pichincha province, Ecuador. Photos by Carl R. Hutter.

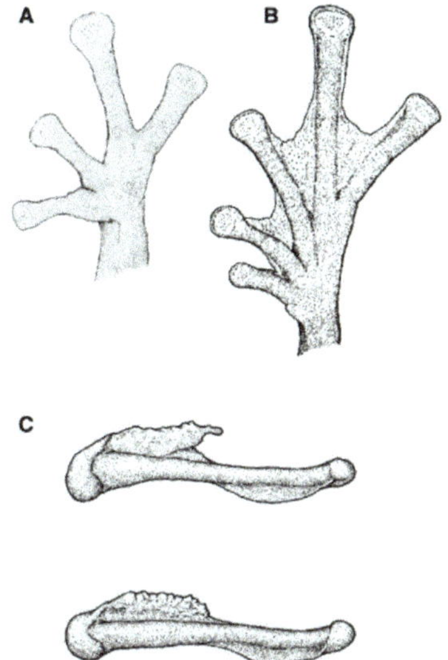

Figure 155. *Nymphargus griffithisi.* (**A**) Hand in ventral view, KU 118040. (**B**) Foot in ventral view, KU 118040. (**C**) Ventral humeral crests ((**Top**): KU 288992; (**Bottom**): KU 188148). (**A,B**) Modified from Lynch and Duellman [22]. (**C**) Modified from Guayasamin et al. [1].

Color in ethanol: Dorsal surfaces lavender with small dark marks. Anterior half to two-thirds of parietal peritoneum white, posterior part translucent; pericardium white; most visceral peritonea clear, except for white peritoneum covering renal capsules.

Biology and ecology: During the rainy season (December–April), at night, *Nymphargus griffithsi* is active on vegetation 50–900 cm above streams, waterfalls, and small rivers in primary and secondary cloud forests [5,87]. Reproductive activity peaks in February–April. Males are highly territorial and engage in aggressive interactions with conspecifics; when fighting, males have been observed dangling upside-down venter to venter holding onto a branch with their hind legs [5]. Females lay clutches of 14–22 eggs on leaves overhanging fast-flowing water. These later expand into a hanging gelatinous mass upon absorption of water; eventually, tadpoles hatch and fall into the stream below [87]. At Reserva Las Gralarias, *N. griffithsi* has been observed at Five-frog Creek, Heloderma Creek, Hercules Creek, and Kathy's Creek [88,92]. Short-term parental care is provided by females; no parental care is provided by males [25].

Call (Figure 156): The information presented below is from Hutter and Guayasamin (2012). The call of *N. griffithsi* is a tonal call, produced at a rate of 1.1–1.9 calls per minute (mean = 1.6 ± 0.4). The call is composed by a single note with a duration of 0.103–0.148 s (mean = 0.122 ± 0.009) and a dominant frequency at 3790–4307 Hz (mean = 4107 Hz; ± 105.5 Hz).

Tadpole: Not described.

Distribution (Figure 157): Although *Nymphargus griffithsi* has been reported in Ecuador and Colombia [22], herein we restrict its distribution to Ecuador, specifically, to localities in Pichincha, Santo Domingo de los Tsáchilas, and Cotopaxi provinces, at elevations between 1220–2430 m. Other populations that have been previously identified as *N. griffithsi* require further taxonomic evaluation.

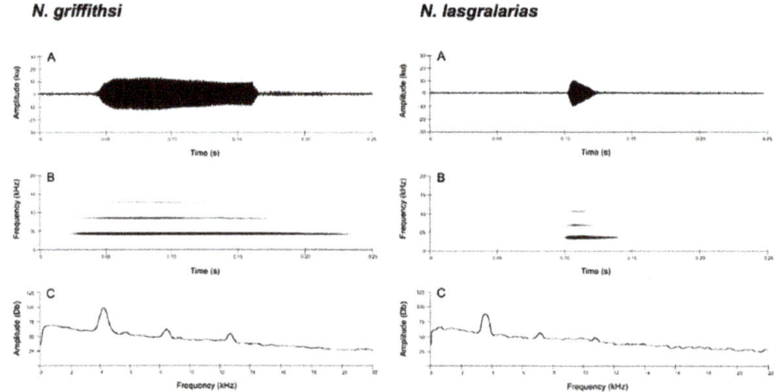

Figure 156. Advertisement call of *Nymphargus griffithisi* and *N. lasgralarias*. (**A**) Oscillogram. (**B**) Audiospectrogram. (**C**) Power spectrum. Modified from Hutter and Guayasamin [88].

Figure 157. Distribution of *Nymphargus griffithsi* in Ecuador (yellow dots).

Conservation status: Globally, *N. griffithsi* is listed by the IUCN as *Least Concern* [263]. Recent records in Ecuador come from Reserva Las Gralarias, Reserva Río Guajalito, Reserva Otonga, La Favorita Station, and near the town of Chiriboga. At Reserva Las Gralarias, the species is not infected by the chytrid fungus *Batrachochytrium dendrobatidis* (*n* = 5) [92]. Given the more restricted distribution of the species, *N. griffithsi* should be considered as *Endangered*, following IUCN criteria B2a, B2(iii).

Evolutionary relationships (Figure 136): *Nymphargus griffithsi* and *N. lasgralarias* are sister taxa.

Taxonomic remarks: Several populations previously assigned to *N. griffithsi* (see Specimens examined) require further evaluation. Vocalization studies on Colombian populations assigned to *N. griffithsi* also suggest that several species are concealed under this name [264]. Paratypes of *Centrolenella scirtetes* seem to correspond to *N. griffithsi* (see Taxonomic remarks of *Centrolene lynchi*).

Specimens examined: *Nymphargus griffithsi*, <u>Ecuador</u>: *Provincia de Cotopaxi:* 11.5 km W of Pilaló (0.94815° S, 78.989633° W; 1500 m), QCAZ 34113; *Provincia de Pichincha:* Río Saloya, 1219 m, BMNH 1940.2.20.4 (holotype), BMNH 1940.2.20.3 (paratype); km 14 on the San Juan de Chillogallo–Chiriboga road (0.275895° S, 78.721647° W; 2120 m), QCAZ 29531; km 16 on the San Juan de Chillogallo–Chiriboga road (0.278161° S, 78.706067° W; 2430 m), QCAZ 29524–30; La Victoria (0.16285° S, 77.909667° W; 2104 m), QCAZ 24801; Tandapi (0.416388° S, 78.7988° W), QCAZ 351, KU 118009–20; Reserva Las Gralarias "Hercules Giant

Tree Frog Creek" (0°01.529′ S, 78°42.243′ W; 2175 m), MZUTI 100, 102, 099; Reserva Las Gralarias "Five Frog Creek" (0°01.870′ S, 78°42.358′ W; 2150 m), MZUTI 101; Reserva Las Gralarias "Heloderma Creek" (0°01.245′ S, 78°42.370′ W; 2200 m), MZUTI 098; 1 km SW San Ignacio (0.44861° S, 78.74777° W; 1920 m), KU 178108–21; 3.5 NE Mindo (0.03222° S, 78.76138° W; 1340 m), KU 164564–76; 5 km ESE Chiriboga (0.245277° S, 78.7261° W; 2010 m), KU 164519–37; 5.6 km SE Tandayapa (0.0333° S, 78.7166° W; 1910 m), KU 202792. *Provincia de Santo Domingo de los Tsáchilas:* 14 km W of Chiriboga (0.26527778° S, 78.847778° W; 1960 m), KU 164544–63; 4 km W Chiriboga (0.24277° S, 78.7855°; 2120 m), KU 142649.

Nymphargus aff. *griffithsi.* <u>Ecuador</u>: *Provincia Imbabura:* 23.2 km W of Apuela, Cordillera de Intag (0.2666° N, 78.6° W; 2190 m), KU 178122–36; San Antonio de Cuellaje, Finca de Estuardo Ayala (0.4775 N, 78.56263 W), QCAZ 42164. *Provincia de Carchi:* Chilma Bajo (0.86472 N, 78.04972 W; 2071 m), QCAZ 40176–77; 9.9. km E of Maldonado on the Maldonado-Tulcán road (0.83472° N, 78.051388° W; 2130 m), QCAZ 12572; km 5 on the Chilma Bajo–El Placer road (0.85705° N, 78.032476° W; 2222 m), QCAZ 39992, 39994; ca. 5 km W La Gruel (0.916667° N, 78.1333° W; 2340 m), KU 202784–91, 202796, 202801.

Nymphargus humboldti new species Guayasamin, Cisneros-Heredia, McDiarmid, Hutter (Figures 158 and 159).

http://zoobank.org/urn:lsid:zoobank.org:act:AD71F7CC-0718-449E-88DC-0A64302D9855.

Common names: English: Humboldt's Glassfrog. Spanish: Rana de cristal de Humboldt.

Etymology: The specific epithet *humboldti* honors Alexander von Humboldt for his unparalleled contributions to biogeography, integrative perspective of the sciences and arts, humanism, and also for his devotion towards mountains.

Holotype: ZSFQ 0388, adult male, collected from Volcán Sumaco (0.61387° S, 77.590° W; 1738 m), Napo Province, Ecuador, by Jose Vieira on 11 September 2018.

Paratopotypes: QCAZ 9402, adult male collected by Juan M. Guayasamin on 19 February 1996; QCAZ 41071, 41073–74, 41077–78, 41081–82, 41150–51, 41314–15, adult males and females, collected by Elicio Tapia and Raúl E. Ruíz on 20–30 March 2009.

Paratypes: QCAZ 45712–16, 45730, 47511–14, adult males and females, collected by Elicio Tapia from Río Yana Challuwa Yaku (1.26764° S, 78.04797° W; 1800–2400 m), Reserva Comunitaria Ankaku, Pastaza Province.

Generic placement: The new species is placed in the clade *Nymphargus* (sensu Guayasamin et al. [1]) based on morphological and molecular data. All species in *Nymphargus* share an absence of webbing between Fingers I–III and absence or reduced webbing between Fingers III and IV; additionally, males lack humeral spines (except *N. grandisonae*). *Nymphargus humboldti* sp. nov. also has these traits. Molecular analyses of mitochondrial genes (12S, 16S, ND1) unambiguously place the new species in the clade *Nymphargus* (Figure 16).

Identification: Among glassfrogs that inhabit the Amazonian slopes of the Andes and the Amazonian lowlands, *Nymphargus humboldti* sp. nov. (Figure 158) is distinguished by having a green dorsum with small yellow spots and moderate body size (male SVL 23.3–25.2 mm; female SVL 24.3–25.9 mm), and lacking hand webbing and humeral spines. *Nymphargus humboldti* sp. nov. is most similar to *N. siren*, which is smaller (SVL in males, 19.8–22.6 mm; in females, 22.5–23.3 mm). Other species with a similar dorsal color pattern include *Espadarana audax*, *Nymphargus cariticommatus*, *N. siren*, *Rulyrana flavopunctata*, *R. mcdiarmidi*, and *Teratohyla midas*. Differences among these species include the presence of hand webbing between Fingers III and IV in *E. audax*, *R. flavopunctata*, *R. mcdiarmidi*, and *T. midas* (absent in *N. humboldti* sp. nov.), white digestive tract in *T. midas* (opaque in *N. humboldti* sp. nov.), and humeral spine in males in *E. audax* (absent in *N. humboldti* sp. nov.). *Nymphargus cariticommatus* differs by having white renal peritoneum and white esophagus (transparent renal pericardium and opaque esophagus in *N. humboldti* sp. nov.) and less foot webbing. Two additional species endemic to Colombia, *N. luminosus* and *N. spilotus*, can be confused with *N. humboldti* sp. nov. Both species are conspicuously larger than *N. humboldti* sp. nov. (*N. luminosus*, male SVL = 27.8–30.0 mm, female SVL = 32.7 mm; *N. spilotus*, male SVL = 25.3–26.4 mm, female SVL = 27.6–28.5 mm); also, *N. luminosus* in only found on the Pacific flank of the Andes in Colombia, whereas *N. spilotus* is only know from its type locality (Rancho Quemado, Corregimiento Florencia, Departamento de Caldas) on the eastern slope of Cordillera Central.

Figure 158. *Nymphargus humboldti* sp. nov. in life. Male from Río Yana Challuwa, Pastaza province, Ecuador, QCAZ 47514, paratype. Photos by Luis A. Coloma.

Diagnosis: (1) Vomers with edentate dentigerous process; (2) snout truncated to bluntly rounded in lateral profile; truncated in dorsal view; (3) tympanum oriented almost vertically, with slight lateral and posterior inclinations, its diameter about 20%–27% of eye diameter; upper fourth of tympanic annulus obscured by supratympanic fold; tympanic membrane pigmented as surrounding skin; (4) dorsal skin shagreen, with minute spicules in males; (5) venter areolate; pair of enlarged subcloacal warts; (6) white parietal peritoneum covering anterior 50%–60% of venter (condition P2); white pericardium; translucent peritonea covering intestines, stomach, kidneys, gall and urinary bladders (condition V1); (7) liver tetralobed, covered by transparent peritoneum (condition H0); (8) humeral spines absent; (9) webbing absent between inner fingers, absent or basal between Fingers III and IV; webbing formula III (3⁻–3)—(2²ᐟ³–3⁻) IV; (10) feet about two-thirds webbed; webbing formula: I (2⁻–2)—(2⁺–2¹ᐟ⁴) II (1⁺–1¹ᐟ²)—(2⁺–2¹ᐟ²) III (1⁺–1¹ᐟ⁴)—(2¹ᐟ⁴–3⁺) IV (2¹ᐟ²–3⁻)—(1²ᐟ³–2⁻) V; (11) ulnar and tarsal folds present, low; (12) concealed prepollex; in males, nuptial pad Type I; (13) Finger I about same length as Finger II (Finger I length 94%–102% of Finger II); (14) disc width of Finger III about 50%–55% of eye diameter; (15) in life, dorsum green with yellow spots (Figure 158); bones green; (16) in preservative, dorsum lavender with small white spots; (18) melanophores usually lacking from dorsal surfaces of fingers and toes, except for few on Toe V; (19) males call from upper surfaces of leaves; call unknown; (20) fighting behavior unknown; (21) egg deposition site unknown; parental care unknown; (22) tadpoles unknown; (23) small body size; in males, SVL 23.3–25.2 mm ($\overline{X}$ = 24.3 ± 0.684, *n* = 13); in females, SVL 25.5–27.4 mm (*n* = 3).

Description of holotype: Adult male, SVL 24.9 mm. Head slightly wider than long (head length 98% of head width); snout truncated in dorsal and lateral profiles; canthus rostralis indistinct, slightly concave; loreal region slightly concave; lips slightly flared; nostril protuberant, closer to tip of snout than to eye, directed frontolaterally; internarial area barely depressed. Eyes large, directed anterolaterally at about a 50° angle; transverse diameter of disc of Finger III 53% eye diameter. Supratympanic fold low, obscuring upper portion of tympanic annulus; tympanum oriented mostly vertically, but with slight posterolateral inclination; tympanic membrane transparent, its upper half pigmented as surrounding skin. Dentigerous processes of vomer low, situated transversely between choanae, lacking teeth; choanae large, longitudinally rectangular; tongue ovoid, with ventral posterior fourth not attached (free) to floor of mouth, posterior margin notched; vocal slits extending posterolaterally from a point about midway along lateral margin of tongue) to angle of jaws. Humeral spine absent. Low ulnar fold evident on external ventrolateral margin of arm; inner ulnar fold absent; relative lengths of fingers: III > IV > II > I; webbing absent between Fingers I, II, and III, basal between Fingers III and IV, webbing formula III 3⁻–2²/³ IV; discs expanded, nearly elliptical; disc pads nearly triangular in shape; subarticular tubercles small, round, simple; supernumerary tubercles absent; palmar tubercle elliptical, simple; nuptial pad large (Type I), ovoid, granular, extending from ventrolateral base to dorsal surface of Finger I, covering proximal half of Finger I. Length of tibia 58% SVL; low inner tarsal fold evident; outer tarsal fold absent; two-thirds webbed; webbing formula of foot: I 2—2⁺ II 1⁺—2¹/⁴ III 1⁺—3⁺ IV 3⁻—2⁻ V; discs on toes round to elliptical; disc on Toe IV narrower than disc on Finger III; disc pads triangular; inner metatarsal tubercle large, ovoid; outer metatarsal tubercle round, barely evident; subarticular tubercles small, round; supernumerary tubercles absent. Skin on dorsal surfaces of head, body, and lateral surface of head and flanks shagreen with numerous minute spinules; throat smooth; belly and lower flanks areolate; cloacal opening directed posteriorly at upper level of thighs; cloacal ornamentation absent. Two enlarged tubercles below the level of vent.

Coloration of the holotype in life (Figure 158): Green dorsum with small yellow spots; upper lip white. Anterior 60% of venter white, posterior portion transparent. Bones green. Iris greyish cream with slight yellow hue and fine black reticulations; yellow circumpupilary ring.

Coloration of the holotype in ethanol: Dorsal surfaces of head, body, and limbs lavender with small white spots; white upper lip. Anterior 60% of ventral parietal peritoneum white. Heart covered by white pericardium; translucent peritonea on gall and urinary bladders; iridophores absent from digestive tract, liver, and kidneys.

Measurements of holotype (mm): *Nymphargus humboldti* sp. nov., adult male, ZSFQ 0388: SVL = 24.9, head length = 8.1, head width = 8.3, eye–nostril = 1.9, nostril–snout = 1.1, IOD = 2.6, upper eyelid width = 2.3, eye diameter = 3.5, tympanum diameter = 0.8, tibia = 14.5, foot length = 11.5, radio–ulna length = 5.5, hand = 8.4, Finger I length = 4.9, Finger II length = 5.2, disc of Finger III width = 0.9

Biology and ecology: At Río Yana Challuwa Yaku, *Nymphargus humboldti* sp. nov. was found in sympatry with *N. anomalus*. During the night, males were calling from the upper surfaces of leaves near a fast-flowing stream. Parental care is unknown.

Call: Not described.

Tadpole: Not described.

Distribution (Figure 159): *Nymphargus humboldti* sp. nov. is known from few localities on the Amazonian slopes of the Ecuadorian Andes at elevations between 1770 and 2400 m.

Figure 159. Distribution of *Nymphargus humboldti* sp. nov. in Ecuador (yellow dots).

Conservation status: Following IUCN criteria, we suggest placing *Nymphargus humboldti* sp. nov. in the *Data Deficient* category.

Evolutionary relationships (Figure 136): *Nymphargus humboldti* sp. nov. is sister to *N. siren*.

Nymphargus lasgralarias Hutter and Guayasamin, 2012 [88] (Figures 154 and 160–162).

Nymphargus lasgralarias Hutter and Guayasamin, 2012 [88]. Holotype: MZUTI 096.
Type locality: "Five Frog Creek" (0°01.870′ S, 78°42.358′ W; 2150 m) at Reserva Las Gralarias, Pichincha province, Ecuador".

Common names: English: Las Gralarias Glassfrog. Spanish: Rana de Cristal de Las Gralarias.

Etymology: The epithet *lasgralarias* refers to the type locality of the species, Reserva Las Gralarias. The species was dedicated to the reserve and the team of people, led by Dr. Jane Lyons, for their efforts to conserve Ecuadorian cloud forests [88].

Identification: Among *Nymphargus* species found on the Pacific versant of the Andes of Ecuador, *N. lasgralarias* can only be confused with *N. buenaventura* and *N. griffithsi*. Dorsal texture and color pattern readily separate *N. buenaventura*, which, in life, has a light-green dorsum with warts corresponding to pale yellow spots, whereas the dorsum of *N. lasgralarias* is homogenously green and lacks warts. Additionally, *N. buenaventura* is smaller (male SVL in *N. lasgralarias* = 24.6–26.5 mm; male SVL in *N. buenaventura* = 20.9–22.4 mm). *Nymphargus lasgralarias* is most similar to *N. griffithsi*. The two species differ in the following traits: (i) Dorsal color pattern: Homogenously green in *N. lasgralarias*; green with very small black spots in *N. griffithsi* (Figure 154); (ii) body size: Male SVL in *N. lasgralarias* = 24.6–26.5 mm, whereas male SVL in *N. griffithsi* = 22.5–24.2 mm; and call: Short, pulsed note lasting 0.016–0.044 s in *N. lasgralarias*; long tonal note lasting 0.103–0.148 s in *N. griffithsi*. Two Colombian species from the Pacific slopes of the Andes, *N. cristinae* [255] and *N. prasinus* [120], can also be confused with *N. lasgralarias*. *Nymphargus lasgralarias* is distinguished from *N. cristinae* by being smaller (male SVL in *N. lasgralarias* = 24.6–26.5 mm; male SVL in *N. cristinae* = 26.0–31.1 mm), having a snout that is truncated in dorsal view and protruding in lateral view (subacuminate in dorsal view, truncated in lateral view in *N. cristinae*; see Ruiz-Carranza and Lynch [255]), and lacking palmar supernumerary tubercles (supernumerary small, abundant in *N. cristinae*). *Nymphargus prasinus* shares the same color pattern as *N. lasgralarias*, but is recognized by having a round snout in dorsal view (truncated *N. lasgralarias*), five to seven teeth on each process of the vomer (teeth absent in *N. gralarias*), and being considerably larger (male SVL 33.0–34.5 mm; *n* = 3) [88].

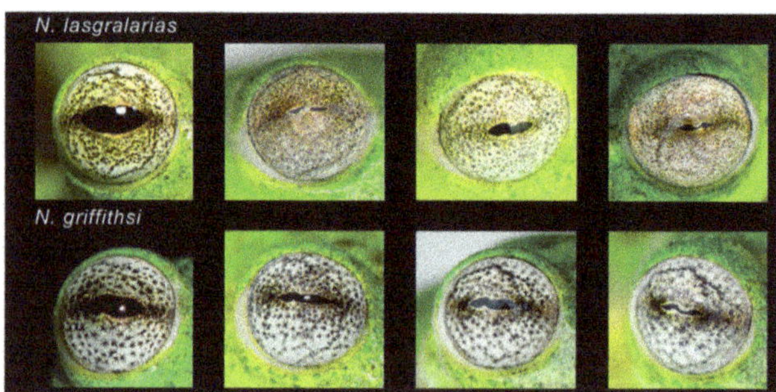

Figure 160. (**Top Row**): Iris of *Nymphargus lasgralarias*. (**Bottom Row**): Iris of *N. griffithsi*. Modified from Hutter and Guayasamin [88].

Characterization. (1) Vomerine teeth absent; (2) snout truncated in dorsal profile, protruding in lateral profile; (3) tympanum small; supratympanic fold present; tympanic membrane translucent, pigmented only on its upper half; (4) skin texture finely shagreen, with microspiculations; (5) ventral skin areolate, with pair of large subcloacal warts; cloaca surrounded by low warts; (6) upper half of ventral parietal peritoneum covered by iridophores (condition P2), all other peritonea translucent, except for thin layer of iridophores covering heart and renal capsules (condition V1); (7) liver lobed, lacking iridophores (condition H0); (8) humeral spines absent; (9) webbing absent between fingers; (10) foot about half webbed; webbing formula: I (2–2⁻)—(2⁺–2^{1/2}) II (2–2⁻)—(3⁻–3) III (2⁻–2)—(3⁻–3) IV (3–3⁺)—2 V; (11) ulnar and tarsal folds low, barely evident; (12) nuptial pad Type I; concealed prepollex; (13) first finger slightly shorter than second; (14) eye diameter larger than width of disc on Finger III; (15) in life: Green dorsum, lacking spots; (16) in preservative: Dorsum pale lavender; (17) iris golden yellow, with numerous small black spots; weakly reticulate; (18) hands and feet yellowish green; melanophores absent from fingers and toes or, when present, restricted to dorsal surfaces of Finger IV and Toes IV and V; (19) males call from upper side of leaves along streams; vocalizations emitted in series of one to five calls; each call pulsed, with a duration of 0.016–0.044 s, and non-modulated dominant frequency at 3618–3963 Hz; (20) fighting behavior unknown; (21) short-term maternal care present; parental care by males absent; (22) egg clutches deposited on upper surface of leaves at terminal margin, transitioning to hanging as eggs develop; (23) tadpoles unknown; (24) SVL in adult males 24.6–26.5 mm (mean = 25.3 ± 0.7368; *n* = 7); in females 26.3–27.2 mm (*n* = 2).

Color in life (Figure 154, Figure 160, Figure 161): Dorsum light green, lacking dark spots; flanks yellowish white; bones green; fingers and toes yellow with a faint green tint. Venter white anteriorly and translucent posteriorly. Iris background golden with numerous dark spots and very light reticulation [88].

Color in ethanol: Dorsal surfaces of head and body are cream; fingers and toes cream. Upper half of ventral parietal peritoneum covered by iridophores (white), all other peritonea translucent, except for thin layer of iridophores covering heart and renal capsules [88].

Biology and ecology: The following information is taken from Hutter and Guayasamin [88]. *Nymphargus lasgralarias* inhabits small, permanent streams (ca. 3 m width) within primary montane forest with minimal disturbance. The species is active during the night and emits advertisement calls from the tops of small-sized ferns, small leaves, and long palm leaves, 1–9 m above the stream. Fighting behavior is unknown, but the description by Duellman and Savitzky [5] could apply to *N. lasgralarias* as well. Small clutches (12–36 eggs per mass; *n* = 23) are deposited on the upper surface of a leaf near its terminal margin, and transition into hanging masses as the eggs develop (Figure 160). Tadpoles are unknown. At Reserva Las Gralarias, the species has been found at Ballux Creek, Five-frog Creek, Heloderma Creek, Hercules Creek, Chalguayacu Creek, Kathy's Creek, and Lucy's Creek. *Nymphargus lasgralarias* occurs sympatrically with six other centrolenid species: *Centrolene ballux*, *C. heloderma*, *C. lynchi*, *C. peristicta*, *Nymphargus grandisonae*, and *N. griffithsi*. Reproductive activity was recorded between April 5 and July 1 2011, but increased dramatically at the end of April and peaked in the middle of May [88]. Females provide short-term parental care; male parental care is absent [25].

Figure 161. *Nymphargus lasgralarias* from Reserva Las Gralarias, Pichincha province, Ecuador. (**Top**): Adult male. (**Bottom**): Spider (Ctenidae) preying on egg clutch. Photos by Jaime Culebras.

Call (Figure 156): Males emit calls during the night, from the tops of small-sized ferns, small leaves, and long palm leaves, 1–6 m above the stream. Calls are pulsed and emitted in series (one to four calls per call series; mean = 2.7 ± 0.7 calls). Five-call series are emitted sporadically. Each series has duration of 0.033–2.541 s (mean = 1.529 ± 0.597 s) and an interval of 8.6–78.6 s (mean 33.8 ± 18.4 s) between series, with an interval of 0.088–1.513 s (mean = 0.873 ± 0.205 s) between calls within a series. The call repetition rate is 2.0–9.9 (5.5 ± 2.7) calls per minute (n = 6 individuals), each call sounding like a "tick" or "click"; call duration is 0.0160–0.0440 s (mean = 0.0257 ± 0.0058; n = 119) and dominant frequency is at 3445–3962 Hz (mean = 3691 ± 131.9 Hz) [88].

Tadpole: Not described.

Distribution (Figure 157): *N lasgralarias* is known from localities on the western slope of the Ecuadorian Andes at elevations between 1850–2300 m. Specifically, it has been found within the limits of the Pichincha and Cotopaxi provinces ([87,88], this work). Individuals that resemble *N. lasgralarias*, but are genetically different, are found in northern Ecuador (see Specimens examined).

Conservation status: *N. lasgralarias* has not been evaluated by the IUCN. Main threats are habitat destruction and fragmentation due to agriculture and cattle, introduced species (trout), infectious diseases (*Batrachochytrium dendrobatidis*), and climate change. At Reserva Las Gralarias, the species is infected by the chytrid fungus *B. dendrobatidis*, but no recent declines have been observed [92]. Following IUCN criteria (B2a, biii, biv), we suggest considering the species as *Endangered*.

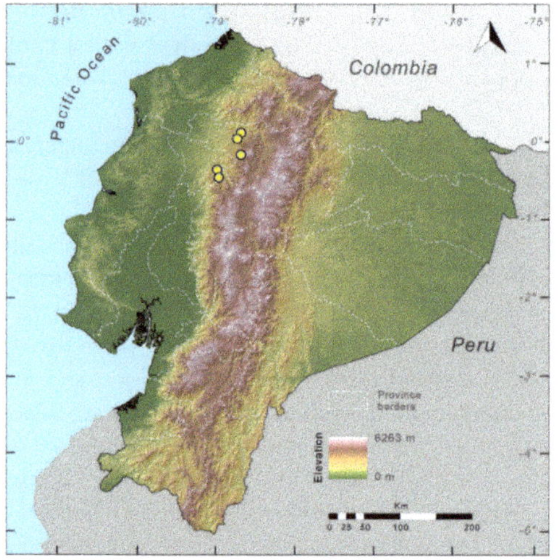

Figure 162. Distribution of *Nymphargus lasgralarias* in Ecuador (yellow dots).

Evolutionary relationships (Figure 136): *N. lasgralarias* and *N. griffithsi* are sister taxa.

Specimens examined: *Nymphargus lasgralarias:* <u>Ecuador</u>: *Provincia de Cotopaxi:* Bosque Integral Otonga (0.55 S, 79.46667 W; 2000 m), QCAZ 13115; Reserva Otonga (0.676 S, 76.397 W; 1950 m), QCAZ 11689–90. *Provincia Imbabura: Provincia de Pichincha:* Five Frog Creek (0°01.870′ S, 78°42.358′ W; 2150 m) at Reserva Las Gralarias, MZUTI 096 (holotype), MZUTI 091–095, 097; Kathy's Creek (0°01.398′ S, 78°43.772′ W; 2000 m), Reserva Las Gralarias, MZUTI 091–095; Hercules Giant Tree Frog Creek (0°01.529′ S, 78°42.243′ W; 2175 m), Reserva Las Gralarias, MZUTI 097; Nanegal Grande (0.1167 N, 78.6667 W; 2300 m), QCAZ 46012; 9 km SE Tandayapa (0.01667 S, 78.6833 W; 2160 m), KU 164577–87.

Nymphargus aff. *lasgralarias:* Ecuador: *Provincia de Imbabura:* 23.2 km W of Apuela, Cordillera de Intag (0.2666 N, 78.6 W; 2190 m), KU 178122–36; San Antonio de Cuellaje, Finca de Estuardo Ayala (0.4775 N, 78.56263 W), QCAZ 42164; Santa Rosa, Reserva Alto Chocó (0.36939 N, 78.44942 W; 2104 m), QCAZ 31768.

Nymphargus laurae Cisneros-Heredia and McDiarmid, 2007 [17] (Figures 163 and 164).

Nymphargus laurae Cisneros-Heredia and McDiarmid, 2007 [17]. Holotype: USNM 288453.
Type locality: "Loreto, Upper Rio Napo (=Loreto region, near the town of Loreto) (ca. 77°20′ S, 00°40′ W, ca. 500 m elevation), lower slopes of the Sumaco Volcano, on the Cordillera Oriental, eastern slopes of the Andes, Provincia de Orellana, República del Ecuador".

Common names: English: Laura's Glassfrog. Spanish: Rana de cristal de Laura.

Etymology: The specific epithet is a patronym for Laura Heredia, DFCH's grandmother, for her support of Diego's interest in animals and science [17].

Identification: *Nymphargus laurae* can be distinguished from most centrolenids by the presence of dorsal ocelli (a trait shared only with *N. anomalus*, *N. lindae* sp. nov., *N. cochranae*, and *N. ignotus*), two papillae on each toe except Toe V (Figure 163), and relatively small size (SVL = 19.9 mm in male holotype). *Nymphargus anomalus* differs by having smaller and more abundant ocelli, and dark spots and punctuations amidst the ocelli. *Nymphargus lindae* sp. nov. differs from *N. laurae* mainly by being larger (in *N. lindae* sp. nov., male SVL = 23.0–26.3 mm; female SVL = 27.2–27.8 mm), having vomerine teeth (absent in *N. laurae*), and lacking papillae on toe discs. *Nymphargus ignotus* differs by having, in life, a tan dorsal coloration, with smaller and more abundant ocelli, and by being larger (22.3–25.4 mm SVL in males of *N. ignota*); further, *N. ignotus* is known only from western Colombia around 1900 m, whereas *N. laurae* is known from the slopes of Volcán Sumaco, on the Amazonian slopes of the Andes. *Nymphargus laurae* is very similar to *N. cochranae*; both species share a similar color pattern and are likely to be sympatric; *N. cochranae* however differs by having much smaller ocelli, no ocelli on forearms and shanks, and larger body size (male *N. cochranae* SVL = 23.8–26.7 mm), and lacking papillae on the toes. See Taxonomic Remarks.

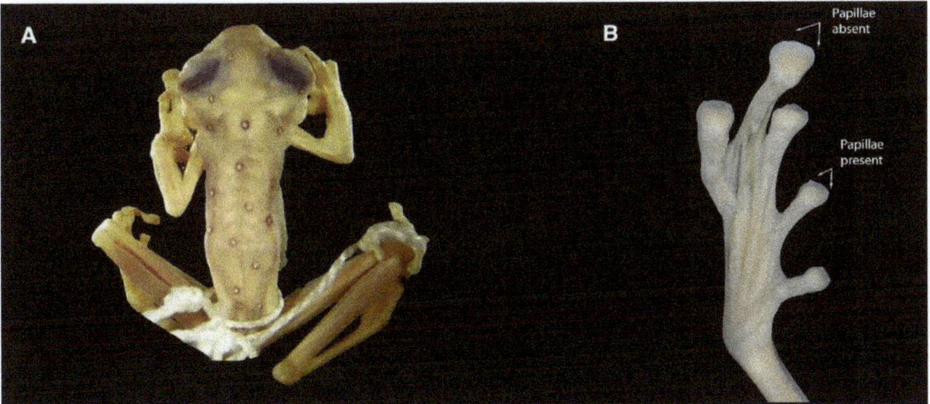

Figure 163. *Nymphargus laurae* in preservative, USNM 288453, from Ecuador, lower slopes of Volcán Sumaco. (**A**) Dorsal view of holotype. (**B**) Ventral view of foot; note presence of two papillae on each toe disc, except Toe IV. Photos by James Poindexter.

Diagnosis: The following traits characterize *N. laurae:* (1) Vomerine teeth absent; (2) snout truncated in dorsal and profile views; nostrils slightly elevated producing a slight depression in the internarial area; loreal region concave; (3) tympanic annulus evident, oriented dorsolaterally with dorsal inclination; weak supratympanic fold from behind eye to insertion of arm; (4) dorsal skin slightly shagreen with elevated warts corresponding to ocelli, and scattered spicules; (5) ventral skin granular; pair of large, round, flat subcloacal warts; other cloacal ornamentation absent; (6) parietal peritoneum white, covering anterior two-thirds of abdomen (condition P2); white pericardium; all other peritonea clear (condition V1); (7) liver lobed, hepatic peritoneum clear (condition H0); (8) humeral spine absent; (9) webbing basal between Fingers I, II and III, outer fingers III $2^{2/3}$—$2^{1/2}$ IV; (10) webbing on feet I 2^{-}—2^{+} II $1^{1/2}$—2^{+} III 1^{+}—$2^{1/2}$ IV $2^{1/2}$—$1^{1/2}$ V; (11) no dermal folds or tubercles on hands, forearms, feet, or tarsi; (12) unpigmented nuptial pad Type I; concealed prepollex; (13) second finger longer than first; (14) eye diameter greater than width of disc on Finger III; (15) color in life, green with yellow spots surrounded by black ocelli; (16) color in preservative, dorsal surfaces tan cream with dark, reddish–lavender ocelli; (17) iris coloration in life unknown; (18) melanophores absent on fingers and toes; (19) calling site and call unknown; (20) fighting behavior unknown; (21) egg deposition site and parental care unknown; (22) tadpole unknown; (23) minute body size; SVL in male holotype 19.9 mm; females unknown.

Color in life: Green dorsum with black ocelli with yellow center [17].

Color in ethanol (Figure 163): Dorsal surfaces cream-colored with minute reddish–lavender melanophores appearing as punctuations and forming a reddish –lavender shadow on sides of body. Dark reddish–lavender ocelli on head and body, the center of each encircles a cream-colored wart. Venter cream. Parietal peritoneum with iridophores; white pericardium; all other peritonea clear [17].

Biology and ecology: Almost no information is available for *Nymphargus laurae*. The type locality is within the Foothill Evergreen forests. Parental care is unknown.

Call: Not described.

Tadpole: Not described.

Distribution (Figure 164): *Nymphargus laurae* only is known from its type locality, near the town of Loreto, lower slopes of the Sumaco Volcano (ca. 77°20′ S, 00°40′ W; ca. 500 m), Ecuador [17], within the Amazonian Tropical Rainforest ecoregion.

Figure 164. Distribution of *Nymphargus laurae* in Ecuador (yellow dot).

Conservation status: *Nymphargus laurae* is classified as *Critically Endangered* by the IUCN [265]; however, given that its type locality has a degree of uncertainty (collected in 1955 from the general area of "Loreto, Upper Rio Napo") and that the species is known from a single specimen, we consider that the category of *Data Deficient* is more appropriate.

Evolutionary relationships: It is likely that *Nymphargus laurae* is a close relative of *N. cochranae* and *N. lindae* sp. nov., given their similar morphologies and color patterns.

Taxonomic Remarks: The description of *Nymphargus laurae* was based on a single specimen and, therefore, intraspecific variation is unknown. Given that this species shares a number of traits (general morphology, webbing, color pattern) with *N. cochranae*, a relatively abundant species on the lower slopes of Volcán Sumaco (nearby the type locality of *N. laurae*), we are presented with two possibilities: (i) The holotype of *N. laurae* represents an anomalous individual of *N. cochranae*, or (ii) *N. laurae* is a valid species, that might occur in sympatry with *N. cochranae*. Also, *N. laurae* is very similar to *N. lindae* sp. nov.; although, the two species are geographically distant (*N. lindae* sp. nov. is only known from Cordillera del Cóndor, whereas *N. laurae* is known from the lower slopes of the Volcán Sumaco); the two taxa differ in body size, vomerine teeth, and papillae on toes (see Diagnosis). As mentioned above, at the moment, it is impossible to determine the extent of intraspecific variation in *N. laurae*.

Specimens examined: *Nymphargus laurae:* <u>Ecuador</u>: *Provincia de Orellana:* Loreto region, near the town of Loreto, lower slopes of Volcán Sumaco (ca. 77°20′ S, 00°40′ W, ca. 500 m), on the Cordillera Oriental, USNM 288453.

Nymphargus lindae new species Guayasamin (Figures 165 and 166).

http://zoobank.org/urn:lsid:zoobank.org:act:A8E08511-149C-4DA3-8F74-57C04C7CC302

Common names: English: Linda's Glassfrog. Spanish: Rana de Cristal de Linda.

Etymology: The specific epithet honors Linda Trueb, one of the most influential amphibian systematist of our days. Linda, as the curator of the herpetological collection of the University of Kansas, has led one of the most prolific and solid research group on amphibian biology, mentoring numerous students (including JMG). Her work on the evolution of skeletal diversity, ontogeny, and scientific illustration is outstanding.

Holotype: QCAZ 41572, adult male, collected from Miazi Alto (4.25044° S, 78.61356° W; 1200 m), Cordillera del Cóndor, Provincia de Zamora Chinchipe, Ecuador, by Juan M. Guayasamin and Elicio Tapia on 8 April 2009.

Paratopotypes: QCAZ 41562–71, 41573–74, 41587, 41594, 41597, 41599, 41644–47, 41654–56, 41658, 42446, adult males and females collected by Juan M. Guayasamin, Elicio Tapia, Silvia Aldás, and Holger Braun on 8–12 April 2009.

Generic placement: The new species is placed in the clade *Nymphargus* (sensu Guayasamin et al. [1]) based on morphological and molecular data. All species in *Nymphargus* share an absence of webbing between Fingers I–III and an absence or reduced webbing between Fingers III and IV; additionally, males lack humeral spines (except *N. grandisonae*). *Nymphargus lindae* sp. nov. has the aforementioned traits. Molecular analyses of mitochondrial genes (12S, 16S, ND1) unambiguously place the new species in the clade *Nymphargus* (Figure 16).

Identification: *Nymphargus lindae* sp. nov. can be distinguished from most other glassfrogs by having, in life, a green dorsum with dark lavender to black ocelli enclosing yellow to orange spots (Figure 165), and lacking webbing between fingers. *Nymphargus lindae* sp. nov. is most similar to other glassfrogs with dorsal ocellated patterns (*N. anomalus*, *N. cochranae*, *N. ignotus*, *N. laurae*, *N. ocellatus*). *Nymphargus anomalus* differs by having, in life, a pale brown dorsum and scattered black and lavender flecks between ocelli (green dorsum lacking flecks in *N. lindae* sp. nov.); *N. ignotus* has a brown dorsum and occurs on the Pacific slopes of the Andes, whereas *N. lindae* sp. nov. has a green dorsum and is restricted to the Amazonian slopes of the Cordillera del Cóndor; *N. ocellatus* has a dorsal color pattern of large dark rings that, in life, have greenish–white centers (smaller ocelli with yellow to orange centers in *N. lindae* sp. nov.); *N. laurae*, known from a single adult male, is smaller (SVL = 19.7 mm in *N. laurae*; male SVL = 23.0–26.3 mm in *N. lindae* sp. nov.), has pointed papillae in Toes I–IV (absent in *N. lindae* sp. nov.), and no teeth on the vomers (present in *N. lindae* sp. nov.; see Taxonomic Remarks). *Nymphargus lindae* sp. nov. differs from its sister species, *N. cochranae*, by the relative size of ocelli (minute *in N. cochranae* and conspicuously larger in *N. lindae* sp. nov.) and female body size (*N. cochranae*, SVL = 27.8–30.3 mm; *N. lindae* sp. nov., SVL 27.2–27.8 mm).

Figure 165. *Nymphargus lindae* sp. nov. in life. (**Left**): Adult female, paratype, QCAZ 41597. (**Right**): Male, also part of the type series. Photos taken at the type locality by Juan M. Guayasamin.

Diagnosis: (1) Dentigerous process of the vomer with two to four teeth; (2) snout truncated to bluntly rounded in lateral profile; truncated in dorsal view; (3) tympanum oriented almost vertically, with slight lateral and posterior inclinations, its diameter about 20%–26% of eye diameter; upper fourth of tympanic annulus obscured by supratympanic fold; tympanic membrane pigmented as surrounding skin; (4) dorsal skin shagreen, with minute spicules in males; (5) venter areolate; pair of enlarged subcloacal warts; (6) white parietal peritoneum covering about anterior 60% of venter (condition P3); white pericardium; translucent peritonea covering intestines, stomach, kidneys, gall and urinary bladders (condition V1); (7) liver tetralobed, covered by transparent peritoneum (condition H0); (8) humeral spines absent; (9) webbing absent between inner fingers, absent or basal between Fingers III and IV (Figure 165); webbing formula III $(2^{4/5}-3^-)$—$(2^{3/4}-3^-)$ IV; (10) feet about two-thirds webbed (Figure 165); webbing formula: I 2^-—$(2^{1/4}-2^{1/3})$II $(1^{1/4}-1^{1/3})$—$(2^{1/4}-2^{1/3})$ III $(1^{1/4}-1^{1/3})$—$(2^{2/3}-3^-)$ IV $(2-2^+)$—2^- V; (11) ulnar and tarsal folds present, low; (12) concealed prepollex; in males, nuptial pad Type I; (13) Finger I about same length as Finger II or slightly shorter (Finger I length 94%–100% of Finger II); (14) disc of Finger III width about 46%–50% of eye diameter; (15) in life, dorsum green with dark lavender to black ocelli enclosing yellow to orange spots (Figure 165); bones green; (16) in preservative, dorsum lavender with black ocelli with white centers; (17) iris white with slight pale yellow hue and thin black reticulation; (18) melanophores usually lacking from dorsal surfaces of fingers and toes, except for few on Toe V; (19) males call from the upper surfaces of leaves; call unknown; (20) fighting behavior unknown; (21) egg deposition site unknown; parental care unknown; (22) tadpoles unknown; (23) small to medium body size; in males, SVL 23.0–26.3 mm ($\overline{X}$ = 25.1 ± 0.848, *n* = 12); in females, SVL 27.2–27.8 mm (*n* = 2).

Description of holotype: Adult male, SVL 25.8 mm. Head slightly wider than long (head length 95% of head width); snout truncated in dorsal and lateral profiles; canthus rostralis indistinct, slightly concave; loreal region slightly concave; lips slightly flared; nostril protuberant, closer to tip of snout than to eye, directed frontolaterally; internarial area barely depressed. Eyes large, directed anterolaterally at an angle ~50°; transverse diameter of disc of Finger III 49% eye diameter. Supratympanic fold low, obscuring upper portion of tympanic annulus; tympanum oriented mostly vertically, but with slight posterolateral inclination; tympanic membrane translucent, pigmented as surrounding skin. Dentigerous process of vomer low, situated transversely between choanae, with three teeth; choanae large, longitudinally rectangular; tongue ovoid, with ventral posterior fourth not attached to floor of mouth and posterior margin notched; vocal slits extending posterolaterally from about the lateral margin of tongue (at about half the length of tongue) to angle of jaws. Humeral spine absent. Low ulnar folds evident on external and internal ventrolateral margins of arm; relative lengths of fingers: III > IV > II > I; webbing absent between Fingers I–III, basal between Fingers III and IV, webbing formula III 3^-–$2^{3/4}$ IV; discs expanded, nearly elliptical; disc pads nearly triangular shaped; subarticular tubercles small, round, simple; low supernumerary tubercles; palmar tubercle elliptical, simple; nuptial pad large (Type I), ovoid, granular, extending from ventrolateral base to dorsal surface of Finger I, covering proximal half of Finger I. Length of tibia 59% SVL; low inner tarsal fold evident; outer tarsal fold absent; foot two-thirds webbed; webbing formula of foot: I 2^-—$2^{1/3}$ II $1^{1/4}$—$2^{1/3}$ III $1^{1/3}$—$2^{2/3}$ IV 3^-—2^- V; discs on toes round to elliptical, lacking papillae; disc on Toe IV narrower that disc on Finger III; disc pads triangular; inner metatarsal tubercle large, ovoid; outer metatarsal tubercle not evident; subarticular tubercles small, round; supernumerary tubercles low. Skin on dorsal surfaces of head, body, and lateral surface of head and flanks shagreen with numerous minute spinules; throat smooth; belly and lower flanks areolate; cloacal opening directed posteriorly at upper level of thighs; cloacal ornamentation absent. Ventral surface of thighs with pair of enlarged tubercles.

Coloration of the holotype in life (Figure 165): Green dorsum with dark lavender to black ocelli enclosing yellow to orange spots. Upper lip white. Anteriorly, about 60% of ventral parietal peritoneum white, posterior portion translucent. Bones green. Iris white with slight pale-yellow hue and thin black reticulation.

Coloration of the holotype in ethanol: Dorsal surfaces of head, body, and limbs lavender with dark lavender ocelli encircling white spots. Anterior 60% of the ventral parietal peritoneum white. Heart covered by white pericardium; translucent peritonea covering gall and urinary bladders; iridophores absent from digestive tract, liver, and kidneys.

Measurements of holotype (mm): *Nymphargus lindae*, adult male, QCAZ 41572: SVL = 25.8, head length = 7.9, head width = 8.3, eye–nostril = 1.9, nostril–snout = 0.8, IOD = 2.4, upper eyelid width = 2.5, eye diameter = 3.5, tympanum diameter = 0.85, tibia = 15.1, foot length = 11.9, radio–ulna length = 5.5, hand length = 8.1, Finger I length = 5.2, Finger II length = 5.4, disc of Finger III width = 1.7.

Biology and ecology: *Nymphargus lindae* sp. nov. is the most abundant species in streams at Miazi Alto, Cordillera del Cóndor. Males were observed calling on leaves and branches about 30–400 cm above a stream. An amplectant pair and two additional males were observed fighting on a branch; males that were not in amplexus continuously approached the amplectant male and tried to push him away with kicks. Parental care is unknown.

Call: Not described.

Tadpole: Not described.

Distribution (Figure 166): *Nymphargus lindae* sp. nov. is only known from the type locality, Miazi Alto (4.25044° S, 78.61356° W; 1200 m), at Cordillera del Cóndor, Provincia de Zamora-Chinchipe, Ecuador. It inhabits the Eastern Foothill Forest ecoregion.

Figure 166. Distribution of *Nymphargus lindae* sp. nov. in Ecuador (yellow dot).

Conservation status: We suggest placing *Nymphargus lindae* sp. nov. in the *Endangered* category, following IUCN criteria B2a, B2(iii). The main threats for the species in Cordillera del Cóndor are habitat destruction and contamination associated with mining activities.

Evolutionary relationships (Figure 136): *Nymphargus lindae* sp. nov. is sister to *N. cochranae*.

Taxonomic Remarks: *Nymphargus lindae* sp. nov. is similar to *N. laurae*. Although the two species are geographically distant (*N. lindae* sp. nov. is endemic to the isolated Cordillera del Cóndor, whereas *N. laurae* is endemic to the lower slopes of Volcán Sumaco, nearby the town of Loreto), differences between these taxa are limited (body size, vomerine teeth, papillae on toes; see Diagnosis). Since *N. laurae* was described based on a single specimen, it is impossible to determine its intraspecific variation. Additional samples from Volcán Sumaco that correspond to the description of *N. laurae* need to be examined to support or refute the validity of the specific status of the two species.

Nymphargus mariae (Duellman and Toft, 1979 [266]; Figures 167–169).

Centrolenella mariae Duellman and Toft, 1979 [266]. Holotype: KU 174713.
Type locality: "Serranía de Sira, ± 1550 m, Departamento Huánuco, Perú".
Cochranella mariae—Ruiz-Carranza and Lynch, 1991 [6]; Cisneros-Heredia and McDiarmid,
    2006 [267].
Centrolene mariae—Duellman and Schulte, 1993 [174].
Nymphargus mariae—Guayasamin, Castroviejo-Fisher, Trueb, Ayarzagüena, Rada, and Vilà,
    2009 [1].
Centrolenella puyoensis Flores and McDiarmid, 1989 [268]. Holotype: MCZ 91187. Type locality:
    "1.0 km W Puyo, Provincia de Pastaza, Ecuador, between 1000–1050 m elevation".
    Synonymy by Cisneros-Heredia & Guayasamin, 2014 [175].
Cochranella puyoensis—Ruiz-Carranza and Lynch, 1991 [6].
Centrolene puyoense—Stuart, Hoffmann, Chanson, Cox, Berridge, Ramani, and Young, 2008 [269].
Nymphargus puyoensis—Guayasamin, Castroviejo-Fisher, Trueb, Ayarzagüena, Rada, and Vilà,
    2009 [1].

Common names: English: María's Glassfrog. Spanish: Rana de Cristal de María.

Etymology: The specific epithet is a patronym for María Koepcke, who devoted her life to biological exploration in Peru [266].

Identification: *Nymphargus mariae* can be distinguished from most glassfrogs by having, in life, a green dorsum with relatively large yellow–green spots, a white parietal peritoneum covering the anterior half of the venter, and by lacking humeral spines and webbing between fingers (Figure 167). *Rulyrana flavopunctata*, *T. midas*, *N. cariticommatus*, *N. humboldti* sp. nov., and *N. siren* have a similar color pattern, but the yellow spots are conspicuously larger in *N. mariae*. Two species, *Sachatamia albomaculata* and *"Centrolene" medemi*, also have large yellow spots on the dorsum, but *S. albomaculata* has webbing between Fingers III and IV and inhabits the Pacific lowlands (*N. mariae* lacks hand webbing and is found on the Amazonian slopes of the Andes and Amazonian lowlands). Also, *"C." medemi* is larger than *N. mariae* (*"C." medemi*: SVL 25.5–30.8 mm in adult males, 34.7–44.3 mm in adult females; *N. mariae*: SVL 23.4–30.2 mm in males, SVL 25.1–32.8 mm in females), has more webbing between Fingers III and IV, and males have humeral spines.

Figure 167. *Nymphargus mariae* in life. Adult female, QCAZ 37923, from a stream tributary of Río Lliquino, Pastaza province, Ecuador. Photos by Martín Bustamante.

Diagnosis: (1) Vomerine teeth present, each vomer with three to four teeth; (2) snout truncated in dorsal aspect, round in lateral profile (Figure 168); (3) tympanum oriented dorsolaterally, with slight posterior inclination, its diameter about 32%–35% of eye diameter; tympanic annulus visible, low supratympanic fold evident, tympanic membrane partially pigmented and clearly differentiated from surrounding skin; (4) dorsum shagreen with spiculated flat warts corresponding to light spots; (5) venter areolate; pair of enlarged subcloacal warts (Figure 6); (6) white lining on the anterior 40% of the ventral parietal peritoneum (condition P2); translucent peritonea covering intestines, stomach, gall bladder, and urinary bladder (condition V1); (7) liver tetralobed, covered by transparent peritoneum (condition H0); (8) humeral spines absent; (9) webbing absent between Fingers I, II, and III, and reduced or absent between outer fingers, III (3^--3)—$(2^{1/3}-2^{2/3})$ IV (Figure 168); (10) webbing formula on foot: I $(2-2^-)$—$(2^{1/3}-2^{1/2})$ II $1^{1/2}$—$2-2^{2/3}$ III $(1^{1/3}-1^{2/3})$—$2^{2/3}$ IV (2^+-3^-)—$(1^{2/3}-2^-)$ V; (11) ulnar fold low; inner and outer tarsal folds low and thin; (12) concealed prepollex; nuptial pad Type I; (13) Finger I about same length as Finger II; (14) disc of Finger III narrow, about 34%–38% of eye diameter; (15) in life, dorsum green to dark green with minute and relatively large yellow–green spots; bones green; (16) in preservative, dorsum dark lavender with minute and large cream spots; (17) iris greyish cream with transverse brown bar and fine dark grey reticulations; pale yellow circumpupilar ring; (18) melanophores absent from Fingers I and II, and Toes I, II, and III; few present on Finger III; numerous on Finger IV and Toes IV and V; (19) males call from the upper side of leaves; call consists of a single note with a duration of 0.015–0.018 s; time between calls is 1.733–1.940 s; dominant frequency is at 3234–4299 Hz; (20) fighting behavior unknown; (21) egg deposition site and parental care unknown; (22) tadpoles unknown; (23) medium body size; in males, SVL 22.4–31.7 mm ($\overline{X}$ = 26.3 ± 2.866, n = 7); in females, SVL 25.1–30.1 mm ($\overline{X}$ = 28.3 ± 2.347, n = 4).

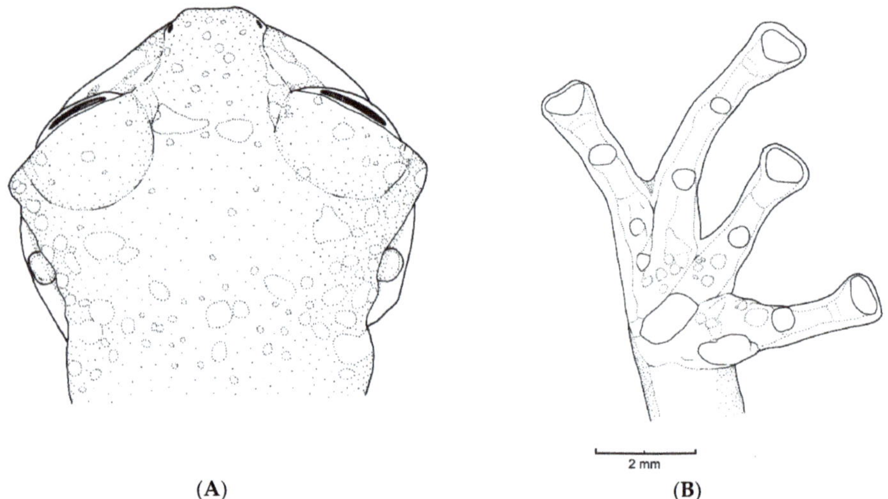

(**A**) (**B**)

Figure 168. *Nymphargus mariae*, holotype, female, KU 174713. (**A**) Head in lateral view. (**B**) Hand in ventral view. Illustrations by Juan M. Guayasamin.

Color in life (Figure 167): Dorsal surfaces green to dark green, with minute and large yellow spots. Anterior half of venter white, posterior part translucent. Iris greyish cream with transverse brown bar and fine dark grey reticulations; pale yellow circumpupilar ring.

Color in ethanol: Dorsum lavender to dark lavender, with minute and large cream spots; iridophores on anterior 40% of the ventral parietal peritoneum; white pericardium; clear hepatic peritoneum; translucent peritoneum covering intestines, stomach, gall bladder, and urinary bladder.

Biology and ecology: Nymphargus mariae is active during the night. Species found at the type locality of N. puyoensis included Boana cinerascens, Pristimantis conspicillatus, P. diadematus, P. lacrimosus, P. lathanites, P. martiae, and P. quaquaversus [268]. Parental care is unknown.

Call: Males call from the upper surfaces of leaves. The following call description is based on a recording of one male of *Nymphargus mariae* made by Diego Paucar on 22 February 2008 at stream tributary of Río Lliquino (1.72553 S, 78.98058 W; 400 m), Provincia de Pastaza, Ecuador. The call consists of a single non-frequency modulated note. Each call has a duration of 0.015–0.018 s ($n = 3$); time between calls is 1.733–1.940 s. Dominant frequency is at 3234–4299 Hz; a first harmonic is visible at 7178–7928 Hz, and a second harmonic is visible at 10767–11714 Hz.

Tadpole: Not described.

Distribution (Figure 169): *Nymphargus mariae* is known from several localities on the Amazonian slopes of the Cordillera Oriental of the Ecuadorian Andes at elevations between 400 and 1078 m ([176,266–268], this work). In Ecuador, the potential distribution of the species is 28,298 km^2.

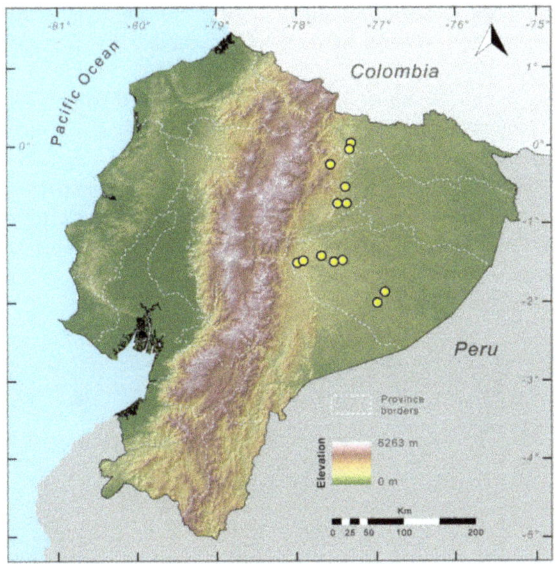

Figure 169. Distribution of *Nymphargus mariae* in Ecuador (yellow spots).

Conservation status: Globally, *Nymphargus mariae* is listed as *Least Concern* by the IUCN [270]. We agree with this conservation status.

Evolutionary relationships (Figure 136): *Nymphargus mariae* is sister to *N. colomai* sp. nov.

Specimens examined: *Nymphargus mariae:* Ecuador: *Provincia de Pastaza:* 1 km W Puyo (1.4833 S, 78.0 W; 1000–1050 m), MCZ 91187; Andean foothills in the Upper Bobonaza River Basin (ca. 2.0 S, 77.0 W), USNM 291298; stream tributary of Río Lliquino (1.72553 S, 78.98058 W; 400 m), QCAZ 37932; near Villano (1.47445 S, 77.53529 W; 440 m), QCAZ 39293; Sacha Yacu (1.39519° S, 77.72946° W; 1078 m), MZUTI 183. *Provincia de Napo:* ca. 45 km E of Narupa (ca. 0.729 S, 77.374 W; ca. 800 m), on the Hollín–Loreto road, DFCH-USFQ D285; *Provincia de Orellana:* Reserva Río Bigal (0.52525 S, 77.41785 W; 930 m), QCAZ 48529; Río Huataracu (ca. 0.729 S, 77.374; ca. 800 m), ca. 70 km E of Hollín, on the Hollín-Loreto road, QCAZ 7104, 7499; *Provincia de Sucumbíos:* Lumbaqui (ca. 0.05 N, 77.333 W; ca. 500 m). Peru: *Departamento de Huánuco:* Serranía de Sira (ca. 9.367 S, 74.75 W; 1550 m), KU 174713.

Localities from the literature: *Nymphargus mariae:* Ecuador: *Provincia de Pastaza:* Río Pucayacu (1.942 S, 77.042 W) [253]; Conambo (1.86197 S, 76.906 W; 337 m) [176]; 1 km W of Puyo (1.493 S, 78.026 W; 1000–1050 m) [253]; Río Lliquino (1.41486 S, 77.54047 W; 380 m) [176]. *Provincia de Napo:* Río Putuyacu, 45 km E of Narupa (0.734 S, 77.49 W; 800 m) [176,253]. *Provincia de Orellana:* Río Huataraco (1.46986 S, 77.92477 W; 347 m) [176]. *Provincia de Sucumbíos:* Río Verde (0.23786 S, 77.576 W; 726 m) [176]; Lumbaqui (0.04675 S, 77. 34358 W; 515 m) [176].

Nymphargus manduriacu Guayasamin, Cisneros-Heredia, Vieira, Kohn, Gavilanes, Lynch, Hamilton, and Maynard, 2019 [21] (Figures 170–173).

Nymphargus manduriacu Guayasamin, Cisneros-Heredia, Vieira, Kohn, Gavilanes, Lynch, Hamilton, and Maynard, 2019 [21]. Holotype: ZSFQ 0466, by original designation.
Type locality: "Reserva Río Manduriacu (0.310755° N, 78.8569° W; 1,215 m), Provincia de Imbabura, República del Ecuador".

Common names: English: Manduriacu glassfrog. Spanish: Rana de Cristal de Manduriacu.

Etymology: The specific epithet *"manduriacu"* refers to the type locality of the species, Río Manduriacu Reserve, Ecuador, a conservation area managed by Fundación EcoMinga (https://ecomingafoundation.wordpress.com/) [21].

Identification: *Nymphargus manduriacu* is easily differentiated from most glassfrogs by lacking webbing between inner fingers (Figure 171) and having, in life, a greyish–green dorsum with numerous yellow spots, which sometimes are surrounded by an ill-defined black ring (i.e., false ocelli; Figure 170). On the Pacific slopes of the Ecuadorian and Colombian Andes, similar species include the following: *N. buenaventura*, *N. ignotus*, *N. spilotus*, and *N. luminosus*. *Nymphargus buenaventura* has a light green dorsum with diffuse pale yellow spots; *N. ignotus* exhibits a pale tan to olive–brown dorsum with black ocelli surrounding orange or yellow spots; *N. luminosus* has a green dorsum with numerous yellow spots; finally, *N. spilotus* has an olive green back with small yellow spots [21].

Figure 170. *Nymphargus manduriacu* in life from Reserva Río Manduriacu, Imbabura province. (A–C) Adult male, ZSFQ 0466. (D–F) Adult female, ZSFQ 0462. Photos by Jose Vieira/Tropical Herping. Obtained from Guayasamin et al. [194].

Diagnosis: *Nymphargus manduriacu* exhibits the following combination of traits: (1) Dentigerous process of vomer low or absent, lacking vomerine teeth; (2) snout truncated in dorsal view, and truncated to slight rounded in lateral view; (3) tympanic annulus barely evident, lower three-fourths visible, tympanic membrane colored as dorsal skin, supratympanic fold present; (4) dorsal skin shagreen, with microspicules in adult males; (5) ventral skin granular, subcloacal area with two large subcloacal warts; (6) parietal peritoneum white, iridophores covering one-third to one-half parietal peritoneum (conditions P2 or P3); pericardium white (i.e., covered by iridophores), all other visceral peritonea clear (condition V1); (7) liver lobed and hepatic peritoneum clear (lacking iridophore layer, condition H0); (8) adult males lacking humeral spines; (9) webbing between Fingers I, II, and III absent, basal between Fingers III and IV (Figure 171); (10) toe webbing basal between Toes I and II, III $1\frac{1}{2}$–($2\frac{1}{2}$–3⁻) III ($1^{1/3}$–$1\frac{1}{2}$)–(3–3⁻) IV (3–3⁻)–($1\frac{1}{2}$–2⁻) V; (11) lacking dermal ornamentations in the form of tubercles, folds, or fringes on hands, arms, feet, or legs; (12) nuptial excrescences Type I and VI; concealed prepollex; (13) Finger I slightly longer than Finger II; (14) diameter of eye larger than width of disc on Finger III; (15) color in life, greyish green to olive green with yellow spots, which, sometimes, are surrounded an ill-defined black ring (i.e., false ocelli); bones green; (16) color in preservative, lavender dorsum with cream spots; (17) iris coloration in life: Light grey with thin grey reticulations and pale yellow hue around pupil; (18) melanophores present and abundant along Fingers III and IV, less dense on Finger II, and rarely present on Finger I; furthermore, present and abundant along Toes IV and V, less dense on Toe III, only at the base of Toes I and II; (19) males call from upper side of leaves; advertisement call is a high-pitched "chirp", with a single, pulsed note with a duration of 0.093–0.118 s ($\overline{X}$ = 0.10 ± 0.007; *n* = 10) and a dominant frequency at 4052–4447 Hz ($\overline{X}$ = 4267.7 ± 118.3); (20) fighting behavior unknown; (21) egg masses deposited on upper side of leaves, clutch size 15–32 (*n* = 4); no long-term parental care provided by either males or females; (22) tadpoles undescribed; (23) SVL in adult males 24.0–25.7 mm (*n* = 3), and in an adult female 28.8 mm.

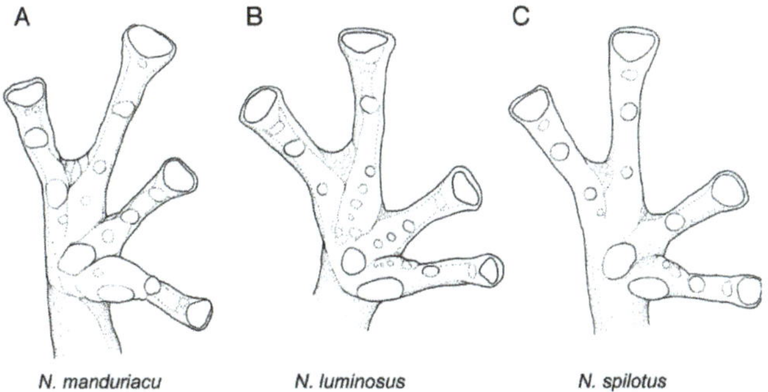

Figure 171. Hand webbing of *Nymphargus manduriacu* and similar species. (**A**) *N. manduriacu*, ZSFQ 0463, adult male, paratype. (**B**) *N. luminosus*, ICN 15930, adult female, holotype. (**C**) *N. spilotus*, ICN 35255, adult female, holotype. Modified from Guayasamin et al. [194].

Color in life (Figure 170): Dorsal surfaces greyish green to olive green with yellow spots, with melanophores concentrated around yellow spots, sometimes looking like false ocelli. Upper lip unpigmented. Inner fingers and toes with yellowish hue. Anterior half of ventral parietal peritoneum white, posterior portion translucent. Green bones. Iris light grey with thin, dark grey reticulations and pale-yellow hue around pupil [21].

Color in ethanol: Dorsal surfaces of body and limbs grey lavender with small white spots. Parietal peritoneum white, iridophores covering one-third to one-half parietal peritoneum. Heart white (covered by iridophores); all other visceral peritonea unpigmented [21].

Biology and ecology: *Nymphargus manduriacu* has only been found at Río Manduriacu Reserve and, although the reserve has been visited several times, *N. manduriacu* was only regularly detected during February 2018, with the site experiencing particularly heavy rains on a daily basis [21].

Call (Figure 172): The advertising call was described by Guayasamin et al. [21], as follows. Each call is a high-pitched "chirp" that consists of a single note with a duration of 0.093–0.118 s ($\overline{X}$ = 0.10 ± 0.007; *n* = 10). Notes are clearly pulsed (8–12 pulses per note; $\overline{X}$ = 10.33 ± 1.366). In each call, there is a slight increase in the dominant frequency with time; the dominant frequency is at 4052–4447 Hz ($\overline{X}$ = 4268 ± 118.3). Time between calls is 3.9–8.6 s ($\overline{X}$ = 5.72 ± 1.82).

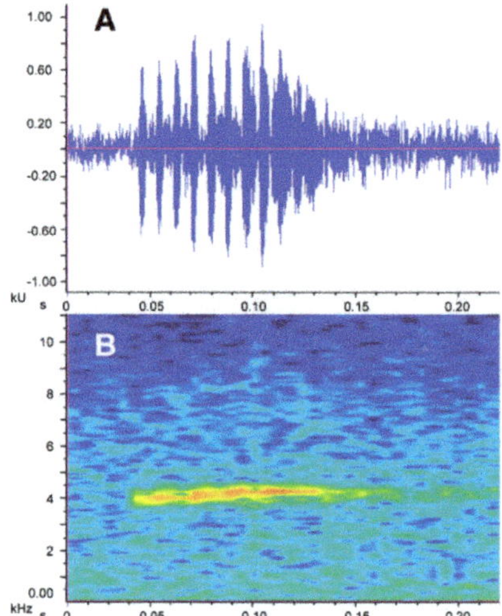

Figure 172. Call of *Nymphargus manduriacu*, LBE-C-042, from Reserva Río Manduriacu. (**A**) Oscillogram. (**B**) Audio-spectrogram. Obtained from Guayasamin et al. [21].

Tadpole: Not described.

Distribution (Figure 173): *Nymphargus manduriacu* is known from a few streams within the Río Manduriacu Reserve (0.31° N, 78.85° W; 1215–1242 m), Imbabura province, Ecuador [21].

Figure 173. Distribution of *Nymphargus manduriacu* in Ecuador (yellow dot).

Conservation status: Guayasamin et al. [21] suggest placing the species in the *Critically Endangered*. At Río Manduriacu Reserve (the only known locality of the species), mining has become one the most dangerous threat to biodiversity, especially to species with restricted distributions.

Evolutionary relationships: *N. manduriacu* was inferred as sister to *N. balionotus* [21].

Specimens examined: <u>Ecuador</u>: *Imbabura province:* Reserva Río Manduriacu (0.310° N, 78.857° W; 1215–1230 m), ZSFQ 0462–66 (type series).

Nymphargus megacheirus (Lynch and Duellman, 1973 [22]; Figures 174–176).

Centrolenella megacheira Lynch and Duellman, 1973 [22]. Holotype: KU 143245.
Type locality: "16.5 km NNE of Santa Rosa, 1700 m, on Quito–Lago Agrio road, Provincia Napo, Ecuador".
Cochranella megacheira—Ruiz-Carranza and Lynch, 1991 [6].
Nymphargus megacheirus—Cisneros-Heredia and McDiarmid, 2007 [17].

Common names: English: Large-handed Glassfrog. Spanish: Rana de Cristal de manos grandes.

Etymology: The specific epithet is from the Greek words *megas*, meaning large, and *cheiros*, meaning hand; the name is used to refer to the exceedingly large hands of the species [22].

Identification: *Nymphargus megacheirus* can be distinguished from other glassfrogs by its green dorsum with small blue to black spots (Figure 174), relatively large size (adult males, SVL 26.8–31.5 mm; adult females, SVL 31.2–32.9 mm), basal or no webbing among fingers (Figure 175), and lack of humeral spines. On the Amazonian slopes of the Ecuadorian Andes, only *N. cochranae* has similar characteristics; however, *N. cochranae* has small dark ocelli enclosing orange dots on the dorsum (Figure 140), whereas *N. megacheirus* has solid dark spots (Figure 174). *Nymphargus megacheirus* has a similar dorsal coloration and hand-webbing pattern as *N. garciae*, but the two species have allopatric distributions with *N. garciae* inhabiting higher elevations (1900–2700 m) than *N. megacheirus* (1300–1740 m). Additionally, *N. megacheirus* is slightly larger than *N. garciae* (*N. garciae*, SVL 25.1–29.9 mm in adult males: SVL 25.9–28.4 mm in adult females).

Figure 174. *Nymphargus megacheirus* in life. Adult male, holotype, KU 143245. Photo by W. E. Duellman.

Diagnosis: (1) Vomers lacking teeth; (2) snout truncated in dorsal and lateral profiles (Figure 175); (3) tympanum relatively small, its diameter 18.8%–25.4% eye diameter, dorsal border of tympanic annulus covered by supratympanic fold, tympanic membrane pigmented as surrounding skin; (4) dorsal skin of males shagreen to pustular, shagreen in females; numerous spicules present in males; in females, spicules present only on head, tympanic region, and limbs; (5) skin of venter areolate; pair of enlarged subcloacal warts; (6) anterior two-thirds to three-fourths of venter covered by white parietal peritoneum, posterior portion transparent (condition P3); white pericardium; translucent peritoneum covering intestines, stomach, testes, kidneys, gall bladder, and urinary bladder (condition V1); (7) liver tetralobed, two large ventral lobes partially covering two smaller lobes; hepatic peritoneum transparent (condition H0); (8) humeral spines absent; (9) hand webbing absent between inner finger, absent or basal between outer fingers (Figure 175), webbing formula: III $(2^{1/2}$–3)—$(2^{1/2}$–3) IV; (10) foot about one-half webbed: I

$(2–2^-)$—$(2^+–2^{1/2})$ II $(1–1^{2/3})$—$(2^{1/2}–2^{2/3})$ III $(1^+–1^{3/4})$—$(2^{1/2}–3^-)$ IV $(2^{1/2}–3^-)$—$(1^{1/2}–2^-)$ V; (11) ulnar fold conspicuous; tarsal folds low; (12) concealed prepollex; in males, nuptial pad Type I; (13) Finger I usually slightly shorter than Finger II (Finger I 88.9%–101.3% length of Finger II); (14) disc of Finger III relatively large, 49.7%–59.0% eye diameter; (15) in life, dorsum green with blue to black dots (Figure 174); bones green; (16) in ethanol, dorsal surfaces lavender with small dark purple dots; (17) iris greyish bronze with thin black reticulation; (18) dorsal surfaces of fingers and toes lacking melanophores; (19) males call from upper side of leaves overhanging streams; calls unknown; (20) fighting behavior unknown; (21) eggs deposition site and parental care unknown; (22) tadpoles unknown; (23) medium body size; in adult males, SVL 26.8–31.5 mm ($\overline{X}$ = 28.3 ± 0.902, *n* = 29); in adult females, SVL 31.2–32.9 mm ($\overline{X}$ = 32.3 ± 0.7805, *n* = 4).

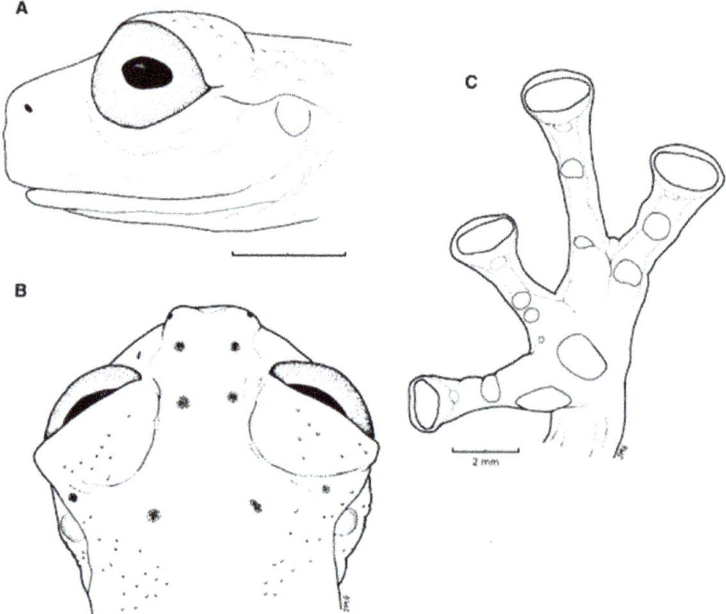

Figure 175. *Nymphargus megacheirus*, KU 143269. (**A**) Head in lateral view. (**B**) Head in dorsal view. (**C**) Hand in ventral view. Illustrations by Juan M. Guayasamin.

Color in ethanol: Dorsal surfaces of head, body, forelimbs, and hind limbs lavender with small, round, black spots [22]. White lining covering pericardium and nearly two-thirds of anterior ventral parietal peritoneum. Liver, intestines, stomach, testes, kidneys, gall bladder, and urinary bladder lack iridophores.

Biology and ecology: *Nymphargus megacheirus* is active at night. During the reproductive season, males were calling from the upper surfaces of leaves overhanging fast-moving streams in cloud forest at the type locality (16.5 km NNE of Santa Rosa) in October 1971; *Espadarana audax*, *Centrolene pipilata*, and *N. siren* were also found there. At the Río Azuela, *N. anomalus*, *N. siren*, *Hyalinobatrachium pellucidum*, and *C. pipilata* occurred along the same streams with *Nymphargus megacheirus* [22]. Parental care is unknown.

Call: Not described.

Tadpole: Not described.

Distribution (Figure 176): *Nymphargus megacheirus* is endemic to the Amazonian slope of the Andes of Ecuador and Colombia at elevations between 1300 and 1750 m ([22,101], this work). In Ecuador, this species has been reported from the provinces of Napo and Sucumbíos (Specimens Examined). The habitat of the species in Ecuador is within the Eastern Montane Forest region.

Figure 176. Distribution of *Nymphargus megacheirus* in Ecuador (yellow dots).

Color in life (Figure 174): Dorsal surfaces green with small blue to black spots. Ventral surfaces mostly white, except for translucent posterior third of venter. Upper lip and ulnar folds white. Bones green. Iris greyish bronze [22].

Conservation status: Globally listed as *Endangered* by the IUCN [271]. The last records of *Nymphargus megacheirus* correspond to specimens collected at Río Azuela and Río Salado on 24 February 1979 (USNM 286700–01, RWM, pers. obs. [17]). Surveys at Río Azuela have failed to find additional individuals [91]. Then, considering its limited distribution and lack of recent records, we suggest that the species should be considered as *Critically Endangered*.

Evolutionary relationships (Figure 136): With the current taxon and gene sampling, *Nymphargus megacheirus* is sister to *N. anomalus*.

Specimens examined: *Nymphargus megacheirus:* <u>Ecuador</u>: *Provincia de Napo:* 16.5 km NNE Santa Rosa (0.21861 S, 77.7319 W, 1700 m), KU 143245–72; 14.7 km (by road) NE of Río Salado (0.12889 S, 77.6083 W, 1300 m), USNM 286701; 2 km SSW Río Reventador (0.1 S, 77.6 W, 1700 m), KU 164614; *Provincia de Sucumbíos:* Río Azuela (0.1167 S, 77.6167 W, 1740 m), KU 143273–77, 166329; Rio Azuela, where river crosses Quito road (0.1166 S, 77.6166 W, 1700 m.), USNM 286700. <u>Colombia</u>: *Departamento de Putumayo:* 10.3 km W El Pepino (1.05 N, 76.9559 W, 1300 m), KU 169664–65.

Nymphargus posadae (Ruiz-Carranza and Lynch, 1995 [255]; Figures 177–179).

Cochranella posadae Ruiz-Carranza and Lynch, 1995 [255]. Holotype: ICN 11307.

<u>Type locality</u>: "Departamento de Cauca, municipio de Inzá, Km 61 carretera Popayán a Inzá, vertiente oriental Cordillera Central, 2°34′ latitud, 76°4′ W de Greenwich, 2800 m", Colombia.

Nymphargus posadae—Cisneros-Heredia and McDiarmid 2007 [17].

Common names: English: Posada's Glassfrog. Spanish: Rana de Cristal de Posada.

Etymology: The specific epithet honors Dr. Andrés Posada Arango, for his work in the fields of zoology, botany, education, and conservation biology in Colombia [255].

Identification: Among glassfrogs that inhabit the Amazonian slopes of the Ecuadorian Andes, *Nymphargus posadae* is unique by having a green dorsum with small greenish–white warts (Figure 177), a slightly sloping snout in lateral profile, and by lacking webbing between fingers. The only species that could be confused with *N. posadae* is *Centrolene buckleyi*, which has humeral spines in adult males (humeral spines absent in *N. posadae*).

Figure 177. *Nymphargus posadae* in life. Adult male from Yanayacu Biological Station, 2100 m, Napo province, Ecuador, QCAZ 25090. Photos by Martín Bustamante.

Diagnosis: (1) Vomers with edentate dentigerous process; (2) snout truncated to round in dorsal aspect, and truncated to slightly sloping in lateral profile (Figure 178); (3) tympanum almost indistinguishable; small when visible, its diameter 21.2%–26.5% of eye diameter; tympanic membrane not differentiated from surrounding skin; supratympanic fold present; (4) dorsal skin covered with numerous small warts and some scattered larger warts; no spicules visible; (5) ventral skin areolate; pair of enlarged subcloacal warts; (6) white parietal peritoneum covering anterior 50%–60% of venter (condition P2); white pericardium; no iridophores in peritonea covering intestines, stomach, and kidneys; translucent peritoneum around gall and urinary bladders (condition V1); (7) liver lobate, covered by transparent peritoneum (condition H0); (8) humeral spines absent; (9) webbing between fingers absent or greatly reduced; webbing formula: III (2^{3/4}–3⁻)—(2^{2/3}–2^{3/4}) IV; (10) feet about two-thirds webbed; webbing formula: I (2⁻–2)—(2⁺–2^{1/4}) II (1^{1/3}–1^{1/2})—(2^{1/3}–2^{3/4}) III 1^{1/2}—(2^{2/3}–2^{3/4}) IV (2^{3/4}–3)—2⁻ V; (11) ulnar and tarsal folds low or absent; (12) concealed prepollex; in males, nuptial pad Type I; (13) Finger II longer than Finger I (Finger I length 92.2%–95.7% of Finger II); (14) disc width of Finger III about 49.3%–52.9% of eye diameter; (15) in life, dorsum green with small greenish–white warts; bones green; (16) in preservative, dorsum lavender with small white dots; (17) iris white with thin dark grey reticulations; (18) dorsal surfaces of fingers and toes lacking melanophores, except for proximal portion of Toes IV and V; (19) males call from upper surfaces of leaves; calls unknown; (20) fighting behavior unknown; (21) eggs deposition site and parental care unknown; (22) tadpole unknown; (23) large body size; male SVL 30.7–34.1 mm ($\overline{X}$ = 32.3, *n* = 6); female SVL 30.2–33.3 mm ($\overline{X}$ = 31.4, *n* = 4).

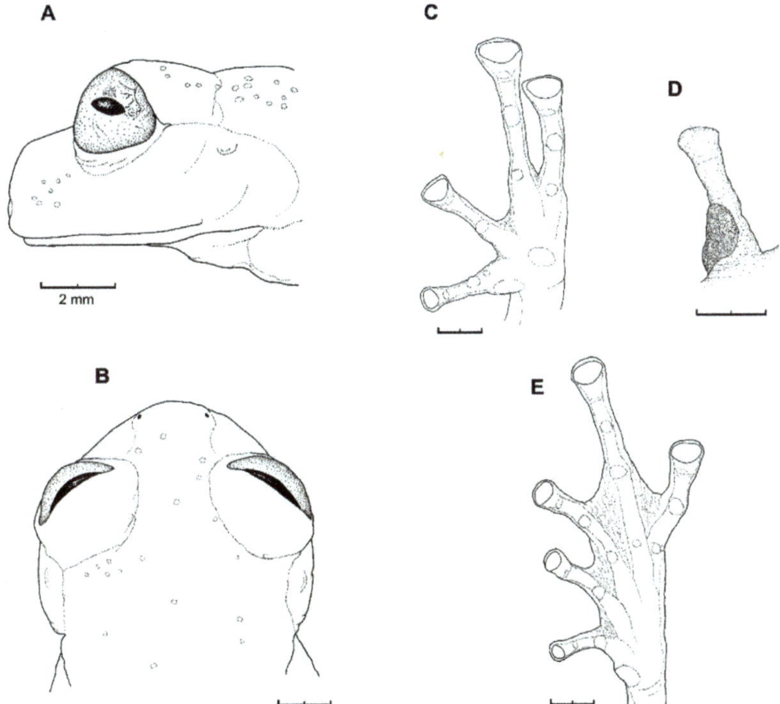

Figure 178. *Nymphargus posadae*, adult males from Yanayacu Biological Station, Napo province, Ecuador. (**A**) Head in lateral view, QCAZ 26023. (**B**) Head in dorsal view, QCAZ 26023. (**C**) Hand in ventral view, QCAZ 25090. (**D**) Finger I in dorsal view, QCAZ 26023. (**E**) Foot in ventral view, QCAZ 26023. Modified from Guayasamin et al. [20].

Color in life (Figure 177): The following description corresponds to individuals from Yanayacu Biological Station [20]. Dorsal surfaces of head, body, and limbs bright green with small, scattered, greenish–white warts. Upper lip white; lower lip with thin white border. Ventrolateral border of arm, Finger IV, tarsus, and Toe V white. Cloacal region with numerous small, white warts. White parietal peritoneum covering about anterior half of venter. Iris white with thin dark grey reticulations.

Color in ethanol: Dorsal surfaces of head, body, and limbs lavender with some larger warts being bluish white. Upper lip white; thin white line evident on lower lip. Ventrolateral border of arm, Finger IV, tarsus, and Toe V white. Ventral surfaces of forearm and tarsus completely covered with white in two specimens (QCAZ 25090 and 26022). Dorsally, Fingers I and II and Toes I, II, and III unpigmented: Some pigmentation visible on Fingers III and IV and Toes IV and V. Cloacal region with several white warts. In adult males, nuptial pad cream (Type I). Parietal peritoneum white anteriorly, covering approximately 50%–60% of venter. Pericardium silver white. No iridophores on the hepatic peritoneum, digestive tract, or kidneys ([20], this work).

Variation: Males from Yanayacu Biological Station are smaller (SVL = 30.7–31.9, $n = 3$) than those from Colombian localities (SVL = 32.7–34.1 mm, $n = 3$) [20,255].

Biology and ecology: In Colombia, the species was observed on vegetation and rocks along a creek [255]. Carranza and Lynch [255] reported that the eggs have a pigmented animal pole (dark brown) and unpigmented vegetal pole (cream). Unfortunately, they did not mention where the eggs were deposited. In Ecuador, three individuals were collected during three years of inventory work at

Yanayacu; all frogs were found calling on the same night (12 June 2003), on ferns 110–220 cm above a stream. Males call from the upper surfaces of leaves [20]. Parental care is unknown.

Call: Not described.

Tadpole: Not described.

Distribution (Figure 179): In Colombia, *Nymphargus posadae* is known from the Caldas, Cauca, and Huila departments on the eastern flank of the Central Cordillera of the Andes, between 1100 and 2800 m [255]. In Ecuador, the species has been reported from localities on the Amazonian slopes of the Andes at elevations of 1750–2100 m ([17,20], this work). *Nymphargus posadae* is also present in Peru (Cordillera del Cóndor) [272].

Figure 179. Distribution of *Nymphargus posadae* in Ecuador (yellow dots).

Conservation status: Globally, *Nymphargus posadae* is considered as *Least Concern* by the IUCN [273]. The species has a relatively large distribution and it is found within several protected areas. Thus, we agree with the current conservation assessment.

Evolutionary relationships (Figure 136): Given the current taxon and gene sampling, *Nymphargus posadae* is sister to *N. pluvialis*.

Specimens examined: *Nymphargus posadae:* <u>Colombia</u>: *Departamento de Huila:* 6.2 km NW of San José de Isnos, 1940 m. <u>Ecuador</u>: *Provincia de Napo:* Yanayacu Biological Station (0°41′ S, 77°53′ W; 2100 m), QCAZ 25090, 26022–23. *Provincia de Zamora Chinchipe:* tributary of Río Jambue, ca. 15 km S from Zamora (ca. 4°14′ S, 78°57′ W; 1750 m). *Provincia de Sucumbíos:* Río Chingual, ca. 3 km N of Sebundoy, ca. 20 km N of La Bonita (ca. 0°26′ S, 77°32′ W; 1890 m), USNM 288464-65. Peru: Cordillera del Cóndor (ca. 05°25′16.5″ S, 78°35′23.2″ W; 1890 m).

Nymphargus siren (Lynch and Duellman, 1973 [22]; Figures 180–182).

Centrolenella siren Lynch and Duellman, 1973 [22]. Holotype: KU 146610.
Type locality: "small tributary of the Río Salado, about 1 km upstream from the Río Coca,
 1410 m, Provincia Napo, Ecuador."
Cochranella siren—Ruiz-Carranza and Lynch, 1991 [6].
emphNymphargus siren—Cisneros-Heredia and McDiarmid, 2007 [17].

Common names: English: Siren Glassfrog. Spanish: Rana de Cristal sirena.

Etymology: In Greek mythology, sirens were bird-women, who by their sweet singing enticed seafarers to destruction; the name is used in allusion to the calls of these frogs that entice biologists to the nocturnal perils of streams [22].

Identification: Among glassfrogs that inhabit the Amazonian slopes of the Ecuadorian Andes, *Nymphargus siren* (Figure 180) is distinguished by having a green dorsum with small yellow spots, a partially white venter, small size (male SVL < 22.6 mm; female SVL < 23.3 mm), and lacking humeral spines. Similar species from eastern Ecuador include *Espadarana audax*, *Nymphargus humboldti* sp. nov., *Rulyrana flavopunctata*, and *Teratohyla midas*. Differences among these species include the presence of humeral spines in adult males of *E. audax* (humeral spines absent in *N. siren*), moderate webbing between Fingers III and IV in *E. audax* and *R. flavopunctata* (webbing absent in *N. siren* and basal in *T. midas*), and the white peritoneal covering of the digestive tract of *T. midas* (digestive tract opaque or translucent in *N. siren*, *E. audax*, and *R. flavopunctata*). *Nymphargus humboldti* sp. nov. and *N. siren* have non-overlapping body sizes (*N. humboldti* sp. nov., male SVL = 23.3–25.2 mm; female SVL = 24.3–25.9 mm).

Figure 180. *Nymphargus siren* in life. Individual (MZUTI 774) from creek on the Oyacachi–El Chaco trail, 1878 m, Napo province, Ecuador. Photos by Luis A. Coloma.

Diagnosis: (1) Vomers with edentate dentigerous process; (2) snout usually truncated in dorsal aspect; truncated to slightly protruding in lateral profile (Figure 181), (3) tympanum oriented almost vertically, with slight lateral and posterior inclinations, its diameter about 20.5%–30.5% of eye diameter; upper fourth of tympanic annulus obscured by supratympanic fold; tympanic membrane clearly differentiated from surrounding skin; (4) dorsal skin shagreen, usually with spicules in males; (5) pair of enlarged subcloacal warts; (6) white parietal peritoneum covering anterior half of venter (condition P2); white pericardium; translucent to opaque peritonea covering intestines, stomach, and kidneys; translucent peritoneum around gall and urinary bladders (condition V1); (7) liver lobate, covered by transparent peritoneum (condition H0); (8) humeral spines absent; (9) webbing absent between inner fingers, absent or basal between outer fingers (Figure 181); webbing formula IV $(2^{2/3}–3)$—$(2^{1/4}–2^{3/4})$ V; (10) feet about two-thirds webbed; webbing formula: I $(2^{-}–2)$—$(2^{+}–2^{1/2})$ II $(1^{1/4}–1^{1/2})$—$(2^{1/2}–3^{-})$ III $(1^{+}–2^{-})$—$(2^{3/4}–3^{-})$ IV $(3^{-}–3)$—$(2^{-}–2)$ V; (11) ulnar fold absent; external tarsal fold absent; low inner tarsal fold evident; (12) concealed prepollex; in males, nuptial pad Type I; (13) Finger II slightly longer than Finger I (Finger I length 91.5%–99.6% of Finger II); (14) disc of Finger III width about 43.3%–58.5% of eye diameter; (15) in life, dorsum green with small yellow spots (Figure 180); bones green; (16) in preservative, dorsum lavender with small white spots; (17) iris, in life, whitish cream, with a yellow hue around pupil and fine, dark grey reticulations; (18) melanophores absent from dorsal surfaces of fingers and toes, except for few on Toe V; (19) males call from upper surface of leaves, call unknown; (20) fighting behavior unknown; (21) egg deposition site and parental care unknown; (22) tadpoles unknown; (23) minute body size; males, SVL 19.8–22.6 mm ($\overline{X}$ = 20.9 ± 0.931, n = 24); females, SVL 22.5–23.3 mm (n = 2).

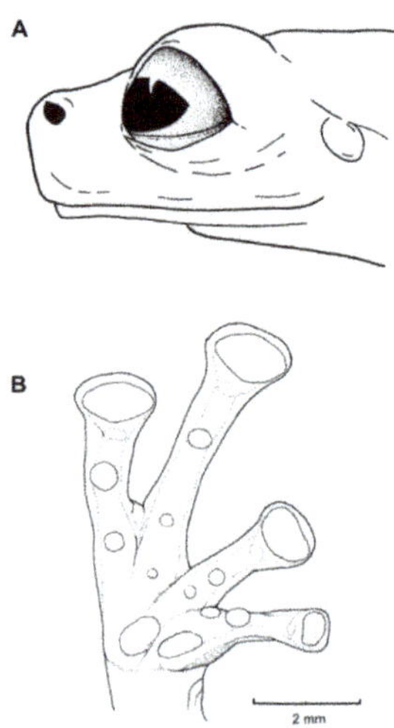

Figure 181. *Nymphargus siren*, KU 146610. (**A**) Head in lateral view. (**B**) Hand in ventral view. Illustrations by Juan M. Guayasamin.

Color in life (Figure 180): Green dorsum with small yellow spots, which are narrowly bordered by black in some individuals [22]. Anterior half of venter white, posterior half transparent. Bones green. Iris whitish cream, with yellow hue around pupil and fine, dark grey reticulations.

Color in ethanol: Dorsal surfaces of head, body, and limbs light to dark lavender with small white spots. Anterior half of ventral parietal peritoneum white; posterior half translucent. Translucent peritonea covering gall and urinary bladders. Iridophores absent from digestive tract, liver, and kidneys.

Variation: Spicules are absent in a few males (KU 164636, 297290). As noted, the presence of spicules may be a reflection of reproductive activity. Two males (KU 178199, 297292) lack yellow spots on the dorsum.

Biology and ecology: Relatively large numbers of individuals of *Nymphargus siren*, including calling males, were observed at the type locality on 7 April 1972 (13 individuals) and again on 18 March 1975 (15 individuals; WED's field notes). Many specimens were also noted at 3.2 km NNE Oritoyacu on 15 July 1977 (21 individuals; John D. Lynch's field notes). Males call from the upper surfaces of leaves (WED field notes, 7 April 1972). Parental care is unknown.

Call (Figure 182): We analyzed 14 notes from two individuals (MZUTI 765, 775). The typical advertisement call is short and is composed by a single note. Note duration is 20–42 (mean = 24, SD = 7) ms. Notes are generally pulsed and have one to three (mean = 2.2, SD = 0.6) amplitude peaks, where the first peak is more pronounced that the others. Notes have their peak amplitude in the first 50% of the note (relative peak time: Range = 0.0595–0.1751, mean = 0.123, SD = 0.032). Pulses within a note have a rate of 24–130 (mean = 86, SD = 23) pulses per second. The dominant frequency of a note measured at peak amplitude is 4737–6029 (mean = 4977, SD = 323) Hz and is contained within the fundamental frequency. The fundamental frequency has a lower limit of 4651–5943 (mean = 4842, SD = 331) Hz and a higher limit of 4823–6115 (mean = 5088, SD = 318) Hz.

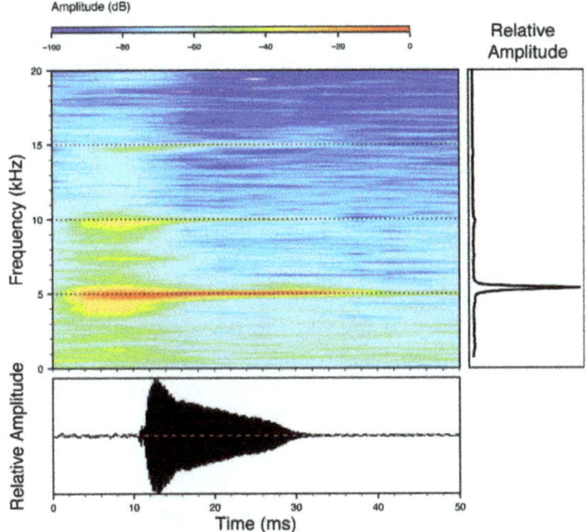

Figure 182. Call of *Nymphargus siren*, MZUTI 765, recorded at trail between Oyacachi and El Chaco, 1645 m, Napo province, Ecuador. Air temperature = 15 °C.

Tadpole: Not described.

Distribution (Figure 183): *Nymphargus siren* inhabits the cloud forests of the Amazonian slopes of the Ecuadorian and Colombian Andes at elevations between 1410 and 2000 m ([22,176], this work). In Ecuador, the species has been reported from the provinces of Napo and Sucumbíos. Rodríguez et al. [274] reported *N. siren* from Peru (Departamento Ayacucho), but we consider this report to be based on a misidentification and restrict the distribution of *N. siren* to Colombia and Ecuador. In Ecuador, the potential distribution of the species is 3881 km^2 within the Eastern Montane Forest region.

Figure 183. Distribution of *Nymphargus siren* in Ecuador (yellow dots).

Conservation status: *Nymphargus siren* is listed as *Vulnerable* by the IUCN because the extent of its occurrence is less than 20,000 km^2, its distribution is severely fragmented, and the extent and quality of its habitat continues to decline [275]. No data are available on the population demography of the species nor on its susceptibility to chytridiomycosis, climate change, and/or changes on UV radiation. Recent records of *N. siren* are from the Río Azuela (March 2000) [91], Yanayacu Biological Station (April 2008) [125], and the Oyacachi–El Chaco trail (May 2012; JMG pers. obs.). The population status of the species at localities where it was historically abundant (e.g., Río Salado; 3.2 km NNE Oritoyacu) is unknown. We maintain its conservation category.

Evolutionary relationships (Figure 136): *Nymphargus siren* and *N. humboldti* sp. nov. are sister species.

Remarks: Although Lynch and Duellman [22] mentioned that some individuals have up to two teeth on the dentigerous process of the vomer, all the specimens examined by us had edentate vomers.

Specimens examined: *Nymphargus siren:* <u>Ecuador</u>: *Provincia de Napo:* Yanayacu Biological Station, 2100 m, QCAZ 37971, 37975; tributary of the Río Salado, about 1 km upstream from the Río Coca (0.19167 S, 77.6997 W; 1410 m), KU 146610 (holotype), KU 146611–23 (paratypes), 164635–49, 178191–206, 146610–23, QCAZ 14425, 30975, 30977–80; Baeza (0.46795 S, 77.567 W; 1650 m), KU 190020–21; 16.5 km NNE Santa Rosa (0.2186 S, 77.732 W; 1700 m), KU 143288–89, 143291, 143293–94; 3.2 km NEE Oritoyacu (0.4597 S, 77.867 W; 1910 m), KU 178170–90; Oyacachi-El Chaco trail at elevations between 1645–1800 m, MZUTI 765–776. *Provincia de Sucumbíos:* Río Azuela (0.11667 S; 77.6167 W, 1740 m), QCAZ 15263, 15266, KU 143295–97, 155499–501; <u>Colombia</u>: *Putumayo:* 35 km SE of San Francisco, 1950 m, KU 169668–69.

Nymphargus sucre Guayasamin, 2013 [276] (Figures 184–186).

Nymphargus sucre Guayasamin, 2013 [276]. Holotype: MZUTI 1421.
<u>Type locality:</u> "creek on the Plan de Milagro–Gualaceo road (3.0077° S, 78.53318° W; 2159 m), Provincia Morona Santiago, Ecuador."

Common names: English: Sucre's glassfrog. Spanish: Rana de Cristal de Sucre.

Etymology: The specific epithet is a noun in apposition and honors Antonio José de Sucre, who, with Simón Bolívar, led the independence of most Andean countries (Bolivia, Colombia, Ecuador, Peru, and Venezuela) from Spain. The epithet also makes reference to the national currency of Ecuador between 1884 and 2000; the Sucre disappeared in the year 2000, when it was replaced by the US dollar after a disastrous economic policy that affected millions of Ecuadorians. As an analogy, the current destruction of habitats in southeastern Ecuador (as in many other regions) is likely to drive many species to extinction if activities such as mining, oil extraction, road building, cattle, and agriculture are promoted irresponsibly, and without assessing their effect on megadiverse areas and endangered species [276].

Identification: *Nymphargus sucre* is distinguished from most glassfrogs by having, in life, a brownish–yellow dorsal surface with yellow spots (Figure 184), and lacking webbing between the fingers. *Nymphargus sucre* is most similar to four other species that lack the typical green dorsal coloration of centrolenids (*N. anomalus*, *N. colomai* sp. nov., *N. ignotus*, and *N. rosada*; Figure 185). *Nymphargus anomalus* and *N. ignotus* differ from *N. sucre* mainly by having, in life, a pale tan to brown dorsum with black ocelli (ocelli absent in *N. sucre*) and lacking yellow spots. *Nymphargus rosada* is distinguished by its pink coloration in life with yellowish orange spots on the dorsum (brownish–yellow dorsum with yellow spots in *N. sucre)* and by being larger with non-overlapping SVL in adult males (SVL = 24.9–28.3 in male *N. rosada* [133]; SVL = 21.6–22.3 mm in male *N. sucre*). Finally, *N. colomai* sp. nov. differs by having numerous diffuse yellow spots on the dorsum (fewer and well-defined spots in

N. sucre), and a white iris with a contrasting horizontal black stripe (iris lacking horizontal stripe and having a yellow hue around the pupil in *N. sucre* [276]; Figure 185).

Other species that may have yellowish–green dorsal patterns and could be confused with *N. sucre* are *N. armatus*, *N. oreonympha*, *N. nephelophila*, and *N. ruizi*. These four species are readily distinguished from *N. sucre* by having black dorsal spots and lacking yellow spots. Additionally, adult males of *N. armatus* have nuptial pads with a Type III morphology, whereas *N. sucre* has a Type I morphology. Adult males of *N. oreonympha*, *N. nephelophila*, *N. armatus*, and *N. ruizi* are larger (SVL 22.6–24.1 mm, SVL 24.0–26.3 mm, SVL 23.3–24.8 mm, SVL 24.3–26.4 mm, respectively [154,277,278] than males of *N. sucre* (SVL 21.6–22.3 mm). Finally, *N. spilotus*, an endemic to the eastern slope of the Cordillera Central of Colombia, has a dorsal pattern that resembles that of *N. sucre* (with yellow spots), but is a considerably larger species (adult male SVL = 25.3–26.4 mm; female SVL = 27.6–28.5 mm [133]) and has prominent vomerine teeth (absent in *N. sucre* [276]).

Figure 184. *Nymphargus sucre* in life. (**A–C**) Adult male, holotype, MZUTI 1421. (**D**) Adult female, MZUTI 1422. Photos by Alejandro Arteaga/Tropical Herping. Obtained from Guayasamin [276].

Figure 185. Glassfrog species similar to *Nymphargus sucre* [276]. (**A**) *Nymphargus anomalus*, QCAZ 47507; photo by Luis A. Coloma. (**B**) *N. colomai* sp. nov., QCAZ 41590; photo by Juan M. Guayasamin. (**C**) *N. rosada*, photo by M. Rivera. (**D**) *N. sucre*, MZUTI 1421; photo by A. Arteaga/Tropical Herping.

Diagnosis: (1) Dentigerous process of the vomer present, but lacking teeth; (2) snout truncated in lateral and dorsal views; (3) tympanum visible without magnification, oriented almost vertically, with slight lateral and posterior inclinations, its diameter about 23%–26% of eye diameter; upper fourth of tympanic annulus obscured by supratympanic fold; tympanic membrane translucent, but with melanophores like those on surrounding skin; (4) dorsal skin shagreen, with numerous minute spicules in males; (5) venter areolate; pair of enlarged subcloacal warts; (6) white parietal peritoneum covering about anterior half of venter (condition P2); white pericardium; translucent peritonea covering intestines, stomach, kidneys, gall and urinary bladders (condition V1); (7) liver trilobed, covered by transparent peritoneum (condition H0); (8) humeral spines absent; (9) webbing absent between inner fingers, basal between Fingers III and IV; webbing formula III ($2^{3/4}$–3^-)—($2^{1/2}$–$2^{3/4}$) IV; (10) feet about two-thirds webbed; webbing formula: I (2–2^-)—2^+ II (1^+–$1^{1/3}$)—(2^+–$2^{1/3}$) III (1–$1^{1/2}$)—($2^{1/3}$–$2^{1/2}$) IV ($2^{2/3}$–3^-)—2^- V; (11) ulnar and tarsal folds present, but low and inconspicuous, lacking pigmentation; (12) concealed prepollex; in males, nuptial pad Type I; (13) Finger I slightly shorter than Finger II (Finger I length 93%–96% of Finger II); (14) disc width of Finger III about 37%–53% of eye diameter; (15) in life, dorsum brownish yellow with yellow spots; color of bones unknown; (16) in preservative, dorsum grey lavender with small white or unpigmented spots; (17) iris white with minute dark spots, thin reticulation, and yellow hue around pupil; (18) dorsal surfaces of fingers and toes lacking melanophores, except for few on Fingers III and IV, and Toes III, IV, and V; (19) males call from the upper surfaces of leaves; call unknown; (20) fighting behavior unknown; (21) egg deposition site and parental care unknown; (22) tadpoles unknown; (23) medium body size; males, SVL 21.6–22.3 mm (n = 2); female, SVL 24.3 mm (n = 1).

Color in life (Figure 184): Brownish–yellow dorsum with small yellow spots; dorsum of the holotype with a greenish hue. Upper lip unpigmented. Anterior half of ventral parietal peritoneum white, posterior portion translucent. Color of bones unknown. Iris white with minute dark spots, thin reticulation, and yellow hue around pupil [276].

Color in ethanol: Dorsal surfaces of head, body, and limbs grey lavender with small white or unpigmented spots. Anterior half of ventral parietal peritoneum white. Heart covered by white pericardium; translucent peritonea covering gall and urinary bladders; iridophores absent from digestive tract, liver, and kidneys. Observations on internal anatomy were based on specimen MZUTI 1421 [276].

Biology and ecology: During the night, males of *Nymphargus sucre* were observed calling on leaves at 90–130 cm above a stream on 6 June 2012. Only one other species of glassfrog (*N. cariticommatus*) was found at the same stream [276]. Parental care is unknown.

Call: Not described.

Tadpole: Not described.

Distribution (Figure 186): *Nymphargus sucre* is only known from the type locality, a creek on the Plan de Milagro–Gualaceo road (3.0077° S, 78.53318° W; 2140–2160 m), Morona Santiago province, Ecuador [276].

Conservation status: *Nymphargus sucre* has not been evaluated by the IUCN. Because of habitat loss and mining, we suggest that the species should be considered as *Critically Endangered*, following IUCN criteria B1, B2a,b(iii).

Evolutionary relationships (Figure 136): *Nymphargus sucre* is sister to *N. cariticommatus*.

Figure 186. Distribution of *Nymphargus sucre* in Ecuador (yellow dot).

Specimens examined: *Nymphargus sucre:* Ecuador: *Morona Santiago province:* Plan de Milagro-Gualaceo road (3.0077° S, 78.53318° W; 2100 m), MZUTI 1421 (holotype), MZUTI 1420, 1421 (paratypes).

Nymphargus wileyi (Guayasamin, Bustamante, Almeida-Reinoso, Funk, 2006 [20];
Figures 187–189).

Cochranella wileyi Guayasamin, Bustamante, Almeida-Reinoso, Funk, 2006 [20]. Holotype: QCAZ 26028.
Type locality: "Yanayacu Biological Station (0°41' S, 77°53' W; 2100 m), Provincia de Napo, Ecuador."
Nymphargus wileyi—Cisneros-Heredia and McDiarmid, 2007 [17].

Common names: English: Wiley's Glassfrog. Spanish: Rana de Cristal de Wiley.
Etymology: The specific name is a noun in the genitive case and a patronym for Edward O. Wiley, for his influential work on the development of phylogenetic systematics and use of the evolutionary species concept [20].

Identification: *Nymphargus wileyi* differs from most species of glassfrogs by having a uniform green dorsum (Figure 187), white renal peritoneum, and by lacking any webbing between the fingers. On the Amazonian slopes of the Andes, only *N. cariticommatus* can be confused with *N. wileyi*. The differences between these two species are subtle; *N. wileyi* has a uniform dorsum (dorsum with small yellow dots in *N. cariticommatus*) and has a faint layer of iridophores covering the anterior portion of the esophagus (white peritoneum covering most of the esophagus in *N. cariticommatus*). The only other species with a uniform green dorsum on the Amazonian slopes of the Andes is *Espadarana durrellorum*, which differs by having webbing between Fingers III and IV, and a humeral spine in males.

Figure 187. *Nymphargus wileyi* in life. (**Left**): Adult male from Yanayacu Biological Station, Napo province, QCAZ 37972. (**Right**): Egg clutches of *N. wileyi* at Yanayacu Biological Station. Photos by Martín Bustamante.

Diagnosis: (1) Vomers lacking teeth; (2) snout truncated in dorsal aspect, truncated to protruding in lateral profile (Figure 188); (3) tympanum oriented almost vertically, with slight posterior and lateral inclinations, its diameter 31.4%–37.8% of eye diameter; tympanic annulus visible anteroventrally, tympanic membrane unpigmented and clearly differentiated from surrounding skin; (4) dorsal surfaces of males and females shagreen, with small spicules evident in males; (5) venter areolate; pair of enlarged subcloacal warts (Figure 15); (6) anterior 40%–60% of ventral parietal peritoneum white, posterior portion transparent (condition P2); white pericardium; translucent to opaque peritoneum covering intestines, stomach, testes, gall bladder, and urinary bladder; peritoneum around kidneys white with unpigmented spots (condition V1); (7) liver tetralobed, covered by transparent peritoneum (condition H0); (8) humeral spines absent; (9) webbing absent between inner fingers; webbing reduced between outer fingers, III 3^-—$2^{2/3}$ IV; (10) concealed prepollex; in males, nuptial pad Type I; (11) Finger II slightly longer than Finger I (Finger I 92%–97% of Finger II); (12) ulnar fold low or absent; inner tarsal low and thin, outer tarsal fold absent; (13) webbing formula on foot I 2—$2^{1/3}$ II ($1^{1/3}$–$1^{2/3}$)—($2^{1/2}$–3^-) III (1^+–2^-)—($2^{2/3}$–3^-) IV (3^-–3)—(2^-–2) V (Figure 188); (14) disc of Finger III of moderate size, about 50%–60% of eye diameter; (15) in life, dorsum pale green; bones green; (16) in preservative, dorsum uniform pale lavender; (17) iris coppery white with black reticulation; (18) fingers and toes lacking melanophores on dorsal surfaces; (19) males call from the upper sides of leaves; call unknown; (20) fighting behavior unknown; (21) females deposit eggs on upper surface of leaves near streams (Figure 187); short-term maternal care unknown; prolonged parental care absent; (22) tadpoles undescribed; (23) small body size; males, SVL 24.0–26.2 mm ($\bar{X}$ = 24.6; *n* = 5); 27.1 mm in one female.

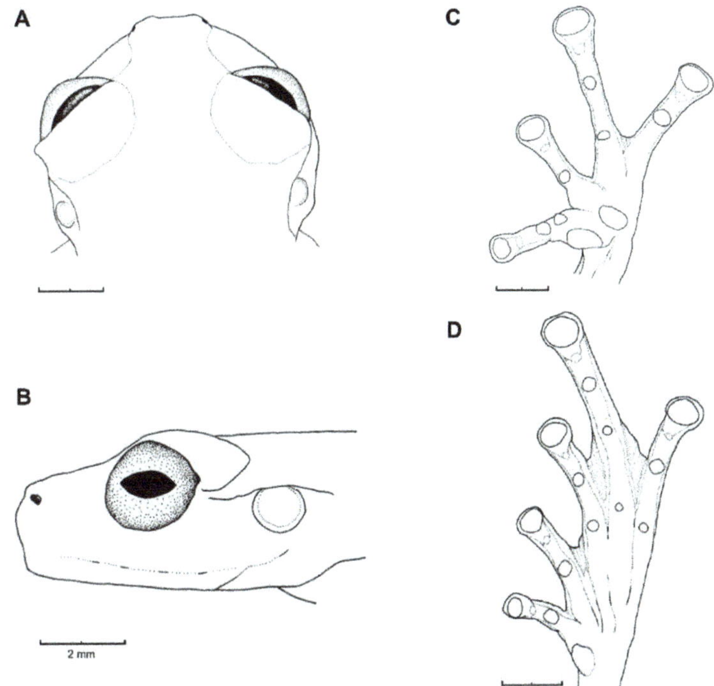

Figure 188. *Nymphargus wileyi*. (**A**) Head in dorsal view, QCAZ 26029. (**B**) Head in lateral view, QCAZ 26029. (**C**) Hand in ventral view, QCAZ 26028. (**D**) Foot in ventral view, QCAZ 26028. Modified from Guayasamin et al. [20].

Color in life (Figure 187): Dorsal surfaces pale green, lacking spots; lower venter transparent; parietal peritoneum white, covering anterior part of abdomen (heart not visible); iris coppery white with black reticulation; bones green [20].

Color in ethanol: Dorsal surfaces of head, body, and limbs pale lavender; upper lip white; dorsally, all fingers, Toes I–III, and most of Toe IV unpigmented; cloacal region mostly unpigmented, except for few minute white flecks. Cream nuptial pad on Finger I. Ventral parietal peritoneum white anteriorly, pericardium white, hepatic peritoneum clear, visceral peritoneum opaque, peritoneum around kidneys white with unpigmented spots (dissected specimens: QCAZ 26029–30) [20].

Biology and ecology: *Nymphargus wileyi* is a nocturnal species that seems to be restricted to primary forest. Individuals have been found at night on leaves 120–220 cm above streams (five males) or above the ground (one female). Males call from the upper surfaces of leaves; males are territorial and are usually found near egg clutches, which are on the tip of leaves; distances between egg clutches and males can be as close as 20 cm (Figure 187). Males were never observed on the same leaf as the egg clutch; also, several egg clutches had no male nearby; these observations suggest that prolonged parental care is absent [20]. Short-term maternal care is unknown. Clutches are deposited on the dorsal surface of leaves near their tips (Figure 187). The number of eggs per clutch varies from 19–28 ($\overline{X}$ = 22; n = 17); eggs and embryos in early developmental stages are whitish [20].

Call: Not described.

Tadpole: Not described.

Distribution (Figure 189): *Nymphargus wileyi* is known only from the type locality (Yanayacu Biological Station, 2100 m) on the Amazonian slopes of the Cordillera Oriental of the Ecuadorian Andes [20]. The habitat of the species is within the Eastern Montane Forest region.

Figure 189. Distribution of *Nymphargus wileyi* in Ecuador (yellow dot).

Conservation status: *Nymphargus wileyi* is listed as *Data Deficient* by the IUCN [279]. We consider this category appropriate.

Evolutionary relationships (Figure 136): *Nymphargus wileyi* is sister to a clade formed by *N. sucre* and *N. cariticommatus*. These three species are found on the Amazonian slopes of the Andes.

Taxonomic Remarks: Only one discrete morphological character separates *Nymphargus wileyi* from *N. cariticommatus*, the absence of pale spots in *N. wileyi*. Also, in preserved material, the esophageal peritoneum of *N. cariticommatus* is more pronounced and extensive than in *N. wileyi*, in which the iridophores are restricted to the anterior portion of the esophageal peritoneum. Surprisingly, the two species, although closely related, are not sister to each other (Figure 136).

Specimens examined: *Nymphargus wileyi:* <u>Ecuador</u>: *Provincia de Napo:* Yanayacu Biological Station (0°41′ S, 77°53′ W; 2100 m), QCAZ 26028 (holotype), 26024, 26029–30, 26057, 27435.

Genus *Rulyrana* Guayasamin, Castroviejo-Fisher, Trueb, Ayarzagüena, Rada, and Vilà, 2009 [1].

Etymology: The name *Rulyrana* honours Pedro Ruiz-Carranza† and John D. Lynch (Instituto de Ciencias Naturales, Universidad Nacional de Colombia), who have contributed enormously to the understanding of anuran diversity, biology, and evolution. Together, Pedro and John made an outstanding effort to describe the amphibian diversity of Colombia, producing a series of glassfrog papers [6,26,27, 101,110,113,129,133,154,246,255,277] that includes the description of the genus *Hyalinobatrachium* and numerous species. The name *Rulyrana* is femine in gender and comes from an arbitrary association of the two first letters of Ruiz and Lynch (Ruly) and the word rana (=frog). *Ruly* also is the nickname of JMG's good friend and colleague Martín Bustamante, whose work on biodiversity, conservation, and even sportive indoctrination, is remarkable ([1], this work).

Rulyrana flavopunctata (Lynch and Duellman, 1973 [22]; Figures 190–194).

Centrolenella flavopunctata Lynch and Duellman, 1973 [22]. Holotype: KU 121048.
Type locality: "Mera, Provincia Pastaza, Ecuador."
Cochranella flavopunctata—Ruiz-Carranza and Lynch, 1991 [6].
Rulyrana flavopunctata—Guayasamin, Castroviejo-Fisher, Trueb, Ayarzagüena, Rada, and Vilà,
2009 [1].

Common names: English: Yellow-spotted Glassfrog. Spanish: Rana de Cristal de
Puntos Amarillos.

Etymology: The specific name is a combination of the Latin *flavus*, meaning golden yellow,
and *punctatus*, meaning dotted, and is used in reference to the dorsal coloration of the species [22].

Identification: Among glassfrogs that inhabit the Amazonian slopes of the Ecuadorian Andes,
Rulyrana flavopunctata is distinguished by having a green to dark green dorsum with small yellow spots
(Figure 190), a partially white venter, small to moderate body size (male SVL 21.0–23.9 mm; female SVL
24.0–27.4 mm), extensive webbing between Fingers III and IV, and lacking humeral spines. Species with
a similar color pattern from eastern Ecuador include *Espadarana audax*, *Nymphargus siren*, *N. humboldti*
sp. nov., *N. mariae*, *R. mcdiarmidi*, and *Teratohyla midas*. None of the species in the genus *Nymphargus*
has webbing between Fingers III and IV; males of *Espadarana audax* have humeral spines; *T. midas* has
very few yellow spots on the upper flanks and a white digestive tract; and *R. mcdiarmidi* has a slightly
larger body size (male SVL 22.8–26.9 mm, female SVL 25.4–30.2 mm) and warts on its dorsum (absent
in *R. flavopunctata*). The differences between *R. flavopunctata* and *R. mcdiarmidi* are subtle and more
work is necessary to verify the validly of the specific status of the latter. See Taxonomic Remarks.

Figure 190. *Rulyrana flavopunctata* in life. Ecuador, Napo province, Cordillera de los Guacamayos,
1564 m, MZUTI 1250 (**left**), 1260 (**right**). Photos by Eduardo Toral.

Diagnosis: (1) Teeth on vomers present or absent; each vomer with zero to three teeth; (2) snout
usually rounded in dorsal and lateral views; (3) tympanum visible, its diameter 27.9%–38.7% of eye
diameter; supratympanic fold low; (4) dorsal skin shagreen; males with spicules; (5) skin of venter
areolate; pair of slightly enlarged subcloacal warts; (6) anterior half of ventral parietal peritoneum
white, posterior half translucent (condition P2); pericardium white; peritonea covering intestines,
stomach, and kidneys lacking iridophores; urinary bladder transparent (condition V1); (7) lobed
liver lacking iridophores (condition H0); (8) humeral spines absent; (9) webbing between Fingers
I and II absent or basal, moderate between Fingers II and III, extensive between Fingers III and
IV; webbing formula for outer fingers: II (1$^{1/2}$–2$^-$)—(3–3$^+$) III (1$^{1/3}$–2$^-$)—(1$^+$–1$^{1/2}$) IV; (10) feet about
three-fourths webbed; webbing formula: I (0$^+$–1)—(1–1$^{3/4}$) II (0$^+$–1)—(0$^+$–2) III (0$^+$–1)—(1$^{1/3}$–2$^+$) IV

$(2^- - 2^+)$—$(0^+ - 1)$ V (Figure 191); (11) ulnar fold low or absent; inner tarsal fold low, short; outer tarsal fold absent; (12) concealed prepollex; in males, nuptial pad Type I; (13) Finger I about same length as Finger II (Finger I length 95.5%–104.0% of Finger II); (14) disc of Finger III relatively large, its width 47.7%–67.0% of eye diameter; (15) in life, dorsum green to dark green with well-defined yellow spots; bones green; (16) in preservative, dorsum dark lavender to slate grey with small white to cream spots; (17) iris pale greyish white, with or without golden tint, with dark grey or brown flecks or fine reticulation; (18) dorsal surfaces of fingers and toes usually with melanophores, but some individuals lack melanophores on Finger I or on Fingers I and II; (19) males call from the upper side of leaves or rocks; call unknown; (20) fighting behavior unknown; (21) egg clutches placed on rock walls; parental unknown; (22) tadpoles unknown; (23) small body size in adult males, SVL 21.0–23.9 mm ($\overline{X}$ = 22.1 ± 0.834, n = 10); in adult females, SVL 24.0–27.4 mm ($\overline{X}$ = 25.3 ± 1.207, n = 6).

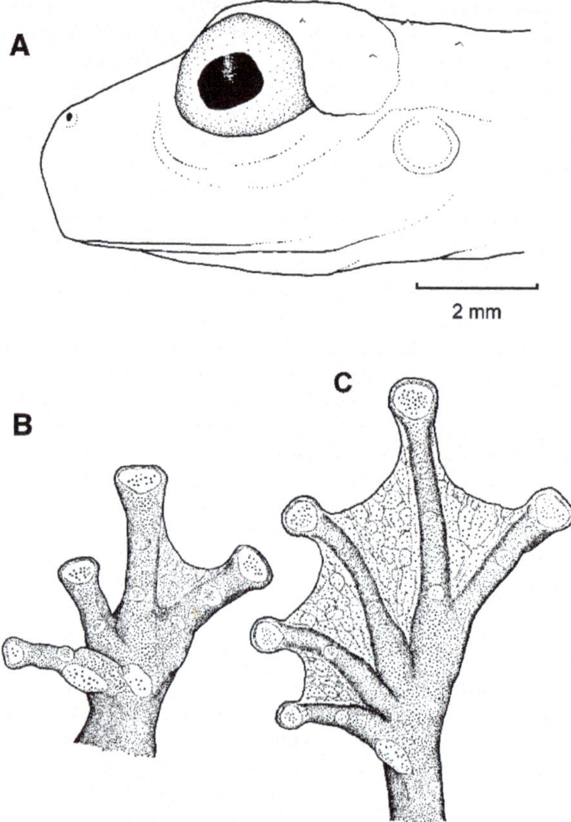

Figure 191. *Rulyrana flavopunctata*. (**A**) Head in lateral view, holotype, adult male, KU 121046. (**B**) Hand in ventral view, KU 121050. (**C**) Foot in ventral view, KU 121048. (**B**,**C**) Modified from Lynch and Duellman [22].

Color in life (Figure 190): Conspicuous variation has been observed in dorsal and ventral coloration. Dorsal surfaces of head and body vary from green to very dark green, with small yellow spots. Anterior half of ventral parietal peritoneum white; posterior half usually translucent, but some individuals have a milky colored peritoneum. Iris pale greyish white to brown, with yellow tint, with dark grey or brown flecks or fine reticulation ([22], this work). At one locality (Río Tayuntza), individuals showed considerable variation in dorsal coloration, from green to almost black, with yellow spots, indicating that intraspecific variation in this species is not necessarily associated with geography.

Color in ethanol: Dorsal surfaces of head, body, and limbs dark lavender to slate grey, with small white or cream spots usually corresponding to warts. Internal organs lacking iridophores, except for heart that is covered by white pericardium.

Biology and ecology: Most individuals were found in a deep ravine where males were perched on small herbs in the spray-zone of a small waterfall. At several small streams between Mera and the Río Alpayacu, males were calling from the upper sides of leaves on 2, 14, and 24 July 1968 [22]. At Río Tayuntza (Morona Santiago province, Ecuador) egg clutches of *R. flavopunctata* were found on rock walls during the rainy season [280] (Figure 192). Parental care is unknown.

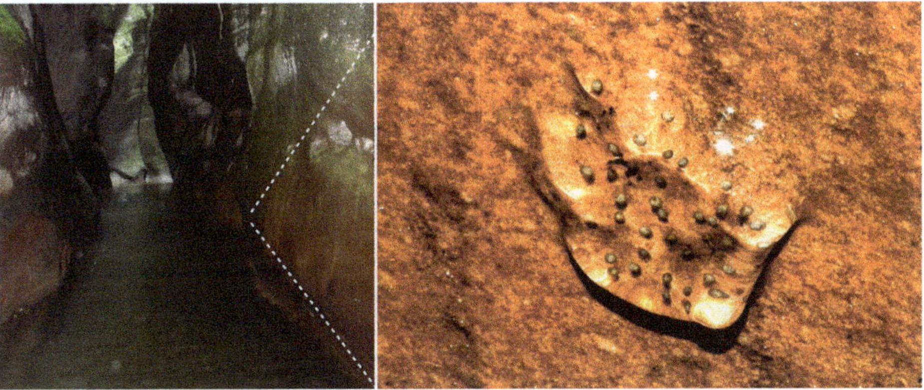

Figure 192. (Left): Habitat of a glassfrog tentatively assigned to *Rulyrana flavopunctata* or *R. mcdiarmidi*. **(Right):** Egg clutch of the rock walls of the river. Locality: Río Tayuntza, Cuevas de los Tayos, Río Tayuntza, Morona Santiago province, Ecuador. Photos by Octavio Jiménez-Robles (**Left**) and Ignacio de la Riva (**Right**). Modified from Jiménez-Robles et al. [280].

Call (Figure 193): We analyzed 85 notes contained within 13 calls from two individuals (MZUTI 1476, 1684). Each call is composed by a single short note that has a duration of 3–8 (mean = 4, SD = 1) ms. Notes are clearly pulsed; the first pulse of each note has the highest amplitude. Pulses within a note have a rate of 125–333 (mean = 262, SD = 55) pulses per second. The dominant frequency of a note measured at peak amplitude is 5857–7580 (mean = 6931, SD = 472) Hz and is contained within the fundamental frequency. The fundamental frequency has a lower limit of 5685–7149 (mean = 6648, SD = 447) Hz and a higher limit of 6115–7924 (mean = 7382, SD = 491) Hz.

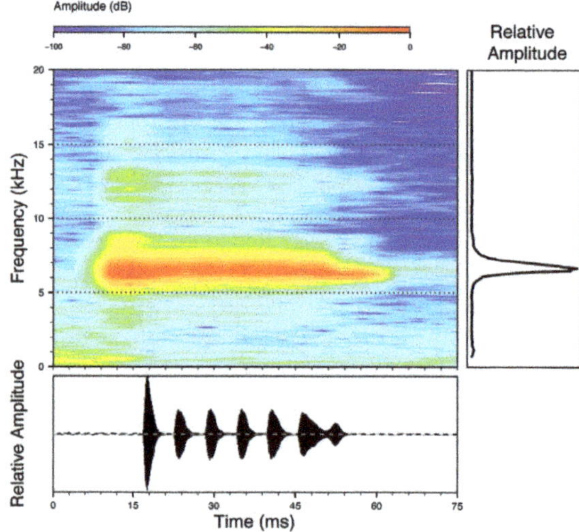

Figure 193. Call of *Rulyrana flavopunctata*, MZUTI 1684, recorded at Quebrada Pangayaku, 1645 m, Napo province, Ecuador. Air temperature = 19.4 °C.

Tadpole: Not described.

Distribution (Figure 194): *Rulyrana flavopunctata* is known from the Amazonian slopes of the Andes of Colombia and Ecuador at elevations between 300 and 1715 m ([22,101,281], this work). In Ecuador, this species has been recorded from localities in the provinces of Napo, Morona Santiago, Pastaza, Sucumbíos, and Tungurahua, at elevations between 720 and 1715 m (Specimens examined). In Ecuador, the potential distribution of the species is 16,369 km^2 within the Eastern Foothill Forest and Eastern Montane Forest regions.

Figure 194. Distribution of *Rulyrana flavopunctata* in Ecuador (yellow dots).

Conservation status: Globally, *Rulyrana flavopunctata* is listed as *Least Concern* by the IUCN [281]. In Ecuador, there are several recent reports of the species, including: Parroquia Teniente Hugo Ortíz (30 March 2007), 7.6 km W of 9 de Octubre (August 2006), 6.8 km N of Limón (21 April 2004), and Río Tayuntza (April 2011), Cordillera de los Guacamayos (April 2012), Quebrada el Plancón (March 2012), Río Hollín (August 2012), Quebrada Pangayacu (August 2012), lower slopes of Volcán Sumaco (February 2012, 2018), Narupa (June 2018). We agree with its current conservation status.

Evolutionary relationships (Figure 195): According to our molecular phylogeny, *Rulyrana flavopunctata* and *R. mcdiarmidi* are genetically indistinguishable.

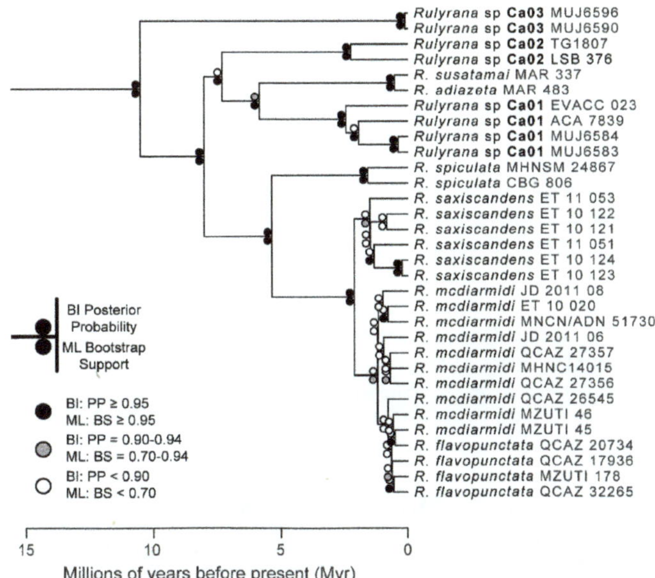

Figure 195. Evolutionary relationships of species in the glassfrog genus *Rulyrana*. The trees were inferred using maximum likelihood and Bayesian criteria.

Taxonomic Remarks: Morphological differences between *Rulyrana flavopunctata* and *R. mcdiarmidi* are subtle (body size, skin texture, color pattern), and the two species are genetically indistinguishable (Figure 195). It is possible that *R. mcdiarmidi* represents a geographic variation of *R. flavopunctata* or that the two species have recently diverged. Information on call variation and other traits are necessary to resolve the species status of *R. mcdiarmidi*.

Specimens examined: Ecuador: *Provincia Morona Santiago:* Río Tayuntza, Cuevas de los Tayos (02.1106° S, 77.75185° W; 700 m). *Provincia Napo:* Río Hollín (0.69583 S; 77.730277 W; 1190 m), QCAZ 22360–62; Reserva Narupa (0.6848 S, 77.742 W, 1170–1228 m), ZSFQ 0374–79. *Provincia Pastaza:* tributary of Río Rivadeneira (1.3604° S, 77.86534° W), SC 34963; 9.5 km NW Mera (1.4 S, 78.166667 W; 1270 m), KU 178094; Mera (1.466667 S, 78.13331 W; 1100 m), KU 121041–46, 121049–51, 178093; Río Alpayacu, 1 km E Mera (1.466667 S, 78.08333 W; 1080 m), KU 121047; near Río Rivadeneira (1.144307 S, 77.99667 W; 982 m), QCAZ 20734–35; km 6 on San Ramón-El Triunfo road (1.35998 S, 77.86564 W; 875 m), QCAZ 37911, 33269; Sacha Yacu (1.39519° S, 77.72946° W; 1078 m), MZUTI 176. *Provincia Sucumbíos:* Bermejo No. 4, 15 km ENE Umbaqui (0.1833 N, 77.366667 W; 720 m), KU 123224.

Localities from the literature: *Rulyrana flavopunctata:* Ecuador: *Provincia de Napo:* San José Abajo, on the eastern slope of Volcán Sumaco (00°32′ S, 77°24′ W; 700–1000 m), AMNH 22187. *Provincia de Pastaza:* 13 km WSW Puyo (ca. 1°34′ S, 78°06′ W; 1000 m), TCWC 24032 [22].

Rulyrana mcdiarmidi (Cisneros-Heredia, Venegas, Rada, and Schulte 2008 [272];
Figures 196 and 197).

Cochranella mcdiarmidi Cisneros-Heredia, Venegas, Rada, and Schulte, 2008 [272]. Holotype:
DFCH-USFQ D132.
Type locality: "small rivulet tributary of the Jambue River, ca. 6 km S from Zamora (ca. 04°03′ S,
78°56′ W, 1150 m), on the western slope of Contrafuerte de Tzunantza, Cordillera Oriental,
eastern slopes of the Andes, Provincia de Zamora Chinchipe, República del Ecuador".
Rulyrana mcdiarmidi—Guayasamin, Castroviejo-Fisher, Ayarzagüena, Trueb, Rada, and Vilà,
2009 [1].

Common names: English: McDiarmid's Glassfrog. Spanish: Rana de Cristal de McDiarmid.
Etymology: The specific epithet honors Dr. Roy McDiarmid, in recognition of his contributions to
the understanding of Neotropical herpetology [272].
Identification: *Rulyrana mcdiarmidi* can be distinguished from all other glassfrogs by having a
moderate-sized body (male SVL 22.8–26.9 mm; female SVL 25.4–30.2 mm), green dorsum with diverse
darker shadows and pale yellow to green spots (Figure 170), dorsal skin with warts, ventral parietal
peritoneum completely white or with posterior portion translucent, thick ulnar folds, and extensive
hand and foot webbing. Among centrolenids from the eastern slopes of the Andes, species with a
similar color pattern include *Espadarana audax*, *Nymphargus siren*, *N. humboldti* sp. nov., *N. mariae*,
R. flavopunctata, and *Teratohyla midas*. None of the species in the genus *Nymphargus* has webbing
between Fingers III and IV; males of *Espadarana audax* have humeral spines; *T. midas* has very few
yellow spots on the upper flanks and a white digestive tract; *R. flavopunctata* differs by being slightly
smaller (male SVL 21.0–23.9 mm; female SVL 24.0–27.4 mm), having yellow dorsal spots, and lacking
dorsal warts. See Taxonomic Remarks.

Figure 196. *Rulyrana mcdiarmidi* in life. Adult (QCAZ 32265) from 7.6 km W of 9 de Octubre, 1715 m,
Morona Santiago province, Ecuador. Photos by Martín Bustamante.

Diagnosis: (1) Vomerine teeth present; (2) snout rounded to subtruncated in dorsal and lateral views; (3) tympanic annulus evident, oriented dorsolaterally; weak supratympanic fold; (4) dorsal skin smooth to shagreen, with numerous warts; (5) ventral skin coarsely granular; subcloacal area coarsely granular, with abundant low, flat warts; other cloacal ornamentation absent; (6) ventral parietal peritoneum white, iridophores covering entirely or almost entirely abdomen to level of groin (condition P4); pericardium white, all other visceral peritonea clear (condition V1); (7) liver lobed, lacking iridophores (condition H0); (8) humeral spine absent; (9) webbing absent between Fingers I and II, basal between II and III, moderate to extensive between outer fingers; webbing formula: II $1^{1/2}$—3^{+} III ($1^{1/2}$–2^{-})—(1^{-}–1^{+}) IV; (10) webbing on feet extensive, webbing formula: I (1^{-}–1)—(1^{+}–$1^{1/3}$) II (0^{+}–1)—($1^{1/3}$–$1^{2/3}$) III (1–1^{-})—($1^{2/3}$–2^{+}) IV 2^{-}—(0^{+}–1) V; (11) thick, non-enameled, non-crenulated ulnar fold; low, short inner tarsal fold; (12) nuptial excrescences Type I in adult males; concealed prepollex; (13) first finger slightly shorter than second; (14) eye diameter larger than width of disc on Finger III; (15) in life, dorsum dark olive green with creamy yellow or light yellowish orange or light green spots, bones green; (16) color in preservative, dorsal surfaces tan grey, pale brown, or greyish lavender with diffuse light tan spots; (17) in life, iris olive brown to brown with thin dark reticulations; (18) abundant melanophores widespread on all fingers and toes; (19) males call from rocks along streams and waterfalls; call unknown; (20) fighting behavior unknown; (21) egg clutches unknown, parental care unknown; (22) tadpoles unknown; (23) small to medium body size; SVL in adult males 22.8–26.9 mm ($\overline{X}$ = 24.5 ± 1.139; n = 11), and in adult females 25.4–30.2 mm ($\overline{X}$ = 28.0 ± 1.384; n = 13).

Color in life (Figure 196): Olive green dorsum with darker suffusions and yellowish cream, orange, or green spots; males with smaller dorsal spots than females. Venter yellowish cream. Iris olive brown to brown with fine dark reticulations. Green bones [272].

Color in ethanol: Dorsal surfaces grey or greyish lavender with light spots. Venter cream. Parietal peritoneum entirely covered by iridophores to level of groin; pericardium white, all other peritonea lack white lining [272].

Biology and ecology: Natural history information on *Rulyrana mcdiarmidi* is scarce. It is nocturnal and males call from rocks in the spray zone of small waterfalls in Foothill Evergreen forests. Parental care is unknown [272].

Call: Not described.

Tadpole: Not described.

Distribution (Figure 197): *Rulyrana mcdiarmidi* is known from localities in southeastern Ecuador (Morona Santiago and Zamora Chinchipe provinces) and one in northeastern Peru (Departamento de Cajamarca) at elevations between 1150 and 1500 m ([272], this work). In Ecuador, the potential distribution of the species is 19,002 km^2.

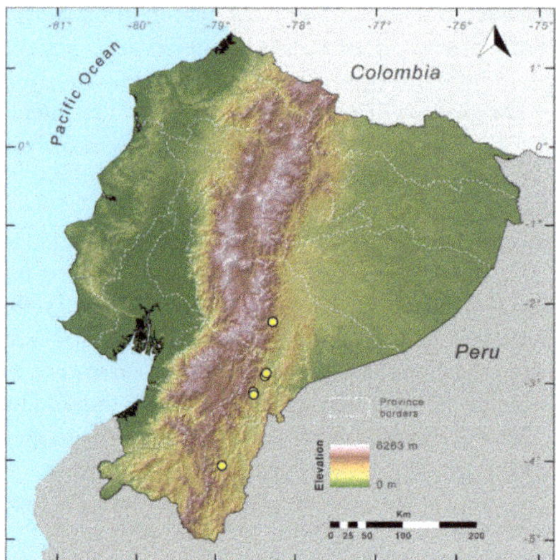

Figure 197. Distribution of *Rulyrana mcdiarmidi* in Ecuador (yellow dots).

Conservation status: *Rulyrana mcdiarmidi* is considered as *Data Deficient* by the IUCN [282]. Given the distribution of the species and the current threats if faces (i.e., habitat fragmentation, contamination by mining), we suggest that it should be considered as *Vulnerable*, following IUCN criteria following IUCN criteria B1, B2a, B2biii.

Evolutionary relationships (Figure 195): *Rulyrana mcdiarmidi* and *R. flavopunctata* are genetically indistinguishable. The clade *mcdiarmidi* + *flavopunctata* is sister to *R. saxiscandens*.

Taxonomic remarks: Morphological differences between *Rulyrana flavopunctata* and *R. mcdiarmidi* are subtle (body size, skin texture, color pattern), and the two species are not separated genetically (Figure 195). Therefore, it is possible that *R. mcdiarmidi* represents a geographic variant of *R. flavopunctata*, or that the two species have recently diverged. Information on call variation and ecological requirements are necessary to resolve the species status of *R. mcdiarmidi*.

Specimens examined: *Rulyrana mcdiarmidi:* <u>Ecuador</u>: *Provincia de Zamora Chinchipe*: small rivulet tributary of the Jambue River, ca. 6 km S from Zamora (ca. 04°03′ S, 78°56′ W, 1150 m), on the western slope of Contrafuerte de Tzunantza, Cordillera Oriental, eastern slopes of the Andes, DFCH-USFQ D132; km 90 of the Gualaceo–Indanza–Cochay road, ca. 1 km SW of Conchay (ca. 03°06′ S, 78°25′ W; 1100 m), DFCH-USFQ AL15. *Provincia de Morona Santiago:* Río Napinaza, 6.8 km N of Limón (2.92278 S, 78.407 W; 985 m), QCAZ 27356–58; 7.6 km W of 9 de Octubre on the 9 de Octubre–Guamote road (2.225 S, 78.2904 W; 1715 m), QCAZ 32265; 2.2 km S of San Juan Bosco (3.10612 S, 79.52515 W; 1013 m), QCAZ 26443; 4.8 km N of Rosario (2.8858 S, 78.38804 W; 841 m), QCAZ 26484; 3.1 km S of San Juan Bosco (3.1467 S, 78.53559 W; 1278 m), QCAZ 26426. <u>Peru</u>: *Departamento de Cajamarca:* Provincia de Jaen: stream tributary of the Río Chinchipe (ca. 05°25′16.5″ S, 78°35′23.2″ W; 1250 m), extreme southwestern slope of the Cordillera del Condor, MUSM 26322–4.

Genus *Sachatamia* Guayasamin, Castroviejo-Fisher, Trueb, Ayarzagüena, Rada, and Vilà, 2009 [1].

Etymology: The generic name *Sachatamia* comes from the Kichwa words *sacha*, meaning forest, and *tamia*, meaning rain, and refers to the tropical rainforest occupied by the species in this genus; *Sachatamia* is feminine in gender [1].

Sachatamia albomaculata (Taylor, 1949 [145]; Figures 198–200).

Centrolenella albomaculata Taylor, 1949 [145]. Holotype: KU 23814.
Type locality: "Los Diamantes, one mile south of Guápiles (Cantón de Pococí, Provincia Limón), Costa Rica."
Cochranella albomaculata—Taylor, 1951 [15].
Sachatamia albomaculata—Guayasamin, Castroviejo-Fisher, Trueb, Ayarzagüena, Rada, and Vilà, 2009 [1].

Common names: English: Cascade Glassfrog [24], White-spotted glassfrog. Spanish: Rana de Cristal de Cascada [24], Rana de cristal de puntos blancos.

Etymology: The specific epithet *albomaculata* is a combination of the Latin words *albus* (=white) and *macula* (=spot, stain, mark) and refers to the dorsal color pattern of the species.

Identification: *Sachatamia albomaculata* is the only glassfrog on the Pacific versant of the Andes that has, in life, a green dorsum with minute and large yellow spots (Figure 198), relative extensive webbing between Fingers III and IV (Figure 199), and lacks humeral spines. *Sachatamia punctulata* seems to fall within the variation of *S. albomaculata* (see Taxonomic Remarks).

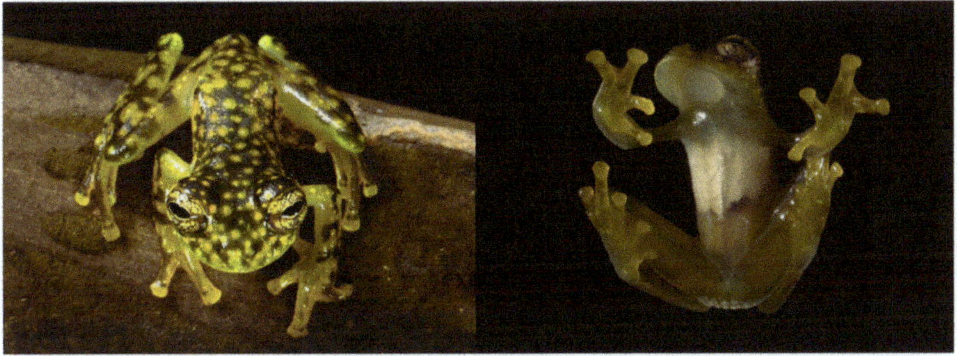

Figure 198. *Sachatamia albomaculata* in life, adult male, QCAZ 40816. Ecuador: Imbabura: near Lita, ca. 500 m. Photos by M. Bustamante (**left**) and Luis A. Coloma (**right**).

Diagnosis: (1) Teeth on dentigerous process of the vomer present, each process bearing four to six teeth; (2) snout round to truncated in dorsal profile, and mostly truncated in lateral profile (Figure 199); (3) tympanum evident, its diameter 25.7%–34.2% of eye diameter, dorsal border of tympanic annulus covered by supratympanic fold, tympanic membrane clearly differentiated from surrounding skin; (4) dorsal skin of males and females shagreen, spicules present in males; (5) skin on venter areolate; pair of enlarged subcloacal warts, which are more evident in females; (6) anterior half of ventral parietal peritoneum covered by white iridophores, posterior half transparent (condition P2); white pericardium; translucent peritoneum covering intestines, stomach, testes, kidneys, gall bladder, and urinary bladder (condition V1); (7) liver tetralobed, two large ventral lobes covering two smaller lobes;

hepatic peritoneum transparent (condition H0); (8) humeral spines absent; (9) no webbing between Fingers I and II; webbing between other fingers as follows: II ($1^{1/3}$–2^+)—(3–$3^{1/2}$) III ($1^{1/3}$–2)—($1^{1/3}$–$1^{3/4}$) IV; (10) foot webbing extensive: I 1—($1^{3/4}$–2) II (0^+–1)—(2–2^+) III (1^-–$1^{1/4}$)—(2^-–2^+) IV (2^-–2^+)—(1–1^+) V; (11) ulnar and tarsal folds low; (12) concealed prepollex; in males, nuptial pad Type I; (13) Finger I about same length as Finger II (Finger I 96.5%–103.9% length of Finger II); (14) disc of Finger III of moderate size, 41.9%–45.0% of eye diameter; (15) in life, dorsum green with small and large yellow spots; bones green; (16) in ethanol, dorsal surfaces of head, body, and limbs dark lavender with small and large whitish cream spots; venter cream; (17) iris silvery white to yellow with black reticulation; (18) melanophores usually present on dorsal surfaces of outer fingers and toes; (19) males call from the upper surfaces of leaves and rocks along streams and in spray zone of waterfalls; calls consists of a rapid high-pitched single note, "tik", with a duration of 0.001–0.002 s; the dominant frequency is at 6.1–7.1 kHz; (20) fighting behavior unknown; (21) eggs deposited on the upper surfaces of leaves or on rocks; short-term maternal care present; prolonged parental care absent; (22) tadpoles with flattened body; dark brown with violet tint dorsally; venter translucent, with few pigment spots laterally; upper jaw smoothly curved, lower jaw V-shaped; labial tooth row formula 1-2/3; (23) small body size, SVL of adult males 22.1–24.7 mm ($\overline{X}$ = 23.6 ± 0.920, *n* = 7); SVL of adult females 26.7–29.0 mm ($\overline{X}$ = 27.9 ± 0.891, *n* = 6).

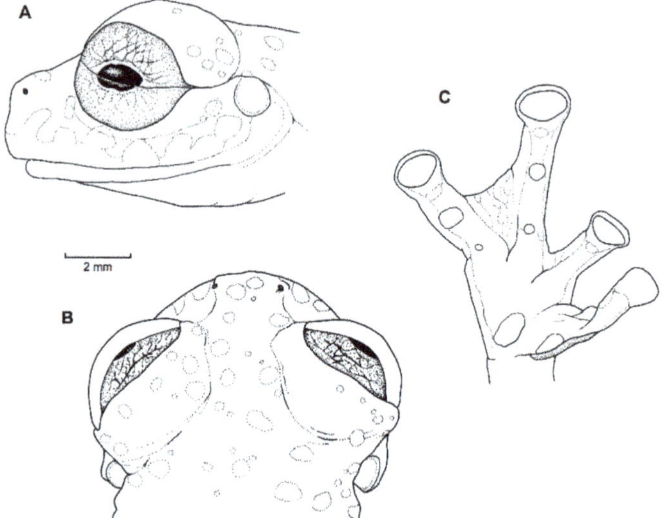

Figure 199. *Sachatamia albomaculata*, adult male, QCAZ 4325. (**A**) Head in lateral view. (**B**) Head in dorsal view. (**C**) Hand in ventral view. Illustrations Juan M. Guayasamin.

Color in life (Figure 198): In Central American, two clearly different color patterns have been reported, one with a green dorsum and small yellow spots; see photographs in AmphibiaWeb [283], McCranie and Wilson (81:plate 10), Savage 2002 (142:plate 198), Kubicki (24:118), and the other with small and large yellow spots (24:114). The holotype of the species, as well as Ecuadorian populations of the species, exhibit the latter coloration pattern. The venter is white on the anterior half and transparent posteriorly. The bones are green. The iris varies from white to yellow with dark reticulations.

Color in ethanol: Dorsal surfaces of head, body, and limbs lavender, with small whitish cream spots or with small and large spots. White lining covers pericardium and anterior half of ventral parietal peritoneum. Liver, intestines, stomach, testes, kidneys, and gall and urinary bladders covered by translucent peritonea.

Biology and ecology: On the night of 1 March 1994, a male and a female (QCAZ 4324–25) were found in amplexus on a leaf about 180 cm above a stream (6 km E of Lita). At a Reserve of the Universidad de la Paz (Costa Rica), *Sachatamia albomaculata* was abundant in most of the smaller creeks that feed into the upper Río Jaris basin [284]. During the wet season, males and egg clutches were found throughout the stream channel and females deposited their eggs on the upper surfaces of leaves hanging over the water. At the beginning of the dry seasons (December 2001 and 2002), they noted a shift in the reproductive behavior and oviposition sites of this species. A large aggregation of males and egg clutches (*n* = 53; December 2002) were found on rocks in the splash area of two waterfalls; calling males and two egg clutches were also seen on a large boulder in the Jaris river. These observations suggested that in some areas *S. albomaculata* uses wet splash zones to prolong breeding during part of the dry season. The karyotype of *Sachatamia albomaculata* is 2N = 20 [153]. Females provide short-term parental care; prolonged parental care absent [25].

Call: Males typically call from the upper surfaces of leaves but have also been seen calling from plants growing on the trunks of fallen trees crossing streams, from vegetation and branches growing along rocky riparian walls and from rocky surfaces along the streams ([24,148], RWM, pers. obs.). The call consists of a rapid high-pitched single note, "tik"; the average note duration of the call is 0.001–0.002 s; the dominant frequency is 6.1–7.1 kHz [24].

Egg masses and tadpoles: In the oviduct and shortly after laying, the eggs are black-and-white [24,148]. In Costa Rica, during the wet season, females deposit eggs on the upper surface of riparian vegetation; during the dry season, egg clutches are deposited in the splash area of waterfalls [284]. The eggs are distinctive in that a space without eggs often occurs in the center of the mass ([24], RWM, pers. obs.). Egg clutches have 37–81 eggs (mean = 57 ± 18) [147]. The tadpole of *S. albomaculata* was described in detail by Hoffmann [147] and we present a summary of his description below. Embryos in stage 22 are tan in color, whereas hatchlings at stage 25 are dark greyish brown, a coloration that extends to the flanks of the body and onto the venter. Ventral coloration is lighter because of the yellow yolk visible through the skin. The larvae of this species are among the most densely pigmented centrolenids in Costa Rica. While hatchlings are consuming their yolk reserve, the oral disc remains incomplete. More developed tadpoles (stages 25–37) have the typical dorsoventrally compressed body shape, which, together with the larvae of *S. ilex* and *T. spinosa*, are the most compressed of Costa Rican centrolenids. The oral disc is very similar to that of *S. ilex*; it is not emarginate and has a single row of 24–40 marginal papillae with a distinctly broad anterior labium (dorsal gap), a trait that is also present in *S. ilex*; broad dorsal gap; papillae are often flattened, irregular, and sometimes absent. The upper jaw is smoothly curved, and the lower jaw is broad and V-shaped. The LTRF probably is 2(2)/3 but most tadpoles lack the A-1 tooth row; if present, the A-2 is short and consists mostly of a few teeth in short rows along the lateral margins of the upper jaw. Three ridges are evident on the lower labium and all or some of them bear teeth [147]. Whether these irregularities in the labial tooth rows result from laboratory rearing or are typical of the species remains to be demonstrated.

Distribution (Figure 200): *Sachatamia albomaculata* is known from humid lowlands and premontane slopes from north-central Honduras to western Colombia and Ecuador at elevations between 20 and 1500 m ([130,148,160], this work). In Ecuador, this species has been reported from the provinces of Esmeraldas and Imbabura at elevations below 700 m. In Ecuador, the potential distribution of the species is 22,608 km^2 within the Chocoan Tropical Forest and the Western Foothill Forest regions.

Figure 200. Distribution of *Sachatamia albomaculata* in Ecuador (yellow dots).

Conservation status: Listed as *Least Concern* by the IUCN in view of its wide distribution and presumed large population [285]. In Ecuador, the species is known mainly from a few reports in the Chocó, an area with the highest deforestation rate (because of oil palm and wood extraction) and spreading mining. We suggest that the species should be considered as *Endangered* in Ecuador, following IUCN criteria A2, B2a, B2(iii).

Evolutionary relationships (Figure 195): *Sachatamia albomaculata* is sister to a clade formed by *S. punctulata* + *S. electrops*.

Taxonomic Remarks: *S. albomaculata*, as currently defined, has two clear dorsal color patterns: (a) Dorsum with small yellow spots, (b) dorsum with small and large yellow spots (see *Color in life* section). Two scenarios can explain the observed polymorphism: Two or more species occur within what is now recognized as *S. albomaculata*, or *S. albomaculata* has a natural polymorphic coloration. To test the two hypotheses, it is necessary to plot the occurrence of the patterns, document variation and possible sympatry, and obtain acoustic data and molecular samples of the different color patterns for comparative analyses. It would also be worth obtaining samples from the Colombian species *S. punctulata*, which has a color pattern identical to pattern (a) of *S. albomaculata*.

Specimens examined: *Sachatamia albomaculata*: <u>Colombia</u>: *Departamento de Antioquia*: Dabeiba, Río Amparradó, Quebrada Iotó, 805 m, ICN 10685–87. Costa Rica: Limón: Los Diamantes, one mile south of Guápiles, KU 23814 (holotype). <u>Ecuador</u>: *Provincia de Imbabura*: 6 km SE of Lita (0°47′41″ N, 78°25′43″ W), QCAZ 4324–25. *Provincia de Esmeraldas*: Estero Vicente, an affluent of the Río San Miguel 0°47′32″ N, 79°11′52″ W, 225–275 m), QCAZ 11369–70; Reserva Biológica Canandé (0°27′4″ N, 79°08′45″ W, 700 m), MECN 2618–19; 2.1 km E of Durango (1.02477 N, 78.61746 W; 284 m), QCAZ 32172–73; 5 km E of Lita, on the Lita–Ibarra road (0.84773 N, 78.42175 W), QCAZ 40816; Reserva Itapoa (0.513° N, 79.134° W, 321 m), MZUTI 3013.

Localities from the literature: *Sachatamia albomaculata*: Ecuador: *Provincia de Esmeraldas*: Estación Biológica Bilsa (0.359 N, 79.701 W) [160].

Sachatamia ilex (Savage, 1967 [202]; Figures 201–203).

Centrolenella ilex Savage, 1967 [202]. Holotype: LACM 25205.

Type locality: "Costa Rica: Provincia de Limón: Cantón de Limón: Alta Talamanca: 16 km SW
    Amubri, on Río Lari, 300 m"; corrected to 14 km SW Amubri, Río Lari, Cantón de
    Talamanca, Provincia de Limón, 300 m by Savage, 1974 [183].

Centrolene ilex—Ruiz-Carranza and Lynch, 1991 [6].

Sachatamia ilex—Guayasamin, Castroviejo-Fisher, Trueb, Ayarzagüena, Rada, and Vilà, 2009 [1].

Common names: English: Holly's Glassfrog. Spanish: Rana de Cristal de Holly, Rana de
Cristal Fantasma.

Etymology: The specific name *ilex* honors Priscilla Hollister "Holly" Starrett (*Ilex* is the genus of
the holly trees), for her contributions to centrolenid systematics [202].

Identification: *Sachatamia ilex* is distinguished from other glassfrogs by having, in life, a uniform
dark green dorsum, a white to pale yellow or greenish–yellow iris with thick and contrasting black
reticulations (Figure 201), and a moderate body size (male SVL 27.0–29.0 mm; female SVL 28.0–34.0 mm).
Adult males have a pointed humeral spine (Figure 201) that is embedded in the arm musculature (not
easily visible externally). *Sachatamia ilex* closely resembles spotless populations of *Espadarana prosoblepon*
(Figure 95) and *E. callistomma* (Figure 89), the males of which have externally visible humeral spines;
most individuals of *E. prosoblepon* lack the thick black reticulation of the iris that is evident in *S. ilex*.
Other species with uniform green dorsal coloration that could be confused with *S. ilex* are "*Cochranella*"
prasina, which lacks webbing between Fingers II, III, and IV (present in *S. ilex*), and *Teratohyla spinosa*,
which is much smaller in body size (SVL < 21 mm) and lacks humeral spines.

Figure 201. *Sachatamia ilex* in life. (**A,B**) Adult male, Alto Tambo, Esmeraldas province, 620 m, QCAZ
47193. (**C**) Adult male, Alto Tambo–El Placer, Esmeraldas province, 706 m, QCAZ 48338. (**D**) Adult
male, Alto Tambo-El Placer, Esmeraldas province, not collected. Photos by L. A. Coloma.

Diagnosis: (1) Each dentigerous process of the vomer bearing three to five teeth; (2) snout
truncated in dorsal and lateral profiles; (3) tympanum oriented almost vertically, relatively small,
its diameter 26.1%–28.7% of eye diameter; supratympanic fold conspicuous, covering dorsal margin of
tympanic annulus; tympanic membrane pigmented, but clearly differentiated from surrounding skin;
(4) dorsal surfaces of males and females smooth to shagreen, lacking spicules; (5) venter areolate; lacking
pair of enlarged subcloacal warts; (6) anterior two-thirds of ventral parietal peritoneum white, posterior
third transparent (condition P3); white pericardium; translucent peritoneum covering intestines,
stomach, kidneys, testes, gall bladder, and urinary bladder (condition V1); (7) liver tetralobed, covered
by transparent peritoneum (condition H0); (8) in males, humeral spines present, but embedded in arm

musculature; (9) webbing absent between Fingers I and II, reduced between Fingers II and III, extensive between III and IV; webbing formula II ($1^{1/2}$–2^-)—(3^+–$3^{1/4}$) III ($1^{1/3}$–2)—(1^+–$1^{1/3}$) IV; (10) webbing formula on foot I (1–$1^{1/4}$)—($1^{1/2}$–2^-) II (1–1^+)—($1^{3/4}$–2) III (1–1^+)—(2^-–2^+) IV ($1^{1/2}$–2^+)—(1–1^+) V; (11) ulnar fold absent or low and inconspicuous; inner tarsal fold low, short; outer tarsal fold absent; (12) concealed prepollex; in males, nuptial pad Type I; (13) Finger I slightly shorter or about same length as Finger II (Finger I 97.2%–100.0% of Finger II); (14) disc of Finger III relatively narrow, its width about 34.0%–45.9% of eye diameter; (15) in life, dorsum uniform green; upper lip white; bones green; (16) in preservative, dorsum uniform lavender; (17) iris, in life, silvery white to greenish yellow with thick, contrasting black reticulation; (18) melanophores usually present on fingers and toes, except sometimes absent from Toes IV and V; (19) males call from upper sides of leaves near streams; each call is frequency modulated and consists of a single note with a duration of 0.077–0.086 s; dominant frequency starts at 6562–6937 Hz, rises to 7406–7500 Hz, and drops to 6562–6843 Hz; (20) males fight upside down, grasping one another venter to venter; (21) brown eggs deposited on upper side of leaves; parental care unknown; (22) tadpoles with flattened body shape; blackish brown dorsal coloration, venter translucent; upper jaw in form of broad arch; lower jaw V-shaped; labial tooth row formula 0/2-3; (23) medium body size; SVL of adult males 27.0–29.0 mm; SVL of adult females 28.0–35.0 mm.

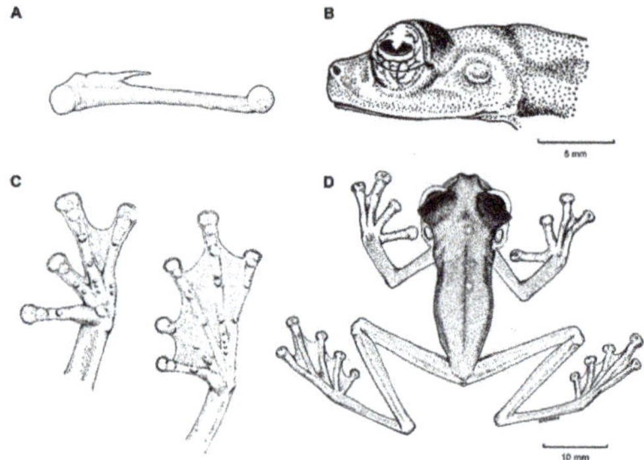

Figure 202. *Sachatamia ilex.* (**A**) Humeral spine of adult male, LACM 72910, not to scale. (**B**) Head in lateral, paratype. (**C**) Hand (left) and foot (right) in ventral view, paratype, not to scale. (**D**) Dorsal view of paratype. (**A**) Modified from Guayasamin et al. [1]. (**B–D**) Modified from Savage [202].

Color in life (Figure 201): Dorsum uniform dark green; throat and venter cream white; dull yellow hands and feet; upper lip with thin white line. White parietal peritoneum covers anterior two-thirds of venter; clear visceral and hepatic peritonea. Bones green. Iris silvery white to greenish yellow with thick, contrasting black reticulations.

Color in ethanol: Dorsal surfaces of head and body uniform lavender, although faint cream spots might be visible in some individuals [202]; limbs creamy lavender dorsally; margin of upper lip white or cream. White parietal peritoneum covering anterior two-thirds of venter; pericardium white; clear peritoneum covering liver, intestines, stomach, kidneys, and urinary bladder. In males, nuptial pad creamy white.

Biology and ecology: Nocturnal and arboreal. On 10 November 1999, at Hacienda La Joya (northwestern Ecuador), one male was calling, and two females were gravid, suggesting reproductive activity [130]. Kubicki [24] described and illustrated the combat behavior of *Sachatamia ilex* as follows: Males hang vertically from vegetation by the tips of their toes and grasp each other chest to chest in a type of arm lock; males may hang in this position for several minutes to hours, and try to dislodge the other male's feet, so that he loses grip of the vegetation. During this behavior, the two males produce softer *preep* calls. Parental care is unknown.

Call: The following call description is based on a recording of one male of *Sachatamia ilex* made by Elicio Tapia on 17 May 2010 at Reserva Otokiki (0.91104° N, 78.57369° W; 706 m), Provincia de Esmeraldas, Ecuador. Males call from the upper surfaces of leaves. Ten calls where emitted during a five-minute interval; the call consists of a single, pulsed note; the note ends with a series of pulses the amplitude of which is conspicuously lower than the pulses at the beginning of the note. Each note has a duration of 0.077–0.086 s ($\overline{X} = 0.08 \pm 0.033$, $n = 6$). The note is frequency modulated; the dominant frequency at the beginning of the note (pulsed section) is at 6562–6937 Hz ($\overline{X} = 6640 \pm 150.2$, $n = 6$); during the tonal section of the note, the dominant frequency increases to 7406–7500 Hz ($\overline{X} = 7422 \pm 38.4$, $n = 6$); finally, it drops to 6562–6843 Hz ($\overline{X} = 6718 \pm 128.0$, $n = 6$). The maximum value of the dominant frequency is always during the first pulsed section of the note.

Egg masses and tadpoles: Brown eggs are deposited as a single layer on the upper leaf surface of plants overhanging streams; sometimes a mass has a slight hole in the center, resembling a doughnut, because they might rotate around from the middle while laying eggs; egg clutches have 12–25 eggs [24]. Species of *Drosophila* have been observed to parasitize egg clutches of *S. ilex* [147]. The tadpole description presented below is a summary obtained from Hoffmann [147]. When hatching, larvae have a dark greyish–brown dorsal coloration; the body is nearly circular in cross-section with a length of about 12 mm; it is longer than other Costa Rican centrolenid hatchlings. In tadpoles that are living off their yolk reserves, oral discs are incomplete. As development continues (Gosner stages 28–40) the body becomes flattened and a dense dorsal pigmentation develops, giving them a blackish–brown ground tone; there is a slightly violet coloration because of the vascularized underlying organs. Lateral body pigmentation decreases ventrally, and the tail coloration is the same as the dorsum, decreasing gradually towards its ventral side. The snout is rounded in dorsal view and acutely angled in lateral view. The anteroventral oral disc has marginal papillae only on the lateral and ventral margins; the anterior labium forms the upper lip and is uniformly broad; the disc is not emarginate. Because of the lack of an anterior disc area, the labial tooth rows A-1 and A-2 are absent. P-1 is usually completely developed, P-2 is shorter not always completely developed, and P-3 lacks teeth or has very few of them [147].

Distribution (Figure 203): *Sachatamia ilex* is known from localities in eastern Nicaragua, Costa Rica, western Panama, western Colombia, and western Ecuador at elevations from sea level up to 1420 m ([130,148,217,284,285], this work). In Ecuador, *Sachatamia ilex* has been found in Esmeraldas and Pichincha provinces at elevations between 150 and 800 m ([130], this work). In Ecuador, the potential distribution of the species is 12,669 km^2 within the Chocoan Tropical Forest and the Western Foothill Forest regions.

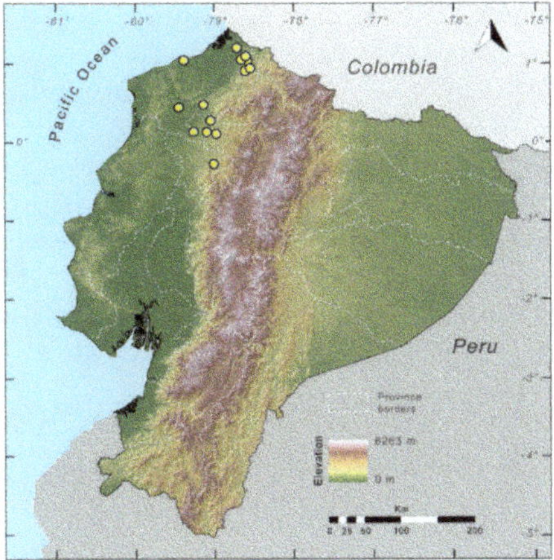

Figure 203. Distribution of *Sachatamia ilex* in Ecuador (yellow dots).

Conservation status: *Sachatamia ilex* is globally listed as *Least Concern* [285]. Kubicki (2007) mentioned that *S. ilex* is fairly common in much of its range along the Caribbean drainage in Costa Rica. In Ecuador, habitat reduction is a serious threat for this and other species that inhabit the Chocó Ecoregion; thus, at the local level, we suggest that the species should be considered as *Endangered*, following IUCN criteria B2a, B2(iii).

Evolutionary relationships (Figure 204): Molecular evidence places *Sachatamia ilex* and *S. orejuela* as sister species.

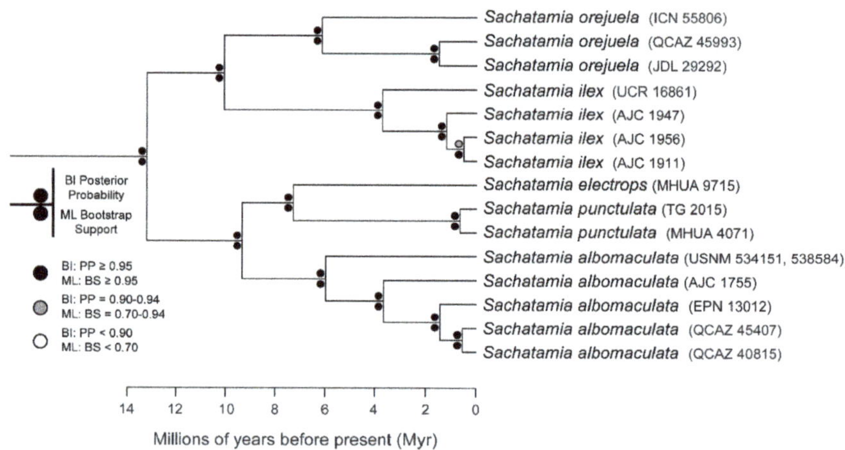

Figure 204. Evolutionary relationships of species in the glassfrog genus *Sachatamia*. The trees were inferred using maximum likelihood and Bayesian criteria.

Specimens examined: *Sachatamia ilex:* <u>Colombia</u>: *Departamento de Antioquia:* Dabeiba, Río Amparradó, Quebrada Iotó, 805 m, ICN 10625–29, 10630 (C&S), 10631–32. Costa Rica: *Provincia de Limón:* Costa Rican Amphibian Research Center, UCR 16861. <u>Ecuador</u>: *Provincia de Esmeraldas:* Reserva Biológica Canandé (0.45° N, 79.14° W, 700 m), MECN 2620–26; Río Tululbí (1.0634° N, 78.59397° W, 189 m), MECN 3199–03; Río Verde (0.92° N, 78.63° W; 300 m), MECN 3204; Río Verde (1.183° N, 78.7166° W; 300 m), MECN 3199; recinto Ventanas (0.89816° N, 78.6175° W. 200 m), MECN 3204; Reserva Otokiki (0.91104° N, 78.57369° W; 706 m), QCAZ 48338; 4 km N of Durango (1.44307° N, 77.99667° W; 253 m), QCAZ 33325; Río La Carolina (0.70449° N, 78.20115° W; 500 m), QCAZ 35363; 2.1 km E Durango (1.02477° N, 78.61746° W; 284 m), QCAZ 32158; 4 km W Durango (1.02348° N, 78.19296° W; 238 m), QCAZ 33057. *Provincia de Santo Domingo de los Tsáchilas:* 4 km NW La Florida, Finca Gloria (0.25694° N, 79.0538° W; 896 m), QCAZ 19881; La Florida, near Alluriquín (0.2836° N, 79.0188° W), QCAZ 13055. *Provincia de Pichincha:* Puerto Quito (0.1167° N, 79.2667° W; 150 m), KU 221613; Hacienda La Joya, km 109 of the Calacalí–Nanegalito–P.V. Maldonado Road, next to the town of San Vicente de Andoas (0.083° N, 78.983° W; 750–800 m), DFCH-USFQ D260–61; near Pedro Vicente Maldonado (0.10421° N, 79.10279° W; 544 m), QCAZ 35429. Panama: *Comarca San Blas:* Camp Summit, 400 m, KU 116464.

Sachatamia orejuela (Duellman and Burrowes, 1989 [86]; Figures 205 and 206).

Centrolenella orejuela Duellman and Burrowes, 1989 [86]. Holotype: KU 145081.
Type locality: "between El Tambo and La Costa, 800 m, Departamento de Cauca, Colombia."
Cochranella orejuela—Ruiz-Carranza and Lynch, 1991 [6].
"Cochranella" orejuela—Guayasamin, Castroviejo-Fisher, Trueb, Ayarzagüena, Rada, and Vilà,
    2009 [1].
Sachatamia orejuela—Twomey, Delia, and Castroviejo-Fisher, 2014 [19].
Teratohyla sornozai—Cisneros-Heredia, Yánez-Muñoz, and Ortega-Andrade, 2009 [286].
    Synonymy by Cisneros-Heredia, Yánez-Muñoz, and Ortega-Andrade, 2010 [287].

Common names: English: Orejuela's Glassfrog. Spanish: Rana de Cristal de Orejuela.

Etymology: The specific name *orejuela* is a patronym for the Orejuela family (Jorge, Anamaría,
and Tomás), who, at the time when the study was conducted, resided, and administered the Reserva
La Planada, Colombia [86].

Identification: *Sachatamia orejuela* is mainly recognized by having a uniformly olive-green dorsum
(Figure 205). In the Pacific versant of the Andes, only the following species have a uniformly green
dorsum: *Nymphargus prasinus*, *Sachatamia ilex*, *S. orejuela*, *Teratohyla spinosa*, and some populations of
E. prosoblepon. From these species, *Sachatamia orejuela* is distinguished by the absence of humeral spines
(spines present in males of *E. prosoblepon* and *S. ilex*), relatively large body size (SVL 27.3–33.8 mm in *S.
orejuela*; SVL < 22.0 mm in *T. spinosa*), and extensive webbing between Fingers III and IV (webbing
absent or basal in *Nymphargus prasinus*). Additionally, *S. orejuela* is found in a very specific microhabitat,
on rocks nearby waterfalls.

Figure 205. *Sachatamia orejuela* in life. Adult male, QCAZ 45993. Locality near Río Aguas Verdes, 670 m,
Imbabura province, Ecuador. Photos by Luis A. Coloma.

Diagnosis: (1) Each vomer with two to five teeth; (2) snout truncated in dorsal aspect, slightly protruding in lateral profile; (3) tympanum small, visible, its diameter 18.5%–25.6% of eye diameter; tympanic membrane pigmented as surrounding skin; supratympanic fold conspicuous; (4) dorsal surfaces finely shagreen, with small spicules evident in sexually active males; (5) ventral surfaces of body areolate; pair of enlarged subcloacal warts absent; in males, cloacal region with minute spinules; (6) anterior half of the parietal peritoneum white (condition P2); white pericardium, translucent visceral peritoneum (condition V1); (7) tetralobed liver covered by transparent peritoneum (condition H0); (8) humeral spines absent; (9) webbing between Fingers I and II basal, moderate between Fingers II and III, and extensive between Fingers III and IV; webbing formula for fingers: II (1–1$^{1/2}$)—(3$^-$–3$^+$) III (1$^+$–2$^-$)—(1–1$^+$) IV; (10) webbing on foot extensive: I 0$^+$—(1$^-$–1) II 0$^+$—(0$^+$–1) III 0$^+$—(0$^+$–1) IV (1$^{1/3}$–1$^{1/2}$)—0 V; (11) ulnar and tarsal folds absent; (12) nuptial pad Type I, concealed prepollex; (13) Finger I slightly longer than Finger II (Finger II about 95%–98% length of Finger I); (14) disc of Finger III large, its width about 62%–70% of eye diameter; (15) in life, dorsum uniform dark olive green; (16) in preservative, dorsum uniform dull grey; (17) iris, in life, dark grey with yellow ring around pupil; (18) melanophores on dorsal surfaces of all fingers and toes; (19) males call from rocks along or within streams; call unknown; (20) fighting behavior unknown; (21) eggs deposition site unknown; parental care unknown; (22) tadpoles unknown; (23) medium body size; SVL 27.3–28.3 mm (*n* = 3) in males; 29.6–33.8 mm (*n* = 6) in females.

Color in life (Figure 205): Uniform dark olive-green dorsum, greyish green flanks; throat translucent, with light green hue; venter whitish cream. Bones green. Iris dark grey to brown ([86], this work).

Color in ethanol: Dorsum, including hands, feet, and webbing dull grey [86]. Anterior half of venter cream white, posterior half translucent; white pericardium, translucent visceral peritonea, and transparent hepatic peritoneum (liver brown).

Biology and ecology: During the night, *Sachatamia orejuela* is mostly found on rocks along steep stream banks or within the stream in the spray zone of cascades ([86,288], this work). As other species adapted to spray zones (e.g., *Centrolene geckoidea*, *C. paezorum*, "*Centrolene*" *petrophilum*, "*Centrolene*" *medemi*), hand and foot webbing in *Sachatamia orejuela* is relatively extensive when compared to glassfrogs not found in this microhabitat. Spiders of the genus *Clubiona* have been observed to prey on juveniles of *S. orejuela* [289]. Parental care is unknown.

Call: Not described.

Tadpole: Not described.

Distribution (Figure 206): *Sachatamia orejuela* is known from four localities in southern Colombia and four localities in northern Ecuador ([86,101,288,290], this work). All localities are on the Pacific flank of the Andean Cordillera Occidental at elevations between 500 and 1250 m, within the Eastern Foothill Forest region. In Ecuador, the potential distribution of the species is 27,514 km^2.

Conservation status: *Sachatamia orejuela* is classified as *Least Concern* by the IUCN [291]. However, in Ecuador, the habitat of the species is fragmented and threatened by logging and mining. We suggest that the species should be considered as *Endangered*, following IUCN criteria B2a, B2(iii).

Evolutionary relationships (Figure 204): *Sachatamia orejuela* and *S. ilex* are inferred as sister species. Herein, we modify the phylogenetic definition of the genus *Sachatamia* [1] as follows: *Sachatamia* is the clade stemming from the most recent common ancestor of *Centrolenella albomaculata* Taylor 1949 [145], and *Cochranella orejuela* Duellman and Burrowes 1989 [86].

Taxonomic Remarks: *Teratohyla sornozai* was synonymized with *Sachatamia orejuela* by Cisneros-Heredia et al. [287]. Examination of the type series of *T. sornozai* showed that all differences that separated it from *S. orejuela* were expressions of ontogenetic and intraspecific variation of the later.

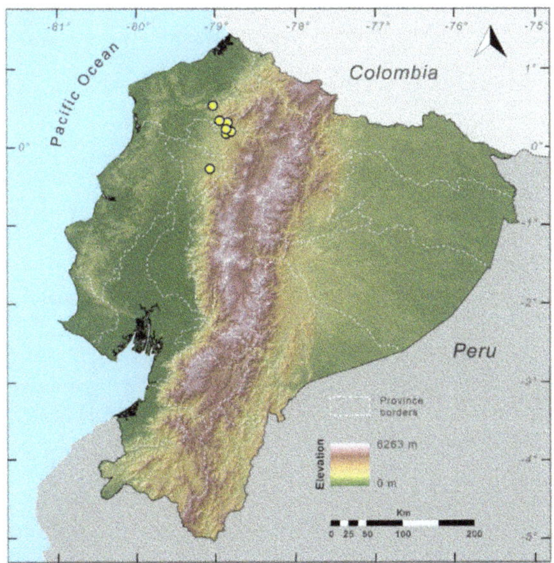

Figure 206. Distribution of *Sachatamia orejuela* in Ecuador (yellow dots).

Specimens examined: *Sachatamia orejuela:* <u>Colombia</u>: *Departamento de Cauca:* between El Tambo and La Costa, 800 m, KU 145081 (holotype), 145080 (paratype). <u>Ecuador</u>: *Provincia de Esmeraldas:* Reserva Biológica Canandé (0.5299 N, 79.0354 S; 550 m), DHMECN 2634. *Provincia de Imbabura:* Zona de amortiguamiento de Reserva Cotacachi Cayapas, near Río Aguas Verdes (0.331010° N, 78.93152° W; 670 m), QCAZ 45993; Reserva Río Manduriacu (0.3108 N, 78.8576 S; 1230 m), JMG 1581. *Provincia de Pichincha:* Bosque Protector Mashpi (00°10′2.34″ N 78°52′2.32″ W; 1200 m), DHMECN 04309; Río Chalpi (00°13′32.38″ N 78°51′28.87″ W; 615 m), DHMECN 04551; Río Anope (00°12′45.54″ N 78°48′58.34″ W; 1080 m), in the surroundings of the town of Saguangal, DHMECN 04552. *Provincia de Santo Domingo de los Tsáchilas:* Trail within Hotel Tinalandia (0.2727 S, 79.079 W), QCAZ 45452.

Localities from the Literature: Colombia: *Departamento de Nariño:* Pialapí, 1250 m, IND-AN 1520–21, LP248 [86]; Departamento Valle del Cauca: Campo Alegre (IUCN, 2010). Ecuador: *Provincia de Esmeraldas:* Reserva Biológica Canandé (0.306 N, 79.138 W; 550 m), DHMECN 2634. *Provincia de Pichincha:* Bosque Protector Mashpi (0.167 N, 78.867 W; 550 m), DHMECN 4308. *Provincia de Imbabura:* Stream tributary of Río Naranjal (0.351 N, 78.917 W; 750 m), DHMECN 3522 ([286] as *Teratohyla sornozai*).

Genus *Teratohyla* Taylor 1951 [15].

Etymology: The generic name *Teratohyla* is derived from the Greek *teras*, meaning monster, marvel or wonder, and the word *Hyla*, traditionally associated with treefrogs. The origin of the frog name *Hyla* is based on the mythological Greek boy *Hylas* and, although the boy's name is masculine, it has been unambiguously treated as feminine by amphibian systematists [292].

Teratohyla amelie (Cisneros-Heredia and Meza-Ramos, 2007 [253]; Figures 207–209).

Cochranella amelie Cisneros-Heredia and Meza-Ramos, 2007 [253]. Holotype: DHMECN 3066.
Type locality: "Comunidad de Oglán, Cantón Arajuno, Provincia de Pastaza, República del
       Ecuador (01°18′65″ S, 77°42′41″ W, 600 m elevation)."
Teratohyla amelie—Guayasamin, Castroviejo-Fisher, Trueb, Ayarzagüena, Rada, and Vilà, 2009 [1].

Common names: English: Amelie's Glassfrog. Spanish: Rana de Cristal de Amelie.

Etymology: The specific name is for *Amelie*, protagonist of the movie "Le Fabuleux Destin d'Amélie Poulain", a movie where little details play an important role in the achievement of *joie de vivre*; like the important role that glassfrogs and all amphibians and reptiles play in the health of our planet. [253].

Identification: *Teratohyla amelie* differs from most species inhabiting the Amazonian lowlands and Amazonian slopes of the Andes by having a completely transparent ventral parietal peritoneum (Figure 207). In Ecuador, only species in the genus *Hyalinobatrachium* and *Chimerella mariaelenae* share the transparent parietal peritoneum character. *Chimerella mariaelenae* differs from *T. amelie* by having a green dorsum with many dark lavender punctuations and scattered larger dark spots in life (dorsum uniform green in *T. amelie*). Also, adult males of *C. mariaelenae* have a small humeral spine (absent in *T. amelie*). *Teratohyla amelie* differs from all species in the genus *Hyalinobatrachium* (characteristics in parentheses) by having a uniform green dorsum in life that turns lavender in preservative (green dorsum with small yellow spots in life; cream in preservative), by depositing its egg clutches on the upper surfaces of leaves (underside of leaves), and by having a Type I nuptial pad (TypeV pad in most *Hyalinobatrachium*). Two other species that are not found in Ecuador, *T. pulverata* and *Vitreorana antisthenesi*, also have a transparent ventral peritoneum and deposit eggs on the upper surfaces of leaves. *Teratohyla pulverata*, however, is restricted to Central America and areas west of the Andes (*T. amelie* is endemic to Amazonian lowlands of Ecuador and Peru), and differs from *T. amelie* by having a sloping snout in lateral view and a green dorsum with white spots in life. The Venezuelan *V. antisthenesi* can be differentiated from *T. amelie* by the presence of vomerine teeth, a green dorsum with light spots, and being larger (21.4–26.2 mm SVL in adult males of *V. antisthenesi* [106]). *Teratohyla midas* is also found on the Amazonian lowlands, but it has yellow dorsal spots and a venter that is white anteriorly and transparent posteriorly.

Diagnosis: (1) Vomerine teeth absent; (2) snout truncated in dorsal view and rounded in lateral view; (3) tympanic annulus evident, oriented dorsolaterally with posterior inclination; very weak supratympanic fold above the tympanum; (4) dorsal skin shagreen; (5) ventral and subcloacal surfaces granular, pair of enlarged subcloacal warts; (6) parietal peritoneum completely transparent (lacking iridophores, condition P0), all visceral peritonea covered with iridophores (condition V5); (7) liver white, bulbous (condition H2); (8) humeral spine absent; (9) webbing on hand absent between Fingers I and II, moderate to extensive between outer fingers: II $(1–1^+)$—$(3–3^+)$ III $(1^{1/2}–2^-)$—$(1–1^{1/2})$ IV; (10) webbing on feet extensive: I $(0–1)$—$(1^{1/2}–2^-)$ II $(1^-–1)$—2^- III $(1^-–1^+)$—$(2^-–2^{1/4})$ IV $(2^-–2^{1/3})$—$(0^-–1^-)$ V; (11) low non-enameled ulnar fold; low non-enameled inner tarsal fold; (12) nuptial excrescences present, Type I; concealed prepollex; (13) first finger longer than second, (14) eye diameter larger than width of disc on Finger III; (15) color in life, uniform green to bluish–green dorsally; (16) in preservative, dorsal surfaces uniform lavender; (17) in life, iris cream brown; (18) numerous melanophores on all fingers and toes; (19) males call from upper sides of leaves; each call consists of single note with three or four pulses; dominant frequency at 4984–7085 Hz, reaching maximum frequency at 5906–6023 Hz; (20) fighting behavior unknown; (21) egg clutches deposited in single layer on upper surfaces of leaves next to streams; males call from tops of leaves next to their clutches; parental care unknown; (22) tadpoles unknown; (23) minute body size; SVL in adult males 18.1–19.3 mm ($\overline{X}$ = 18.8 ± 0.47; n = 6); females unknown.

Figure 207. *Teratohyla amelie* in life. (**Left**): Adult male, QCAZ 37920, Ecuador, Pastaza province, stream tributary of Río Lliquino, 350 m (photo by Martín Bustamante). (**Right**): Adult male, QCAZ 47204, Ecuador, Pastaza province, Reserva Ecológica Río Anzu, 1200 m. Photo by Lucas Bustamante/Tropical Herping.

Color in life (Figure 207): Uniform green to bluish–green dorsum, lacking flecks or spots. Greenish throat, other ventral surfaces transparent. White heart and viscera visible ventrally. Green bones. Iris cream brown.

Color in ethanol: Dorsal surfaces uniformly lavender. Ventral surfaces cream. Ventral parietal peritoneum without iridophores. Iridophores covering heart, digestive tract, and liver.

Biology and ecology: *Teratohyla amelie* is a recently described species and little is known about its natural history. It is nocturnal and males call from and pairs deposit single layer clutches on the upper surfaces of *Heliconia* leaves about 80 cm above small streams in pristine Lowland Evergreen forests. Parental care is unknown.

Call (Figure 208): The description provided below is based on a recording obtained by Diego Paucar at a tributary of Río Lliquino (1.45295° S; 77.443° W, 350 m; QCAZ 38779), Provincia Pastaza, Ecuador, on 15 May 2008. Calls are emitted every 10–30 s ($\overline{X}$ = 17.8 s ± 7.55, n = 8). Each call consists of a single pulsed note that lasts 0.011–0.014 s ($\overline{X}$ = 0.013 ± 0.001, n = 10); each call has three to four pulses ($\overline{X}$ = 3.5 ± 0.515, n = 10). The dominant frequency is between 4984–7085 Hz (n = 10), with its maximum frequency at 5906–6023 Hz; a first harmonic is visible at 10,923–12,474 Hz (n = 10); a second harmonic is visible at 16,815–19,120 Hz (n = 10).

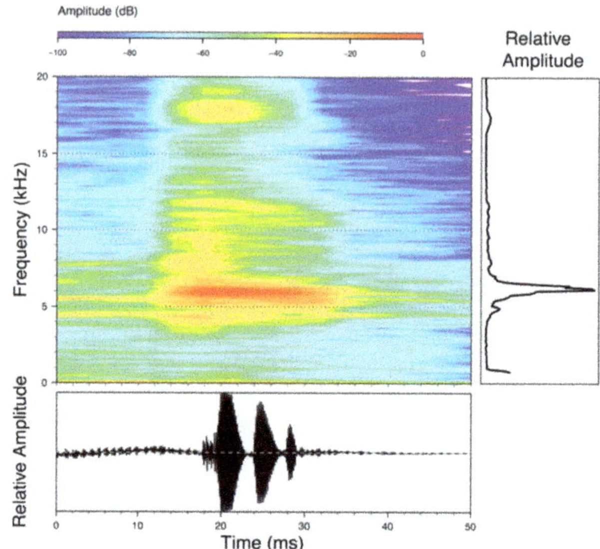

Figure 208. Call of *Teratohyla amelie* (QCAZ 38779) from a tributary of Río Lliquino (1.45295° S; 77.443° W, 350 m; QCAZ 38779), Provincia Pastaza, Ecuador.

Tadpole: Not described.

Distribution (Figure 209): *Teratohyla amelie* is known from a few localities in the foothills and Amazonian lowlands of Ecuador and Peru at elevations between 350 and 1037 m ([19,176,253], this work). In Ecuador, it has been found at only five localities, all within Pastaza Province. In Peru, *T. amelie* is known from a single locality: Km 10 from Quincemil towards Puerto Maldonado (13°12′03.6″ S; 70°40′28.9″ W; 572 m) [19]. In Ecuador, the potential distribution of the species is 24,852 km^2.

Figure 209. Distribution of *Teratohyla amelie* in Ecuador (yellow dots).

Conservation status: *Teratohyla amelie* is classified as *Data Deficient* by the IUCN [293]. However, the distribution of the species is much broader than previously recorded. Thus, we suggest that this species should considered as *Least Concern*.

Evolutionary relationships (Figures 16 and 210): *Teratohyla amelie* is the sister species to *T. pulverata* [2]. Given the lowland distributions of these two species—*T. amelie* east of the Andes and *T. pulverata* from west of the Andes—the most likely mechanism of speciation is vicariance through the uplift of the Andes.

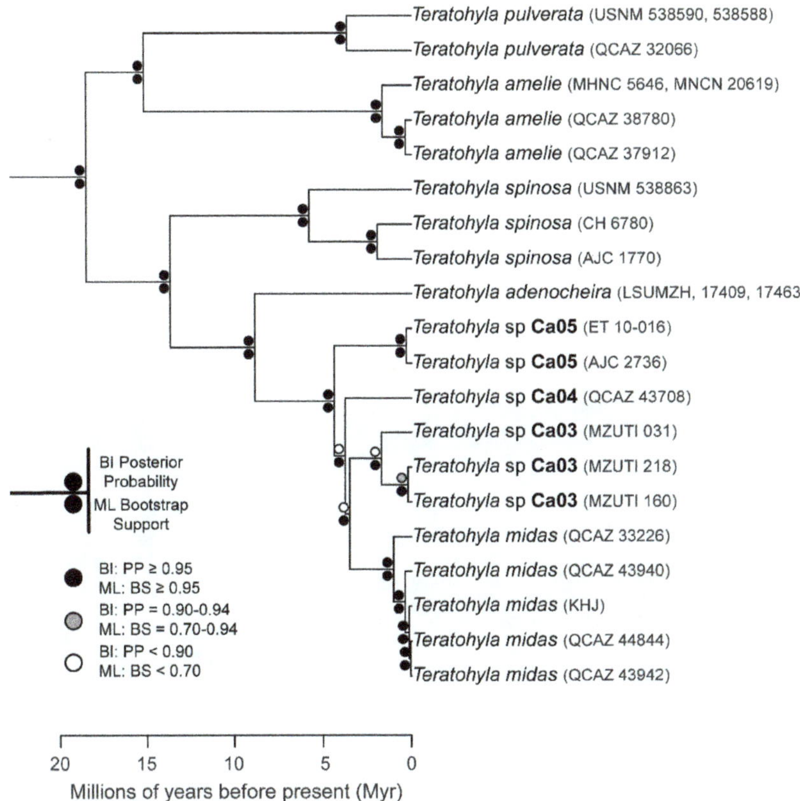

Figure 210. Evolutionary relationships of species in the glassfrog genus *Teratohyla*. The trees were inferred using maximum likelihood and Bayesian criteria.

Specimens examined: *Teratohyla amelie:* <u>Ecuador</u>: *Provincia de Pastaza:* Comunidad de Oglán (1.318° S, 77.711° W, 600 m), DHMECN 3066, 3591; Tributary of Río Lliquino (1.45295° S; 77.443° W; 350 m), QCAZ 37920–21, 38779; tributary of Río Rivadeneira (1,3604° S, 77.86534° W), QCAZ 48734; Reserva Ecológica Río Anzu (1.40608° S; 78.0479° W; 1037 m), QCAZ 47204; Reserva Ecológica Shanca Arajuno, Río Shanca Arajuno (1.35998° S; 77.86564° W; 850–875 m), QCAZ 37912–13; Curintza (2.05747° S, 76.751° W), DHMECN 4372.

Photographic records: <u>Ecuador</u>: *Zamora Chinchipe province:* Reserva Natural Maycu (4.24859° S, 78.6574° W), QCAZ 66821. *Morona Santiago province:* Kimm (3.0146° S, 78.0354° W), QCAZ 73545 [294].

Teratohyla midas (Lynch and Duellman, 1973 [22]; Figures 211–214).

Centrolenella midas Lynch and Duellman, 1973 [22]. Holotype: KU 123219.
Type locality: "Santa Cecilia, 340 m, Provincia Napo (Sucumbíos), Ecuador".
Cochranella midas—Ruiz-Carranza and Lynch, 1991 [6].
Teratohyla midas—Guayasamin, Castroviejo-Fisher, Trueb, Ayarzagüena, Rada, and Vilà, 2009 [1].

Common names: English: Midas' Glassfrog. Spanish: Rana de Cristal Midas.

Etymology: The specific name *Midas* is that of a king in Greek mythology, at whose touch everything turned to gold. The epithet is associated with Río Aguarico (=rich water), near the type locality of the species, in reference to gold found in the river, and in allusion to the gold flecks on the frogs [22].

Identification: Among glassfrogs that inhabit the Amazonian lowlands of Ecuador, *Teratohyla midas* is unique by having a green dorsum with a few small yellow dots that are usually concentrated on the upper flanks (Figure 211); other important diagnostic traits include a white digestive tract, small body size (male SVL < 22.5 mm), and the absence of humeral spines. Other species with similar dorsal coloration that inhabit the Amazonian slopes of the Andes are *Nymphargus cariticommatus*, *N. siren*, *N. humboldti* sp. nov., and *Rulyrana flavopunctata*. All these species have numerous yellow spots uniformly distributed on the dorsum (limited to the upper flanks of the body in *T. midas*). Also, none of these species have a white digestive tract, and species of *Nymphargus* have less hand webbing than *T. midas*.

Figure 211. *Teratohyla midas* in life. Adult male from stream near Tena, 708 m, MZUTI 1621. Photos by Eduardo Toral.

Diagnosis: (1) Vomers with dentigerous process bearing zero to four teeth; (2) snout truncated in dorsal aspect; truncated to bluntly rounded in lateral profile (Figure 212); (3) tympanum moderate, oriented almost vertically, with slight lateral and posterior inclinations, its diameter 29.1%–36.3% of eye diameter; only upper border of tympanic annulus obscured by supratympanic fold; tympanic membrane clearly differentiated from surrounding skin; (4) dorsal skin smooth; males with minute spicules visible only under magnification (×250); (5) venter areolate; pair of enlarged subcloacal warts; (6) white parietal peritoneum covering anterior 40%–50% of venter (condition P2); white pericardium and gastrointestinal peritoneum; translucent peritonea on liver and gall and urinary bladders (condition V2); (7) liver lobate, covered by transparent peritoneum (condition H0); (8) humeral spines absent; (9) webbing absent between Fingers I and II, reduced between Fingers II and III, and moderate between outer fingers (Figure 212); formula: II (2⁻–2)—(3–3$^{1/3}$) III (2⁻–2⁺)—(1$^{3/4}$–2) IV; (10) feet about two-thirds webbed; webbing formula: I (1–1⁺)—(2–2⁺) II (1–1⁺)—(2–2⁺) III (1⁻–1⁺)—(2–2⁺) IV (2–2⁺)—1 V; (11) ulnar and tarsal folds absent; (12) concealed prepollex; in males, nuptial pad Type I; (13) Finger I slightly longer than Finger II (Finger II length 94.1%–98.4% of Finger I); (14) width of disc of Finger

III about 34.4%–39.7% of eye diameter; (15) in life, dorsum green with few small yellow dots usually concentrated on flanks; bones green; (16) in preservative, dorsum lavender with small white dots; (17) iris, in life, silvery white with pale yellow hue, faint brown horizontal stripe, and grey reticulations; yellow ring around pupil; (18) melanophores usually covering fingers and toes; (19) males call from the upper surfaces of leaves; call consists of a series of three notes; each note is 0.15–0.16 s in length, with a dominant frequency of 7030–7060 Hz; (20) fighting behavior unknown; (21) eggs deposited on the upper side of leaves along streams; short-term maternal care present; prolonged parental care absent; (22) tadpoles unknown; (23) minute body size; in males, SVL 19.0–22.5 mm ($\overline{X}$ = 19.9 ± 1.346, n = 8); in females, SVL 20.9–26.8 mm ($\overline{X}$ = 23.0 ± 2.135, n = 8).

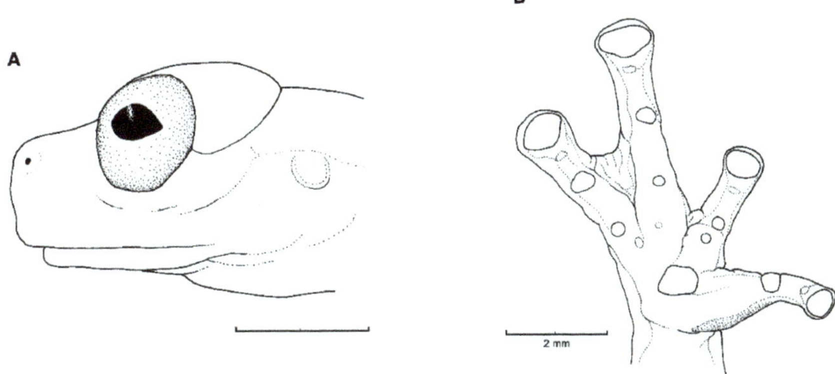

Figure 212. *Teratohyla midas*, QCAZ 33226. (**A**) Head in lateral view. (**B**) Hand in ventral view. Illustrations by Juan M. Guayasamin.

Color in life (Figure 211): Dorsum green with few small yellow spots on upper flanks. White upper lip. Venter white anteriorly and transparent posteriorly. Iris silvery white with pale yellow hue, faint brown stripe, and thin grey reticulations. Green bones.

Color in ethanol: Dorsal surfaces of head, body, and limbs cream lavender to lavender. Anterior half of the venter cream white, posterior half cream with some translucence. Digestive tract white (with iridophores).

Variation: Two paratypes (females KU 107026, 150622) lack yellow spots on the dorsum. Additionally, these individuals were the largest females examined (SVL = 26.8 and 25.4 mm, respectively). When excluding them from the summary statistics for the female body size, the values change as following: SVL = 20.9–23.8 mm ($\overline{X}$ = 21.95 ± 1.009, n = 6). Teeth on the dentigerous process of the vomer are always present in females; vomerine teeth are less evident or absent in males.

Biology and ecology: This species has been found throughout the year on leaves of herbs, shrubs, and trees along small rivulets in the Amazonian rainforest ([22,139], this work). At Santa Cecilia, *Teratohyla midas* is sympatric with *Hyalinobatrachium munozorum* and *Cochranella resplendens* [22]. Diaz-Ricaurte et al. [295] provide a description of the amplexus, oviposition, and parental care behaviors in the species. The observed female laid 27 eggs that, after fertilization, were covered by the female for almost an hour (brooding).

Call (Figure 213): The following description is based on a call of *Teratohyla midas* (LBE-C-036), recorded at the Boanamo community, 240 m, Orellana province, Ecuador, by Morley Read. Calls are emitted in series; 16 calls were recorded in a 3-min period. Time between calls is 10.2–15.1 s (mean = 13.02 ± 1.732). Each call is short and composed by a single note, which has a duration of 0.073–0.092 s (mean = 0.082 ± 0.007). Notes usually have three or four well-differentiated pulses (mean = 3.14 ± 0.378). The dominant frequency measure at its peak is 7210–7390 Hz (mean = 7285 ± 60.4).

Calls are similar to those reported by Twomey et al. [19] for Peruvian populations of the species, although they report a dominant frequency at 6760–7060 Hz.

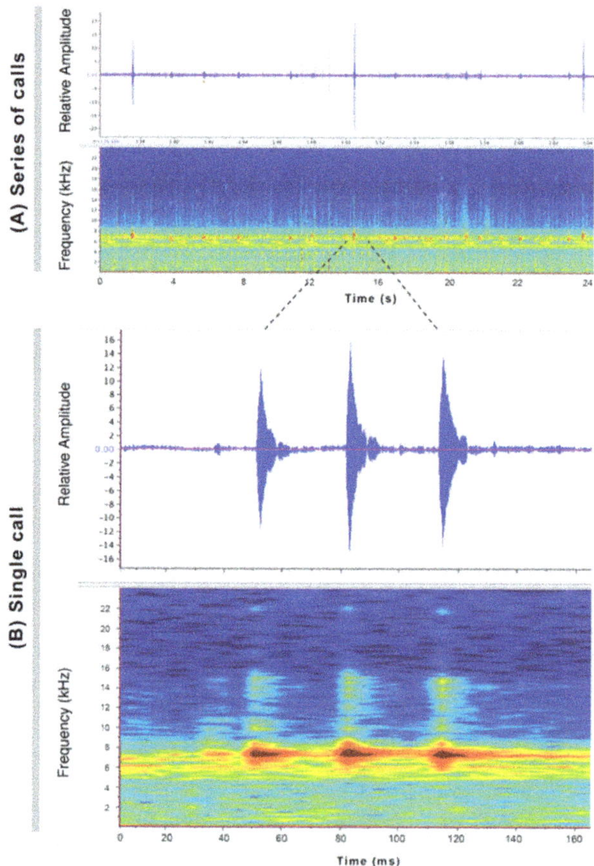

Figure 213. Calls of *Teratohyla midas*, LBE-C-036, recorded at Boanamo community, 240 m, Orellana province, Ecuador, by Morley Read. (**A**) Call series: fragment showing 3 calls. (**B**) Single call.

Tadpole: Not described.

Distribution (Figure 214): *T. midas* has been reported from the Amazonian lowlands (<940 m) of Ecuador, Peru, and Brazil, and is likely to be present in Colombia and Bolivia ([17,19,22,139,229,296], this work). In Ecuador, *T. midas* localities are at elevation of 190–930 m, with a potential distribution of 51,794 km^2 within the Amazonian Tropical Rainforest and the Eastern Foothill Forest regions.

Figure 214. Distribution of *Teratohyla midas* in Ecuador (yellow dots).

Conservation status: Globally, *Teratohyla midas* is listed as *Least Concern* by the IUCN [297]. In Ecuador, the species has a broad distribution; thus, we agree with the current conservation status.

Evolutionary relationships (Figure 210): *Teratohyla midas*, as currently recognized, is the sister species of *T. adenocheira*. However, there is considerable genetic structure within *Teratohyla midas* and it might represent a species a complex.

Specimens examined: *Teratohyla midas:* <u>Ecuador</u>: *Provincia de Morona Santiago:* Río Kusutka (02.111° S, 77.73897° W; 650 m), nearby Wisui Biological Station, MZUTI 031. *Provincia de Napo:* Puerto Misahualli (1.033 S, 77.667 W; 400 m), QCAZ 20001–02; S side of Río Napo, 6.5 km ESE Puerto Misahualli at La Cruz Blanca in Jatun Sacha Biological Reserve (1.05 S, 77.6 W; 395 m), QCAZ 9038–44; Tena, near airport (0.985 S, 77.825 W; 550 m), QCAZ 8819–21. *Provincia de Orellana:* stream nearEstación Científica Yasuní (0.667 S, 76.4 W; 230–240 m), QCAZ 22876, 23895; Estación Científica Yasuní, near the bird observation tower (0.67491° S, 76.39834° W), QCAZ 19316; Río Napo, Añangu (0.52492° S, 76.38445° W; 255 m), QCAZ 43939–44; Tiputini Biodiversity Station (0.6167° S, 76.167° W; 190–270 m), DFCH-D102; km 37–38 on the Pompeya–Iro road (ca. 0.6 S, 76.45 W), QCAZ 17311–12, 17314–16. *Provincia de Pastaza:* Pomona (1.6833 S, 77.883 W; 930 m), QCAZ 33225–26; Estación Hola Vida near Pomona (1.6284° S, 77.9095° W; 837 m), QCAZ 33225–26; Parroquia Teniente Hugo Ortíz (1.37 S, 77.955 W), QCAZ 33252–53; Río Oglán, Curaray (01°19′ S, 77°35′ W; 600 m), USNM 288437; tributary of Río Lliquino (1.18183°S, 77.47879° W; 400 m), QCAZ 37933–34; tributary of Río Lliquino, Villano B Camp (1.45295° S, 77.443° W; 350 m), QCAZ 37916–19; Río Pucayacu (1.374148°S, 77.90955° W; 940 m), QCAZ 33252–53. *Provincia de Sucumbíos:* Río Pañayacu (0.40667 S, 76.113 W; 210 m), QCAZ 20329; Monte Tour (0.0315 S, 76.321 W; 290 m); near Lago Agrio (0.0847 S, 76.8828 W; 370 m), KU 125334; Pozo Peña Blanca, E of Dureno and near Pacayacu (0.0001 S, 76.648 W; 260 m), QCAZ 6385; Puerto Bolívar (0.0886 S, 76.142 W; 240 m), QCAZ 28134, 28137; Puerto Libre (0.061 N, 76.75 W; 280 m), KU 123220–23; Estación Científica PUCE at the Reserva de Producción Faunística Cuyabeno (0.083 S, 76.166 W; 220 m), QCAZ 6015; Río Conejo, 2 km N of Santa Cecilia (0.097778 N, 76.99 W), KU 153256; Santa Cecilia (0.05 N, 76.9667 W; 340 m), KU 105283, 107026, 123219, 146625, 150622–23, 152487, 158518.

Photographic records: *Teratohyla midas:* <u>Ecuador</u>: *Morona Santiago province:* Kimm, 5 km NW airline distance from Santiago (3.0149 S, 78.0356 W), QCAZ 73554 [294]. Road Peñas-Shaimi, 2.8 Km E of Río Yaupi by road (2.9663 S, 77.84682 W), QCAZ 73546–47 [294]. *Orellana province:* Reserva Río Bigal (0.532913° S, 77.423228° W; 981 m), photo by Morley Read. *Pastaza province:* Lorocachi (1.625259 S, 75.99035 W), QCAZ 56024 [294].

Teratohyla pulverata (Peters, 1873 [298]; Figures 215–217).

Hyla pulverata Peters, 1873 [298]. Holotype: ZMB 7842, according to Duellman, 1977 [118].
Type locality: "Chiriqui", Panama; at the time of the description "Chiriqui" included both
 Atlantic and Pacific versants of extreme western Panama, according to Myers, 1982 [299].
Centrolene pulveratum—Dunn, 1931 [236].
Centrolenella pulveratum—Taylor, 1949 [145].
Cochranella pulverata—Taylor, 1951 [15].
Cochranella petersi—Goin, 1961 [97]. Type locality: "Rio Durango, N. W. Ecuador". Placed in
 synonymy by Guayasamin, Cisneros-Heredia, and Castroviejo-Fisher, 2008 [300].
Centrolenella petersi—Goin, 1964 [187].
Centrolenella pulverata—Savage, 1967 [202].
Hyalinobatrachium pulveratum—Ruiz-Carranza and Lynch, 1991 [6].
Hyalinobatrachium petersi—Ruiz-Carranza and Lynch, 1998 [27].
Teratohyla pulverata—Guayasamin, Castroviejo-Fisher, Trueb, Ayarzagüena, Rada, and Vilà,
 2009 [1].

Common names: English: Dusty Glassfrog [24]. Spanish: Rana de Cristal polvosa [24].

Etymology: The specific name comes from the Latin *pulvereus*, meaning dusty, and refers to the minute white spots on the dorsum of the species.

Identification: Among glassfrogs that inhabit the Chocoan forest of Ecuador, *Teratohyla pulverata* is unique by having a green dorsum with small white spots, completely transparent ventral parietal peritoneum (Figure 215), white hepatic and gastrointestinal peritonea, a sloping snout in lateral profile, and by lacking humeral spines. Only species in the genus *Hyalinobatrachium* could be confused with *T. pulverata*; however, *T. pulverata* has green bones (white in *Hyalinobatrachium*), deposits eggs on the upper sides of leaves (undersides of leaves in *Hyalinobatrachium*), and has conspicuous teeth on the vomers (teeth absent in *Hyalinobatrachium*).

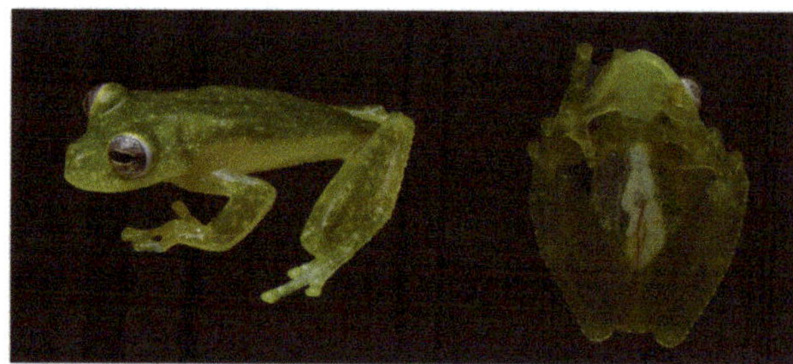

Figure 215. *Teratohyla pulverata* in life. Individual (QCAZ 32224) from Silanche, Pichincha province. Photos by Luis A. Coloma.

Diagnosis: *Teratohyla pulverata* exhibits the following traits: (1) Dentigerous process of the vomer with two to four teeth; (2) snout rounded in dorsal view, sloping in lateral profile (Figure 216); (3) tympanum visible, relatively small, its diameter 20.2%–23.3% of eye diameter; tympanic annulus visible except for upper border covered by supratympanic fold; tympanic membrane differentiated and translucent, pigmented as surrounding skin; (4) dorsal surfaces shagreen; males with small spicules (only visible under magnification); (5) ventral surfaces granular, lacking pair of enlarged subcloacal warts; (6) ventral parietal peritoneum completely transparent (condition P0); pericardium and gastrointestinal peritoneum white (condition V2); (7) white, bulbous liver (i.e., covered by iridophores; condition H2); (8) humeral spines absent; (9) webbing between Fingers I and II absent or basal; webbing formula: II $(1^{+}–1^{1/3})$—$(2^{4/5}–3^{-})$ III $(1^{1/3}–1^{2/3})$—$(1^{+}–2^{-})$ IV; (10) feet about two-thirds webbed; webbing formula:

I (1^--1)—$(1^{2/3}-2^-)$ II (1^--1)—$(1^{3/4}-2^-)$ III $(1-1^+)$—$(1^{2/3}-2^+)$ IV (2^--2^+)—(1^--1^+) V; (11) metacarpal, ulnar, metatarsal, and tarsal enameled folds present, with low, enameled tubercles; (12) nuptial pad Type I in adult males; concealed prepollex; (13) Fingers I and II about same length (Finger II length 98%–103% of Finger I length); (14) disc of Finger III small, its width 20.1%–23.5% of eye diameter; (15) in life, dorsum green with white flecks and dots; bones green; (16) in preservative, dorsum cream to light lavender with unpigmented or white flecks and dots; (17) iris greyish white with thin dark grey reticulations and minute yellow flecks; yellow to cream circumpupilary ring borders the pupil; (18) melanophores partially covering dorsal surface of Finger IV, absent from Fingers I–III; (19) males call from upper side of leaves; call emitted as series of three notes (each note = 0.05 s), dominant frequency at 5600–6200 Hz; (20) fighting behavior unknown; (21) egg masses deposited on upper side of leaves; short-term maternal care present; prolonged parental care absent; (22) in tadpoles, tooth row formula 2/3; A-2 tooth row is widely separated in the center; (23) small body size; adult males, SVL 22.0–24.5 mm (n = 13); adult females, SVL 25.3–28.3 mm (n = 5).

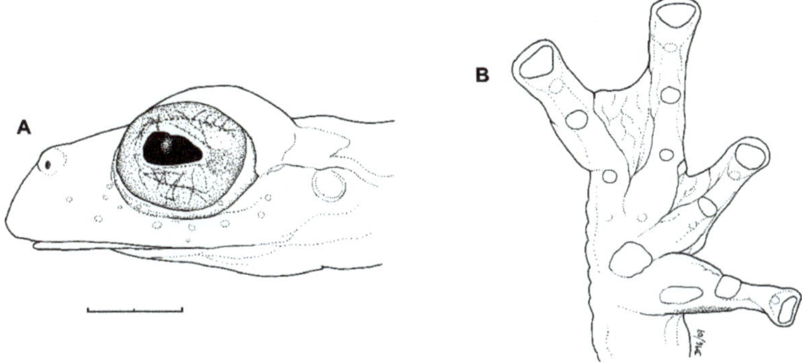

Figure 216. *Teratohyla pulverata*, QCAZ 32066. (**A**) Head in lateral view. (**B**) Hand in ventral view. Scale bar = 2 mm. Illustrations by Juan M. Guayasamin.

Color in life (Figure 215): Dorsal surfaces of head and body green with small, white dots. Dorsally, limbs green with pale yellow spots, with white centers. Upper lip white; small white dots visible below eye. Ulnar and tarsal folds with low, white tubercles. Cloacal region with numerous small, white warts. Parietal peritoneum transparent (all internal organs visible ventrally). White peritonea covering heart, liver, digestive tract, testes, and gall bladder. Transparent peritonea covering urinary bladder and kidneys. Iris grey white with thin dark grey reticulations and minute yellow flecks; thin yellow line around pupil.

Color in ethanol: Dorsal surfaces of head, body, and limbs cream with numerous minute lavender flecks and lacking white iridophores, which are visible in life. Upper lip white; small white spots visible below eye. Low tubercles on ulnar and tarsal folds white, although this coloration is often lost after years of preservation (e.g., KU 116493). Cloacal region with several white warts. Internal organs that are covered by white peritonea in life remain white in preservative (i.e., heart, liver, digestive tract, testes, and gall bladder).

Biology and ecology: As in all centrolenid frogs, the reproductive activity of *Teratohyla pulverata* is in vegetation along streams. In Ecuador, *T. pulverata* was found sympatrically with *Espadarana prosoblepon* and *H. fleischmanni* in Estero Aguacate, and with *Sachatamia ilex*, *S. albomaculata*, *E. prosoblepon*, and *H. aureoguttatum* in Río Bogotá [199]. Delia et al. [25] experimentally demonstrated that short-term brooding provided by females of *T. pulverata* greatly impacts the survival of clutches. This form of parental care reduces mortality associated to dehydration, predation, and fungal infection [25].

Call: Males of *Teratohyla pulverata* call from the upper surfaces of leaves [24]. The call of is a quick high-pitched "tik", normally repeated several times with roughly a second pause between calls [24]. The dominant frequency is at 5.7–6.2 kHz [149,150].

Egg masses: The following information was mostly obtained from Hawley [301]. Unlike other glassfrog egg clutches, the jelly surface is crenulated [17]. The eggs have a yellowish–green coloration and are deposited in a single layer on the upper surface of leaves. The coloration persists for the first five days of development, then changes to pink over the next five days, and lastly to darker shades of red during the remainder of development. The number of eggs per clutch is 23–57 eggs ($n = 62$). Development from oviposition until the last egg hatched was 6–24 days. There was a positive correlation between hatching and nocturnal rainfall. Survivorship per clutch was 85% ± 4% ($n = 59$); 42% percent of clutches hatched without embryonic mortality. Mortality in embryos was caused mainly by flooding (5% or 128 eggs). Desiccation caused mortality in July and August, whereas flooding produced mortality in September. Hatching in embryos in clutches with high mortality is earlier than embryos in clutches with low mortality, perhaps as a response to avoid fungal infection on decaying embryos or predation. Delia et al. [25] reported that clutches contain 59 eggs (±7.5, $n = 50$).

Tadpole: The tadpole of *Teratohyla pulverata* was described by Hoffmann [302]; herein, we present a summary of his description. The ecomorphological guild of the tadpole is exotroph, lotic, and burrower. When hatching (Gosner Stage 24–25) tadpoles have a dark reddish coloration due to the dorsal star-like pigmentation spots and the reddish visceral and muscular system. Young larvae have a nearly circular shape in cross-section, whereas later larval stages change to a typical flat shape. The LTRF is 2/3; the A-2 row had a wide gap; P-3 is slightly shorter than the equal sized P-1 and P-2. The upper jaw sheath forms a smooth arc and has a pointed serration. The lower jaw sheath is smooth and V-shaped. Marginal papillae are present on both sides of the oral disc; the papillae are arranged in a single row but have a wide dorsal gap on the anterior labium. The eye position is dorsal.

Distribution (Figure 217): *Teratohyla pulverata* is known from the humid lowlands on the Atlantic versant from north-central Honduras, and on the Pacific slope from southwestern Costa Rica up to 960 m; it is also known from the lowland forest on the western slopes of the Andes of Colombia and Ecuador at elevations below 400 m ([84,101,148,198,199,300,303], this work). In Ecuador, this species is known from localities in the provinces of Esmeraldas and Pichincha at elevations below 400 m (Specimens Examined). In Ecuador, the potential distribution of the species is 25,146 km^2 within the Chocoan Tropical Forest region.

Conservation status: Globally, *Teratohyla pulverata* is listed as *Least Concern* by the IUCN [304]. Major threats for species inhabiting the Chocó Ecoregion are contamination associated to mining and deforestation because of agriculture (mainly oil palm) and pasture lands. In Ecuador, because of the mentioned threats, we suggest that the species should be considered as *Endangered*, following IUCN criteria B2a, B2(iii).

Evolutionary relationships (Figure 210): Guayasamin et al. [1] mentioned that the placement of *Teratohyla pulverata* in the genus *Teratohyla* is supported by a dataset that includes nuclear and mitochondrial genes, and the combined nuclear genes. They also mentioned that the topology recovered by one particular mitochondrial gene (ND1) strongly supports a clade composed of *Vitreorana castroviejoi* + *V. antisthenesi* + *T. pulverata*. Most genetic data at hand favor a sister relationship between *T. pulverata* and *T. amelie* ([2,3,19], this work). Given the lowland distributions of these two species—*T. amelie* east of the Andes and *T. pulverata* from west of the Andes—the most likely mechanism of speciation is vicariance through the uplift of the Andes.

Figure 217. Distribution of *Teratohyla pulverata* in Ecuador (yellow dots).

Remarks: From our observations, it is evident that iridophores (white iridophores) are usually maintained in preserved specimens during long periods of time. For example, the specimen KU 85476 was collected on August 1964 and still maintains the white colorations in most of its internal organs (only a portion of the digestive tract has lost the iridophores and now is cream). However, small amounts of iridophores are diffused in ethanol. In *Teratohyla pulverata*, the white dots on the dorsal surfaces and ulnar and tarsal folds are not visible in some preserved specimens. Given the importance of the presence/absence of iridophores in centrolenid taxonomy, it is important that researchers take the time to carefully record and photograph the coloration in life.

Specimens examined: *Teratohyla pulverata:* <u>Costa Rica</u>: *Puntarenas:* Rincón de Osa, UCR 17417, USNM 219379–87. <u>Ecuador</u>: *Provincia de Esmeraldas:* Río Durango (1.04186° N, 78.62405° W; 243 m), BM 1902.5.27.24 (holotype of *H. petersi*), BM 1902.5.27.25, QCAZ 32066, QCAZ 45414, DHMECN 2612, 3194–3195; Río Zapayo (0.78333° N, 78.9333° W), BM 1902.7.29.36–37; Río Bogotá, DHMECN 3194–95; Reserva Ecológica Cotacachi-Cayapas, Charco Vicente, 60 m, QCAZ 11367–68; Estero Aguacate (00°39′ N, 80°03′ W; 10–64 m), Parroquia San Francisco del Cabo, DHMECN 3194–95. *Provincia de Pichincha:* Silanche (0.1333° N, 79.1333° W; 400 m), QCAZ 32224. Nicaragua: Matagalpa: Finca Tepeyac, 10.5 km N and 9.0 km E of Matagalpa, 960 m, KU 85476. <u>Panama</u>: Coclé: Quebrada Guabalito, Palmarazo, Parque Nacional Omar Torrijos, CH 5122; Darién: Río Jaque, 1.5 km above Río Imamado, 50 m, KU 116493. Honduras: Olancho, USNM 342214–21.

Teratohyla spinosa (Taylor 1949 [145]; Figures 218–221).

Centrolenella spinosa Taylor, 1949 [145]. Holotype: KU 23809.
Type locality: "Los Diamantes, one mile south of Guápiles, (Cantón de Pococí, Provincia de
  Limón,) Costa Rica". Savage, 1974 [183], commented on the type locality.
Teratohyla spinosa—Taylor, 1951 [15].
Cochranella siren—Ruiz-Carranza and Lynch, 1991 [6].
Teratohyla spinosa—Guayasamin, Castroviejo-Fisher, Trueb, Ayarzagüena, Rada, Vilà, 2009 [1].

Common names: English: Dwarf Glassfrog [24], Minute Spiny Glassfrog. Spanish: Rana de
Cristal Enana [24], Rana de Cristal espinosa.
Etymology: The specific name comes from the Latin *spina*, meaning thorn, and refers to the
spine-like prepollex that is evident next to the thumb (Figure 12).
Identification: Among glassfrogs that inhabit the lowlands of western Ecuador (Chocó Ecoregion),
Teratohyla spinosa is unique in having a uniform green dorsum, a white to yellowish–white iris with
contrasting black reticulations (Figure 218), small size (SVL < 21.0 mm), and lacking humeral spines.
Also, this species is exceptional by having a spine (prepollex) that is clearly separated from Finger I
(Figures 12 and 219).

Figure 218. *Teratohyla spinosa* in life. (**Top row**): Adult male from Durango, Esmeraldas province,
QCAZ 45410. (**Bottom row**): Amplectant pair from Reserva Otokiki, Esmeraldas province, Ecuador.
Photos by Luis A. Coloma.

Diagnosis: (1) Vomers with dentigerous process that bear or lack teeth; (2) snout truncated in
dorsal aspect; truncated to slightly protruding in lateral profile (Figure 219); (3) tympanum with
pronounced lateral inclination, its diameter about 23.5%–32.0% of eye diameter; upper border of

tympanic annulus obscured by supratympanic fold; tympanic membrane clearly differentiated from surrounding skin; (4) dorsal skin smooth, lacking spicules; (5) ventral skin smooth to slightly areolate; lacking pair of enlarged subcloacal warts; (6) white parietal peritoneum covering anterior 50%–60% of venter (condition P3); white pericardium; no iridophores in peritonea covering intestines, stomach, and kidneys; translucent peritoneum around gall and urinary bladders (condition V1); (7) liver lobate, covered by transparent peritoneum (condition H0); (8) humeral spines absent; (9) webbing absent between Fingers I and II, reduced between Fingers II and III, and extensive between Fingers III and IV (Figure 219); formula: II ($1^{2/3}$–2^-)—(3^+–$3^{1/4}$) III (2^-–2)—(1^+–$1^{2/3}$) IV; (10) distinct prepollex; in males, nuptial pad Type I (see Remarks); (11) Finger I slightly longer than Finger II (Finger II length 90.1%–92.5% of Finger I); (12) ulnar fold low; tarsal folds absents; (13) feet with relatively extensive webbing; formula: I 1—(2^-–2) II (1^-–1)—(2^-–2^+) III (1^+–$1^{1/2}$)—(2^+–$2^{1/2}$) IV (2^-–$2^{1/2}$)—(1^-–1) V; (14) width of Finger III disc about 33.1%–40.4% of eye diameter; (15) in life, dorsum uniform green; bones green; (16) in preservative, dorsum creamy lavender; (17) in life, iris white to yellowish white, with black contrasting reticulations; (18) melanophores present only on dorsal surfaces of Finger IV (and sometimes Finger III) and Toes IV and V; (19) males call from the upper surfaces of leaves; each call consists of one to three pulsed notes; the dominant frequency is at 6469–6937 Hz; (20) fighting behavior unknown; (21) eggs deposited on the underside of leaves; short-term maternal care present; prolonged parental care absent; (22) oral apparatus of tadpoles with tooth row formula 0/0; (23) minute body size; in males, SVL 18.0–19.7 mm ($\overline{X}$ = 19.0 ± 0.525, n = 10); in females, SVL 19.7–20.7 mm (n = 3).

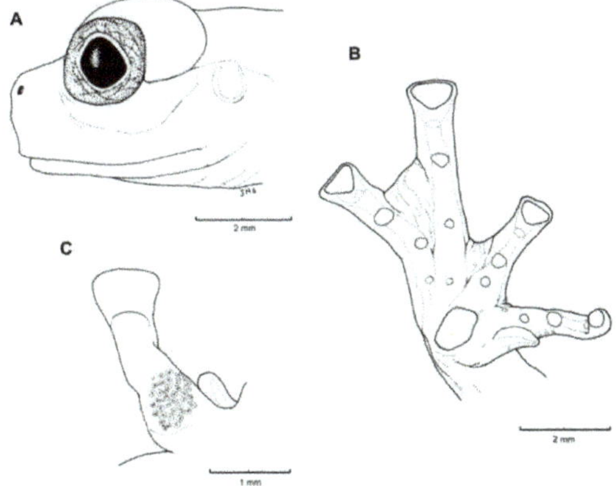

Figure 219. *Teratohyla spinosa*, adult male. (**A**) Head in lateral view, QCAZ 31321. (**B**) Hand in ventral view, KU 164668. (**C**) Finger I in dorsal view, QCAZ 10450; note nuptial pad and prepollical spine. Illustrations by Juan M. Guayasamin.

Color in life (Figure 218): Yellowish–green dorsum. Anterior half of venter white, posterior part transparent. Green bones. Iris silvery white to yellowish white, with marked black reticulations.

Color in ethanol: Dorsal surfaces of head, body, and limbs cream with light lavender hue. Anterior half of venter white, posterior part translucent. Translucent peritonea covering gall and urinary bladders. Iridophores absent from digestive tract, liver, and kidneys.

Biology and ecology: In Ecuador, *Teratohyla spinosa* has been found breeding along streams near San Francisco de Bogotá (25 May 2006) and Alto Tambo (April 2010). Individuals were active during the night; several males were calling and amplectant pairs were observed. Females provide short-term parental care, whereas any form of prolonged parental care is absent [25].

Call (Figure 220): Males call from the upper surfaces of leaves. Ibáñez et al. [197] described the call as a series of harsh "creep-creeps" followed by a considerable pause, with a dominant frequency at 6.8–7.2 kHz. The description provided below is based on a recording obtained by JMG at Reserva

Otokiki (0.91104° N, 78.57369° W; 706 m), Esmeraldas province, Ecuador. Calls are emitted every 12–35 s; each call usually has two notes (range = 1–3 notes, $\overline{X}$ = 1.92 ± 0.515, n = 12; Figure 220); calls with one note have a duration of 0.124–0.139 s (n = 2); calls with two notes last 0.431–0.594 s ($\overline{X}$ = 0.545 ± 0.067, n = 9); a single three-note call had a duration of 0.793 s. Notes are pulsed, with 5–15 pulses per note ($\overline{X}$ = 7.87 ± 2.096, n = 23). The dominant frequency is at 6469–6937 Hz ($\overline{X}$ = 6664 ± 129.2, n = 12); a first harmonic is visible at 12,843–13,687 Hz ($\overline{X}$ = 13351 ± 305.6, n = 12); a second harmonic is sometimes visible at 19,312–20,812 Hz ($\overline{X}$ = 20039 ± 9885, n = 8).

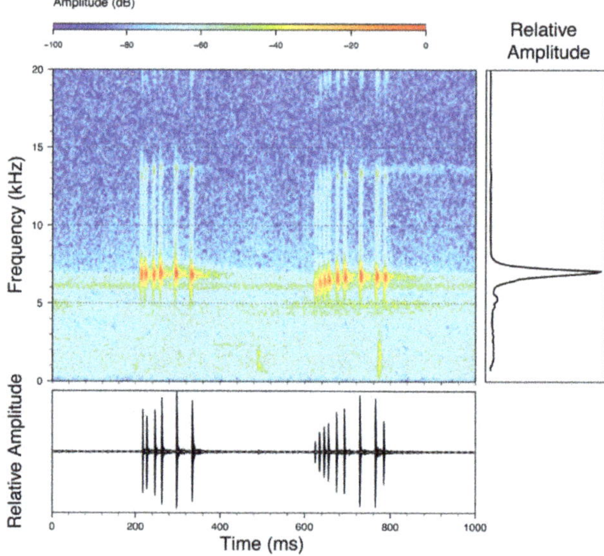

Figure 220. Call of *Teratohyla spinosa*, LBE-C-002, recorded at Reserva Otokiki, 706 m, Esmeraldas province, Ecuador.

Egg masses and tadpoles: Egg masses are attached on the underside of the leaves near their margins and 0.5 to 2.5 m above the water and sometimes over the shore [24,146,147]. When laid, the eggs of *Teratohyla spinosa* are greenish white; the embryos are initially yellow and become darker during development, reaching a dark grey, brownish, or reddish–tan color at hatching [24,146,147]. Egg masses contain 15–25 eggs, with each egg and associated capsule measuring 5–7 mm in diameter; clutch mass is small, measuring about 21 × 25 × 7 mm [24,146,147]. The following description of the tadpole is primarily derived from Savage [148], but with some modifications (RWM, pers. obs.). Tiny on hatching, total length about 11 mm at stage 25; body elongated, depressed; mouth ventral; nares and crescent-shaped eyes dorsal; midlateral, posterior, sinistral spiracle; vent tube median; tail long with rounded tip; caudal fins reduced. Oral disc complete, moderate, with single row of marginal papillae (21–45) laterally and ventrally, wide dorsal gap above mouth; LTRF probably 2/3, but when rows are present, rows of teeth are weaker and sometimes irregular; A-1 row often not evident [147]; A-2 usually present, but often visible only as traces of teeth on the lateral extremes of the tooth ridge; jaw sheaths robust and well developed, upper broadly arched and lower one v-shaped, both with sharp serrations along their entire lengths. Body and tail greyish brown dorsally, lighter ventrally; bright yellow yolk mass colors posterior half of the body and covers the liver and intestine; tail musculature mostly brown except pale area anteroventrally; fins are opaquely transparent with scattered dark spots. Starrett [146] and Hoffmann [147] also provided descriptions of the tadpoles. Starrett mentioned that her specimens lacked labial teeth and suggested that possibly this species does not develop labial teeth. Savage [148] considered her specimens to be anomalous and possibly the result of raising the tadpoles in captivity. Hoffmann agreed, but also noted how variable the tooth rows were in his specimens, which also were lab-raised. A better understanding of the nature of the labial teeth will only become available with the examination of field collected larvae of *T. spinosa*.

Distribution (Figure 221): *Teratohyla spinosa* is distributed from eastern Honduras, through Nicaragua, Costa Rica, and Panama in Central America [84,134,145,148,305–308]. In South America, it is found in the Chocó Ecoregion of Colombia south to Río Palenque in Ecuador [86,101,134]. In Ecuador, the species has been reported from the provinces of Esmeraldas, Pichincha, and Los Ríos at elevations below 700 m. In Ecuador, the potential distribution of the species is 38,671 km^2, within the Chocoan Tropical Forest and the Western Foothill Forest regions.

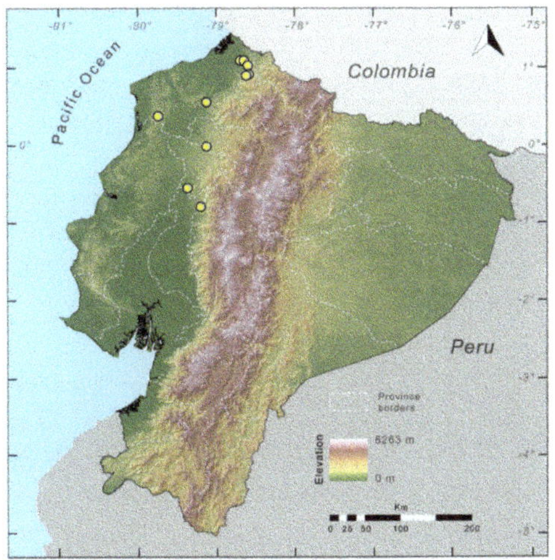

Figure 221. Distribution of *Teratohyla spinosa* in Ecuador (yellow dots).

Conservation status: Globally, *Teratohyla spinosa* is listed as *Least Concern* by the IUCN [309]. In Ecuador, the habitat of the species (Chocó) is continuously being fragmented by logging activity and habitat transformation towards oil palm plantations; therefore, we suggest that the species should be considered as *Endangered*, following IUCN criteria B2a, B2(iii).

Evolutionary relationships (Figure 210): *Teratohyla spinosa* is sister species to a clade formed by *T. midas* + *T. adenocheira*. *Teratohyla spinosa* is found in the Chocó, whereas *T. midas* and *T. adenocheira* are endemic to the Amazon basin. The most like speciation mechanism that separated *T. spinosa* from its closest relatives is vicariance through the uplift of the Andes.

Specimens examined: *Teratohyla spinosa:* <u>Ecuador</u>: *Provincia de Cotopaxi:* 6 km W of Guasaganda (0.797° S, 79.212° W), QCAZ 6809. *Provincia de Esmeraldas:* 2 km E of San Francisco de Bogotá, on the San Francisco–Durango road (1.08585° N, 78.68575° W; 77 m), QCAZ 32124–25; 1.3 km E of San Francisco de Bogotá, on the San Francisco–Durango road (1.0888° N, 78.69563° W), QCAZ 32138; near Durango (1.07934° N, 78.66954° W; 74 m), QCAZ 31321; 4 km N of Durango (1.44307° N, 77.99667° W; 253 m), QCAZ 33319; Bosque Integral Otokiki (0.91241° N, 78.58092° W; 700 m), QCAZ 48283; 7 km N of Durango in the Durango–San Lorenzo road (1.07934° N, 78.66954° W; 74 m), QCAZ 31321. *Provincia Los Ríos:* Estación Biológica Río Palenque (0.55° S, 79.3667° W; 220 m), KU 164663–68. *Provincia Pichincha:* Río Blanco, Río Yambi (00°01′ S, 79°08′ W; ca. 700 m), USNM 288443.

Localities from the literature: <u>Ecuador</u>: *Provincia de Esmeraldas:* Estación Biológica Bilsa (0.36667 N, 79.750 W), Reserva Ecológica Mache Chindul [310].

Photographic record: <u>Ecuador</u>: *Provincia de Esmeraldas:* Reserva Tesoro Escondido (0.5337 N, 79.1445 W), QCAZ 65398 [294].

Vitreorana ritae (Lutz in Lutz and Kloss, 1952 [311]; Figures 222–226).

Centrolene ritae Lutz in Lutz and Kloss, 1952 [311]. Holotype at MNRJ, now lost. Neotype: MCZ
    A96522.
Type locality: "headwaters of Río Caiwima, a tributary of the Río Amayaca-Yacu, ca. 70 km NNE
    Puerto Nariño, Amazonas, Colombia (approximately 3°20′ S, 70°20′ W)" Cisneros-Heredia,
    2013 [312].
Centrolenella oyampiensis Lescure, 1975 [313]. Holotype: MNHNP 1973.1673. Type locality:
    "village Zidok (Haut-Oyapock), Guyane française". Placed in synonymy by Cisneros-
    Heredia, 2013 [312].
Cochranella oyampiensis—Ruiz-Carranza and Lynch 1991 [6].
Centrolenella ametarsia Flores, 1987 [314]. Holotype: MCZ 96522. Type locality: "the headwaters
    of Río Caiwima, a tributary of the Río Amaca-Yacu, ca. 70 km NNE Puerto Nariño,
    Amazonas, Colombia (approximately 3°20′ S, 70°20′ W)". Placed in synonymy by Cisneros-
    Heredia, 2013 [312].
Cochranella ametarsia—Ruiz-Carranza and Lynch 1991 [6].
Vitreorana oyampiensis—Guayasamin, Castroviejo-Fisher, Trueb, Ayarzagüena, Rada, and Vilà,
    2009 [1].
"*Cochranella*" *ritae*—Guayasamin, Castroviejo-Fisher, Trueb, Ayarzagüena, Rada, and Vilà,
    2009 [1].
Vitreorana ritae—Cisneros-Heredia, 2013 [312].

Common names: English: Rita's Glassfrog. Spanish: Rana de Cristal de Rita.

Etymology: The name *Vitreorana* has its origin in the Latin words *vitreum*, meaning glass, and
rana, meaning frog. The name refers to the total or partial ventral transparency of these frogs [1]. The
specific name *ritae* is a patronym to Rita Kloss, assistant of Bertha Lutz.

Identification: Among glassfrogs, *Vitreorana ritae* is unique by having a green dorsum with small
black spots (Figure 222), a small size (≤24 mm), white gastrointestinal peritoneum, and a distinct
prepollex (Figure 12). The only glassfrog that could be confused with *V. ritae* is *V. helenae*, which differs
by having a yellow iris (greyish white with a fine dark reticulation in *V. ritae*), light greenish–yellow
dorsum with dark punctuations (green with dark punctuations in *V. ritae*; Figure 222), and a mostly
white hepatic peritoneum (hepatic peritoneum mostly transparent, showing the brown liver, except for
some iridophores on the upper border in *V. ritae* [300].

Diagnosis: *Vitreorana ritae* has the following combination of traits: (1) Dentigerous process of
vomer with one tooth or lacking teeth; (2) snout rounded in dorsal and lateral views (Figure 223);
(3) tympanum visible, moderate in size, its diameter 25.8%–35.4% of eye diameter; tympanic annulus
visible except for upper border covered by supratympanic fold; tympanic membrane differentiated and
translucent, pigmented as surrounding skin; (4) dorsal surfaces shagreen; spinules absent; (5) ventral
surfaces granular, pair of enlarged subcloacal warts; (6) anterior 25%–40% of ventral parietal peritoneum
white, posterior portion transparent (condition P2–P3); pericardium and gastrointestinal peritoneum
white (condition V2); (7) lobed liver covered by transparent peritoneum, except for anterior border that
may be covered by thin layer of iridophores (condition H0); (8) humeral spines absent; (9) webbing
between Fingers I–III absent, moderate between Fingers III and IV (Figure 223); webbing formula:
III $(2^{-}-2^{1/3})$—$(1^{+}-2^{-})$ IV; (10) webbing between toes moderate; webbing formula: I 1—$(2^{-}-2)$ II
$(1-1^{+})$—$(2-2^{1/4})$ III $(1^{+}-1^{1/2})$—2^{+} IV $(2-2^{1/3})$—1 V; (11) low ulnar fold, lacking iridophores; low inner
tarsal fold present, lacking iridophores; outer tarsal fold absent; (12) nuptial pad of males Type I;
distinct prepollex; (13) Finger I slightly longer than Finger II (Finger II length 84%–92% of Finger I);
(14) disc of Finger III moderate, its width 31.0%–42.3% of eye diameter; (15) in life, dorsum green
with small dark spots (Figure 222); bones green; (16) in preservative, dorsum lavender with dark
spots; (17) iris greyish white with thin dark reticulation; (18) melanophores covering dorsal surfaces of

Fingers III and IV; (19) males call from the upper side of leaves; single and double note advertisement call of 0.10–0.15 s duration, dominant frequency at 4640–5160 Hz; (20) fighting behavior unknown; (21) eggs deposited on upper or underside of leaves; short-term maternal care present; prolonged parental care absent; (22) tadpoles at stage 25 with labial tooth row formula 0/1–2; oral disc small and ventral with one row of large marginal papillae laterally and posteriorly; upper jaw sheath wide and robust, lower jaw sheath wide, V-shaped; dorsum reddish brown, venter whitish, tail muscle reddish and tail fins transparent (Figure 224); (23) minute body size; adult male SVL 17.1–20.1 mm ($\overline{X}$ = 18.8 ± 1.250, *n* = 6); two adult females SVL 19.8–19.9 mm; Lima et al. [315] provided the following data for the species in central Amazonia: SVL in males 17–21 mm, in females 21–24 mm.

Figure 222. *Vitreorana ritae* from Yasuní National Park, Ecuador. Photo by Jaime Culebras.

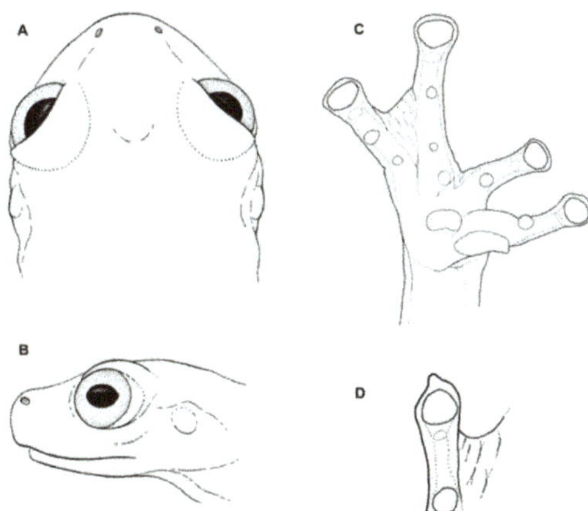

Figure 223. *Vitreorana ritae.* (**A**) Head in dorsal view, MCZ 96522. (**B**) Head in lateral view, MCZ 96522. (**C**) Hand in ventral view, with exposed prepollex, ICN 50847. (**D**) Toe I in ventral view, with papilla at tip. Not drawn to scale. (**A,B**) Modified from Flores [314]; (**C,D**) by Juan M. Guayasamin.

Color in life (Figure 222): Green dorsum with small dark blue to black spots. Green bones. Iris greyish white with thin dark reticulation. Venter white anteriorly, turning transparent posteriorly; white digestive tract.

Color in ethanol: Dorsum lavender with small dark lavender spots. Color of parietal and visceral peritonea variable; holotype and one Colombian specimen (ICN 50847) with transparent parietal and visceral peritonea. Two specimens (ICN 50846, QCAZ 28138) with white parietal peritoneum on anterior half of venter, and white peritoneum on intestines and stomach. We consider that the white lining has been dissolved in the holotype and ICN 50847, but it is equally possible that *Vitreorana ritae* is polymorphic for the presence of white lining on the parietal and visceral peritonea. It may also be possible that the species is a composite of more than one species.

Biology and ecology: During the night, *Vitreorana ritae* has been found on vegetation, about 1–5 m above streams, and on the trunk of a *Ceiba*, about 7 m above the ground [300,314]. At Reserva Florestal Adolpho Ducke in Manaus, Amazonas, Brazil, breeding individuals and clutches of *V. ritae* were observed on vegetation above small streams in forests [300]. Adults were found throughout the year, and vocalizations were heard throughout the night and were more common between January and May (from the middle to the end of the rainy season). Short-term maternal care is present, whereas any form of prolonged parental care is absent [25].

Egg masses and tadpoles (Figure 224): The information shown below is from Menin et al. [316]. At Reserva Florestal Adolpho Ducke, Amazonas, Brazil, clutches were deposited on the upper (*n* = 4) or undersides (*n* = 2) of leaves hanging over a stream, from 0.5 to 1 m above water surface. Mean clutch size was 16 eggs (*n* = 4; range 12–18). The eggs were green and deposited in a single layer with adhered jelly capsules. Tadpoles were green while in the egg capsules but became reddish brown with transparent tail after hatching. Tadpoles from a single clutch raised in the laboratory hatched at Gosner's [191] stage 24. Tadpoles were found buried in the leaf litter along the edges of slow-flowing streams. The description below is based on eleven tadpoles at stage 25. In life, dorsum reddish brown, venter whitish, tail muscle reddish and tail fins transparent. In preserved specimens, dorsum transparent with dark melanophores, venter whitish with melanophores on posterior part and tail translucent with melanophores on tail musculature. Body elongate and depressed in lateral view, and rectangular and elongate in dorsal view. Body and tail 27%–28% and 72%–73% of total length, respectively. Body wider than deep; maximum body width behind eyes. Snout broad and truncated in dorsal and ventral views and flattened in lateral view. Eyes small, close together, and facing dorsolaterally. Interorbital distance six times larger than maximum eye diameter. Narial openings small, nearer snout than eyes. Spiracle single, short, sinistral, positioned on longitudinal axis of body and directed posteriorly. Vent tube short, positioned along ventral midline and attached directly to ventral fin. Tail musculature robust; caudal fins low, similar in height, deeper than the robust caudal musculature only on the posterior one-third of the tail. Dorsal fin originates at tail–body junction and increases in depth throughout the first third of the tail, and then gradually decreases to a rounded tip. Ventral fin originating on the tail musculature, slightly arched, and maintaining the same height throughout the distal two-thirds of the tail. Tail tip rounded. Oral disc small and ventral (Figure 224), bordered laterally and posteriorly by one row of large marginal papillae with pointed tips. Marginal papillae on posterior border larger than lateral ones. Labial tooth row formula (LTRF) 0/1–2. All labial tooth rows nearly the same length. Upper jaw sheath wide and robust; lower jaw sheath wide, V-shaped; both hardly serrated.

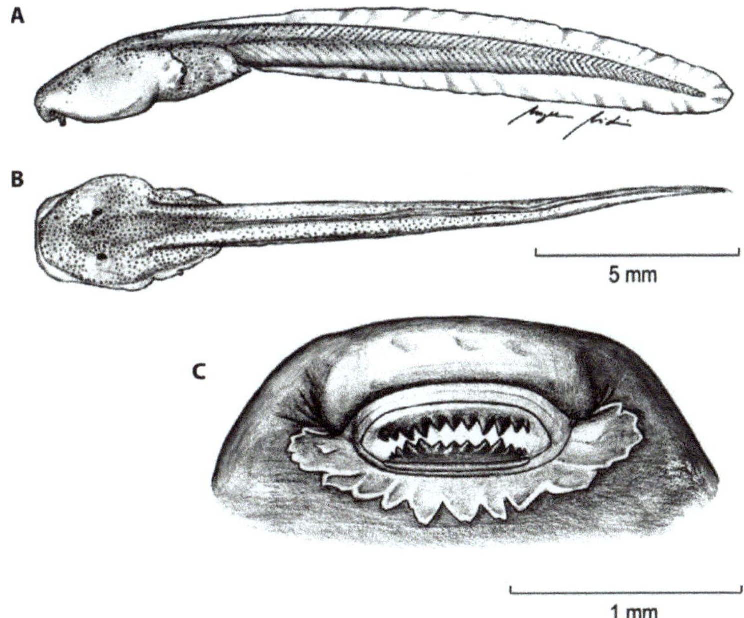

Figure 224. Tadpole of *Vitreorana ritae* at Gosner stage 25. (**A**) Lateral view. (**B**) Dorsal view. (**C**) Oral disc. Modified from Menin et al. [316].

Call (Figure 225): At central Amazonia of Brazil, males of *Vitreorana ritae* were observed calling on vegetation above streams, about 0.5 to 3 m high on leaves of aquatic or stream side plants (*n* = 20) [316]. Herein we report a call obtained by Morley Read at Boanamo community (LBE-C-039; 240 m) and Shiripuno Lodge (LBE-C-039; 250 m), Orellana province, Ecuador. We analyzed 31 notes contained within 10 calls from two individuals. The typical advertisement call is relatively short (range = 288–1095 ms, mean = 685.2 ms, SD = 280.1 ms) and contains two to five notes per call (mean = 3.3, SD = 1.1). Each note has a duration of 61–122 (mean = 76, SD = 16.3) ms. Notes are strongly pulsed and have 10–19 (mean = 13.7, SD = 2) amplitude peaks throughout the note, whereby some pulses are difficult to delimit because of a lower amplitude. Pulses within a note have a rate of 131–224 (mean = 171, SD = 24) pulses per second. Notes have their peak amplitude in the last 50% of the note (relative peak time: Range = 0.4425–0.8815, mean = 0.719, SD = 0.106), where the amplitude increases from a lower to a higher amplitude at the end of the note. The dominant frequency of a note measured at peak amplitude is 4688–5344 (mean = 5044, SD = 182) Hz and is contained within the fundamental frequency. The fundamental frequency has a lower limit of 4312–4875 (mean = 4605, SD = 164) Hz and a higher limit of 4969–6029 (mean = 5284, SD = 229) Hz. The call described for the Ecuadorian populations matches the general information provided by Zimmerman and Bogart [317] for Brazilian populations.

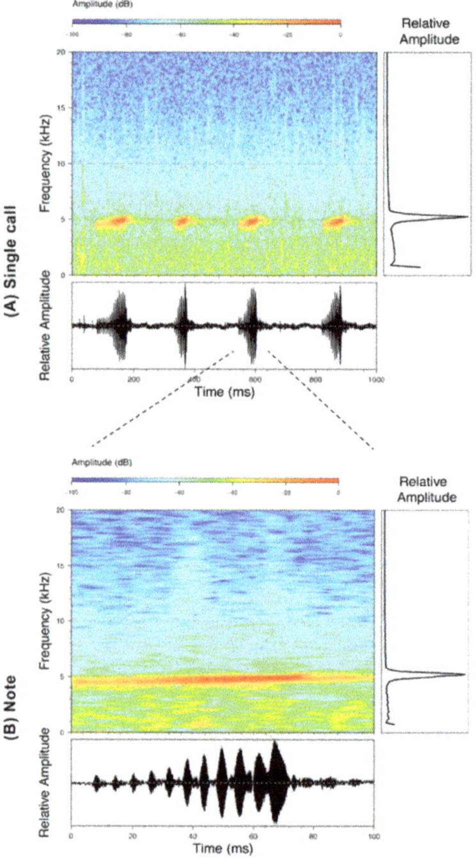

Figure 225. Calls of *Vitreorana ritae*, LBE-C-039, recorded by Morley Read at Boanamo Community, 240 m, Orellana province, Ecuador. (**A**) A single call with four notes. (**B**) A single note.

Distribution (Figure 226): *Vitreorana ritae* is found from the Guianas across western Amazonia [17,222,229,300,313–316,318]. In Ecuador, the species has been reported from five localities in the provinces of Orellana and Sucumbíos at elevations below 260 m (Specimens Examined). In Ecuador, the potential distribution of the species is 42,600 km^2 within the Amazonian Tropical Rainforest region.

Conservation status: *Vitreorana ritae* is listed as *Data Deficient* by the IUCN [297]. This species is known only from a handful of localities but is likely to occur more widely in the Amazon basin. Further surveys with emphasis on sampling bromeliads may provide a better understanding of the population dynamics of this species [166,319]. Given the broad distribution of the species in the Amazon basin, we suggest that it should be placed in the *Least Concern* category.

Evolutionary relationships (Figure 227): *Vitreorana ritae* is sister to *V. helenae*.

Taxonomic remarks: Cisneros-Heredia [312] placed *C. ametarsia* and *C. oyampiensis* under the synonymy of *C. ritae*, arguing that several of the traits described by Lutz [311] were misinterpreted and that Lutz's description of apparently "large size" of the discs of *C. ritae* was misrepresentative. We adopt the change proposed by Cisneros-Heredia [312] but note that new material of *ritae* from its type locality ("Benjamin Constant, Alto Solimões", Estado do Amazonas, Brazil) will provided a definitive solution to any taxonomic uncertainty.

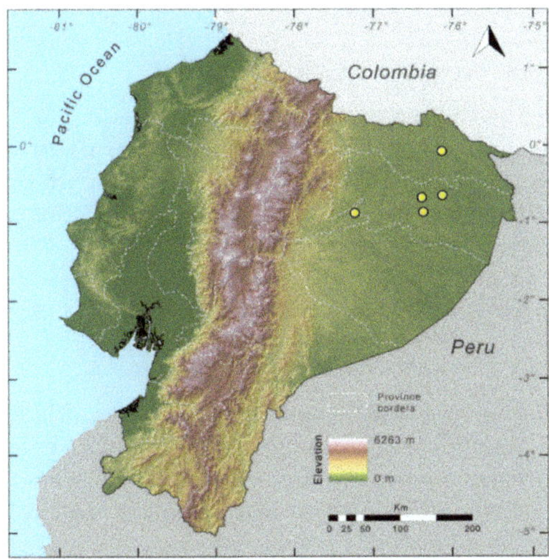

Figure 226. Distribution of *Teratohyla ritae* in Ecuador (yellow dots).

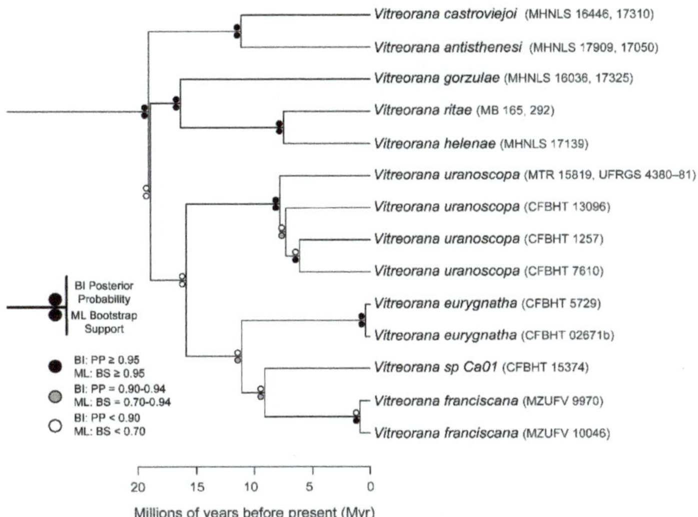

Figure 227. Evolutionary relationships of species in the glassfrog genus *Vitreorana*. The tree was inferred using maximum likelihood and Bayesian criteria.

Specimens examined: *Teratohyla ritae:* <u>Colombia</u>: *Departamento de Amazonas:* headwaters of Río Caiwima, tributary of the Río Amayaca-Yacu, MCZ A-96522 (holotype of *C. ametarsia*, neotype of *V. ritae*); Leticia, ICN 50846–47, ICN (JDL 24472). <u>Ecuador</u>: *Provincia de Orellana:* Río Yasuní, 200 km upstream from Río Napo (0.85° S, 76.383° W; 200 m), KU 175216; Tiputini Biodiversity Station (0.65° S, 76.13° W; 210 m), DFCH-USFQ D162; Estación Científica Yasuní PUCE, 240 m, QCAZ 16652; Yasuní, km 8 on the Pompeya-Iro road, 260 m, QCAZ 22709. *Provincia de Sucumbíos:* Puerto Bolívar (0.09° S, 76.14° W; 240 m), 260 m, QCAZ 28138.

5. Discussion

5.1. Evolutionary Relationships of Glassfrogs and Other Anurans

Historically, glassfrogs have been thought to be closely related to hylids [93,94,298,320–324], although Jiménez de la Espada [83] considered *Centrolene geckoidea* to be a Neotropical representative of *Rhacophorus*, and Taylor [15] believed the South African heleophrynid frogs and glassfrogs were closely related. Noble, as early as 1931 [321], suggested a close affinity between glassfrogs and *Allophryne ruthveni*, which he described as a "toothless *Centrolenella*". The hypothesis by Noble has been supported by molecular and morphological studies [2,8,11–14,325–328], but see Haas [329], and by now it is well established that Allophrynidae is the sister family of Centrolenidae.

5.2. Evolutionary Relationships and Generic Placement within Centrolenidae

The molecular evolutionary trees of glassfrogs, constructed since 2008, have been accompanied by a reinterpretation of the systematics, biogeography, richness patterns, behavior, and speciation of the group ([1,3,19], this work). Continuous and collaborative efforts have allowed a better taxon and gene sampling, increasing the accuracy of the inference of evolutionary relationships [330–340].

Our phylogenetic sampling includes 75% of the described diversity of Centrolenidae and all but three (*"Centrolene" medemi*, *Nymphargus buenaventura*, *N. laurae*) of the known Ecuadorian species. All genera are supported and congruent with the taxonomy proposed by Guayasamin et al. [1]. Thus, recognized genera within Centrolenidae are stable and meet the stability criteria described by Guayasamin et al. [1] and Vences et al. [341]. However, relationships among genera are not fully resolved and further genetic data are necessary. At the species level, relationships are stable and well supported in most cases, allowing straightforward generic placement and facilitating recognition of cryptic species. Among Ecuadorian species, only one species, *"Centrolene" medemi*, still has an uncertain generic placement. A full species list of Centrolenidae and their current generic placement is provided in Table S1.

5.3. Biogeography

Our divergence dating results show that Centrolenidae and Allophrynidae separated from each other 25–55 Mya, and centrolenids started to diversify about 25–41 Mya, which are slightly older ages than previous estimates [3]. The most recent estimate of divergence times found an age of ~45 Mya for the split between Centrolenidae and Allophrynidae [328]. The geographic origin of glassfrogs is placed, unambiguously, in South America, originating at mid-elevations or climatically similar habitats [3,28]. From there, glassfrogs have dispersed into Central America multiple times [2,3]. Notably, the lowland clades (e.g., *Hyalinobatrachium*, *Teratohyla*) have colonized Central America, whereas Andean clades with high levels of species richness (e.g., *Nymphargus*, *Centrolene*) are completely absent from Central America [3]. This is likely in part due to climatic-niche conservatism constraining centrolenids from dispersing to lowland regions from the Andes [3,28].

Even though glassfrog diversity thrives in the Andes (Figures 228 and 229), the specific ancestral biogeographic region of centrolenid origins is ambiguous [3]; what we do know, however, is that the most common recent ancestor inhabited mid-elevation mountains (1000–2000 m), and that lower and higher elevation habitats were colonized more recently [3,28]. Also, the Andes facilitated an explosion of species that is best explained by greater time (ancient cradle or montane museum hypothesis; or time-for-speciation effect more generally) rather than faster diversification at mid-elevations (montane species pump), despite the recently of the major Andean uplift [3,28,342]. At a broad scale, speciation within the Andes is produced, mainly, by a combination of landscape heterogeneity (e.g., with both longitudinal and transversal barriers) and climatic-niche conservatism in species [28]. There are several examples of niche conservatism at the generic level; most species of *Centrolene*, *Nymphargus*, *Celsiella*, and *Ikakogi* are restricted to mountains, whereas others such as *Hyalinobatrachium*, *Vitreorana*, and *Cochranella* are more prevalent in lowlands.

Elevational distribution of glassfrogs in the Pacific lowlands and Andean slopes of Ecuador

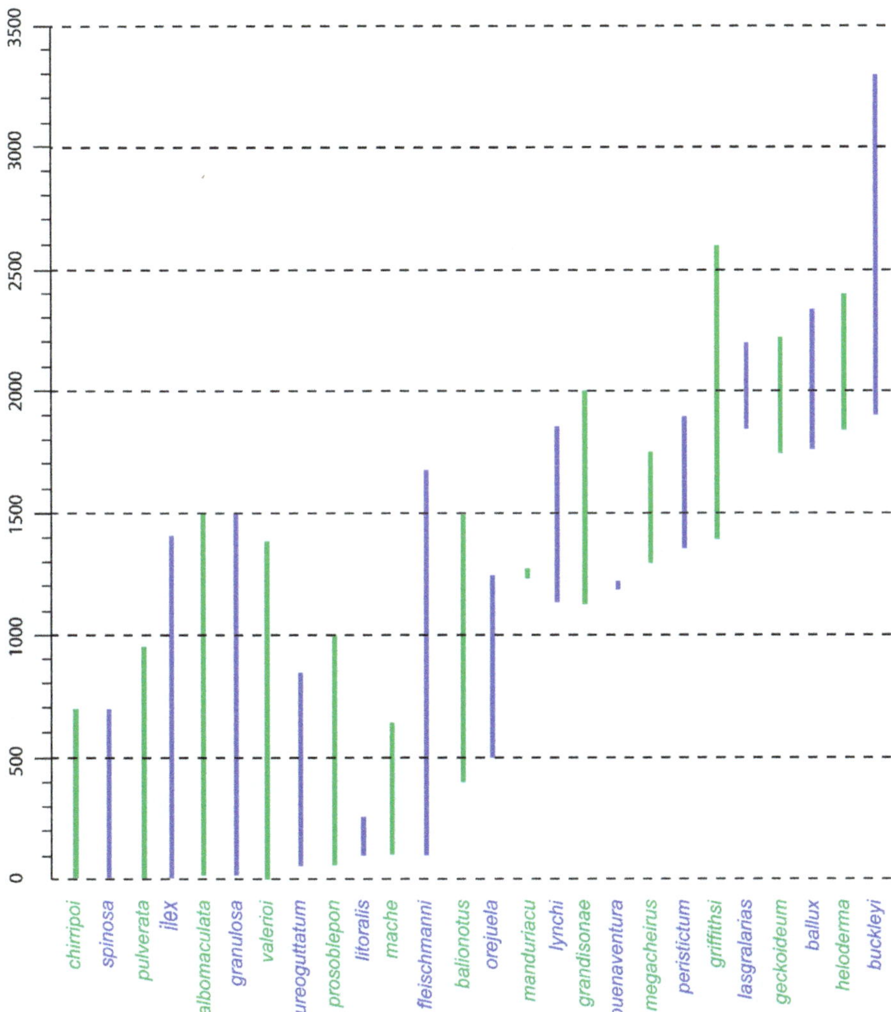

Figure 228. Elevational distribution of glassfrogs found on the Pacific lowlands and western slopes of the Ecuadorian Andes.

Elevational distribution of glassfrogs in the Amazonian lowlands and slopes of Ecuador

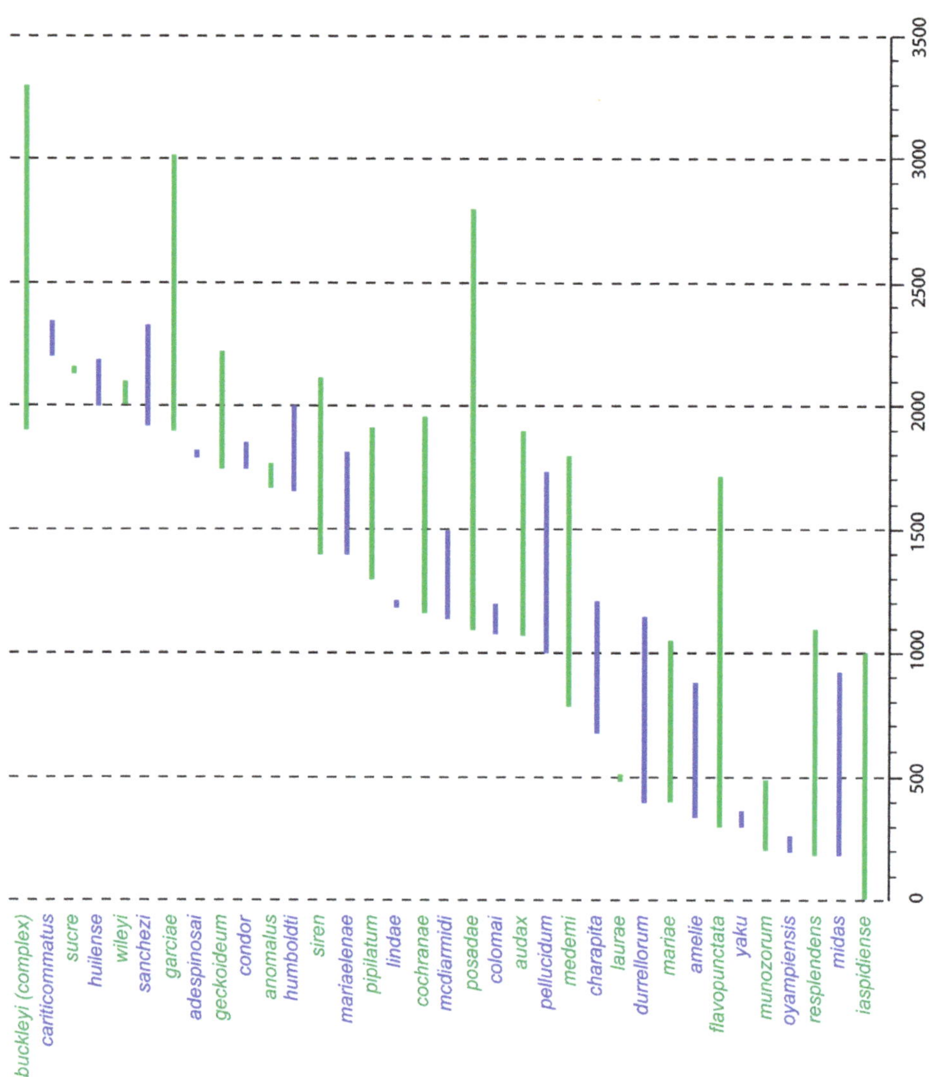

Figure 229. Elevational distribution of glassfrogs on the Amazonian lowlands and Amazonian slopes of the Ecuadorian Andes.

Glassfrogs have shown to be successful, in terms of species numbers, mostly in tropical montane areas (Figures 230 and 231). There is a clear reduction of species richness as latitude increases, with low temperature and low precipitation being the most likely limiting factors. These two variables also explain the absence (or near absence) of glassfrogs in dry or highly seasonal areas (Cerrado, Llanos, Chaco) and high-elevation regions within the tropics (e.g., Páramo; Figures 230–232).

In Ecuador, glassfrogs also reach their diversity peak in cloud and montane forests (Figure 232). They have a moderate species richness in the Choco and Amazonian Rainforest and are completely absent from xeric and dry areas of southwestern Ecuador, although they can be found in semideciduous forests (Figure 232). Below, we describe the patterns and diversification mechanisms that are prevalent in Centrolenidae.

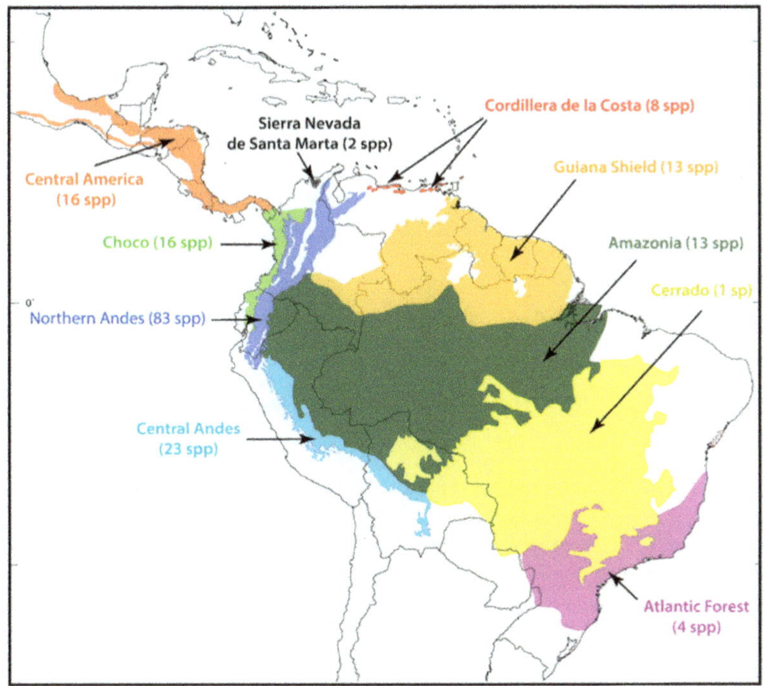

Figure 230. Biogeographical regions in the Neotropics (sensu Duellman [343]). Each region is followed by its number of glassfrog species. Note that one particular species can occur in more than one biogeographic area. Modified from Castroviejo-Fisher et al. [3].

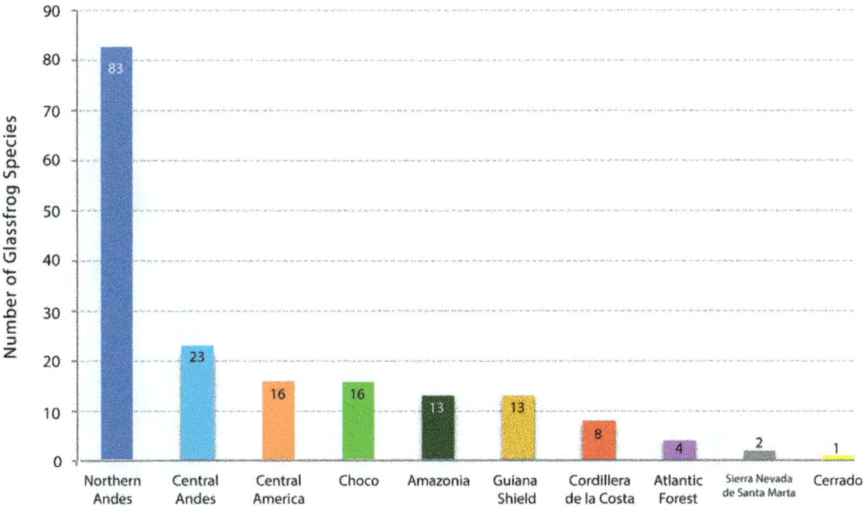

Figure 231. Species richness of glassfrogs per biogeographic region in the Neotropics.

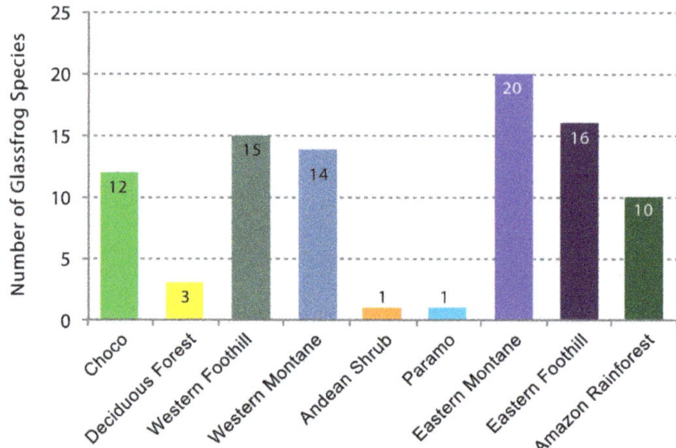

Figure 232. Species richness of glassfrogs per biogeographic region in Ecuador.

5.4. Speciation

Speciation reflects the interactions among several biotic, abiotic, and historical events that have operated over various temporal and spatial scales. Most models that attempt to explain population divergence assume a role of geography (e.g., allopatry, parapatry) combined with niche conservatism or divergence to generate species [344–346]. For example, the linearity of the Andes results in elongate geographical ranges and reduces potential contact and gene flow among parapatric populations [347,348] (Figure 233). Pleistocene glaciations have produced recurrent fragmentation, isolation, and reconnection of montane forests and their faunas [349–353], but see Bush and Oliveira [354]. Species that today occupy the lowlands of the Chocó and Amazon ecoregions may had been separated by the uplift of the Andes, creating the opportunity for large-scale, co-occurring allopatric speciation.

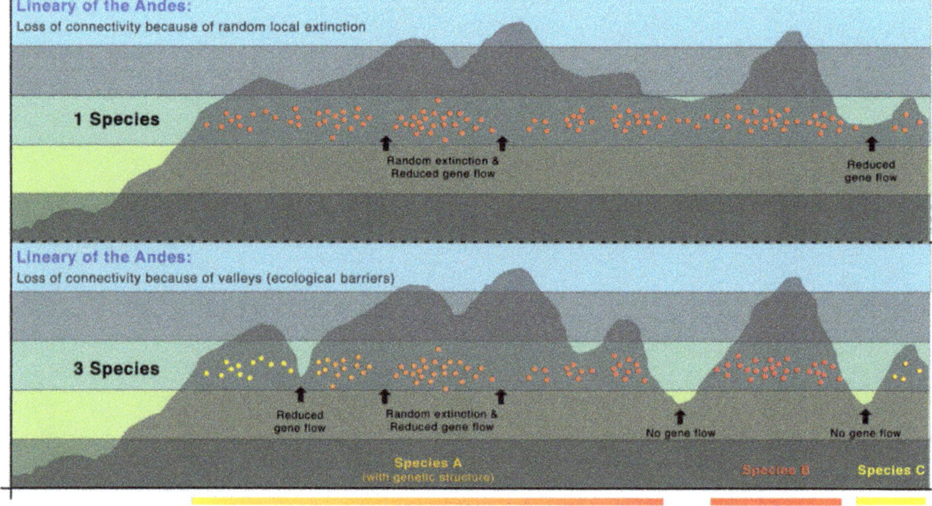

Figure 233. Speciation modes. Effect of the linearity of the Andes in diversification processes. Modified from Guayasamin et al. [194].

Speciation may or may not be coupled with an ecological shift (Figure 234). The tendency of populations to retain their ancestral ecological niche and failing to adapt to the new environmental conditions facilitates lineage divergence when ecological barriers are present [344–356]. Recent studies demonstrate that closely related centrolenid species have, in most cases, similar abiotic ecological niches and that elevational shifts are rare in the family [3,28] (Figure 233). Then, lineage divergence in glassfrogs seems to be driven mostly by allopatric speciation coupled with niche conservatism [28]. There are, however, clear exceptions where clades are adapted to very different climates, meaning that niche divergence has occurred at least a few times (e.g., *Hyalinobatrachium* is mostly a lowland clade, whereas *Centrolene* and *Nymphargus* are montane clades). The relevance of niche conservatism in glassfrog diversification is at odds with studies that predicted and showed that, in amphibians, the primary causes of speciation were adaptation to climates (elevated regions vs. lowland regions) coupled with fragmentation of the once contiguous lowlands; in other words, allopatric speciation with ecological evolution [357,358] (Figure 234).

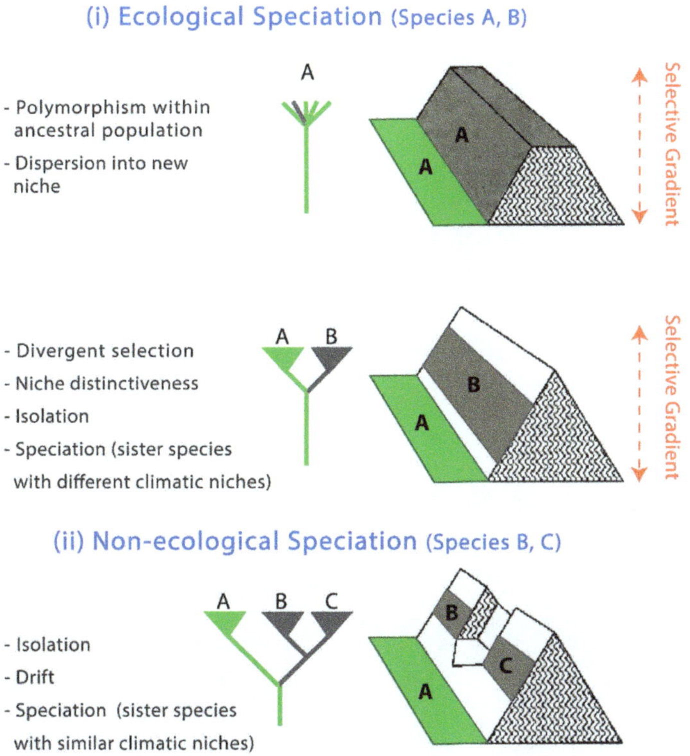

Figure 234. Speciation modes. (**i**) Ecological speciation in elevational gradients. (**ii**) Non-ecological speciation in the same elevational band.

Giving that our phylogeny includes the most complete taxon sampling so far, we are presented with the opportunity to search for specific lineages that match the different scenarios that explain tropical diversity. We acknowledge that many more species of glassfrogs remain to be discovered, changing the topology that is the backbone of our interpretations; however, we expect that these variations will be relatively minor and that the general patterns of speciation will hold. Also, we work under the assumption that geography is a primary player in the process of speciation and that closely related species are likely to be geographically nearby; this assumption is supported by the relatively low vagility that amphibians have in relation to other vertebrates [259,359–362], and that speciation is likely to produce sister species that are found nearby (allopatry), adjacent (parapatry), or in the same place (sympatry) [345,363]. Thus, below, we summarize the speciation hypotheses that are relevant for the Ecuadorian Andes and lowlands and include the examples that fit these scenarios.

- *Allopatric speciation of lowland species mediated by the uplift of the Andes.* Within glassfrogs, there are some examples where sister species inhabit the Chocoan and Amazonian lowlands: (i) *Teratohyla amelie* + *T. pulverata*, (ii) *T. spinosa* + (*T. midas* + *T. adenocheira*), (iii) *Cochranella granulosa* and *C. resplendens*, (iv) *Hyalinobatrachium muiraquitan* + (*H. fleischmanni* + *H. tatayoi*). The most parsimonious speciation mechanism that explains this pattern is allopatric (vicariant) speciation coupled with niche conservatism (i.e., non-ecological speciation), the vicariant event being the uplift of the Andes.

- *Allopatric speciation facilitated by the linearity of mountain ranges* (Figures 233 and 234). This is the most prevalent speciation mode in centrolenids. In this case, allopatric (vicariant) speciation is coupled with niche conservatism (i.e., non-ecological speciation), but the geographic barriers are, usually, river valleys or depressions. The linearity and complexity of the Andes is, likely, one of the major allopatric speciation forces behind the diversity of *Nymphargus* and *Centrolene*, genera that present conspicuous niche conservatism [28] (Figure 235). The linearity of the Andes results in species with elongated geographical ranges that reduces opportunities for contact and gene flow among parapatric populations [347,348,364]. Also, contact between populations is further restricted by the presence of deep river canyons that transect the Andes and thereby facilitate allopatric speciation [347]. On the Amazonian slope of the Ecuadorian Andes, such river canyons include, potentially, the following: Quijos, Pastaza, Paute, Zamora, and Nangaritza (Figure 236). On the Pacific slopes of the Andes, rivers that potentially disrupt gene flow are Mira, Guayllabamba, Chanchán, and Jubones [365–370] (Figure 236). Most sister species within the Andean *Centrolene* and *Nymphargus* fit into this speciation pattern. Similarly, sister glassfrog species in *Celsiella* (*C. revocata* + *C. vozmedianoi*) and *Vitreorana* (*V. antisthenesi* + *V. castroviejoi*), distributed in the elongated Cordillera de la Costa in Venezuela, are another example of speciation facilitated by the linearity of a mountain chain and, in this particular case, the effect of the Unare Depression.

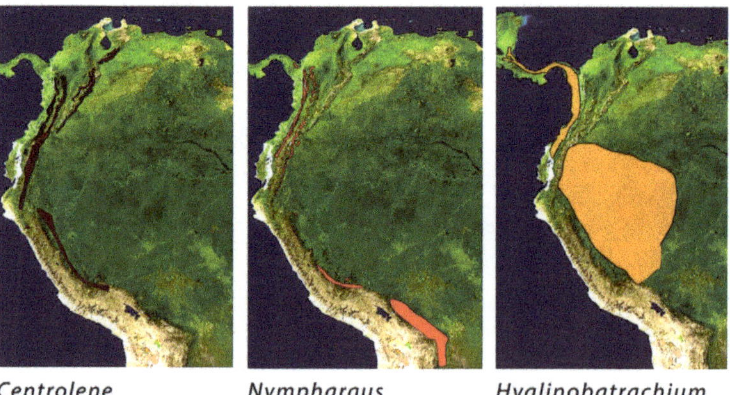

Figure 235. Niche conservatism in glassfrogs at the generic level (modified from Guayasamin et al. [1]). Note that distribution of genera is restricted to particular biogeographic regions, suggesting that closely related species have similar climatic requirements. Number of species per genera are as follows: *Centrolene* = 24, *Nymphargus* = 41, *Hyalinobatrachium* = 32.

Figure 236. Potential river valleys that promote speciation in Ecuador.

- *Speciation along elevational gradients* (Figure 234). Speciation involving adaptation to new climates (i.e., ecological speciation) has been hypothesized as a major generator of diversity in amphibians [357] (Figure 234). Ancestral reconstructions suggest that glassfrogs originated at mid-elevations or climatically similar habitats, and that the lowlands and higher elevation habitats were colonized more recently [28]. Within Andean clades, few species have shifted towards the lowlands (e.g., *Nymphargus mariae*, *Centrolene durrellorum*), and only two species, *Centrolene buckleyi* and *C. venezuelensis*, reach elevations above 3000 m. Similarly, within mainly lowland clades, few taxa have been able to speciate in the Andes (e.g., *Hyalinobatrachium pellucidum*). Thus, the data at hand strongly suggest that niche shifts in centrolenids are rare [28] and that clades tend to radiate within the same climatic regimes (e.g., *Centrolene* and *Nymphargus* in the highlands; *Teratohyla* and *Hyalinobatrachium* in the lowlands).

- *Sympatric speciation.* Examples of sympatric speciation are rare and, in most cases, difficult to test [345]. Within Centrolenidae, there are very few cases of sister species that are found in sympatry; some such examples include: (i) *Nymphargus griffithsi* + *N. lasgralarias*, and (ii) *N. cariticommatus* + *N. sucre*. The possibility of sympatric speciation in glassfrogs remains to be tested and contrasted with other plausible scenarios (i.e., that species originated in allopatry, and that the current sympatry is a consequence of subsequent range expansions and secondary contact).

5.5. Pending Taxonomic Problems and Candidate Species

We have identified a total of 24 candidate species (Table S6), as well as numerous pending taxonomic problems, which we describe below:

- *Centrolene buckleyi:* Different sources of evidence (i.e., genetic, acoustic) suggest that *C. buckleyi*, as currently defined, is a species complex [2,20,98]. We find three lineages (*Centrolene* sp. *Ca02*, *C.* sp. *Ca04*, and *C.* sp. *Ca05*) that are morphologically similar to *C. buckleyi*. None of these candidate species are sister to populations from the neotype locality of *C. buckleyi*. The extensive distribution of *C. buckleyi* in the high Andes of Colombia and Ecuador provide multiple opportunities for isolation and speciation. A taxonomic evaluation of this species complex is greatly needed, especially in the face of population declines that seem to have affected this species [91,103].

- *Centrolene peristicta:* Giving the striking morphological similarity and phylogenetic proximity between *C. peristicta* and *C. antioquiensis*, it is possible that they represent a single evolutionary lineage. Call analysis is critical to assess this possibility.

- *Espadarana prosoblepon:* Within Centrolenidae, this species has the broadest latitudinal range. Also, it has considerable variation in terms of color pattern and body size. Historically, several species (*Hylella puncticrus, Hylella parabambae, Hyla ocellifera*) have been recognized in what is now accepted as a single species, *E. prosoblepon*. The most similar species in Ecuador, *E. callistomma*, is not reciprocally monophyletic with specimens identified as *E. prosoblepon*. This might be the result of incorrect identification, hybridization, or introgression. However, as mentioned above, calls of *E. prosoblepon* and *E. callistomma* are different. *Espadarana prosoblepon* may also represent a species complex that requires subdivision.

- *Hyalinobatrachium munozorum:* We identify one candidate species related to *H. munozorum* in Ecuador (*H.* sp. *Ca02*). Species in the genus *Hyalinobatrachium* share several morphological traits, sometimes making species recognition difficult. Accordingly, more life history information including data on calls, patterns of parental care, and tadpole morphology, together with denser sampling for genetic data are critical for species identifications and to uncover any cryptic diversity [231,232].

- *Nymphargus garciae, N. nephelophilus,* and *N. oreonympha:* Given the morphological similarity among these three species, we recommend their revision and consideration of the possibility that they represent a single species. Currently, molecular data are available only for *N. garciae*.

- *Nymphargus griffithsi:* Cryptic species are likely hidden in *N. griffithsi* as we identify three candidate species with strong morphological similarity (*N.* sp. *Ca01, N.* sp. *Ca02, N.* sp. *Ca03*). We recommend an evaluation of *N. griffithsi* across its entire range, including populations of *N. buenaventura*.

- *Nymphargus laurae: Nymphargus laurae* is morphologically similar to *N. lindae* sp. nov. and *N. cochranae*. Differences among these taxa are difficult to assess because *N. laurae* was described based on a single specimen; additionally, some of the diagnostic traits could be the result of prolonged preservation (e.g., reduction of body size) [371]. The taxonomic resolution of this problem requires collecting more specimens from the type locality of *N. laurae* (slopes of Volcán Sumaco).

- *Rulyrana flavopunctata* and *R. mcdiarmidi:* Genetic data show that these two species are not reciprocally monophyletic. This is a pattern expected in early diverging species but could also indicate that the two species actually represent a single evolutionary species with geographic variation. Analyses of calls and other life history data are needed to validate or reject the validity of *R. mcdiarmidi*.

- *Sachatamia albomaculata:* As currently defined, this species has two clearly differentiated dorsal color patterns: (i) Dorsum with small yellow spots and (ii) dorsum with small and large yellow spots. This variation could correspond to two taxa or represent natural intraspecific polymorphism. Additionally, *S. punctulata* has a color pattern identical to pattern (i) of *S. albomaculata*. Clearly, more detailed work is needed.

- *Teratohyla midas:* We identify a candidate species in *T. midas* (*T.* sp. *Ca03*), which has substantial genetic differentiation from *T. midas* from near the type locality. However, the species may have substantial genetic differentiation or population structure. In terms of morphology and DNA data, *T. midas* is very similar to *T. adenocheira*. Both species have a small body size, similar color pattern, and inhabit the Amazonian lowlands (<1000 m). The most conspicuous difference between these species is the density of dorsal yellow spots; *T. adenocheira* has numerous dots, whereas *T. midas* has very few that are restricted to the upper flanks [372]. We recommend a reassessment of the geographic variation of *T. midas* and the validity of *T. adenocheira*.

5.6. Evolution of Translucency, Parental Care, and Humeral Spines

Translucency: The widespread occurrence of translucency in glassfrogs is puzzling. A recent study (Barnett et al. 2020) [373] found that perceived luminance of glassfrogs changes depending on the immediate background, a change that is more pronounced on the legs, suggesting that camouflage is through edge diffusion. The strategy of disrupting the typical frog body outline might be even more efficient in species that exhibit complete ventral translucency and where internal organs are covered by reflective iridophores (Figures 13A and 102). It has also been shown that the dorsal green coloration in glassfrogs has similar reflective properties as photosynthetic leaves [374], also supporting the relevance of camouflage as an antipredatory mechanism. The venters of all glassfrogs are partially or completely transparent; therefore, it is reasonable to assume that this feature appeared in the ancestor of the clade. Also, complete ventral transparency has evolved multiple independent times within Centrolenidae (e.g., *Hyalinobatrachium*, *Chimerella*, *Vitreorana*).

Parental care: Until very recently, parental care in glassfrogs was considered to be rare and, when present, provided exclusively by males [4,17]. However, Delia et al. [25], based on detailed observations of 40 species, demonstrated that parental care is widespread in Centrolenidae (Table 8). In species thought to lack parental care, Delia et al. [25] observed that, just after oviposition and fertilization, females exhibit a short brooding behavior; this behavior significantly reduces embryonic mortality (experimentally tested in *Cochranella granulosa* and *Teratohyla pulverata*). Even though we still lack information on parental care for most glassfrogs species, the results by Delia et al. [25] have produced a major shift in what we thought we knew about parental behavior in this frog family. Ancestral reconstructions suggest that the most recent common ancestor of glassfrogs exhibited a short, female-only parental care, from which some species (mostly *Hyalinobatrachium* and some *Centrolene*) have evolved prolonged, male-only parental care [25]. Additionally, the repeated evolution of complex male care is always associated with reductions in egg jelly and changes in oviposition sites [375]. Female mate choice and the evolution of parental care is an area of glassfrog biology that still needs further research.

Table 8. Parental care in glassfrogs. **Short-term maternal care** = immediately after oviposition, female provides brooding to egg clutch for several hours; after this initial brooding, clutches remain unattended. **Prolonged male care** = male provides parental care to egg clutch for several days. **Prolonged maternal care** = female provides parental care to egg clutch for several days. Each type of parental care is coded as Absent (0), Present (1), or Unknown (?). Terminology and data are summarized from Delia et al. [25].

Species	Short-Term Maternal Care	Prolonged Male Care	Prolonged Maternal Care	Source
	Subfamily: Centroleninae			
Centrolene antioquiensis	?	1	0	Delia et al. [25]
Centrolene ballux	?	0	0	Delia et al. [25]
Centrolene buckleyi	1	0	0	Delia et al. [25]
Centrolene daidalea	0	1	0	Cardozo-Urdaneta & Señaris [376], Delia et al. [25]
Centrolene geckoidea	?	1	0	Lynch et al. [113], Grant et al. [112]
Centrolene hesperia	1	0	0	Cadle & McDiarmid [377]
Centrolene hybrida	?	0	0	Delia et al. [25]
Centrolene lynchi	1	0	0	Delia et al. [25]
Centrolene peristicta	0	1	0	Delia et al. [25]

Table 8. *Cont.*

Species	Short-Term Maternal Care	Prolonged Male Care	Prolonged Maternal Care	Source
Subfamily: Centroleninae				
Centrolene sanchezi	?	0	0	Delia et al. [25]
Centrolene savagei	0	1	0	Vargas-Salinas et al. [378], Delia et al. [25]
Chimerella corleone	?	0	0	Delia et al. [25]
Chimerella mariaelenae	1	0	0	Delia et al. [25]
Cochranella erminea	?	0	0	Delia et al. [25]
Cochranella euknemos	1	0	0	Delia et al. [25]
Cochranella granulosa	1	0	0	Delia et al. [25]
Cochranella guayasamini	?	0	0	Delia et al. [25]
Cochranella resplendens	1	0	0	Delia et al. [25]
Espadarana andina	?	0	0	Cabanzo-Olarte et al. [379]
Espadarana audax	1	0	0	Delia et al. [25]
Espadarana prosoblepon	1	0	0	Jacobson [188], Delia et al. [25]
Nymphargus grandisonae	1	0	0	Delia et al. [25]
Nymphargus griffithsi	1	0	0	Delia et al. [25]
Nymphargus lasgralarias	1	0	0	Delia et al. [25]
Nymphargus ignotus	?	0	0	Restrepo & Naranjo [380]
Nymphargus wileyi	?	0	0	Delia et al. [25]
Sachatamia albomaculata	1	0	0	Delia et al. [25]
Sachatamia punctulata	?	0	0	Delia et al. [25]
Teratohyla midas	1	0	0	Diaz-Ricaurte et al. [295], Delia et al. [25]
Teratohyla pulverata	1	0	0	Delia et al. [25]
Teratohyla spinosa	1	0	0	Delia et al. [25]
Vitreorana aff. *eurygnatha*	?	0	0	Gouveia et al. [381]
Vitreorana ritae	1	0	0	Delia et al. [25]
Subfamily: Hyalinobatrachinae				
Celsiella vozmedianoi	?	1	0	Señaris & Ayarzagüena [106]
Hyalinobatrachium aureoguttatum	0	1	0	Valencia-Aguilar et al. [196], Delia et al. [25]
Hyalinobatrachium chirripoi	0	1	0	Delia et al. [25]
Hyalinobatrachium colymbiphyllum	0	1	0	McDiarmid [4], Delia et al. [25]
Hyalinobatrachium fleischmanni	0	1	0	Greer & Wells [213], Jacobson [195], Hayes [189], Delia et al. [25]
Hyalinobatrachium mondolfii	?	1	0	Delia et al. [25]
Hyalinobatrachium orientale	0	1	0	Lehtinen et al. [382]
Hyalinobatrachium pallidum	?	1	0	Cardozo-Urdaneta & Señaris [376]
Hyalinobatrachium pellucidum	0	1	0	Delia et al. [25]
Hyalinobatrachium talamancae	0	1	0	Delia et al. [25]

Table 8. *Cont.*

Species	Short-Term Maternal Care	Prolonged Male Care	Prolonged Maternal Care	Source
Subfamily: Centroleninae				
Hyalinobatrachium valerioi	0	1	0	McDiarmid [4] Vockenhuber et al. [239,383], Delia et al. [25]
Hyalinobatrachium vireovittatum	?	1	0	Hayes [189]
Ikakogi tayrona	0	0	1	Bravo-Valencia & Delia [47], Delia et al. [25]

Humeral spines: Male glassfrogs are highly territorial and they have evolved very unique behaviors and structures that are specifically associated to male-to-male combats (Table 9). The most conspicuous morphological trait that males use during fights are humeral spines [52,103], which are bony processes that project from each humeral bone (Figure 14), and that are present only in males of several glassfrog species (see Guayasamin et al. [1]). Although humeral spines are a rarity among amphibians, they have evolved multiple independent times in Centrolenidae [2], suggesting that they provide a selective advantage. Armaments (i.e., humeral spines) in glassfrogs probably only allow males to obtain or defend a territory [52,103], and most likely have no direct role in attracting females, which presumably choose a mate based on the quality of his territory, and acoustic and behavioral displays.

Table 9. Centrolenid species in which combat behavior has been documented, coded by the type of combat behaviors: **Primitive**: Combat in axillary amplexus-like position or wrestling on the leaves; **derived**: Combat dangling by hind limbs; both: Primitive and derived. Modified from Rojas-Runjaic and Cabello [384].

Species	Combat Behavior Type	Observed Duration	References
Subfamily Centroleninae			
Centrolene acanthidiocephalum	Derived	N/A	Ruiz-Carranza (pers. comm. in: Bolívar et al. [103])
Centrolene buckleyi	Derived	37 min	Bolívar et al. [103]
Centrolene daidalea	Both	39 min	Rojas-Runjaic & Cabello [385]
Centrolene hesperia	Derived/Head-to-Vent	N/A	Cadle & McDiarmid [378]
Centrolene lynchi	Derived/Head-to-Vent	5 min	Dautel et al. [51]
Cochranella granulosa	Derived	2 h 30 min	Kubicki [24]
Espadarana andina	Derived	N/A	Guayasamin & Barrio-Amorós [386]
Espadarana prosoblepon	Both	45 min	Derived: Jacobson [189]; Kubicki [24] Primitive & derived: Rojas-Runjaic & Cabello [384]
Nymphargus griffithsi	Derived	10 min	Duellman & Savitzky [5]
Nymphargus grandisonae	Derived/Head-to-Vent	5 h 2 min *	Hutter et al. [52]
Nymphargus ignotus	Derived		Restrepo-Toro [387]
Sachatamia ilex	Both	N/A	Derived: Kubicki [24] Primitive & derived: Rojas-Runjaic & Cabello [384]
Subfamily Hyalinobatrachinae			
Hyalinobatrachium colymbiphyllum	Primitive	N/A	Savage [149]
Hyalinobatrachium valerioi	Primitive	5 min	McDiarmid & Adler [239]
Hyalinobatrachium fleischmanni	Both	26 min * [a]	Primitive: McDiarmid & Adler [239]; Greer & Wells [214]; Jacobson [189]; Kubicki [24]; Derived and primitive: Delia et al. [216]

[a] the only complete fight observed lasted 3 min, but the longest observed fight lasted 26 min, beginning mid-conflict.
* Complete observation of combat behavior from the beginning to the end.

6. Conservation

Evaluating the conservation status and conservation needs of amphibian species is extremely challenging. Traditional threats (e.g., habitat destruction, water contamination, pesticides) and the emergence of novel factors (e.g., climate change, introduced species, and emerging diseases) need to be combined when developing conservation strategies [23,76,124,270]. With all these considerations in mind, we present our evaluation of the conservation of each Ecuadorian glassfrog in Table S7. The most endangered species are: *Centrolene buckleyi, C. charapita, C. geckoidea, C. medemi, C. pipilata, Cochranella mache, Nymphargus balionotus, N. manduriacu, N. megacheirus,* and *N. sucre* (Table S7) and, therefore, are most likely to go extinct because of any of the aforementioned variables. Additionally, numerous species are *Data Deficient* (Table S7) and urgently require additional research.

The most conspicuous and immediate threat to glassfrog conservation is habitat destruction, which is rampant in the Chocó ecoregion (Esmeraldas Province) and also on the northwestern slopes of the Andes (Figure 18). The Chocó suffers the highest deforestation rate in Ecuador, thus, affecting all the species that occur there. Five Chocoan glassfrogs (*H. fleischmanni, C. geckoidea, T. pulverata, T. spinosa, Cochranella mache*) have more than half of their potential distribution affected by human activities (Table S4).

Habitat preservation is the most effective mechanism to protect diversity. However, we first need to identify the areas that should be prioritized for such actions. As a first approach, it is clear that both at the South American and Ecuadorian scales, mountain ranges harbor the highest density of species and, also, species that tend to have small distributions. Also, as shown in Figure 230, the Northern Andes are, by far, the most important biogeographic regions in terms of glassfrog species richness. Then, any effort directed towards the conservation of Andean forests will greatly benefit glassfrogs.

In Ecuador, more specifically, when the potential distributions of all glassfrog species are combined, we are able to identify areas with both high diversity and low human disturbance that are not included in the current system of protected areas of Ecuador (Figure 237). Based on this information, specifically, we recommend: (i) The implementation of a biological corridor between the National Parks of Cayambe-Coca and Sumaco, (ii) creation of a new protected area in the lower montane evergreen forest of the western Andes in Pichincha Province, and (iii) protection of the endemic Chocoan forest (threatened mainly by wood extraction and palm plantations). The aforementioned conservation actions agree with the recommendations produced by broader studies [76,388], in terms of the areas that need to be protected in Ecuador given the current threats and biodiversity patterns (Figures 238 and 239).

Other conservation actions should be directed towards species that are known from few localities that lie within areas that are (or will be) affected by human activities; this is the case, for example, for *Centrolene condor*, species endemic to the Cordillera del Cóndor, where hundreds of hectares have been conceded for mining activities [389]. A similar situation applies to *N. manduriacu*, known from a single locality (Reserva Río Manduriacu) that is concessioned for mining [21].

One of the novel threats to amphibian diversity worldwide is the disease caused by the fungus *Batrachochytrium dendrobatidis* (*Bd*), known as chytridiomycosis [23,390,391]. Chytridiomycosis has been implicated in the extinction of numerous Andean species, mainly harlequin toads (*Atelopus*) [31,33,392]. In Ecuador, ecological modeling predicts that the highest suitability for chytridiomycosis is in the Andes, at elevations above 2000 m [393]. The impact of the disease on Ecuadorian glassfrogs has not been assessed, but preliminary studies show that *Bd* has a relatively high prevalence in several species of Andean glassfrogs [92]. Because of the absence of long-term monitoring of amphibian populations in Ecuador, there is no certainty of the effect of *Bd* on most amphibians, but in places where *Bd* is present, it is possible that all vulnerable species are already extinct. Glassfrogs that persist in spite of infections are probably resistant to the disease or are exposed to a less-pathogenic strains of the fungus [93]. The disappearance of species such as *Nymphargus balionotus* and *Centrolene geckoidea* from relatively pristine areas may be related to chytridiomycosis.

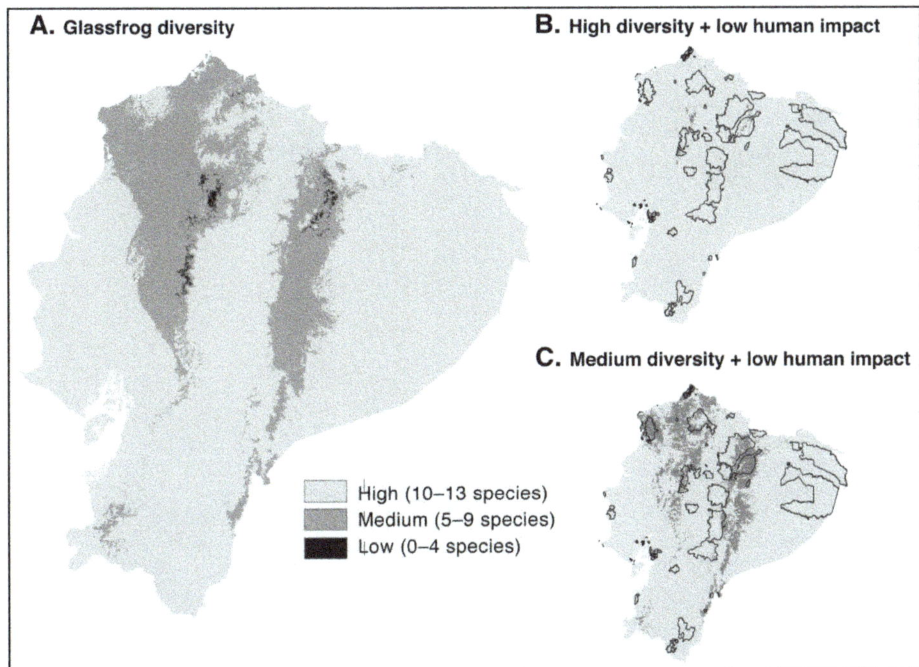

Figure 237. Glassfrog species diversity in Ecuador. (**A**) Species diversity simplified into three categories: High, medium, and low. (**B**) Map showing protected areas in Ecuador and, in dark, regions that combine high glassfrog diversity and low impact by human activities. (**C**) Map showing protected areas in Ecuador and, in dark, regions with both medium diversity and low impact by human activities.

Another specific conservation threat to Andean species that depend on rivers and streams is the introduced rainbow trout, *Oncorhynchus mykiss* [124,394]. Laboratory experiments on the interaction of this exotic fish and tadpoles of the red-spotted glassfrog (*Nymphargus grandisonae*) show that trout prey on tadpoles and that its presence results in an increased mortality and phenotypic change [124]. Most likely, similar effects are produced on other amphibian species with aquatic larvae. Given that the rainbow trout has been introduced into most Ecuadorian Andean rivers and lakes, the only option to reduce its impact is to start eradication programs, at least in protected areas and wherever the trout might be threatening endangered amphibians [124]. Other studies suggest that climate change, chytridiomycosis, and their synergetic effects, likely represent the major threats that amphibians face in Ecuador [393] and worldwide [395].

Finally, climate change represents one of the most challenging conservations phenomenon for biodiversity, given its global scale and potentially large effect even on species that are found in relatively pristine areas. It has been shown that the same variables that explain the high levels of tropical diversity (e.g., narrow thermal tolerance and low dispersal) also make tropical species more vulnerable to rapid thermal change [396]. The thermal breath of tropical amphibians is poorly known, but it can be correlated with their elevational distribution; glassfrogs show, in general, a narrow elevational range (mean = 653 m ± 526; *n* = 150 species; Table S1), meaning that their optimal thermal niches are only available at very restricted elevations. The survival of a species depends, then, on its ability to disperse at a pace fast enough to find itself in a suitable climatic environment. Dispersion will also depend on factors that are extrinsic to the organism, such as habitat continuity. In areas where fragmentation is severe, species will not be able to shift their distributional ranges. From a conservation perspective, it is critical to have areas with elevational gradients and both terrestrial and riparian corridors to allow species' movements. Alternatively, in fragmented landscapes, assisted dispersal might represent a necessary conservation action [396].

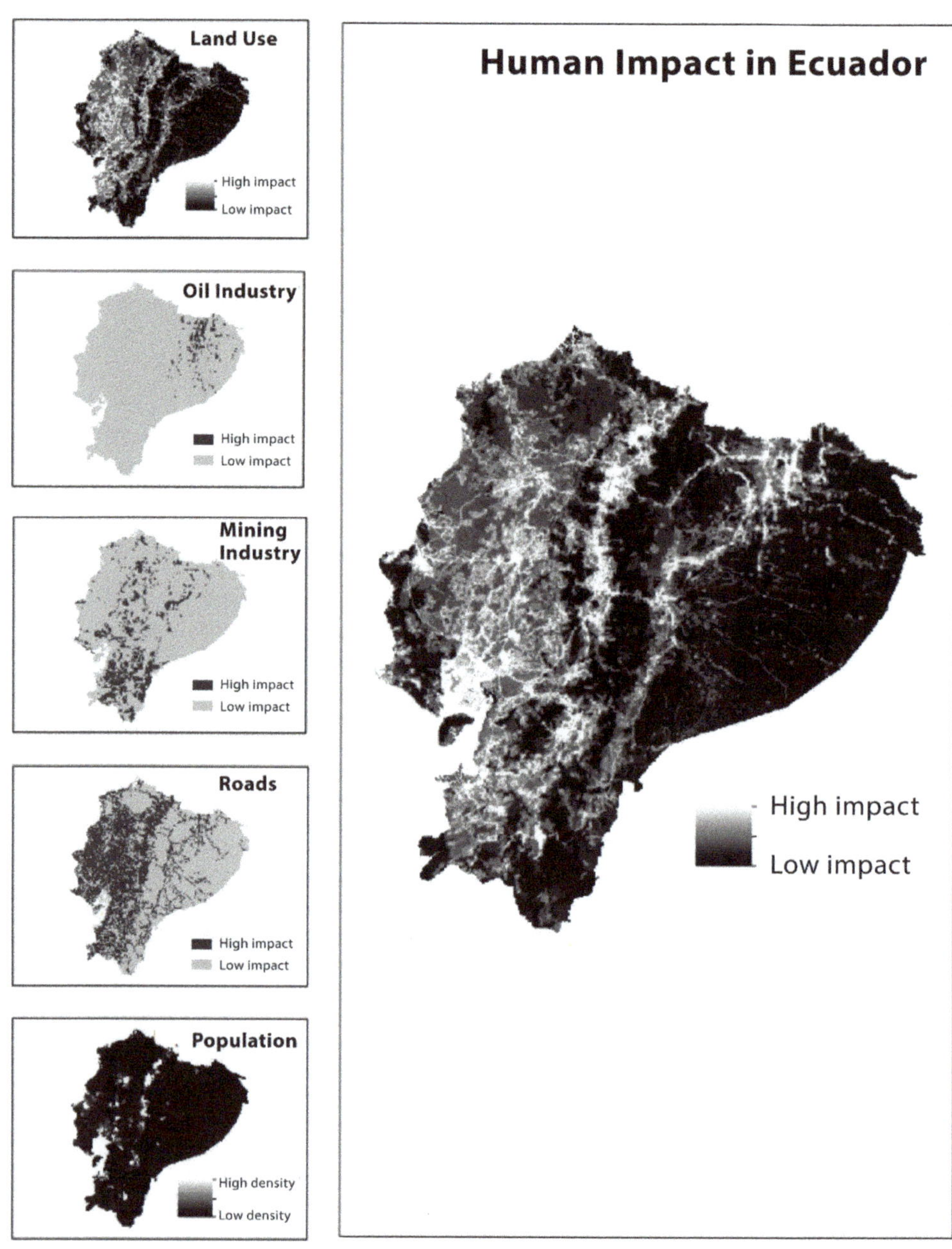

Figure 238. Human impact on ecosystems of Ecuador.

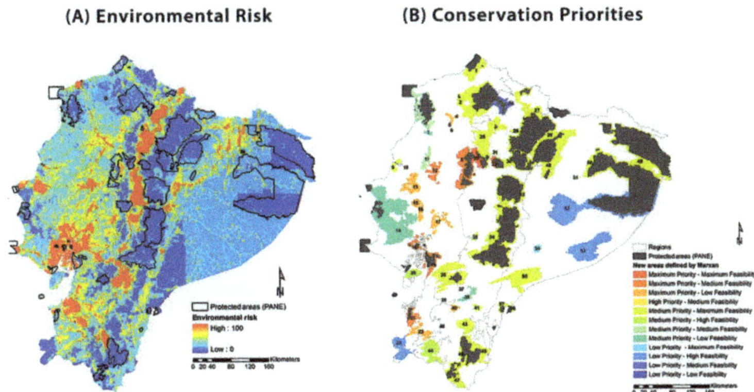

Figure 239. Environmental risk and conservation priorities in Ecuador. (**A**) Environmental risk surface for continental Ecuador. This surface takes into account information on roads, human population density, airports, dams, agriculture and husbandry, and oil and mining industry [73]. (**B**) Conservation priorities in Ecuador; this map was constructed using the potential distribution of 809 species (amphibians, birds, mammals, plants), combined with feasibility of conservation [73].

The alternative solution for species is to adapt to the novel climatic conditions associated with anthropogenic climate change; however, rates of climatic niche change among populations of plants and animals are dramatically slower than projected rates of future climate change. This means that, most likely, populations may not be able to change their climatic niches rapidly enough to keep pace with changing conditions as global climate warms, with dispersal being the only venue to avoid extinction [397,398].

Supplementary Materials: The following are available online at http://www.mdpi.com/1424-2818/12/6/222/s1, Table S1: Taxonomy, biogeographic distribution, and elevation of all recognized species of glassfrogs, Table S2: Taxon and genetic markers used in this study. Sequences generated in previous studies were downloaded from GenBank. Newly generated sequences are in bold blue, Table S3: Summary statistics for each marker used in the phylogenetic analyses, Table S4: Potential distribution of glassfrog species in Ecuador, with percentage affected by human activities, Table S5: Area Under Curve (AUC) values for potential distribution models, Table S6: Candidate species recognized in this study, Table S7: Conservation status of Ecuadorian glassfrogs.

Author Contributions: J.M.G. and D.F.C.-H. conceived the study; J.M.G. wrote the manuscript, with contributions from all authors; J.M.G., D.F.C.-H., P.P., R.W.M., and C.R.H. collected the data; J.M.G., C.R.H., and P.P. analyzed the data; J.M.G. and C.R.H. rendered all figures and tables. All authors have read and agreed to the published version of the manuscript.

Funding: JMG's research was supported by the National Science Foundation (DEB-1046408, DEB-1045960, DEB-1045991DEB–0608011, EF–0334928, DEB-1046408); Secretaría Nacional de Educación Superior, Ciencia, Tecnología e Innovación de Ecuador (PIC-08-0000470); Partnerships for Enhanced Engagement in Research Science [grant number P1-108]; JRS Biodiversity Foundation; the American Philosophical Society through the Lewis and Clark Fund for Exploration and Field Research; IUCN-Save Our Species; Pontificia Universidad Católica del Ecuador; Mashpi Biodiversity Reserve, Panorama Society Grant and Harris Scholarship Award of the University of Kansas Natural History Museum; Universidad Tecnológica Indoamérica; and Universidad San Francisco de Quito (Collaboration Grants 11164, 16871; COCIBA Grants 5521, 5467, 16808). DFCH's research was supported by María Elena Heredia, Laura Heredia, the Smithsonian Women's Committee, the 2002 Research Training Program, National Museum of Natural History, Smithsonian Institution, Tiputini Biodiversity Station, Instituto de Ecología Aplicada ECOLAP-USFQ, Río Guajalito Protected Forest, Mashpi Protected Forest/Metropolitan Tourists, Fundación Futuro, the Global Amphibian Assessment, Conservation International, the Atelopus Initiative, the Research Analysis Network for Neotropical Amphibians (RANA, supported by the National Science Foundation DEB-0130273), the Russel E. Train Education for Nature Program, World Wildlife Fund WWF, Programa "Becas de Excelencia", Secretaría de Educación Superior, Ciencia, Tecnología e Innovación (SENESCYT), Ecuador, and Universidad San Francisco de Quito USFQ (Chancellor grants, COCIBA grants, Collaboration grants, projects HUBI ID 34, 36, 39, 48, 1057, 7703, 12253, 13524, 16953). Work by JMG and DFCH was supported by "Proyecto Descubre Napo", an initiative of Universidad San Francisco de Quito in association with Wildlife Conservation Society and funded by the Gordon and Betty Moore Foundation as part of the project: WCS Consolidating Conservation of Critical Landscapes (mosaics) in the Andes.

Acknowledgments: This study took a very long time to write and review; it was supposed to be part of JMG's PhD dissertation but, instead, it took 12 additional years of punctuated work and the creative input of all coauthors, mainly CRH. There are numerous people to thank and, for sure, we will forget to mention some (we apologize in advance). This article was greatly improved by discussions with Santiago Castroviejo-Fisher, Marco Rada, Jesse Delia, Alessandro Catenazzi, Rudolf von May, Annabelle Wang, Elisa Bonaccorso, and three anonymous reviewers. We are grateful to the many individuals and institutions who provided specimens, tissues, calls, photos, and information included in this study: Luis A. Coloma, Ítalo Tapia (Centro Jambatu), Linda Trueb, William E. Duellman, John E. Simmons, Rafe Brown (University of Kansas), Alejandro Arteaga, Lucas Bustamante, Jose Vieira (Tropical Herping), John D. Lynch (ICN), James A. Poindexter, Ron Heyer, Addison Wynn (USNM), Darrel Frost, Charles W. Myers, Charles J. Cole, Linda S. Ford (AMNH), Jeff Streicher, David Gower and Mark Wilkinson (BMNH), James Hanken and José Rosado (MCZ), Chris Funk (Colorado State University), Ivan Ineich, Roger Bour, and Jean-Christophe de Massary (MNHN); Janalee Caldwell and Laurie Vitt (OMNH), Santiago Castroviejo-Fisher, Marco Rada (Pontificia Universidad Católica do Rio Grande do Sul), Taran Grant (Universidade de São Paulo), Jane Lyons (Reserva Las Gralarias), Marco M. Reyes (Fundación Óscar Efrén Reyes), Kelly Zamudio (Cornell University), Mario Yánez-Muñoz (INABIO), Brian Kubicki (Costa Rican Amphibian Research Center), Santiago Ron (QCAZ), Ana Almendáriz (EPN), Jean-Marc Touzet and Ana María Velasco (FHGO), Carolina Reyes-Puig, Emilia Peñaherrera and David Brito (ZSFQ), Mauricio Ortega-Andrade (Ikiam) and Andrew J. Crawford (Smithsonian Research Tropical Institute, Panama). For assistance and comments during the development of this project, we extend our sincere gratitude mainly to Luis A. Coloma, Martín R. Bustamante, Elisa Bonaccorso, Santiago Castroviejo-Fisher, and Marco Rada. We also thank Alejandro Arteaga, Lucas Bustamante, Eduardo Toral, Gabriela Gavilanes, Nathalia Valencia, Carolina Reyes-Puig, Diego Páez-Moscoso, Andrea Encalada, Ítalo Tapia, Nicolás Peñafiel, Diana Flores, Gabriela Nicholls, Gene Schupp, Henry Imba, Esteban Suárez, César Barrio-Amorós, Mario Yánez-Muñoz, Jonathan Guillemot, Mark Mulligan, Susana Cárdenas, Jesse Delia, Janeth Lessmann, Jeffrey Arellano, David Brito, Emilia Peñaherrera, Ana Nicole Acosta, Gabriel Muñoz, Ernesto Villacís, Katiuska Valarezo, María Olga Borja, Sebastián Cruz, María Elena Heredia, Andrés León-Reyes, Wendy Loor, Pablo Melo, Carolina Proaño, Daniel Proaño, Geovanna Robayo, Javier Robayo, Mayer Rodríguez, and Tomi Sugahara for their help in numerous ways (discussions, fieldwork, labwork, photography). Several friends and colleagues contributed photographs for this article; their names are listed in the figure captions, but we would like to explicitly thank Luis Coloma, Alejandro Arteaga, Jose Vieira, Lucas Bustamante, Jesse Delia, and Jaime Culebras. Research permits in Ecuador were issued by the Ministerio del Ambiente (#033-IC-FAU-DNBAPVS/MA, #56-IC-FAU/FLO-DPN/MA, #05-2013-IC-FAU-DPAP-MA, MAE-DNB-CM-2015-0017). JMG thanks John D. Lynch, Angela Suárez, Celsa Señaris, Chris Funk, Cameron Ghalambor, and Cesar Barrio-Amorós for their hospitality during visits to Colombia, Venezuela, and United States. DFCH is grateful to George R. Zug, W. Ronald Heyer, Robert P. Reynolds, Kenneth A. Tighe, Steve W. Gotte, Carole C. Baldwin, and Mary Sangrey for their support at the USNM, to Julian Faivovich, Taran Grant, Anita Peñaherrera, Jean-Marc Touzet, Claudia Torres-Gastello, Juana Suarez, Jeff Streicher, and Leonardo Zurita for their support during his visits to New York, Paris, Lima, and London; and to David Romo, Consuelo Barriga de Romo, Kelly Swing, Vlastimil Zak, Francisco Vintimilla, Carlos Burneo, Carolina Proaño, Ana Sevilla, and Roque Sevilla for their support for fieldwork. *Juan Manuel dedicates this study to Nina and Elisa.*

Conflicts of Interest: The authors declere no conflict of interest.

References

1. Guayasamin, J.M.; Castroviejo-Fisher, S.; Trueb, L.; Ayarzagüena, J.; Rada, M.; Vilà, C. Phylogenetic systematics of glassfrogs (Amphibia: Centrolenidae) and their sister taxon *Allophryne ruthveni*. *Zootaxa* **2009**, *2100*, 1–97. [CrossRef]

2. Guayasamin, J.M.; Castroviejo-Fisher, S.; Ayarzagüena, J.; Trueb, L.; Vilà, C. Phylogenetic relationships of glassfrogs (Centrolenidae) based on mitochondrial and nuclear genes. *Mol. Phylogenet. Evol.* **2008**, *48*, 574–595. [CrossRef]

3. Castroviejo-Fisher, S.; Guayasamin, J.M.; Gonzalez-Voyer, A.; Vilà, C. Neotropical diversification seen through glassfrogs. *J. Biogeogr.* **2014**, *41*, 66–80. [CrossRef]

4. McDiarmid, R.W. Evolution of parental care in frogs. In *The Development of Behavior: Comparative and Evolutionary Aspects*; Burghardt, G.M., Bekoff, M., Eds.; Garland STPM Press: New York, NY, USA, 1978; pp. 127–147.

5. Duellman, W.E.; Savitzky, A.H. Aggressive behavior in a centrolenid frog, with comments on territoriality in anurans. *Herpetologica* **1976**, *32*, 401–404.

6. Ruiz-Carranza, P.M.; Lynch, J.D. Ranas Centrolenidae de Colombia I. Propuesta de una nueva clasificación genérica. *Lozania* **1991**, *57*, 1–30.

7. Hayes, M.P.; Starrett, P.H. Notes on a collection of centrolenid frogs from the Colombian Chocó. *Bull. South. Calif. Acad. Sci.* **1980**, *79*, 89–96.

8. Wiens, J.J.; Fetzner, J.W.; Parkinson, C.L.; Reeder, T.W. Hylid frog phylogeny and sampling strategies for speciose clades. *Syst. Biol.* **2005**, *54*, 719–748. [CrossRef] [PubMed]

9. Guayasamin, J.M.; Trueb, L. A new species of glassfrog (Anura: Centrolenidae) from the lowlands of northwestern Ecuador, with comments on centrolenid osteology. *Zootaxa* **2007**, 27–45. [CrossRef]

10. Darst, C.R.; Cannatella, D.C. Novel relationships among hyloid frogs inferred from 12S and 16S mitochondrial DNA sequences. *Mol. Phylogenet. Evol.* **2004**, *31*, 462–475. [CrossRef] [PubMed]

11. Faivovich, J.; Haddad, C.F.B.; Garcia, P.C.A.; Frost, D.R.; Campbell, J.A.; Wheeler, W.C. Systematic review of the frog family Hylidae, with special reference to Hylinae: Phylogenetic analysis and taxonomic revision. *Bull. Am. Mus. Nat. Hist.* **2005**, *294*, 1–240. [CrossRef]

12. Frost, D.R.; Grant, T.; Faivovich, J.; Bain, R.H.; Haas, A.; Haddad, C.F.B.; de Sa, R.O.; Channing, A.; Wilkinson, M.; Donnellan, S.C.; et al. The amphibian tree of life. *Bull. Am. Mus. Nat. Hist.* **2006**, *297*, 1–370. [CrossRef]

13. Grant, T.; Frost, D.R.; Caldwell, J.P.; Gagliardo, R.; Haddad, C.F.B.; Kok, P.J.R.; Means, D.B.; Noonan, B.P.; Schargel, W.F.; Wheeler, W.C. Phylogenetic systematics of Dart-poison frogs and their relatives (Amphibia: Athesphatanura: Dendrobatidae). *Bull. Am. Mus. Nat. Hist.* **2006**, *299*, 1–262. [CrossRef]

14. Pyron, R.A.; Wiens, J.J. A large–scale phylogeny of Amphibia including over 2800 species, and a revised classification of advanced frogs, salamanders, and caecilians. *Mol. Phylogenet. Evol.* **2011**, *61*, 543–583. [CrossRef]

15. Taylor, E.H. Two new genera and a new family of tropical American frogs. *Proc. Biol. Soc. Wash.* **1951**, *64*, 33–40.

16. Sanchiz, B.; De la Riva, I. Remarks on the tarsus of centrolenid frogs (Amphibia: Anura). *Graellsia* **1993**, *49*, 115–117.

17. Cisneros-Heredia, D.F.; McDiarmid, R.W. Revision of the characters of Centrolenidae (Amphibia: Anura: Athesphatanura), with comments on its taxonomy and the description of new taxa of glassfrogs. *Zootaxa* **2007**, *1572*, 1–82. [CrossRef]

18. Castroviejo-Fisher, S.; Vilà, C.; Ayarzagüena, J.; Blanc, M.; Ernst, R. Species diversity of *Hyalinobatrachium* glassfrogs (Amphibia: Centrolenidae) from the Guiana Shield, with the description of two new species. *Zootaxa* **2011**, *3132*, 1–55. [CrossRef]

19. Twomey, E.M.; Delia, J.; Castroviejo-Fisher, S. A review of northern Peruvian glassfrogs (Centrolenidae), with the description of four new remarkable species. *Zootaxa* **2014**, *3851*, 1–87. [CrossRef]

20. Guayasamin, J.M.; Bustamante, M.R.; Almeida-Reinoso, D.; Funk, W.C. Glass frogs (Centrolenidae) of Yanayacu Biological Station, Ecuador, with the description of a new species and comments on centrolenid systematics. *Zool. J. Linn. Soc.* **2006**, *147*, 489–513. [CrossRef]

21. Guayasamin, J.M.; Cisneros-Heredia, D.F.; Vieira, J.; Kohn, S.; Gavilanes, G.; Lynch, R.L.; Hamilton, P.S.; Maynard, R.J. A new glassfrog (Centrolenidae) from the Chocó-Andean Río Manduriacu Reserve, Ecuador, endangered by mining. *PeerJ* **2019**, *7*, e6400. [CrossRef]

22. Lynch, J.D.; Duellman, W.E. A review of the centrolenid frogs of Ecuador, with descriptions of new species. *Occas. Pap. Mus. Nat. Hist.* **1973**, *16*, 1–66.

23. Lips, K.R.; Brem, F.; Brenes, R.; Reeve, J.D.; Alford, R.A.; Voyles, J.; Collins, J.P. Emerging infectious disease and the loss of biodiversity in a Neotropical amphibian community. *Proc. Natl. Acad. Sci. USA* **2006**, *103*, 3165–3170. [CrossRef]

24. Kubicki, B. *Ranas de vidrio de Costa Rica/Glass frogs of Costa Rica*; Editorial INBio: Santo Domingo de Heredia, Costa Rica, 2007; 304p.

25. Delia, J.; Bravo-Valencia, L.; Warkentin, K.M. Patterns of parental care in Neotropical glassfrogs: Fieldwork alters hypotheses of sex-role evolution. *J. Evol. Biol.* **2017**, *30*, 898–914. [CrossRef]

26. Ruiz-Carranza, P.M.; Lynch, J.D. Ranas Centrolenidae de Colombia VIII. Cuatro nuevas especies de *Centrolene* de la Cordillera Central. *Lozania* **1995**, *65*, 1–16.

27. Ruiz-Carranza, P.M.; Lynch, J.D. Ranas Centrolenidae de Colombia XI. Nuevas especies de ranas de cristal del genero *Hyalinobatrachium*. *Rev. Acad. Colomb. Cienc. Exact. Fís. Nat.* **1998**, *22*, 571–586.

28. Hutter, C.R.; Guayasamin, J.M.; Wiens, J.J. Explaining Andean megadiversity: The evolutionary and ecological causes of glassfrog elevational richness patterns. *Biol. Lett.* **2013**, *16*, 1135–1144. [CrossRef]

29. Vieites, D.R.; Wollenberg, K.C.; Andreone, F.; Köhler, J.; Glaw, F.; Vences, M. Vast underestimation of Madagascar's biodiversity evidenced by an integrative amphibian inventory. *Proc. Natl. Acad. Sci. USA* **2009**, *106*, 8267–8272. [CrossRef]

30. Stuart, S.N.; Chanson, J.S.; Cox, N.A.; Young, B.E.; Rodrigues, A.S.; Fischman, D.L.; Waller, R.W. Status and trends of amphibian declines and extinctions worldwide. *Science* **2004**, *306*, 1783–1786. [CrossRef]

31. La Marca, E.; Lips, K.R.; Lötters, S.; Puschendorf, R.; Ibañez, R.; Rueda-Almonacid, J.V.; Schulte, R.; Marty, C.; Castro, F.; Manzanilla-Puppo, J.; et al. Catastrophic population declines and extinctions in neotropical harlequin frogs (Bufonidae: *Atelopus*). *Biotropica* **2005**, *37*, 190–201. [CrossRef]

32. Wake, D.B.; Vradenburg, V.T. Are we in the midst of the sixth mass extinction? A view from the world of amphibians. *Proc. Natl. Acad. Sci. USA* **2009**, *105*, 11466–11473. [CrossRef]

33. Scheele, B.C.; Pasmans, F.; Berger, L.; Skerratt, L.F.; Martel, A.; Beukema, W.; Acevedo, A.A.; Burrowes, P.A.; Carvalho, T.; Catenazzi, A.; et al. Amphibian fungal panzootic causes catastrophic and ongoing loss of biodiversity. *Science* **2019**, *363*, 1459–1463. [CrossRef] [PubMed]

34. Simpson, G.G. The species concept. *Evolution* **1951**, *5*, 285–298. [CrossRef]

35. Simpson, G.G. *Principles of Animal Taxonomy*; Columbia University Press: New York, NY, USA, 1961.

36. Wiley, E.O. The evolutionary species concept reconsidered. *Syst. Zool.* **1978**, *27*, 17–26. [CrossRef]

37. De Queiroz, K. Ernst Mayr and the modern concept of species. *Proc. Natl. Acad. Sci. USA* **2005**, *102*, 6600–6607. [CrossRef]

38. De Queiroz, K. Species concepts and species delimitation. *Syst. Biol.* **2007**, *56*, 879–886. [CrossRef]

39. Dayrat, B. Towards integrative taxonomy. *Biol. J. Linn. Soc.* **2005**, *85*, 407–415. [CrossRef]

40. Padial, J.M.; Miralles, A.; De la Riva, I.; Vences, M. The integrative future of taxonomy. *Front. Zool.* **2010**, *7*, 16. [CrossRef]

41. Alberch, P.; Gale, E. A developmental analysis of an evolutionary trend: Digital reduction in amphibians. *Evolution* **1985**, *39*, 8–23. [CrossRef]

42. Fabrezi, M.; Alberch, P. The carpal elements of anurans. *Herpetologica* **1996**, *52*, 188–204.

43. Shubin, N.; Alberch, P. A Morphogenetic approach on the origin and basic organization of the tetrapod limb. In *Evolutionary Biology*; Hecht, M., Wallace, B., Prance, G., Eds.; Plenum Press: New York, NY, USA, 1986; pp. 319–387.

44. Savage, J.M.; Heyer, W.R. Variation and distibution in the tree-frog genus *Phyllomedusa* in Costa Rica, Central America. *Beiträge Zur Neotropischen Fauna* **1967**, *5*, 111–131. [CrossRef]

45. McDiarmid, R.W.; Altig, R. (Eds.) *Tadpoles: The Biology of Anuran Larvae*; The University of Chicago Press: Chicago, IL, USA, 1999; 444p.

46. Flores, G. A new *Centrolenella* (Anura) from Ecuador, with comments on nuptial pads and prepollical spines in *Centrolenella*. *J. Herpetol.* **1985**, *19*, 313–320. [CrossRef]

47. Bravo-Valencia, L.; Delia, J. Maternal care in a glassfrog: Care function and commitment to offspring in *Ikakogi tayrona*. *Behav. Ecol. Sociobiol.* **2016**, *70*, 41–48. [CrossRef]

48. Sueur, J.; Aubin, T.; Simonis, C. Seewave, a free modular tool for sound analysis and synthesis. *Bioacoustics* **2008**, *18*, 213–226. [CrossRef]

49. Cocroft, R.B.; Ryan, M.J. Patterns of advertisement call evolution in toads and chorus frogs. *Anim. Behav.* **1995**, *49*, 283–303. [CrossRef]

50. Dautel, N.; Salgado Maldonado, A.L.; Abuza, R.; Imba, H.; Griffin, K.; Guayasamin, J.M. Advertisement and combat calls of the glass frog *Centrolene lynchi* (Anura: Centrolenidae), with notes on combat and reproductive behavior. *Phyllomedusa* **2011**, *10*, 31–43. [CrossRef]

51. Hutter, C.R.; Escobar-Lasso, S.; Rojas-Morales, J.A.; Gutiérrez-Cárdenas, P.D.A.; Imba, H.; Guayasamin, J.M. The territoriality, vocalizations and aggressive interactions of the red-spotted glassfrog, *Nymphargus grandisonae*, Cochran and Goin, 1970 (Anura: Centrolenidae). *J. Nat. Hist.* **2013**, *47*, 3011–3032. [CrossRef]

52. Köhler, J.; Jansen, M.; Rodríguez, A.; Kok, P.R.J.; Toledo, L.F.; Emmrich, E.; Glaw, F.; Haddad, C.F.B.; Rödel, M.R.; Vences, M. The use of bioacoustics in anuran taxonomy: Theory, terminology, methods and recommendations for best practice. *Zootaxa* **2017**, *4251*, 1–124. [CrossRef]

53. Perl, R.B.; Nagy, Z.T.; Sonet, G.; Glaw, F.; Wollenberg, K.C.; Vences, M. DNA barcoding Madagascar's amphibian fauna. *Amphib. Reptil.* **2014**, *10*, 15685381-00002942. [CrossRef]

54. Felsenstein, J. Cases in which parsimony or compatibility methods will be positively misleading. *Syst. Zool.* **1978**, *27*, 401–410. [CrossRef]

55. Felsenstein, J. *Inferring Phylogenies*; Sinauer Associates: Sunderland, MA, USA, 2004.

56. Jin, L.; Nei, M. Limitations of the evolutionary parsimony method of phylogenetic analysis. *Mol. Biol. Evol.* **1990**, *7*, 82–102.

57. Katoh, K.; Standley, D.M. MAFFT multiple sequence alignment software version 7, improvements in performance and usability. *Mol. Biol. Evol.* **2013**, *30*, 772–780. [CrossRef] [PubMed]

58. Nguyen, L.T.; Schmidt, H.A.; von Haeseler, A.; Minh, B.Q. IQ-TREE: A fast and effective stochastic algorithm for estimating maximum-likelihood phylogenies. *Mol. Biol. Evol.* **2015**, *32*, 268–274. [CrossRef] [PubMed]

59. Kalyaanamoorthy, S.; Minh, B.Q.; Wong, T.K.; von Haeseler, A.; Jermiin, L.S. ModelFinder: Fast model selection for accurate phylogenetic estimates. *Nat. Methods* **2017**, *14*, 587. [CrossRef] [PubMed]

60. Minh, B.Q.; Nguyen, M.A.T.; von Haeseler, A. Ultrafast approximation for phylogenetic bootstrap. *Mol. Biol. Evol.* **2013**, *30*, 1188–1195. [CrossRef]

61. Bouckaert, R.; Heled, J.; Kühnert, D.; Vaughan, T.; Wu, C.H.; Xie, D.; Drummond, A.J. BEAST 2, a software platform for Bayesian evolutionary analysis. *Plos Comput. Biol.* **2014**, *10*, e1003537. [CrossRef]

62. Hutter, C.R.; Lambert, S.M.; Wiens, J.J. Rapid diversification and time explain amphibian richness at different scales in the Tropical Andes, Earth's most biodiverse hotspot. *Am. Nat.* **2017**, *190*, 828–843. [CrossRef]

63. Rambaut, A.; Drummond, A.J. Tracer v1.4. 2007. Available online: http://beast.bio.ed.ac.uk/Tracer (accessed on 14 August 2019).

64. IUCN. *IUCN Red List Categories and Criteria: Version 3.1*; IUCN Species Survival Commission; IUCN: Gland, Switzerland; Cambridge, UK, 2001.

65. Sierra, R.; Cerón, C.; Palacios, W.; Valencia, R. *Propuesta Preliminar de un Sistema de Clasificación de Vegetación Para el Ecuador Continental*; Proyecto INEFAN/GEF-BIRF y ECOCIENCIA: Quito, Ecuador, 1999.

66. Ron, S.R.; Guayasamin, J.M.; Menéndez-Guerrero, P. Biodiversity and Conservation Status of Ecuadorian Amphibians. In *Amphibian Biology*; Part 2; Heatwole, H., Barrio-Amoros, C.L., Wilkinson, H.W., Eds.; Surrey Beatty & Soons PTY Limited: Baulkham Hills, Australia, 2011; Volume 9, pp. 129–170.

67. Cerón, C.; Palacios, W.; Valencia, R.; Sierra, R. Las formaciones naturales de la Costa del Ecuador. In *Propuesta Preliminar de un Sistema de Clasificación de Vegetación para el Ecuador Continental*; Sierra, R., Ed.; Proyecto INEFAN/GEF-BIRF y EcoCiencia: Quito, Ecuador, 1999; pp. 55–73.

68. Palacios, W.; Cerón, C.; Valencia, R.; Sierra, R. Las formaciones naturales de la Amazonía del Ecuador. In *Propuesta Preliminar de un Sistema de Clasificación de Vegetación para el Ecuador Continental*; Sierra, R., Ed.; Proyecto INEFAN/GEF-BIRF y EcoCiencia: Quito, Ecuador, 1999; pp. 109–117.

69. Valencia, R.; Cerón, C.; Palacios, W.; Sierra, R. Las formaciones naturales de la Sierra del Ecuador. In *Propuesta Preliminar de un Sistema de Clasificación de la Vegetación para el Ecuador Continental*; Sierra, R., Ed.; Proyecto INEFAN/GEF-BIRF y EcoCiencia: Quito, Ecuador, 1999; pp. 79–108. 194p.

70. Valencia, R.; Foster, R.B.; Villa, G.; Condit, R.; Svenning, J.C.; Hernández, C.; Romoleroux, K.; Losos, E.; Magård, E.; Balslev, H. The especies distributions and local hábitat variation in the Amazon: Large forest plot in Eastern Ecuador. *J. Ecol.* **2004**, *92*, 214–229. [CrossRef]

71. Ridgely, R.S.; Greenfield, P.J. *The Birds of Ecuador*; Cornell University Press: Ithaca, NY, USA, 2001.

72. Bass, M.S.; Finer, M.; Jenkins, C.N.; Kreft, H.; Cisneros-Heredia, D.F.; McCracken, S.F.; Pitman, N.C.A.; English, P.H.; Swing, K.; Villa, G.; et al. Global conservation significance of Ecuador's Yasuní National Park. *PLoS ONE* **2010**, *5*, e8767. [CrossRef]

73. Lessmann, J.; Muñoz, J.; Bonaccorso, E. Maximizing species conservation in continental Ecuador: A case of systematic conservation planning for biodiverse regions. *Ecol. Evol.* **2014**, *4*, 2410–2422. [CrossRef]

74. Jarnevich, C.S.; Stohlgren, T.J.; Kumar, S.; Morisette, J.T.; Holcombe, T.R. Caveats for correlative species distribution modeling. *Ecol. Inform.* **2015**, *29*, 6–15. [CrossRef]

75. Radosavljevic, A.; Anderson, R. Making better Maxent models of species distributions: Complexity, overfitting and evaluation. *J. Biogeogr.* **2014**, *41*, 629–643. [CrossRef]

76. Hijmans, R.J.; Cameron, S.E.; Parra, J.L.; Jones, P.G.; Jarvis, A. Very high resolution interpolated climate surfaces for global land areas. *Int. J. Climatol.* **2005**, *25*, 1965–1978. [CrossRef]

77. Phillips, S.J.; Anderson, R.P.; Schapire, R.E. Maximum entropy modeling of species geographic distributions. *Ecol. Model.* **2006**, *190*, 231–259. [CrossRef]

78. Phillips, S.J.; Dudík, M. Modeling of species distributions with Maxent: New extensions and a comprehensive evaluation. *Ecography* **2008**, *31*, 161–175. [CrossRef]

79. Graham, C.H.; Hijmans, R.J. A comparison of methods for mapping species ranges and species richness. *Glob. Ecol. Biogeogr.* **2006**, *15*, 578–587. [CrossRef]

80. Elith, J.; Leathwick, J. Predicting species distributions from museum and herbarium records using multiresponse models fitted with multivariate adaptive regression splines. *Divers. Distrib.* **2007**, *13*, 265–275. [CrossRef]

81. Liu, C.; Berry, P.M.; Dawson, T.P.; Pearson, R.G. Selecting thresholds of occurrence in the prediction of species distributions. *Ecography* **2005**, *28*, 385–393. [CrossRef]

82. Jiménez-Valverde, A.; Lobo, J.M. Threshold criteria for conversion of species presence to either-or presence-absence. *Acta Oecologica* **2007**, *31*, 361–369. [CrossRef]

83. Jiménez de la Espada, M. Nuevos batracios americanos. *An. De La Soc. Española De Hist. Nat.* **1872**, *1*, 85–88.

84. McCranie, J.R.; Wilson, L.D. *The Amphibians of Honduras*; Society for the Study of Amphibians and Reptiles: Ithaca, NY, USA, 2002; 625p.

85. Barrio-Amorós, C.L.; Rojas-Runjaic, F.J.M.; Señaris, J.C. Catalogue of the amphibians of Venezuela: Illustrated and annotated species list, distribution, and conservation. *Amphib. Reptile Conserv.* **2019**, *13*, 1–198.

86. Duellman, W.E.; Burrowes, P.A. New species of frogs, *Centrolenella*, from the Pacific Versant of Ecuador and Southern Colombia. *Univ. Kans. Mus. Nat. Hist. Occas. Pap.* **1989**, *132*, 1–14.

87. Arteaga, A.; Bustamante, L.; Guayasamin, J.M. *The Amphibians and Reptiles of Mindo: Life in the Cloudforest*; Universidad Tecnológica Indoamérica: Quito, Ecuador, 2013.

88. Hutter, C.R.; Guayasamin, J.M. A new cryptic species of glassfrog (Centrolenidae: *Nymphargus*) from Reserva Las Gralarias, Ecuador. *Zootaxa* **2012**, *3257*, 1–21. [CrossRef]

89. Márquez, R.; De la Riva, I.; Bosch, J. Advertisement calls of three glass frogs from the Andean forests (Amphibia: Anura: Centrolenidae). *Herpetol. J.* **1996**, *6*, 97–99.

90. Cisneros-Heredia, D.F.; Wild, E.; Reyes, J.P.; Coloma, L.A.; Gutierrez, P.; Ron, S.R.; Bolívar, W. The IUCN Red List of Threatened Species. *Cent. Ballux.* **2019**, e.T54907A85877221. [CrossRef]

91. Bustamante, M.R.; Ron, S.R.; Coloma, L.A. Cambios en la diversidad en siete comunidades de anuros en los Andes de Ecuador. *Biotropica* **2005**, *37*, 180–189. [CrossRef]

92. Guayasamin, J.M.; Mendoza, A.M.; Longo, A.V.; Zamudio, K.R.; Bonaccorso, E. High prevalence of *Batrachochytrium dendrobatidis* in an Andean frog community (Reserva Las Gralarias, Ecuador). *Amphib. Reptile Conserv.* **2014**, *8*, 33–44.

93. Boulenger, G.A. *Catalogue of the Batrachia Salientia s. Ecaudata in the collection of the British Museum*; Trustees, British Museum (Natural History): London, UK, 1882; 495p.

94. Noble, G.K. Two new batrachians from Colombia. *Bull. Am. Mus. Nat. Hist.* **1920**, *42*, 441–446.

95. Rivero, J.A. Salientia of Venezuela. *Bull. Mus. Comp. Zool.* **1961**, *126*, 1–207.

96. Cochran, D.M.; Goin, C.J. *Frogs of Colombia*; United States National Museum Bulletin, Smithsonian Institution Press: Washington, DC, USA, 1970; 655p.

97. Goin, C.J. Three new centrolenid frogs from Ecuador. *Zool. Anz.* **1961**, *166*, 95–104.

98. Amador, L.; Parada, A.; D'Elía, G.; Guayasamin, J.M. Uncovering hidden specific diversity of Andean glassfrogs of the *Centrolene buckleyi* species complex (Anura: Centrolenidae). *PeerJ* **2018**, *6*, e5856. [CrossRef]

99. ICZN. International Code of Zoological Nomenclature. In *The International Trust for Zoological Nomenclature*, 4th ed.; ICZN: London, UK, 1999; 306p.

100. Bolívar-G, W.; Grant, T.; Osorio, L.A. Combat behavior in *Centrolene buckleyi* and other centrolenid frogs. *Alytes* **1999**, *16*, 77–83.

101. Ruiz-Carranza, P.M.; Ardila-Robayo, M.C.; Lynch, J.D. Lista actualizada de la fauna de Amphibia de Colombia. *Rev. Acad. Colomb. Cienc. Exact. Fís. Nat.* **1996**, *20*, 365–415.

102. Duellman, W.E.; Wild, E.R. Anuran amphibians from the Cordillera de Huancabamba, northern Peru: Systematics, ecology, and biogeography. *Occas. Pap. Mus. Nat. Hist. Univ. Kans.* **1993**, *157*, 1–53.

103. Guayasamin, J.M. The IUCN Red List of Threatened Species. *Cent. Buckleyi.* **2010**, e.T54908A11220443. [CrossRef]

104. Myers, C.W.; Donnelly, M.A. A tepui herpetofauna on a granitic mountain (Tamacuari) in the borderland between Venezuela and Brazil: Report from the Phipps Tapirapecó Expedition. *Am. Mus. Novit.* **1997**, *3213*, 1–71.

105. Rivero, J.A. Los Centrolénidos de Venezuela. *Mem. De La Soc. De Cienc. Nat. La Salle* **1968**, *28*, 301–334.

106. Señaris, J.C.; Ayarzegüena, J. *Revisión taxonómica de la Familia Centrolenidae (Amphibia; Anura) de Venezuela*; Publicaciones del Comité Español del Programa Hombre y Biosfera—Red IberoMaB de la UNESCO: Sevilla, Spain, 2005; 337p.

107. Cisneros-Heredia, D.F.; Morales-Mite, M. A new species of glassfrog from elfin forests of the Cordillera del Cóndor, southeastern Ecuador. *Herpetozoa* **2008**, *21*, 49–56.

108. Almendáriz, A.; Batallas, D. Nuevos datos sobre la distribución, historia natural y el canto de *Centrolene condor* Cisneros-Heredia y Morales-Mite 2008 (Amphibia: Anura: Centrolenidae). *Rev. Politécnica* **2012**, *30*, 42–53.

109. Cisneros-Heredia, D.; Angulo, A. The IUCN Red List of Threatened Species 2009. *Cent. Condor* **2009**, e.T158474A5200524. [CrossRef]

110. Ruiz-Carranza, P.M.; Hernández-Camacho, J.; Ardila-Robayo, M.C. Una nueva especie colombiana del género *Centrolene* Jiménez de la Espada 1872 (Amphibia: Anura) y redefinición del género. *Caldasia* **1986**, *15*, 431–444.

111. Rueda-Almonacid, J.V. Estudio anatómico y relaciones sistemáticas de *Centrolene geckoideum* (Salientia: Anura: Centrolenidae). *Trianea* **1994**, *5*, 133–187.

112. Grant, T.; Bolívar, G.W.; Castro, F. The advertisement call of *Centrolene geckoideum*. *J. Herpetol.* **1998**, *32*, 452–455. [CrossRef]

113. Lynch, J.D.; Ruiz-Carranza, P.M.; Rueda, J.V. Notes on the distribution and reproductive biology of *Centrolene geckoideum* Jimenez de la Espada in Colombia and Ecuador (Amphibia: Centrolenidae). *Stud. Neotrop. Fauna Environ.* **1983**, *18*, 239–243. [CrossRef]

114. Goodman, D.E.; Goin, C.J. The Habitat of *Centrolene geckoideum* in Ecuador. *Herpetologica* **1970**, *26*, 276.

115. Cisneros-Heredia, D.F.; Yánez-Muñoz, M.H. A new species of glassfrog (Centrolenidae) from the southern Andean foothills on the west Ecuadorian region. *South Am. J. Herpetol.* **2007**, *2*, 1–10. [CrossRef]

116. Cisneros-Heredia, D.F.; Yánez-Muñoz, M. Amphibia, Anura, Centrolenidae, *Centrolene balionotum*, *Centrolene geckoideum*, and *Cochranella cariticommata*: Distribution extension, new provincial records, Ecuador. *Check List* **2007**, *3*, 39–42. [CrossRef]

117. Cisneros-Heredia, D.F.; Wild, E.; Gonzalez Duran, G.A.; Reyes, J.P.; Coloma, L.A.; Yánez-Muñoz, M.; Rivera, M.; Gutierrez, P.; Ron, S.R.; Bolívar, W. The IUCN Red List of Threatened Species 2019. *Cent. Geckoideum* **2019**, e.T54911A85877414. [CrossRef]

118. Duellman, W.E. Liste der rezenten Amphibien und Reptilien: Hylidae, Centrolenidae, Pseudidae. *Das Tierreich* **1977**, *95*, 1–225.

119. González-Fernández, J.E. Anfibios colectados por la Comisión Científica del Pacífico (entre 1862 y 1865) conservados en el Museo Nacional de Ciencias Naturales de Madrid. *Graellsia* **2006**, *62*, 111–158. [CrossRef]

120. Duellman, W.E. Three new species of centrolenid frogs from the Pacific versant of Ecuador and Colombia. *Univ. Kans. Mus. Nat. Hist. Occas. Pap.* **1981**, *88*, 1–9.

121. Krynak, K.L.; Wessels, D.G.; Imba, S.M.; Lyons, J.A.; Guayasamin, J.M. Newly discovered population of Bumpy Glassfrog, *Centrolene heloderma* (Duellman, 1981), with discussion of threats to population persistence. *Check List* **2018**, *14*, 261. [CrossRef]

122. Mejía, D.; Cisneros-Heredia, D.F.; Gómez, D.; Vargas-Salinas, F.; Chaves Portilla, G.; Gonzalez Duran, G.A.; Coloma, L.A.; Rada, M.; Gutierrez, P.; Bernal, M.H.; et al. The IUCN Red List of Threatened Species 2019. *Cent. Heloderma* **2019**, e.T54916A85878225. [CrossRef]

123. Martín-Torrijos, L.; Sandoval-Sierra, J.V.; Diéguez-Uribeondo, J.; Bosch, J.; Muñoz, J.; Guayasamin, J.M. Rainbow Trout (*Oncorhynchus mykiss*) threaten Andean amphibians. *Neotrop. Biodivers.* **2016**, *2*, 26–36. [CrossRef]

124. Mendoza, Á.; Wild, E.; Delia, J.; Lynch, J.; Bravo, L.; Rada, M. The IUCN Red List of Threatened Species 2019. *Cent. Huilense* **2019**, e.T54918A49363441. [CrossRef]

125. Guayasamin, J.M.; Funk, W.C. The amphibian community at Yanayacu Biological Station, Ecuador, with a comparison of vertical microhabitat use among *Pristimantis* species and the description of a new species of the *Pristimantis myersi* group. *Zootaxa* **2009**, *2220*, 41–66. [CrossRef]

126. Coloma, L.A.; Ron, S.; Wild, E.; Cisneros-Heredia, D. The IUCN Red List of Threatened Species 2004. *Cent. Lynchi* **2004**, e.T54924A11225650. [CrossRef]

127. Duellman, W.E. The identity of *Centrolenella grandisonae* Cochran and Goin (Anura, Centrolenidae). *Trans. Kans. Acad. Sci.* **1980**, *83*, 26–32. [CrossRef]

128. Suárez-Mayorga, A.M. Lista preliminar de la fauna Amphibia presente en el transecto la Montañita-Alto de Gabinete, Caqutá, Colombia. *Rev. Acad. Colomb. Cienc. Exact. Fís. Nat.* **1999**, *23*, 395–405.

129. Ruiz-Carranza, P.M.; Lynch, J.D. Ranas Centrolenidae de Colombia II. Nuevas especies de *Centrolene* de la Cordillera Oriental y Sierra Nevada de Santa Marta. *Lozania* **1991**, *58*, 1–26.

130. Guayasamin, J.M.; Cisneros-Heredia, D.F.; Yánez-Muñoz, M.H.; Bustamante, M.R. Notes on geographic distribution. Amphibia, Centrolenidae, *Centrolene ilex, Centrolene litorale, Centrolene medemi, Cochranella albomaculata, Cochranella ametarsia*: Range extensions and new country records. *Check List* **2006**, *2*, 24–25. [CrossRef]

131. Lynch, J.; Renjifo, J.; Rada, M.; Bolívar, W. The IUCN Red List of Threatened Species 2019. *Cent. Medemi* **2019**, e.T54926A49364088. [CrossRef]

132. Salgado, A.L.; Guayasamin, J.M. Parental care and reproductive behavior of the minute dappled glassfrog (Centrolenidae: *Centrolene peristictum*). *South Am. J. Herpetol.* **2018**, *13*, 211–220. [CrossRef]

133. Ruiz-Carranza, P.M.; Lynch, J.D. Ranas Centrolenidae de Colombia, X. Los centrolénidos de un perfil del flanco oriental de la Cordillera Central en el Departamento de Caldas. *Rev. Acad. Colomb. Cienc. Exact. Fís. Nat.* **1997**, *21*, 541–553.

134. Cisneros-Heredia, D.F.; McDiarmid, R.W. Amphibia, Centrolenidae, Centrolene peristictum, Centrolene prosoblepon, Cochranella cochranae, Cochranella midas, Cochranella resplendens, Cochranella spinosa, Hyalinobatrachium munozorum: Range extensions and new provincial records. *Check List* **2005**, *1*, 18–22. [CrossRef]

135. Cisneros-Heredia, D.F.; Wild, E.; Lynch, J.; Mueses-Cisneros, J.J.; Reyes, J.P.; Coloma, L.A.; Gutierrez, P.; Ron, S.R. The IUCN Red List of Threatened Species 2019. *Cent. Peristictum* **2019**, e.T54931A85879425. [CrossRef]

136. Cisneros-Heredia, D.F.; Reyes, J.P.; Coloma, L.A.; Yánez-Muñoz, M.; Ron, S.R. The IUCN Red List of Threatened Species 2019. *Cent. Pipilatum* **2019**, e.T54933A98643479. [CrossRef]

137. Wild, E.R. Two new species of centrolenid frogs from the Amazonian slope of the Cordillera-Oriental, Ecuador. *J. Herpetol.* **1994**, *28*, 299–310. [CrossRef]

138. Lynch, J.; Rueda-Almonacid, J.V. The IUCN Red List of Threatened Species 2019. *Cent. Sanchezi* **2019**, e.T54939A49365296. [CrossRef]

139. Cisneros-Heredia, D.F.; McDiarmid, R.W. A new species of the genus *Centrolene* (Amphibia: Anura: Centrolenidae) from Ecuador with comments on the taxonomy and biogeography of Glassfrogs. *Zootaxa* **2006**, *1344*, 1–32.

140. Terán-Valdez, A.; Guayasamin, J.M. The tadpole of the glassfrog *Chimerella mariaelenae* (Anura: Centrolenidae). *CienciAmérica* **2014**, *3*, 17–22.

141. Cisneros-Heredia, D.F. Amphibia, Anura, Centrolenidae, *Chimerella mariaelenae* (Cisneros-Heredia and McDiarmid, 2006), *Rulyrana flavopunctata* (Lynch and Duellman, 1973), *Teratohyla pulverata* (Peters, 1873), and *Teratohyla spinosa* (Taylor, 1949): Historical records, distribution extension and new provincial record in Ecuador. *Check List* **2009**, *5*, 912–916.

142. Cisneros-Heredia, D.F.; Guayasamin, J.M. Amphibia, Anura: Centrolenidae, *Centrolene mariaelenae*: Distribution extension, Ecuador. *Check List* **2006**, *2*, 93–95. [CrossRef]

143. Catenazzi, A.; Venegas, P.J. Anfibios y reptiles/Amphibians and reptiles. *Rapid Biol. Soc. Invent. Rep.* **2012**, *24*, 106–117. (in Spanish); 260–271 (in English).

144. Catenazzi, A.; Cisneros-Heredia, D.F.; Sánchez, J.; Brito, J.; Szekely, P.; Betancourt, R. The IUCN Red List of Threatened Species 2018. *Chimerella Mariaelenae* **2018**, e.T135968A516500. [CrossRef]

145. Taylor, E.H. Costa Rican frogs of the genera *Centrolene* and *Centrolenella. Univ. Kans. Sci. Bull.* **1949**, *33*, 257–270. [CrossRef]

146. Starrett, P.H. Descriptions of tadpoles of Middle American frogs. *Misc. Publ. Mus. Zool. Univ. Mich.* **1960**, *110*, 1–37.

147. Hoffmann, H. The glass frog tadpoles of Costa Rica (Anura: Centrolenidae): A study of morphology. *Abh. Der Senckenberg Ges. Für Nat.* **2010**, *567*, 1–78.

148. Savage, J.M. *The Amphibians and Reptiles of Costa Rica: A Herpetofauna Between Two Continents, Between Two Seas*; The University of Chicago Press: Chicago, IL, USA; London, UK, 2002; 934p.

149. Savage, J.M.; Starrett, P.H. A new fringe-limbed tree-frog (family Centrolenidae) from lower Central America. *Copeia* **1967**, 604–609. [CrossRef]

150. Ibáñez, D.R.; Rand, A.S.; Jaramillo, A. *Los Anfibios del Monumento Natural Barro Colorado, Parque Nacional Soberanía y Áreas Adyacentes*; Mizrachi, E., Pujol, S.A., Eds.; Fundación Natura, Smithsonian Tropical Research Institute: Santa Fe, Bogota, Colombia, 1999.

151. Culebras, J.; Angiolani-Larrea, F.N.; Tinajero-Romero, J.; Pellet, C.; Yeager, J. First record and notable range extension of the glass frog *Cochranella granulosa* (Taylor, 1949) (Anura, Centrolenidae) found in Ecuador. *Herpetol. Notes* **2020**, *13*, 353–355.

152. Solís, F.; Ibáñez, R.; Jaramillo, C.; Chaves, G.; Savage, J.; Cruz, G.; Wilson, L.D.; Köhler, G.; Kubicki, B.; Sunyer, J. The IUCN Red List of Threatened Species 2010. *Cochranella Granulosa* **2010**, e.T54964A11232691. [CrossRef]

153. Duellman, W.E. Additional studies of chromosomes of anurans amphibians. *Syst. Zool.* **1967**, *16*, 38–43. [CrossRef] [PubMed]

154. Ruiz-Carranza, P.M.; Lynch, J.D. Ranas centrolenidos de Colombia. IX. Dos nuevas especies del suroeste de Colombia. *Lozania* **1996**, *68*, 1–11.

155. Frost, D.R. *Amphibian Species of the World: An Online Reference, Version 3.0*; Museum of Natural History: New York, NY, USA, 2004. Available online: http://research.amnh.org/vz/herpetology/amphibia/American (accessed on 12 November 2019).

156. Cisneros-Heredia, D.F.; Morales, M.; Rada, M.; Pinto-Erazo, M.; Yánez-Muñoz, M.; Calderón, M.; Ortega-Andrade, M.; Bejarano-Muñoz, P.; Grant, T. The IUCN Red List of Threatened Species 2019. *Cochranella Litoralis* **2019**, e.T54923A49367704. [CrossRef]

157. Guayasamin, J.M.; Bonaccorso, E. A new species of glass frog (Centrolenidae: *Cochranella*) from the lowlands of northwestern Ecuador, with comments on the *Cochranella granulosa* group. *Herpetologica* **2004**, *60*, 485–494. [CrossRef]

158. Cisneros-Heredia, D.F.; Delia, J.; Yánez-Muñoz, M.H.; Ortega-Andrade, H.M. Natural history and intraspecific variation of the Ecuadorian Blue Glassfrog *Cochranella mache* Guayasamin & Bonaccorso, 2004 (Anura: Centrolenidae). *Herpetozoa* **2008**, *2*, 57–66.

159. Ortega-Andrade, H.M.; Rojas-Soto, O.; Paucar, C. Novel data on the ecology of *Cochranella mache* (Anura: Centrolenidae) and the importance of protected Areas for this Critically Endangered glassfrog in the Neotropics. *PLoS ONE* **2013**, *8*. [CrossRef]

160. Ortega-Andrade, H.M.; Bermingham, J.; Aulestia, C.; Paucar, C. Herpetofauna of the Bilsa Biological Station, province of Esmeraldas, Ecuador. *Check List* **2010**, *6*, 119–154. [CrossRef]

161. Jaramillo-Martinez, A.F.; Valencia-Z, A.; Cardona, V.E.; Castro-Herrera, F.; Cisneros-Heredia, D.F. Range extension of *Cochranella mache* Guayasamin and Bonaccorso, 2004 (Anura: Centrolenidae) with comments on the distribution of *C. euknemos* (Savage and Starrett, 1967) in Colombia. *Herpetol. Notes* **2015**, *8*, 161–163.

162. Guayasamin, J.M. *Cochranella mache*. In IUCN 2013. IUCN Red List of Threatened Species. Version 2013.1. 2006. Available online: www.iucnredlist.org (accessed on 26 September 2013).

163. Cisneros-Heredia, D.F.; Delia, J.; Yánez-Muñoz, M.H.; Ortega-Andrade, H.M. Endemic Ecuadorian glassfrog *Cochranella mache* is critically endangered because of habitat loss. *Oryx* **2009**, *44*, 114–117. [CrossRef]

164. Aguayo, R.; Harvey, M.B. A new glassfrog of the *Cochranella granulosa* group (Anura: Centrolenidae) from a Bolivian cloud forest. *Herpetologica* **2006**, *62*, 323–330. [CrossRef]

165. Duellman, W.E. The biology of an Equatorial herpetofauna in Amazonian Ecuador. *Misc. Publ. Mus. Nat. Hist. Univ. Kans.* **1978**, *65*, 1–352.

166. Guayasamin, J.M.; Ron, S.R.; Cisneros-Heredia, D.F.; Lamar, W. A new species of frog of the *Eleutherodactylus lacrimosus* assemblage (Leptodactylidae) from the Western Amazon Basin, with comments on the utility of canopy surveys in lowland rainforest. *Herpetologica* **2006**, *62*, 191–202. [CrossRef]

167. Terán-Valdez, A.; Guayasamin, J.M.; Coloma, L.A. Description of the tadpole of *Cochranella resplendens* and redescription of the tadpole of *Hyalinobatrachium aureoguttatum* (Anura, Centrolenidae). *Phyllomedusa* **2009**, *8*, 105–124. [CrossRef]

168. Mijares-Urrutia, A. Una nueva especie de rana arborícola (Amphibia: Hylidae) de un bosque nublado del oeste de Venezuela. *Rev. Bras. de Biol.* **1998**, *58*, 659–663. [CrossRef]

169. Rada, M.; Rueda-Almonacid, J.V.; Velásquez-Álvarez, A.A.; Sánchez-Pacheco, S.J. Descripción de las larvas de dos centrolénidos (Anura: Centrolenidae) del noroccidente de la Cordillera Oriental, Colombia. *Pap. Avulsos Zool.* **2007**, *47*, 259–272. [CrossRef]

170. Noonan, B.; Bonett, R. A new species of *Hyalinobatrachium* (Anura: Centrolenidae) from the highlands of Guyana. *J. Herpetol.* **2003**, *37*, 91–97. [CrossRef]

171. Twomey, E.M.; Brown, J.L. Range extension: *Cochranella resplendens*. *Herpetol. Rev.* **2005**, *36*, 1.

172. Molina-Zuluaga, C.; Cano, E.; Restrepo, A.; Rada, M.; Daza, J.M. Out of Amazonia: The unexpected trans-Andean distribution of *Cochranella resplendens* (Lynchy and Duellman, 1978) (Anura: Centrolenidae). *Zootaxa* **2017**, *4238*, 268–274. [CrossRef]

173. Cisneros-Heredia, D.F.; Brown, J. The IUCN Red List of Threatened Species 2018. *Cochranella Resplendens* **2018**, e.T54985A54458254. [CrossRef]

174. Duellman, W.E.; Schulte, R. New species of centrolenid frogs from northern Peru. *Univ. Kans. Mus. Nat. Hist. Occas. Pap.* **1993**, *155*, 1–33.

175. Cisneros-Heredia, D.F.; Guayasamin, J.M. Notes on the taxonomy of some glassfrogs from the Andes of Peru and Ecuador (Amphibia: Centrolenidae). *Pap. Avulsos Zool.* **2014**, *54*, 161–168. [CrossRef]

176. Yánez-Muñoz, M.H.; Meza-Ramos, P.; Ortega-Andrade, H.M.; Mueses-Cisneros, J.J.; Reyes-Puig, M.; Reyes-Puig, J.P.; Durán, J.C. Nuevos datos de distribución de ranas de cristal (Amphibia: Centrolenidae) en el oriente de Ecuador, con comentarios sobre la diversidad en la región. *Avances en Ciencias e Ingenierías* **2010**, *3*, 33–40. [CrossRef]

177. Mueses-Cisneros, J.J.; Brito, J.; Sánchez, J.; Szekely, P.; Betancourt, R. The IUCN Red List of Threatened Species 2018. *Espadarana Audax* **2018**, e.T78163103A78162747. [CrossRef]

178. Ospina-Sarria, J.J.; Bolívar-G, W.; Mendez-Narvaez, J. Amphibia, Anura, Centrolenidae, *Espadarana callistomma* (Guayasamin and Trueb, 2007): First country records from Colombia. *Check List* **2010**, *6*, 244–245. [CrossRef]

179. Cisneros-Heredia, D.F.; Guayasamin, J.; Rada, M. The IUCN Red List of Threatened Species 2019. *Espadarana Callistomma* **2019**, e.T135738A85908349. [CrossRef]

180. Cisneros-Heredia, D.F. A new species of glassfrog of the genus *Centrolene* from the foothills of Cordillera Oriental of Ecuador (Anura: Centrolenidae). *Herpetozoa* **2007**, *20*, 27–34.

181. Cisneros-Heredia, D.F.; Brito, J.; Sánchez, J.; Yánez-Muñoz, M.; Szekely, P.; Betancourt, R. The IUCN Red List of Threatened Species 2019. *Espadarana Durrellorum* **2019**, e.T135835A98657065. [CrossRef]

182. Boettger, O. *Katalog der Batrachier-Sammlung im Museum der Naturforschenden Gesellschaft in Frankfurt am Main*; Knauer: Frankfurt am Main, Germany, 1892.

183. Savage, J.M. Type localities for species of amphibians and reptiles described from Costa Rica. *Rev. De Biol. Trop. San José* **1974**, *22*, 71–122.

184. Boulenger, G.A. Descriptions of new batrachians collected by Mr. C.F. Underwood in Costa Rica. *Ann. Mag. Nat. Hist. Lond.* **1896**, *18*, 340–342. [CrossRef]

185. Günther, A.C.L.G. Reptilia and Batrachia. Part 166. *Biol. Cent. Am.* **1901**, *7*, 269–292.

186. Noble, G.K. Some Neotropical batrachians preserved in the United States National Museum with a note on the secondary sexual characters of these and other amphibians. *Proc. Biol. Soc. Wash.* **1924**, *37*, 65–72.

187. Goin, C.J. Distribution and synonymy of *Centrolenella fleischmanni* in northern South America. *Herpetologica* **1964**, *20*, 1–8.

188. Jacobson, S.K. Reproductive behavior and male mating success in two species of glass frogs (Centrolenidae). *Herpetologica* **1985**, *41*, 396–404.

189. Hayes, M.P. A study of clutch attendance in the Neotropical frog *Centrolenella fleischmanni* (Anura: Centrolenidae). Ph.D. Thesis, University of Miami, Coral Gables, FL, USA, 1991, Unpublished.

190. McCaffery, R.; Lips, K. Survival and abundance in males of the glass frog *Espadarana (Centrolene) prosoblepon* in central Panama. *J. Herpetol.* **2013**, *47*, 162–168. [CrossRef]

191. Gosner, K.L. A simplified table for staging anuran embryos and larvae with notes on identification. *Herpetologica* **1960**, *16*, 183–190.

192. Kubicki, B.; Bolaños, F.; Chaves, G.; Solís, F.; Ibáñez, R.; Coloma, L.A.; Ron, S.R.; Wild, E.; Cisneros-Heredia, D.F.; Renjifo, J. The IUCN Red List of Threatened Species 2010. *Espadarana Prosoblepon* **2010**, e.T54934A11228804. [CrossRef]

193. Eaton, T.H. An anatomical study of a neotropical tree frog, *Centrolene prosoblepon* (Salientia: Centrolenidae). *Univ. Kans. Sci. Bull.* **1958**, *39*, 459–472.

194. Guayasamin, J.M.; Vieira, J.; Glor, R.E.; Hutter, C.R. A new glassfrog (Centrolenidae: *Hyalinobatrachium*) from the Topo river basin, Amazonian slopes of the Andes of Ecuador. *Amphib. Reptile Conserv.* **2019**, *13*, 133–144 e194.

195. Barrera-Rodríguez, M.; Ruiz-Carranza, P.M. Una nueva especie del género *Centrolenella* Noble 1920 (Amphibia: Anura: Centrolenidae) de la Cordillera Occidental de Colombia. *Trianea* **1989**, *3*, 77–84.

196. Valencia-Aguilar, A.; Castro-Herrera, F.; Ramírez-Pinilla, M.P. Microhabitats for oviposition and male clutch attendance in *Hyalinobatrachium aureoguttatum* (Anura: Centrolenidae). *Copeia* **2012**, *2012*, 722–731. [CrossRef]

197. Ibáñez, D.R.; Jaramillo, F.E.; Jaramillo, A. Ampliación del ámbito de distribución y descripción del renacuajo de la rana de cristal *Hyalinobatrachium aureoguttatum* (Anura: Centrolenidae). *Rev. Acad. Colomb. Cienc. Exact. Fís. Nat.* **1999**, *23*, 293–298.

198. Acosta-Galvis, A.R. Ranas, Salamandras y Caecilias (Tetrapoda: Amphibia) de Colombia. *Biota Colomb.* **2000**, *1*, 289–319.

199. Bustamante, M.R.; Cisneros-Heredia, D.F.; Yánez-Muñoz, M.H.; Ortega-Andrade, H.M.; Guayasamin, J.M. Amphibia, Centrolenidae, *Cochranella pulverata*, *Hyalinobatrachium aureoguttatum*: Distribution extension, Ecuador. *Check List* **2007**, *3*, 271–276. [CrossRef]

200. Solís, F.; Ibáñez, R.; Jaramillo, C.; Fuenmayor, Q.; Castro, F.; Grant, T. The IUCN Red List of Threatened Species 2010. *Hyalinobatrachium Aureoguttatum* **2010**, e.T55003A11234997. [CrossRef]

201. Taylor, E.H. Notes on Costa Rican Centrolenidae with descriptions of new forms. *Univ. Kans. Sci. Bull.* **1958**, *39*, 41–68.

202. Savage, J.M. A new tree-frog (Centrolenidae) from Costa Rica. *Copeia* **1967**, *1967*, 325–331. [CrossRef]

203. McCranie, J.R.; Wilson, L.D. Two new species of centrolenid frogs of the genus *Hyalinobatrachium* from eastern Honduras. *J. Herpetol.* **1997**, *31*, 10–16. [CrossRef]

204. Kubicki, B. Rediscovery of *Hyalinobatrachium chirripoi* (Anura: Centrolenidae) in southeastern Costa Rica. *Rev. Biol. Trop.* **2004**, 215–218. [CrossRef]

205. Solís, F.; Ibáñez, R.; Chaves, G.; Savage, J.M.; Jaramillo, C.; Fuenmayor, Q.; Castro, F.; Grant, T.; Wild, E.R.; Kubicki, B. The IUCN Red List of Threatened Species 2008. *Hyalinobatrachium Chirripoi* **2008**, e.T55006A11235906. [CrossRef]

206. Boettger, O. Ein neuer Laubfrosch aus Costa Rica. Berichte der Senckenbergischen Naturforschenden Gesellschaft in Frankfurt am Main. 1893, Volume 1893, pp. 251–252. Available online: https://www.biodiversitylibrary.org/itemdetails/33793 (accessed on 29 December 2019).

207. Mertens, R. Die herpetologische Sektion des Natur-Museums und Forschungs-Institutes Senckenberg in Frankfurt a. M. nebst einem Verzeichnis ihrer Typen. *Senckenbergiana Biol.* **1967**, *48*, 1–106.

208. Cope, E.D. Third addition to a knowledge of the Batrachia and Reptilia of Costa Rica. *Proc. Acad. Nat. Sci. USA* **1894**, *46*, 194–206.

209. Starrett, P.H.; Savage, J.M. The systematic status and distribution of Costa Rican glass-frogs, genus *Centrolenella* (Family Centrolenidae), with description of a new species. *Bull. South. Calif. Acad. Sci.* **1973**, *72*, 57–78.

210. Boulenger, G.A. Reptilia and Batrachia (1894). *Zool. Rec.* **1895**, *31*, 1–44.

211. Taylor, E.H. New tailless Amphibia from Mexico. *Univ. Kans. Sci. Bull.* **1942**, *28*, 67–89.

212. Mendoza, A.M.; Bolívar-García, W.; Vásquez-Domínguez, E.; Ibánez, R.; Parra-Olea, G. The role of Central American barriers in shaping the evolutionary history of the northernmost glassfrog, *Hyalinobatrachium fleischmanni* (Anura: Centrolenidae). *PeerJ* **2019**, *7*, e6115. [CrossRef]

213. Greer, B.J.; Wells, K.D. Territorial and reproductive behavior of the tropical American frog *Centrolenella fleischmanni*. *Herpetologica* **1980**, *36*, 318–326.

214. Villa, J. Biology of a Neotropical glass frog, *Centrolenella fleischmanni* (Boettger), with special references to its frogfly associates. *Milwaukee Public Mus. Contrib. Biol. Geol.* **1984**, *21*, 1–17.

215. Delia, J.; Cisneros-Heredia, D.F.; Whitney, J.; Murrieta-Galindo, R. Observations on the reproductive behavior of a Neotropical Glassfrog, *Hyalinobatrachium fleischmanni* (Anura: Centrolenidae). *South Am. J. Herpetol.* **2010**, *5*, 1–12. [CrossRef]

216. McDiarmid, R.W. Glass frog romance along a tropical stream. *TerraLos Angeles Cty. Mus. Nat. Hist.* **1975**, *13*, 14–18.

217. Frost, D.R. *Amphibian Species of the World: An Online Reference, Version 6.0*; Museum of Natural History: New York, NY, USA, 2019. Available online: http://research.amnh.org/vz/herpetology/amphibia/American (accessed on 12 November 2019).

218. Cruz, F.K.; Urgiles, V.L.; Sánchez-Nivicela, J.C.; Siddons, D.C.; Cisneros-Heredia, D.F. Southernmost records of *Hyalinobatrachium fleischmanni* (Anura: Centrolenidae). *Check List* **2017**, *13*, 67–70. [CrossRef]

219. Coloma, L.A.; Ron, S.R.; Wild, E.R.; Cisneros-Heredia, D.F.; Solís, F.; Ibáñez, R.; Santos-Barrera, G.; Kubicki, B. The IUCN Red List of Threatened Species 2010. *Hyalinobatrachium Fleischmanni* **2010**, e.T55014A11238651. [CrossRef]

220. Ayarzagüena, J. *Los Centrolenidos de la Guayana Venezolana*; Publicaciones de la Asociación de Amigos de Doñana: Sevilla, Spain, 1992; Volume 1, pp. 1–46.

221. Duellman, W.E. Amphibian species of the world: Additions and corrections. *Univ. Kans. Publ. Mus. Nat. Hist.* **1993**, *21*, 1–372.

222. Lescure, J.; Marty, C. *Atlas des Amphibiens de Guyane*; Muséum National d'Histore Naturelle: Paris, France, 2001; 388p.

223. Yánez-Muñoz, M.H.; Pérez-Peña, P.; Cisneros-Heredia, D.F. New country records of *Hyalinobatrachium iaspidiense* (Amphibia, Anura, Centrolenidae) from the Amazonian lowlands of Ecuador and Peru. *Herpetol. Notes* **2009**, *2*, 49–52.

224. Barrio-Amorós, C.L.; Brewer-Carias, C. Herpetological results of the 2002 expedition to Sarisariñama, a tepui in Venezuelan Guayana, with the description of five new species. *Zootaxa* **2008**, *1942*, 1–68. [CrossRef]

225. Guayasamin, J.M.; North, S. Amphibia, Centrolenidae, *Hyalinobatrachium iaspidiense*: Distribution extension. *Check List* **2009**, *5*, 526–529. [CrossRef]

226. Cordeiro-Duarte, A.C.; Sanaiotti, T.M.; Duarte, E.L.C.; Pereira, O. Geographic Distribution. *Hyalinobatrachium nouraguensis. Herpetol. Rev.* **2002**, *33*, 220.

227. Caldwell, J.P.; Shepard, D.B. Frogs of the Lower Cristalino River Area. In *Amphibians and Reptiles of the Lower Cristalino River Region of the Southern Amazon*; Vitt, L.J., Caldwell, J.P., Colli, G.R., França, F.G.R., Shepard, D.B., Eds.; Norman, Sam Noble Oklahoma Museum of Natural History: Norman, OK, USA, 2005. Available online: http://www.omnh.ou.edu/personnel/herpetology/vitt/Cerrado/Cristalino/index.htm (accessed on 12 September 2019).

228. Ernst, R.; Rödel, M.O.; Arjoon, D. On the cutting edge–The anuran fauna of the Mabura Hill Forest Reserve, Central Guyana. *Salamandra* **2005**, *41*, 179–194.

229. Kok, P.J.R.; Castroviejo-Fisher, S. Glassfrogs (Anura: Centrolenidae) of Kaieteur National Park, Guyana, with notes on the distribution and taxonomy of some species of the family in the Guiana Shield. *Zootaxa* **2008**, *1680*, 25–53. [CrossRef]

230. La Marca, E.; Señaris, C. The IUCN Red List of Threatened Species. *Hyalinobatrachium Iaspidiense* **2004**, e.T55018A11239792. [CrossRef]

231. Castroviejo-Fisher, S.; Padial, J.M.; Chaparro, J.C.; Aguayo, R.; De la Riva, I. A new species of *Hyalinobatrachium* (Anura: Centrolenidae) from the Amazonian slopes of the central Andes, with comments on the diversity of the genus in the area. *Zootaxa* **2009**, *2143*, 24–44. [CrossRef]

232. Castroviejo-Fisher, S.; Moravec, J.; Aparicio, J.; Guerrero-Reinhards, M.; Calderón, G. DNA taxonomy reveals two new species records of *Hyalinobatrachium* (Anura: Centrolenidae) for Bolivia. *Zootaxa* **2011**, *2798*, 64–68. [CrossRef]

233. Cisneros-Heredia, D.F.; McDiarmid, R.W. Primer registro de *Hyalinobatrachium ruedai* (Amphibia: Centrolenidae) en Ecuador, con notas sobre otras especies congenéricas. *Herpetotropicos* **2006**, *3*, 21–28.

234. Rodríguez, L.; Martinez, J.L.; Coloma, L.A.; Ron, S.R.; Cisneros-Heredia, D.F.; Angulo, A.; Icochea, J.; Acosta, A. The IUCN Red List of Threatened Species 2010. *Hyalinobatrachium Munozorum* **2010**, e.T195860A115339384. [CrossRef]

235. Angulo, A.; Cisneros-Heredia, D.F.; Icochea, M.J.; Martinez, J.L.; Jungfer, K.; Rodriguez, L.; Coloma, L.A.; Ron, S.R.; Sinsch, U.; Arizabal, W. The IUCN Red List of Threatened Species 2014. *Hyalinobatrachium Pellucidum* **2014**, e.T47255219A47255232. [CrossRef]

236. Dunn, E.R. New frogs from Panama and Costa Rica. *Occas. Pap. Boston Soc. Nat. Hist.* **1931**, *5*, 385–401.

237. Hayes, M.P. Predation on the adults and prehatching stages of glass frogs (Centrolenidae). *Biotropica* **1989**, *15*, 74–76. [CrossRef]

238. McDiarmid, R.W.; Adler, K. Notes on the territorial and vocal behaviour of Neotropical frogs of the genus *Centrolenella. Herpetologica* **1974**, *30*, 75–78.

239. Vockenhuber, E.A.; Hödl, W.; Amézquita, A. Glassy fathers do matter: Egg attendance enhances embryonic survivorship in the glass frog *Hyalinobatrachium valerioi. J. Herpetol.* **2009**, *43*, 340–344. [CrossRef]

240. Solís, F.; Ibáñez, R.; Chaves, G.; Savage, J.M.; Bolaños, F.; Kubicki, B.; Jaramillo, C.; Fuenmayor, Q.; Coloma, L.A.; Ron, S.R.; et al. The IUCN Red List of Threatened Species 2008. *Hyalinobatrachium Valerioi* **2008**, e.T55036A11243874. [CrossRef]

241. Guayasamin, J.M.; Cisneros-Heredia, D.F.; Maynard, R.J.; Lynch, R.L.; Culebras, J.; Hamilton, P.S. A marvelous new glassfrog (Centrolenidae, *Hyalinobatrachium*) from Amazonian Ecuador. *ZooKeys* **2017**, *673*, 1–20. [CrossRef] [PubMed]

242. Maynard, R.J.; Aall, N.C.; Saenz, D.; Hamilton, P.S.; Kwiatkowski, M.A. Road-edge effects on herpetofauna in a lowland Amazonian rainforest. *Trop. Conserv. Sci.* **2016**, *9*, 264–290. [CrossRef]

243. Coloma, L.A.; Ron, S. The IUCN Red List of Threatened Species. *Nymphargus Anomalus* **2004**, e.T54946A11230785. [CrossRef]

244. Lynch, J.D. A new ocellated frog (Centrolenidae) from western Colombia. *Proc. Biol. Soc. Wash.* **1990**, *103*, 35–38.

245. Bolívar-G, W.; Coloma, L.A.; Ron, S.R.; Cisneros-Heredia, D.F.; Lynch, J.D.; Wild, E.R. The IUCN Red List of Threatened Species. *Cochranella Balionota* **2004**, e.T54948A11231013. [CrossRef]

246. Lynch, J.D.; Ruiz-Carranza, P.M. A remarkable new centrolenid frog from Colombia with a review of nuptial excrescences in the family. *Herpetologica* **1996**, *52*, 525–535.

247. Yánez-Muñoz, M.H.; Sánchez, J.C.; López, K.; Rea, S.E.; Meza-Ramos, P.; Oyagata, C.L.A.; Guerrero, P. Ampliaciones del rango de distribución de algunas especies de anfibios y reptiles en el suroccidente de Ecuador. *Av. En Cienc. E Ing.* **2014**, *6*, B2–B5.

248. Cisneros-Heredia, D.F. The IUCN Red List of Threatened Species. *Nymphargus Buenaventura* **2008**, e.T135767A4198831. [CrossRef]

249. Coloma, L.A.; Ron, S.R.; Almeida, D. The IUCN Red List of Threatened Species 2004. *Nymphargus Cariticommatus* **2004**, e.T54950A11231356. [CrossRef]

250. Salaman, P.G.W.; Donegan, T.M. (Eds.) Colombia '98 expedition to Serranía de los Churumbelos: Preliminary Report. *Fund. Proaves* **1998**, *1*, 1–46.

251. Coloma, L.A.; Ron, S.R.; Almeida, D. The IUCN Red List of Threatened Species 2010. *Nymphargus Cochranae* **2010**, e.T54954A11231957. [CrossRef]

252. Guayasamin, J.M.; Tapia, E.; Aldas, S.; Deichmann, J. Anfibios y Reptiles de los Tepuyes de la Cuenca Alta del Río Nangaritza, Cordillera del Cóndor. In *Evaluación Ecológica Rápida de la Biodiversidad de los Tepuyes de la Cuenca Alta del Río Nangaritza, Cordillera del Cóndor, Ecuador*; Guayasamin, J.M., Bonaccorso, E., Eds.; Conservación Internacional: Quito, Ecuador, 2011; pp. 56–61.

253. Cisneros-Heredia, D.F.; Meza-Ramos, P. An enigmatic new species of glassfrog (Amphibia: Anura: Centrolenidae) from the Amazonian Andean slopes of Ecuador. *Zootaxa* **2007**, *1485*, 33–41. [CrossRef]

254. Almendáriz, A.; Simmons, J.E.; Vaca-Guerrero, J.; Brito, J. Overview of the herpetofauna of the unexplored Cordillera del Cóndor of Ecuador. *Amphibian Reptile Conserv.* **2014**, *8*, 45–64.

255. Ruiz-Carranza, P.M.; Lynch, J.D. Ranas Centrolenidae de Colombia, V. Cuatro nuevas especies de *Cochranella* de la Cordillera Central. *Lozania* **1995**, *62*, 1–24.

256. Quevedo, A.; Mejía, D.; Gómez, D.; Vargas-Salinas, F.; Gonzalez Duran, G.A.; Lynch, J.; Bernal, M.H.; Gutierrez, P.; Bolívar, W. The IUCN Red List of Threatened Species 2019. *Nymphargus Garciae* **2019**, e.T54962A85873982. [CrossRef]

257. Sorokin, A.; Steigerwald, E. Interspecific combat between *Nymphargus* aff. *grandisonae* and *Espadarana prosoblepon* (Anura, Centrolenidae). *Herpetol. Notes* **2017**, *10*, 283–285.

258. Bogert, C.M. The influence of sound on the behavior of amphibians and reptiles. In *Animal Sounds and Communication. American Institute of Biological Sciences Publication No. 7*; Lanyon, W.E., Tavolga, W.N., Eds.; Intelligencer Printing Company: Washington, DC, USA, 1960; pp. 137–320.

259. Duellman, W.E.; Trueb, L. *Biology of Amphibians*; The Johns Hopkins University Press: Baltimore, Maryland; London, UK, 1994.

260. Ospina-Sarria, J.J.; Bolívar, W.; Mendez-Narvaez, J.; Burbano-Yandi, C. The tadpole of *Nymphargus grandisonae* (Anura: Centrolenidae) from Valle del Cauca, Colombia. *South Am. J. Herpetol.* **2011**, *6*, 79–86. [CrossRef]

261. Bolívar-G, W.; Coloma, L.A.; Ron, S.R.; Cisneros-Heredia, D.F.; Wild, E.R. The IUCN Red List of Threatened Species. *Nymphargus Grandisonae* **2004**, e.T54914A11222520. [CrossRef]

262. Vanegas-Guerrero, J.; Ramírez-Castaño, V.A.; Guevara-Molina, S.C. *Nymphargus grandisonae* (Cochran y Goin 1970). *Catálogo de Anfibios y Reptiles de Colombia* **2014**, *2*, 51–55.

263. Mejía, D.; Gómez, D.; Vargas-Salinas, F.; Gonzalez Duran, G.A.; Coloma, L.A.; Bernal, M.H.; Rada, M.; Yánez-Muñoz, M.; Gutierrez, P.; Ron, S.R.; et al. The IUCN Red List of Threatened Species 2018. *Nymphargus Griffithsi* **2018**, e.T54965A85873493. [CrossRef]

264. Arcila-Pérez, L.F.; Rios-Soto, J.A.; Montilla, S.O.; Londoño-Guarnizo, C.A.; Vargas-Salinas, F. Vocalization and natural history in populations of a glassfrog assigned to *Nymphargus griffithsi* in the Central Andes of Colombia. *Herpetol. Rev.* **2017**, *48*, 275–280.

265. Cisneros-Heredia, D.F. The IUCN Red List of Threatened Species. *Nymphargus Laurae* **2008**, e.T135996A4223953. [CrossRef]

266. Duellman, W.E.; Toft, C.A. Anurans from Serranía de Sira, Amazonian Perú: Taxonomy and biogeography. *Herpetologica* **1979**, *35*, 60–70.

267. Cisneros-Heredia, D.F.; McDiarmid, R.W. Review of the taxonomy and conservation status of the Ecuadorian glassfrog *Centrolenella puyoensis* Flores & McDiarmid (Amphibia: Anura: Centrolenidae). *Zootaxa* **2006**, *1361*, 21–31.

268. Flores, G.; McDiarmid, R.W. Two new species of South American *Centrolenella* (Anura: Centrolenidae) related to *C. Mariae. Herpetol.* **1989**, *45*, 401–411.

269. Stuart, S.N.; Hoffmann, M.; Chanson, J.S.; Cox, N.A.; Berridge, R.J.; Ramani, P.; Young, B.E. (Eds.) *Threatened Amphibians of the World*; Lynx Editions: Barcelona, Spain; International Union for the Conservation of Nature: Gland, Switzerland; Conservation International: Arlington, VA, USA, 2008.

270. Cisneros-Heredia, D.F.; Ortega-Andrade, M.; Bejarano-Muñoz, P. The IUCN Red List of Threatened Species. *Nymphargus Mariae* **2018**, e.T88380390A89226232. [CrossRef]

271. Bolívar, W.; Coloma, L.A.; Ron, S.; Almendáriz, A.; Grant, T. The IUCN Red List of Threatened Species 2004. *Nymphargus Megacheirus* **2004**, e.T54969A11233360. [CrossRef]

272. Cisneros-Heredia, D.F.; Venegas, P.J.; Rada, M.; Schulte, R. A new species of glassfrog (Anura: Centrolenidae) from the foothill Andean forests of Ecuador and Peru. *Herpetologica* **2008**, *64*, 341–353. [CrossRef]

273. Guerrero, J.A.; Mueses-Cisneros, J.J.; Brito, J.; Guayasamin, J.; Sánchez, J.; Szekely, P.; Betancourt, R.; Castroviejo-Fisher, S. The IUCN Red List of Threatened Species. *Nymphargus Posadae* **2018**, e.T54981A85874598. [CrossRef]

274. Rodríguez, L.O.; Córdova, J.; Icochea, J. Lista preliminar de los anfibios del Perú. *Publicaciones Museo Historia Natural Universidad Nacional Mayor San Marcos* **1993**, *45*, 1–22.

275. Coloma, L.A.; Ron, S.R.; Cisneros-Heredia, D.F. The IUCN Red List of Threatened Species 2010. *Nymphargus Siren* **2010**, e.T54992A11226688. [CrossRef]

276. Guayasamin, J.M. A new yellow species of glassfrog (Centrolenidae: *Nymphargus*) from the Amazonian slopes of the Ecuadorian Andes. *Zootaxa* **2013**, *3651*, 193–200. [CrossRef] [PubMed]

277. Ruiz-Carranza, P.M.; Lynch, J.D. Ranas Centrolenidae de Colombia IV. Nuevas especies de *Cochranella* del grupo *ocellata* de la Cordillera Oriental. *Lozania* **1991**, *60*, 1–13.

278. Lynch, J.D. A new centrolenid frog from the Andes of western Colombia. *Rev. Acad. Colomb. Cienc. Exact. Fís. Nat.* **1993**, *18*, 567–569.

279. Guayasamin, J.M. The IUCN Red List of Threatened Species. *Nymphargus Wileyi* **2008**, e.T135864A4213232. [CrossRef]

280. Jiménez-Robles, O.; de la Riva, I.; Guayasamin, J.M. *Rulyrana flavopunctata* (Yellow-spotted Cochran Frog): Reproduction. *Herpetol. Rev.* **2015**, *46*, 238–239.

281. Castro, F.; Rueda, J.V.; Bolívar-G, W.; Estupinan, R.A.; Almendáriz, A. The IUCN Red List of Threatened Species. *Rulyrana Flavopunctata* **2004**, e.T54961A11219097. [CrossRef]

282. Cisneros-Heredia, D.; Angulo, A. The IUCN Red List of Threatened Species. *Rulyrana Mcdiarmidi* **2009**, e.T158475A5200698. [CrossRef]

283. AmphibiaWeb. Information on Amphibian Biology and ConservationBerkeley, California: AmphibiaWeb. Available online: http://amphibiaweb.org/ (accessed on 30 September 2019).

284. Puschendorf, R.; Kubicki, B.; Ryan, M.; Vaughan, C. *Cochranella albomaculata* (NCN). Reproduction. *Herpetol. Rev.* **2004**, *35*, 52.

285. Solís, F.; Ibáñez, R.; Chaves, G.; Savage, J.M.; Jaramillo, C.; Fuenmayor, Q.; Castro, F.; Grant, T.; Wild, E.R.; Kubiki, B.; et al. Iucn Red List Threat. Species. *Sachatamia Albomaculata* **2010**, e.T54944A11230494. [CrossRef]

286. Cisneros-Heredia, D.F.; Yánez-Muñoz, M.H.; Ortega-Andrade, H.M. Description of a new species of *Teratohyla* Taylor (Amphibia: Athesphatanura: Centrolenidae) from north-western Ecuador. *Zootaxa* **2009**, *2227*, 53–62. [CrossRef]

287. Cisneros-Heredia, D.F.; Yánez-Muñoz, H.M.; Ortega-Andrade, H.M. *Teratohyla sornozai* Cisneros-Heredia, Yánez-Muñoz y Ortega-Andrade es un sinónimo junior de *Rulyrana orejuela* Duellman y Burrowes (Amphibia, Anura, Centrolenidae). *Av. En Cienc. E Ing.* **2010**, *3*, B3–B4. [CrossRef]

288. Yánez-Muñoz, M.H.; Cisneros-Heredia, D.F. Amphibia, Anura, Centrolenidae, *Cochranella orejuela*, first country record from Ecuador. *Check List* **2008**, *4*, 50–54. [CrossRef]

289. Almeida-Reinoso, D.; Coloma, L.A. Natural History Notes: *Rulyrana orejuela* (Orejuela glass frog). Predation. *Herpetol. Rev.* **2012**, *43*, 126.

290. Guevara-Molina, S.C.; Rada, M. Sachatamia orejuela. *Catálogo de Anfibios y Reptiles de Colombia* **2018**, *4*, 68–72.

291. Castro, F.; Gonzalez Duran, G.A.; Lynch, J.; Grant, T. The IUCN Red List of Threatened Species 2019. *Sachatamia Orejuela* **2019**, e.T54976A49367802. [CrossRef]

292. Myers, C.W.; Stothers, R.B. The myth of Hylas revisited: The frog name *Hyla* and other commentary on *Specimen medicum* (1768) of J.N. Laurenti, the "father of herpetology". *Arch. Nat. History Lond.* **2006**, *33*, 241–266. [CrossRef]

293. Cisneros-Heredia, D.F. The IUCN Red List of Threatened Species 2008. *Teratohyla Amelie* **2008**, e.T135771A4199328. [CrossRef]

294. Ron, S.R.; Merino-Viteri, A.; Ortiz, D.A. *Anfibios del Ecuador*; Version 2019.0.; Pontificia Universidad Católica del Ecuador: Museo de Zoología, Córdoba, Argentina, 2019. Available online: https://bioweb.bio/faunaweb/amphibiaweb (accessed on 5 March 2020).

295. Diaz-Ricaurte, J.; Molina, E.; Guevara, C.; Díaz-Morales, R. *Teratohyla midas* (Santa Cecilia Cochran Frog). Reproductive behavior. *Herpetol. Rev.* **2016**, *47*, 12–13.

296. Araújo, K.d.C.; Pansonato, A.; de Oliveira, R.H.; Morais, D.H.; de Carvalho, V.T.; Ávila, R.W. Advertisement call and new distribution records from Brazil of *Teratohyla midas* (Lynch & Duellman, 1973) (Anura, Centrolenidae). *Check List* **2018**, *14*, 303–308.

297. Rodríguez, L.; Martinez, J.L.; Azevedo-Ramos, C.; Coloma, L.A.; Ron, S.R. The IUCN Red List of Threatened Species 2004. *Teratohyla Midas* **2004**, e.T54971A11219321. [CrossRef]

298. Peters, W.C.H. Über eine neue Schildkrötenart, Kinosternon Effeldtii und einige andere neue oder weniger bekannte Amphibien. *Mon. der Königlich Preuss. Akad. der Wiss. zu Berl.* **1873**, 603–618.

299. Myers, C.W. Spotted poison frogs: Descriptions of three new Dendrobates from western Amazonia, and resurrection of a lost species from "Chiriqui". *Am. Mus. Novit.* **1982**, *2721*, 1–23.

300. Guayasamin, J.M.; Cisneros-Heredia, D.F.; Castroviejo-Fisher, S. Taxonomic identity of *Cochranella petersi* Goin, 1961 and *Centrolenella ametarsia* Flores, 1987. *Zootaxa* **2008**, *1815*, 25–34. [CrossRef]

301. Hawley, T.J. Embryonic development and mortality in *Hyalinobatrachium pulveratum* (Anura: Centrolenidae) of south-western Costa Rica. *J. Trop. Ecol.* **2006**, *22*, 731–734. [CrossRef]

302. Hoffmann, H. Description of the previously unknown tadpole of *Hyalinobatrachium pulveratum* (Anura: Centrolenidae). *Rev. De Biol. Trop.* **2004**, *52*, 219–228. [CrossRef]

303. Köhler, G. *Anfibios y Reptiles de Nicaragua*; Herpeton: Offenbach, Germany, 2001.

304. Solís, F.; Ibáñez, R.; Castro, F.; Grant, T.; Acosta-Galvis, A.; Kubicki, B. The IUCN Red List of Threatened Species. *Teratohyla Pulverata* **2010**, e.T55030A11242870. [CrossRef]

305. Ibáñez, D.R.; Solís, F.A.; Jaramillo, C.A.; Rand, A.S. An overview of the herpetology of Panama. In *Mesoamerican Herpetology: Systematics, Zoogeography and Conservation*; Johnson, J.D., Webb, R.G., Flores-Villela, O.A., Eds.; The University of Texas at El Paso: El Paso, TX, USA, 2000; pp. 159–170.

306. Sunyer, J.; Köhler, G. New country and departmental records of herpetofauna in Nicaragua. *Salamandra* **2007**, *43*, 57–62.

307. Sunyer, J.; Páiz, G.; Dehling, D.M.; Köhler, G. A collection of amphibians from Río San Juan, southeastern Nicaragua. *Herpetol. Notes* **2009**, *2*, 189–202.

308. Köhler, G. *Amphibians of Central America*; Herpeton: Offenbach, Germany, 2011.

309. Coloma, L.A.; Ron, S.R.; Wild, E.; Cisneros-Heredia, D.; Solís, F.; Ibáñez, R.; Jaramillo, C.; Chaves, G.; Savage, J.; Cruz, G.; et al. The IUCN Red List of Threatened Species 2010. *Teratohyla Spinosa* **2010**, e.T54996A11227658. [CrossRef]

310. Jongsma, G.F.M.; Hedley, R.W.; Duraes, R.; Karubian, J. Amphibian diversity and species composition in relation to habitat type and alteration in the Mache-Chindul Reserve, northwest Ecuador. *Herpetologica* **2014**, *70*, 34–46. [CrossRef]

311. Lutz, B.; Kloss, G.R. Anfíbios anuros do alto Solimões e Rio Negro. Apontamentos sôbre algumas formas e suas vicariantes. *Memórias do Instituto Oswaldo Cruz. Rio de Jan.* **1952**, *50*, 625–678. [CrossRef]

312. Cisneros-Heredia, D.F. *Centrolene ritae* Lutz is a senior synonym of *Cochranella oyampiensis* Lescure and *Cochranella ametarsia* Flores (Anura: Centrolenidae). *ACI Avances en Ciencias e Ingenierías* **2013**, *5*, B1–B4. [CrossRef]

313. Lescure, J. Contribution a l'étude des amphibiens de Guyane Francaise. V. Les centrolenidae. *Bull. Soc. Zool. Fr.* **1975**, *100*, 385–394.

314. Flores, G. A new *Centrolenella* from the Amazonian Lowlands of Colombia. *J. Herpetol.* **1987**, *21*, 185–190. [CrossRef]

315. Lima, A.P.; Magnusson, W.E.; Menin, M.; Erdtmann, L.K.; Rodrigues, D.J.; Keller, C.; Hödl, W. *Guia de Sapos da Reserva Adolfdo Ducke, Amazônia Central*; Áttema Design Editorial: Manaus, Brazil, 2006.

316. Menin, M.; Lima, A.; Rodriguez, D. The tadpole of *Vitreorana oyampiensis* (Anura: Centrolenidae) in Central Amazonia, Brazil. *Zootaxa* **2009**, *2203*, 65–68. [CrossRef]

317. Zimmerman, B.L.; Bogart, J.P. Vocalizations of primary forest frog species in the central Amazon. *Acta Amaz.* **1984**, *14*, 473–519. [CrossRef]

318. Muñoz-Saravia, A.; Aguayo-Vedia, C.R. Sobre la presencia de *Vitreorana oyampiensis* (Lescure, 1975) (Amphibia: Centrolenidae) en Bolivia. *Cuad. Herpetol.* **2009**, *23*, 97–99.

319. McCracken, S.F.; Forstner, M.R.J. Bromeliad patch sampling technique for canopy herpetofauna in neotropical forests. *Herpetol. Rev.* **2008**, *39*, 170–174.

320. Lynch, J.D. The transition from archaic to advanced frogs. In *Evolutionary Biology of the Anurans: Contemporary Research on Major Problems*; Vial, J.L., Ed.; University of Missouri Press: Columbia, NY, USA, 1973.

321. Noble, G.K. *The Biology of the Amphibia*; McGraw-Hill: New York, NY, USA, 1931; pp. 133–182, 577p.

322. Duellman, W.E. On the classification of frogs. *Occas. Pap. Mus. Nat. Hist. Univ. Kans.* **1975**, *42*, 1–14.

323. Duellman, W.E. *Hylid Frogs of Middle America*, 2nd ed.; Society for the Study of Amphibians and Reptiles, Contributions to Herpetology: Lawrence, KS, USA, 2001; No. 18.

324. Ford, L.S.; Cannatella, D.C. The major clades of frogs. *Herpetol. Monogr.* **1993**, *7*, 94–117. [CrossRef]

325. Austin, J.D.; Lougheed, S.C.; Tanner, K.; Chek, A.A.; Bogart, J.P.; Boag, P.T. A molecular perspective on the evolutionary affinities of an enigmatic neotropical frog, *Allophryne ruthveni*. *Zool. J. Linn. Soc.* **2002**, *134*, 335–346. [CrossRef]

326. Burton, T. Muscles of the pes of hylid frogs. *J. Morphol.* **2004**, *260*, 209–233. [CrossRef] [PubMed]

327. Pyron, R.A. Biogeographic analysis reveals ancient continental vicariance and recent oceanic dispersal in amphibians. *Syst. Biol.* **2014**, *63*, 779–797. [CrossRef] [PubMed]

328. Feng, Y.J.; Blackburn, D.C.; Liang, D.; Hillis, D.M.; Wake, D.B.; Cannatella, D.C.; Zhang, P. Phylogenomics reveals rapid, simultaneous diversification of three major clades of Gondwanan frogs at the Cretaceous–Paleogene boundary. *Proc. Natl. Acad. Sci. USA* **2017**, *114*, E5864–E5870. [CrossRef] [PubMed]

329. Haas, A. The phylogeny of frogs as inferred from primarily larval characters. *Cladistics* **2003**, *19*, 23–89.

330. Wiley, E.O. *Phylogenetics: The Theory and Practice of Phylogenetic Systematics*; John Wiley and Sons: New York, NY, USA, 1981.

331. Lecointre, G.; Philippe, H.; Van Le, H.L.; Le Guyader, H. Species sampling has a major impact on phylogenetic inference. *Mol. Phylogenetics Evol.* **1993**, *2*, 205–224. [CrossRef]

332. Hillis, D.M. Inferring complex phylogenies. *Nature* **1996**, *383*, 130–131. [CrossRef]

333. Hillis, D.M. Taxonomic sampling, phylogenetic accuracy, and investigator bias. *Syst. Biol.* **1998**, *47*, 3–8. [CrossRef]

334. Rannala, B.; Huelsenbeck, J.P.; Yang, Z.; Nielsen, R. Taxon sampling and the accuracy of large phylogenies. *Syst. Biol.* **1998**, *47*, 702–710. [CrossRef]

335. Zwickl, D.J.; Hillis, D.M. Increased taxon sampling greatly reduces phylogenetic error. *Syst. Biol.* **2002**, *51*, 588–598. [CrossRef] [PubMed]

336. Pollock, D.D.; Zwickl, D.J.; McGuire, J.A.; Hillis, D.M. Increased taxon sampling is advantageous for phylogenetic inference. *Syst. Biol.* **2002**, *51*, 664–671. [CrossRef] [PubMed]

337. Poe, S. Evaluation of the strategy of long-branch subdivision to improve the accuracy of phylogenetic methods. *Syst. Biol.* **2003**, *52*, 423–428. [CrossRef] [PubMed]

338. Rokas, A.; Carroll, S.B. More genes or more taxa? The relative contribution of gene number and taxon number to phylogenetic accuracy. *Mol. Biol. Evol.* **2005**, *22*, 1337–1344. [CrossRef]

339. Heath, T.A.; Zwickl, D.J.; Kim, J.; Hillis, D.M. Taxon sampling affects inferences of macroevolutionary processes from phylogenetic trees. *Syst. Biol.* **2008**, *57*, 160–166. [CrossRef]

340. Heath, T.A.; Hedtke, S.M.; Hillis, D.M. Taxon sampling and the accuracy of phylogenetic analyses. *J. Syst. Evol.* **2008**, *46*, 239–257.

341. Vences, M.; Guayasamin, J.M.; Miralles, A.; de La Riva, I. To name or not to name: Criteria to promote economy of change in Linnaean classification schemes. *Zootaxa* **2013**, *3636*, 201–244. [CrossRef]

342. Hoorn, C.; Wesselingh, F.P.; Ter Steege, H.; Bermudez, M.A.; Mora, A.; Sevink, J.; Jaramillo, C. Amazonia through time: Andean uplift, climate change, landscape evolution, and biodiversity. *Science* **2010**, *330*, 927–931. [CrossRef]

343. Duellman, W.E. Distribution Patterns of Amphibians in South America. In *Patterns of Distribution of Amphibians: A Global Perspective*; Duellman, W.E., Ed.; The Johns Hopkins University Press: Baltimore, MD, USA, 1999; pp. 255–328.

344. Peterson, A.T.; Soberon, J.; Sanchez-Cordero, V. Conservatism of ecological niches in evolutionary time. *Science* **1999**, *285*, 1265–1267. [CrossRef]

345. Coyne, J.A.; Orr, H.A. *Speciation*; Sinauer Associates, Inc.: Sunderland, MA, USA, 2004.

346. Wiens, J.J. Speciation and ecology revisited: Phylogenetic niche conservatism and the origin of species. *Evolution* **2004**, *58*, 193–197. [CrossRef]

347. Remsen, J.V. High incidence of "leapfrog" pattern of geographic variation in Andean birds: Implications for the speciation process. *Science* **1984**, *224*, 171–173. [CrossRef] [PubMed]

348. Graves, G.R. Linearity of geographic range and its possible effect on the population structure of Andean birds. *Auk* **1988**, *105*, 47–52. [CrossRef]

349. Vuilleumier, F. Pleistocene speciation in birds living in the high tropical Andes. *Nature* **1969**, *223*, 1179–1180. [CrossRef]

350. Haffer, J. *Avian Speciation in Tropical South America*; Nutall Ornithological Club: Cambridge, MA, USA, 1974.

351. Hackett, S.J. Molecular systematics and zoogeography of flowerpiercers in the *Diglossa baritula* complex. *Auk* **1995**, *112*, 156–170. [CrossRef]

352. Behling, H.; Hooghiemstra, H. Holocene Amazon rainforest-savanna dynamics and climatic implications: High-resolution pollen record from Laguna Loma Linda in eastern Colombia. *J. Quat. Sci.* **2000**, *15*, 687–695. [CrossRef]

353. Bonaccorso, E.; Koch, I.; Peterson, A.T. Pleistocene fragmentation of Amazonian species' ranges. *Divers. Distrib.* **2006**, *8*, 157–164. [CrossRef]

354. Bush, M.B.; de Oliveira, P.E. The rise and fall of the refugial hypothesis of Amazonian speciation: A paleoecological perspective. *Biota Neotrop.* **2006**, *6*, 1–17. [CrossRef]

355. Wiens, J.J.; Graham, C.H. Niche conservatism: Integrating evolution, ecology, and conservation biology. *Ann. Rev. Ecol. Evol. Syst.* **2005**, *36*, 519–539. [CrossRef]

356. Losos, J.B. Phylogenetic niche conservatism, phylogenetic signal and the relationship between phylogenetic relatedness and ecological similarity among species. *Ecol. Lett.* **2009**, *11*, 995–1003. [CrossRef]

357. Lynch, J.D.; Duellman, W.E. Frogs of the genus *Eleutherodactylus* in western Ecuador: Systematics, ecology, and biogeography. *Univ. Kans. Nat. Hist. Mus. Spec. Publ.* **1997**, *23*, 1–236.

358. Graham, C.H.; Ron, S.R.; Santos, J.C.; Schneider, C.J.; Moritz, C. Integrating phylogenetics and environmental niche models to explore speciation mechanisms in dendrobatid frogs. *Evolution* **2004**, *58*, 1781–1793. [CrossRef] [PubMed]

359. Martof, B.S. Home range and movements of the green frog, *Rana clamitans*. *Ecology* **1953**, *34*, 529–543. [CrossRef]

360. Dole, J.W. Spring movements of leopard frogs, *Rana pipiens* Schreber, in Northern Michigan. *Am. Midl. Nat.* **1967**, *78*, 167–181. [CrossRef]

361. Dole, J.W. Dispersal of recently metamorphosed leopard frogs, *Rana pipiens*. *Copeia* **1971**, *2*, 221–228. [CrossRef]

362. Berven, K.A.; Grudzien, T.A. Dispersal in the wood frog (*Rana sylvatica*): Implications for genetic population structure. *Evolution* **1990**, *22*, 2047–2056.

363. Lynch, J.D. The gauge of speciation: On the frequencies of modes of speciation. In *Speciation and Its Consequences*; Otte, D., Endler, J.A., Eds.; Sinauer Associates: Sunderland, MA, USA, 1989; pp. 527–556.

364. Graves, G.R. Elevational correlates of speciation and intraspecific geographic variation in plumage in Andean forest birds. *Auk* **1985**, *102*, 556–579. [CrossRef]

365. Krabbe, N. Arid valleys as dispersal barriers to high-Andean forest birds in Ecuador. *Cotinga* **2008**, *29*, 28–30.

366. Guayasamin, J.M.; Bonaccorso, E.; Duellman, W.E.; Coloma, L.A. Genetic differentiation in the nearly extinct harlequin toads (Bufonidae: *Atelopus*), with emphasis on the *Atelopus ignescens* and *A. bomolochos* species complexes. *Zootaxa* **2010**, *2574*, 55–68. [CrossRef]

367. Guayasamin, J.M.; Krynak, T.; Krynak, K.; Culebras, J.; Hutter, C.R. Phenotypic plasticity raises questions for taxonomically important traits: A remarkable new Andean rainfrog (*Pristimantis*) with the ability to change skin texture. *Zool. J. Linn. Soc.* **2015**, *173*, 913–928. [CrossRef]

368. Guayasamin, J.M.; Hutter, C.R.; Tapia, E.E.; Culebras, J.; Peñafiel, N.; Pyron, R.A.; Arteaga, A. Diversification of the rainfrog *Pristimantis ornatissimus* in the lowlands and Andean foothills of Ecuador. *PLoS ONE* **2017**, *12*, e0172615. [CrossRef]

369. Hutter, C.R.; Guayasamin, J.M. Cryptic diversity concealed in the Andean cloud forests: Two new species of rainfrogs (*Pristimantis*) uncovered by molecular and bioacoustic data. *Neotrop. Biodivers.* **2015**, *1*, 36–59. [CrossRef]

370. Prieto-Torres, D.A.; Cuervo, A.M.; Bonaccorso, E. On geographic barriers and Pleistocene glaciations: Tracing the diversification of the Russet-crowned Warbler (*Myiothlypis coronata*) along the Andes. *PLoS ONE* **2018**, *13*, e0191598. [CrossRef] [PubMed]

371. Deichmann, J.; Boundy, J.; Williamson, G.B. Anuran artifacts of preservation: 27 years later. *Phyllomedusa* **2009**, *8*, 51–58. [CrossRef]

372. Harvey, M.B.; Noonan, B.P. Bolivian glass frogs (Anura: Centrolenidae) with a descrition of a new species from Amazonia. *Proc. Biol. Soc. Wash.* **2005**, *118*, 428–441. [CrossRef]

373. Barnett, J.B.; Michalis, C.; Anderson, H.M.; McEwen, B.L.; Yeager, J.; Pruitt, J.N.; Scott-Samuel, N.E.; Cuthill, I.C. Imperfect transparency and camouflage in glass frogs. *Proc. Natl. Acad. Sci. USA* **2020**, 201919417. [CrossRef]

374. Schwalm, P.A.; Starrett, P.H.; McDiarmid, R.W. Infrared reflectance in leaf-sitting neotropical frogs. *Science* **1977**, *196*, 1225–1226. [CrossRef]

375. Delia, J.; Bravo-Valencia, L.; Warkentin, K.M. The evolution of extended parental care in glassfrogs: Do egg-clutch phenotypes mediate coevolution between the sexes? *Ecol. Monogr.* **2020**. [CrossRef]

376. Cardozo-Urdaneta, A.; Señaris, J.C. Vocalización y biología reproductiva de las ranas de cristal *Hyalinobatrachium pallidum* y *Centrolene daidaleum* (Anura, Centrolenidae) en la sierra de Perijá, Venezuela. *Mem. De La Fund. La Salle De Cienc. Nat.* **2012**, *70*, 87–105.

377. Cadle, J.E.; McDiarmid, R.W. Two new species of *Centrolenella* (Anura: Centrolenidae) from northwestern Peru. *Proc. Biol. Soc. Wash.* **1990**, *103*, 746–768.

378. Vargas-Salinas, F.; Quintero-Ángel, A.; Osorio-Domínguez, D.; Rojas-Morales, J.A.; Escobar-Lasso, S.; Gutiérrez-Cárdenas, P.D.A.; Amézquita, A. Breeding and parental behaviour in the glass frog *Centrolene savagei* (Anura: Centrolenidae). *J. Nat. Hist.* **2014**, *48*, 1689–1705. [CrossRef]

379. Cabanzo-Olarte, L.C.; Ramírez-Pinilla, M.P.; Serrano-Cardozo, V.H. Oviposition, site preference, and evaluation of male clutch attendance in *Espadarana andina* (Anura: Centrolenidae). *J. Herpetol.* **2013**, *47*, 314–320. [CrossRef]

380. Restrepo, J.H.; Naranjo, L.G. Ecología reproductiva de una población de *Cochranella ignota* (Anura: Centrolenidae). *Rev. Acad. Colomb. Cienc. Exact. Fís. Nat.* **1999**, *23*, 49–60.

381. Gouveia, S.F.; Faria, R.G.; da Rocha, P.A. Local distribution and notes on reproduction of *Vitreorana* aff. *eurygnatha* (Anura: Centrolenidae) from Sergipe, Northeastern Brazil. *Herpetol. Bull.* **2012**, *120*, 16–21.

382. Lehtinen, R.M.; Green, S.E.; Pringle, J.L. Impacts of paternal care and seasonal change on offspring survival: A multiseason experimental study of a Caribbean frog. *Ethology* **2014**, *120*, 400–409. [CrossRef]

383. Vockenhuber, E.A.; Hödl, W.; Karpfen, U. Reproductive behavior of the glass frog *Hyalinobatrachium valerioi* (Anura: Centrolenidae) at the tropical stream Quebrada Negra (La Gamba, Costa Rica). *Stapfia* **2008**, *88*, 335–348.

384. Guayasamin, J.M.; Barrio-Amorós, C. Combat behavior in *Centrolene andinum* (Rivero, 1968) (Anura: Centrolenidae). *Salamandra* **2005**, *41*, 153–155.

385. Rojas-Runjaic, F.J.; Cabello, P. *Centrolene daidaleum* (Ruiz-Carranza & Lynch, 1991)(Anura, Centrolenidae): A glassfrog with primitive and derived combat behavior. *Zootaxa* **2011**, *2833*, 60–64.

386. Restrepo-Toro, J.H. Ecología conductual de una Rana arbórea Neotropical. Bachelor's Thesis, Universidad del Valle, Cali, Colombia, 1996.

387. Cuesta, F.; Peralvo, M.; Merino-Viteri, A.; Bustamante, M.; Baquero, F.; Freile, J.F.; Muriel, P.; Torres-Carvajal, O. Priority areas for biodiversity conservation in mainland Ecuador. *Neotrop. Biodivers.* **2017**, *3*, 93–106. [CrossRef]

388. Vandegrift, R.; Thomas, D.C.; Roy, B.A.; Levy, M. *The Extent of Recent Mining Concessions in Ecuador*; Rainforest Information Center: Nimbin, NSW, Australia, 2018.

389. Berger, L.; Speare, R.; Daszak, P.; Green, D.E.; Cunningham, A.A.; Goggin, C.L.; Slocombe, R.; Ragan, M.A.; Hyatt, A.D.; McDonald, K.R. Chytridiomycosis causes amphibian mortality associated with population declines in the rain forests of Australia and Central America. *Proc. Natl. Acad. Sci. USA* **1998**, *95*, 9031–9036. [CrossRef]

390. Bonaccorso, E.; Guayasamin, J.M.; Méndez, D.; Speare, R. Chytridomycosis as a possible cause of population declines in *Atelopus cruciger* (Anura: Bufonidae). *Herpetol. Rev.* **2003**, *34*, 331–334.

391. Scheele, B.C.; Pasmans, F.; Skerratt, L.F.; Berger, L.; Martel, A.; Beukema, W.; Acevedo, A.A.; Burrowes, P.A.; Carvalho, T.; Catenazzi, A.; et al. Response to comment on "Amphibian fungal panzootic causes catastrophic and ongoing loss of biodiversity". *Science* **2020**, *367*, eaay2905. [CrossRef] [PubMed]

392. Menéndez-Guerrero, P.A.; Graham, C.H. Evaluating multiple causes of amphibian declines of Ecuador using geographical quantitative analyses. *Ecography* **2013**, *36*, 756–769. [CrossRef]

393. Vimos, D.J.; Encalada, A.C.; Ríos-Touma, B.; Suárez, E.; Prat, N. Effects of exotic trout on benthic communities in high-Andean tropical streams. *Freshw. Sci.* **2015**, *34*, 770–783. [CrossRef]

394. Krynak, K.L.; Wessels, D.G.; Imba, S.M.; Krynak, T.J.; Snyder, E.B.; Lyons, J.A.; Guayasamin, J.M. Call survey indicates rainbow trout farming alters glassfrog community composition in the Andes of Ecuador. *Amphib. Reptile Conserv.* **2020**, *14*, e234.

395. Polato, N.R.; Gill, B.A.; Shah, A.A.; Gray, M.M.; Casner, K.L.; Barthelet, A.; Messer, P.W.; Simmons, M.P.; Guayasamin, J.M.; Encalada, A.C.; et al. Narrow thermal tolerance and low dispersal drive higher speciation in tropical mountains. *Proc. Natl. Acad. Sci. USA* **2018**, *115*, 12471–12476. [CrossRef] [PubMed]

396. Hoegh-Guldberg, O.; Hughes, L.; McIntyre, S.; Lindenmayer, D.B.; Parmesan, C.; Possingham, H.P.; Thomas, C.D. Assisted colonization and rapid climate change. *Science* **2008**, *321*, 345–346. [CrossRef] [PubMed]

397. Quintero, I.; Wiens, J.J. Rates of projected climate change dramatically exceed past rates of climatic niche evolution among vertebrate species. *Ecol. Lett.* **2013**, *16*, 1095–1103. [CrossRef]

398. Jezkova, T.; Wiens, J.J. Rates of change in climatic niches in plant and animal populations are much slower than projected climate change. *Proc. R. Soc. B* **2016**, *283*, 20162104. [CrossRef]

Article

Recent and Rapid Radiation of the Highly Endangered Harlequin Frogs (*Atelopus*) into Central America Inferred from Mitochondrial DNA Sequences

Juan P. Ramírez [1,2], César A. Jaramillo [3,4,5], Erik D. Lindquist [6], Andrew J. Crawford [1,3,5] and Roberto Ibáñez [3,5,7,8,*]

[1] Department of Biological Sciences, Universidad de los Andes, Calle 19 No. 1-60, Bogotá 111711, Colombia; jp.ramirez10@uniandes.edu.co (J.P.R.); andrew@dna.ac (A.J.C.)
[2] Department of Biology, San Diego State University, San Diego, CA 92182, USA
[3] Círculo Herpetológico de Panamá, Apartado 0824-00122, Panama; jaramilc@si.edu
[4] Departamento de Histología y Neuroanatomía Humana, Facultad de Medicina, Universidad de Panamá, Apartado 0824-03366, Panama
[5] Smithsonian Tropical Research Institute, Apartado 0843-03092, Panama
[6] Department of Biological Sciences, Messiah University, Mechanicsburg, PA 17055, USA; elindquist@messiah.edu
[7] Sistema Nacional de Investigación, SENACYT, Apartado 0816-02852, Panama
[8] Departamento de Zoología, Universidad de Panamá, Apartado 0824-03366, Panama
* Correspondence: ibanezr@si.edu; Tel.: +507-6650-6544

Received: 21 August 2020; Accepted: 15 September 2020; Published: 18 September 2020

Abstract: Populations of amphibians are experiencing severe declines worldwide. One group with the most catastrophic declines is the Neotropical genus *Atelopus* (Anura: Bufonidae). Many species of *Atelopus* have not been seen for decades and all eight Central American species are considered "Critically Endangered", three of them very likely extinct. Nonetheless, the taxonomy, phylogeny, and biogeographic history of Central American *Atelopus* are still poorly known. In this study, the phylogenetic relationships among seven of the eight described species in Central America were inferred based on mitochondrial DNA sequences from 103 individuals, including decades-old museum samples and two likely extinct species, plus ten South American species. Among Central American samples, we discovered two candidate species that should be incorporated into conservation programs. Phylogenetic inference revealed a ladderized topology, placing species geographically furthest from South America more nested in the tree. Model-based ancestral area estimation supported either one or two colonization events from South America. Relaxed-clock analysis of divergence times indicated that *Atelopus* colonized Central America prior to 4 million years ago (Ma), supporting a slightly older than traditional date for the closure of the Isthmus. This study highlights the invaluable role of museum collections in documenting past biodiversity, and these results could guide future conservation efforts. An abstract in Spanish (Resumen) is available as supplementary material.

Keywords: Bufonidae; cryptic species; forensic taxonomy; Great American Biotic Interchange; historical biogeography; Isthmus of Panama; Middle America; molecular phylogenetics; phylogeography

---

1. Introduction

Amphibians are experiencing a global conservation crisis, with an estimated 41 to 50% of species suffering population declines [1–3]. These declines are likely the result of interactions among several factors of primarily anthropic origin, including overexploitation, habitat destruction, pollution, and the effects of emerging epizootic pathogens [4]. The most alarming cause of global declines and extinctions of amphibians is chytridiomycosis, a disease caused by the chytrid fungus, *Batrachochytrium dendrobatidis* (*Bd*) [5,6].

Harlequin frogs of the genus *Atelopus* (Anura: Bufonidae) have arguably suffered the most dramatic population declines and extinctions of any diverse genus of amphibians. Because of their diurnal habits, bright coloration, and previously high local abundance, species of this genus were a prominent element of many Neotropical communities until about 35 years ago [7–9]. Since the late 1980's, the majority of species of *Atelopus* have not been seen in their historic localities [10]. These declines were likely driven by *Bd*, and *Atelopus* are known to be highly susceptible to chytridiomycosis [11,12]. Declines of *Atelopus* and other amphibian species have been considered to be most severe in populations at higher elevations (above 1000 m above sea level) [13], hypothetically because the lower temperatures present at higher altitudes are associated with higher levels of infectivity and fecundity in *Bd* [14].

The eight species of Central American *Atelopus* are no exception to these patterns of endangerment and decline. These forest species have a very similar natural history, in which reproduction is associated with fast-flowing streams where eggs are laid in the water and tadpoles develop ([15], pers. obs.). The only exception is the scarcely known *A. chirripoensis* from the Costa Rican *páramo* which breeds in small ponds [16]. Central American *Atelopus* were once abundant members of amphibian communities at certain localities, particularly during their breeding season ([15,16], pers. obs.). Currently, all eight are listed as Critically Endangered by the International Union for Conservation of Nature (IUCN) and the three species restricted to elevations above 1000 m (*A. chiriquiensis*, *A. chirripoensis*, and *A. senex* [17]) are considered possibly extinct [18]. The much reduced populations of the remaining five species have been prioritized for *ex situ* conservation through captive breeding programs in Panama and abroad [19,20]. *Atelopus zeteki* is endemic to central Panama and it has not been seen in the wild since 2009. Individuals were collected before the epizootic declines, however, and this species is maintained in a well-managed captive breeding program [21]. The four species still found in the wild, *A. varius* of Costa Rica and Panama, plus *A. certus*, *A. glyphus*, and *A. limosus* of eastern Panama, are being reared in captivity by the Panama Amphibian Rescue and Conservation Project [22,23].

The success of *ex situ* conservation programs depends on having a complete and robust taxonomy, because correct delimitation and identification of species (1) allows a more efficient use of resources by avoiding the unnecessary protection of taxonomically invalid species [24], (2) can prevent neglecting previously unrecognized species in conservation programs [25], and (3) helps to avoid the *ex situ* generation of hybrids possibly maladapted to the environments inhabited by their parental species [26]. However, the taxonomy of Central American *Atelopus* is still incompletely resolved. Most species have been evaluated only based on characteristics of adult external morphology, and coloration pattern has been given special importance, even though it has been found that in *Atelopus* this character is not always concordant with genetic differentiation [27,28]. Additionally, the phylogenetic relationships of the species of *Atelopus* in Central America are drastically understudied, because only three species have been included in any published molecular phylogenies of the genus [10,28,29].

Given that Central American *Atelopus* seem to have a South American origin [10], understanding their phylogenetic relationships and historical biogeography could shed light on the timing of the closure of the Isthmus of Panama and its effect on animal lineages moving between continents. The term Great American Biotic Interchange (GABI) is sometimes used to refer to the massive interchange of mammalian lineages starting around 2.7 million years ago (Ma) [30]. Here, however, we use the term more broadly to refer to any exchange of fauna or flora between North and South America during the Neogene and Quaternary periods (cf. [31]). How the GABI relates temporally to the final geological completion of the Isthmus of Panama is a controversial topic, because the traditional date of roughly 3 Ma [32,33] has been challenged recently by Montes et al. [34] on geological evidence, and by Bacon et al. [35] on DNA sequence data, who propose a much older closing of the Isthmus around 15–10 Ma. In turn, O'Dea et al. [36] charged both studies with biased data collection and erroneous interpretation, and the controversy continues. Nonetheless, DNA sequence studies on amphibians and reptiles are often supportive of a closure date older than 3 Ma (e.g., [37–39]). In this work, the phylogenetic relationships of the species of *Atelopus* of Central America are estimated from mitochondrial DNA sequence data in order to evaluate the current taxonomic status of populations and species, as well as to investigate the number and timing of colonization events between South America and Central America.

2. Materials and Methods

2.1. Sampling

The character data generated for this study consisted of mitochondrial DNA sequences from 103 individuals of Central American *Atelopus* (Figure 1) plus congeneric samples of individuals from Colombia, representing *A. elegans*, and *A. laetissimus*. Samples included toe-clips and tissue samples from vouchered specimens, including many tissue loans from natural history museum collections. DNA sequence data from an additional eight South American species of *Atelopus* were obtained from GenBank to evaluate the possible monophyly of Central American samples, including data from *A.* cf. *spumarius*, *A. flavescens*, *A.* cf. *hoogmoedi*, *A. nanay*, *A. peruensis*, *A. pulcher*, *A. epikheistos*, and *A. bomolochos*. As outgroup to the genus *Atelopus*, we included additional GenBank sequence data representing the bufonids *Bufo gargarizans minshanicus*, *B. japonicus*, *B. stejnegeri*, *Duttaphrynus melanostictus*, *Anaxyrus boreas*, and *Bufotes viridis*. See Supplementary Table S1.

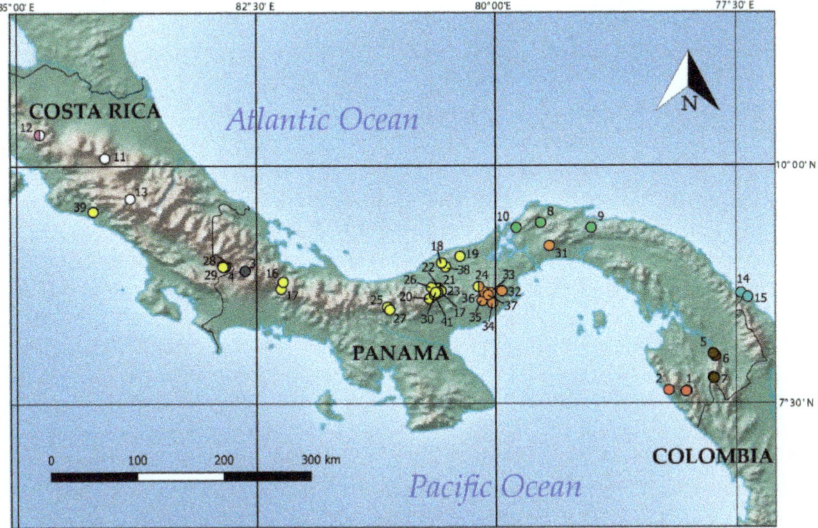

Figure 1. Map of Isthmian Central America showing the collecting localities of the samples used for this study. Each species is indicated with a different color, as follows: *Atelopus chiriquiensis*, grey; *A. senex*, white; *A. varius*, yellow; *A. zeteki*, orange; *A. limosus*, green; *A. certus*, red; *A. glyphus*, brown; *A.* sp. "Puerto Obaldía-Capurganá", light blue; *A.* sp. "Monteverde", violet. The yellow|orange and violet|white circles indicate the localities where *A. varius* and *A. zeteki* or *A.* sp. "Monteverde" and *A. senex*, respectively, were found in sympatry.

2.2. Molecular Laboratory Protocols

Genomic DNA was extracted using a standard phenol-chloroform procedure. A 714-base pair (bp) fragment of cytochrome *b* (cyt *b*) was amplified with the primers CB1 (5′-CCATCCAACATCTCAGCATGATGAAA-3′) and CB3 (5′-GGCGAATAGGAAGTATCATTC-3′ [40]). A 639-bp fragment of the 3′ end of cytochrome oxidase I (COI) was amplified using the primers

COIf (5′-CCTGCAGGAGGAGGAGAYCC-3′) and COIa (5′-AGTATAAGCGTCTGGGTAGTC-3′ [41]). PCR, DNA sequencing, and contig assembly protocols followed those of Crawford et al. [42]. As no internal gaps were inferred, sequences were aligned by eye and no premature codons were found when translating inferred codons with the software Mesquite version 3.10 [43].

2.3. Phylogenetic Analyses

Phylogenetic inference was performed on the concatenated 2-gene data set totaling 1353 bp using maximum parsimony (MP), maximum likelihood (ML), and Bayesian criteria. The software PAUP*, version 4.0a149 for Windows [44] was used for heuristic tree searching under MP inference, based on 5000 replicate searches (each from random starting trees) with both the MaxTrees command and the rearrangement limit on each tree set to 100,000. These search conditions were repeated five additional times to evaluate the completeness of each search by counting the number of novel most-parsimonious trees obtained per each additional run. In total, 97 equally parsimonious trees were obtained as a result, which differed only slightly in terms of branch lengths and phylogenetic position of some highly similar samples of *Atelopus senex*, *A. varius*, *A. zeteki*, and *A. glyphus*.

The software PartitionFinder 2 [45] was used to simultaneously select the best-fit partition scheme and corresponding substitution models according to the Bayesian Information Criterion (BIC) and using the greedy algorithm of Lanfear et al. [46]. We assumed a maximum of six possible partitions across the concatenated data matrix (i.e., by gene and by codon). The resulting partition scheme and models are presented in Table S2. Posteriorly, a total of 300 independent tree searches (i.e., three runs with hundred replicates each) were conducted using GARLI version 2.01 [47] on the Cyberinfrastructure for Phylogenetic Research (CIPRES) online platform [48], maintaining the search parameter values as default.

For MP and ML analyses we assessed statistical support of branches via 1000 non-parametric bootstrap replicates [49]. MP bootstrapping was done in PAUP*, with 30 tree searches performed on each character matrix generated by sampling with replacement, and with other search parameters as above. For ML bootstrapping in GARLI, 1000 pseudo-replicates distributed in twenty runs of 50 independent searches each were run on CIPRES, again maintaining the parameter values as default.

Bayesian phylogenetic analysis (BA) using reversible-jump Markov chain Monte Carlo (rjMCMC) was performed with BEAST 2 version 4.8 [50], as implemented in the CIPRES portal, and using the based substitution model implemented in the RBS package of BEAST 2 [50,51]. The posterior distribution of trees including dates and rates was estimated by assuming a relaxed normal clock model of evolution and a birth-death tree prior (a diversification model that considers both speciation and extinction [52]). The prior on substitution rate assumed a normal distribution with a mean of 0.00957 (i.e., 0.957%) and standard deviation of 0.0010 per lineage per million years. The mean was based on the mitochondrial substitution rate of Macey et al. [53], obtained from mitochondrial protein-coding genes in Eurasian toads of the genus *Bufo*, as corrected by Crawford [54]. The prior distribution of other parameters were not modified from their default values. Bayesian rjMCMC analyses were run for 100 million generations, saving one sampled tree per 1000 generations and with the first 10% of trees discarded as burn-in. Stationarity and mixing of the log-likelihood values and parameter estimates were evaluated with Tracer version 1.7 [55]. We confirmed that all parameters estimated from the post-burn-in set of trees had effective sample sizes > 200. The resulting set of trees was summarized with TreeAnnotator 2.4.8, part of the BEAST 2 package.

2.4. Biogeographic Analyses

To determine the number and direction of colonizations between South America and Central America, we inferred ancestral areas using a model-based approach assuming the BA tree trimmed to one sample per species (including unconfirmed candidate species, see below). The R-package 'BioGeoBEARS' [56,57] was used to evaluate the fit of six biogeographical models (DEC, DEC-J, DIVA, DIVA-J, BAYAREA-LIKE, BAYAREALIKE-J) to our data based on the corrected Akaike information criterion (AICc). The model recovered as having the best fit was then used for conducting the ancestral area estimation, again using 'BioGeoBEARS'. Species of *Atelopus* were coded as being from either Central America or South America, with the dividing line assumed to be the Uramita fault that runs parallel to the Atrato River in Chocó, Colombia [34].

2.5. Pairwise Genetic Distances

The genetic distances between sequences of each gene among all samples were estimated with the software MEGA version 7 [58], assuming a Kimura 2-parameter model of sequence evolution (K2P [59]). Default values were retained for all parameters. Sites with ambiguous or missing bases were removed only in pairwise comparisons. An unconfirmed candidate species was identified as a reciprocally monophyletic clade recovered in the BA consensus tree that had a genetic distance from its sister clades larger than the minimum genetic distances between other named sister species (2.08% or 2.44% in COI or cyt *b*, respectively). Even though the genetic distance cut-offs used herein may seem low in comparison to those used for other frogs and for different loci (e.g., 3% for the case of the 16S ribosomal RNA gene and 10% for the barcode fragment of COI [60,61]), they allow defining clades that mostly correspond to the species recognized by the current taxonomy of the genus in Central America.

3. Results

3.1. Phylogenetic Analyses

The complete, concatenated DNA sequence alignment consisted of a total of 1353 bp from 121 individuals of *Atelopus* plus samples from six individuals of outgroup bufonids. The majority of ingroup samples had sequences of the COI fragment (unavailable for two samples, or 1.6% of the total), whereas six ingroup terminals (4.7% of the total) lack DNA sequence data from cyt *b*. Premature stop codons were not observed in any of the sequences. Taking into account the ingroup only, of the 714 sites in cyt *b*, 142 were parsimony-informative and 35 were singletons, and of the 639 aligned sites of COI, 136 were parsimony-informative and 10 were singletons (Table 1).

Table 1. Number of parsimony-informative sites, singletons, and invariant sites (in base pairs) for each gene region among the ingroup samples.

Gene	Length of Sequence (bp)	Invariant Sites	Singletons	Parsimony-Informative Sites
cyt *b*	714	537	35	142
COI	639	493	10	136

The trees with the highest likelihood and parsimony scores (one of the 97 more parsimonious and very similar trees was chosen randomly) are shown in Figure 2 and Supplementary Figure S1, respectively, but outgroup non-*Atelopus* bufonid samples were removed to aid in visualization of the ingroup. The Bayesian consensus phylogeny of ingroup samples is shown in Figure 3, while the complete BA tree with the outgroup is presented in Supplementary Figure S3. The topologies obtained by the three inference methods (MP, ML and BA) were highly congruent with respect to the Central American samples, differing only in the phylogenetic position of *A. senex* and *A. chiriquiensis*. The South American *Atelopus* samples consistently formed three well supported clades and the phylogenetic relationships among the three were distinct among the different inference methods (see below). In general, the majority of nodes in the three phylogenies were highly supported (e.g., bootstrap values > 75% or posterior probabilities > 0.95), with the exception of most internal nodes within *A. zeteki* and *A. varius*, among some nodes within Central American *Atelopus* in the ML topology (see Figure 2), and regarding the phylogenetic relationships between *A. senex* and *A. chiriquiensis*.

In all topologies, the genus *Atelopus* was recovered as monophyletic. The South American congeners, with the exception of *A. elegans* (see below), comprised three clades, all with significant support: (1) the species of the Amazonian foothills of Peru and Ecuador, plus the Guianan lowlands

(*Atelopus* cf. *spumarius*, *A. pulcher*, *A. flavescens*, and *A.* cf. *hoogmoedi*), corresponding to the 'Amazonian-Guianan clade' of Lötters et al. (2011); (2) species from the high Andes of Ecuador and Peru (*A. epikheistos*, *A. nanay*, *A. bomolochos*, and *A. peruensis*); and (3) *A. laetissimus* from the Sierra Nevada de Santa Marta of north-western Colombia. In the MP topology, clade 1 was the sister to all other *Atelopus*, whereas in the ML and BA topologies, *A. laetissimus* was the sister to all other sampled congeners. Clade 2 was the sister to our focal Central American lineage of *Atelopus* in both MP and BA topologies, with relatively high support in the BA consensus tree (posterior probability of 0.93; Figure 3 and Supplementary Figure S2). In the ML tree, clades 1 and 2 were sisters and formed the sister lineage to the Central American clade of *Atelopus*.

The remaining species of *Atelopus* included in this work comprised a monophyletic group containing eight clades of mostly Central American distribution. All but two of these clades were assignable to a previously described name, because we included samples from near the type localities of all known Central American species of the genus. The two exceptions were that we had no samples of *A. chirripoensis*, and the group comprised of Caribbean Coast samples from Puerto Obaldía, Panama, and Capurganá, Colombia may represent an unnamed species, referred to as *A.* aff. *limosus* in Flechas et al. [62] and Lewis et al. [23]. This latter clade plus *A. elegans* from the Pacific coast of Colombia (Figure 4D5) together formed the sister lineage of all other Central American species in all analyses (Figures 2 and 3). The first split within this clade of Central American endemics separated (from all remaining samples) two species from eastern Panama, *A. certus* and *A. glyphus*, which were recovered as sisters in all analyses. The next split separated from all remaining samples either *A. senex* plus *A. chiriquiensis* (BA) or a clade containing *A. senex*, *A. chiriquiensis*, plus the central Panama species, *A. limosus*, *A. varius*, and *A. zeteki* (ML and MP), although neither of these hypotheses received high statistical support. In trees inferred by all three methods, *A. limosus* of central Panama was the sister group to a clade containing samples identified as *A. varius* and *A. zeteki*. *Atelopus varius* (e.g., Figure 4B3,B4) ranges from central Costa Rica to central Panama, and *A. zeteki* (e.g., Figure 4B5,C1,C2) is found near and east of Penonomé in central Panama. Specimens of these two species were found in sympatry in Juan Lana, near San Miguel Abajo, Panama. The identification of the *Atelopus* from the Monteverde Cloud Forest Reserve in northern Costa Rica (Figure 4A1,A2), as *A. varius*, was not supported by the phylogenetic analyses. Therefore, *A. varius* seems to have a smaller distribution range in Costa Rica, whose type locality is "Veragoa", western Panama [63,64].

For both mitochondrial genes, the genetic distances between the eight clades of Central American *Atelopus* were generally much larger than within any clade (Table 2), with the exception of the mtDNA clades designated as *A. varius* and *A. senex*. However, this pattern is driven by the individual with institutional code MVZ 164818 (from Monteverde, Costa Rica) which has genetic distances comparable to, or even larger than, those found among the eight clades previously mentioned (Table 2). Each of the eight clades had genetic distances greater than 2.08 at COI (and greater than 2.44 at cyt *b*) from their sister clade (Table 2), except for the case of *A. glyphus*, for which its estimated COI genetic distance (1.43–2.07) from its sister species is lower than the cut-off value.

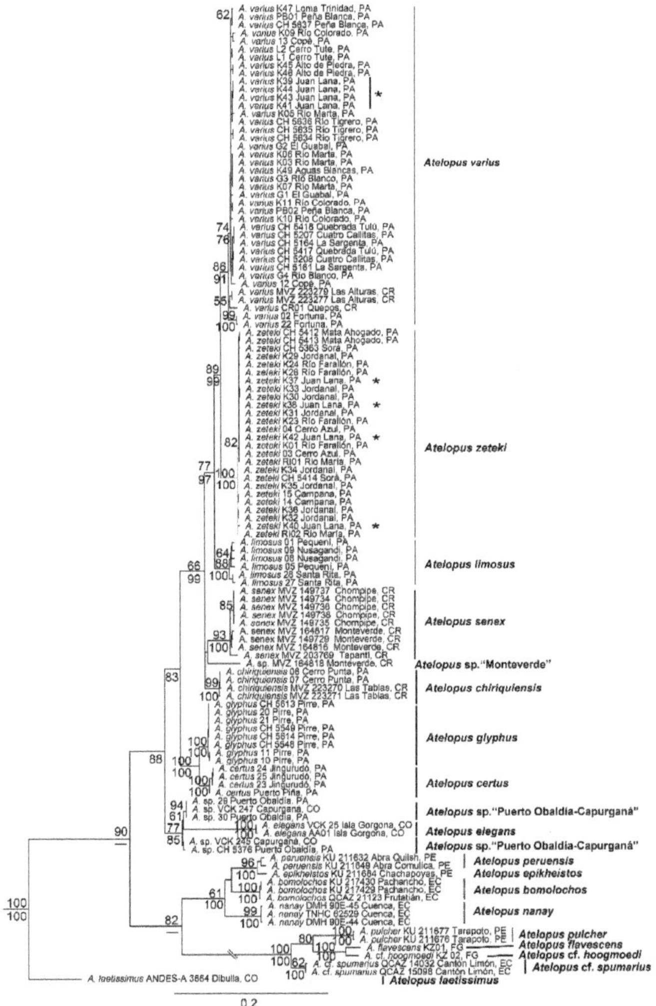

Figure 2. Maximum likelihood (ML) tree inferred from DNA sequences of two mitochondrial genes (COI and cyt *b*) from the species of *Atelopus* of Central and South America. Outgroup samples consisting of non-*Atelopus* bufonids were removed for better visualization of the ingroup and are shown in Supplementary Figure S2. Branch support for all major clades is indicated as non-parametric bootstrap values under ML and maximum parsimony (MP) above and below each branch, respectively. However, to improve visualization of the tree, support values for only the earliest-diverging lineages of within-population samples are shown. A dash indicates MP bootstrap support below 50%. Specimens are indicated by their field or museum voucher number, or if not available, by the name of the locality where they were obtained. Specimens from the locality in which *A. varius* and *A. zeteki* are found in sympatry (Juan Lana, Panama) are marked with asterisks.

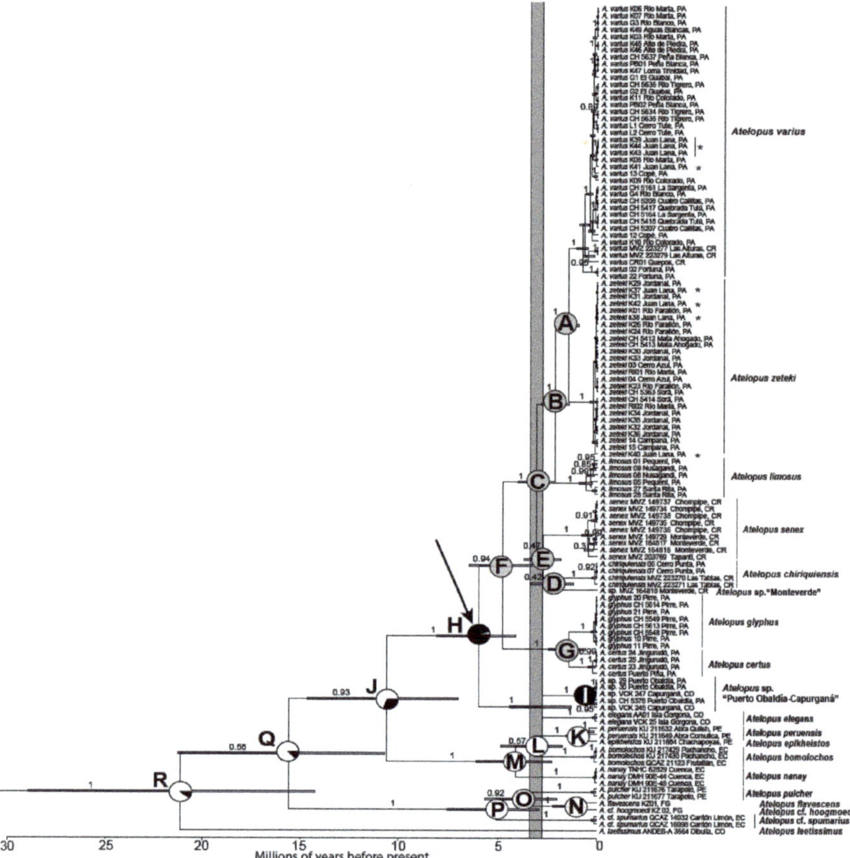

Figure 3. A timetree of *Atelopus* diversification from a relaxed-clock MCMC Bayesian analysis using the software BEAST 2, calibrated using a mitochondrial DNA substitution rates of mitochondrial genes of bufonid frogs of Macey et al. [53], as corrected by Crawford [54]. Outgroup samples of non-*Atelopus* bufonid genera were trimmed for improved visualization of the ingroup and are shown in Supplementary Figure S3. Numbers above branches are the posterior probabilities of adjacent nodes. 95% credibility interval of the age of each node is indicated by horizontal bars. The grey vertical shading shows an interval between 3.2 to 2.76 million years ago (Ma) indicating the estimated time when gene flow among shallow marine animals and movement of surface water across the Panamanian Isthmus ended, according to O'Dea et al. [36]. Note that the minimum age of node H, the oldest node reconstructed as being entirely Central American, is still older than 3.2 Ma. Shading in the circles on each node correspond to the ancestral area reconstructions under the DEC model (white refers to South America, pale grey to Central America, and black to ancestors recovered as inhabiting both continents), and the letters in each circle correspond to rows in Table 3. Specimens are indicated by their field or museum voucher number, or if not available, with an arbitrary number and the name of the locality where they were obtained (sample details in Supplementary Table S1). Specimens from the locality in which *A. varius* and *A. zeteki* are found in sympatry (Juan Lana, Panamá) are marked with asterisks.

Figure 4. Geographic color pattern variation of some harlequin frogs from central to eastern Panama and adjacent Colombia. Rows are arranged from west (**1**) to east (**5**) for columns (**A–C**), and north (**1**) to south (**5**) for column (**D**). Columns represent regions from Central America to Colombia: (**A**) Costa Rica; (**B**) western to central Panama; (**C**) central to eastern Panama and adjacent Colombia; (**D**) eastern Panama and Darién Province, Panama (**1–4**) to adjacent Colombia (**5**). Unscaled photographic images of: (**A1,A2**) *Atelopus* sp. "Monteverde" from the Monteverde Cloud Forest Reserve, Punta Arenas-Alajuela Provinces, Costa Rica (Figure 1, locality 12); (**A3**) male and (**A4**) female *Atelopus senex* from Volcán Barba, Parque Nacional Braulio Carrillo, Heredia Province, Costa Rica; (**A5**) male *Atelopus senex*, MVZ 203769, from the Refugio Nacional Tapantí, Costa Rica (Figure 1, locality 13); (**B1**) male and (**B2**) female *Atelopus chiriquiensis* from Río Candelaria, Parque Internacional La Amistad, Panama; (**B3**) *Atelopus varius* from Cerro Tute, Parque Nacional Santa Fe, Panama (Figure 1, locality 27); (**B4**) *Atelopus varius* from Río Tigrero, Coclé Province, Panama (Figure 1, locality 20); (**B5**) *Atelopus zeteki*, sample K37, Quebrada Juan Lana, San Miguel Abajo, Coclé Province, Panama (Figure 1, locality 24); (**C1**) *Atelopus zeteki* from the headwaters of Río María, Sorá, Panamá Oeste Province, Panama (Figure 1, locality 37); (**C2**) *Atelopus zeteki* from Cerro Azul (Cerro La Victoria), Panamá Province, Panama (Figure 1, locality 31); (**C3**) *Atelopus limosus* from type locality, San Juan de Pequení, Panamá Province, Panama (Figure 1, locality 8); (**C4**) *Atelopus limosus* from Nusagandi, Comarca Guna Yala, Panama (Figure 1, locality 9); (**C5**) *Atelopus* sp. "Puerto Obaldía-Capurganá" from Capurganá, Chocó Department, Colombia (Figure 1, locality 15); (**D1**) *Atelopus glyphus*, sample no. 10, from north-western Serranía de Pirre, Panama (Figure 1, locality 5); (**D2**) *Atelopus glyphus*, CH 5613, from southern Serranía de Pirre, Panama (Figure 1, locality 7); (**D3**) *Atelopus certus* from Cerro Sapo, Darién Province, Panama; (**D4**) *Atelopus certus*, CH 4665, from Serranía de Jingurudó, Darién Province, Panama (Figure 1, locality 1); (**D5**) *Atelopus elegans* from Isla Gorgona, Cauca Department, Colombia. Some of Central American *Atelopus* have bright colors with contrasting patterns and are known to have potent skin toxins [65] and are, therefore, considered to be aposematic [66]. Photo credits: David Cannatella (**A5**), David M. Dennis (**A1,B1,B2,C2,C4**), Sandra V. Flechas (**C5,D5**), Michael and Patricia Fogden (**A2,A3,A4**), Marcos A. Guerra (**C3,D1,D2**), Roberto Ibáñez (**D4**), César A. Jaramillo (**D3**), Erik D. Lindquist (**B3,B4**), and Guido Sterkendries (**B5,C1**).

Table 2. Pairwise genetic distances obtained using a Kimura 2-parameter model of sequence evolution (i.e., K2P distance × 100) for the mitochondrial fragments of the COI and cyt *b* genes of the species of *Atelopus* from Central America. Ranges represent minimum and maximum observed distances. Intraspecific genetic distances are shown along the diagonal of the matrix. Values above and below the diagonal correspond to the interspecific genetic distances of the cyt *b* and COI genes, respectively.

	Taxon	1	2	3	4	5	6	7	8	9
1	*A.* sp. "Puerto Obaldía-Capurganá"	0–0.79 0–0.86	7.94–8.59	7.45–7.93	9.17–10.60	10.9–11.4	8.58–8.91	9.23–10.80	9.57–11.80	10.1–10.80
2	*A. certus*	4.55–5.40	0–0.63 0–0.28	2.91–3.16	7.58–9.60	9.78–9.96	7.77–7.80	7.95–9.11	7.31–8.26	8.12–8.45
3	*A. glyphus*	4.89–5.22	1.43–2.07	0 0–0.42	7.58–9.11	9.93–10.10	7.62–7.79	8.28–9.10	7.93–8.90	8.44–8.77
4	*A. senex*	6.09–7.14	6.78–8.22	6.59–7.32	0.16–1.33 0–1.92	7.00–8.68	4.06–5.19	5.90–8.42	5.12–8.44	6.86–8.18
5	*A.* sp. "Monteverde" (MVZ 164818)	6.80–7.13	6.61–7.31	6.43	4.62–4.91	0 0	5.27	6.36–7.00	6.67–7.46	7.00–7.32
6	*A. chiriquiensis*	5.76–6.27	6.80–7.33	6.62–6.80	4.91–5.53	4.73–4.90	0–0.16 0–0.28	3.46–4.37	4.20–5.58	5.13–5.75
7	*A. limosus*	5.92–6.79	6.09–7.49	5.92–6.62	4.72–5.93	5.93–6.28	5.08–5.43	0.31–1.26 0.28–1.42	3.75–5.27	4.36–5.91
8	*A. varius*	6.12–7.00	6.28–8.25	5.92–6.98	4.81–5.96	5.56–6.29	4.24–4.92	3.89–4.74	0–2.08 0–2.44	2.86–4.66
9	*A. zeteki*	6.44–7.13	6.63–7.69	6.09–6.44	5.15–6.45	5.92–6.63	4.73–5.24	3.71–4.89	2.72–3.72	0–0.79 0–0.28

3.2. Divergence Times and Ancestral Area Estimation

The Dispersal and Local Extinction and Cladogenesis biogeographical model (DEC [67]) was found to have the best fit to the geographic distribution of species and the BA phylogeny (Figure 3, Supplementary Table S3). The ancestral area estimation was mostly well-supported with probabilities > 0.95 across nodes (Table 3 and Supplementary Table S4). Combining these results with the BA timetree suggested that the most recent common ancestor (MRCA) of the genus *Atelopus* existed in the early Miocene of South America, 21.1 Ma (95% posterior credibility interval, CI: 14.21 to 28.95 Ma). In addition, the most recent common ancestor (MRCA) of the Central American species of the genus (plus the Colombian *A. elegans*) was recovered as diverging from the other species of *Atelopus* during the late Miocene, (i.e., stem age at 10.6 Ma, CI: 7.02 to 14.64 Ma; Figure 3, node J), with a high probability (0.958) that it was distributed in both Central and South America at the time. *Atelopus* sp. "Puerto Obaldía-Capurganá" (aka, *A.* aff. *limosus*) plus *A. elegans* shared a MRCA 2.72 Ma (late Pliocene; CI: 1.8 to 3.69 Ma) which was inferred to inhabit both Central and South America, while all members of the sister lineage (Figure 3, node F) to this clade were recovered as being firmly of Central American origin. The timetree allowed us to bound the timing of dispersal into Central America as follows. The hypothesis of just one colonization of Central America (followed by one back dispersal event by *Atelopus elegans*) implies that the colonization took place between the stem age (Figure 3, node J; 10.6 Ma) and crown age (node H at 6.01 Ma, CI: 4.11 to 8.1 Ma) of the MRCA of all Central America samples, i.e., between roughly 10.6 Ma and 6.01 Ma. The alternative scenario of two independent invasions implies the age of first and more successful (in terms of descendants) took place between the stem age (node H at 6.01 Ma) and crown age (node F at 4.79 Ma, CI: 3.29 to 6.47 Ma) of the oldest clade of exclusively (i.e., monophyletic) Central American samples (Figure 3).

Table 3. Estimated crown ages (in millions of years ago, Ma) and results of the ancestral area estimation using the Dispersal-Extinction-Cladogenesis (DEC) model for the species of *Atelopus* of Central America and other selected nodes within the genus. Ages were estimated by concatenated MCMC Bayesian analyses using the software BEAST 2. See text for details. Node labels are indicated on the tree in Figure 3. CA = Central America, SA = South America.

Node Label	Mean Crown Age (Ma)	95% Posterior Credibility Interval	DEC Estimation
A	1.49	0.92–2.09	CA (100%)
B	2.12	1.41–2.86	CA (100%)
C	3.06	2.15–4.02	CA (100%)
D	2.27	1.22–3.35	CA (100%)
E	2.72	1.80–3.69	CA (100%)
F	4.79	3.29–6.47	CA (100%)
G	1.41	0.59–2.47	CA (100%)
H	6.01	4.11–8.10	CA-SA (95.8%), CA (4.2%)
I	2.75	1.30–4.41	CA-SA (100%)
J	10.6	7.02–14.64	SA (69.7%), CA-SA (30.3%)
Q	15.55	10.68–21.23	CA-SA (14.5%), SA (85.5%)
R	21.12	14.21–28.95	CA-SA (11.3%), SA (88.7%)

4. Discussion

4.1. Phylogenetic Systematics and Biogeography

The phylogenetic relationships inferred here among *Atelopus* are mostly in agreement with those found by Lötters et al. [10], a study also based solely on mitochondrial DNA sequences (12S and 16S genes), but which included only three of the Central American species, *A. varius*, *A. zeteki*, and *A. chiriquiensis*. The current study and Lötters et al. [10] found that *Atelopus* is monophyletic and of South American origin. Additionally, both studies recovered a clade of Amazonian-Guianan species which is sister to a clade containing the Andean, Chocoan, and Central American species. In contrast with Lötters et al. [10], however, in this work we included sequences of one species (*A. laetissimus*) from the Sierra Nevada de Santa Marta (SNSM), an isolated massif in north-western Colombia well known for its endemicity. The phylogenetic position of *A. laetissimus* is noteworthy, as it diverged very early in the history of the genus, being either sister to the rest of *Atelopus* (Figures 2 and 3), or the sister to a clade formed by Andean, Chocoan, and Central American species (Supplementary Figure S1). A similar pattern was found in the frogs of the clade Allocentroleniae, in which the basal split separates the Guianas and Amazonian regions from a clade containing a species from the SNSM that is sister to all other known species [68].

While a single colonization event can explain the presence of *Atelopus* in Central America, our data are also compatible with two dispersal events (Figure 3). Further sampling of other Chocoan *Atelopus*, such as *A. spurrelli*, *A. longibrachius*, *A. balios*, and the mainland populations of *A. elegans*, may be needed to resolve this ambiguity. Savage and Bolaños [16] proposed that the enigmatic *A. chirripoensis* of Costa Rica is more closely related to the *A. ignescens* species complex of Ecuador and southern Colombia than to other Central American species of *Atelopus*, which would imply yet another potential colonization event. However, testing this hypothesis is difficult given that *A. chirripoensis* has not been seen since 1980, and it is now probably extinct [16,18]. The remarkable external similarity between *A. chirripoensis* and the *A. ignescens* complex might also be explained by morphological convergence, given that both are found in similar *páramo* environments [16].

Our analyses suggest that the ancestral lineage of *Atelopus* reached Central America between 14.64 and 3.29 Ma with point estimates ranging from 10.6 to 4.79 Ma, depending on whether one postulates one or two invasions. In all cases, the estimated time frame is consistent with recent molecular biogeographic studies (e.g., [38,69]) supporting a non-traditional, older closure date for the Panamanian Isthmus by 10 Ma [34]. One could also argue that *Atelopus* dispersed over water prior to the completion of the Isthmus, although amphibians are sensitive to dehydration and to salt water [66,70,71]. Assuming an old date for the formation of the Isthmus, one recent hypothesis predicts that the Isthmus was forested during the Miocene and Pliocene epochs until 2.5 Ma when savanna-like environments appeared [72]. Our proposed time-frame for colonization by *Atelopus* of the Isthmus is temporally consistent with this prediction, as *Atelopus* species are affiliated with moist forests. Studies of additional groups of terrestrial organisms with low dispersal capabilities (such as insects, fossorial reptiles, or understory birds) or that are sensitive to salt water (such as amphibians) are needed to confirm whether there may be a tendency of inferring early crossings between continents by wet-forest lineages [31].

4.2. Geographic Color Pattern Variation

Most Central American species of *Atelopus* are strikingly variable in their color patterns, not only between but also within populations (e.g., [63]). In some species, such as *A. chiriquiensis* (Figure 4B1,B2), *A. senex* (Figure 4A3,A4) and populations of *A. varius*, a noticeable sexual dimorphism in coloration is sometimes present. On the other hand, color patterns can be very similar between species, resulting in species misidentification [28,73], especially in species that lack other clear diagnostic morphological characters. Our molecular data revealed the presence of a distinct species in the Monteverde Cloud Forest Reserve, northern Costa Rica, that has until now been confused with *A. varius*, very likely due to their similarity in color pattern.

In general, recently metamorphosed and juvenile individuals of *A. certus*, *A. chiriquiensis*, *A. glyphus*, *A. limosus*, *A. varius*, and *A. zeteki* have a barred dorsal color pattern, consisting of a series of dark and light colored chevrons, which usually changes or fades ontogenetically as they grow older ([74], E.D.L. and R.I. pers. obs.). This barred pattern sometimes remains in adult individuals within certain populations, and may even be the predominant color pattern, as observed in some populations of *A. glyphus* (Figure 4D1), *A. limosus* (Figure 4C4), *A. varius* (Figure 4B3,B4), and *A. zeteki*. Therefore, natural and/or sexual selection pressures might be behind this ontogenetic change in coloration, resulting in color pattern differences among geographic areas.

Geographic variation in color pattern has been documented for *A. varius* and *A. zeteki* [63,73], and we found distinct color pattern variants among populations of *A. glyphus* and *A. limosus* along their distribution ranges. In the northern region of the Serranía de Pirre (e.g., Figure 1, localities 5 and 6), adults of *A. glyphus* have a barred dorsal color pattern (Figure 4D1), while in the highlands of the southern portion of Serranía de Pirre above Cana (e.g., Figure 1, locality 7), adults have a predominantly uniform brown dorsum with small, light markings (Figure 4D2). In the western area of the distribution range of *A. limosus* (e.g., Figure 1, localities 8 and 10), adults have a uniform olive green dorsum ([63,75]; Figure 4C3), whereas populations to the East (e.g., Figure 1, locality 9) have a barred color pattern on the dorsum (Figure 4C4). The mtDNA data presented here supports our contention that these color pattern differences indeed represent geographic variation between conspecific populations. In contrast, the undescribed *Atelopus* sp. "Puerto Obaldía-Capurganá" from eastern Panama seems to exclusively have a barred dorsal color pattern (Figure 4C5).

4.3. Taxonomic and Conservation Implications

The phylogenetic results obtained here are broadly concordant with the current taxonomic arrangement of the species of *Atelopus*. Our data reveal two Central American lineages whose taxonomic status needs further attention, however. First, the specimens from Puerto Obaldía, Panama, and Capurganá, Colombia, were already of uncertain taxonomy. The names *A. varius glyphus* [76,77], *A. varius* [78] and *A.* aff. *limosus* [23,62] have been applied to this binational population, but our molecular phylogenetic results are not compatible with these proposed names. We cannot, however, reject the hypothesis that these Caribbean samples might correspond to a previously unreported population of *A. spurrelli* from the Pacific coast of Colombia, since we were unable to sample this latter taxon. More likely, however, these specimens represent an undescribed species.

Second, a specimen from the Monteverde Cloud Forest in Costa Rica (MVZ 164818) is most likely a member of an undescribed (and probably extinct) species, given its large genetic distance from all other specimens, including several sympatric individuals of its sister lineage, *A. senex*. The specimens from the Monteverde, Chompipe, and Tapantí areas in Costa Rica highlight a taxonomic problem that requires further study. Specimens from populations at Chompipe (the slopes of Volcán Barba) and Tapantí are traditionally considered to be *A. senex* (Figure 4A3–A5), while the ones from around Monteverde are assigned to *A. varius* ([15,63,79]; Figure 4A1,A2). The specimen (MVZ 164818) from the Monteverde Cloud Forest Reserve stood as a member of an undescribed (and probably extinct) species, although its phylogenetic position had no significant statistical support. This specimen was inferred to be the sister clade to either *A. senex*, *A. chiriquiensis*, or to a larger clade containing both. Richards and Knowles [28] also reported on the taxonomic confusion around Monteverde specimens and noted the low genetic divergence (i.e., 0.5–1.2%) between samples from Monteverde and *A. senex* specimens. However, we confirm that the specimen MVZ 164818 from Monteverde is certainly not *A. varius*, and it should likely be assigned to an undescribed species. Furthermore, the specimens from Monteverde MVZ 164816 and MVZ 164818 share the same morphology and color pattern based on our examination, but here we show they are genetically distinct at the level of species, as *A. senex* and *A.* sp. "Monteverde", respectively. Assuming no errors on the original source of the samples and DNA sequencing, possible explanations for the difference between morphological features and our mtDNA analysis are that genetic introgression between species has occurred or both species and/or their hybrids were present at Monteverde. To resolve this taxonomic problem, we suggest including more DNA data in future analyses, since further sampling of individuals would be difficult given the catastrophic population declines. Finally, the poorly known *A. spurrelli*, *A. certus* (Figure 4D3,D4), and *A. glyphus* (Figure 4D1,D2) are in need of further taxonomic study.

We confirm the results of Richards and Knowles [28] regarding the distinctiveness of *A. zeteki* (Figure 4B5) from *A. varius*, and we uncovered the locality (Juan Lana, Panama) in which these species are found in sympatry, as was hinted by Zippel et al. [73]. As previously suggested [28,73], these two species could have been hybridizing in sympatry at this locality, a hypothesis that deserves further exploration given their morphological similarity and recent divergence (2.12 Ma, CI: 1.41 to 2.86 Ma). Additionally, we established the identity of individuals from a population in Cerro Azul (Cerro La

Victoria), Panama (Figure 4C2). Savage [63] suggested this to be an introduced population of *A. varius* that resembled those from El Valle de Antón and Cerro Campana. According to our phylogenetic analyses, the individuals from this population are *A. zeteki*.

Finding support for the traditional taxonomy of Central American *Atelopus* brings hope to the success of the *ex situ* conservation programs being undertaken with the described species, as it indicates that resources are being used efficiently in frog conservation, given that all surviving species are being included by these programs. In addition, since each named species appears to be a distinct genetic entity, species ranges are not being underestimated, and no species is being protected unnecessarily. However, this is not the case for the two candidate species uncovered in this study, as they are not currently part of any *ex situ* or *in situ* conservation program. Furthermore, finding that the original identification of most specimens (except some specimens of *A. varius* and *A. zeteki*, Supplementary Table S1) is concordant with the results of the phylogenetic analyses, suggests that cryptic diversity is low among Central American *Atelopus*, and that hybridization resulting from misidentifications of specimens in captive programs should be rare. At least in the case of *A. limosus* this has been shown to be the case [80]. Further study is needed using multilocus nuclear markers or genomic data, especially on the likely case of hybridization between *A. varius* and *A. zeteki*. Additional genotyping should be conducted on captive specimens currently housed in *ex situ* programs. Finally, recognizing the taxonomic distinctiveness of the candidate species found herein is urgently needed in order to initiate the corresponding conservation measures to guarantee their long-term survival.

5. Conclusions

In this work, the phylogenetic relationships among seven of the eight described species of Central American *Atelopus*, plus eight South American congeneric species, were inferred based on mitochondrial DNA sequence data. The phylogenetic analyses revealed a ladderized topology showing the Central American species as a monophyletic group, and placing the species geographically furthest from South America more nested in the tree. We detected two previously unrecognized candidate species, including an undescribed species from Costa Rica involved in a taxonomic confusion that requires further study. We showed that species in eastern Panama, also show geographic variation in their color patterns, and clarify the species identity of individuals from some of these populations. Biogeographic models supported either one or two colonization events from South America, indicating that *Atelopus* reached Central America prior to 4 million years ago (Ma), a timing slightly older than the traditional date estimated for the closure of the Isthmus. Furthermore, this study underscores the invaluable role of museum collections in documenting biodiversity, and the relevance of genetic analyses for guiding conservation efforts.

Supplementary Materials: The following are available online at http://www.mdpi.com/1424-2818/12/9/360/s1; Resumen (Abstract in Spanish); Figure S1: One of the 97 most-parsimonious trees (selected arbitrarily) inferred from DNA sequences of two mitochondrial genes (COI and cyt *b*) for samples of *Atelopus* of Central and South America; Figure S2: Maximum likelihood phylogenetic tree inferred from DNA sequences of two mitochondrial genes (COI and cyt *b*) from species of *Atelopus* of Central and South America, including the extended non-*Atelopus* bufonid outgroup; Figure S3: Timetree of *Atelopus* based on a relaxed-clock MCMC Bayesian analysis of DNA sequences of two mitochondrial genes, COI and cyt *b*, calibrated using the estimated substitution rate for bufonid frogs of Macey et al. [53], as corrected by Crawford [54], and including the extended non-*Atelopus* bufonid outgroup; Table S1: Specimens of *Atelopus* and other bufonid frogs used for the present study, including museum voucher, locality, geographic coordinates, elevation and GenBank accession numbers; Table S2: Best-fit partitioning scheme for the two protein-coding mitochondrial genes, cytochrome *b* (cyt *b*) and cytochrome oxidase subunit I (COI) and models of nucleotide substitution for each partition, based on the Bayesian information criterion (BIC) as implemented in PartitionFinder 2 [45]; Table S3: Comparison among biogeographic models evaluated with the software BioGeoBEARS; Table S4: Estimated crown ages (in millions of years ago, Ma) and results of the ancestral area estimates under the Dispersal-Extinction-Cladogenesis (DEC) model, assuming the MCMC Bayesian consensus tree obtained using the software BEAST 2 and concatenating the two mitochondrial genes.

Author Contributions: J.P.R., C.A.J., E.D.L., A.J.C. and R.I. designed the study; C.A.J., E.D.L. and R.I. collected and procured tissues; C.A.J. and R.I. examined specimens; C.A.J. conducted laboratory work to extract and sequence DNA; J.P.R., C.J.A. and A.J.C. performed the bioinformatic analyses on DNA sequences; J.P.R., C.A.J. and A.J.C.

performed data curation; J.P.R., A.J.C. and R.I. prepared the manuscript with the contributions and edits from all authors. All authors have read and agreed to the published version of the manuscript.

Funding: This research was partially funded by Jegg Ettling, Field Research for Conservation (FRC) Program, Saint Louis Zoological Park. A.J.C. was supported by Research Program INV-2017-51-1432 from the School of Sciences, Universidad de los Andes. R.I. was funded by the Panama Amphibian Rescue and Conservation project, Sistema Nacional de Investigación (Secretaría Nacional de Ciencia, Tecnología e Innovación–SENACYT) and Minera Panamá (First Quantum Minerals Ltd.), during the preparation of the manuscript.

Acknowledgments: We thank the following persons and museums for providing access to tissue samples and/or specimens: David B. Wake and Tami Mott (Museum of Vertebrate Zoology, University of California at Berkeley), David C. Cannatella, Travis LaDuc, and Santiago Ron (Texas Natural History Collection, Texas Memorial Museum, University of Texas at Austin), Luis A. Coloma (Museo de Zoología, Pontificia Universidad Católica del Ecuador), William E. Duellman and Juan M. Guayasamin (Natural History Museum and Biodiversity Research Center, University of Kansas), Federico Bolaños (Museo de Zoología, Universidad de Costa Rica), and Yiselle Cano (Museo de Historia Natural C. J. Marinkelle, Universidad de los Andes, Bogotá). Thanks to Kevin Zippel, Anthony P. Wisniewski, Edgardo J. Griffith, Vicky Poole, Peter B. Johantgen, Allan Gillogly, Fidel E. Jaramillo, José Luis Atehortúa, Carlos A. Navas, Shyh-Hwang Chen, Victoriano González, and Karla Aparicio for donating specimens or tissues. Field assistance in Panama was provided by Frank A. Solís, James Coronado, and G. Urrieta. Alessandro Catenazzi and Juan C. Chaparro helped with permits in Peru. This work was conducted under the following permits: Instituto Nacional de Recursos Naturales Renovables de Panamá (scientific permits No. 27-92, No. 27-93, No. 46-94, No. 58-95, No. 03-97, No. 10-98, No. 20-2000, No. 39-2000 and No. 41-2000), Autoridad Nacional del Ambiente de Panamá (scientific permits No. SC/A 013-2001, No. SE/A-083-2001, No. SE/A-104-2001, No. SE/A-2-03, No. SE/A-78-03 and No. SE/A-07902, scientific import permits No. SIM/A 007-02, No. SIM/A 011-02, No. SIM/A-14-02, No. SIM/A-14-03 and No. SIM/A-16-03, scientific re-exportation permit No. SEX/A 059-02 and No. SER/A-1-04), and the U.S. Fish and Wildlife Service (declaration for export). We owe thanks to Eldredge Bermingham for opening the doors of his molecular genetics laboratory to us, to Lee A. Weigt for preliminary molecular work of initial tissues, and to Nimiadina Herrera and Maribel González for support and guidance in the laboratory. Thanks also to the members of the @CrawLab, especially Paola Montoya and Luis A. Rueda, for their feedback on this project. We acknowledge David Cannatella, David M. Dennis, Sandra V. Flechas, Michael and Patricia Fogden, Marcos A. Guerra, and Guido Sterkendries for photographic images of frogs in the graphical abstract and Figure 4. J.P.R. thanks Viviana Romero-Alarcón for constructive discussions about phylogenetic analyses and Martha Ramírez-Pinilla for support.

Conflicts of Interest: The authors declare no conflict of interest.

References

1. Stuart, S.; Chanson, J.S.; Cox, N.A.; Young, B.E.; Rodrigues, A.S.L.; Fishman, D.L.; Waller, R.W. Status and trends of amphibian declines and extinctions worldwide. *Science* **2004**, *306*, 1783–1786. [CrossRef] [PubMed]

2. Hoffmann, M.; Hilton-Taylor, C.; Angulo, A.; Böhm, M.; Brooks, T.M.; Butchart, S.H.M.; Carpenter, K.E.; Chanson, J.; Collen, B.; Cox, N.A.; et al. The impact of conservation on the status of the world's vertebrates. *Science* **2010**, *330*, 1503–1509. [CrossRef] [PubMed]

3. González-del-Pliego, P.; Freckleton, R.P.; Edwards, D.P.; Koo, M.S.; Scheffers, B.R.; Pyron, R.A.; Jetz, W. Phylogenetic and trait-based prediction of extinction risk for data-deficient amphibians. *Curr. Biol.* **2019**, *29*, 1557–1563. [CrossRef]

4. Hayes, T.B.; Falso, P.; Gallipeau, S.; Stice, M. The cause of global amphibian declines: A developmental endocrinologist's perspective. *J. Exp. Biol.* **2010**, *213*, 921–933. [CrossRef] [PubMed]

5. Skerratt, L.F.; Berger, L.; Speare, R.; Cashins, S.; McDonald, K.R.; Phillott, A.D.; Hines, H.B.; Kenyon, N. Spread of chytridiomycosis has caused the rapid global decline and extinction of frogs. *EcoHealth* **2007**, *4*, 125–134. [CrossRef]

6. Scheele, B.C.; Pasmans, F.; Skerratt, L.F.; Berger, L.; Martel, A.; Beukema, W.; Acevedo, A.A.; Burrowes, P.A.; Carvalho, T.; Catenazzi, A.; et al. Amphibian fungal panzootic causes catastrophic and ongoing loss of biodiversity. *Science* **2019**, *363*, 1459–1463. [CrossRef]

7. Ron, S.R.; Duellman, W.E.; Coloma, L.A.; Bustamante, M.R. Population decline of the Jambato toad *Atelopus ignescens* (Anura: Bufonidae) in the Andes of Ecuador. *J. Herpetol.* **2003**, *37*, 116–126. [CrossRef]

8. Coloma, L.A.; Duellman, W.E.; Almendáriz C., A.; Ron, S.R.; Terán-Valdez, A.; Guayasamin, J.M. Five new (extinct?) species of *Atelopus* (Anura: Bufonidae) from Andean Colombia, Ecuador, and Peru. *Zootaxa* **2010**, *2574*, 1–54. [CrossRef]

9. Guayasamin, J.M.; Bonaccorso, E.; Duellman, W.E.; Coloma, L.A. Genetic differentiation in the nearly extinct harlequin frogs (Bufonidae: *Atelopus*), with emphasis on the Andean *Atelopus ignescens* and *A. bomolochos* species complexes. *Zootaxa* **2010**, *2574*, 55–68. [CrossRef]

10. Lötters, S.; Meijden, A.; Coloma, L.A.; Boistel, R.; Cloetens, P.; Ernst, R.; Lehr, E.; Veith, M. Assessing the molecular phylogeny of a near extinct group of vertebrates: The Neotropical harlequin frogs (Bufonidae: *Atelopus*). *Syst. Biodivers.* **2011**, *9*, 45–57. [CrossRef]

11. DiRenzo, G.V.; Langhammer, P.F.; Zamudio, K.R.; Lips, K.R. Fungal infection intensity and zoospore output of *Atelopus zeteki*, a potential acute chytrid supershedder. *PLoS ONE* **2014**, *9*, e93356. [CrossRef] [PubMed]

12. DiRenzo, G.V.; Tunstall, T.S.; Ibáñez, R.; deVries, M.S.; Longo, A.V.; Zamudio, K.R.; Lips, K.R. External reinfection of a fungal pathogen does not contribute to pathogen growth. *EcoHealth* **2018**, *15*, 815–826. [CrossRef] [PubMed]

13. La Marca, E.; Lötters, S.; Puschendorf, R.; Ibáñez, R.; Rueda-Almonacid, J.V.; Schulte, R.; Marty, C.; Castro, F.; Manzanilla-Puppo, J.; García-Pérez, J.E.; et al. Catastrophic population declines and extinctions in neotropical harlequin frogs (Bufonidae: *Atelopus*). *Biotropica* **2005**, *37*, 190–201. [CrossRef]

14. Woodhams, D.C.; Alford, R.A.; Briggs, C.J.; Johnson, M.; Rollins-Smith, L.A. Life-history trade-offs influence disease in changing climates: Strategies of an amphibian pathogen. *Ecology* **2008**, *89*, 1627–1639. [CrossRef]

15. Savage, J.M. *The Amphibians and Reptiles of Costa Rica: A Herpetofauna between Two Continents, between Two Seas*; University of Chicago Press: Chicago, IL, USA, 2002.

16. Savage, J.M.; Bolaños, F. An enigmatic frog of the genus *Atelopus* (Family Bufonidae) from Parque Nacional Chirripó, Cordillera de Talamanca, Costa Rica. *Rev. Biol. Trop.* **2009**, *57*, 381–386. [CrossRef]

17. Köhler, G. *Amphibians of Central America*; Herpeton Verlag: Offenbach, Germany, 2011.

18. IUCN The IUCN Red List of Threatened Species. Available online: www.iucnredlist.org (accessed on 10 January 2019).

19. Gagliardo, R.; Crump, P.; Griffith, E.; Mendelson, J.; Ross, H.; Zippel, K. The principles of rapid response for amphibian conservation, using the programmes in Panama as an example. *Int. Zoo Yearb.* **2008**, *42*, 125–135. [CrossRef]

20. Zippel, K.; Johnson, K.; Gagliardo, R.; Gibson, R.; McFadden, M.; Browne, R.; Martinez, C.; Townsend, E. The Amphibian Ark: A global community for *ex situ* conservation of amphibians. *Herpetol. Conserv. Biol.* **2011**, *6*, 340–352.

21. Estrada, A.; Gratwicke, B.; Benedetti, A.; DellaTogna, G.; Garrelle, G.; Griffith, E.; Ibáñez, R.; Ryan, S.; Miller, P.S. *The Golden Frogs of Panama (Atelopus zeteki, A. varius): A Conservation Planning Workshop*; IUSN/SSC Conservation Breeding Specialist Group: Apple Valley, MN, USA, 2014.

22. Cikanek, S.J.; Nockold, S.; Brown, J.L.; Carpenter, J.W.; Estrada, A.; Guerrel, J.; Hope, K.; Ibáñez, R.; Putman, S.B.; Gratwicke, B. Evaluating group housing strategies for the *ex-situ* conservation of harlequin frogs (*Atelopus* spp.) using behavioral and physiological indicators. *PLoS ONE* **2014**, *9*, e90218. [CrossRef]

23. Lewis, C.H.R.; Richards-Zawacki, C.L.; Ibáñez, R.; Luedtke, J.; Voyles, J.; Houser, P.; Gratwicke, B. Conserving Panamanian harlequin frogs by integrating captive-breeding and research programs. *Biol. Conserv.* **2019**, *236*, 180–187. [CrossRef]

24. Allendorf, F.W.; Luikart, G.H.; Aitken, S.N. *Conservation and the Genetics of Populations*, 2nd ed.; John Wiley & Sons, Ltd.: West Sussex, UK, 2012.

25. Daugherty, C.H.; Cree, A.; Hay, J.M.; Thompson, M.B. Neglected taxonomy and continuing extinctions of tuatara (*Sphenodon*). *Nature* **1990**, *347*, 177–179. [CrossRef]

26. Frankham, R.; Briscoe, D.A.; Ballou, J.D. *Introduction to Conservation Genetics*; Cambridge University Press: Cambridge, UK, 2002.

27. Noonan, B.P.; Gaucher, P. Phylogeography and demography of Guianan harlequin toads (*Atelopus*): Diversification within a refuge. *Mol. Ecol.* **2005**, *14*, 3017–3031. [CrossRef] [PubMed]

28. Richards, C.L.; Knowles, L.L. Tests of phenotypic and genetic concordance and their application to the conservation of Panamanian golden frogs (Anura, Bufonidae). *Mol. Ecol.* **2007**, *16*, 3119–3133. [CrossRef] [PubMed]

29. Lötters, S.; van der Meijden, A.; Rödder, D.; Köster, T.E.; Kraus, T.; La Marca, E.; Haddad, C.F.B.; Veith, M. Reinforcing and expanding the predictions of the disturbance vicariance hypothesis in Amazonian harlequin frogs: A molecular phylogenetic and climate envelope modelling approach. *Biodivers. Conserv.* **2010**, *19*, 2125–2146. [CrossRef]

30. Stehli, F.G.; Webb, S.D. *The Great American Biotic Interchange*; Plenum Press: New York, NY, USA, 1985.
31. Cody, S.; Richardson, J.E.; Rull, V.; Ellis, C.; Pennington, R.T. The Great American Biotic Interchange revisited. *Ecography* **2010**, *33*, 326–332. [CrossRef]
32. Keigwin, J.D., Jr. Pliocene closing of the Isthmus of Panama, based on biostratigraphic evidence from nearby Pacific Ocean and Caribbean Sea cores. *Geology* **1978**, *6*, 630–634. [CrossRef]
33. Coates, A.G.; Obando, J.A. The geologic evolution of the Central American Isthmus. In *Evolution and Environment in Tropical America*; Jackson, J.B.C., Budd, A.F., Coates, A.G., Eds.; University of Chicago Press: Chicago, IL, USA, 1996; pp. 21–56.
34. Montes, C.; Cardona, A.; Jaramillo, C.; Pardo, A.; Silva, J.C.; Valencia, V.; Ayala, C.; Pérez-Angel, L.C.; Ramírez, V.; Niño, H. Middle Miocene closure of the Central American seaway. *Science* **2015**, *348*, 226–229. [CrossRef]
35. Bacon, C.D.; Silvestro, D.; Jaramillo, C.; Smith, B.T.; Chakrabarty, P.; Antonelli, A. Biological evidence supports an early and complex emergence of the Isthmus of Panama. *Proc. Natl. Acad. Sci. USA* **2015**, *112*, 6110–6115. [CrossRef]
36. O'Dea, A.; Lessios, H.A.; Coates, A.G.; Eytan, R.I.; Restrepo-Moreno, S.A.; Cione, A.L.; Collins, L.; de Queiroz, K.; Farris, D.W.; Norris, R.D.; et al. Formation of the Isthmus of Panama. *Sci. Adv.* **2016**, *2*, e1600883. [CrossRef]
37. Daza, J.M.; Smith, E.N.; Páez, V.P.; Parkinson, C.L. Complex evolution in the Neotropics: The origin and diversification of the widespread genus *Leptodeira* (Serpentes: Colubridae). *Mol. Phylogenet. Evol.* **2009**, *53*, 653–667. [CrossRef]
38. Pinto-Sánchez, N.R.; Ibáñez, R.; Madriñán, S.; Sanjur, O.I.; Bermingham, E.; Crawford, A.J. The Great American Biotic Interchange in frogs: Multiple and early colonization of Central America by the South American genus *Pristimantis* (Anura: Craugastoridae). *Mol. Phylogenet. Evol.* **2012**, *62*, 954–972. [CrossRef]
39. Elmer, K.R.; Bonett, R.M.; Wake, D.B.; Lougheed, S.C. Early Miocene origin and cryptic diversification of South American salamanders. *BMC Evol. Biol.* **2013**, *13*, 59. [CrossRef]
40. Palumbi, S.R. Nucleic acids II: The polymerase chain reaction. In *Molecular Systematics*; Hillis, D.M., Moritz, C., Mable, B.K., Eds.; Sinauer Associates: Sunderland, MA, USA, 1996; pp. 205–247.
41. Kessing, B.; Croom, H.; Martin, A.; McIntosh, C.; McMillan, W.O.; Palumbi, S. *The Simple Fool's Guide to PCR, Version 1.0*; University of Hawaii: Honolulu, HI, USA, 1989.
42. Crawford, A.J.; Bermingham, E.; Polanía, C. The role of tropical dry forest as a long-term barrier to dispersal: A comparative phylogeographical analysis of dry forest tolerant and intolerant frogs. *Mol. Ecol.* **2007**, *16*, 4789–4807. [CrossRef] [PubMed]
43. Maddison, W.P.; Maddison, D.R. Mesquite: A Modular System for Evolutionary Analysis. Version 3.10. Available online: http://mesquiteproject.org (accessed on 1 September 2016).
44. Swofford, D.L. *PAUP*. Phylogenetic Analysis Using Parsimony (*and Other Methods)*; Version 4; Sinauer Associates: Sunderland, MA, USA, 2003.
45. Lanfear, R.; Frandsen, P.B.; Wright, A.M.; Senfeld, T.; Calcott, B. PartitionFinder 2: New methods for selecting partitioned models of evolution for molecular and morphological phylogenetic analyses. *Mol. Biol. Evol.* **2016**, *34*, 772–773. [CrossRef]
46. Lanfear, R.; Calcott, B.; Ho, S.Y.; Guindon, S. PartitionFinder: Combined selection of partitioning schemes and substitution models for phylogenetic analyses. *Mol. Biol. Evol.* **2012**, *29*, 1695–1701. [CrossRef]
47. Zwickl, D.J. Genetic Algorithm Approaches for the Phylogenetic Analysis of Large Biological Sequence Datasets under the Maximum Likelihood Criterion. Ph.D. Thesis, The University of Texas at Austin, Austin, TX, USA, 2006.
48. Miller, M.A.; Pfeiffer, W.; Schwartz, T. Creating the CIPRES Science Gateway for inference of large phylogenetic trees. In Proceedings of the Gateway Computing Environments Workshop (GCE), New Orleans, LA, USA, 14 November 2010; pp. 1–8.
49. Felsenstein, J. Confidence limits on phylogenies: An approach using the bootstrap. *Evolution* **1985**, *39*, 783–791. [CrossRef] [PubMed]
50. Bouckaert, R.; Heled, J.; Kühnert, D.; Vaughan, T.; Wu, C.H.; Xie, D.; Suchard, M.A.; Rambaut, A.; Drummond, A.J. BEAST 2: A software platform for Bayesian evolutionary analysis. *PLoS Comput. Biol.* **2014**, *10*, e1003537. [CrossRef] [PubMed]

51. Bouckaert, R.; Alvarado-Mora, M.V.; Pinho, J.R. Evolutionary rates and HBV: Issues of rate estimation with Bayesian molecular methods. *Antivir. Ther.* **2013**, *18*, 497–503. [CrossRef] [PubMed]

52. Gernhard, T. The conditioned reconstructed process. *J. Theor. Biol.* **2008**, *253*, 769–778. [CrossRef]

53. Macey, J.R.; Schulte, J.A., II; Larson, A.; Fang, Z.; Wang, Y.; Tuniyev, B.S.; Papenfuss, T.J. Phylogenetic relationships of toads in the *Bufo bufo* species group from the eastern escarpment of the Tibetan Plateau: A case of vicariance and dispersal. *Mol. Phylogenet. Evol.* **1998**, *9*, 80–87. [CrossRef]

54. Crawford, A.J. Relative rates of nucleotide substitution in frogs. *J. Mol. Evol.* **2003**, *57*, 636–641. [CrossRef]

55. Rambaut, A.; Drummond, A.J.; Xie, D.; Baele, G.; Suchard, M.A. Posterior summarization in Bayesian phylogenetics using Tracer 1.7. *Syst. Biol.* **2018**, *67*, 901–904. [CrossRef]

56. Matzke, N.J. BioGeoBEARS': Biogeography with Bayesian (and Likelihood) Evolutionary Analysis in R Scripts. Available online: http://CRAN.R-project.org/package=BioGeoBEARS (accessed on 21 February 2018).

57. Matzke, N.J. Model selection in historical biogeography reveals that founder-event speciation is a crucial process in island clades. *Syst. Biol.* **2014**, *63*, 951–970. [CrossRef]

58. Kumar, S.; Stecher, G.; Tamura, K. MEGA7: Molecular Evolutionary Genetics Analysis version 7.0 for bigger datasets. *Mol. Biol. Evol.* **2016**, *33*, 1870–1874. [CrossRef]

59. Kimura, M. A simple method for estimating evolutionary rate of base substitutions through comparative studies of nucleotide sequences. *J. Mol. Evol.* **1980**, *16*, 111–120. [CrossRef] [PubMed]

60. Fouquet, A.; Gilles, A.; Vences, M.; Marty, C.; Blanc, M.; Gemmel, N.J. Underestimation of species richness in Neotropical frogs revealed by mtDNA analyses. *PLoS ONE* **2007**, *2*, e1109. [CrossRef] [PubMed]

61. Vences, M.; Thomas, M.; Bonett, R.M.; Vieites, D.R. Deciphering amphibian diversity through DNA barcoding: Chances and challenges. *Philos. Trans. R. Soc. B Biol. Sci.* **2005**, *360*, 1859–1868. [CrossRef]

62. Flechas, S.V.; Sarmiento, C.; Cárdenas, M.E.; Medina, E.M.; Restrepo, S.; Amézquita, A. Surviving chytridiomycosis: Differential anti-*Batrachochytrium dendrobatidis* activity in bacterial isolates from three lowland species of *Atelopus*. *PLoS ONE* **2012**, *7*, e44832. [CrossRef]

63. Savage, J.M. The harlequin frogs, genus *Atelopus*, of Costa Rica and western Panama. *Herpetologica* **1972**, *28*, 77–94.

64. Lötters, S.; Böhme, W.; Günther, R. Notes on the type material of the neotropical harlequin frogs *Atelopus varius* (Lichtenstein & Martens, 1856) and *Atelopus cruciger* (Lichtenstein & Martens, 1856) deposited in the Museum für Naturkunde of Berlin (Anura, Bufonidae). *Mitteilungen Aus Dem Mus. Für Naturkunde Berl. Zool. Reihe* **1998**, *74*, 173–184.

65. Rodríguez, C.; Rollins-Smith, L.; Ibáñez, R.; Durant-Archibold, A.A.; Gutiérrez, M. Toxins and pharmacologically active compounds from species of the family Bufonidae (Amphibia, Anura). *J. Ethnopharmacol.* **2017**, *198*, 235–254. [CrossRef]

66. Wells, K.D. *The Ecology and Behavior of Amphibians*; University of Chicago Press: Chicago, IL, USA, 2007.

67. Ree, R.H.; Smith, S.A. Maximum likelihood inference of geographic range evolution by dispersal, local extinction, and cladogenesis. *Syst. Biol.* **2008**, *57*, 4–14. [CrossRef]

68. Castroviejo-Fisher, S.; Guayasamin, J.M.; Gonzalez-Voyer, A.; Vila, C. Neotropical diversification seen through glassfrogs. *J. Biogeogr.* **2014**, *4*, 66–80. [CrossRef]

69. Winston, M.E.; Kronaure, D.J.C.; Moreau, C.S. Early and dynamic colonization of Central America drives speciation in Neotropical army ants. *Mol. Ecol.* **2017**, *26*, 859–870. [CrossRef]

70. Warburg, M.R. *Ecophysiology of Amphibians Inhabiting Xeric Environments*; Springer: Berlin, Germany, 1997.

71. Cruz-Piedrahita, C.; Navas, C.A.; Crawford, A.J. Life on the edge: A comparative study of ecophysiological adaptations of frogs to tropical semiarid environments. *Physiol. Biochem. Zool.* **2018**, *91*, 740–756. [CrossRef]

72. Bacon, C.D.; Molnar, P.; Antonelli, A.; Crawford, A.J.; Montes, C.; Vallejo-Pareja, M.C. Quaternary glaciation and the Great American Biotic Interchange. *Geology* **2016**, *44*, 375–378. [CrossRef]

73. Zippel, K.; Ibáñez, D.R.; Lindquist, E.D.; Richards, C.L.; Jaramillo, A.C.A.; Griffith, E.J. Implicaciones en la conservación de las ranas doradas de Panamá, asociadas con su revisión taxonómica. *Herpetotropicos* **2006**, *3*, 29–39.

74. Lindquist, E.D.; Hetherington, T.E. Tadpoles and juveniles of the Panamanian golden frog, *Atelopus zeteki* (Bufonidae), with information on development of coloration and patterning. *Herpetologica* **1998**, *54*, 370–376.

75. Ibáñez, D.R.; Jaramillo, C.A.; Solís, F.A. Una especie nueva de *Atelopus* (Amphibia: Bufonidae) de Panamá. *Caribb. J. Sci.* **1995**, *31*, 57–64.

76. Dunn, E.R. New frogs from Panama and Costa Rica. *Occas. Pap. Boston Soc. Nat. Hist.* **1931**, *5*, 385–401.

77. Cochran, D.M.; Goin, C.J. Frogs of Colombia. *U. S. Natl. Mus. Bull.* **1970**, *288*, 1–655. [CrossRef]
78. Lynch, J.D.; Suárez-Mayorga, A. Catálogo de anfibios en el Chocó Biogeográfico. In *Colombia Diversidad Biótica IV, El Chocó Biogeográfico/Costa Pacífica*; Rangel, O., Ed.; Universidad Nacional de Colombia: Bogotá, Colombia, 2004; Volume I, pp. 654–668.
79. Pounds, J.A. Amphibians and reptiles. In *Monteverde: Ecology and Conservation of a Tropical Cloud Forest*; Nadkarni, N.M., Wheelwright, N.T., Eds.; Oxford University Press: New York, NY, USA, 2000; pp. 149–177.
80. Crawford, A.J.; Cruz, C.; Griffith, E.; Ross, H.; Ibáñez, R.; Lips, K.R.; Driskell, A.C.; Bermingham, E.; Crump, P. DNA barcoding applied to *ex situ* tropical amphibian conservation programme reveals cryptic diversity in captive populations. *Mol. Ecol. Resour.* **2013**, *13*, 1005–1018. [CrossRef]

 *diversity*

Article

Barcoding Analysis of Paraguayan Squamata

Pier Cacciali [1,2,]* **Emilio Buongermini** [3,]* **and Gunther Köhler** [4,5,]*

1   Instituto de Investigación Biológica del Paraguay, Del Escudo 1607, 1425 Asunción, Paraguay
2   Guyra Paraguay, Av. Cnel. Carlos Bóveda, Parque Ecológico Capital Verde—Viñas Cué,
     1719 Asunción, Paraguay
3   Subtropica, Teniente Rivas 741, 1732 Asunción, Paraguay
4   Senckenberg Forschungsinstitut und Naturmuseum Frankfurt, Senckenberganlage 25,
     60325 Frankfurt a.M., Germany
5   Institute for Ecology, Evolution & Diversity, Goethe-University, Biologicum, Building C,
     Max-von-Laue-Str. 13, 60438 Frankfurt am Main, Germany
*   Correspondence: pier_cacciali@yahoo.com (P.C.); ebuongermini@gmail.com (E.B.);
     gkoehler@senckenberg.de (G.K.); Tel.: +595-994-624-017 (P.C.); +595-983-467-983 (E.B.);
     +49-(0)69-7542-1232 (G.K.)

Received: 26 June 2019; Accepted: 23 August 2019; Published: 30 August 2019

Abstract: Paraguay is a key spot in the central region of South America where several ecoregions converge. Its fauna (and specifically its herpetofauna) is getting better studied than years before, but still there is a lack of information regarding molecular genetics, and barcoding analyses have proven to be an excellent tool in this matter. Here, we present results of a barcoding analysis based on 16S rRNA gene sequences, providing valuable data for the scientific community in the region. We based our fieldwork in several areas of Paraguay. We analyzed 249 samples (142 sequenced by us) with a final alignment of 615 bp length. We identified some taxonomic incongruences that can be addressed based on our results. Furthermore, we identify groups, where collecting efforts and research activities should be reinforced. Even though we have some blanks in the geographical coverage of our analysis—and there is still a lot to do towards a better understanding of the taxonomy of the Paraguayan herpetofauna—here, we present the largest genetic dataset for the mitochondrial DNA gene 16S of reptiles (particularly, Squamata) from Paraguay, which can be used to solve taxonomic problems in the region.

Keywords: amphisbaenians; lizards; snakes; South America; taxonomy

---

1. Introduction

Ideally, profound knowledge of biodiversity is the first step before any conservation action, sustainable management or biological study is carried out in a given area [1,2]. The scientific community is not only aware that the world is facing a major extinction event [3], but also that many lineages are disappearing even before becoming known to science [4,5]. In the last decades, the use of molecular data has helped to improve our knowledge about biological diversity and the use of genetics as a tool for species recognition is now routine. Molecular data are more often used every day and are applied to the species identification [6–8] and species delimitation [9–12] of all kinds of living organisms, with applications even in food control quality [13].

In this context DNA barcoding analysis is the examination and comparison of short and stable fragments of DNA (DNA barcodes), usually mitochondrial, that represent genetic identifiers for a species [14], and have proven to be a reliable technique for taxonomy [15–17]. Nevertheless, researchers have to be cautious because despite being a power tool, barcoding analysis potentially can lead to misinterpretations if sequences used for comparison were generated from misidentified specimens [18]. Therefore, it is strongly recommended that molecular genetic tools are complemented with morphological, bioacoustics, and ecological data [19]. In conclusion, molecular genetics open a path for more detailed taxonomic studies [20].

For biological works that are concerned with the central portion of South America, Paraguay is critical since the country is located in a confluence zone of different ecoregions, such as Cerrado, Pantanal, Atlantic Forest, Chaco (Humid and Dry), and Southern Cone Mesopotamian Savanna, each of them having its own distinct origin and evolutionary history [21,22]. Additionally, Paraguay is key for works in northern Argentina, Uruguay, southwestern Brazil, and Bolivia. In spite of this biogeographical importance, Paraguay has been poorly explored, and in the current era of molecular genetics, the investigations that include genetic samples from Paraguay are extremely rare in herpetology. For instance, the natural history museum of Paraguay (Museo Nacional de Historia Natural del Paraguay) started its tissue collection for genetic analyses in this decade, whereas other neighbor countries began cryo tissue collections already some decades ago.

The herpetofauna from Paraguay, and specifically the squamate diversity, is still poorly known, evidenced by the fact that even in the last decade, and without the help of molecular tools, several new records for the country were made (e.g., *Ophiodes fragilis, Epictia vellardi, Chironius exoletus, Lygophis paucidens, Philodryas livida,* and *Micrurus silviae*) [23–28] and some species new to science were described (*Tropidurus lagunablanca* Carvalho, 2016; *Tropidurus tarara* Carvalho, 2016; *Tropidurus teyumirim* Carvalho, 2016; *Ophiodes luciae* Cacciali & Scott, 2015; *Phalotris normanscotti* Cabral & Cacciali, 2015) [29–31]. The incorporation of molecular genetics in taxonomy opened new pathways leading to a higher resolution in species delimitation, identifying several cryptic species. Some herpetologists in the region included genetic samples of Squamata from Paraguay (Gamble et al., 2012, Werneck et al., 2012, Morando et al., 2014, Recoder et al., 2014) although only very occasionally [32–35].

In 2015 we started a project of barcoding the reptile fauna (Squamata specifically) of Paraguay. It is important to mention here that the two genetic markers commonly used for barcoding analyses are the mtDNA genes 16S rRNA and Cytochrome Oxidase Subunit I (COI) [15,16,36–38] and both markers seem to work rather equally good for species identification. However, for the South American herpetofauna, 16S was more used than COI, and it is therefore better represented in GenBank for comparison. In addition, studies with 16S have shown not only good results in species recognition but also the systematic relationships among related species [37,39,40]. Thus, we decided to use sequences of the mtDNA gene 16S in our barcoding analysis.

During the project we gathered a lot of information about squamate diversity, and as a product of this work, some papers were published [41–45]. The use of DNA barcoding offers a starting point for recording the number of species that occur in a given region. Our results show how the use of DNA barcode data can augment and increase the accuracy of herpetological inventory surveys. Our barcoding study of the Paraguayan Squamata reveals the depth of taxonomic diversity in this country. Furthermore, our DNA barcode data represent the so far most comprehensive DNA barcode reference library for lizards and snakes of Paraguay. These reference data provide the scientific community with resources of numerous possibilities, ranging from species inventories, species identifications, taxonomic studies to wildlife trafficking.

2. Materials and Methods

2.1. Study Area

Paraguay is located in the center of South America (Figure 1) between parallels 18°18′ and 27°30′ S, and the meridians 54°19′ and 62°38′ W; with a total surface of 406,752 km². The country is divided by the Paraguay River into two portions: the Occidental Region (commonly known as "Chaco") with an area of 246,925 km² (60.7% of the country), and the Eastern Region (or Oriental Region) with a surface of 159,827 km² (39.9% of the country).

Figure 1. Administrative divisions of Paraguay, showing the two regions known as "Occidental Region" or Chaco (light brown) and "Oriental Region" (light green), divided by the Paraguay River (highlighted in blue in the map).

The topography of the Chaco region is a flat savanna with few small isolate hills in the center/north and east. Bad drainage creates vast flooded areas especially south and east, north-west has some dunes formation and is extremely dry. The oriental region is undulated with hills and the highest point in the center is well irrigated by tributaries of the Paraguayan and Parana rivers basins.

The climatic conditions vary in a northwestern–southeastern gradient, being more humid and cooler in the southeast. The mean annual temperature in the whole country is about 23 °C, being 24.5 °C in the western region and 22.5 °C in the eastern region. It is important to note that there are two seasons, the wet season, in which it rains frequently, coincide with the warm period from October to April and the dry season from May to September, where rain is less frequent and coincides with the coldest period.

There is a big difference with respect to the variation in temperature, given that the mean maximum is 25 °C, but the absolute maximum temperature could reach around 50 °C, especially in the northwestern portion of the country in January or February. The coldest month is July, and the absolute minimum temperature can be −6 °C in the south. Nevertheless, in Paraguay the "true" winter usually does not last longer than 16 days each year. Thus, Paraguay has a warm/hot climate during most parts of the year.

2.2. Data Collection

Even though there are some blanks in the areas sampled, the coverage of collecting sites in this study is rather vast (Figure 2). Nevertheless, there are two ecoregions in the Occidental Region (Cerrado Chaqueño and Médanos del Chaco), from where we have no samples. The methods used in the field were the traditional techniques for herpetology: Active searching at different times of the day and night, examining potential shelters (e.g., barks, logs, caves, mud, leaf litter, etc.) [46] (Figure 3). Fast moving lizards (e.g., *Ameiva* and *Teius*) were collected using compressed air rifles [47]. Additionally, some habitats, such as ant nests and swamps, were dug looking for hypogeal organisms [48], and floating vegetation was sampled using a trawl net (Figure 3). In total, 147 days of fieldwork were accounted for this project, and about 400 specimens collected.

It is important to highlight that the exotic lizard *Hemidactylus mabouia* is currently widely distributed in the country and now is part of the Paraguayan herpetofauna, and thus also included in the study. These genetic data may help in future studies about the colonization of the species, which in Paraguay has been recorded in the Concepción, San Pedro, Central, Alto Paraná, and Itapúa departments [49].

Reptiles that were captured alive were euthanized with a pericardial injection of a solution of embutramide, mebezonium iodide, and tetracaine hydrochloride (T-61®, Intervet International GmbH, Unterschleissheim, Germany) or Sodium Thiopental (Tiopental Sódico®, Biosano, Chile). The Secretaría del Ambiente from Paraguay (Currently "Ministerio del Ambiente y Desarrollo Sustentable") authorized the collecting of specimens though permits SEAM [Secretaría del Ambiente] N° 004/11 and 009/2014. Exportation permits for tissues and specimens were also issued by the same authority through the permits SEAM N° 002/14, 016/2016, and 084/2016.

Figure 2. Ecoregions of Paraguay, showing the collecting sites (brown dots) for this project.

Figure 3. Sampling methods during fieldwork included diverse techniques to search in different environments.

After euthanasia, tissue samples were taken either from the muscle of the thigh, tongue, finger clips, tail (when regenerated), or liver. Tissues were preserved in vials containing 98% non-denatured ethanol, and stored at −20 °C as soon as possible.

Hemipenes of Squamata were everted after euthanasia, with an injection of 70% ethanol after manually everting the organs. All specimens were fixed with a solution of 36% formalin and 96% ethanol in the proportion of 5:1000 (e.g., 5 mL formalin in 1 L ethanol), injected in the body cavity, thighs, and thickest part of the tail. Following fixation, the specimens were maintained in 70% ethanol.

2.3. Molecular Protocols

We used two different methods of DNA extraction. For sets containing few samples (usually eight or fewer), we used the DNeasy® Blood & Tissue Kit of Qiagen® (Hilden, Germany), whereas for sets of 96 samples we used the fiberglass plate [50]. Both methods are detailed below. The DNA was isolated from tissues whenever possible, or taken from preserved specimens that had been stored for a considerable time in 70% ethanol at room temperature in some cases.

For the DNeasy® method, we used tissue fragments of ~2 mm^2. When buffers formed precipitates, they were warmed up at 56 °C before use. All reagents for this protocol are included in the kit. Tissues were digested adding 180 μL (all values are for individual samples) of ATL Buffer and 20 μL of proteinase K. Samples in that mix were incubated in a rocking platform at 56 °C for 4 to 12 h until the tissue was completely lysed.

Following digestion, 200 μL of AL lysis Buffer + 200 μL of ethanol (98%) were added. This mix was centrifuged (8000 rpm) in DNeasy® Mini spin columns, discarding all the flow-through. Then, 500 μL of AW1 washing Buffer was added and centrifuged (8000 rpm) discarding the flow-through. Finally, 500 μL of AW2 washing Buffer was added and centrifuged (14,000 rpm) discarding the flow-through. The final elution was made with 200 μL of AE Buffer, after an incubation of one minute, followed by centrifugation (8000 rpm).

For the fiberglass extraction method, we used tissue fragments of ~1 mm^2. Specifications of reagents used in this protocol, are detailed in Table S1. Initially, the samples were washed with 50 μL (values per sample) of a solution of 1× Tris-Ethylenediaminetetraacetic acid (TE) Buffer to remove the remaining ethanol, for ~15 h. Following, the samples were digested with 50 μL of a solution of Vertebrate lysis Buffer and proteinase K (10:1), and incubated in a rocking platform at 56 °C for 12–24 h.

Once the samples were digested, the DNA extraction was made adding 100 μL of Binding Buffer and centrifuging at 2800 rpm. These products were transferred to a Pall® (Cortland, NY, USA) AcroPrep® filter plate, where the plate was vacuumed for 2 min. Then it was added 180 μL of Washing Buffer 1 and vacuumed again for 2 min. Posteriorly, it was added 750 μL of the Washing Buffer 2 and vacuumed for 2. Then TE Buffer was used to elute the DNA, adding 50 μL and incubating it for 2 min at 56 °C.

We amplified fragments of the mtDNA 16S gene with forward and reverse reactions using the following primers respectively: F: L2510 (5′-CGCCTGTTTAACAAAAACAT-3′) and R: H3056 (5′-CGGTCTGAACTCAGATCACGT-3′) [6]. The master mix cocktail used for amplification was of 1 μL of the DNA template, 2.5 μL of Y Buffer, 4 μL of dNTPs, 0.5 μL of TaqPol, 1 μL of MgCl$_2$, 1 μL of the forward and reverse primers, and 14 μL of distilled water, reaching a final volume of 25 μL. Amplification reactions were performed in an Eppendorf Mastercycler® pro (Hamburg, Germany) thermocycler using the following PCR conditions: initial denaturation 2 min (94 °C)—[denaturation 35 s (94 °C)—hybridization 35 s (48.5 °C)—elongation 60 s (72 °C)] × 40—final elongation 10 min (72 °C) [51]. Sequencing was performed using a BigDye® Terminator (ThermoFisher Scientific®, Waltham, MA, USA) with the following cycling conditions: 1 min at 95 °C, 30 × [10 s at 95 °C, 10 s at 50 °C, 2 min at 60 °C], with 10 μL of reaction volume.

Additionally to our own samples, we included non-Paraguayan data from sequences downloaded from GenBank, selecting preferably those sequences associated with museum vouchers, to avoid common problems of misidentifications in that repository [52–54]. In most cases, we downloaded only sequences from species represented in our samples, except for *Bothrops* and *Amphisbaena*. In these two cases, we downloaded samples from all the species present in Paraguay, because of the difficulty of these taxa for morphological identification. In the case of *Micrurus*, there is only one sequence of a species present in Paraguay available in GenBank (JQ627286). A particular case was the only available sample of *Vanzosaura rubricauda* (AF420716) uploaded in the framework of a lizards' phylogeny [55]. That specimen (MRT 05059) from Vacaria, Estado de Bahia, Brazil, actually is *V. multiscutata* [35]. Nevertheless, it was included in the analysis to evaluate the clustering with the genetic sample from Paraguay.

Codes of sequences downloaded from GenBank, plus accession numbers of sequences generated in this work, are available in the Table S2.

2.4. Data Analysis

Chromatograms of forward and reverse sequences were assessed, and a consensus sequence for each sample generated in SeqTrace 0.9.0 [56]. For sequences alignment, we employed MAFFT2 [57,58] through the webserver [59], which includes a special search strategy (Q-INS-i) for the secondary structure of the rRNA 16S [60]. No later manual edition was introduced. Results of MAFFT2 were visualized in MSA Viewer [61] and exported as fasta files.

The best scheme for substitution model was explored in PartitionFinder 2.1.1 [62], using linked branch lengths (supported by most of the phylogenetic programs) using a PhyML 3.0 analysis [63]. Given that the analysis is based on Squamata, which can show highly divergent clades, we estimated the relative quality of the statistical models using the Bayesian Information Criterion (BIC) [64] since it penalizes more the number of parameters in the model and then is better for a large degree of heterogeneity [65].

Given that it is not recommended to use both +I (significant proportion of invariable sites) plus +G (rate of variation among sites follows a gamma distribution) together in the same substitution model, we chose the best suggested model using +I or +G in the partition schemes, but never both together [66].

The phylogenetic hypothesis was performed under a Maximum Likelihood (ML) approach, using IQ-Tree [67] through its webserver [68], setting 10,000 non-parametric bootstrap replicates plus 10,000 replicates of Shimodaira-Hasegawa approximate likelihood ratio (SH-aLRT) [69] and 10,000 ultrafast bootstrap (UFBoot) approximation replicates [70]. We used a sequence of *Sphenodon punctatus* to root the tree [71], which has been proposed as the sister clade to Squamata [72]. Here, it is important to highlight that the phylogenetic hypothesis is used to sorting groups, and we do not seek for a comprehensive evolutionary reconstruction.

For visualization and edition (branch arrangement, colors, font sizes, etc.) of the tree generated through ML analysis, we used FigTree 1. 4.3 [73]. The final alignment plus the ML tree are stored in TreeBASE repository (https://treebase.org/) under the submission number 24616. To do this, we first managed the alignments and trees in nexus format and combined them in a single file (containing one alignment and the ML tree) using Mesquite 3.31 [74].

3. Results

We generated a total of 142 sequences of 64 species of Squamata from Paraguay, including one exotic species: *Hemidactylus mabouia*. In the Table S2, we present a list of specimens used for genetic analyses based on the field work for this project. For comparison we added 107 sequences from GenBank (Table S2). The final alignment constituted of a dataset of 249 samples of 615 bp length. Sequences are available in GenBank.

The best substitution model for the Barcoding dataset was GTR+G, according to the BIC. The sample of *Sphenodon punctatus* was retrieved as the sister clade to the Squamata (Figure 4). Deep nodes have low bootstrap values, meaning that the phylogenetic relationships are weakly supported. Nevertheless, the shallowest divergences have higher support values, recovering most of the genera included in the analysis as monophyletic, with the exception of *Manciola* (Scincidae) and the tribe Xenodontini (Colubridae).

The tribe Xenodontini (Figure S1) of the Subfamily Dipsadinae (Colubridae) contains the samples of *Erythrolamprus aesculapii* in a monophyletic clade, whereas *E. poecilogyrus* appears as paraphyletic. *Erythrolamprus reginae* clusters sister to the above-mentioned taxa. The genus *Xenodon* seems to be paraphyletic, given that two samples of *Xenodon pulcher* are sister to *Erythrolamprus*, whereas *Xenodon merremi* is sister to *Xenodon pulcher* + *Erythrolamprus*. Finally, in this clade a sample of *Erythrolamprus sagittifer* is sister to a sample of *Lygophis dilepis*.

Sister to Xenodintini is a clade composed of *Phalotris* + *Philodryas* (Figure S2). Both genera are monophyletic in the tree. The genera *Psomophis* and *Dipsas* are clustered together, and nested as sister to the above-mentioned snakes (Figure S3). Other genera of Dipsadinae that are rendered as monophyletic are *Hydrodynastes*, *Helicops*, and *Thamnodynastes* (Figure S4). A clade grouping members of the Colubrinae subfamily is composed of three genera (*Chironius*, *Leptophis*, and *Palusophis*) that also show monophyly (Figure S5). The *Pseudoboini* (Dipsadinae: Colubridae) is shown in its own clade (Figure S6) with four of the five genera used in the analysis being monophyletic, whereas the two species of the genus *Phimophis* appear in different positions of the gene tree.

The genus *Micrurus* (Elapidae) is the sister clade of the Colubridae, whereas the genus *Bothrops* (Viperidae) seems to be the sister to Elapidae + Colubridae (Figure S7). Located in a most basal position among snakes are the two species of *Epicrates* (Boidae), with *Amerotyphlops* (Typhlopidae) as the sister clade of the remaining snakes (Figure S8). The genus *Amphisbaena* (Amphisbaenidae) is monophyletic, where *A. alba* and *A. bolivica* are in their own clades, and *A. mertensii* shows also monophyly (Figure S9). *Amphisbaena angustifrons* is the sister taxon of the other *Amphisbaena*, with a sample of *Amphisbaena* sp. (PCS 314) as the most basal taxon of the clade (Figure S9). In our analysis, *Amphisbaena* is sister to Teiidae + Gymnophthalmidae. Gymnophthalmidae appears as a monophyletic clade, and the four genera show monophyly as well (Figure S10).

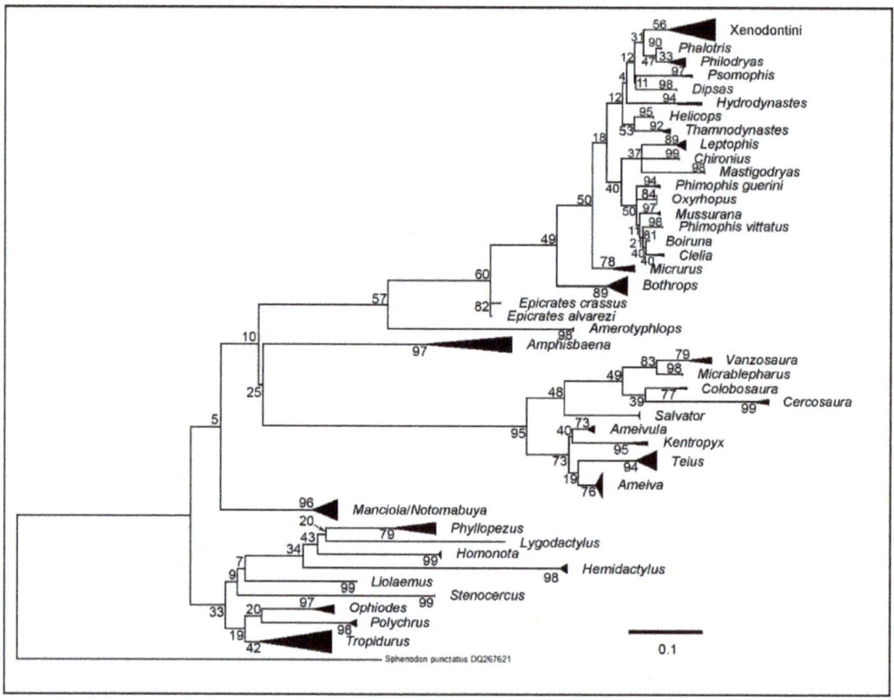

Figure 4. General view of the barcoding tree performed with a Maximum Likelihood approach. Clades are collapsed (in genera or tribes) for a better visualization. Numbers on nodes represent Bootstrap values. The scale bar represents branch length (substitutions/site).

The Family Teiidae is shown as paraphyletic. The only Tupinambinae in our samples was *Salvator*, which clusters as sister to Gymnophthalmidae (Figure S11). Samples of Teiinae are clustered together showing monophyly, where *Ameivula* and *Kentropyx* are sister clades (Figure S11), as are *Teius* and *Ameiva* (Figure S12). The Family Scincidae is sister to the all above-mentioned clades (Figure S13). Samples of *Manciola* show paraphyly (Figure S13). The remaining cluster contains members of the Anguidae, Gekkonidae, Phyllodactylidae, Liolaemidae, Polychrotidae, and Tropiduridae families. The clade composed by geckos shows monophyly in the genera, but not in the families given that *Phyllopezus* and *Homonota* are currently placed in Phyllodactylidae, whereas *Hemidactylus* and *Lygodactylus* are Gekkonidae (Figure S14). Sister to the Gekkota (Gekkonidae + Phyllodactylidae) is *Liolaemus*, and *Stenocercus* is rendered as sister to the Gekkota + *Liolaemus*.

The last cluster, sister to Gekkota + *Liolaemus* + *Stenocercus*, is represented by *Ophiodes* (Anguidae), *Polychrus* (Polychrotidae) (Figure S15), and *Tropidurus* (Tropiduridae) (Figure S16). In this case, the Family Tropiduridae is polyphyletic since the other member of the family (*Stenocercus*) is sister clade to Gekkota + *Liolaemuus*.

4. Discussion

Molecular genetics, and in particular barcoding analyses, proved to be a powerful tool to generate preliminary information about the taxonomic status of problematic taxa [20,75]. We present here the most comprehensive analysis of genetic samples of Squamata from Paraguay. The results obtained here will be useful to help identify questionable specimens and in some cases also to clarify some taxonomic issues of the Squamata fauna from the central region of South America. Thus, the data generated here will have a positive impact in a larger geographic context, beyond Paraguay's borders.

As said before, genetics alone will not yield a well-founded taxonomy. Nevertheless, molecular genetics open a path for defining operational taxonomic units (OTUs), identifying potential undescribed species and pointing to taxonomic problems, and thus have to be seen as a first informative step and a complementary evidence line in the framework of the modern integrative taxonomic approach [20,76].

Some taxonomic results of this project were already published. For instance, the samples of *Colobosaura* exhibit large genetic distances, and then *Colobosaura kraepelini* was revalidated [41]. The *Tropidurus* samples show monophyly in the species of the *torquatus* group (*T. catalanensis* and *T. etheridgei*), but indicate several uncertainties within the *spinulosus* group (formerly *T. guarani*, *T. lagunablanca*, *T. spinulosus*, *T. tarara*, and *T. teyumirim*), that resulted in the synonymization of *T. guarani* with *T. spinulosus*, and *T. tarara* and *T. teyumirim* with *T. lagunablanca* [43]. Regarding the Family Phyllodactylidae, there is strong evidence for the recognition of two different *Homonota* species in the Chaco [42,44] and a highly distinctive *Phyllopezus* clade, separated from populations from Cerrado and Chaco [45].

The samples of *Vanzosaura rubricauda* from Cerrado (field number "ALA") show a high branch distance compared with *Vanzosaura rubricauda* from Chaco (GK 3801) which is even larger than the distance from *V. multiscutata* (Figure S10). Integrating molecular and morphological data, a new species of *Vanzosaura* (*V. savanicola*) was previously described, and *Gymnodactylus multiscutatus* was transferred to the genus *Vanzosaura* [35]. Nevertheless, their genetic tree [35] included only a single sample from Paraguay and none from Argentina. In their map, obviously two divergent populations of *Vanzosaura rubriauda* are recognized: One west of the Paraguay River in the Dry Chaco, and another east of the Paraguay River in the Cerrado. Keeping a conservative approach, the authors maintained *V. rubricauda* as a single taxonomic unit, but with our additional samples it might be possible to generate new taxonomic hypotheses.

Furthermore, we recommend further studies on Amphisbaenidae, because one of the major and latest revisions of Amphisbaenidae in the Neotropics concluded that *A. mertensi* and *A. cunhai* (not recorded in Paraguay) are the most basal lineages of the genus [77]. Our analysis showed that the most basal sample (*Amphisbaena* sp. PCS 314) seems to be a different species as those within the remaining clade. Additional analyses, including more samples and a detailed morphological revision, are necessary to assess the specific status of that specimen.

The weakest part of this work was the analysis of snakes. These animals are usually the harder ones to sample. Compared to the actual diversity of Colubridae, our dataset had fewer samples of this family and therefore it was not possible to draw detailed taxonomic conclusions. However, the presence of the genus *Xenodon* in two different clusters suggests that more taxonomic work with this group of snakes is needed. Several taxonomic modifications occurred within the Colubridae in the last decade, where the genera *Lystrophis* and *Waglerophis* were synonymized with *Xenodon*, based on the analysis of gene sequences 12S and 16S for the genus *Lystrophis*, and Cytb and bdnf for one sample of *Waglerophis merremii* [78]. In our analysis, we found the samples of *X. pulcher* (previously *Lystrophis pulcher*) separated from *X. merremii* (previously *Waglerophis merremii*). It is desirable to perform phylogenies in this group using more nuclear data to get more robust relationships in the deep nodes.

In the clade of the genus *Thamnodynastes*, the two species used in our analysis (*T. chaquensis* and *T. hypoconia*) are nested in the same node. A more specific genetic study of the genus is highly recommendable. Furthermore, the revision of the phylogenetic status of *Phimophis* is advisable, since here it appears polyphyletic. In a former study, two samples of *Phimophis* were used: *P. guerini* (GQ457761) and *P. iglesiasi* (JQ598891) and due to polyphyly the authors described the genus *Rodriguesophis* to include the latter species [78]. This genus is characterized by the absence of the loreal scale. Both *P. guerini* and *P. vittatus* have a loreal scale, so they cannot be assigned to *Rodriguesophis*. Thus, a deeper integrative (morphological and molecular) analysis is needed to understand their relationships.

The genus *Micrurus* is scarcely represented in GenBank, and comparisons are not possible. The only sample from GenBank is *M. altirostris*, which is differentiated from Paraguayan samples by a rather long branch distance. There is a polytomy with three samples (PCS 310, 334, and 337), and the only identified specimen is *Micrurus pyrrhocryptus* (PCS 310) from Pantanal (northern part of Paraguay). The other samples are from Concepción at the other side of the river, and with a body color different from the pattern of *M. pyrrhocryptus*. It is important to highlight here that some of the specimens that we have of *Micrurus*, were decapitated (killed by rural farmers) and therefore, without genetic samples for comparison (in GenBank) and without cephalic data (which contains important diagnostic characters) [79], its specific taxonomic allocation becomes difficult.

Some of our Paraguayan samples of *Bothrops* of the *neuwiedi* complex, from distant parts of Paraguay, are clustered with a sample of *B. diporus* from GenBank (Samples PCS 302, PCS 318, PCS 331, PCS 504, Figure S7). A thorough analysis of this complex of *Bothrops* is needed to understand the true diversity in the group.

Regarding the Scincidae, a sample from GenBank of *Manciola guaporicola* (KX364960) from Mbaracayú Reserve (Paraguay) appears out of the clade of the remaining *M. guaporicola* from Paraguay and Brazil. This is also a topic that should be further investigated.

Regarding conservation, one of the major problems in Paraguay for several years was habitat loss due to extensive soybean crops in the eastern part of Paraguay [80]. Nevertheless, habitat fragmentation is currently also affecting the landscapes of the Occidental Region of Paraguay [81,82]. Thus, currently, the protected areas are the best strategy for conservation of biodiversity in Paraguay, although many conservation units face legal problems (e.g., lack of official measurements, management plans, forest guards, infrastructure, etc.) and then the long-term maintenance of their biodiversity is not guaranteed [80]. There are some reptile species absent from protected areas in Paraguay; therefore, monitoring and conservation efforts should be intensified for these taxa [83]. One species recently revalidated is *Colobosaura kraepelini*. This lizard is known only from the holotype from the locality of Puerto Max (San Pedro Department), the neotype from Altos and an additional specimen from San Bernardino, both localities in Cordillera Department [41]. There are no protected areas in the Cordillera Department, but there are some in the northern portion of Central Department (border with Cordillera), located less than 10 km from the known localities of *C. kraepelini*. The presence of this species in a conservation unit should be confirmed, but is possible that it is protected by "Monumento Natural Cerro Chororí" and "Monumento Natural Cerro Kõi". It is important to note that the conservation unit closer to the distribution of *C. kraepelini* is the "Parque Nacional Lago Ypacaraí", although only the lagoon is protected and not the surroundings. The species *Homonota septentrionalis* was described from the driest part of Paraguay (northwestern Chaco) and is abundant in the "Parque Nacional Teniente Enciso" [44]. The four species of *Tropidurus* found in Paraguay [43] are well represented in several protected areas. The last herpetofaunal conservation assessment was published in 2009 [84], and thus a new conservation assessment of Paraguayan reptile fauna including the new taxa, is necessary to provide a sound basis for conservation planning for those species that require special attention.

Finally, given that the sequenced specimens are yet a small portion of the actual diversity of Paraguay (Figure 5), it will be of the utmost importance to continue and expand these studies that will further improve our taxonomic knowledge.

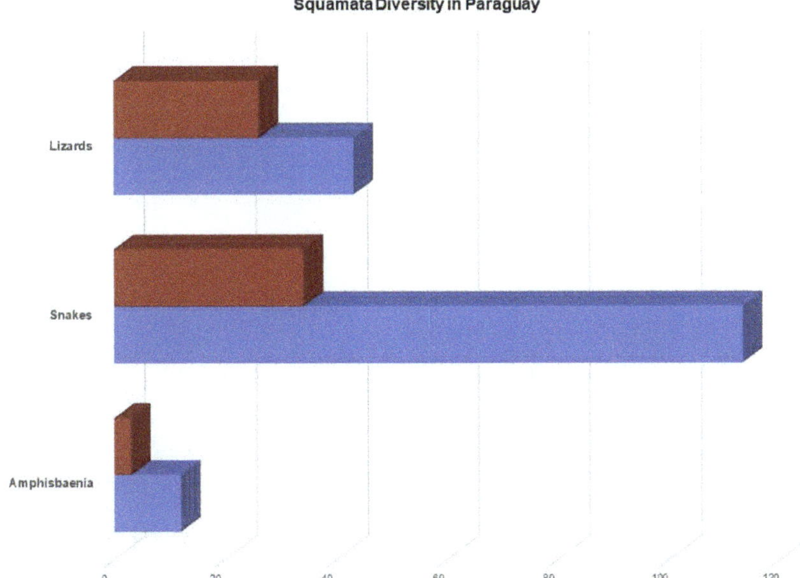

Figure 5. Comparison between known diversity of Paraguayan Squamata (blue bars) vs. diversity of sampled taxa for this study (red bars).

Supplementary Materials: The following are available online at http://www.mdpi.com/1424-2818/11/9/152/s1, Table S1: Reagents used for molecular procedures; Table S2: Specimens used in the barcoding analysis; Figures S1–S16: Details of sub-trees sections from the general barcoding analysis.

Author Contributions: Conceptualization, P.C., E.B., and G.K.; data curation, P.C., E.B., and G.K.; methodology, P.C.; formal analysis, P.C.; supervision, G.K.; writing—original draft preparation, P.C.; writing—review and editing, G.K. and E.B.; funding acquisition, P.C. and E.B.

Funding: This research was developed in the framework of the Ph.D. thesis project "Lizards of Paraguay: an integrative approach to solve taxonomic problems in central South America" presented by P.C. in the Faculty of Biosciences of the Johann Wolfgang Goethe-Universität, which had a strong financial support from the Deutscher Akademischer Austauschdienst (DAAD). Additionally, P.C. received a subsidy from the Consejo Nacional de Ciencia y Tecnología (CONACYT) through the Programa Nacional de Incentivo a los Investigadores (PRONII), field equipment provided by Idea Wild, and a grant to perform fieldwork activities in the Mbaracayú Reserve by funded by the GNB Bank through the Fundación Moisés Bertoni (FMB). Additionally, we received financial support from PRESIDENT ENERGY to pay the APC of this work.

Acknowledgments: This was a major project that had the help of many people. For this, we owe gratitude to Norman Scott, Luciano Avila, Mariana Morando, Tony Gamble, Sebastian Lotzkat, Martin Jansen, Arne Schultze, Christian Printzen, Raúl Maneyro, and Abel Batista, who provided advice and technical support at different stages of the project. This project would not have been achieved without the valuable help of many colleagues and friends who offered their help during field work. Hence, thank you so much Jorge Ayala, Frederick Bauer, Aníbal Bogado, Enrique Bragayrac, Diego Bueno, Emilio Buongermini, "Huguitus" Cabral, Paulo Campos Filho, Gloria "Lolex" Céspedes, Jorge A. Céspedez, Julia Coda, Sixto Fernández, Andrea Ferreira, Marcela Ferreira, Celeste "Tita" Gauto, Kevin Guest, Hugo del Castillo, Stefan Harrison (you rock man!), Monica Kozykariski, Victoria Kuntz, Arne Lestheruis, Pamela Marchi, Nicolás Martínez, José Méndez, Martha Motte, Cristian F. Pérez, Pastor Pérez, Rachel Pitts, Sergio Ríos, Lía Romero, Mirtha Ruiz Díaz, Humberto "Ka'umber" Sánchez, Rebecca Sheehan, Nelson Silva, María E. Tedesco, Dulcy Vázquez, Thomas and Sabine Vinke, Akira Yoshikawa, and Víctor Zaracho. Furthermore, we have to acknowledge the hospitality of persons (Rubén Ávila and Nilda Torres de Ávila, Massimo and Angela Coda, Rosario Gabaglio, Ana María Macedo, and Goli Stroessner) and organizations (Fundación Moisés Bertoni, Guyra Paraguay, and Para La Tierra) that provided accommodations during field wok. We also want to express our gratitude to some friends and colleagues that provided tissue samples, literature, or valuable information. They are Luciano J. Avila, Francisco Brusquetti, Arley Camargo, Santiago Carreira, André de Carvalho, Mariana Morando, and Paul Smith. Additionally, we are grateful to the staff (especially Heike Kappes) of the Grunelius-Möllgaard Laboratory (Senckenberg Forschungsinstitut und Naturmuseum Frankfurt - SMF), and to Linda Mogk (SMF) for lab support. P.C. also wants to thank to Tachi,

Dante and Rafa for patience and support. Finally, to the CONACYT for providing important tools for scientific research in the country.

Conflicts of Interest: The authors declare no conflict of interest. The funders had no role in the design of the study; in the collection, analyses, or interpretation of data; in the writing of the manuscript, or in the decision to publish the results.

References

1. Pullin, A.; Sutherland, W.; Gardner, T.; Kapos, V.; Fa, J.E. Conservation priorities: Identifying need, taking action and evaluating success. In *Key Topics in Conservation Biology 2*; Macdonald, D.W., Willis, K.J., Eds.; John Wiley & Sons, Ltd.: Hoboken, NJ, USA, 2013; pp. 3–22. [CrossRef]
2. Baumgartner, J.; Esselman, R.; Salzer, D.; Young, J. *Conservation Action Planning Handbook*; The Nature Conservancy: Arlington, VA, USA, 2007.
3. Ripple, W.J.; Wolf, C.; Newsome, T.M.; Galetti, M.; Alamgir, M.; Crist, E.; Mahmoud, M.I.; Laurance, W.F. World scientists' warning to humanity: A second warning. *Bioscience* **2017**, *67*, 1026–1028. [CrossRef]
4. Young, B.E.; Stuart, S.N.; Chanson, J.S.; Cox, N.A.; Boucher, T.M. *Disappearing Jewels: The Status of New World Amphibians*; NatureServe: Arlington, VA, USA, 2004.
5. Johnson, C.N.; Balmford, A.; Brook, B.W.; Buettel, J.C.; Galetti, M.; Guangchun, L.; Wilmshurst, J.M. Biodiversity losses and conservation responses in the Anthropocene. *Science* **2017**, *356*, 270–275. [CrossRef] [PubMed]
6. Palumbi, S.R.; Martin, A.; Romano, S.; Mcmillan, W.O.; Stice, L.; Grabowski, G. *The Simple Fool's Guide to PCR*; University of Hawaii Press: Honolulu, HI, USA, 1991.
7. Rudnick, J.A.; Katzner, T.E.; Bragin, E.A.; DeWoody, J.A. Species identification of 548 birds through genetic analysis of naturally shed feathers. *Mol. Ecol. Notes* **2007**, *7*, 757–762. [CrossRef]
8. Yang, L.; Tan, Z.; Wang, D.; Xue, L.; Guan, M.X.; Huang, T.; Li, R. Species identification through mitochondrial rRNA genetic analysis. *Sci. Rep.* **2014**, *4*, 1–11. [CrossRef] [PubMed]
9. Sites, J.; Marshall, J.C. Delimiting species: A renaissance issue in systematic biology. *Trends. Ecol. Evol.* **2003**, *18*, 462–470. [CrossRef]
10. Pinzón, J.H.; LaJaunesse, T.C. Species delimitation of common reef corals in the genus *Pocillopora* using nucleotide sequence phylogenies, population genetics and symbiosis ecology. *Mol. Ecol.* **2011**, *20*, 311–325. [CrossRef] [PubMed]
11. Khodami, S.; Martínez Arbizu, P.; Stöhr, S.; Laakmann, S. Molecular species delimitation of Icelandic brittle stars (Ophiuroidea). *Pol. Polar Res.* **2014**, *35*, 243–260. [CrossRef]
12. Leliaert, F.; Verbruggenc, H.; Vanormelingend, P.; Steena, F.; López-Bautistab, J.M.; Zuccarelloe, G.C.; De Clercka, O. DNA-based species delimitation in algae. *Eur. J. Phycol.* **2014**, *49*, 179–196. [CrossRef]
13. Barcaccia, G.; Lucchini, M.; Cassandro, M. DNA Barcoding as a molecular tool to track down mislabeling and food piracy. *Diversity* **2016**, *8*, 2. [CrossRef]
14. Lane, N. On the origin of bar codes. *Nature* **2009**, *462*, 272–274. [CrossRef]
15. Hebert, P.D.N.; Cywinska, A.; Ball, S.; de Waard, J.R. Biological identifications through DNA barcodes. *Proc. R. Soc. B* **2003**, *270*, 313–321. [CrossRef] [PubMed]
16. Murphy, R.W.; Crawford, A.J.; Bauer, A.M.; Che, J.; Donnellan, S.C.; Fritz, U.; Haddad, C.F.B.; Nagy, Z.T.; Poyarkov, N.A.; Vences, M.; et al. Cold Code: The global initiative to DNA barcode amphibians and nonavian reptiles. *Mol. Ecol. Resour.* **2013**, *13*, 161–167. [CrossRef]
17. Perl, R.G.B.; Nagy, Z.T.; Sonet, G.; Glaw, F.; Wollenberg, K.C.; Vences, M. DNA barcoding Madagascar's amphibian fauna. *Amphib.-Reptil.* **2014**, *35*, 197–206. [CrossRef]
18. Shen, Y.Y.; Chen, X.; Murphy, R.W. Assessing DNA barcoding as a tool for species identification and data quality control. *PLoS ONE* **2013**, *8*, e57125. [CrossRef] [PubMed]
19. Meiri, S.; Mace, G.M. New taxonomy and the origin of species. *PLoS Biol.* **2007**, *5*, e194. [CrossRef]
20. Hajibabaei, M.; Singer, G.A.C.; Hebert, P.D.N.; Hickey, D.A. DNA barcoding: How it complements taxonomy, molecular phylogenetics and population genetics. *Trends Genet.* **2007**, *23*, 167–172. [CrossRef] [PubMed]
21. Spichiger, R.; Palese, R.; Chautems, A.; Ramella, L. Origin, affinities and diversity hot spots of the Paraguayan dendrofloras. *Candollea* **1995**, *50*, 515–537.

22. Morrone, J.J. *Biogeografía de América Latina y el Caribe*; Manuales & Tesis; SEA/UNESCO/CYTED: Zaragoza, España, 2001.

23. Cacciali, P.; Scott, N. Revisión del Género Ophiodes de Paraguay (Squamata: Anguidae). *Bol. Soc. Zool. Urug.* **2012**, *21*, 1–8.

24. Cabral, H.; Netto, F. *Epictia vellardi*. Geographic distribution. *Herp. Rev.* **2016**, *47*, 83.

25. Cacciali, P.; Cabral, H. The genus *Chironius* (Serpentes, Colubridae) in Paraguay: Composition, distribution, and morphology. *Bas. App. Herp.* **2015**, *29*, 51–60. [CrossRef]

26. Cacciali, P.; Smith, P.; Källberg, A.; Pheasey, H.; Atkinson, K. Reptilia, Squamata, Serpentes, *Lygophis paucidens* Hoge, 1952: First records for Paraguay. *Check List* **2013**, *9*, 131–132. [CrossRef]

27. Smith, P.; Cacciali, P.; Scott, N.; del Castillo, H.; Pheasey, H.; Atkinson, K. First record of the globally-threatened Cerrado endemic snake *Philodryas livida* (Amaral, 1923) (Serpentes, Dipsadidae) from Paraguay, and the importance of the Reserva Natural Laguna Blanca to its conservation. *Cuad. Herpetol.* **2014**, *28*, 169–171.

28. Cacciali, P.; Espínola, D.; Centrón Viñales, S.; Gauto Espínola, I.; Cabral, H. Squamata, Serpentes, *Micrurus silviae* Di-Bernardo, Borges-Martins and Silva, 2007: Presence confirmation in Paraguay. *Check List* **2011**, *7*, 809–810. [CrossRef]

29. Carvalho, A.L.G. Three new species of the *Tropidurus spinulosus* group (Squamata: Tropiduridae) from Eastern Paraguay. *Am. Mus. Nov.* **2016**, *3853*, 1–44. [CrossRef]

30. Cacciali, P.; Scott, N. Key to the *Ophiodes* (Squamata: Sauria: Diploglossidae) of Paraguay with the description of a new species. *Zootaxa* **2015**, *3980*, 42–50. [CrossRef]

31. Cabral, H.; Cacciali, P. A new species of *Phalotris* (Serpentes: Dipsadidae) from the Paraguayan Chaco. *Herpetologica* **2015**, *71*, 72–77. [CrossRef]

32. Gamble, T.; Colli, G.R.; Rodrigues, M.T.; Werneck, F.P.; Simons, A.M. Phylogeny and cryptic diversity in geckos (*Phyllopezus*; Phyllodactylidae; Gekkota) from South America's open biomes. *Mol. Phyl. Evol.* **2012**, *62*, 943–953. [CrossRef]

33. Werneck, F.P.; Gamble, T.; Colli, G.R.; Rodrigues, M.T.; Sites, J. Deep diversification and long-term persistence in the South American "Dry Diagonal": Integrating continent-wide phylogeography and distribution modeling of geckos. *Evolution* **2012**, *66*, 3014–3034. [CrossRef]

34. Morando, M.; Medina, C.D.; Ávila, L.J.; Pérez, C.H.F.; Buxton, A.; Sites, J.W. Molecular phylogeny of the New World gecko genus *Homonota* (Squamata: Phyllodactylidae). *Zool. Scr.* **2014**, *43*, 249–260. [CrossRef]

35. Recoder, R.S.; Werneck, F.; Teixeira, M.; Colli, G.R.; Sites, J.W.; Rodrigues, M.T. Geographic variation and systematic review of the lizard genus *Vanzosaura* (Squamata, Gymnophthalmidae), with the description of a new species. *Zool. J. Lin. Soc.* **2014**, *171*, 206–225. [CrossRef]

36. Sacchi, C.T.; Whitney, A.M.; Mayer, L.W.; Morey, R.; Steigerwalt, A.; Boras, A.; Weyant, R.S.; Popovic, T. Sequencing of 16S rRNA gene: A rapid tool for identification of *Bacillus anthracis*. *Emerg. Infect. Dis.* **2002**, *8*, 1117–1123. [CrossRef]

37. Jansen, M.; Schultze, A. Molecular, morphology and bioacoustic data suggest Bolivian distribution of a large species of the *Leptodactylus pentadactylus* group (Amphibia: Anura: Leptodactylidae). *Zootaxa* **2012**, *3307*, 35–47. [CrossRef]

38. Scherz, M.D.; Vences, M.; Borrell, J.; Ball, L.; Nomenjanahary, D.H.; Parker, D.; Rakotondratsima, M.; Razafimandimby, E.; Starnes, T.; Rabearivony, J.; et al. A new frog species of the subgenus *Asperomantis* (Anura, Mantellidae, *Gephyromantis*) from the Bealanana District of northern Madagascar. *Zoosyst. Evol.* **2017**, *93*, 451–466. [CrossRef]

39. Batista, A.; Hertz, A.; Köhler, G.; Mebert, K.; Veselý, M. Morphological variation and phylogeography of frogs related to *Pristimantis caryophyllaceus* (Anura: Terra-rana: Craugastoridae) in Panama. *Salamandra* **2014**, *50*, 155–171.

40. Köhler, G.; Townsend, J.H.; Petersen, C.B. A taxonomic revision of the *Norops tropidonotus* complex (Squamata, Dactyloidae), with the resurrection of *N. spilorhipis* (Álvarez del Toro and Smith, 1956) and the description of two new species. *Mesoam. Herpetol.* **2016**, *3*, 8–41.

41. Cacciali, P.; Martínez, N.; Köhler, G. Revision of the phylogeny and chorology of the tribe Iphisini with the revalidation of *Colobosaura kraepelini* Werner, 1910 (Reptilia, Squamata, Gymnophthalmidae). *ZooKeys* **2017**, *669*, 89–105. [CrossRef]

42. Cacciali, P.; Morando, M.; Medina, C.D.; Köhler, G.; Motte, M.; Avila, L.J. Taxonomic analysis of Paraguayan samples of *Homonota fasciata* Duméril & Bibron (1836) with the revalidation of *Homonota horrida* Burmeister (1861) (Reptilia: Squamata: Phyllodactylidae) and the description of a new species. *PeerJ* **2017**, *5*, e3523. [CrossRef]

43. Cacciali, P.; Köhler, G. Diversity of *Tropidurus* (Squamata: Tropiduridae) in Paraguay—An integrative taxonomic approach based on morphological and molecular genetic evidence. *Zootaxa* **2018**, *4375*, 511–536. [CrossRef]

44. Cacciali, P.; Morando, M.; Avila, L.J.; Köhler, G. Description of a new species of *Homonota* (Reptilia, Squamata, Phyllodactylidae) from the central region of northern Paraguay. *Zoosyst. Evol.* **2018**, *94*, 147–161. [CrossRef]

45. Cacciali, P.; Lotzkat, S.; Gamble, T.; Köhler, G. Cryptic diversity in the Neotropical gecko genus *Phyllopezus* Peters, 1878 (Reptilia: Squamata: Phyllodactylidae): A new species from Paraguay. *Int. J. Zool* **2018**. [CrossRef]

46. Cacciali, P. *Colecta y Preparación de Anfibios y Reptiles: Manual para Colecta Científica*; Editorial Académica Española: Saarbrücken, Germany, 2013.

47. Scrocchi, G.; Kretzschmar, S. *Guía de Métodos de Captura y Preparación de Anfibios y Reptiles para Estudios Científicos y Manejo de Colecciones Herpetológicas*; Miscelánea N° 102; Fundación Miguel Lillo: San Miguel de Tucumán, Argentina, 1996.

48. Simmons, J.E. *Herpetological Collecting and Collections Management*; Society for the Study of Amphibians and Reptiles: Salt Lake, UT, USA, 2002.

49. Cacciali, P.; Scott, N.; Aquino, A.L.; Fitzgerald, L.A.; Smith, P. The Reptiles of Paraguay: Literature, distribution, and an annotated taxonomic checklist. *Spec. Publ. Mus. Southwest. Biol.* **2016**, *11*, 1–373.

50. Ivanova, N.V.; Dewaard, J.R.; Hebert, P.D. An inexpensive, automation-friendly protocol for recovering high-quality DNA. *Mol. Ecol. Notes* **2006**, *6*, 998–1002. [CrossRef]

51. Lotzkat, S.; Hertz, A.; Bienentreu, J.F.; Köhler, G. Distribution and variation of the giant alpha anoles (Squamata: Dactyloidae) of the genus *Dactyloa* in the highlands of western Panama, with the description of a new species formerly referred to as *D. microtus*. *Zootaxa* **2013**, *3626*, 1–54. [CrossRef]

52. Vilgalys, R. Taxonomic misidentification in public DNA databases. *New Phytol.* **2003**, *160*, 4–5. [CrossRef]

53. Zhang, G. Specimens versus sequences. *Science* **2009**, *323*, 1672. [CrossRef]

54. Park, K.S.; Ki, C.S.; Kang, C.I.; Kim, Y.J.; Chung, D.R.; Peck, K.R.; Song, J.H.; Lee, N.Y. Evaluation of the GenBank, EzTaxon, and BIBI services for molecular identification of clinical blood culture isolates that were unidentifiable or misidentified by conventional methods. *J. Clin. Microbiol.* **2012**, *50*, 1792–1795. [CrossRef]

55. Pellegrino, K.C.M.; Rodrigues, M.T.; Yonenaga-Yassuda, Y.; Sites, J.W. A molecular perspective on the evolution of microteiid lizards (Squamata, Gymnophthalmidae), and a new classification for the family. *Biol. J. Lin. Soc.* **2001**, *74*, 315–338. [CrossRef]

56. Stucky, B.J. SeqTrace: A graphical tool for rapidly processing DNA sequencing chromatograms. *J. Biomol. Tech.* **2012**, *23*, 90–93. [CrossRef]

57. Katoh, K.; Misawa, K.; Kuma, K.; Miyata, T. MAFFT: A novel method for rapid multiple sequence alignment based on fast Fourier transform. *Nucleic Acids Res.* **2002**, *30*, 3059–3066. [CrossRef]

58. Katoh, K.; Standley, D.M. MAFFT multiple sequence alignment software version 7: Improvements in performance and usability. *Mol. Biol. Evol.* **2013**, *30*, 772–780. [CrossRef]

59. Katoh, K.; Rozewicki, J.; Yamada, K.D. MAFFT online service: Multiple sequence alignment, interactive sequence choice and visualization. *Brief. Bioinform.* **2017**, bbx108. [CrossRef]

60. Katoh, K.; Toh, H. Recent developments in the MAFFT multiple sequence alignment program. *Brief. Bioinform.* **2008**, *9*, 286–298. [CrossRef]

61. Yachdav, G.; Wilzbach, S.; Rauscher, B.; Sheridan, R.; Sillitoe, I.; Procter, J.; Lewis, S.E.; Rost, B.; Goldberg, T. MSAViewer: Interactive JavaScript visualization of multiple sequence alignments. *Bioinformatics* **2016**, *32*, 3501–3503. [CrossRef]

62. Lanfear, R.; Frandsen, P.B.; Wright, A.M.; Senfeld, T.; Calcott, B. PartitionFinder 2: New methods for selecting partitioned models of evolution for molecular and morphological phylogenetic analyses. *Mol. Biol. Evol.* **2016**, *34*, 772–773. [CrossRef]

63. Guindon, S.; Dufayard, J.F.; Lefort, V.; Anisimova, M.; Hordijk, W.; Gascuel, O. New algorithms and methods to estimate maximum-likelihood phylogenies: Assessing the performance of PhyML 3.0. *Syst. Biol.* **2010**, *59*, 307–321. [CrossRef]

64. Schwarz, G.E. Estimating the dimension of a model. *Ann. Stat.* **1978**, *6*, 461–464. [CrossRef]
65. Brewer, M.J.; Butler, A.; Cooksley, S. The relative performance of AIC, AICc and BIC in the presence of unobserved heterogeneity. *Met. Ecol. Evol.* **2016**, *7*, 679–692. [CrossRef]
66. Yang, Z.; Landry, J.F.; Hebert, P.D.N. A DNA Barcode Library for North American Pyraustinae (Lepidoptera: Pyraloidea: Crambidae). *PLoS ONE* **2016**, *11*, e0161449. [CrossRef]
67. Nguyen, L.T.; Schmidt, H.A.; von Haeseler, A.; Minh, B.Q. IQ-TREE: A fast and effective stochastic algorithm for estimating Maximum Likelihood phylogenies. *Mol. Biol. Evol.* **2015**, *32*, 268–274. [CrossRef]
68. Trifinopoulos, J.; Nguyen, L.T.; von Haeseler, A.; Minh, B.Q. W-IQ-TREE: A fast online phylogenetic tool for Maximum Likelihood analysis. *Nucleic Acid Res.* **2016**, *44*, W232–W235. [CrossRef]
69. Anisimova, M.; Gil, M.; Dufayard, J.F.; Dessimoz, C.; Gascuel, O. Survey of branch support methods demonstrates accuracy, power, and robustness of fast Likelihood-based approximation schemes. *Syst. Biol.* **2011**, *60*, 685–699. [CrossRef]
70. Minh, B.Q.; Thi Nguyen, M.A.; von Haeseler, A. Ultrafast approximation for phylogenetic bootstrap. *Mol. Biol. Evol.* **2013**, *30*, 1188–1195. [CrossRef]
71. Miller, H.C. Cloacal and buccal swabs are a reliable source of DNA for microsatellite genotyping of reptiles. *Conserv. Genet.* **2006**, *7*, 1001–1003. [CrossRef]
72. Pyron, R.A.; Burbrink, F.T.; Wiens, J.J. A phylogeny and revised classification of Squamata, including 4161 species of lizards and snakes. *BMC Evol. Biol.* **2013**, *13*, 93. [CrossRef]
73. Bogaardt, C.; Carvalho, L.; Hill, V.; O'Toole, A.; Rambaut, A. FigTree, Version 1.4.3, Molecular Evolution, Phylogenetics and Epidemiology, United Kingdom. 2018. Available online: http://tree.bio.ed.ac.uk/software/figtree/ (accessed on 14 November 2018).
74. Madison, W.P.; Madison, D.R. Mesquite: A Modular System for Evolutionary Analysis, Version 3.2, Mesquite Project, United States. 2017. Available online: http://mesquiteproject.org (accessed on 15 June 2018).
75. Chapple, D.G.; Ritchie, P.A. A retrospective approach to testing DNA barcoding method. *PLoS ONE* **2013**, *8*, e77882. [CrossRef]
76. Padial, J.M.; Miralles, A.; De la Riva, I.; Vences, M. The integrative future of taxonomy. *Front. Zool.* **2010**, *7*, 16. [CrossRef]
77. Mott, T.; Vieites, D.R. Molecular phylogenetics reveals extreme morphological homoplasy in Brazilian worm lizards challenging current taxonomy. *Mol. Phylogenet. Evol.* **2009**, *51*, 190–200. [CrossRef]
78. Grazziotin, F.G.; Zaher, H.; Murphy, R.W.; Scrocchi, G.; Benavides, M.A.; Zhang, Y.P.; Bonatto, S.L. Molecular phylogeny of the New World Dipsadidae (Serpentes: Colubroidea): A reappraisal. *Cladistics* **2012**, *28*, 437–459. [CrossRef]
79. Da Silva, N.J.; Sites, J.W. Revision of the *Micrurus frontalis* complex (Serpentes: Elapidae). *Herpetol. Monogr.* **1999**, *13*, 142–194. [CrossRef]
80. Cartes, J.L. Brief history of conservation in the Interior Atlantic Forest. In *The Atlantic Forest of South America*; Galindo-Leal, C., Gusmão Câmara, I., Eds.; Island Press: London, UK, 2003; pp. 269–287.
81. Huang, C.; Kim, S.; Song, K.; Townshend, J.; Davis, P.; Altstatt, A.; Rodas, O.; Yanosky, A.A.; Clay, R.; Tucker, C.J.; et al. Assessment of Paraguay's forest cover change using Landsat observations. *Glob. Planet. Chang.* **2009**, *67*, 1–12. [CrossRef]
82. Yanosky, A.A. Paraguay's challenge of conserving natural habitats and biodiversity with global markets demanding for products. In *Conservation Biology: Voices from the Tropics*; Sodhi, N.S., Gibson, L., Raven, P.H., Eds.; Wiley-Blackwell: Oxford, UK, 2013; pp. 113–119.
83. Cacciali, P.; Cabral, H.; Yanosky, A.A. Conservation implications of protected areas' coverage for Paraguay's reptiles. *Parks* **2015**, *21*, 101–119. [CrossRef]
84. Motte, M.; Núñez, K.; Cacciali, P.; Brusquetti, F.; Scott, N.; Aquino, A.L. Categorización del estado de conservación de los anfibios y reptiles de Paraguay. *Cuad. Herpetol.* **2009**, *23*, 5–18.

 *diversity*

Article

The Western Amazonian Richness Gradient for Squamate Reptiles: Are There Really Fewer Snakes and Lizards in Southwestern Amazonian Lowlands?

Daniel L. Rabosky [1,*], Rudolf von May [1,2], Michael C. Grundler [1] and Alison R. Davis Rabosky [1]

[1]   Museum of Zoology & Department of Ecology and Evolutionary Biology, University of Michigan, Ann Arbor, MI 48109, USA; rvonmay@gmail.com (R.v.M.); mgru@umich.edu (M.C.G.); ardr@umich.edu (A.R.D.R.)
[2]   Biology Program, California State University Channel Islands, Camarillo, CA 93012, USA
*   Correspondence: drabosky@umich.edu

Received: 15 September 2019; Accepted: 14 October 2019; Published: 18 October 2019

Abstract: The lowland rainforests of the Amazon basin harbor some of the most species-rich reptile communities on Earth. However, there is considerable heterogeneity among climatically-similar sites across the Amazon basin, and faunal surveys for southwestern Amazonia in particular have revealed lower species diversity relative to sites in the northwestern and central Amazon. Here, we report a herpetofaunal inventory for Los Amigos Biological Station (LABS), a lowland site located in the Madre de Dios watershed of southern Peru. By combining active search and passive trapping methods with prior records for the site, we provide a comprehensive species list for squamate reptiles from LABS. We also estimate an "expected" list for LABS by tabulating additional taxa known from the regional species pool that we consider to have a high probability of detection with further sampling. The LABS total of 60 snake and 26 lizard taxa is perhaps the highest for any single site in the southern Amazon. Our estimate of the regional species pool for LABS suggests that the southwestern Amazonian lowlands harbor at least 25% fewer species of snakes relative to the western equatorial Amazon, a diversity reduction that is consistent with patterns observed in several other taxonomic groups. We discuss potential causes of this western Amazonian richness gradient and comment on the relationship between spatial diversity patterns in squamates and other taxa in the Amazon basin.

Keywords: species richness; diversity gradient; community structure; reptiles; neotropics; Amazon; rainforest; lizard; snake

---

1. Introduction

Perhaps most the fundamental challenge in biogeography is to explain why species richness varies across the surface of the Earth. Regardless of spatial grain of sampling, the most prominent biodiversity pattern on our planet is the extent to which the number of species differs between sites. The most famous example of this pattern is the latitudinal diversity gradient, whereby species richness peaks in the tropics and decreases with latitude as one moves towards increasingly temperate or polar regions. Importantly, species richness can vary substantially even among climatically-matched sites, for reasons that remain difficult to explain [1]. For example, species richness of rainforest trees is far higher in the Neotropics and southeast Asian tropics than in climatically-matched sites from the African tropics [2]. Similarly, the number of broadly-sympatric lizard taxa in the spinifex deserts of arid Australia greatly exceeds the number of lizard species that occur in any other region on Earth, including both climatically-matched desert regions and wet tropical regions alike [3–6].

Australian lizards aside, the Amazon basin and eastern Andes represent the most biodiverse region on Earth for the majority of terrestrial organisms [7,8]. The rainforests of western Amazonia are characterized by extreme species richness, and represent one of the largest remaining wilderness areas on the planet. In spite of their high overall diversity, sites within the Amazon can vary substantially in species richness. Pitman et al. [9] observed that tree species richness at area-matched lowland sites varied along a gradient from north to south across the western Amazon. At Yasuni National Park, in the Ecuadorian Amazon (1° S), standardized survey plots (1 ha [hectare]) contain an average of 239 species, versus 174 species for sites in the Madre de Dios watershed of southern Peru (12° S). Pitman et al. [9] explored several possible explanations for this western Amazonian richness gradient, including the influence of biotic and abiotic differences between regions, concluding that regional climatic factors play an important role in mediating differences in species richness.

Squamate reptiles (lizards and snakes) show an intriguing pattern of species richness variation across the Amazon basin, and especially along a north-to-south gradient that extends from eastern Ecuador to southeastern Peru (Figure 1). Although snakes are phylogenetically nested within squamates, we nonetheless use the word "lizard" throughout this article to refer to all squamates that are not snakes, owing to major differences in ecology, abundance, and detectability between snakes and non-snake squamates [10]. Species lists for sites from the northwest Amazon (e.g., Rio Amazonas and Rio Napo of Peru and Ecuador) are markedly higher than those for the southwestern Amazonian lowlands, and include a number of genera that are not represented in the south [11]. More recently, estimates of species richness from reconstructed geographic ranges and museum samples suggest that richness of snakes in particular is considerably higher in the northwestern Amazon (= western equatorial Amazon) relative to the south [12–14]. A casual inspection of published species lists would suggest that sites in southern Peru harbor fewer species of both snakes and lizards relative to sites approximately 1000–1500 km further north. For example, 48–52 species of snakes and 24 species of lizards are known from Cusco Amazonico [15], an intensively-studied site along the Madre de Dios River in southern Peru. In contrast, approximately 94 species of squamates are known from a similar-sized area in the Ecuadorian Amazon [7,16]. These and other results [11,15] imply that alpha diversity for squamates in the northern Amazon is 30% higher relative than communities in the south, a ratio that approaches that observed for tree assemblages from the same regions [9].

However, it is challenging to draw general conclusions about variation in species richness and turnover in community composition for Amazonian squamate reptiles. These difficulties emerge for at least three reasons. First, the geographic scale of sampling for published squamate inventories is highly variable. The lack of standardization has necessitated that researchers perform biogeographic comparisons on communities that differ by three to four orders of magnitude in spatial extent (e.g., 300 ha to 4,500,000 ha; [11]). Second, squamate reptiles—and especially snakes—are notoriously difficult to sample; many species have detectabilities so low as to render them effectively invisible in the communities where they occur [10,17,18]. Finally, there is no single standardized protocol for sampling rainforest squamate communities: The low detectabilities of many species lead researchers to adopt a variety of methods for sampling communities, towards the goal of providing the most complete species list for an area. For these reasons, there remains a tremendous amount of uncertainty in community composition across Amazonian squamate communities.

Here, we report a comprehensive list of squamate taxa from a targeted sampling of a lowland rainforest site from southeastern Peru. From 2001–2018, we used a variety of methods to survey approximately 10 km^2 (1000 ha) of primary and secondary forest at Los Amigos Biological Station (hereafter, LABS) in the Madre de Dios region of southeastern Peru. We provide lists of those species documented explicitly at LABS, as well as an expanded list that includes other taxa from the region whose occurrence at LABS is highly probable with further sampling. Our work builds on previous surveys and compilations for this general region [19–21], and provides one of the most thorough and spatially-explicit inventories of squamates for a single Amazonian site. We contrast species richness at LABS to other Amazonian sites, and we compare these results to patterns documented for trees, lianas, birds, and other taxa. Our ultimate goal is to determine whether a true north-to-south gradient in species richness exists for squamate reptile communities in the western Amazon basin.

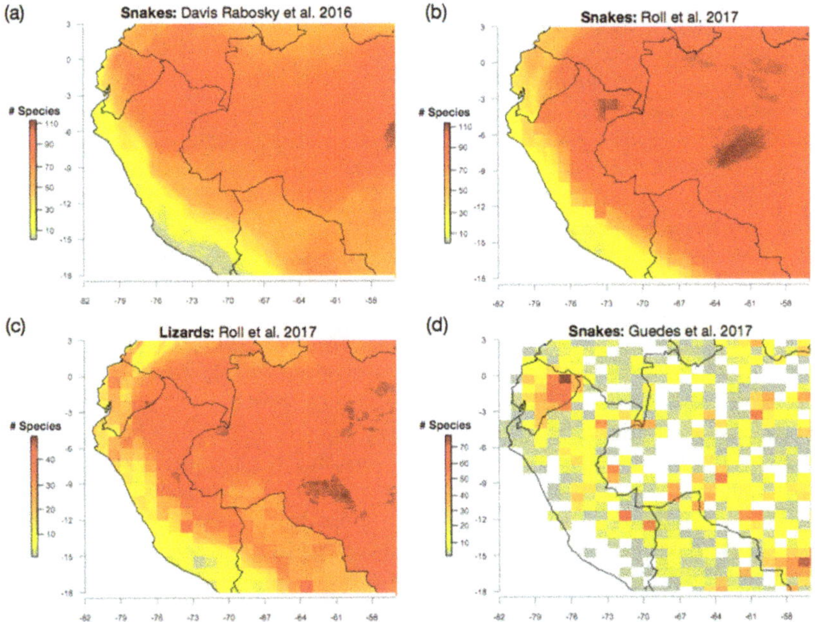

Figure 1. Species richness for squamate reptiles in the western Amazon basin as inferred from range reconstructions and primary occurrence data. Estimates of snake (**a,b**) and lizard (**c**) species richness from published range maps [13,14] suggests higher diversity in the western equatorial Amazon (northern Peru, Ecuador) relative to southern Peru. (**d**) Snake species richness from georeferenced museum occurrence data [12] is consistent with higher richness in the Ecuadorian Amazon relative to southern Peru. Results in (**d**) are heavily affected by biases in sampling and data availability; the data compilation included proportionately fewer occurrence records maintained by Peruvian institutions. Both (**a,b**) imply that some sites in the north contain 20–30 more species of snakes than sites from southern Peru.

2. Materials and Methods

2.1. Study Site

Research was performed at Los Amigos Biological Station (LABS; 12°34′07′′S, 70°05′57′′W, 250 m elev; also known as CICRA in previous literature), located in Manu province, Madre de Dios region, southeast Peru. In addition, we carried out field surveys at Centro de Monitoreo 1 (CM1; 12°34′17′′ S, 70°04′29′′ W), located ca. 3.5 km east from LABS, and Centro de Monitoreo 2 (CM2; 12°26′57′′ S, 70°15′06′′ W, 260 m), located ca. 25 km northwest from LABS. The annual rainfall recorded at LABS ranges between 2700 and 3000 mm [22]. The dry season (June–September) has markedly lower rainfall and is slightly cooler than the wet season. The mean annual temperature ranges between 21 °C and 26 °C (N. Pitman, pers. comm.). Most of our surveys took place in four forest types: mature floodplain forest (hereafter referred to as floodplain forest), terra firme forest, bamboo forest dominated by two native woody bamboo species (*Guadua sarcocarpa* and *G. weberbaueri*), and palm swamp dominated by a native palm (*Mauritia flexuosa*). We also surveyed other terrestrial habitats including secondary forest, riparian forest, clearings, and river banks; and various types of aquatic habitats including oxbow lakes, streams, permanent ponds (e.g., "Pozo Don Pedro"), seasonal ponds, and other wetlands (e.g., small swamps dominated by other palm species). Further details regarding primary forest types and aquatic habitats is available in earlier references [20,21,23–25].

2.2. Data Collection

From 2001 to 2018, we surveyed reptile and amphibian communities at LABS using a variety of sampling techniques. Survey methodologies included (1) leaf-litter plot surveys, (2) nocturnal visual encounter surveys, (3) passive trapping using pitfall traps and funnel traps, and (4) a variety of active and targeted search strategies to increase encounter probabilities for specific taxa and/or within particular microhabitats. Leaf-litter sampling protocols used at this location have been described previously [20] and involved exhaustive search through randomly-selected 100 m^2 or 25 m^2 square plots by teams of trained observers at night. Nocturnal visual encounter surveys were performed by actively search for reptiles and amphibians along the official LABS trail network (trail surveys), and by searching along 50 m or 100 m transects (transect surveys) that we established away from major trails [20]. Both trail and transect surveys typically involved groups of 2–3 researchers, but survey procedures differed, with trail surveys conducted at a more rapid walking pace (mean pace: 0.36 km/hr) relative to transect surveys (mean pace: 0.10 km/hr). Use of vertebrate animals was approved by the Animal Care and Use Committee of the University of Michigan (protocol PRO00008306).

In general, fieldwork from 2001–2015 emphasized transect surveys and leaf-litter plot surveys, and work from 2016–2018 was focused primarily on trail surveys and passive trapping. However, pitfall trapping and trail surveys occurred prior to 2016, and some transect/leaf-litter sampling occurred between 2016–2018. From 2016–2018, we conducted approximately 340 km and 14,500 min of nocturnal trail surveys. We also performed at least 32.5 km and 20,500 min of transect surveys (~650 transects), along with approximately 375 leaf-litter plot surveys.

Preliminary surveys with drift fences and pitfall traps took place in the dry seasons of 2005 and 2006. In four forest types (floodplain forest, terra firme forest, bamboo forest, and palm swamp), we set up three arrays of drift fences and pitfall traps. Each array had three fences of 5 m each in length. We used plastic sheets for making the fences and we buried each fence 10 cm in the ground, leaving at least 50 cm of fence above ground. We used four 19 L plastic buckets as pitfall traps, following a lineal construction design. The opening of each bucket was flush with the ground surface. The arrays remained in the same location up to four weeks, and were checked at least once per day.

From 2016–2018, we established a regular grid of pitfall and funnel traps in multiple habitats; sites were surveyed in March–April 2016, November–December 2016, November–December 2017, and November–December 2018. Each trapline involved either three pitfall traps or four funnel traps. Funnel trap [26] dimensions were 18 cm × 18 cm × 79 cm, and were purchased from Terrestrial Ecosystems (Mt Claremont, Australia). Traps were connected by 12 m of plastic drift fencing. All traps were checked at least once per day. From 2016–2018, we conducted a total of 9181 trap-days of funnel trap sampling, and 6670 trap-days of pitfall sampling. Many records were also obtained through opportunistic sampling, particularly medium-to-large diurnal snakes that were most commonly encountered on trails while researchers checked traps or performed other activities.

In general, we collected multiple voucher specimens per species, to facilitate taxonomic work, to generate a permanent and curated record of squamate biodiversity for LABS and to enable us to create and curate community resources that can only be provided by physical specimens (e.g., samples of venom, Duvernoy's gland, glandular skin, and internal parasites). Multiple tissue types were taken from all vouchered animals and preserved in RNAlater. All vouchered specimens were deposited in the herpetological collections at the Museo de Historia Natural, Universidad Nacional Mayor de San Marcos (MUSM; Lima, Peru), or in the University of Michigan Museum of Zoology (UMMZ). Records for several species were obtained through opportunistic sampling by researchers or visitors to LABS who were unaffiliated with our research group. In these cases, we required that the records be accompanied by photographs and specific sublocality information.

2.3. Data Tabulation and Analysis

We compiled three lists of squamate reptiles for LABS. First, we consider only the set of species documented explicitly by our group from the immediate vicinity of the LABS station. The area of this survey region is approximately 10 km², and thus provides a useful "local scale" counterpart to other spatially explicit, localized surveys in Amazonia (e.g., surveys at Cusco Amazonica, Peru; Tiputini Biological Station, Ecuador; Reserva Ducke, Brazil). The second list includes all species documented from the Los Amigos conservation concession, including parts of the reserve or its buffer zone that were located across the Rio Los Amigos (CM1) or some distance upriver from LABS (CM2). Finally, we generated an expected list that included all species known from lowland Amazonian rainforest with documented occurrences in the broader Los Amigos / Madre de Dios region; this includes all species known from other lowland sites within roughly 150 km of LABS, and which we consider to have a high likelihood of detection with continued work at the site. Where possible, we used recent taxonomic works to update the identifications of many specimens in our dataset (e.g., *Erythrolamprus*; [27]).

We analyzed the tempo of species accumulation through time for our 2016–2018 fieldwork, where much of our efforts were focused on squamate sampling, and where we believe we had a reasonably consistent sampling effort from day-to-day. We used randomized sampling from the full set of survey days from 2016–2018 (112 days) to generate an expected species accumulation curve as a function of survey effort. Using nonlinear regression, we fitted a simple hyperbolic model for species accumulation through time (equation 2 in [28]), and used it to predict both the total richness of the LABS squamate fauna and to estimate the effort required to achieve a comprehensive inventory. In this model [28], the estimated species richness is given by $S(n) = S_{max} \, n/(B + n)$, where B and S_{max} are fitted constants, and n is the number of samples. Here, a sample is equated to a "survey day", and S_{max} is the asymptotic richness as n becomes large.

For comparative purposes, we also compiled data on squamate richness from a number of previous studies. Where possible, we updated records from those previous lists to reflect recent taxonomic changes. However, we did not attempt to comprehensively address taxonomic changes to species lists, because most taxonomic changes involved updates to specific epithets, and thus had little impact on either the total or generic species richness at a given site. We also obtained estimates of species richness at the local-to-regional scale for a range of other taxa (e.g., trees, birds, lianas), focusing in particular on northern and southern lowland sites in the western Amazon basin. This generally involved comparing sites from the Rio Napo drainage (Yasuní, Tiputini) to those from southern Peru (Madre de Dios watershed). These and other locations discussed in text are shown in Figure 2.

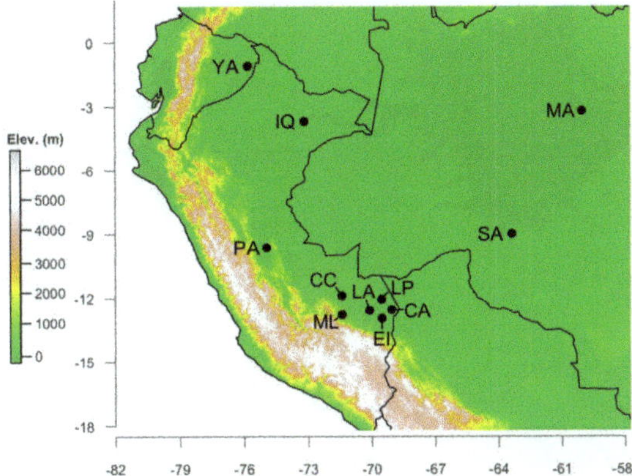

Figure 2. Localities in the western and central Amazon basin discussed in this article. YA: Yasuní National Park and Tiputini Biodiversity Station; IQ: Iquitos (Loreto, Peru); PA: Panguana Biological Station; CC: Cocha Cashu Biological Station; LA: Los Amigos Biological Station (LABS); ML: Manu Learning Centre; EI: Explorer's Inn; CA: Cusco Amazónico; LP: Las Piedras Biological Station; SA: Samuel Hydroelectric Project (Rondônia, Brazil); MA: Manaus, Brazil. All localities are lowland rainforest sites that receive approximately 2200–3200 mm of precipitation annually, although southern sites are markedly more seasonal than those in the north.

3. Results

The documented squamate fauna from LABS includes at least 26 species of lizards and 60 species of snakes (Appendix A: Table A1), an increase of 19 species relative to the most recently-published list for the site [16]. Several additional snake and lizard species are known from the Los Amigos Conservation Concession and are shown in Table A1 (LABS-R). Finally, the expected Los Amigos (LABS-X) list includes taxa that are known from geographically-nearby and climatically-similar sites in southern Peru and includes a total of 36 lizard and 68 snake taxa (Table A1: LABS-X). For the LABS-X list, we did not include taxa known only from foothill sites in the region (e.g., *Dipsas pavonina*; [16]). We also did not include taxa whose nearest documented occurrences are in adjacent regions of Brazil (e.g., *Bothrocophias hyoprora*; [29]). Table A1 provides, where available, representative catalog numbers for several reference specimens deposited in the herpetological collections at the UMMZ and MUSM. For the 2016–2018 survey period, mean numbers of captures per day ranged from 0–12 (snakes) and 0–19 (lizards), with resulting in averages of 2.9 (snakes) and 3.6 (lizards) species per day. The number of species detected per day throughout this survey period is shown in Figure 3. Overall, capture rates fluctuate through time and show considerable temporal autocorrelation, as expected if capture rates are responding to temporally-correlated drivers (e.g., weather; lunar cycle). The empirical species accumulation curve for our 2016–2018 work is shown in Figure 4. Expected richness as a function of the number of survey-days is shown in Figure 5. The fitted hyperbolic function provided a good fit to the overall tempo of species accumulation as a function of sampling (Figure 5), and the predicted maximum richness is 58 and 28 species of snakes and lizards, respectively. For lizards, the total reported species richness from LABS is similar to the predicted richness as extrapolated from the species accumulation curve (N = 26 versus N = 27 species; Figure 4). For snakes, the observed total is slightly higher than the predicted total, but somewhat lower than the number recorded during the 2016–2018 sampling period (N = 49 species). Using the fitted accumulation curve (Figure 5) and assuming a homogeneous sampling process, we estimate that an additional 260 days of sampling (e.g., more than twice the sampling effort already expended) would be needed to record 95% of the predicted snake richness in this system.

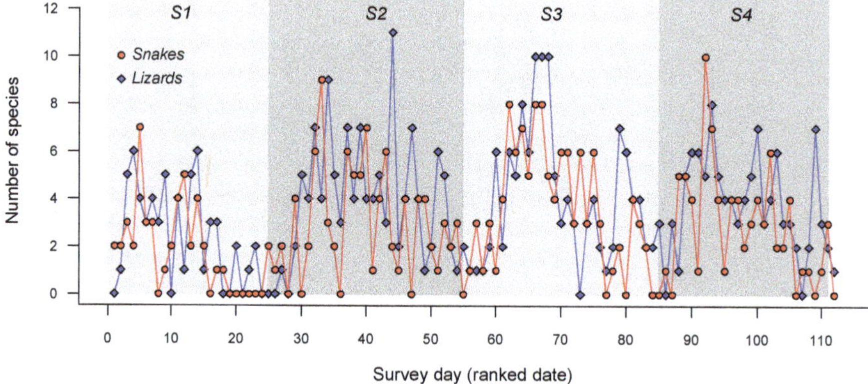

Figure 3. Numbers of species of snakes and lizards detected per sampling day during 2016–2018 fieldwork. Sampling periods: S1, March–April 2016; S2, November–December 2016; S3, November–December 2017; S4, November–December 2018.

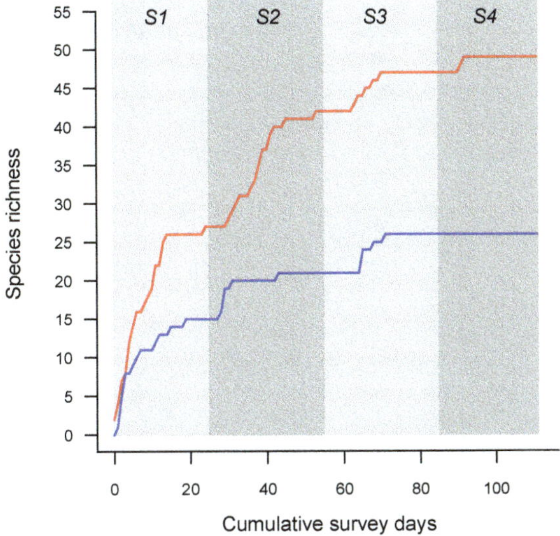

Figure 4. Cumulative numbers of species of snakes (red) and lizards (blue) during 2016–2018 fieldwork at Los Amigos Biological Station (LABS). Sampling periods: S1, March–April 2016; S2, November–December 2016; S3, November–December 2017; S4, November–December 2018.

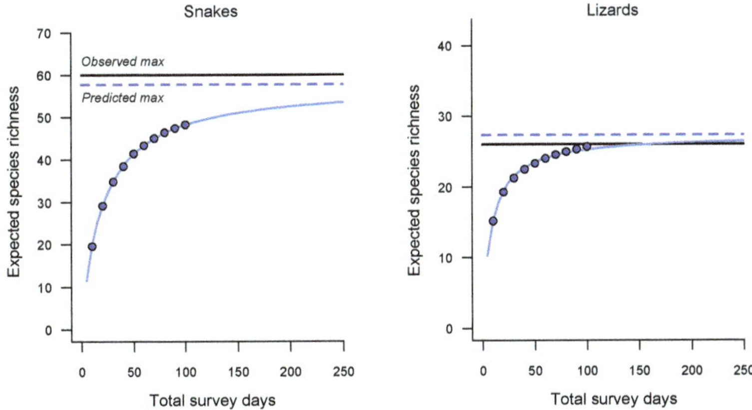

Figure 5. Expected species accumulation curves as a function of survey days for snakes (**left**) and lizards (**right**), using survey data from 2016–2018. Points show mean species richness as a function of survey days obtained from resampling the observed data; underlying blue line corresponds to fitted hyperbolic function [30]. Observed maximum (solid horizontal line) is the total number of species documented from LABS (N = 60), and predicted max (blue dashed line) is the expectation from the resampled survey data.

4. Discussion

With at least 60 species of snakes and 26 species of lizards documented, the squamate reptile diversity at LABS is the highest local-scale diversity reported for the southwestern Amazon basin. These numbers are comparable to those reported for Tiputini Biodiversity Station (Table 1), which is claimed to hold the global alpha diversity record for reptiles [7]. These numbers are based on broadly-comparable sampling regions; as defined here, the LABS total includes species detected in an area of approximately 10 km^2 in the vicinity of the main research facilities, versus 6.5 km^2 for Tiputini. However, the comparisons offered in Table 1 do not address gross discrepancies in survey effort and should nonetheless be treated provisionally. LABS has been the subject of considerable long-term research, and we expect that the site has been more intensively surveyed than Tiputini. The "regional" list for LABS (LABS-X, Table A1) includes 68 snake species and 36 lizard species, and is thus considerably richer than faunas typically reported for the southwestern Amazon basin (Table 1 [11,15]).

Nonetheless, it is clear that the snake fauna of central/northwestern Amazonia is richer than for LABS and other southwestern sites. At least 89 species of snakes are known from the Iquitos region [31], which represents a 29% increase in richness relative to the LABS regional list (Table A1: LABS-X). The regional list for Iquitos encompasses a large spatial scale [31], but it is unlikely that the discrepancies in snake richness between these regions can be explained as an artifact of scale. For example, the Iquitos region includes multiple genera of snakes and lizards that have not been reported from the lowlands of southern Peru (e.g., *Arthrosaura*, *Bothrocophias*, *Iguana*, *Loxopholis*, *Thamnodynastes*), and several other genera are considerably more species-rich in the Iquitos region (e.g., *Amerotyphlops*, *Atractus*, *Helicops*, *Micrurus*, *Erythrolamprus*). Given the high richness of the Iquitos region overall, we predict that lowland sites in the Ecuadorian Amazon (Tiputini, Yasuní; Table 1) are more diverse than reported previously [7]. Southern Peru has been subject to greater sampling effort, and the discrepancies between local-scale tabulations for southern and northern sites (Table 1) seem likely to increase with additional sampling in the Rio Napo basin and adjacent regions.

Table 1. Species richness for lizards and snakes for selected sites in the western Amazon basin. Localities are shown in Figure 2. LABS, LABS-R, and LABS-X are documented and estimated richness totals for Los Amigos Biological Station, as in Table A1.

Locality	Scale	Latitude	Precipitation	Snakes	Lizards	Reference
Manu Learning Center (MLC) [1]	local	−12.8	3000	40	21	[16]
Los Amigos: LABS	local	−12.6	2600	60	27	This paper
Los Amigos: LABS-R	regional	−12.6	2600	62	28	This paper
Los Amigos: LABS-X	regional	−12.6	2600	68	36	This paper
Cusco Amazonico (CA)	local	−12.5	2400	52	28	[15,32]
Tambopata (EI)	local	−12.8	2500	50	26	[32]
Cocha Cashu	local	−11.9	2500	32	25	[16]
Panguana	local	−9.6	2500	44	27	[33]
Yasuni [2]	regional	−0.6	3100	78	33	[7]
Tiputini	local	−0.6	3100	63	31	[7]
Iquitos [3]	regional	−3.7	2900	89	39	[11]

[1] Higher elevation than other sites (500 m, vs 220–280 for others), presumably accounting for higher precipitation. [2] Bass et al. (2010) list does not include several taxa that are almost certainly present: *Ameiva ameiva*, *Epictia diaplocia*, and *Xenopholis scalaris*; listed totals include these taxa. [3] Added *Epictia diaplocia* to Iquitos list; vouchered specimens known from region [34].

4.1. A Western Amazonian Richness Gradient: Does It Exist?

For squamate reptiles, and especially for snakes, our results imply that there is a clear gradient in species richness from north to south in the western Amazon basin. Results parallel those found in a wide variety of taxa, including trees, vascular epiphytes, lianas, and frogs (Table 2). For example, regardless of spatial scale, tropical tree communities in Yasuní contain roughly 40–50% more species than sites at Cocha Cashu or LABS [9]. The causes of this western Amazonian richness gradient remain poorly understood. Pitman et al. [9] noted several structural differences between northern (Yasuní) and southern (Cocha Cashu) tree communities, including tree density, average height, leaf size, and seed size. Furthermore, a number of macroecological descriptors differ between these communities, with species from northern Amazonian sites having smaller geographic and altitudinal ranges. Interestingly, more mobile taxa—birds and primates in particular—show no evidence for increased diversity in the north: Communities from Cocha Cashu and Los Amigos have effectively the same diversity as those from Yasuní, Tiputini, and the Iquitos region (Table 2).

Table 2. Species richness for different taxonomic groups in lowland rainforest in the northern equatorial Amazon (Loreto, Peru; Ecuador) and the southwestern Amazon (southern Peru, northwestern Bolivia). Diversity inflation is the proportional increase in species richness at northern sites relative to southern sites.

Taxon	Richness (North)	Richness (South)	Diversity Inflation (%)	Scale	Reference
Trees	239	174	40	local	[9]
Trees	1017	693	50	regional [1]	[9]
Trees	1356	1004	40	regional [2]	[9]
Vascular epiphytes [3]	313	152	110	regional	[35,36]
Lianas	203	115	80	local	[37]; Burnham, pers comm
Birds	550–600	550–600	0	regional	[7,38,39]
Snakes	90	68	30	regional	This study
Lizards	39	36	10	regional	This study
Frogs	141	100	40	regional	[7,32]
Primates	12–14	12–14	0	regional	[7,40–42]

[1] Directly estimated from plot data. [2] Estimated from regional compendia of tree diversity. [3] Estimated from surveys at Parque Nacional Madidi, northern Bolivia (border with Peru)

Pitman et al. [9] reviewed several candidate mechanisms for the western Amazonian richness gradient in trees. They considered it unlikely that higher tree richness in the north could be attributed to greater disturbance, increased biogeographic mixing, and intraspecific density-dependence (e.g., Janzen-Connell effects). Nonetheless, they noted that modern climate could play an important role in facilitating higher northern richness, because northern sites are simultaneously wetter and less seasonal than southern sites. Relative to the north, the southern Amazon is characterized by a greatly exacerbated dry season, with little precipitation during the months of June, July, and August. One possibility is that the generally wetter year-round climate in the north facilitates the persistence of water-limited taxa in the understory [9]; this model might also explain why amphibian diversity is substantially higher in the north (Table 2). However, it is unclear how water availability per se contributes to higher squamate richness; lizards in particular have diversity patterns that are effectively decoupled from annual precipitation, and exceptionally diverse assemblages can be found in regions with both high (e.g., Amazonia) and low (arid Australia) rainfall.

At present, many putative ecological drivers of the richness gradient for squamate reptiles cannot be assessed, because we lack basic information on population structure and dynamics for nearly all species of Amazonian squamates. For starters, we are not aware of any comparative data on the relative density of squamates at northern and southern sites. In addition, "local" communities (Table 1) might not be comparable in some ways. Some fraction of taxa within a particular local assemblage might be transient or otherwise non-coexisting species, maintained by mass effects that occur over larger spatial and temporal scales. Because we have so little information about the processes that generate local assemblages, it is all the more difficult to understand how those processes might vary in space. Furthermore, the lack of standardization in sampling methodology and scale makes it difficult to compare species abundance distributions among sites. We suggest that future studies should also consider historical explanations for the richness gradient, in addition to contemporary ecological processes. The history of landscape-level change in the Amazon basin remains controversial [43,44], but it is nonetheless possible that expansion and contraction of savannah habitats in southwestern Amazonia might have resulted in the loss of some species that have not yet recolonized the region. Some widespread Amazonian taxa (e.g., *Iguana iguana*, *Bothrocophias hyoprora*, *Bothrops taeniautus*) apparently fail to occur in suitable habitat in southern Peru, even as they occur at similar latitudes and similarly seasonal climates in Brazil [29,45].

One site from the south-central Brazilian Amazon, "Samuel", purportedly hosts perhaps the highest snake diversity in the entire Amazon basin [11,46], despite a moderately southern latitude (–8.9) and marked seasonality in rainfall. At first glance, the high richness reported for Samuel would appear to reject local climatic explanations for the north-to-south richness gradient for squamates. However, this site cannot be compared to others listed in Table 1, for several reasons. First, the total richness frequently reported for Samuel (92+ species) is in fact a richness estimate for the Brazilian state of Rondonia [47]. The list includes the results of surveys through two major neotropical biomes (lowland Amazonian rainforest and cerrado) and with a survey area that greatly exceeds 100,000 km^2. A total of 68 snake species were reported from the principal survey in the vicinity of Samuel hydroelectric project (near Porto Velho, Brazil), with a survey area of roughly 200 km^2 [47]. Sampling intensity for this site was considerably greater than for nearly any other comparable site in the Amazon, and involved 1507 individual snake captures. Hence, we view the Samuel richness totals as comparable to the regional estimates for the Los Amigos snake fauna (68 species; Tables 1 and 2) and markedly lower than the regional snake fauna for the northwestern and equatorial western Amazon.

4.2. Taxonomic Issues

The taxonomy of many neotropical squamates is in flux: cryptic species appear to be present in many groups [48–51], and truly new species remain to be formally described. We present a species list for LABS with the caveat that nomenclatural boundaries for some taxa are likely to change in the near future. Any future use of the list presented here in a management context should exercise caution with respect to names assigned to particular taxa. For this reason, we have provided a list of vouchered specimens from LABS that can be cross-referenced by future studies. Potentially problematic groups include the following snake genera: (1) *Chironius*, including the relationship between *C. carinatus* and *C. exoletus* in southern Peru (listed as *C. exoletus/carinatus* in Table A1) as well the potential for cryptic diversity in several other taxa [48,52]; (2) all species in the genus *Erythrolamprus* (= *Liophis*), which appears to consist of five species at LABS, but where—in light of recent taxonomic work [27]—the connections between the taxa we list and *Erythrolamprus/Liophis* taxa reported for other Amazonian sites is unclear; (3) *Atractus*, a megadiverse snake genus that contains at least three species at LABS, but which is in need of comprehensive revision for southern Amazonia; species of this genus are frequently misidentified in both museum and field collections [53,54].

5. Conclusions

We have presented a taxonomic inventory of squamate reptiles for a well-sampled lowland rainforest site in southern Peru. Our results provide further support for the existence of a western Amazonian richness gradient, whereby species richness for some groups of plants and animals declines as one moves from the northwestern Amazon basin (Rio Amazonas, Peru; Ecuadorian Amazon) to the southwest (southern Peru). Further studies of this gradient (Table 2) can provide insights into the evolutionary and ecological processes that have shaped Amazonian biodiversity more generally. However, numerous challenges remain, especially for groups such as squamate reptiles, which are difficult and costly to sample. For example, de Fraga et al. [17] estimate the cost of finding a single snake in the Amazon to be approximately $120, a number that accords well with our own field costs. Martins and Oliveira [55], in one of the most comprehensive ecological studies of Amazonian snakes, estimated that approximately 3.25 h of survey effort were required per individual snake capture. To obtain a reasonably comprehensive taxonomic inventory of reptiles at any single site in the

Amazon is a daunting task, requiring many thousands of person-days of field time. Consequently, our understanding of the ecology of tropical reptile communities lags far behind that of many other taxa. Despite numerous human impacts on Amazonian forest communities, it is not yet possible to develop conservation and management strategies for most squamate reptile taxa, owing to the lack of basic information on the distribution and abundance of species.

Author Contributions: Conceptualization, D.L.R. and R.v.M.; methodology, D.L.R., R.v.M., M.C.G. and A.R.D.R.; formal analysis, D.L.R. and A.R.D.R.; investigation, D.L.R., R.v.M., M.C.G. and A.R.D.R.; resources, D.L.R., M.C.G., A.R.D.R. and R.v.M.; data curation, M.C.G., R.v.M. and D.L.R.; writing—original draft preparation, D.L.R.; writing—review and editing, D.L.R., R.v.M., A.R.D.R., M.C.G.; visualization, A.R.D.R. and D.L.R.; project administration, D.L.R. and R.v.M.; funding acquisition, D.L.R., A.R.D.R. and R.v.M.

Funding: This research was supported by a fellowship from the David and Lucile Packard Foundation to D.L.R., and by the University of Michigan Museum of Zoology. RvM thanks the Amazon Conservation Association, Conservación Amazónica-ACCA, Wildlife Conservation Society, National Science Foundation, and National Geographic Society.

Acknowledgments: For field assistance, we thank E.S. Vargas Laura, C. Macahuache Díaz, R. Villarcorta Díaz, E. Durand Salazar, E.M. Iglesias Antonio, N. Tafur Olortegui, O.L. Huacarpuma Aguilar, Y. Casanca Leon, R. Santa Cruz, C. Sanchez Paredes, H. Cárdenas, V. Herrera, A. Guzmán, C. Alarcón, J.C. Cusi, P. Cerda, M. A. Cowan, D. Nondorf, M. R. Grundler, I. A. Holmes, J. G. Larson, G. Pandelis, I. Russell, B. Sealey, P. O. Title, E. P. Westeen, M. Medina-Müller, N. Carrillo, R. Jennings, K. E. Reider, R. McCracken, Z. Lange, P. Campbell, M. Semeniuk, A. Smith, J. Vollmar, M. Vollmar, C. Pinnell, J. M. Jacobs, S. Kiriakopolos, M. Roscheisen, J. Valdivia, J. Huamán, L. Flores, V. Quispe, J. Martinez, R. Thupa, H. Collado, J. Perez, and park rangers of the Los Amigos Conservation Concession. We especially acknowledge A. Basto, C. Castañeda, J. Valdez, C. Quispe, other members of the supporting staff at LABS, and support from the Amazon Conservation Association. We thank G. Schneider, C. Aguilar, J. Cordova Santa Gadea, J. C. Cusi, M. Fernandez, and C. Whitcher for assistance with specimen logistics at UMMZ and MUSM, and R. Murrell for assistance with financial logistics at the UMMZ. M. Finer generously shared reptile species lists for Yasuní and Tiputini. G. Gagliardi, N. Hidalgo, and T. Paine shared several new squamate records for Los Amigos with us. H. Braz, T. Guedes, N. R. de Albuquerque, and P. Melo Sampaio generously shared information about several snake records from Brazil. Research and collecting permits were issued by the Instituto Nacional de Recursos Naturales (INRENA), the Dirección General Forestal y de Fauna Silvestre (DGFFS), and the Servicio Nacional Forestal y de Fauna Silvestre (SERFOR), Peru (R.D. 11-2008-INRENA-IFFS- DCB, 120-2012-AG-DGFFS-DGEFFS, 064-2013-AG- DGFFS-DGEFFS, 292-2014-AG-DGFFS-DGEFFS, R.D.G. 029-2016-SERFOR-DGGSPFFS, R.D.G. 405-2016-SERFOR-DGGSPFFS).

Appendix A

Table A1. Squamate reptiles of Los Amigos Biological Station (LABS) and adjacent Los Amigos Conservation Concession. LABS column includes only those species documented explicitly from the vicinity of the biological station. LABS-R denotes the full list of documented occurrences from the Los Amigos Conservation Concession, including LABS, CM1, and CM2. LABS-X is the documented and expected regional list for LABS and includes taxa not yet documented from Los Amigos, but which are likely to be detected with further sampling. MUSM: Museo de Historia Natural, Universidad Nacional Mayor de San Marcos (Lima, Peru); UMMZ: University of Michigan Museum of Zoology; RAB: Rabosky Lab field series (formal accession pending; all specimens to be assigned UMMZ catalog numbers).

Clade	Species	LABS	LABS-R	LABS-X	Material Examined
Amphisbaenidae (lizard)	*Amphisbaena alba*			X [1]	
	Amphisbaena fuliginosa	X	X	X	
Anguidae (lizard)	*Diploglossus fasciatus*			X [2]	
Gekkota (lizard)	*Gonatodes hasemani*	X	X	X	RAB 2799, MUSM 38982

Table A1. *Cont.*

Clade	Species	LABS	LABS-R	LABS-X	Material Examined
	Gonatodes humeralis	X	X	X	RAB 3434, MUSM 38984
	Pseudogonatodes guianensis	X	X	X	RAB 2508, MUSM 39869
	Thecadactylus solimoensis	X	X	X	RAB 3277, MUSM 39869
Scincidae (lizard)	*Copeoglossum nigropunctatum*		X	X	
	Exila nigropalmata	X	X	X [3]	RAB 3514, MUSM 39156
	Varzea altamazonica		X	X	
Teioidea (lizard)	*Alopoglossus angulatus*	X	X	X	UMMZ 246710, MUSM 36813
	Ameiva ameiva	X	X	X	RAB 2924, MUSM 39781
	Bachia dorbignyi	X	X	X	RAB 3497, MUSM 39788
	Bachia trisanale	X	X	X [3]	RAB 3469, MUSM 38919
	Cercosaura argulus	X	X	X	RAB 2701, MUSM 38930
	Cercosaura bassleri	X	X	X	UMMZ 245038, MUSM 38937
	Cercosaura eigenmanni	X	X	X	UMMZ 246735, MUSM 36933
	Dracaena guianensis			X [1]	
	Iphisa elegans	X	X	X [3]	RAB 2931, MUSM 39020
	Kentropyx altamazonica			X [1,2]	
	Kentropyx pelviceps	X	X	X	UMMZ 246751, MUSM 37269
	Potamites ecpleopus			X [1,2]	
	Ptychoglossus brevifrontalis	X	X	X	UMMZ 246762, MUSM 39120
	Tupinambis teguixin	X	X	X	RAB 3430
Iguania (lizard)	*Anolis fuscoauratus*	X	X	X	RAB 2868, MUSM 36859
	Anolis ortonii	X	X	X	UMMZ 248371
	Anolis punctatus	X	X	X	RAB 2537, MUSM 39782
	Anolis tandai	X	X	X	RAB 2660, MUSM 38909
	Enyalioides palpebralis	X	X	X	MUSM 38981
	Plica plica	X	X	X	RAB 2704, MUSM 39860

Table A1. *Cont.*

Clade	Species	LABS	LABS-R	LABS-X	Material Examined
	Plica umbra	X	X	X	RAB 2682, MUSM 39099
	Polychrus liogaster			X [1]	
	Stenocercus fimbriatus			X [1,2]	
	Stenocercus roseiventris			X [1,2]	
	Uracentron azureum	X	X	X	
	Uracentron flaviceps	X	X	X	MUSM 39154
Typhlopidae (snake)	*Amerotyphlops reticulatus*	X	X	X	UMMZ 248453, MUSM 36160
Leptotyphlopidae (snake)	*Epictia diaplocia*	X	X	X	RAB 2932, MUSM 39046
Aniliidae (snake)	*Anilius scytale*	X	X	X [3]	RAB 2499, MUSM 36848
Boidae (snake)	*Boa constrictor*	X	X	X	
	Corallus batesii	X	X	X	
	Corallus hortulanus	X	X	X	UMMZ 248448, MUSM 36965
	Epicrates cenchria	X	X	X	RAB 2903, MUSM 37120
	Eunectes murinus	X	X	X	RAB 3417
Colubrinae (snake)	*Chironius carinatus / exoletus*	X	X	X [3]	UMMZ 246087, MUSM 38943
	Chironius fuscus	X	X	X	UMMZ 245047, MUSM 36957
	Chironius multiventris	X	X	X	UMMZ 245049, MUSM 36962
	Chironius scurrulus			X [1,7]	
	Chironius sp.	X	X	X [3,7]	UMMZ 248360
	Dendrophidion dendrophis	X	X	X	MUSM 36979
	Drymarchon corais	X	X	X [3]	RAB 2739, MUSM 36145
	Drymobius rhombifer	X	X	X [3]	UMMZ 245052, MUSM 39824
	Drymoluber dichrous	X	X	X	UMMZ 246799, MUSM 38977
	Leptophis ahaetulla	X	X	X	UMMZ 246823, MUSM 37345
	Oxybelis aeneus			X [1]	
	Oxybelis fulgidus			X [1]	
	Phrynonax poecilonotus	X	X	X [3]	RAB 3532, MUSM 39115
	Rhinobothryum lentiginosum	X	X	X	
	Spilotes pullatus	X	X	X	

Table A1. *Cont.*

Clade	Species	LABS	LABS-R	LABS-X	Material Examined
	Spilotes sulphureus			X [1]	
	Tantilla melanocephala	X	X	X [3]	UMMZ 246845, MUSM 39145
Dipsadinae (snake)	*Apostolepis nigroterminata*	X	X	X [3]	RAB 3393, MUSM 39783
	Atractus elaps	X	X	X [3]	RAB 3414, MUSM 39784
	Atractus major			X [1]	
	Atractus snethlageae	X	X	X [4]	RAB 2666, MUSM 35703
	Atractus sp.	X	X	X [3]	RAB 2924, MUSM 38913
	Clelia clelia	X	X	X	
	Dipsas catesbyi	X	X	X	RAB 2802, MUSM 38970
	Dipsas indica	X	X	X [3]	RAB 3561, MUSM 38973
	Drepanoides anomalus	X	X	X	RAB 3408, MUSM 38974
	Erythrolampus dorsocorallinus [6]	X	X	X [3]	MUSM 36049
	Erythrolampus oligolepis [6]	X	X	X [3]	RAB 3411, MUSM 39049
	Erythrolampus reginae [6]	X	X	X	RAB 3436, MUSM 39837
	Erythrolampus sp.	X	X	X	RAB 2711
	Erythrolampus taeniogaster [6]	X	X	X	
	Helicops angulatus	X	X	X	UMMZ 245053, MUSM 38989
	Helicops leopardinus	X	X	X	
	Helicops polylepis		X	X	
	Hydrops triangularis	X	X	X [3]	RAB 3500, MUSM 38992
	Imantodes cenchoa	X	X	X	UMMZ 246814, MUSM 35857
	Imantodes lentiferus	X	X	X	RAB 2900, MUSM 39834
	Leptodeira annulata	X	X	X	UMMZ 246822, MUSM 39043
	Oxyrhopus formosus	X	X	X	UMMZ 248365, MUSM 39077
	Oxyrhopus melanogenys	X	X	X	UMMZ 246832, MUSM 39855
	Oxyrhopus petolarius	X	X	X	UMMZ 245072, MUSM 39081
	Philodryas argentea			X	

Table A1. *Cont.*

Clade	Species	LABS	LABS-R	LABS-X	Material Examined
	Philodryas viridissima		X	X	
	Pseudoboa coronata			X	
	Pseudoeryx plicatilis	X	X	X	MUSM 24359
	Siphlophis cervinus	X	X	X [3]	MUSM 39141
	Siphlophis compressus	X	X	X	MUSM 39142
	Taeniophallus brevirostris	X	X	X [3]	RAB 2462, MUSM 39143
	Taeniophallus occipitalis	X	X	X	RAB 3391, MUSM 37708
	Xenodon rhabdocephalus	X	X	X [3]	UMMZ 246850, MUSM 39158
	Xenodon severus	X	X	X	RAB 2712
	Xenopholis scalaris	X	X	X	UMMZ 245077, MUSM 39889
Elapidae (snake)	*Micrurus annellatus*	X	X	X	UMMZ 248451, MUSM 39056
	Micrurus hemprichii	X	X	X [3,5]	
	Micrurus lemniscatus	X	X	X	UMMZ 248456, MUSM 39057
	Micrurus obscurus	X	X	X	UMMZ 248458, MUSM 39846
	Micrurus surinamensis	X	X	X [3]	UMMZ 246861, MUSM 37353
Viperidae (snake)	*Bothrops atrox*	X	X	X	RAB 2814, MUSM 35721
	Bothrops bilineatus	X	X	X	UMMZ 246865, MUSM 36919
	Bothrops brazili	X	X	X	MUSM 36922
	Lachesis muta	X	X	X	UMMZ 248454, MUSM 36149

[1] Known from multiple lowland sites in Departatmento de Madre de Dios, Peru [15,19,32]. [2] Known from adjacent Rio Las Piedras watershed; Margarita Medina-Müller and Emma Hume, pers. comm. [3] New LABS records documented as part of this study; compare to Whitworth et al. [16]. [4] *Atractus cf. flammigerus* previously (listed as A. flammigerus in [16]). [5] T. Paine, submitted manuscript; photographed. [6] Taxonomy given here follows recent revision of *Erythrolamprus / Liophis* [27]. [7] Reference [16] lists *Chironius cf. scurrulus* for Los Amigos, which we provisionally interpret as *Chironius sp.* based on more recently collected material.

References

1. Rosenzweig, M.L. *Species Diversity in Space and Time*; Cambridge University Press: Cambridge, UK, 1995.
2. Couvreur, T.L.P. Odd man out: Why are there fewer plant species in African rain forests? *Plant Syst. Evol.* **2015**, *301*, 1299–1313. [CrossRef]
3. Pianka, E.R. Zoogeography andspeciation of Australian desert lizards. *Copeia* **1972**, 127–145. [CrossRef]
4. Pianka, E.R. *Ecology and Natural History of Desert Lizards*; Princeton University Press: Princeton, NJ, USA, 1986.
5. Morton, S.R.; James, C.D. The diversity and abundance of lizards in arid Australia—A new hypothesis. *Am. Nat.* **1988**, *132*, 237–256. [CrossRef]

6. Rabosky, D.L.; Cowan, M.A.; Talaba, A.L.; Lovette, I.J. Species interactions mediate phylogenetic community structure in a hyperdiverse lizard assemblage from arid Australia. *Am. Nat.* **2011**, *178*, 579–595. [CrossRef] [PubMed]

7. Bass, M.S.; Finer, M.; Jenkins, C.N.; Kreft, H.; Cisneros-Heredia, D.F.; McCracken, S.F.; Pitman, N.C.A.; English, P.H.; Swing, K.; Villa, G.; et al. Global Conservation Significance of Ecuador's Yasuni National Park. *PLoS ONE* **2010**, *5*, e8767. [CrossRef]

8. Jenkins, C.N.; Pimm, S.L.; Joppa, L.N. Global patterns of terrestrial vertebrate diversity and conservation. *Proc. Natl. Acad. Sci. USA* **2013**, *110*, E2602–E2610. [CrossRef] [PubMed]

9. Pitman, N.C.A.; Terborgh, J.W.; Silman, M.R.; Nunez, P.; Neill, D.A.; Ceron, C.E.; Palacios, W.A.; Aulestia, M. A comparison of tree species diversity in two upper Amazonian forests. *Ecology* **2002**, *83*, 3210–3224. [CrossRef]

10. Magnusson, W.E. *Snakes and Other Lizards*; Open Science Publishers: New York, NY, USA, 2019.

11. Da Silva, N.J.; Sites, J.W. Patterns of diversity of neotropical squamate reptile species with emphasis on the brazilian amazon and the conservation potential of indigenous reserves. *Conserv. Biol.* **1995**, *9*, 873–901. [CrossRef]

12. Guedes, T.B.; Sawaya, R.J.; Zizka, A.; Laffan, S.; Faurby, S.; Pyron, R.A.; Bernils, R.S.; Jansen, M.; Passos, P.; Prudente, A.L.C.; et al. Patterns, biases and prospects in the distribution and diversity of Neotropical snakes. *Glob. Ecol. Biogeogr.* **2018**, *27*, 14–21. [CrossRef]

13. Rabosky, A.R.D.; Cox, C.L.; Rabosky, D.L.; Title, P.O.; Holmes, I.A.; Feldman, A.; McGuire, J.A. Coral snakes predict the evolution of mimicry across New World snakes. *Nat. Commun.* **2016**, *7*, 11484. [CrossRef]

14. Roll, U.; Feldman, A.; Novosolov, M.; Allison, A.; Bauer, A.M.; Bernard, R.; Bohm, M.; Castro-Herrera, F.; Chirio, L.; Collen, B.; et al. The global distribution of tetrapods reveals a need for targeted reptile conservation. *Nat. Ecol. Evol.* **2017**, *1*, 1677–1682. [CrossRef] [PubMed]

15. Duellman, W. *Cusco Amazonico*; Comstock Publishing: Ithaca, NY, USA, 2005.

16. Whitworth, A.; Downie, R.; von May, R.; Villacampa, J.; MacLeod, R. How much potential biodiversity and conservation value can a regenerating rainforest provide? A 'best-case' approach from the Peruvian Amazon. *Trop. Conserv. Sci.* **2016**, *9*, 224–245. [CrossRef]

17. De Fraga, R.; Stow, A.J.; Magnusson, W.E.; Lima, A.P. The Costs of Evaluating Species Densities and Composition of Snakes to Assess Development Impacts in Amazonia. *PLoS ONE* **2014**, *9*, e105453. [CrossRef] [PubMed]

18. Greene, H.W. *Snakes: The Evolution of Mystery in Nature*; University of California Press: Berkeley, CA, USA, 1997.

19. Catenazzi, A.; Lehr, E.; von May, R. The amphibians and reptiles of Manu National Park and its buffer zone, Amazon basin and eastern slopes of the Andes, Peru. *Biota Neotrop.* **2013**, *13*, 269–283. [CrossRef]

20. Von May, R.; Donnelly, M.A. Do trails affect relative abundance estimates of rainforest frogs and lizards? *Austral Ecol.* **2009**, *34*, 613–620. [CrossRef]

21. Von May, R.; Jacobs, J.M.; Santa-Cruz, R.; Valdivia, J.; Huaman, J.M.; Donnelly, M.A. Amphibian community structure as a function of forest type in Amazonian Peru. *J. Trop. Ecol.* **2010**, *26*, 509–519. [CrossRef]

22. Pitman, N.C.A. *An Overview of the Los Amigos Watershed, Madre de Dios, Southeastern Peru*; (Unpublished Report); Amazon Conservation Association: Washington, DC, USA, 2010.

23. Silman, M.R.; Terborgh, J.W.; Kiltie, R.A. Population regulation of a dominant-rain forest tree by a major seed-predator. *Ecology* **2003**, *84*, 431–438. [CrossRef]

24. Griscom, B.W.; Ashton, P.M.S. A self-perpetuating bamboo disturbance cycle in a neotropical forest. *J. Trop. Ecol.* **2006**, *22*, 587–597. [CrossRef]

25. Olivier, J. Bambusiform grasses (Poaceae) from the Los Amigos River, Madre de Dios, Peru. *Rev. Peru. Biol.* **2008**, *15*, 121–126.

26. Thompson, G.G.; Thompson, S.A. Comparative temperature in funnel and pit traps. *Aust. J. Zool.* **2009**, *57*, 311–316. [CrossRef]

27. Ascenso, A.C.; Costa, J.C.L.; Prudente, A.L.C. Taxonomic revision of the *Erythrolamprus reginae* species group, with description of a new species from Guiana Shield (Serpentes: Xenodontinae). *Zootaxa* **2019**, *4586*, 65–97. [CrossRef]

28. Colwell, R.K.; Coddington, J.A. Estimating terrestrial biodiversity through extrapolation. *Philos. Trans. R. Soc. B* **1994**, *345*, 101–118.

29. De Oliveira, L.; Araujo, I.; Prudente, A.L.C.; de Fraga, R.; de Almeida, A.; Ascenso, A.C. New distributional records of the toad-headed pitviper *Bothrocophias hyoprora* (Amaral, 1935) in Brazil. *Amphib. Reptile Conserv.* **2018**, *12*, 1–4.

30. Colwell, R.K.; Winkler, D.W. A null model for null models in biogeography. In *Ecological Communities: Conceptual Issues and the Evidence*; Strong, D.R., Simberloff, D., Abele, L.G., Thistle, A.B., Eds.; Princeton University Press: Princeton, NJ, USA, 1984; pp. 344–359.

31. Dixon, J.R.; Soini, P. *The Reptiles of the Upper Amazon Basin, Iquitos Region, Peru I–VII*; Milwaukee Public Museum: Milwaukee, WI, USA, 1986.

32. Doan, T.M.; Arriaga, W. Microgeographic variation in species composition of the herpetofaunal communities of Tambopata Region, Peru. *Biotropica* **2002**, *34*, 101–117. [CrossRef]

33. Schluter, A.; Icochea, J.; Perez, J.M. Amphibians and reptiles of the lower Rio Llullapichis, Amazonian Peru: Updated species list with ecological and biogeographic notes. *Salamandra* **2004**, *40*, 141–146.

34. Francisco, B.C.S.; Pinto, R.R.; Fernandes, D.S. Taxonomy of *Epictia munoai* (Orejas-Miranda, 1961) (Squamata: Serpentes: Leptotyphlopidae). *Zootaxa* **2012**, *3512*, 42–52. [CrossRef]

35. Kreft, H.; Koster, N.; Kuper, W.; Nieder, J.; Barthlott, W. Diversity and biogeography of vascular epiphytes in western Amazonia, Yasuní, Ecuador. *J. Biogeogr.* **2004**, *31*, 1463–1476. [CrossRef]

36. Aceby, A.; Kromer, T. Diversidad y distribución vertical de epífitas en los alrededores del campamento Río Eslabón y de la laguna Chalalán, Parque Nacional Madidi, Dpto. La Paz, Bolivia. *Rev. Soc. Boliv. Botánica* **2001**, *3*, 104–123.

37. Burnham, R.J.; Romero-Saltos, H.G. Diversity and distribution of lianas in Yasuní, Ecuador. In *Ecology of Lianas*; Schnitzer, S.A., Bongers, F., Burnham, R.J., Putz, F.E., Eds.; Wiley: West Sussex, UK, 2015; pp. 50–64.

38. Karr, J.R.; Robinson, S.K.; Blake, J.G.; Bierregaard, R.O.J. Birds of four neotropical forests. In *Four Neotropical Rainforests*; Gentry, A.H., Ed.; Yale University Press: New Haven, CT, USA, 1990.

39. Sullivan, B.L.; Wood, C.L.; Ilif, M.J.; Bonney, R.E.; Fink, D.; Kelling, S. eBird: A citizen-based bird observation network in the biological sciences. *Biol. Conserv.* **2009**, *142*, 2288–2292. [CrossRef]

40. Janson, C.H.; Emmons, L.H. Ecological structure of the nonflying mammal community at Cocha Cashu Biological Station, Manu National Park, Peru. In *Four Neotropical Rainforests*; Gentry, A.H., Ed.; Yale University Press: New Haven, CT, USA, 1990; pp. 314–338.

41. Peres, C.A. Primate community structure at twenty western Amazonian flooded and unflooded forests. *J. Trop. Ecol.* **1997**, *13*, 381–405. [CrossRef]

42. Pitman, R.L. Checklist de los Mamíferos no Voladores de Cocha Cashu y Pakitza. 2008. Available online: https://cochacashu.sandiegozooglobal.org/species-lists/ (accessed on 5 September 2019).

43. Hoorn, C.; Wesselingh, F.P.; ter Steege, H.; Bermudez, M.A.; Mora, A.; Sevink, J.; Sanmartin, I.; Sanchez-Meseguer, A.; Anderson, C.L.; Figueiredo, J.P.; et al. Amazonia through time: Andean uplift, climate change, landscape evolution, and biodiversity. *Science* **2010**, *330*, 927–931. [CrossRef] [PubMed]

44. Rull, V. Neotropical biodiversity: Timing and potential drivers. *Trends Ecol. Evol.* **2011**, *26*, 508–513. [CrossRef] [PubMed]

45. De Souza, J.R.D.; Venancio, N.M.; de freitas, M.A.; de Souza, M.B.; Verissimo, D. First record of *Bothrops taeniatus* Wagler, 1824 (Reptilia, Viperidae) for the state of Acre, Brazil. *Check List* **2013**, *9*, 430–431. [CrossRef]

46. Avila-Pires, T.C.S.; Vitt, L.J.; Sartorius, S.S.; Zani, P.A. Squamata (Reptilia) from four sites in southern Amazonia, with a biogeographic analysis of Amazonian lizards. *Bol. Mus. Para. Emílio Goeldi. Cienc. Nat. Belém* **2009**, *4*, 99–118.

47. Da Silva, N. The snakes from Samuel hydroelectric power plant and vicinity, Rondonia, Brazil. *Herpetol. Nat. Hist.* **1993**, *1*, 37–86.

48. Hamdan, B.; Pereira, A.G.; Loss-Oliveira, L.; Rodder, D.; Schrago, C.G. Evolutionary analysis of *Chironius* snakes unveils cryptic diversity and provides clues to diversification in the Neotropics. *Mol. Phylogenet. Evol.* **2017**, *116*, 108–119. [CrossRef]

49. Nunes, P.M.S.; Fouquet, A.; Curcio, F.F.; Kok, P.J.R.; Rodrigues, M.T. Cryptic species in *Iphisa elegans* Gray, 1851 (Squamata: Gymnophthalmidae) revealed by hemipenial morphology and molecular data. *Zool. J. Linn. Soc.* **2012**, *166*, 361–376. [CrossRef]

50. Prudente, A.L.C.; Passos, P. New Cryptic Species of *Atractus* (Serpentes: Dipsadidae) from Brazilian Amazonia. *Copeia* **2010**, *010*, 397–404. [CrossRef]

51. Sturaro, M.J.; Avila-Pires, T.C.S.; Rodrigues, M.T. Molecular phylogenetic diversity in the widespread lizard *Cercosaura ocellata* (Reptilia: Gymnophthalmidae) in South America. *Syst. Biodivers.* **2017**, *15*, 532–540. [CrossRef]

52. Torres-Carvajal, O.; Echevarria, L.Y.; Lobos, S.E.; Venegas, P.J.; Kok, P.J.R. Phylogeny, diversity and biogeography of Neotropical sipo snakes (Serpentes: Colubrinae: *Chironius*). *Mol. Phylogenet. Evol.* **2019**, *130*, 315–329. [CrossRef]

53. Meneses-Pelayo, E.; Passos, P. New Polychromatic Species of *Atractus* (Serpentes: Dipsadidae) from the Eastern Portion of the Colombian Andes. *Copeia* **2019**, *107*, 250–261. [CrossRef]

54. Passos, P.; Scanferla, A.; Melo-Sampaio, P.R.; Brito, J.; Almendariz, A. A giant on the ground: Another large-bodied *Atractus* (Serpentes: Dipsadinae) from Ecuadorian Andes, with comments on the dietary specializations of the goo-eaters snakes. *An. Acad. Bras. Cienc.* **2019**, *91*. [CrossRef] [PubMed]

55. Martins, M.; Oliveira, M.E. Natural history of snakes in forests of the Manaus region, central Amazonia, Brazil. *Herpetol. Nat. Hist.* **1999**, *6*, 78–150.

Article

Ecological and Conservation Correlates of Rarity in New World Pitvipers

Irina Birskis-Barros [1,2], Laura R. V. Alencar [2], Paulo I. Prado [2], Monika Böhm [3] and Marcio Martins [2,*]

1   School of Natural Sciences, University of California, Merced, CA 95340, USA
2   Departamento de Ecologia, Instituto de Biociências, Universidade de São Paulo,
    05508-09 São Paulo, SP, Brazil
3   Institute of Zoology, Zoological Society of London, Regent's Park, London NW1 4RY, UK
*   Correspondence: martinsmrc@usp.br; Tel.: +55-11-30918470

Received: 30 June 2019; Accepted: 23 August 2019; Published: 27 August 2019

Abstract: Rare species tend to be especially sensitive to habitat disturbance, making them important conservation targets. Thus, rarity patterns might be an important guide to conservation efforts. Rabinowitz's approach defines rarity using a combination of geographical range, habitat specificity, and local abundance, and is frequently used in conservation prioritization. Herein, we use Rabinowitz's approach to classify the New World (NW) pitvipers (family Viperidae) regarding rarity. We tested whether body size and latitude could predict rarity, and we compared rarity patterns with extinction risk assessments and other prioritization methods in order to detect rare species not classified as threatened or prioritized. Most NW pitvipers have large geographical ranges, high local abundances, and narrow habitat breadths. There are 11.8% of NW pitviper species in the rarest category and they occur along the Pacific coast of Mexico, in southern Central America, in the Andean region of Ecuador, and in eastern Brazil. Rarity in NW pitvipers is inversely related to latitude but is not related to body size. Our results indicate that additional species of NW pitvipers are threatened and/or should be prioritized for conservation. Combining complementary approaches to detect rare and threatened species may substantially improve our knowledge on the conservation needs of NW pitvipers.

Keywords: geographical range; habitat breadth; local abundance; threatened species; extinction risk; Viperidae

1. Introduction

There is growing evidence that rare species play key ecological roles in the functioning and structuring of communities [1–3]. Additionally, rare species tend to be more affected by habitat disturbance compared to common species [4–6], and should thus comprise important targets of conservation programs [6]. A better understanding of rarity patterns is therefore necessary to help with guiding conservation efforts [6].

In the context of populations and communities, ecologists tend to define rare species as those showing low abundance and/or a small distribution area (see reviews in [7,8]). In conservation biology, the meaning of rare is usually associated with extinction risk [6], given that low abundance and small distribution areas are linked to heightened extinction risk of species [9]. Although the term rarity has been broadly used, a more precise concept of rarity was proposed by Rabinowitz [10] as a combination of three components: Small geographical range, high habitat specificity, and low local abundance. The so-called "seven forms of rarity" of Rabinowitz [10] is often suggested to have some degree of subjectivity in the classification of species (e.g., [11–13]). However, although several other strategies for the classification of rarity have been proposed (see review [6]), the "seven forms of rarity" approach is still a simple, effective, and widely used framework to characterize rare species from ecological and conservation points of view (e.g., [14–22]).

Rarity patterns resulting from the "seven forms of rarity" approach tend to be strongly correlated with extinction risk assessed by traditional methods like the International Union for Conservation of Nature (IUCN) categories and criteria [14,22]. In fact, this correlation is expected because most methods aimed to detect extinction risk also use two of the criteria used to define rarity, geographical range and population size, among their main criteria [23]. However, using just one aspect of rarity, such as small geographical range, could lead to an underestimation of extinction risk of a species. On the other hand, the interplay between processes at different scales that make species common or rare can create associations between different components of rarity [24,25]. The investigation of such associations is thus necessary for a better understanding of rarity and also has practical consequences like the use of one aspect as a proxy for the other aspects when data are incomplete. Furthermore, detecting intrinsic factors that could predict rarity and its different aspects from species trait data would also help in conservation efforts by accelerating assessments of rarity and subsequent prioritization exercises. Rarity patterns in tropical woody plants, for example, have a strong phylogenetic component, indicating that related species tend to be similar regarding rarity [26]. In birds, abundance is negatively related to body size, and range sizes are better predicted by habitat specificity, migratory status, and clutch size of different species [27]. Thus, identifying correlations among the different aspects of rarity and possible intrinsic factors that can predict rarity would help us to increase the efficacy of conservation strategies.

Vipers (family Viperidae) comprise about 350 snake species distributed in most continents. New World (NW) vipers are all in the subfamily Crotalinae (pitvipers), comprising about 12 genera and 150 species (~ 43% of all vipers [28]) distributed throughout the Americas [29]. NW pitvipers diverged from their Old World relatives around 28 million years ago and arrived in the New World from a single invasion into North America [30–32]. Once their ancestor reached the NW, pitvipers rapidly occupied virtually all terrestrial habitats available [29,30,32,33] and are today conspicuous predators in most NW ecosystems. There is a large body of literature on the biology of NW pitvipers, especially focusing on their medical importance [34,35], complex venom apparatus [36–38], and feeding habits [29,39,40]. Additionally, compared to other snake lineages and reptiles in general, vipers are significantly more threatened, mainly due to habitat loss and persecution by humans [41–43]. Given their high extinction risk but substantial availability of biological information, NW pitvipers comprise an interesting snake lineage to explore rarity patterns, the intrinsic factors affecting these patterns, and the possible associations between rarity and extinction risk.

Herein, we compiled a dataset of NW pitvipers to categorize species according to Rabinowitz [10]. Additionally, we explored how rarity is distributed across space and among lineages. If rare species are located in a more restricted area and/or lineage, this could guide conservation efforts. We also investigated whether intrinsic factors predict rarity patterns. We expected that body size and latitude could predict rarity and thus could be used as proxies in circumstances where data on distribution, habitat breadth or population abundance are not available. We chose these variables because body size is frequently related to life history traits and range size, and abundance was found to be related to latitude in some clades (see additional details in *Predicting rarity* below). Finally, we compared the rarity patterns found with assessments of extinction risk and other prioritization methods in order to detect rare species which may have so far been overlooked in conservation efforts. As Rabinowitz's classification uses three different criteria (geographical ranges, local abundances, and habitat breadths) to define rare species, we expect that some rare species would not be listed as threatened and/or prioritized for conservation, and thus, this would demonstrate that assessments of rarity could provide an additional tool to guide conservation efforts for these key ecological species.

2. Materials and Methods

2.1. Rarity Patterns

We used literature and unpublished information to characterize the geographical range (GR), habitat breadth (HB), and local abundance (LA) of 143 species of NW pitvipers (see Tables S1–S3). We follow the taxonomy of the Reptile Database as of August, 2015 [28], except for some island populations of *Crotalus* (see [43,44]). Each variable considered was dichotomized (see details below). As a result, we built a table with the eight combinations of these variables in the table cells (cf. Rabinowitz [10]). We then categorized species in four rarity categories (see [14]): Not rare (NR), which includes those species that are not rare in any of the three aspects of rarity; low intermediate (LI), with species that are rare in one aspect; high intermediate (HI), with species that are rare in two aspects; and rarest (RT), with species that are rare in all three aspects.

Locality data used to calculate GR of each species were gathered mostly from literature (scientific papers, theses, dissertations, books, reports etc.), but also from museum databases (including VertNet [45]), and from an unpublished database for species that occur in Brazil [46] (see Table S1). Each resulting map was compared to the maps provided by Campbell and Lamar [29] to exclude possible erroneous localities. In a few cases in which locality data were not available, we obtained data directly from published distribution maps.

We estimated GR using a map depicting the point occurrences of a given species and drawing the smallest convex polygon connecting external points and that encompasses the remaining points (cf. extent of occurrence, EOO [47]). We then excluded areas that were clearly unsuitable habitat (using literature and unpublished information on degree of habitat specialization) and that represented over 1% of the total polygon. In order to make very small distributions more realistic, we considered a buffer of five km radius centered at each point of occurrence. For island endemics, we used the area of the island as the species' GR. For most species, the area of our GR was the same as the EOO [47], but in some, the GR obtained was up to one fourth the EOO (e.g., *Bothrops asper*; see Table S4). We then dichotomized the distribution of GRs using its median: GR > 82,900 km^2 = wide distribution; GR < 82,900 km^2 = restricted distribution). All spatial analyses were performed in the software Quantum GIS [48].

We estimated HB of each species using literature data to obtain the number of major habitat types in which the species can be found, considering gross vegetation types: Desert, non-desert open habitats (grassland, shrubland, and savanna), and forests (Table S2). We then dichotomized the distribution of number of habitat types using its median: > one habitat type = wide habitat breadth; one habitat type = narrow habitat breadth (Table S2). To estimate LA of each species, we used a simplification of the ACFOR (abundant, common, frequent, occasional, rare) scale [49] with three categories (abundant, common, and rare), which refer only to the apparent abundance found in the literature (see Table S3). For each species we assigned a value from one to three (rare to abundant). For dichotomization, we used the mean of these values (1.98; one = rare; two and three = not rare) instead of its median (2), as this would make dichotomization impossible. We failed to find LA data for 16 species. Thus, of the 143 species of New World pitvipers considered herein, we had enough data to analyze rarity for 127 species (88% of the species).

We analyzed the associations between the three rarity variables (GR, HB, and LA) using log-linear models [50,51]. We fitted alternative log-models to the frequencies of species in each combination of the three rarity variables. The models expressed all combinations of associations among variables in a multi-way contingency table. We started with a full model (all possible associations between two variables plus the association of the three variables) and then we tested sequentially the loss of fit caused by the removal of each one of these associations [51]. The loss of fit in these sequential comparisons was gauged by the log-likelihood ratio test (LRT) with the Chi-square test. A significant loss of fit when an association was removed was considered support for such association. The first step in this forward selection procedure was to compare through LRT the model with all interactions with the model without the three-way interaction. If the test showed a non-significant loss of fit, we inferred that the data did not support the three-way association between the three forms of rarity. In this case, we proceeded by testing the significance of the three two-way associations (that is, the association between GR and HB, GR and LA, HB and LA). In each of these tests we compared through LRT the model with the target two-way interaction (but with the other remaining interactions) with the model with all two-way interactions. Each of these three tests then allowed us to identify if there was evidence for one of the associations among pairs of rarity forms (for instance, if species with small geographic range and small populations were more frequent than expected if these two states were independent, and so on). We performed these analyses in R 3.1.3 (R Core Team, Vienna, Austria) [52].

2.2. Predicting Rarity

We explored if body size and latitude can predict rarity in NW pitvipers. Body size is frequently related to several life history aspects in animals [53], and might affect abundance (see [54]) and home range size (e.g., [55–57]). In snakes, body size can also be tied to aspects often predicting generation time, such as time of sexual maturity, fecundity, and lifespan (e.g., [58,59]). Thus, body size could potentially predict rarity in snakes. We collected data for body size (total length, i.e., snout-vent length plus tail length) through an extensive literature review (see Table S5).

Latitude could also predict rarity patterns among NW pitvipers. According to Rapoport's rule, in temperate zones species tend to have larger geographical ranges than in the tropics [60–64] and species occurring in tropical regions tend to have narrower habitat breadth than those in temperate zones [65–67]. Additionally, there is evidence that species abundance is positively related to latitude [68,69]. To evaluate if latitude predicts rarity patterns across NW pitvipers, we used the latitude of the centroid of the geographical range of each species.

To investigate if body size and latitude predict rarity, we used Bayesian phylogenetic generalized mixed models implemented in the R package MCMCglmm [70]. Under this approach phylogenetic relatedness is incorporated as a covariance matrix. We followed the approach of Verde Arregoitia et al. [71] and used body size and latitude as fixed effects and the rarity ranking categories as an ordinal dependent variable. We assigned a rank ranging from 1 to 4 to our rarity categories (cf. [14]), in which the lower value represents the rarest category and the higher value indicates the not rare category. We implemented our models using the codes by Verde Arregoitia et al. [71] and the phylogeny generated in Alencar et al. [31]. We ran the analyses for at least 10,000,000 steps to achieve convergence, discarding the first 10% iterations as burn-in. We verified convergence using the Heidelberger and Welch's convergence diagnostic [72], checking for the autocorrelation of successive samples and through visual inspection of trace plots. We considered associations as significant when the 95% credible intervals did not include zero.

2.3. Rarity and Other Assessments of Extinction Risk or Prioritization

Categorizations under the IUCN criteria [47] were gathered from the IUCN Red List [42], from assessments by the IUCN Snake and Lizard Red List Authority, and from the recent assessment of Brazilian snakes for species endemic to Brazil (see [43]). We assessed the remaining species for which previous IUCN categorizations were not available using the IUCN categories and criteria [47] (see Table S4). Continuing declines in habitat quality under criterion B were inferred visually using maps of human influence index [73], superimposed on distribution maps, combined with information on the ability of species to persist in disturbed habitats (obtained in the literature or by specialist consultation).

We also compared our rarity categorization of New World pitvipers of this study (only the rarest, RT category) with those of the Mexican red list [74], the Threat Index of Maritz et al. ([43] for the World vipers), and conservation status assessments using the Environmental Vulnerability Score (EVS; [75]) for Central American [76] and Mexican species [77].

3. Results

3.1. Rarity Patterns

We found marked rarity patterns within NW pitvipers (Figure 1 and Table S4). Considering each component separately, most species have a narrow habitat breadth (78.7% of the species inhabit a single habitat type) or are locally common (79.5%), and around half of them (55.9%) are widely distributed (GR > 82,900 km^2).

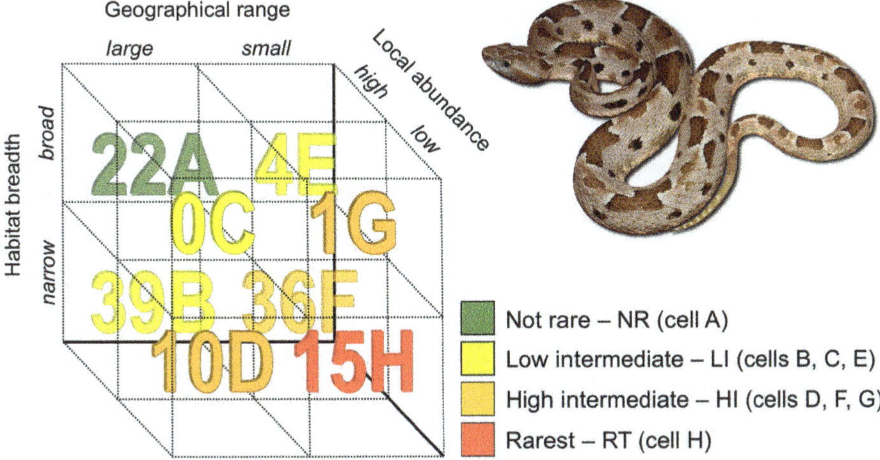

Figure 1. Number of species of New World pitvipers in each rarity cell (letters A to F after each number) of the method proposed by Rabinowitz [10]. The species depicted is *Bothrops pauloensis*, from southeastern Brazil.

Regarding the rarity types of Rabinowitz ([10]; cells A to H in Figure 1), one type does not occur in NW pitvipers (cell C in Figure 1; large GR, broad HB, and low LA; Table S4) and two types comprise less than 5% of the species, with 3.15% of species with small GR, broad HB, and high LA (cell E in Figure 1; Table S4) and 0.8% of species with small GR, broad HB, and low LA (cell G in Figure 1; Table S4). A great number of the NW pitvipers (30.7%) has large GR, narrow HB, and high LA (cell B in Figure 1; Table S4) and 28.3% of the species have small GR, narrow HB, and high LA (cell F in Figure 1; Table S4). Thus, cells B and F represent the majority (59%) of the rarity patterns characterizing the NW pitvipers (Figure 1; Table S4). Besides, 11.8% of the species are in cell H (small GR, narrow HB, and low LA) and 7.9% in cell D (large GR, narrow HB, low LA; Figure 1; Table S4). As for our rarity categories, 11.8% of the species are in the rarest category (cell H), 37.0% are in high intermediate (cells D, F, and G), 33.8% are in low intermediate (cells B, C, and E), and 17.3% are in the not rare category (cell A). Using log-linear models, we found significant, positive associations between GR and HB ($\chi^2 = 7.75$, $p = 0.005$, 1 *df*) and between HB and LA ($\chi^2 = 5.66$, $p = 0.017$, 1 *df*). This indicates that the most parsimonious model predicting the number of species in each rarity category includes the association between GR and HB and the association between HB and LA.

Species in the rarest category occur along the Pacific coast of Mexico, in southern Central America, especially in Costa Rica and Panamá, in the Andean region of Ecuador, and in a narrow area in eastern Brazil (Figure 2A). When both higher rarity categories (high intermediate and rarest) are considered together, rare pitvipers are found in the southwestern United States, throughout Central America from southwestern Mexico to Panamá, and in central and northern South America (Figure 2B). When the distribution of rare pitvipers (high intermediate and rarest; Figure 2A,B) is compared to pitviper richness hotspots throughout the Americas ("hottest" areas in Figure 2D), there is a relatively good congruence between them in North and Central America, but a generally low congruence in South America.

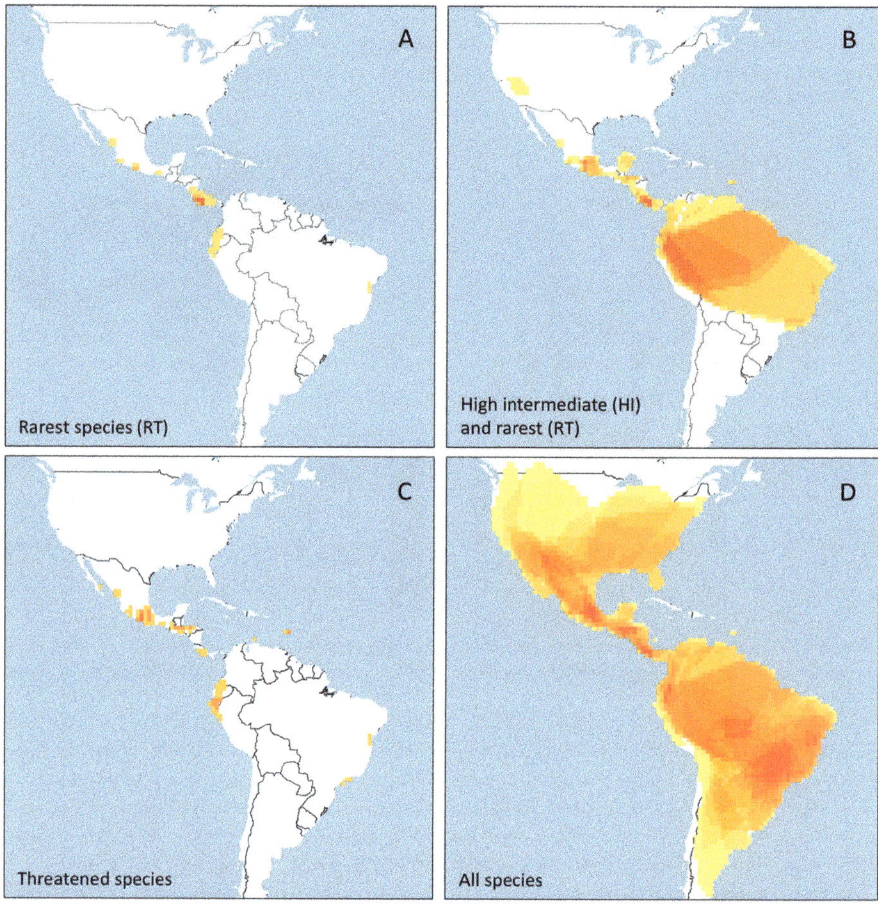

Figure 2. Maps depicting the species richness of New World pitvipers, according to different rarity and threat status categories. (**A**) Distribution of the rarest species (category RT); (**B**) distribution of rare species considering both higher rarity categories (HI and RT); (**C**) distribution of threatened species (IUCN (International Union for Conservation of Nature) categories critically endangered, endangered, and vulnerable); (**D**) species richness of all New World pitvipers.

In general, rarity patterns do not seem to be phylogenetically conserved across NW pitvipers (Figure 3) with the exception of Central American lineages (*Bothriechis, Cerrophidion, Mixcoatlus,* and *Ophryacus*), which show high concentration of rare species (High Intermediate and Rarest; Figure 3). In other lineages (e.g., *Bothrops, Crotalus*), rare species are frequently endemic to islands or to Central and North American mountains (Figure 3 and Figure S1).

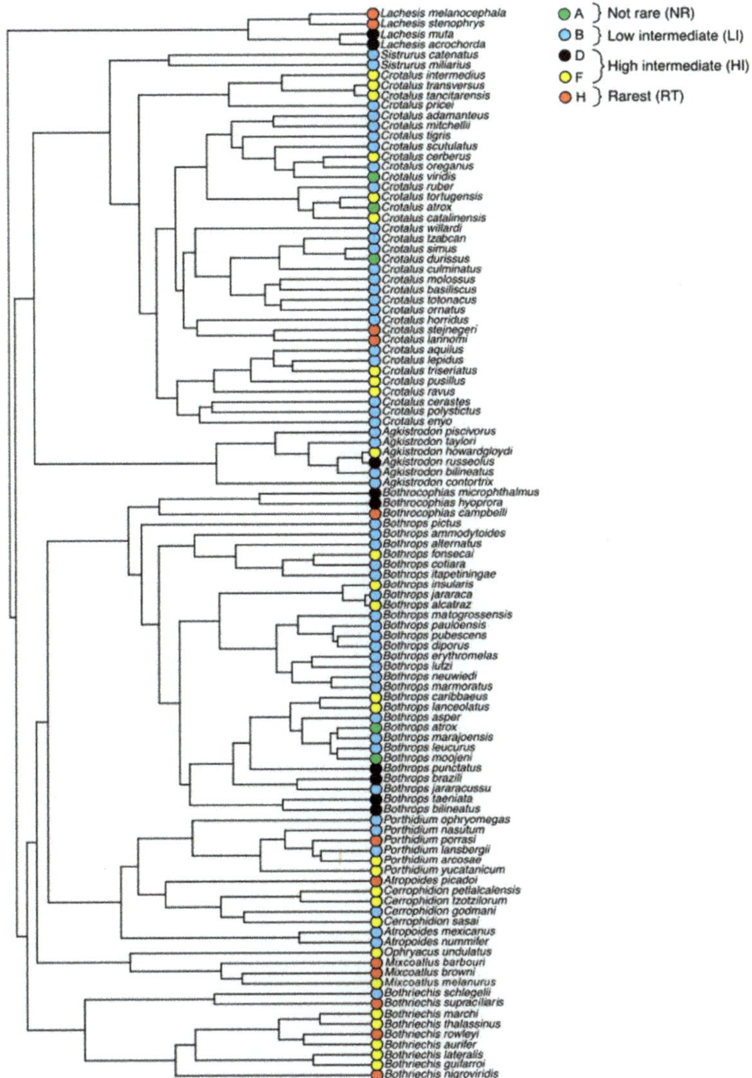

Figure 3. Phylogeny of the New World pitvipers, with colors representing different rarity categories.

3.2. Predicting Rarity

Latitude but not body size predicted rarity in NW pitvipers. Bayesian phylogenetic models suggested an effect of latitude on rarity, i.e., the probability of a lineage being rare decreases with increasing latitude (Table S6).

3.3. Rarity and Other Assessments of Extinction Risk or Prioritization

When rarity patterns are compared to categorizations of extinction risk using IUCN categories and criteria (Table 1; Table S4), 20 species (91.0%) of the 22 threatened species (vulnerable, VU; endangered, EN; and critically endangered, CR) are in the higher rarity categories (high intermediate and rarest) and two (9.1%) are in the low intermediate category (specifically in the category E). For near threatened

(NT) species, two (33.3%) are in low intermediate, three (50.0%) are in high intermediate, and one (16.7%) in the rarest category. Just 22 (23.7%) of least concern (LC) species are in not rare, 38 (40.9%) are in low intermediate, 29 (31.2%) are in high intermediate, and four (4.3%) are in rarest. Finally, one (16.7%) data deficient (DD) species is in low intermediate, three (50%) species are in high intermediate and two (33.3%) in rarest. Additionally, there is a relatively high congruence in the distribution of the species in the rarest category (RT; Figure 2A) and threatened species (Figure 2C).

Table 1. Categories of rarity (based on Rabinowitz [10]) of New World pitvipers in relation to the categories of the IUCN. CR = Critically endangered; DD = data deficient; EX = extinct; EW = extinct in the wild; EN = endangered; LC = least concern; NT = near threatened; VU = vulnerable. Numbers for threatened categories (CR, EN, and VU) are in bold.

Rarity Category	IUCN Categories								Total
	EX	EW	CR	EN	VU	NT	LC	DD	
Not Rare	0	0	**0**	**0**	**0**	0	22	0	22
Low Intermediate	0	0	**0**	**2**	**0**	2	38	1	43
High Intermediate	0	0	**6**	**2**	**4**	3	29	3	47
Rarest	0	0	**1**	**3**	**4**	1	4	2	15
Total	0	0	**7**	**7**	**8**	6	93	6	127

Furthermore, among our 15 rarest species, four are listed as endangered or threatened in the Mexican red list of threatened species [74], two were assessed as having a high threat index in Maritz et al. [43] prioritization, and six and three species were already suggested to show high vulnerability to extinction through EVS in Mexico [77] and in Central America [76], respectively (Table 2).

Table 2. Comparison of the categorization of New World pitvipers of this study (only the rarest, RT category) with those of extinction risk assessments by IUCN ([42] for the World), Secretaría de Medio Ambiente y Recursos Naturales ([78] for Mexico), Johnson et al. ([76] for Central America), Wilson et al. ([77] for Mexico), and the Threat Index of Maritz et al. ([43] for the World). Categories in the Mexican red list are: A = Amenazada (threatened) and P = En Peligro de Extinción (endangered). Category H (high vulnerability species) in the studies using the Environmental Vulnerability Score (EVS) system indicates a score ≥ 14 (see [75]). An empty cell indicates that the species was not assessed.

Rarest Species in this Study (Category RT)	IUCN	Mexican Red List	EVS Mexico	EVS Central America	Among top 30 Threat Index in Maritz et al. [43]
Bothriechis thalassinus	CR			H	no
Bothrocophias campbelli	EN				no
Bothrops ayerbei	DD				no
Bothrops venezuelensis	DD				no
Cerrophidion sasai	LC			H	no
Cerrophidion tzotzilorum	LC				yes
Cerrphidion wilsoni	EN			H	no
Crotalus caliginis	EN				no
Crotalus catalinensis	VU	A	H		no
Crotalus intermedius	VU	A	H		no
Crotalus muertensis	VU		H		no
Crotalus stejnegeri	LC	A	H		no
Crotalus transversus	LC	P	H		yes
Mixcoatlus browni	VU		H		no
Porthidium arcosae	NT				no

4. Discussion

Our results show that most New World pitvipers have a narrow habitat breadth or are locally common, and around half of them are widely distributed. Twenty-two species (17.3%) are in the not rare category and fifteen (11.8%) are in the highest rarity category (Rarest). Most of the latter species are distributed in Central America, especially in highlands and islands, and seven of these rarest species are not classified as threatened by IUCN.

Pitvipers tend to be locally common and to have narrow habitat breadth (see Tables S2 and S3). As a consequence, two of the rarity types in the Rabinowitz cross-classification (cells E and G in Figure 1) together comprise less than 4% of species and one of the rarity types (cell C in Figure 1) was not occupied by any species. This result contrasts with those observed in other groups, like plants [74], mammals [14], and frogs [22], for which empty cells were not recorded, perhaps because the number of species considered herein is smaller than those of these studies. Further rarity studies involving other taxonomically restricted snake lineages (e.g., families, subfamilies) would show whether the patterns described herein are widespread amongst snakes.

A high frequency of habitat specialization and high local abundance was also detected in birds and plants [17,27]. Although few studies have explored patterns of habitat specialization in snakes (e.g., [79,80]), Luiselli [81], using a different approach, found that rare species (including vipers) tend to have narrow realized ecological niches, which can be related to habitat breadth (see below). Additionally, in most studies of NW snake communities, pitvipers tend to appear among the most frequently encountered species (e.g., [82–87]), suggesting a general trend to be locally common. Overall, these results are in accordance with our findings that many NW pitvipers are habitat specialists and locally common.

We found an association between geographical range and habitat breadth in NW pitvipers (see also [88] for snakes in general). Indeed, there seems to be a general trend of association between these traits in different taxa. In a review of the effects of niche breadth on range size in several groups of organisms (diatoms, algae, plants, and animals), Slatyer et al. [89] found that habitat breadth is a good predictor of range size (see also [90,91]). Indeed, many recent studies indicate that populations of specialist species tend to decline faster than populations of generalist species [89], no matter why this association occurs (see discussion in [89], see also [92]). Thus, the high degree of habitat specialization associated with small ranges in NW pitvipers (e.g., 50 species with ranges < 20,000 km^2 have narrow habitat breadth; see also [88]) indicates that this group may be especially vulnerable to human disturbance in their habitats. The association between geographical range and habitat breadth also indicates that there is some redundancy within the rarity aspects included in the "seven forms of rarity" framework, and that it might be possible to use one of these aspects as a proxy for the other when data are incomplete. Additionally, when redundancy within the rarity aspects is high, a simplified framework could be used (e.g., with only two aspects).

The rarest species of NW pitvipers occur in Central America and Andean and Trans-Andean South America (see Figure 2A). Actually, many of the rare pitvipers are restricted to mountains or islands, which translates into small geographical ranges (Figure S2; Table S4) and narrow habitat breadth. In fact, Böhm et al. [88] showed that snakes with a narrow habitat breadth tend to have small geographical ranges, and that snakes have smaller ranges at higher altitudes. Additionally, when the geographical distribution of rare pitvipers (Rarest category; Figure 2A) is compared to pitviper richness hotspots throughout the Americas (Figure 2D), there is a relatively high congruence between them in North and Central America, but a generally lower congruence in South America. This lack of congruence in South America perhaps reflects the fact that most hotspots of pitviper richness in this region are dominated by widespread species, whereas in Central America these hotspots are determined mainly by species with small geographical ranges (see Table S4 and Figure S2).

As predicted, our results show that NW pitvipers occurring at lower latitudes tend to be rare. Indeed, there is evidence of Rapoport's rule among these snakes (correlation between latitudinal midpoint and GR, this latter log-transformed, Pearson's $r = 0.245$, $p < 0.05$, but see [57]), indicating that GR may have contributed to this result. There is also a significant correlation between latitude and number of habitats used by species (Spearman's $r = 0.474$, $p < 0.05$), indicating that HB has also contributed to this trend. Indeed, analyzing snake communities worldwide, Luiselli [81] found that the percentage of rare species in a community is higher in the tropics than in temperate regions. Similar results were found for other groups, such as primates [93], Australian honeyeaters [69], and Australian mammals [68]. However, latitude is just an indicator of geographical location [94] and further studies should search for the factors that are actually explaining higher levels of rarity in the tropics. Letcher and Harvey [95], for instance, tested if climatic variability could explain Rapoports's rule for mammals of the Palearctic region and their results showed that only annual temperature range could predict range size. Thus, an interesting future approach would be to test whether environmental variables, such as temperature, rainfall, and resource distribution, can predict the variation of rarity in New World pitvipers and other groups along the latitudinal gradient.

Although rarity (sensu Rabinowitz [10]) is not directly incorporated in the Red List classification using IUCN categories and criteria [96], there is some congruence between rarity and extinction risk [14,22]. This congruence may occur mainly because certain aspects of rarity, such as small geographical range and small population size, are captured in the IUCN Red List Criteria as symptoms of high extinction risk; thus, extinction risk and rarity may covary [96]. Indeed, there is a high congruence between rare and threatened New World pitvipers (Figure 2A,C, respectively; Table S4). This high congruence is somewhat surprising considering that nearly half of our Rare species are not threatened (four LC, two DD, and one NT) in the IUCN red list. This congruence probably reflects the fact that the extent of occurrence of three quarters of our Rare species falls below the threshold of IUCN Red List criterion B (20,000 km^2; Table S4, [47]). This latter criterion is the most used for reptiles in general [41] and was used in the assessment of almost all threatened NW pitvipers (21 out of 22 species; [42]; this study). Furthermore, although not explicitly included as a specific criterion in the IUCN categories and criteria, local abundance is often taken into account during conservation assessments [97]. Additionally, the high congruence in the geographical distribution of rare and threatened species indicates that rarity is a useful spatial surrogate of endangerment in NW pitvipers. In fact, using a global database of fossil marine animals, Harnik et al. [19] showed that geographical range was the main factor determining extinction risk in the seven forms of rarity, followed by habitat breadth, while local abundance had a minor effect.

Even though we found an overall congruence between rare and threatened NW pitvipers, four species (*Cerrophidion sasai*, *C. tzotzilorum*, *Crotalus stejnegeri*, and *C. transversus*) classified in the rarest category are categorized as "least concern" on the IUCN Red List ([42]). Thus, special attention should be given to these species, as they may become threatened very easily in the near future because they may be disproportionately affected by habitat disturbance [4,6,68]. By the same token, both DD species that appeared in the rarest category (*Bothrops ayerbei* and *B. venezuelensis*) may prove to be threatened or NT when additional information becomes available for them.

Besides the congruence with the IUCN Red List, four of our rarest species are listed as endangered or threatened in the Mexican red list of threatened species ([78]) and additional species appear in Mexican and Central American extinction risk assessments using EVS [75–77] and with a high threat index in Maritz et al. [43] prioritization. Based on the complementary results presented herein and in the IUCN [42] and Mexican [43] red lists and EVS assessments [76–78], species that appear as high priorities in these lists and studies (see Table 2) should receive special attention in future conservation planning. Additionally, one species in the rarest category (*Mixcoatlus browni*) had a high ranking in Maritz et al.'s [43] Ecological and Evolutionary Distinctiveness index (EED). Thus, this species may also be ecologically and evolutionarily unique and therefore deserves special attention for conservation.

5. Conclusions

In this study, we used Rabinowitz's rarity categories to classify New World pitvipers and our results show that although most NW pitvipers are not rare, several species have a combination of features (narrow habitat breadth and small geographical range) that may make them vulnerable to extinction in the near future. We detected regions where a concentration of rare pitvipers occurs (e.g., mountainous regions of Mexico, southern Central America, and Transandean South America), which may become target regions for new protected areas. Our results also show that rarity in NW pitvipers tends to decrease with increasing latitude and hence latitude can be used as a proxy in the lack of data, helping the efforts of conservation in some situations. Lastly, we compared rarity patterns with IUCN assessments and showed that among species in our rarest category, some were not listed as threatened. Thus, we propose the use of our rarity analyses as additional criteria for setting conservation priorities, which can contribute effectively to the conservation of NW pitvipers. Indeed, combining complementary approaches like Rabinowitz's rarity method (this study) with those presented in the previous paragraphs (Table 2), may substantially improve our knowledge on the conservation needs of NW pitvipers.

Supplementary Materials: The following are available online at http://www.mdpi.com/1424-2818/11/9/147/s1, Figure S1: Relief maps showing the distribution of rare vipers in the New World (North and Central America above, South America below). Note that most rare pitvipers occur in mountainous areas. Grid details and richness as in Figure 1. Figure S2: Comparison of the size of viper geographical distribution among regions of the New World. Table S1: Sources of locality data used to calculate geographical range for New World vipers. Table S2: Habitats of New World vipers. Major habitats cited in the literature, number of major habitats, and the category considered in our analyses for habitat breadth. Table S3: Species abundance of New World pitvipers. The column "Abundance" (rare = 1, frequent = 2 and abundant = 3) is based on literature information for each species and LA is the category considered in our analyses based on the previous column (low local abundance for one in previous column, high local abundance for two and three). Table S4: Geographical range (GR_1, in km^2), extent of occurrence (EOO, cf. IUCN), region of the New World (CEAM = Central America; CISA = Cisandean South America; ISLA = island endemics; TRSA = Transandean South America; USNM = United States and northern Mexico), IUCN category (LC = least concern; DD = data deficient; NT = near threatened; VU = vulnerable; EN = endangered; CR = critically endangered), source of IUCN category (BRZ = Brazilian assessment by ICMBio; IUCN = IUCN Red List; SIS = IUCN Species Information System; TS = this study), rarity variables (GR_2 = geographical range; HB = habitat breadth; LA = local abundance), rarity cell in Rabinowitz "seven forms of rarity" (see Figure 1), and rarity category (NR = not rare; LI = low intermediate; HI = high intermediate; RT = rarest) for all species of New World pitvipers. Table S5: Body size (maximum length, in milimeters) and absolute latitude used in the analysis for predicting rarity. The column "references" refers to the data on maximum length. Table S6: Parameter estimates of Bayesian phylogenetic mixed-models analyses. An asterisk indicates a statistically significant result ($p < 0.01$). CI = Confidence intervals.

Author Contributions: Conception of the work, I.B.-B., M.M., and L.R.V.A.; acquisition, analysis, and interpretation of data, I.B.-B., M.M., L.R.V.A., M.B., and P.I.P.; writing—original draft preparation, I.B.-B. and M.M.; writing—review and editing, I.B.-B., M.M., L.R.V.A., M.B., and P.I.P.; acquisition of funding, M.M.

Funding: I.B.-B. L.R.V.A. and M.M. thank São Paulo Research Foundation (FAPESP) (fellowships #2012/15398-7 and #2012/22197-8 to I.B.-B., #2012/02038-2 and #2016/14292-1 to L.R.V.A., grants #2011/50206-9, #2015/21259-8, and #2018/14091-1 to M.M.). M.M. and P.I.P. thank the Brazilian Research Council (CNPq) for research fellowships (306961/2015-6 and 310885/2017-5). M.B. receives funding from The Rufford Foundation.

Acknowledgments: We thank two anonymous reviewers for providing useful suggestions, C. Spencer for providing unpublished data on the distribution of *Crotalus atrox* and C. Nogueira and the group of collaborators of the database on the distribution of Brazilian snakes for providing updated unpublished distribution data.

Conflicts of Interest: The authors declare no conflict of interest.

References

1. Fisher, R.A.; Corbet, A.S.; Williams, C.B. The relation between the number of species and the number of individuals in a random sample from an animal population. *J. Anim. Ecol.* **1943**, *12*, 42–58. [CrossRef]
2. Hubbell, S.P. *A Unified Theory of Biodiversity and Biogeography*; Princeton University Press: Princeton, NJ, USA, 2001; pp. 1–392.
3. Raphael, M.G.; Marcot, B.G. Introduction. In *Conservation of Rare or Little-Known Species: Biological, Social and Economic Considerations*; Raphael, M.T., Moline, R., Eds.; Island Press: Washington, DC, USA, 2007; pp. 1–16.
4. Pimm, S.L.; Jones, H.L.; Diamond, J. On the risk of extinction. *Am. Nat.* **1988**, *132*, 757–785. [CrossRef]
5. Johnson, C.N. Species extinction and the relationship between distribution and abundance. *Nature* **1998**, *394*, 272–274. [CrossRef]
6. Flather, C.H.; Sieg, C.H. Species Rarity: Definition, Causes, and Classification. In *Conservation of Rare or Little-Known Species: Biological, Social and Economic Considerations*; Raphael, M.T., Moline, R., Eds.; Island Press: Washington, DC, USA, 2007; pp. 40–66.
7. Gaston, K.J. *Rarity*; Chapman and Hall Press: London, UK, 1994; pp. 1–205.

8.   Gaston, K.J. What is rarity? In *The Biology of Rarity*; Kunin, W.E., Gaston, K.J., Eds.; Population and Community Biology Series; Springer: Dordrecht, The Netherlands, 1997; Volume 17, pp. 1–21.

9.   Mace, G.M.; Collar, N.J.; Gaston, K.J.; Hilton-Taylor, C.; Akçakaya, R.; Leader-Williams, N.; Milner-Gulland, E.J.; Stuart, S.N. Quantification of extinction risk: IUCN's System for Classifying Threatened Species. *Conserv. Biol.* **2008**, *22*, 1424–1442. [CrossRef] [PubMed]

10. Rabinowitz, D. Seven Forms of Rarity. In *The Biological Aspects of Rare Plant Conservation*; Synge, H., Ed.; John Wiley & Sons: Chichester, UK, 1981; pp. 205–217.

11. Gaston, K.J.; Blackburn, T.M. Evolutionary age and risk of extinction in the global avifauna. *Evol. Ecol.* **1997**, *11*, 557–565. [CrossRef]

12. Pitman, N.C.A.; Terborgh, J.; Silman, M.R.; Nuñez, P.V. Tree species distributions in an upper Amazonian forest. *Ecology* **1999**, *80*, 2651–2661. [CrossRef]

13. Ricklefs, R.E. Rarity and diversity in Amazonian forest trees. *Trends Ecol. Evol.* **2000**, *15*, 83–84. [CrossRef]

14. Yu, J.; Dobson, F.S. Seven forms of rarity in mammals. *J. Biogeogr.* **2000**, *27*, 131–139. [CrossRef]

15. Casazza, G.; Barberis, G.; Minuto, L. Ecological characteristics and rarity of endemic plants of the Italian Maritime Alps. *Biol. Conserv.* **2005**, *123*, 361–371. [CrossRef]

16. Esparza-Olguín, L.; Valverde, T.; Mandujano, M.C. Comparative demographic analysis of three *Neobuxbaumia* species (Cactaceae) with differing degree of rarity. *Popul. Ecol.* **2005**, *47*, 229–245. [CrossRef]

17. Söderström, L.; Séneca, A.; Santos, M. Rarity patterns in members of the Lophoziaceae/Scapaniaceae complex occurring North of the Tropics—Implications for conservation. *Biol. Conserv.* **2007**, *135*, 352–359. [CrossRef]

18. Espeland, E.K.; Emam, T.M. The value of structuring rarity: The seven types and links to reproductive ecology. *Biodivers. Conserv.* **2011**, *20*, 963–985. [CrossRef]

19. Harnik, P.G.; Simpson, C.; Payne, J.L. Long-term differences in extinction risk among the seven forms of rarity. *Proc. R. Soc. Lond B Biol. Sci.* **2012**, *279*, 4969–4976. [CrossRef] [PubMed]

20. Anacker, B.L.; Gogol-Prokurat, M.; Leidholm, K.; Schoenig, S. Climate change vulnerability assessment of rare plants in California. *Madroño* **2013**, *60*, 193–211. [CrossRef]

21. Bennett, J.; Vellend, M.; Lilley, P.; Cornwell, W.; Arcese, P. Abundance, rarity and invasion debt among exotic species in a patchy ecosystem. *Biol. Invasions* **2013**, *15*, 707–716. [CrossRef]

22. Toledo, L.F.; Becker, C.G.; Haddad, C.F.B.; Zamudio, K.R. Rarity as indicator of endangerment in neotropical frogs. *Biol. Conserv.* **2014**, *179*, 54–62. [CrossRef]

23. Andelman, S.J.; Groves, C.; Regan, H.M. A review of protocols for selecting species at risk in the context of US Forest Service viability assessments. *Acta Oecol.* **2004**, *26*, 75–83. [CrossRef]

24. Hanski, I. Dynamics of regional distribution: The core and satellite species hypothesis. *Oikos* **1982**, *38*, 210–221. [CrossRef]

25. Giam, X.; Olden, J.D. Drivers and interrelationships among multiple dimensions of rarity for freshwater fishes. *Ecography* **2018**, *41*, 331–344. [CrossRef]

26. Loza, M.I.; Jiménez, I.; Jorgenses, P.M.; Arellano, G.; Macía, M.J.; Torrez, V.W.; Ricklefs, R.E. Phylogenetic patterns of rarity in a regional species pool of tropical woody plants. *Glob. Ecol. Biogeogr.* **2017**, *26*, 1043–1054. [CrossRef]

27. Cofre, H.L.; Böhning-Gaese, K.; Marquet, P.A. Rarity in Chilean forest birds: Which ecological and life-history traits matter? *Divers. Distrib.* **2007**, *13*, 203–212. [CrossRef]

28. Uetz, P. (Ed.) The Reptile Database. Available online: http://www.reptile-database.org (accessed on 1 September 2015).

29. Campbell, J.A.; Lamar, W.W. *The Venomous Reptiles of the Western Hemisphere*; Cornell University Press: Ithaca, NY, USA, 2004; pp. 1–528.

30. Wüster, W.; Peppin, L.; Pook, C.E.; Walker, D.E. A nesting of vipers: Phylogeny and historical biogeography of the Viperidae (Squamata: Serpentes). *Mol. Phylogenetics Evol.* **2008**, *49*, 445–459. [CrossRef]

31. Alencar, L.R.V.; Quental, T.B.; Grazziotin, F.G.; Alfaro, M.L.; Martins, M.; Venzon, M.; Zaher, H. Diversification in vipers: Phylogenetic relationships, time of divergence and shifts in speciation rates. *Mol. Phylogenetics Evol.* **2016**, *105*, 50–62. [CrossRef]

32. Alencar, L.R.V.; Martins, M.; Greene, H.W. Evolutionary History of Vipers. In *Encyclopedia of Life Sciences*; John Wiley & Sons Ltd.: Chichester, UK, 2018.

33. Castoe, T.A.; Parkinson, C.L. Bayesian mixed models and the phylogeny of pitvipers (Viperidae: Serpentes). *Mol. Phylogenetics Evol.* **2006**, *39*, 91–110. [CrossRef]

34. Swaroop, S.; Grab, B. Snakebite mortality in the world. *Bull. World Health Organ.* **1954**, *10*, 35–76.
35. Chippaux, J.P. Snake-bites: Appraisal of the global situation. *Bull. World Health Organ.* **1998**, *75*, 515–524.
36. Galán, J.A.; Sánches, E.E.; Rodríguez-Acosta, A.; Pérez, J.C. Neutralization of venoms from two Southern Pacific Rattlesnakes (*Crotalus helleri*) with commercial antivenoms and endothermic animal sera. *Toxicon* **2004**, *43*, 791–799. [CrossRef]
37. McCue, M.D. Cost of producing venom in three North American Pitvipers Species. *Copeia* **2006**, *2006*, 818–825. [CrossRef]
38. Gibbs, H.L.; Mackessy, S.P. Functional basis of a molecular adaptation: Prey-specific toxic effects of venom from Sistrurus rattlesnakes. *Toxicon* **2009**, *53*, 672–679. [CrossRef]
39. Klauber, L.M. *Rattlesnakes: Their Habits, Life Histories, and Influence on Mankind*, 2nd ed.; University of California Press: Berkeley, CA, USA, 1972; pp. 1–1533.
40. Martins, M.; Marques, M.; Marques, O.; Sazima, I. Ecological and phylogenetic correlates of feeding habits in neotropical pitvipers of the genus *Bothrops*. In *Biology of the Vipers*; Schuett, G.W., Höggren, M., Douglas, M.E., Greene, H.W., Eds.; Eagle Mountain Publishing: Eagle Mountain, UT, USA, 2002; pp. 307–328.
41. Böhm, M.; Collen, B.; Baillie, J.E.M.; Bowles, P.; Chanson, J.; Cox, N.; Hammerson, G.; Hoffmann, M.; Livingstone, S.R.; Rama, M.; et al. The conservation status of the world's reptiles. *Biol. Conserv.* **2013**, *157*, 372–385. [CrossRef]
42. IUCN. The IUCN Red List of Threatened Species. Version 2017-2. Available online: https://www.iucnredlist.org/ (accessed on 1 July 2017).
43. Maritz, B.; Penner, J.; Martins, M.; Crnobrnja-Isailović, J.; Spear, S.; Alencar, L.R.V.; Sigala-Rodriguez, J.; Messenger, K.; Clark, R.; Jenkins, C.; et al. Identifying global priorities for the conservation of vipers. *Biol. Conserv.* **2016**, *204*, 94–102. [CrossRef]
44. Grismer, L.L. An evolutionary classification of reptiles on islands in the Gulf of California, Mexico. *Herpetologica* **1999**, *55*, 446–469.
45. Constable, H.; Guralnick, R.; Wieczorek, J.; Spencer, C.; Peterson, A.T. VertNet Steering Committee. VertNet: A New Model for Biodiversity Data Sharing. *PLoS Biol.* **2010**, *8*, e1000309. [CrossRef]
46. Nogueira, C.C. Atlas of Brazilian Snakes. Manuscript in preparation.
47. IUCN. *IUCN Red List Categories and Criteria: Version 3.1*, 2nd ed.; IUCN Species Survival Commission: Gland, Switzerland; IUCN: Cambridge, UK, 2012; pp. 1–32.
48. QGIS Development Team. QGIS Geographic Information System. Version 2.8. Open Source Geospatial Foundation Project. Available online: http://www.qgis.org/ (accessed on 1 March 2016).
49. Crisp, D.J.; Southward, A.J. The distribution of intertidal organisms along the coasts of the English Channel. *J. Mar. Biol. Assoc. UK* **1958**, *37*, 157–203. [CrossRef]
50. Dobson, A.J.; Barnett, A. *An Introduction to Generalized Linear Models*; Chapman and Hall/CRC Press: Boca Raton, FL, USA, 2008; pp. 1–320.
51. Agresti, A.; Kateri, M. Categorical data analysis. In *International Encyclopedia of Statistical Science: 206–208*; Lovric, M., Ed.; Springer: Berlin, Germany, 2011.
52. R Core Team. *R: A Language and Environment for Statistical Computing*; R Foundation for Statistical Computing: Vienna, Austria, 2016; pp. 206–208.
53. Bonner, J.T. *Why Size Matters: From Bacteria to Blue Whales*; Princeton University Press: Princeton, NJ, USA, 2011; pp. 1–176.
54. White, E.P.; Ernest, S.M.; Kerkhoff, A.J.; Enquist, B.J. Relationships between body size and abundance in ecology. *Trends Ecol. Evol.* **2007**, *22*, 323–330. [CrossRef]
55. Whitaker, P.B.; Shine, R. A radiotelemetric study of movements and shelter-site selection by free-randing brownsnakes (*Pseudonaja textilis*, Elapidae). *Herpetol. Monogr.* **2003**, *17*, 130–144. [CrossRef]
56. Roth, E.D. Spatial ecology of a cottonmouth (*Agkistrodon piscivorus*) population in East Texas. *J. Herpetol.* **2005**, *39*, 308–313. [CrossRef]
57. Reed, R.N. Interspecific patterns of species richness, geographic range size, and body size among New World venomous snakes. *Ecography* **2003**, *26*, 107–117. [CrossRef]
58. Greene, H.W. *Snakes: The Evolution of Mystery in Nature*; University of California Press: Berkeley, CA, USA, 1997; pp. 1–366.
59. Feldman, A.; Sabath, N.; Pyron, A.R.; Mayrose, I.; Meiri, S. Body sizes and diversification rates of lizards, snakes, amphisbaenians and the tuatara. *Glob. Ecol. Biogeogr.* **2016**, *25*, 187–197. [CrossRef]

60. Rapoport, E.H. *Areography: Geographical Strategies of Species*; Pergamon Press: Oxford, UK, 1982; pp. 1–286.
61. Stevens, G.C. The latitudinal gradient in geographical range: How so many species coexist in the tropics. *Am. Nat.* **1989**, *133*, 240–256. [CrossRef]
62. Meliadou, A.; Troumbis, A.Y. Aspects of heterogeneity in the distribution of diversity of the European herpetofauna. *Acta Oecol.* **1997**, *18*, 393–412. [CrossRef]
63. Arita, H.T.; Rodríguez, P.; Vázquez-Domínguez, E. Continental and regional ranges of North American mammals: Rapoport's rule in real and null worlds. *J. Biogeogr.* **2005**, *32*, 961–971. [CrossRef]
64. Whitton, F.J.S.; Purvis, A.; Orme, C.D.L.; Olalla-Tarraga, M.A. Understanding global patterns in amphibian geographic range size: Does Rapoport rule? *Glob. Ecol. Biogeogr.* **2012**, *21*, 179–190. [CrossRef]
65. Dobzhansky, T. Evolution in the tropics. *Am. Sci.* **1950**, *38*, 209–221.
66. Pagel, M.D.; May, R.M.; Collie, A.R. Ecological aspects of the geographical distribution and diversity of mammalian species. *Am. Nat.* **1991**, *137*, 791–815. [CrossRef]
67. Eeley, H.A.C.; Foley, R.A. Species richness, species range size and ecological specialisation among African primates: Geographical patterns and conservation implications. *Biodivers. Conserv.* **1999**, *8*, 1033–1056. [CrossRef]
68. Johnson, C.N. Rarity in the tropics: Latitudinal gradients in the distribution and abundance in Australian mammals. *J. Anim. Ecol.* **1998**, *67*, 689–698. [CrossRef]
69. Symonds, M.R.R.; Christidis, L.; Johnson, C.N. Latitudinal gradients in abundance, and the causes of rarity in the tropics: A test using Australian honeyeaters (Aves: Meliphagidae). *Oecologia* **2006**, *149*, 406–417. [CrossRef]
70. Hadfield, J.D. MCMC Methods for Multi-Response Generalized Linear Mixed Models: The MCMCglmm R Package. *J. Stat. Softw.* **2010**, *33*, 1–22. [CrossRef]
71. Verde Arregoitia, L.D.; Leach, K.; Reid, N.; Fisher, D.O. Diversity, extinction, and threat status in Lagomorphs. *Ecography* **2015**, *38*, 1155–1165. [CrossRef]
72. Heidelberger, P.; Welch, P.D. Simulation run length control in the presence of an initial transient. *Oper. Res.* **1983**, *31*, 1109–1144. [CrossRef]
73. WCS/CIESIN (Wildlife Conservation Society/Center for International Earth Science Information Network, Columbia University). *Last of the Wild Project, Version 2, 2005 (LWP-2): Global Human Influence Index (HII) Dataset (Geographic)*; NASA Socioeconomic Data and Applications Center (SEDAC): Palisades, NY, USA, 2005. Available online: http://sedac.ciesin.columbia.edu/data/set/wildareas-v2-human-footprint-geographic (accessed on 2 February 2014).
74. Médail, F.; Verlaque, R. Ecological characteristics and rarity of endemic plants from southeast France and Corsica: Implication for biodiversity conservation. *Biol. Conserv.* **1997**, *80*, 269–281. [CrossRef]
75. Wilson, L.D.; McCranie, J.R. The conservation status of the herpetofauna of Honduras. *Amphib. Reptil. Conserv.* **2004**, *3*, 6–33.
76. Johnson, J.D.; Mata-Silva, V.; Wilson, L.D. A conservation reassessment of the Central American herpetofauna based on the EVS measure. *Amphib. Reptil. Conserv.* **2015**, *9*, 1–94.
77. Wilson, L.D.; Mata-Silva, V.; Johnson, J.D. A conservation reassessment of the reptiles of Mexico based on the EVS measure. *Amphib. Reptil. Conserv.* **2013**, *7*, 1–47.
78. SEMARNAT (Secretaría de Medio Ambiente y Recursos Naturales). Norma Oficial Mexicana NOM-059-SEMARNAT-2010-Protección Ambiental-Especies Nativas de México de Flora y Fauna Silvestres-Categorías de Riesgo y Especificaciones Para su Inclusión, Exclusión o Cambio-Lista de Especies en Riesgo. *Diario Oficial de la Federación*, 30 December 2010; 1–78.
79. Foufopoulos, J.; Ives, A.R. Reptiles extinctions on Land-Bridge Islands: Life-history atributes and vulnerability to extinction. *Am. Nat.* **1999**, *153*, 1–25. [CrossRef]
80. Reed, R.N.; Shine, R. Lying in wait for extinction: Ecological correlates of conservation status among Australian Elapid Snakes. *Conserv. Biol.* **2002**, *16*, 451–461. [CrossRef]
81. Luiselli, L. Testing hypotheses on the ecological patterns of rarity using a novel model of study: Snake communities worldwide. *Ecology* **2006**, *6*, 44–58. [CrossRef]
82. Dunn, E.R. Relative abundance of some Panamanian snakes. *Ecology* **1949**, *30*, 39–57. [CrossRef]
83. Duellman, W.E. The Biology of An Equatorial Herpetofauna in Amazonian Ecuador. *Misc. Publ. Mus. Nat. Hist. Univ. Kansas* **1978**, *65*, 1–352.

84. Martins, M.; Oliveira, M.E. Natural history of snakes in forests of the Manaus region, Central Amazonia, Brazil. *Herpetol. Nat. Hist.* **1998**, *6*, 78–150.

85. Fitch, H.S. *A Kansas Snake Community: Composition and Changes Over 50 Years*; Krieger Publishing Company: Malabar, FL, USA, 1999; pp. 1–165.

86. Ford, B.N.; Lancaster, D.L. The species-abundance distribution of snakes in a bottomland hardwood forest of the Southern United States. *J. Herpetol.* **2007**, *41*, 385–394. [CrossRef]

87. Sawaya, R.J.; Marques, O.A.V.; Martins, M. Composition and natural history of a Cerrado snake assemblage at Itirapina, São Paulo state, southeastern Brazil. *Biota Neotrop.* **2008**, *8*, 129–151. [CrossRef]

88. Böhm, M.; Kemp, R.; Williams, R.; Davidson, A.D.; Garcia, A.; McMillan, K.M.; Bramhall, H.R.; Collen, B. Rapoport's rule and determinants of species range size in snakes. *Divers. Distrib.* **2017**, *23*, 1472–1481. [CrossRef]

89. Slatyer, R.A.; Hirst, M.; Sexton, J.P. Niche breadth predicts geographical range size: A general ecological pattern. *Ecol. Lett.* **2013**, *16*, 1104–1114. [CrossRef]

90. Laube, I.; Korntheuer, H.; Schwager, M.; Trautmann, S.; Rahbek, C.; Böhning-Gaese, K. Towards a more mechanistic understanding of traits and range sizes. *Glob. Ecol. Biogeogr.* **2013**, *22*, 233–241. [CrossRef]

91. Estrada, A.; Meireles, C.; Morales-Castilla, I.; Poschlod, P.; Vieites, D.; Araújo, M.B.; Early, R. Species' intrinsic traits inform their range limitations and vulnerability under environmental change. *Glob. Ecol. Biogeogr.* **2015**, *24*, 849–858. [CrossRef]

92. Holt, R.D. Rarity and evolution: Some theoretical considerations. In *The Biology of Rarity*; Kunin, W.E., Gaston, K.J., Eds.; Population and Community Biology Series; Springer: Dordrecht, The Netherlands, 1997; Volume 17, pp. 209–234.

93. Harcourt, A.H. Rarity in the tropics: Biogeography and macroecology of the primates. *J. Biogeogr.* **2006**, *33*, 2077–2087. [CrossRef]

94. Hawkins, B.A.; Diniz-Filho, J.A.F. 'Latitude' and geographical patterns in species richness. *Ecography* **2004**, *27*, 268–272. [CrossRef]

95. Letcher, A.J.; Harvey, P.H. Variation in Geographical Range Size Among Mammals of the Palearctic. *Am. Nat.* **1994**, *144*, 30–42. [CrossRef]

96. Collen, B.; Dulvy, N.K.; Gaston, K.J.; Gärdenfors, U.; Keith, D.A.; Punt, A.E.; Regan, H.M.; Böhm, M.; Hedges, S.; Seddon, M.; et al. Clarifying misconceptions of extinction risk assessment with the IUCN Red List. *Biol. Lett.* **2016**, *12*, 20150843. [CrossRef]

97. Martins, M.; (University of São Paulo, São Paulo, São Paulo, Brazil). Personal communication, 2018.

Article

Biogeography, Systematics, and Ecomorphology of Pacific Island Anoles

John G. Phillips [1,2,*], **Sarah E. Burton** [1,3], **Margarita M. Womack** [4], **Evan Pulver** [1,5] and **Kirsten E. Nicholson** [1]

[1] Department of Biology, Central Michigan University, 2100 Biosciences, Mount Pleasant, MI 48859, USA
[2] Department of Biological Sciences, University of Idaho, 875 Perimeter Dr., Moscow, ID 83844, USA
[3] Department of Fisheries and Wildlife, Michigan State University, 480 Wilson Rd., East Lansing, MI 48824, USA
[4] 3612 Woodley Rd NW, District of Columbia, Washington, DC 20016, USA
[5] 3759 Farm to Market Rd. 1488 #175, The Woodlands, TX 77382, USA
[*] Correspondence: jphillips@uidaho.edu

Received: 1 July 2019; Accepted: 15 August 2019; Published: 21 August 2019

Abstract: Anoles are regarded as important models for understanding dynamic processes in ecology and evolution. Most work on this group has focused on species in the Caribbean Sea, and recently in mainland South and Central America. However, the Eastern Tropical Pacific (ETP) is home to seven species of anoles from three unique islands (Islas Cocos, Gorgona, and Malpelo) that have been largely overlooked. Four of these species are endemic to single islands (*Norops townsendi* on Isla Cocos, *Dactyloa agassizi* on Isla Malpelo, *D. gorgonae* and *N. medemi* on Isla Gorgona). Herein, we present a phylogenetic analysis of anoles from these islands in light of the greater anole phylogeny to estimate the timing of divergence from mainland lineages for each species. We find that two species of solitary anoles (*D. agassizi* and *N. townsendi*) diverged from mainland ancestors prior to the emergence of their respective islands. We also present population-wide morphological data suggesting that both display sexual size dimorphism, similar to single-island endemics in the Caribbean. All lineages on Isla Gorgona likely arose during past connections with South America, and ecologically partition their habitat. Finally, we highlight the importance of conservation of these species and island fauna in general.

Keywords: Dactyloidae; ecomorphology; Iguania; Isla Cocos; Isla Gorgona; Isla Malpelo; island biogeography; lizards; neotropics; overwater dispersal

---

1. Introduction

Island ecosystems often display highly diverse communities, distinct from their continental counterparts. This has led to a rich body of work examining communities on islands to understand their ecological dynamics and the processes that promoted their formation [1–3]. As a result, islands have served as model systems for examining evolutionary processes like community assembly and speciation. Particularly in taxonomic groups with reduced lability, an island–mainland divide can promote allopatric speciation, leading to island endemism [4–6]. Many archipelagos that have featured prominently in the annals of evolutionary biology are of volcanic origin (e.g., the Canary, Galápagos, and Hawaiian Islands), where the nature of their formation necessitates any inhabitants to have almost certainly arrived via overwater dispersal [7], but the process of community assembly on continental islands can be different. While dispersal is a viable explanation for communities on continental islands, any historical connection to the mainland yields the possibility of a vicariant origin for those species. In some cases, support for both options has been presented for a variety of taxa inhabiting the same island or archipelago, with disagreement among biologists [8] and sources within. These studies on island biogeography have produced important, although sometimes controversial ideas [9,10] and additional studies may continue to influence our current understanding of island communities.

One group of organisms that has been influential in our understanding of island biogeography and community assembly therein, are anole lizards (family Dactyloidae). Anoles are highly diverse with 429 species, [11,12] and display a variety of morphological and ecological differences. They inhabit a wide breadth of microhabitats from forest floor to canopy, exhibit a concomitant array of morphological features adapted to exploiting those niches, and display equally diverse behaviors and ecological characteristics [13]. Studies of Caribbean anoles have figured prominently in the literature with respect to novel ecological and evolutionary patterns [14–16] and have played a role in testing hypotheses of island colonization [17,18]. Caribbean anole research has supported overwater dispersal as the generally accepted paradigm for the biogeographic origin Caribbean anoles [19,20]. However, this line of inquiry has been limited to the Caribbean islands, and studies of anoles in Pacific Islands have been lacking.

The paucity of work on Pacific anoles may be attributable to a lack of taxa and islands as only a combined seven species of anoles (Figure 1) are known from the Eastern Tropical Pacific (ETP) across three islands (Islas Cocos, Gorgona, and Malpelo; Figures 2 and 3). The islands are variable in ecology and geology, which may impact the evolution of their biotic communities. Anoles occupying these islands span the extreme ends of the dactyloid phylogeny [20,21], and are highly variable in ecology and natural history (reviewed below). Genetic analyses of these species may shed light on the evolutionary patterns and processes in the Pacific Islands, as well as provide a framework for the comparison of evolutionary patterns to the Caribbean anoles to better understand island evolution. Among the Pacific islands inhabited by anoles, only two contain a single species of anole or "solitary anoles" (*sensu* Schoener [22]): Isla Malpelo, Colombia is occupied by *Dactyloa agassizi,* and Isla Cocos, Costa Rica is inhabited by *Norops townsendi.* The third Pacific anole island, Isla Gorgona off Colombia, is occupied by five species: *D. chocorum, D. gorgonae, D. princeps, N. medemi,* and *N. parvauritus* (Figure 1). Below we briefly review each island in terms of geologic origin and known ecomorphology of their anole communities.

Figure 1. Seven species of anoles found on the islands in the Eastern Tropical Pacific (ETP). Top row, left to right: *Dactyloa agassizi* (Isla Malpelo), *Dactyloa gorgonae* (Isla Gorgona), *Norops townsendi* (Isla Cocos). Bottom row: *Norops medemi, Norops parvauritus, Dactyloa princeps, Dactyloa chocorum* (all Isla Gorgona).

Figure 2. Three islands that are home to the anoles in the ETP. (**a**) Isla Malpelo, (**b**) Isla Cocos, (**c**) Isla Gorgona.

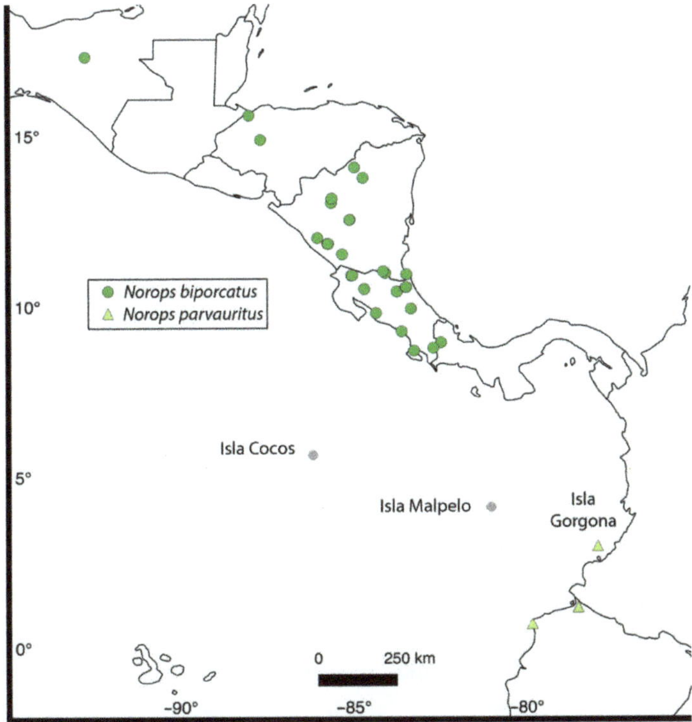

Figure 3. Map of the Eastern Tropical Pacific (ETP) with the three islands in this study. Mainland sampling localities for *Norops biporcatus* and *N. parvauritus* included in the molecular phylogeny are also shown.

While there is debate within the anole community as to the generic level nomenclature of anoles, herein we opt for the eight-genus model proposed by Nicholson et al. [11,18], highlighting the evolutionary distinctness among anoles. Our study illustrates the utility of treating *Norops* and *Dactyloa* as the distinct non-sister genera separated by as much as 31–87 million years ([11,23]; although Román-Palacios et al. [24] performed more rigorous comparative analyses on the dating methods suggesting this basal divergence likely occurred 51–65 Mya). If considered the same genus, a reader who is unfamiliar with the intricacies of anole taxonomy may erroneously assume that solitary anoles *D. agassizi* and *N. townsendi* are comparable in morphology and ecology, which is a misleading assumption. Each species is more closely related to separate clades of solitary Caribbean anoles which make for more appropriate comparisons of ecomorphological evolution on islands (*D. extrema, D. luciae,* and *D. roquet* for *D. agassizi* and *N. concolor, N. lineatus,* and *N. pinchoti* for *N. townsendi;* [25]).

1.1. Isla Malpelo

Isla Malpelo (3.5 km^2, Figure 2a) is a remnant volcanic plug (hardened magma within the vent of a volcano) and is estimated to be 15–17 million years old [26], representing an older stage in the life of a volcanic island. A younger Malpelo was considerably larger and most likely vegetated [27], but now the floral communities of Malpelo's steep slopes are sparse, which likely contributes to a species-poor ecosystem [28–31]. No terrestrial vertebrates are found on Malpelo except for three endemic species of lizards: an anole (*Dactyloa agassizi*), a galliwasp (*Diploglossus millepunctatus*), and a gecko (*Phyllodactylus transversalis*). Without sufficient primary producers to anchor the trophic pyramid of Isla Malpelo's ecosystem, all three lizards are dependent upon the colonies of a unique keystone species, the Nazca Booby (*Sula granti*), which seasonally nests on the island [32]. Lizards and crabs (*Geocarcinus malpilensis*) compete inter- and intra-specifically for booby fecal matter, unprotected eggs, and dropped or regurgitated fish. *Dactyloa agassizi* is the most abundant lizard species on the island with an estimated population of 140,000–206,000 [33]. Its natural history and behavior are unique among anoles, in that individual males do not defend territories or display aggression toward one other [34], (pers. obs. KEN and MMW). This may be due to a lack of importance of territory, given the stochastic nature of nutrient availability and the lack of habitat heterogeneity. They are also very curious and unafraid of humans, readily approaching and investigating them, presumably in search of food [34], (pers. obs. KEN and MMW). *Dactyloa agassizi* belongs to a clade of anoles (*Dactyloa*, sensu Nicholson et al. [21]) ranging from northern Costa Rica to central South America [35–37]. A recent study examined the divergence of *D. agassizi* from mainland *D. insignis*, finding this split to predate the origin of Isla Malpelo [24]. *Dactyloa insignis* is one of the northernmost-ranging *Dactyloa* species, so the mainland ancestor of *D. agassizi* may have originated in northern Costa Rica.

1.2. Isla Cocos

Isla Cocos (24 km^2, Figure 2b) also has a single endemic anole, *Norops townsendi*. The environment of Isla Cocos is dramatically different from Malpelo [38], (pers. obs. KEN and MMW), being a highly vegetated and lush tropical island with cloud forest around its highest peak. Similar to Isla Malpelo, Cocos is volcanic in origin, although its origin is estimated to be much younger at 1.9–2.4 Mya [39]. Isla Cocos likely represents how Malpelo might have appeared earlier in its geologic life, while Cocos will likely more closely resemble Malpelo over time. *Norops townsendi* is found throughout the island from sea level to its peak (Cerro Yglesias, 575 m) and is very abundant [38], (pers. obs. KEN and MMW). The species is most often observed on the trunks of the trees with no obvious preference for tree species on which to perch but has been observed very high up in the canopy, as well as occasionally capturing food items on the ground (pers. obs. KEN and MMW). These observations suggest *N. townsendi* to be an ecological generalist, although we did not collect sufficient data to specifically address the question of its trophic ecology. *Norops townsendi* was previously thought to be morphologically allied to *N. polylepis* from Costa Rica [40], but phylogenetic analyses have since suggested it is allied closer with *N. poecilopus*, a semi-aquatic anole from Panama [21,41–43], but see [20].

1.3. Isla Gorgona

Isla Gorgona (13.3 km^2, Figure 2c) is only 35 km from the coast of Colombia and, like Isla Cocos, is a small tropical island that is separated from the mainland by a deep marine channel (~270 m; [44]). The island falls entirely within Gorgona National Natural Park which covers ~600 km^2 [45]. It has a complex geologic history perhaps dating back to 90 Mya [46–49]. While Isla Gorgona formed separately from the mainland, there is evidence to suggest multiple connections and subsequent separations from the mainland [50]. However, the timing of the connections is unclear. Most recently, Gorgona was connected to mainland South America from 17,000–11,000 years ago during the Pleistocene glaciation [51,52], which may explain some of the biotic similarities between the island and mainland communities [53,54]. The dynamic geologic history sets the stage for different components of Isla Gorgona's biotic community to have originated via different modes: dispersal or vicariance. However, without more refined dates of the connection history we cannot test hypotheses regarding these alternatives, but we can provide interpretations of divergence dates estimated from our data.

Five anole species inhabit the island, representing a diverse range of morphological and ecological modalities as well as phylogenetic affinities (Figure 4). *Dactyloa gorgonae* and *Norops medemi* are endemic to Isla Gorgona, while *D. chocorum*, *D. princeps*, and *N. parvauritus* have populations in mainland South America in addition to the island. The two species of *Norops* (*N. parvauritus* and *N. medemi*) differ greatly from each other in ecomorphology in that *Norops medemi* is smaller and found low on the trunks of trees and is darkly colored and patterned, similar to the tree trunks on which they are found. On the other hand, one of the largest members of *Norops*, the brilliant green *N. parvauritus* has a robust body with slender limbs. This species often goes unobserved as it frequents the forest canopy and is ecologically similar to the *Dactyloa* species. This species was previously recognized as a subspecies of the widespread *N. biporcatus* (ranging from Mexico to central South America) until it was recently elevated to species status [55]. In addition to examining the timing of *N. parvauritus* on Isla Gorgona, we also used this opportunity to evaluate this proposed taxonomic change. Armstead et al. [55] had limited genetic sampling of both *N. biporcatus* (*N* = 5) and *N. parvauritus* (*N* = 3), with no individuals from Isla Gorgona, the type locality for *Norops biporcatus parvauritus*. Herein we address both deficiencies by presenting a broad genetic data across the range including from Isla Gorgona. While Armstead et al. (2017) also presented a morphological analysis to support the species elevation, past work in mainland anoles has shown that delineating species using morphology without range-wide genetic sampling can lead to the violation of a monophyletic species concept (e.g., *N. quaggulus*, [43,56]).

The other three species are in the genus *Dactyloa* (*D. chocorum*, *D. gorgonae*, and *D. princeps*) and are all highly arboreal, but differ in size, coloration, and perch preferences (pers. obs., this study). *Dactyloa gorgonae* is unique among all anoles in that it is bright blue in color, smaller, and slimmer than the other two *Dactyloa*, and believed to be most closely related to a mainland species (*D. chloris*), to which it is morphologically similar except in coloration. Locals on Isla Gorgona believe that *D. gorgonae* is only found within the small inhabited area on the island, where it is observed in the mostly open branches of trees, but that assertion is likely due to frequency of observations and insufficient sampling effort elsewhere on the island. The forest on the island is dense, and their blue coloration is quite cryptic within the foliage of the trees (pers. obs. KEN and MMW). *Dactyloa princeps* was observed frequently on the trunks of trees while *D. chocorum* was observed rarely (only once during our visit). Both *D. princeps* and *D. chocorum* have populations on the mainland, but no studies have examined the morphological differences among island populations.

Understanding the biogeographic history of these Pacific anoles is a crucial initial step in setting the stage for future ecological and evolutionary studies of these species. Investigating the timing and route(s) of colonization for each island, combined with knowledge of each island's likely environment at the time of colonization, will establish the historical context for the unique extant communities. We investigated the biogeography of these Pacific Island anoles using molecular data to address the timing of divergence from mainland ancestors and to evaluate the route of colonization of each island by each of the seven species. This information was used to compare divergence of island lineages to geologic models for each island and to investigate hypotheses of overwater dispersal to each island. Islas Cocos and Malpelo are both volcanic in origin and could only have been colonized via over water dispersal, but we sought to investigate the timing and route of potential pathways. We also investigated the timing of divergence from mainland populations for all five species on Isla Gorgona and compared them to evaluate if their colonization was temporally proximate.

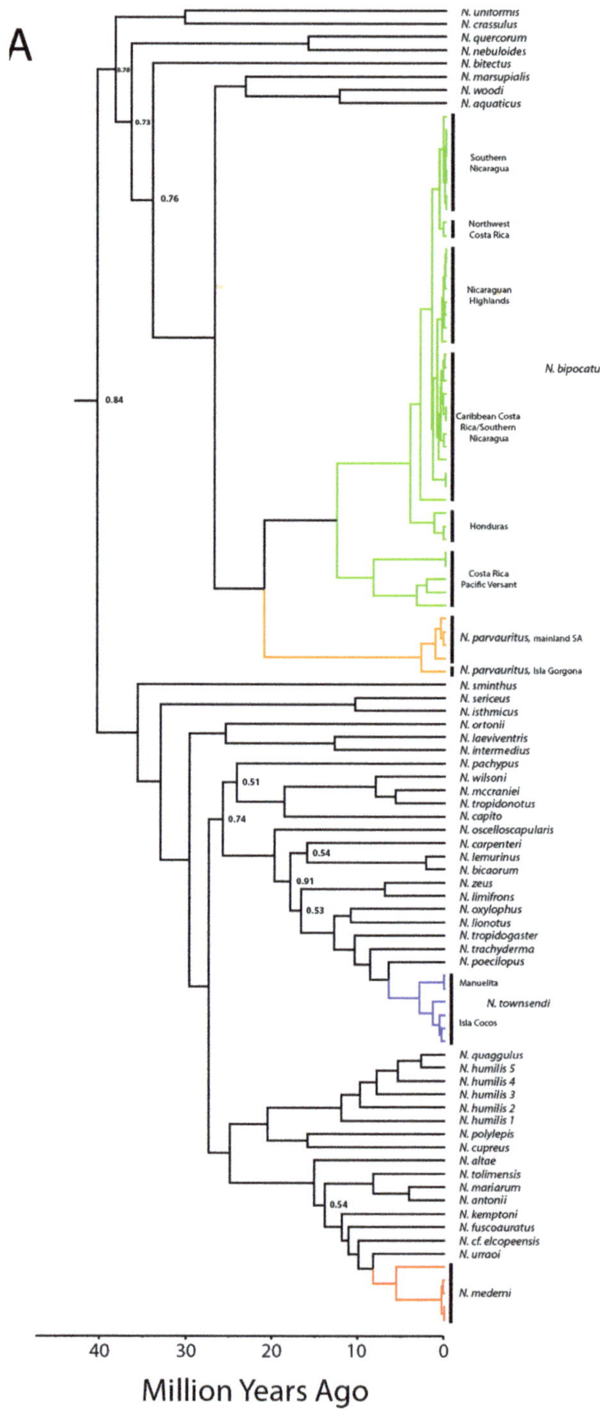

Figure 4. *Cont.*

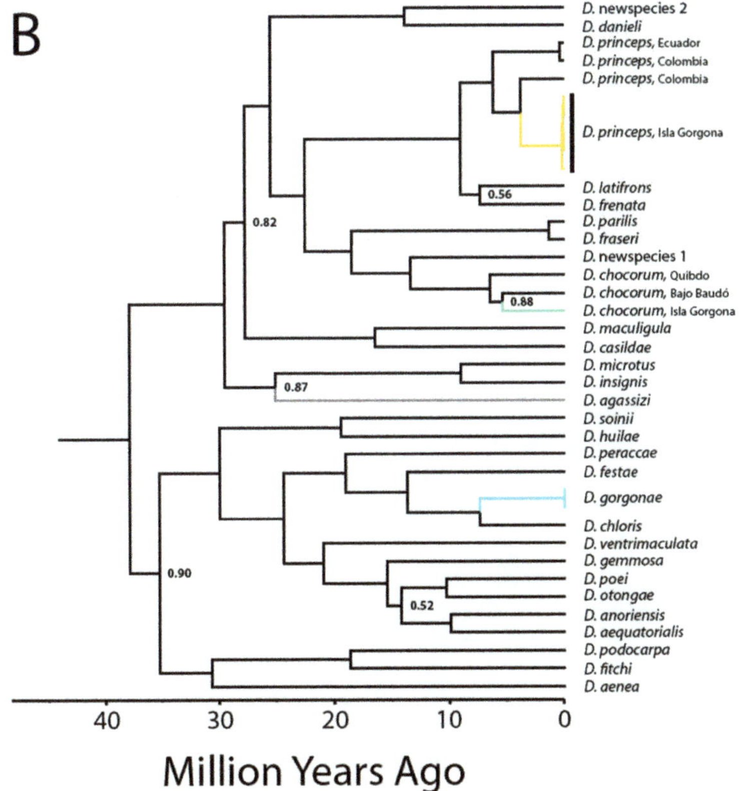

Figure 4. Bayesian reconstruction of *Dactyloa* and *Norops* anoles dated using a combination of fossil calibrations and a mtDNA mutation rate. All lineages of Pacific Island anoles are colored, and posterior values are present on all nodes with <0.95 probability. (**A**) *Norops* clades pertinent to this study. (**B**) *Dactyloa* clades pertinent to this study.

Finally, we collected observations and ecomorphological measurements for comparing to those recorded for Caribbean anoles to examine the relationship in ecomorphological patterns between solitary anoles. Sexual size dimorphism is evident in many solitary species of birds [57,58], mammals [59], and lizards [22,60,61]. Solitary anoles in the Caribbean typically display sexual size dimorphism, presumably to reduce intraspecific competition [22], a trend that likewise holds in mainland populations with species-poor anole communities [62]. Given that sexual size dimorphism has not been fully evaluated in Pacific species, we seek to determine if *D. agassizi* and *N. townsendi* also adhere to this pattern.

2. Methods

Each island was visited in July or August 2004, for 3–10 days. Because of permit restrictions, collection of samples for genetic analysis was restricted to only five individuals per species for all islands. In the case of *Dactyloa gorgonae*, collection was further limited to $N = 2$ by station officials on Isla Gorgona. Cryptic and arboreal niches allowed for collection of only one *D. chocorum* and one *N. parvauritus*, but *D. gorgonae*, *D. princeps*, and *N. medemi* were more common and useful for population-wide morphological data (see Supplementary Table S1 for specimen data). For both *N. townsendi* (Isla Cocos) and *D. agassizi* (Isla Malpelo) full populations were observed to assess habitat use and captured and measured when possible (see Morphology and Ecology section below).

Additionally, to evaluate the taxonomic status of *N. parvauritus*, samples of *N. biporcatus* were collected throughout Central America (Mexico through Panama, Figure 3, Supplementary Table S2). We complied with all applicable Animal Care guidelines (CMU-IACUC # 10–02).

2.1. DNA Sequencing and Phylogenetic Reconstruction

Tissue samples for each species were collected and preserved in 95% EtOH. Voucher specimens were fixed in formalin, stored in EtOH, and deposited with the Universidad de los Andes, Colombia. Qiagen DNEasy kits were used for DNA extraction. PCR amplification of several contiguous mitochondrial genes (ND2, tRNA^{Trp}, tRNA^{Ala}, tRNA^{Asn}, tRNA^{Cys}, tRNA^{Tyr}, and partial CO1) used previously in the literature for phylogenetic reconstruction of mainland anoles [42,43] was conducted. Genes were amplified using Empirical Bioscience Mean Green Master Mix and the same cycling conditions and primers as Nicholson et al. [21]. PCR products were purified using Qiagen QIAquick purification kits, followed by sequencing at Michigan State University's (MSU, East Lansing, MI, USA) sequencing core. Sequence data were edited and aligned using MUSCLE, then combined with published sequences for individuals of the same species from mainland Central or South America and closely related species (Supplementary Table S1). Our newly collected data were combined with published data for 75 *Dactyloa* and *Norops* species plus 10 non-Dactyloid outgroup taxa (Table 1 and Supplementary Table S1) for a total of 1480 aligned base pairs of sequence data (Genbank vouchers in Supplementary Table S1).

Table 1. Mean dates and 95% confidence intervals for select nodes relevant to Pacific Island anoles, using multiple dating methods (fossil-fossil calibrations + mutation rate; rate = mutation rate alone).

Node	Fossil	95%	Rate	95%
Dactyloa agassizi + *Dactyloa insignis/Dactyloa microtus*	29.2	21.6–37.2	33.4	22.4–36.4
Dactyloa chocorum from Isla Gorgona vs. mainland	6.3	3.8–9.9	6.7	4.1–9.1
All *Dactyloa chocorum* coalescence	7.6	4.8–11.2	7.3	5.3–10.5
Dactyloa princeps from Isla Gorgona vs. mainland	4.5	2.7–7.1	3.0	2.8–6.5
All *Dactyloa princeps* coalescence	7.3	5.1–10.5	7.4	5.2–9.4
Dactyloa chloris + *Dactyloa gorgonae*	8.6	4.9–13.2	12.8	5.3–12.2
Norops medemi coalescence (within Isla Gorgona)	6.6	5.2–9.6	6.1	5.0–8.0
Norops medemi + *Norops urraoi*	9.6	10.2–16.3	8.1	7.7–11.8
Norops townsendi coalescence (Isla Cocos vs. Manuelita)	3.5	2.2–4.7	3.1	2.4–4.9
Norops poecilopus + *Norops townsendi*	7.6	5.1–10.1	6.7	5.6–10.0
Norops parvauritus coalescence	3.2	2.4–6.6	3.8	2.0–4.8
Norops biporcatus coalescence	14.7	11.6–20.7	12.5	11.4–18.3
Norops biporcatus + *Norops parvauritus*	24.4	20.2–32.7	21.3	19.4–29.7

Phylogenetic analyses were conducted to investigate the placement of taxa among established species of anoles for which comparative data were available. Phylogenetic reconstructions were conducted using BEAST2 for Bayesian analyses under a lognormal relaxed clock model and a Yule Speciation prior [63]. PARTITIONFINDER was used to select models of evolution as well as to examine the suitability of partitioning. In all cases each gene (including tRNAs) was entered as a potential partition for the first, second, and third codon positions. Branch lengths were unlinked, all models of evolution available in BEAST2 were tested, and a BIC information criterion and greedy algorithm were used. The PARTITIONFINDER analysis recommended a GTR + Γ model of evolution for the mtDNA segment without partitioning. The same alignment for each dating method was run for 20 million generations with 20% burn in removed via TreeAnnotator [63] after evaluating that the analysis had reached stationarity using Tracer v.1.6.0 [64].

We estimated the date of origin for each island species by analyzing our data using BEAST2. We used two approaches to estimating the age of divergence between each island lineage and its mainland relatives: (1) using the Macey et al. [65] calibration rate of 0.65 mutations per 100 bp per million years, used by several mtDNA studies of anoles and related taxa [66–71] and (2) implementing the same mutation rate in addition to three calibration dates for Iguanian fossils used by Prates et al. [37]

and sources within. We used this approach following Román-Palacios et al.'s [24] findings that implementing multiple dating strategies can avoid the pitfalls of trees based only on a single mutation rate [72].

From our phylogenetic reconstruction we compared divergence dates between the island and mainland lineages to the estimated timing of origin for each island; our null hypothesis was that each species' origin would post-date the emergence of each island but had no a priori expectations or hypotheses regarding when each species may have originated. For some species (*N. medemi, N. townsendi*) for which we had multiple samples within an island, we were also able to estimate the coalescence of species on each island, making the assumption that this date represents the youngest possible age of colonization.

2.2. Morphology and Ecology

On each island, observations were made for all species during diurnal and nocturnal survey periods (except *N. agassizi*, as we were not allowed on Isla Malpelo at night). Individuals were captured with collapsible fishing poles fitted with string lassos in order to record population-wide morphological data. Measurements were taken on all captured individuals (SVL, mass, fore- and hind-limb length, head width, depth and length, number of lamellae rows) as well as recording ecological data (perch height, perch diameter, perch type; Supplementary Tables S3–S5).

Student's t-test was used to compare differences between the sexes for each species (although fore- and hind-limb length, head width, depth and length, were all determined to covary with SVL and were omitted from the sexual dimorphism analysis). Additionally, ecological and morphological data for *N. townsendi* were collected at four different locations on the island, including the highest point (Cerro Yglesias) and a small islet (Manuelita) that is disconnected from Isla Cocos. One-way ANOVA assuming unequal variance was implemented to test for differences among sites, which may suggest separate populations. This was followed by a *t*-test for any variables that displayed a significant difference ($p < 0.05$) to evaluate pairwise relationships between populations. To account for multiple tests in all cases, we applied a sequential Bonferroni correction. The same process was used to evaluate the differences in habitat use (perch height, perch diameter) among three of the Isla Gorgona species (*D. gorgonae, D. princeps, N. medemi*).

3. Results

The results of our phylogenetic reconstruction are depicted in Figure 4. All island taxa were reconstructed within their previously assigned genera (*Dactyloa* and *Norops*), and all relationships described herein have high support (posterior probability > 0.88). For Isla Gorgona, *D. chocorum* and *D. princeps* were both nested within clades of mainland conspecifics. The *D. princeps* clade was recovered to be sister to *D. frenatus* and *D. chocorum* sister to *D. fraseri* in all analyses. As previously hypothesized, *D. gorgonae* was placed sister to *D. chloris*. Within *Norops*, the island *N. parvauritus* was placed within a clade of mainland *N. parvauritus* from Ecuador, all of which is sister to *N. biporcatus*. *Norops townsendi* was recovered as sister to *N. poecilopus* and a clade of semiaquatic anoles that also include non-aquatic *N. trachyderma* and *N. tropidogaster*, and endemic *N. medemi* was placed sister to *N. urraoi* (a newly described species from the northwestern Andes) as suggested by Grisales-Martínez et al. [41], and nested within a clade of species from northern Colombia, Panama, and Costa Rica (*N. fuscoauratus, N. kemptoni*, and *N.* "newspecies," (believed to be *N.* cf. *elcopeensis* per [73]). There was also a deeper divergence within *N. medemi*, suggesting the species has been on Isla Gorgona at least for 6.1 Mya (5–8 Mya, 95% CI).

Node dating analyses were similar when using fossil dates + mutation rate versus the molecular calibration rate alone (Figure 4, Dactyloa chloris 1). These estimated dates of *D. agassizi* (as suggested in Román-Palacios et al. [24]) and *N. townsendi* predate the predicted origin for Islas Malpelo and Cocos respectively, while divergence dates for the Isla Gorgona taxa from their mainland counterparts are all younger than the island (all estimates within the past 10 My; Table 1, Figure 5).

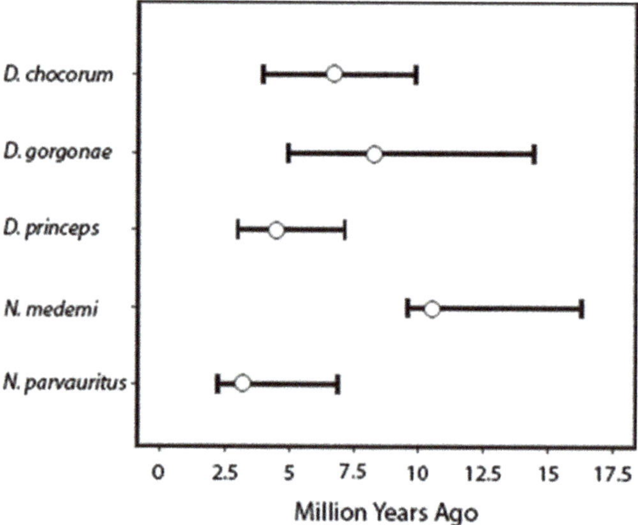

Figure 5. Range of divergence dates (95% CI) for each of the five anole lineages on Isla Gorgona from a mainland ancestor. Mean node age is represented by an open circle.

Morphological Data

Males were significantly larger than females in *D. agassizi*, *D. princeps*, and *N. townsendi* (Table 2). Most of the morphological variables (SVL, head, and limb measurements) are likely the products of having a larger body size and therefore cannot be characterized as independently evolving traits, all three species unquestionably display sexual size dimorphism, similar to solitary anoles in the Caribbean (Table 3). *Norops medemi* on the other hand displayed no significant differences between sexes with the exception of males having one more row of toe lamellae on average. However, sample sizes for this species were very small ($N = 5$ for each sex). Because of small sample sizes for *D. chocorum* and *N. parvauritus*, and only one female measured for *D. gorgonae*, those three species are omitted from further morphological analysis.

Table 2. Average morphological measurements by sex for four species of Pacific Island anoles. (SVL = snout-vent length, # lamelle = # of lamellar rows). *P*-values from pairwise t-tests between the sexes are listed below the means. Standard deviation is included for each mean value in parentheses. Significant values after Bonferroni correction are bolded and italicized.

Sex/Species	N	SVL (mm)	Mass (g)	# Lamelle	PH (m)	PD (m)	# Mites
Isla Malpelo							
Dactyloa agassizi, female	18	67.3 (± 5.9)	7.6 (± 2.2)	36.8 (± 1.7)	1.4 (± 1.2)	0.3 (± 0.3)	
Dactyloa agassizi, male	29	92.9 (± 9.0)	24.1 (± 6.9)	37.1 (± 1.7)	1.5 (± 1.1)	0.3 (± 0.4)	
p		***<0.0001***	***<0.0001***	0.53	0.76	0.88	
Isla Cocos							
Norops townsendi, female	16	46.6 (± 4.2)	2.1 (± 0.7)	18.5 (± 1.6)	0.9 (± 0.6)	0.07 (± 0.06)	5.2 (± 6.4)
Norops townsendi, male	61	50.5 (± 4.0)	2.6 (± 0.8)	18.5 (± 1.2)	1.4 (± 1.2)	0.09 (± 1.0)	7.9 (± 13.3)
p		***0.002***	0.05	0.83	0.12	0.53	0.52
Isla Gorgona							
Dactyloa princeps, female	10	102.1 (± 14.0)	21.1 (± 8.4)	22.9 (± 0.9)	2.4 (± 1.3)	17.3 (± 9.6)	
Dactyloa princeps, male	9	139.8 (± 7.9)	49.1 (± 7.3)	24.0 (± 1.1)	3.6 (± 1.3)	11.4 (± 4.9)	
p		***<0.0001***	***<0.0001***	0.04	0.07	0.12	
Norops medemi, female	5	51.0 (± 2.0)	2.3 (± 0.2)	15.0 (± 0.0)	3.1 (± 2.3)	5.2 (± 4.5)	
Norops medemi, male	5	48.8 (± 3.9)	1.9 (± 0.3)	15.8 (± 0.5)	2.0 (± 0.2)	5.9 (± 7.7)	
p		0.30	0.02	***0.01***	0.37	0.87	

Table 3. Average SVL by sex for solitary species of *Dactyloa* and *Norops*. Standard deviation is provided when available.

Species	Male SVL (mm)	Female SVL (mm)	F/M Ratio	Reference
Dactyloa agassizi	92.9 (± 9.0)	67.3 (± 5.9)	0.72	This study
Dactyloa extrema	65.9	52	0.79	[74]
Dactyloa luciae	72.6	52.1	0.72	[75]
Dactyloa roquet	67.3 (± 7.5)	52.1 (± 6.3)	0.77	[22]
Norops concolor	69.3 (± 0.7)	50.1 (± 0.3)	0.72	[76]
Norops lineatus	69.1	58.1	0.84	[77]
Norops pinchoti	47.7 (± 0.1)	40.2 (± 0.2)	0.84	[76]
Norops townsendi	50.5 (± 4.0)	46.6 (± 4.2)	0.92	This study

Comparative sample sizes from the Caribbean were small for solitary relatives to *D. agassizi* and *N. townsendi* without wide-scale population data available but serve as an exploratory indication of variation within and between species. Both sexes of *D. agassizi* are considerably larger than their Caribbean counterparts, although the magnitude of sexual size dimorphism is similar to *D. luciae* (and also the distantly related *N. concolor*). *Norops townsendi* displayed a lower magnitude of sexual size dimorphism than any other species of solitary *Norops*, although the difference was still significant (Tables 2 and 3).

The four populations of *N. townsendi* displayed significant differences among each other. Two of these populations (Chatham Trail and Playa) showed few differences except for Playa individuals being more massive with longer hindlimbs. Mass could be attributed to differences in prey availability between the sites, so we also re-analyzed our populations with these two sites combined to treat as a single population (hereafter referred to as the "main" population). In all morphological measurements, Cerro Yglesias individuals were significantly smaller than the main population, while Manuelita individuals were significantly larger with a greater number of toe lamellae (Table 4). The main population occupied perches >1 m higher than either the cerro or the islet, which may provide an explanation for the divergence in morphological evolution among the populations.

Perch height, diameter, type, and position were difficult to evaluate objectively for *D. agassizi* because the surface of the island consists of bare rock with nothing to perch upon except occasional boulders and pebbles. We observed some individuals perching on boulders, but most were on the ground or on pebbles, rendering a perch height mean of 1.5 m and perch diameter that was not quantifiable (often individuals were on the ground); perch type could only be described as rock. Alternatively, perch height, diameter, type, and position were measurable for *N. townsendi* because Isla Cocos had a variety of available perch heights and types. *Norops townsendi* was observed throughout the island on everything from sand to branches well within the canopy, making averages or generalizations difficult. Canopy height extended well over 20 m (the upper limit of our measuring abilities) and animals were often observed as high, or within the canopy. The aforementioned lack of samples for two of the Isla Gorgona species, inhibits our ability to evaluate habitat use across the local anole community, especially since the low number (1) of *D. chocorum* and *N. parvauritus* observed is likely due to their preferred niche in the canopy. The forest of Isla Cocos is similar to non-climax lowland rainforest habitats of Central America in having a range of tree species of varying heights and diameters with substantial undercover shrubs providing a variety of perches and niches. However, we found that *D. gorgonae* occupies higher perches than either *D. princeps* or *N. medemi*, while *D. princeps* occupied significantly larger perches than *N. medemi* (Table 5), likely owing to a much larger body size and longer limbs (Table 2). This habitat partitioning aligns with Nicholson et al.'s [21] ecomode classification, where each species occupies a different ecomode: *D. gorgonae* = trunk-crown, *D. princeps* = crown giant, *N. medemi* = trunk ground (Figure 6). *Dactyloa chocorum* and *N. parvauritus* are also both considered crown giants, but *D. princeps* clearly utilizes habitat lower down in the forest.

Table 4. (**A**) Average morphological and ecological measurements for four populations of *Norops townsendi* on Isla Cocos. (**B**) Results for pairwise comparisons of all variables between each population. Significant differences following Bonferroni correction are bolded. (**C**) Results for pairwise comparisons of all variables between each population, while treating the Playa and Trail sites as a single population. Significant values after Bonferroni correction are bolded and italicized. (SVL = snout-vent length, Fore = forelimb length, Hind = hindlimb length, HW = head width, HH = head height, HL = head length, PH = perch height, PD = perch diameter).

Population	SVL (mm)	mass (g)	Fore (mm)	Hind (mm)	HW (mm)	HH (mm)	HL (mm)	# lamellae	PH (m)	PD (cm)	# mites
Cerro Yglesias	45.9 (± 3.0)	1.6 (± 0.4)	21.5 (± 1.4)	40.4 (± 3.5)	7.0 (± 0.5)	5.4 (± 0.6)	12.4 (± 1.2)	17.7 (± 1.2)	0.8 (± 0.1)	11.2 (± 9.9)	7.7 (± 8.9)
Playa by Station	48 (± 2.4)	2.4 (± 0.4)	22.9 (± 1.8)	40.1 (± 2.8)	7.6 (± 0.4)	5.7 (± 0.4)	13.3 (± 0.8)	18.4 (± 1.2)	1.8 (± 0.9)	8.3 (± 12.1)	N/A
Chatham Trail	48.9 (± 3.3)	2.1 (± 0.3)	22.8 (± 1.3)	40.8 (± 1.6)	7.5 (± 0.4)	5.6 (± 0.3)	13.2 (± 0.5)	18.2 (± 1.3)	1.7 (± 1.5)	8.2 (± 6.8)	10.4 (± 15.8)
Playa + Trail	*48.3 (± 2.8)*	*2.3 (± 0.4)*	*22.8 (± 1.6)*	*40.3 (± 2.4)*	*7.5 (± 0.4)*	*5.7 (± 0.4)*	*13.3 (± 0.7)*	*18.3 (± 1.2)*	*1.8 (± 1.2)*	*8.3 (± 9.9)*	*10.4 (± 15.8)*
Manuelita Islet	55.2 (± 3.2)	3.5 (± 0.7)	26.4 (± 2.3)	45.9 (± 3.9)	8.3 (± 0.6)	6.3 (± 0.4)	14.5 (± 1.0)	19.5 (± 0.9)	0.4 (± 0.1)	6.4 (± 6.6)	3.4 (± 6.5)
Population	**SVL**	**mass**	**Fore**	**Hind**	**HW**	**HH**	**HL**	**lamellae**	**PH**	**PD**	**mites**
Playa v Trail	0.32	**0.004**	0.10	**0.002**	0.23	0.61	0.05	0.69	0.34	0.17	N/A
Playa v Cerro	**0.003**	**0.0003**	**0.002**	0.30	**0.001**	0.06	**0.001**	0.21	0.03	0.69	N/A
Cerro v Trail	**0.003**	**0.00001**	0.02	0.31	0.01	0.19	0.04	0.15	**0.0002**	0.14	0.61
Playa v Islet	**< 0.00001**	**< 0.00001**	**< 0.00001**	**0.00001**	**0.0002**	**< 0.00001**	**0.0003**	**0.0006**	**0.0001**	0.20	N/A
Trail v Islet	**< 0.00001**	**< 0.00001**	**< 0.00001**	**< 0.00001**	**0.00001**	**< 0.00001**	**0.00002**	**0.004**	**< 0.00001**	0.77	0.09
Cerro v Islet	**< 0.00001**	**< 0.00001**	**< 0.00001**	**0.001**	**0.00001**	**0.00005**	**0.00003**	**0.0001**	**< 0.00001**	0.12	0.15
Population	**SVL**	**mass**	**Fore**	**Hind**	**HW**	**HH**	**HL**	**lamellae**	**PH**	**PD**	**mites**
Main v Cerro	0.02	**0.00001**	0.02	0.97	**0.0004**	0.07	**0.001**	0.14	**0.01**	0.36	0.60
Main v Islet	**< 0.00001**	**< 0.00001**	**< 0.00001**	**< 0.00001**	**< 0.00001**	**< 0.00001**	**< 0.00001**	**0.0004**	**< 0.00001**	0.46	0.08
Cerro v Islet	**< 0.00001**	**< 0.00001**	**< 0.00001**	**0.001**	**0.00001**	**0.00005**	**0.00003**	**0.0001**	**< 0.00001**	0.12	0.15

Table 5. Mean values and pairwise comparisons of habitat use in Isla Gorgona anoles. Bold values are statistically significant. * Denotes that only one individual was observed and is not believed to reflect typical habitat use.

Species	N	Perch Height (m)	Perch Diameter (cm)
Dactyloa chocorum	1 *	2.2	12.0
Dactyloa gorgonae	8	15.6 (± 7.3)	18.2 (± 7.1)
Dactyloa princeps	21	3.1 (± 1.5)	14.5 (± 7.9)
Norops medemi	30	3.4 (± 2.6)	7.9 (± 5.3)
Norops parvauritus	1 *	4.4	93.0
D. gorgonae vs. D. princeps		**< 0.00001**	0.26
D. gorgonae vs. N. medemi		**< 0.00001**	0.10
D. princeps vs. N. medemi		0.66	**0.04**

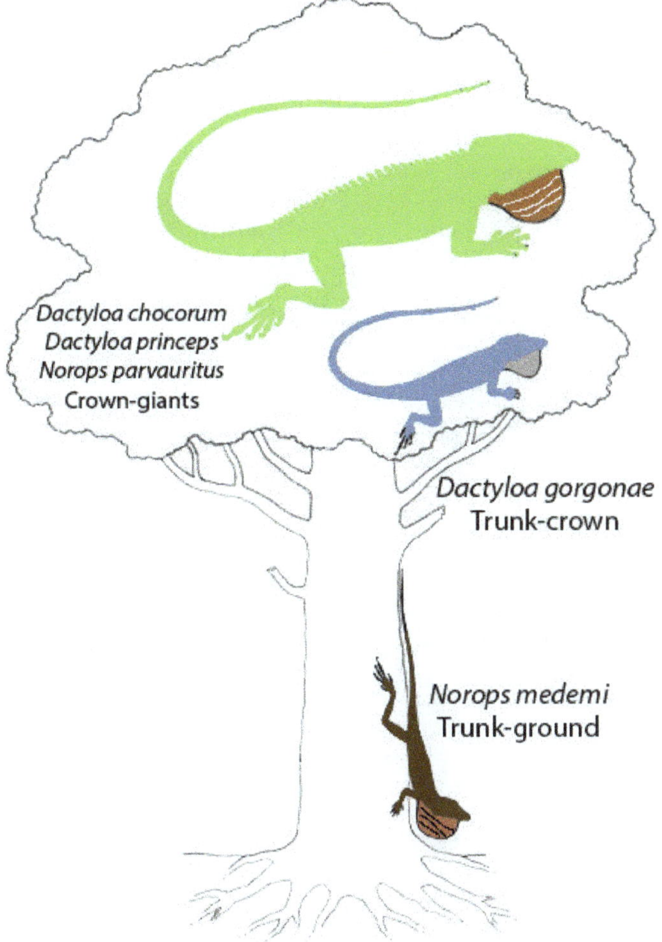

Figure 6. The Isla Gorgona anole community depicted by habitat in accordance with ecomode designation by Nicholson et al. [21]. Figure modified from [13], which was modified from [78].

4. Discussion

4.1. Biogeographic and Phylogenetic Interpretations

We estimated the dates of divergence from mainland populations for Pacific Island anoles from molecular data. The range of estimated dates is consistent with past findings: the ancestors of the included species have a long history on the mainland [20,21,24], are derived from clades with disparate physical and temporal origins. The genus *Dactyloa* represents one of the first splits in the anole tree and arose in the northern South America, while *Norops* (+*Ctenonotus*) is one of the last clades to diverge and occurred approximately 20 million years later [24] in the Caribbean, followed by dispersal to Central America before dispersing south into South America [21].

We interpret our data as follows, beginning with the multi-species island of Gorgona. Isla Gorgona has a complex geologic history leading to its aerial emergence potentially by 27 Mya [50]. The island is close to the mainland (although currently separated by a deep channel) and because of previous contact with the mainland, it could have been easily colonized early by overland dispersal prior to its separation from the mainland, or via overwater dispersal when the sea levels were lower, causing the landmasses to be more easily reachable. Three of the five Isla Gorgona species do not appear to differ morphologically from their mainland counterparts. However, denser sampling in mainland Colombia and Ecuador might be able to elucidate closer relationships than our present sampling allows (see *N. urraoi* example, [41]). *Dactyloa chocorum* occurs both on mainland South America and on Isla Gorgona and there have been no identified morphological distinctions recorded to date. *Dactyloa gorgonae* only differs from mainland *D. chloris* in being solid blue. *Dactyloa princeps* was always believed to be closely related to *D. frenata* with some investigators debating whether any features distinguish the two, unpubl. per [35,79,80], although our phylogenetic estimates support evolutionary distinctness. The lack of obvious morphological differences from mainland relatives may be due to low selection pressure, suggesting the environment on the island is strongly similar to the mainland environment from which it is derived, or to a recent arrival on the island that has yet to produce substantial change.

The two *Norops* species' colonization of Isla Gorgona are even more interesting considering the lack of solid hypotheses regarding the origin and biogeography of *Norops* as a whole. Early hypotheses suggested that South American *Norops* couldn't have invaded from Central America until after the closure of the Panamanian Portal which was believed to have occurred ~3.5 Mya [10]. However, the timing of this event is believed to have occurred multiple times with several estimates at 23–25 Mya ([81,82] with some controversy on this date reviewed in Elmer et al. [83]), 15 Mya [84], and 3.5 Mya [85,86]. More recent work has postulated that perhaps the *Norops* clade was widespread across Central America and South America before the continents were separated by the eastward movement of the proto-Greater Antillean land arc [21]. The estimated dates of divergence for *N. parvauritus* (3.2 Mya) and coalescence for *N. medemi* (6.6 Mya, Table 1) would lend support for the Nicholson et al. [21] biogeographic hypothesis, although more testing is required to investigate that hypothesis. Our sampling of Isla Gorgona was low because of the cryptic nature of some species combined with government permitting restrictions, so these dating estimates are tentative and should be tested in future with additional data. However, our findings of habitat partitioning among three of the Isla Gorgona species is of interest and warrants further study.

The biogeographic scenarios for the two solitary anoles (*D. agassizi*—Isla Malpelo; *N. townsendi*—Isla Cocos) are more difficult to interpret. In both cases, the estimated timing of divergence from mainland relatives (*D. agassizi*: 22.8 Mya; *N. townsendi*: 7.6 Mya) considerably predate the islands' estimated origin (Malpelo: 15–17 Mya; Cocos: 1.9–2.4 Mya). One explanation for both solitary anoles could be that each species originated on the mainland prior to dispersal to each island, followed by subsequent extinction of the species on the mainland, a conclusion also reached by Román-Palacios et al. [24]. The route of dispersal to the islands from the mainland appears to be fairly straightforward from the standpoint of currents, as currents move virtually straight to each island from the mainland at regular intervals throughout the year [29,50], making over-water dispersal plausible. For many over-water dispersal scenarios it may seem improbable for anoles to disperse over great expanses of water, but other species made the same trip and colonized the same islands, and similar dispersal events have been observed and hypothesized for anoles and other lizard fauna in the Caribbean [19,67,87].

Our calculated divergence estimate for *D. agassizi* is similar to Román-Palacios et al. [24], who found that *D. agassizi* and its sister species, *D. insignis*, diverged during the Oligocene, 22–26 Ma. We recovered *D. agassizi* sister to a sister-pair of *D. insignis* and *D. microtus* with a mean divergence of 31 (fossil 95% 22–38) or 29 Ma (rate 95% 23–34). Román-Palacios et al. [24] did not include *D. microtus* in their phylogeny, so this is not a contradiction of their topology, although our dates suggest an older divergence, which is concurrent with the assessment that rate-based methods can overestimate the ages of shallower nodes if gene saturation is in effect [72]. Poe et al. [20] placed *D. agassizi* as sister to a clade of 14 other species of *Dactyloa*, but did not include *D. insignis* in their analysis. They also included *N. townsendi* within a group of anoles aligned with *N. trachyderma*. We note that Poe et al. [20] also included other loci in their analysis, which might account for the differences seen here.

Prior to this study, no one had published an estimated age of colonization for *N. townsendi*. Our analyses suggest a split within *N. townsendi* around 3.5 Mya that separates individuals on Isla Cocos from those on Manuelita Islet (Figure 4). This is problematic in that the divergence predates the origin of the island, but our limitations on sample collection might be influencing our results. A more fine-scale sampling effort could provide a more reliable estimate of the coalescence between the two lineages (incidentally, the Manuelita population was significantly larger than any of the three populations on Isla Cocos, Table 4). The phylogenetic placement of *N. townsendi* was unknown until recently but was presumed to be within the mainland *Norops* clade [88,89]. Our analyses also placed it sister to *N. poecilopus*, a semi-aquatic anole from Panamá (100% posterior probability support), concurring with other recent topologies [21]. This clade also contains two additional semi-aquatic species (*N. lionotus* and *N. oxylophus*). If *N. townsendi* evolved from a semi-aquatic ancestor, it may have had more opportunities for rafting, and also makes biogeographic sense, as the currents could have carried an ancestor from Panama to Isla Cocos. Other semi-aquatic anoles (*N. lynchi*, *N. macrolepis*, *N. rivalis*) occur in coastal South America that are missing from our phylogeny, providing other possible relatives. If one of these shares a more recent common ancestor with *N. townsendi*, it may result in a more recent estimate of colonization closer to the origin of Isla Cocos.

The oceanic currents may have played a significant role in the colonization of the islands in the eastern Pacific region by anoles and other taxa [35,90,91]. El Niño Southern Oscillation (ENSO) cycles are believed to have played an important role in island colonization in the past three to five million years [92,93], which could explain dispersal events to Isla Cocos and even the Galápagos Archipelago [3]. As suggested by Román-Palacios et al. [24], this is too recent to influence the divergence of *D. agassizi*, but the Colombian Current as part of the Humboldt Current has been active for over 100 million years [94], and likely was an important factor in the colonization of Isla Malpelo by all inhabitants. For both *D. agassizi* and *N. townsendi*, divergences predating island emergence strongly suggest dispersal from the mainland (likely lower Central America given their closest ancestor as suggested by our phylogenetic reconstruction). This could have taken place via island hopping, possibly using seamounts that were previously aerial [24,95–97]. Recent work in the ETP suggests such a scenario for previously emergent landmasses that are part of the Galápagos archipelago [98], therefore, we cannot discount the possibility that there are other seamounts in the ETP that were previously emergent and could have served as the "stepping-stones" to Isla Cocos and/or Malpelo.

4.2. Taxonomic Implications

Our findings support Armstead et al. [55] in their recognition of *Norops parvauritus* as a distinct species from *N. biporcatus*. The age of the divide between the two species is older than many other species pairs, so we conclude this taxonomic change was warranted. Our inclusion of a specimen from

Isla Gorgona (the type locality) was highly important to confirm the specific status of *N. parvauritus*. Fine-scale sampling throughout Mesoamerica also suggests geographic structure in *N. biporcatus*, with some division aligning with the continental divide in Costa Rica and the Nicaraguan highlands (Figure 4, Supplementary Figures S1 and S2).

Given the placement of the Isla Gorgona *N. parvauritus* as sister to the rest of its species cannot rule out the possibility of *N. parvauritus* first colonizing Isla Gorgona and then back-colonizing the mainland after ancestral *N. biporcatus* populations had been extirpated. In this way, Isla Gorgona may have served as a refugium for at least one species, but perhaps also for *N. medemi*. While *N. biporcatus* is still wide-ranging for Neotropical anoles, the recently elevated *N. parvauritus* represents what would have been a disjunct portion of its range. This pattern may carry over to other lineages of Neotropical reptiles as we note that there are also disjunct populations of the colubrid snake *Leptophis ahaetulla* in Northwest Ecuador and Isla Gorgona [99], which are yet to be analyzed in a phylogeographic context.

4.3. Natural History Observations

Both *D. agassizi* and *N. townsendi* can be described as ecological generalists unlike solitary Lesser Antillean anoles [100]. *Norops townsendi* appears to clearly conform to this generalization, given that is was found ubiquitously on Isla Cocos in terms of perch height, diameter, and type, as briefly noted by Carpenter [38]. *Dactyloa agassizi* has distinctly unique behavioral characteristics that challenge the original meaning of generalist as applied to the Lesser Antillean anoles. The food web of Isla Malpelo appears to be very tight given the lack of vegetation and scarcity of obvious food resources for all the island inhabitants. Everything seems to revolve around the birds that visit the island, particularly the Nazca Booby (*Sula granti*). Any food dropped by the birds, and all waste products were immediately consumed by all *D. agassizi*, as well as *Diploglossus millepunctatus* lizards and *Geocarcinus malpilensis* crabs which congregate en masse around these resources. The *Geocarcinus* and *Diploglossus* attempt to capture any *D. agassizi* that venture too close [34], pers. obs., and the crabs will also consume eggs of all species [34]. Rand et al. [34] and Wolda [31] reported that *D. agassizi* ate both insects and seeds, and that individuals would quickly consume potential food items revealed when rocks were turned over. All *D. agassizi* departed from typical anole behavior in being very curious, climbing on, and licking the observers (clearly investigating food possibilities), never displaying any defensive behaviors or territoriality. Therefore, insofar as their environment allowed them, *D. agassizi* appeared to be an ecological generalist. Isla Malpelo is unlike most other islands that are home to anoles, so it is not surprising that *D. agassizi* departs from the general patterns observed for solitary Caribbean anoles.

4.4. Conservation of Island Fauna

While these Pacific Island anoles represent novel diversity unseen elsewhere, it is fortunate that all three islands are strictly controlled and monitored, virtually ensuring the conservation of the species. Tourists are somewhat rare to both Islas Cocos and Malpelo, and mostly consist of individuals seeking to scuba dive in the waters around the islands. It requires special permission to step onto the islands and is virtually impossible to be allowed onto Malpelo unless undertaking research. More tourists visit Isla Gorgona, which is also considerably closer to the mainland and faster to visit. Even then, they are only allowed to walk along trails accompanied by a guide and walk within the grounds of the former prison which is slowly being taken over by the forest. These strict controls are highly beneficial to the conservation of each island's flora and fauna and no development of any kind is allowed. Thus, the future seems cautiously bright for the conservation status of these islands.

Only one species (*D. agassizi*) has estimates of population size [32], suggesting 140,000–206,000 individuals. None of the seven species of anoles discussed in this study has received any IUCN designation. However, we currently have no reason to believe any of these species/populations is currently at a high risk of extinction/extirpation because of the restricted access of these islands by the Colombian and Costa Rican governments. It is important that these policies continue, as island species are generally more predisposed to extirpation because of high endemism and smaller population sizes. [6]. In particular, tropical islands should be a priority in conservation since they serve as hotspots for biodiversity in many clades [101,102]. Anthropogenic disturbances have contributed to the extinction of other oceanic taxa [103,104], so the protection of these islands and their native ecosystems needs to continue for the conservation of not just these anoles but all flora and fauna on these islands.

5. Conclusions

Herein, we have estimated the divergence from the mainland for seven species of anole lizards in the ETP. *Dactyloa agassizi* and *N. townsendi* both diverge prior to the emergence of their islands (Islas Malpelo and Cocos respectively), and display sexual size dimorphism, similar to Caribbean relatives. *Norops townsendi* additionally displayed divergent patterns of morphological evolution in different populations across Isla Cocos. All lineages on Isla Gorgona appear to originate within the past 10 My, but geologic history of past connections with the mainland is unclear. Three of the species *D. gorgonae*, *D. princeps*, and *N. medemi* display ecological partitioning. This biogeographic and ecomorphological work set the stage for future questions in the realm of island biogeography and community ecology.

Supplementary Materials: The following are available online at http://www.mdpi.com/1424-2818/11/9/141/s1, Figure S1: Bayesian phylogeny as calculated by using a mutation rate only; Figure S2: Bayesian phylogeny as calculated by both mutation rate and fossil calibrations, displaying node bars for 95% confidence intervals; Table S1: Species used in phylogenetic analyses with Genbank numbers; Table S2: Locality data for all *Norops biporcatus* and *N. parvauritus* samples used in this study; Table S3: Ecological and morphological data from *Dactyloa agassizi* used in this study; Table S4: Ecological and morphological data from *Norops townsendi* used in this study. Table S5: Ecological and morphological data from five species of anoles on Isla Gorgona used in this study.

Author Contributions: K.E.N. designed the study. J.G.P., S.E.B., M.M.W., and K.E.N. collected specimens. J.G.P., S.E.B., E.P., and K.E.N. conducted labwork. J.G.P. and K.E.N. conducted analysis. J.G.P. and K.E.N. led the writing with contributions from all authors.

Funding: This work was supported in part by a grant K.E.N. (labwork, analysis and writing: NSF DEB 0949359 and internal funding from Central Michigan University).

Acknowledgments: The authors would like to thank Andrew C. Kramer and A. Scott McNaught for statistical advice. We would like to thank Mahmood Sasa for logistical and permitting support in Costa Rica, particularly for Isla Cocos. We also thank our Colombian colleagues for logistical support and permitting, especially in accessing Islas Gorgona and Malpelo. We thank Natasha Bloch and Daniel Medina for field assistance on Isla Gorgona, as well as Jeremy Gibson-Brown and the Nicholson lab for editorial comments. Mainland sampling of *Norops biporcatus* was assisted by Sarah Burton, Jenny Gubler, David Laurencio, Lenin Obando, and Javier Sunyer. Additional tissues were provided/collected by Carlos A. Andino, Atanasio Baldonado, Cesar A Cerrato-Mendoza, Sebastian Charley, Wes Chun, Ignacio Cruz, Gabriela Diaz, Stephen Doucette-Riise, Sergio Gonzalez, Lorraine P Ketzler, Francisco Lopez, Ileana R Luque-Montes, Melissa Medina-Flores, Aaron Mendoza, Wendy Naira-Mejia, Ronald Picado, Jay M. Savage, Josiah H Townsend, Scott L. Travers, Laurie J. Vitt and Larry David Wilson. This study conformed to all IACUC regulations and specimens were collected with the appropriate permits (in Costa Rica: permit number ACMIC-002-003, Resolucíones 239-2008-SINAC, 040–2009-SINAC, 017-2010-SINAC and 005–2011-SINAC and MINAET Permit no. 029–2011-SINAC; in Colombia: resolución número 2388, in Nicaragua under DGPN/DB-10–2010 and in Panama under ANAM Scientific Permit No. SEX/A-50-12).

Conflicts of Interest: The authors declare no conflict of interest.

References

1. Chiba, S. Ecological and morphological patterns in communities of land snails of the genus Mandarina from the Bonin Islands. *J. Evol. Biol.* **2004**, *17*, 131–143. [CrossRef] [PubMed]
2. Gillespie, R. Community assembly through adaptive radiation in Hawaiian spiders. *Science* **2004**, *303*, 356–359. [CrossRef] [PubMed]
3. Parent, C.E.; Caccone, A.; Petren, K. Colonization and diversification of Galápagos terrestrial fauna: A phylogenetic and biogeographical synthesis. *Philos. Trans. R. Soc. B Biol. Sci.* **2008**, *363*, 3347–3361. [CrossRef] [PubMed]
4. Carlquist, S.J. *Island Biology*; Columbia University Press: New York, NY, USA, 1974.
5. Hendriks, K.P.; Alciatore, G.; Schilthuizen, M.; Etienne, R.S. Phylogeography of Bornean land snails suggests long-distance dispersal as a cause of endemism. *J. Biogeogr.* **2019**, *46*, 932–944. [CrossRef]

6.	Whittaker, R.J.; Fernández-Palacios, J.M. *Island Biogeography: Ecology, Evolution, and Conservation*, 2nd ed.; Oxford University Press: Oxford, UK, 2007.

7.	Ali, J.R.; Aitchison, J.C. Exploring the combined role of eustasy and oceanic island thermal subsidence in shaping biodiversity on the Galápagos. *J. Biogeogr.* **2014**, *41*, 1227–1241. [CrossRef]

8.	Williams, E.E. A critique of Guyer and Savage (1986): Cladistic relationships among anoles (Sauria: Iguanidae): Are the data available to reclassify the anoles? In *Biogeography of the West Indies: Past, Present, and Future*; Sandhill Crane Press: Gainesville, FL, USA, 1989; pp. 286–341.

9.	Crother, B.I.; Parenti, L.R. *Assumptions Inhibiting Progress in Comparative Biology*; CRC Press: Boca Raton, FL, USA, 2016.

10.	Guyer, C.; Savage, J.M. Cladistic relationships among anoles (Sauria: Iguanidae). *Syst. Zool.* **1986**, *35*, 509–531. [CrossRef]

11.	Nicholson, K.E.; Crother, B.I.; Guyer, C.; Savage, J.M. Translating a clade-based classification into one that is valid under the international code of zoological nomenclature: The case of the lizards of the family Dactyloidae (Order Squamata). *Zootaxa* **2018**, *4461*, 573–586. [CrossRef] [PubMed]

12.	Uetz, P.; Freed, P.; Hošek, J. The Reptile Database. Available online: http://www.reptile-database.org (accessed on 29 June 2019).

13.	Losos, J.B. *Lizards in an Evolutionary Tree: Ecology and Adaptive Radiation of Anoles*; University of California Press: Berkeley, CA, USA, 2009.

14.	Irschick, D.L.; Vitt, L.J.; Zani, P.A.; Losos, J.B. A comparison of evolutionary radiations in mainland and West Indian *Anolis* lizards. *Ecology* **1997**, *78*, 2191–2203. [CrossRef]

15.	Irschick, D.L.; Losos, J.B. A comparative analysis of the ecological significance of maximal locomotor performance in Caribbean *Anolis* lizards. *Evolution* **1998**, *52*, 219–226. [CrossRef] [PubMed]

16.	Losos, J.B. The evolution of convergent community structure in Caribbean *Anolis* communities. *Syst. Biol.* **1992**, *41*, 403–420. [CrossRef]

17.	Losos, J.B.; Warheit, K.I.; Schoener, T.W. Adaptive differentiation following experimental island colonization in *Anolis* lizards. *Nature* **1997**, *387*, 70–73. [CrossRef]

18.	Williams, E.E. The ecology of colonization as seen in the zoogeography of *Anolis* lizards on small islands. *Q. Rev. Biol.* **1969**, *44*, 345–389. [CrossRef]

19.	Hedges, S.B.; Hass, C.A.; Maxson, L.R. Caribbean biogeography: Molecular evidence for dispersal in West Indian terrestrial vertebrates. *Proc. Natl. Acad. Sci. USA* **1992**, *89*, 1909–1913. [CrossRef]

20.	Poe, S.; Nieto-Montes de Oca, A.; Torres-Carvajal, O.; de Queiroz, K.; Velasco, J.A.; Truett, B.; Gray, L.N.; Ryan, M.J.; Köhler, G.; Ayala-Varela, F.; et al. A phylogenetic, biogeographic, and taxonomic study of all extant species of *Anolis* (Squamata; Iguanidae). *Syst. Biol.* **2017**, *66*, 663–697. [CrossRef]

21.	Nicholson, K.E.; Crother, B.I.; Guyer, C.; Savage, J.M. It is time for a new classification of anoles. *Zootaxa* **2012**, *3477*, 1–108. [CrossRef]

22.	Schoener, T.W. Size patterns in West Indian *Anolis* lizards: I. size and species diversity. *Syst. Biol.* **1969**, *18*, 386–401. [CrossRef]

23.	Blankers, T.; Townsend, T.M.; Pepe, K.; Reeder, T.W.; Wiens, J.J. Contrasting global-scale evolutionary radiations: Phylogeny, diversification, and morphological evolution in the major clades of iguanian lizards. *Biol. J. Linn. Soc.* **2013**, *108*, 127–143. [CrossRef]

24.	Román-Palacios, C.; Tavera, J.; Castañeda, D.R. When did anoles diverge? An analysis of multiple dating strategies. *Mol. Phylogenet. Evol.* **2018**, *127*, 655–668. [CrossRef]

25.	Poe, S.; Goheen, J.R.; Hulebak, E.P. Convergent exaptation and adaptation in solitary island lizards. *Proc. R. Soc. B Biol. Sci.* **2007**, *274*, 2231–2237. [CrossRef]

26.	Hoernle, K.; van den Bogaard, P.; Werner, R.; Lissinna, B.; Hauff, F.; Alvarado, G.; Garbe-Schönberg, D. Missing history (16–17 Ma) of the Galápagos hotspot: Implications for the tectonic and biological evolution of the Americas. *Geology* **2002**, *30*, 795–798. [CrossRef]

27.	Stead, J.A. Field observations of the geology of Malpelo Island. *Smithson. Contrib. Zool.* **1975**, *176*, 17–26.

28.	Gorman, G.C.; Chorba, T.L. Terrestrial biology of Malpelo Island: A historical review. *Smithson. Contrib. Zool.* **1975**, *176*, 9–12.

29.	Hertlein, L.G. Contribution to the biogeography of Cocos Island, including a bibliography. *Proc. Calif. Acad. Sci. USA* **1963**, *32*, 219–289.

30. Kiester, A.R.; Hoffman, J.A. Reconnaissance and mapping of Malpelo Island. *Smithson. Contrib. Zool.* **1975**, *176*, 13–20.

31. Wolda, H. The ecosystem on Malpelo Island. *Smithson. Contrib. Zool.* **1975**, *176*, 21–26.

32. López-Victoria, M.; Wolters, V.; Werding, B. Nazca Booby (*Sula granti*) inputs maintain the terrestrial food web of Malpelo Island. *J. Ornithol.* **2009**, *150*, 865–870. [CrossRef]

33. López-Victoria, M. The lizards of Malpelo (Colombia): Some topics on their ecology and threats. *Caldesia* **2006**, *28*, 129–134.

34. Rand, A.S.; Gorman, G.C.; Rand, W.M. Natural history, behavior, and ecology and *Anolis agassizi*. *Smithson. Contrib. Zool.* **1975**, *176*, 27–38.

35. Castañeda, M.D.R.; de Queiroz, K. Phylogenetic relationships of the *Dactyloa* clade of *Anolis* lizards based on nuclear and mitochondrial DNA sequence data. *Mol. Phylogenetics Evol.* **2011**, *61*, 784–800. [CrossRef]

36. Castañeda, M.D.R.; de Queiroz, K. Phylogeny of the *Dactyloa* clade of *Anolis* lizards: New insights from combining morphological and molecular data. *Bull. Mus. Comp. Zool.* **2013**, *160*, 345–398. [CrossRef]

37. Prates, I.; Rodrigues, M.T.; Melo-Sampaio, P.R.; Carnaval, A.C. Phylogenetic relationships of Amazonian anole lizards (*Dactyloa*): Taxonomic implications, new insights about phenotypic evolution and the timing of diversification. *Mol. Phylogenetics Evol.* **2015**, *82*, 258–268. [CrossRef]

38. Carpenter, C.C. The display of the Cocos Island anole. *Herpetologica* **1965**, *21*, 256–260.

39. Castillo, P.; Batiza, R.; Vanko, D.; Malavassi, E.; Barquero, J.; Fernandez, E. Anomalously young volcanoes on old hot-spot traces. I. Geology and petrology of Cocos Island. Geology and petrology of Cocos Island. *Geol. Soc. Am. Bull.* **1998**, *100*, 1400–1414. [CrossRef]

40. Poe, S. Phylogeny of anoles. *Herpetol. Monogr.* **2004**, *18*, 37–89. [CrossRef]

41. Grisales-Martínez, F.A.; Velasco, J.A.; Bolívar, W.; Williams, E.E.; Daza, J.M. The taxonomic and phylogenetic status of some poorly known *Anolis* species from the Andes of Colombia with the description of a *nomen nudum* taxon. *Zootaxa* **2017**, *4303*, 213–230. [CrossRef]

42. Nicholson, K.E.; Guyer, C.; Phillips, J.G. Biogeographic origin of mainland *Norops* (Squamata: Dactyloidae). In *Assumptions Inhibiting Progress in Comparative Biology*; CRC Press: Boca Raton, FL, USA, 2017; pp. 169–184.

43. Phillips, J.G.; Deitloff, J.; Guyer, C.; Hutteman, S.; Nicholson, K.E. Biogeography and evolution of a widespread Central American lizard species complex: *Norops humilis*, (Squamata: Dactyloidae). *BMC Evol. Biol.* **2015**, *15*, 143. [CrossRef]

44. Castro, F.; Ayala, S.; Carvajal, H. Los saurios de las islas Gorgona y Gorgonilla. In *Gorgona*; Universidad de los Andes: Cundinamarca, Colombia, 1979; pp. 189–218.

45. UASPNN. *El Sistema de Parques Nacionales Naturales de Colombia*; Unidad Administrativa Especial del Sistema de Parques Nacionales Naturales (UASPNN), Ministerio del Medio Ambiente. Editorial Nomos: Bogotá, Colombia, 1998.

46. Echeverría, L.M.; Aitken, B. Pyroclastic rocks: Another manifestation of ultramafic volcanism of Gorgona Island. Colombia. *Contrib. Mineral. Petrol.* **1986**, *92*, 428–436. [CrossRef]

47. Gomez, H. Algunos aspectos neotectónicos hacia el suroeste del Litoral Pacífico colombiano. *Rev. CIAF* **1986**, *11*, 281–296.

48. Kerr, A.C. La Isla de Gorgona, Colombia: A petrological enigma? *Lithos* **2005**, *84*, 77–101. [CrossRef]

49. Serrano, L.; Ferrari, L.; Martínez, M.L.; Petrone, C.M.; Jaramillo, C. An integrative geologic, geochronologic and geochemical study of Gorgona Island, Colombia: Implications for the formation of the Caribbean Large Igneous Province. *Earth Planet. Sci. Lett.* **2011**, *309*, 324–336. [CrossRef]

50. Von Prahl, H.; Guhl, F.; Grögl, M. *Gorgona*; Futura Groupo Editorial Ltd.: Bogota, Colombia, 1979.

51. Aguirre, J.; Rangel, O. La isla Gorgona y sus ecosistemas. In *Colombia Pacífico*; Fondo para la Protección del Medio Ambiente (FEN): Bogotá, Colombia, 1993; Volume I, pp. 106–170.

52. Alberico, M. Biogeografía terrestre. Capitulo XI. In *Isla de Gorgona*; Banco Popular: Bogotá, Colombia, 1986; pp. 223–244.

53. Murillo, M.T.; Lozano, G. Hacia la realización de una flórula del Parque Nacional Natural Islas de Gorgona y Gorgonilla, Cauca, Colombia. *Rev. Acad. Col. Cie. Exactas Físicas y Naturales* **1989**, *12*, 277–304.

54. Yockteng, R.; Cavelier, J. Diversidad y mecanismos de dispersión de árboles de la Isla Gorgona y de los bosques húmedos tropicales del Pacífico colombo-ecuatoriano. *Rev. Biol. Trop.* **1998**, *46*, 45–53.

55. Armstead, J.V.; Ayala-Varela, F.; Torres-Carvajal, O.; Ryan, M.J.; Poe, S. Systematics and ecology of *Anolis biporcatus* (Squamata: Iguanidae). *Salamandra* **2017**, *53*, 285–293.

56. Köhler, G.; McCranie, J.R.; Nicholson, K.E.; Kreutz, J. Geographic variation in hemipenial morphology in *Norops humilis* (Peters 1863), and the systematic status of *Norops quaggulus* (Cope 1885) (Reptilia, Squamata, Polychrotidae). *Senckenbergiana Biol.* **2003**, *82*, 213–222.

57. Selander, R.K. Sexual dimorphism and differential niche utilization in birds. *Condor* **1966**, *68*, 113–151. [CrossRef]

58. Clegg, S.M.; Owens, P.F. The 'island rule' in birds: Medium body size and its ecological explanation. *Proc. R. Soc. Lond. Ser. B Biol. Sci.* **2002**, *269*, 1359–1365. [CrossRef]

59. Dayan, T.; Simberloff, D. Size patterns among competitors: Ecological character displacement and character release in mammals, with special reference to island populations. *Mammal Rev.* **1998**, *28*, 99–124. [CrossRef]

60. Schoener, T.W. The ecological significance of sexual dimorphism in size in the lizard *Anolis conspersus*. *Science* **1967**, *155*, 474–477. [CrossRef]

61. Schoener, T.W. Size patterns in West Indian Anolis lizards. II. Correlations with the sizes of particular sympatric species-displacement and convergence. *Am. Nat.* **1970**, *104*, 155–174. [CrossRef]

62. Velasco, J.A.; Poe, S.; González-Salazar, C.; Flores-Villela, O. Solitary ecology as a phenomenon extending beyond insular systems: Exaptive evolution in *Anolis* lizards. *Biol. Lett.* **2019**, *15*, 20190056. [CrossRef]

63. Bouckaert, R.; Heled, J.; Kühnert, D.; Vaughan, T.; Wu, C.; Xie, D.; Suchard, M.A.; Rambaut, A.; Drummond, A.J. BEAST 2: A software platform for Bayesian evolutionary analysis. *PLoS Comput. Biol.* **2014**, *10*, e1003537.

64. Rambaut, A.; Suchard, M.A.; Xie, D.; Drummond, A.J. Tracer v.1.6. 2014. Available online: http://tree.bio.ed.ac.uk/software/tracer/ (accessed on 9 August 2018).

65. Macey, J.R.; Schulte II, J.A.; Ananjeva, N.B.; Larson, A.; Rastegar-Pouyani, N.; Shammakov, S.M.; Papenfuss, T.J. Phylogenetic relationships among Agamid Lizards of the *Laudakia caucasia* species group: Testing hypotheses of biogeographic fragmentation and an area cladogram for the Iranian Plateau. *Mol. Phylogenetics Evol.* **1998**, *10*, 118–131. [CrossRef]

66. Glor, R.E.; Kolbe, J.J.; Powell, R.; Larson, A.; Losos, J.B. Phylogenetic analysis of ecological and morphological diversification in Hispaniolan trunk-ground anoles (*Anolis cybotes* group). *Evolution* **2003**, *57*, 2383–2397. [CrossRef]

67. Glor, R.E.; Losos, J.B.; Larson, A. Out of Cuba: Overwater dispersal and speciation among lizards in the *Anolis carolinensis* subgroup. *Mol. Ecol.* **2005**, *14*, 2419–2432. [CrossRef]

68. Guarnizo, C.E.; Werneck, F.P.; Giugliano, L.G.; Santos, M.G.; Fenker, J.; Sousa, L.; D'Angiolella, A.B.; dos Santos, A.R.; Strüssmann, C.; Rodrigues, M.T.; et al. Cryptic lineages and diversification of an endemic lizard (Squamata, Dactyloidae) of the Cerrado hotspot. *Mol. Phylogenetics Evol.* **2016**, *94*, 279–289. [CrossRef]

69. Ng, J.; Glor, R.E. Genetic differentiation among populations of a Hispaniolan trunk anole that exhibit geographical variation in dewlap color. *Mol. Ecol.* **2011**, *20*, 4302–4317. [CrossRef] [PubMed]

70. Reynolds, R.G.; Strickland, T.R.; Kolbe, J.J.; Falk, B.G.; Perry, G.; Revell, L.J.; Losos, J.B. Archipelagic genetics in a widespread Caribbean anole. *J. Biogeogr.* **2017**, *44*, 2631–2647. [CrossRef]

71. Torres-Carvajal, O.; de Queiroz, K. Phylogeny of hoplocercine lizards (Squamata: Iguania) with estimates of relative divergence times. *Mol. Phylogenetics Evol.* **2009**, *50*, 31–43. [CrossRef]

72. Zheng, Y.; Peng, R.; Kuro-o, M.; Zeng, X. Exploring patterns and extent of bias in estimating divergence time from mitochondrial DNA sequence data in a particular lineage: A case study of salamanders (Order Caudata). *Mol. Biol. Evol.* **2011**, *28*, 2521–2535. [CrossRef]

73. Poe, S.; Scarpetta, S.; Schaad, E.W. A new species of *Anolis* (Squamata: Iguanidae) from Panama. *Amphib. Reptile Conserv.* **2015**, *9*, 1–13.

74. Macedonia, J.M.; Clark, D.L. Headbob display structure in the naturalized *Anolis* lizards of Bermuda: Sex, context, and population effects. *J. Herpetol.* **2003**, *37*, 266–277. [CrossRef]

75. Thorpe, R.S.; Barlow, A.; Malhotra, A.; Surget-Groba, Y. Widespread parallel population adaptation to climate variation across a radiation: Implications for adaptation to climate change. *Mol Ecol.* **2015**, *24*, 1019–1030. [CrossRef] [PubMed]

76. Calderón-Espinosa, M.L.; Forero, A.B. Morphological diversification in solitary endemic anoles: *Anolis concolor* and *Anolis pinchoti* from San Andrés and Providence Islands, Colombia. *South Am. J. Herpetol.* **2011**, *6*, 205–210. [CrossRef]

77. Gartner, G.E.A.; Gamble, T.; Jaffe, A.L.; Harrison, A.; Losos, J.B. Left-right dewlap asymmetry and phylogeography of *Anolis lineatus* on Aruba and Curaçao. *Biol. J. Linn. Soc.* **2013**, *110*, 409–426. [CrossRef]

78. Williams, E.E. Ecomorphs, faunas, island size, and diverse end points in island radiations of *Anolis*. In *Lizard Ecology: Studies of a Model Organism*; Harvard University Press: Cambridge, MA, USA, 1983; pp. 326–370.

79. Savage, J.M.; Talbot, J.J. The giant anoline lizards of Costa Rica and western Panama. *Copeia* **1978**, *1978*, 480–492. [CrossRef]

80. Williams, E.E. New or problematic Anolis from Colombia. V. *Anolis danieli*, a new species of the latifrons species group and a reassessment of *Anolis apollinaris* Boulenger, 1919. *Breviora* **1988**, *489*, 1–25.

81. Farris, D.W.; Jaramillo, C.; Bayona, G.; Restrepo-Moreno, S.A.; Montes, C.; Cardona, A.; Mora, A.; Speakman, R.J.; Glascock, M.D.; Valencia, V. Fracturing of the Panamanian Isthmus during initial collision with South America. *Geology* **2011**, *39*, 1007–1010. [CrossRef]

82. Montes, C.; Cardona, A.; Jaramillo, C.; Pardo, A.; Silva, J.C.; Valencia, V.; Ayala, C.; Pérez-Angel, L.C.; Rodriguez-Parra, L.A.; Ramirez, V.; et al. Middle Miocene closure of the Central American seaway. *Science* **2012**, *348*, 226–229. [CrossRef]

83. Elmer, K.R.; Bonett, R.M.; Wake, D.B.; Lougheed, S.C. Early Miocene origin and cryptic diversification of South American salamanders. *BMC Evol. Biol.* **2013**, *13*, 59. [CrossRef]

84. Montes, C.; Cardona, A.; McFadden, R.; Morón, S.E.; Silva, C.A.; Restrepo-Moreno, S.; Bayona, G.A. Evidence for middle Eocene and younger land emergence in central Panama: Implications for Isthmus closure. *Geol. Soc. Am. Bull.* **2012**, *124*, 780–799. [CrossRef]

85. Coates, A.G.; Obando, J.A. The geologic evolution of the Central American isthmus. In *Evolution and Environment in Tropical America*; University of Chicago Press: Chicago, IL, USA, 1996; pp. 21–56.

86. MacFadden, B.J. Extinct mammalian biodiversity of the ancient New World tropics. *Trends Ecol. Evol.* **2006**, *21*, 157–165. [CrossRef] [PubMed]

87. Censky, E.J.; Hodge, K.; Dudley, J. Over-water dispersal of lizards due to hurricanes. *Nature* **1988**, *395*, 556. [CrossRef]

88. Etheridge, R.E. The Relationships of the Anoles (Reptilia: Sauria: Iguanidae): An Interpretation Based on Skeletal Morphology. Ph.D. Thesis, University Microfilms, Ann Arbor, MI, USA, 1959.

89. Savage, J.M.; Guyer, C. Infrageneric classification and species composition of the anole genera, *Anolis, Ctenonotus, Dactyloa, Norops,* and *Semiurus* (Sauria: Iguanidae). *Amphib. Reptil.* **1989**, *10*, 105–116. [CrossRef]

90. Caccone, A.; Gibbs, J.P.; Ketmaier, V.; Suatoni, E.; Powell, J.R. Origin and evolutionary relationships of giant Galapagos tortoises. *Proc. Natl. Acad. Sci. USA* **1999**, *96*, 13223–13228. [CrossRef] [PubMed]

91. Caccone, A.; Gentile, G.; Gibbs, J.P.; Fritts, T.H.; Snell, H.L.; Betts, J.; Powell, J.R. Phylogeography and history of giant Galapagos tortoises. *Evolution* **2002**, *56*, 2052–2066.

92. Huber, M.; Caballero, R. Eocene El Nino: Evidence for robust tropical dynamics in the "hothouse". *Science* **2003**, *299*, 877–881. [CrossRef] [PubMed]

93. Fedorov, A.V.; Dekens, P.S.; McCarthy, M.; Ravelo, A.C.; Barreiro, M.; Pacanowski, R.C.; Philander, S.G. The Pliocene paradox (mechanisms for a permanent El Niño). *Science* **2006**, *312*, 1485–1489. [CrossRef] [PubMed]

94. Hartley, A.J.; Chong, G.; Houston, J.; Mather, A.E. 150 million years of climate stability: Evidence from the Atacama Desert, northern Chile. *J. Geol. Soc.* **2005**, *162*, 421–444. [CrossRef]

95. Bessudo, S.; Soler, G.A.; Klimley, A.P.; Ketchum, J.T.; Hearn, A.; Arauz, R. Residency of the scalloped hammerhead shark (*Sphyrna lewini*) at Malpelo Island and evidence of migration to other islands in the Eastern Tropical Pacific. *Environ. Biol. Fishes* **2011**, *91*, 165–176. [CrossRef]

96. Jiménez, L.; Acosta, A.; Chong, N. Population structure of *Megabalanus peninsularis* in Malpelo Island, Colombia. *Rev. Biol. Trop.* **2016**, *51*, 461–468.

97. O'Connor, M.; Bruno, J.; Gaines, S. Temperature control of larval dispersal and the implications for marine ecology, evolution, and conservation. *Proc. Natl. Acad. Sci. USA* **2007**, *104*, 1266–1271. [CrossRef] [PubMed]

98. Orellana-Rovirosa, F.; Richards, M. Emergence/subsidence histories along the Carnegie and Cocos Ridges and their bearing upon biological speciation in the Galápagos. *Geochem. Geophys. Geosystems* **2018**, *19*, 4099–4129. [CrossRef]

99. Savage, J.M. *The Amphibians and Reptiles of Costa Rica: A Herpetofauna Between Two Continents, Between Two Seas*; University of Chicago Press: Chicago, IL, USA, 2002; p. 954.

100. Losos, J.B.; de Queiroz, K. Evolutionary consequences of ecological release in Caribbean *Anolis* lizards. *Biol. J. Linn. Soc.* **1997**, *61*, 459–483. [CrossRef]

101. Myers, N.; Mittermeier, R.A.; Mittermeier, C.G.; da Fonseca, G.A.; Kent, J. Biodiversity hotspots for conservation priorities. *Nature* **2000**, *403*, 853–858. [CrossRef] [PubMed]
102. Cubas, J.; Irl, S.D.; Villafuerte, R.; Bello-Rodríguez, V.; Rodríguez-Luengo, J.L.; del Arco, M.; Martín-Esquivel, J.L.; González-Mancebo, J.M. Endemic plant species are more palatable to introduced herbivores than non-endemics. *Proc. R. Soc. Lond. Ser. B Biol. Sci.* **2019**, *286*, 20190136. [CrossRef]
103. Regnier, C.; Achaz, G.; Lambert, A.; Cowie, R.H.; Bouchet, P.; Fontaine, B. Mass extinction in poorly known taxa. *Proc. Natl. Acad. Sci. USA* **2015**, *112*, 7761–7766. [CrossRef]
104. Solem, A. How many Hawaiian land snail species are left and what can we do for them? *Bish. Mus. Occas. Pap.* **1990**, *30*, 27–40.

Article

Endemic Infection of *Batrachochytrium dendrobatidis* in Costa Rica: Implications for Amphibian Conservation at Regional and Species Level

Héctor Zumbado-Ulate [1,*], Kiersten N. Nelson [1], Adrián García-Rodríguez [2,3], Gerardo Chaves [2], Erick Arias [2,3], Federico Bolaños [2], Steven M. Whitfield [4] and Catherine L. Searle [1]

[1] Department of Biological Sciences, Purdue University, West Lafayette, IN 47907, USA
[2] Escuela de Biología, Universidad de Costa Rica, San Pedro, 11501–2060 San José, Costa Rica
[3] Departamento de Zoología, Instituto de Biología, Universidad Nacional Autónoma de México, 04510 Ciudad de México, México
[4] Conservation and Research Department, Zoo Miami, Miami, FL 33177, USA
* Correspondence: hzumbado@purdue.edu

Received: 28 June 2019; Accepted: 7 August 2019; Published: 9 August 2019

Abstract: *Batrachochytrium dendrobatidis* (*Bd*) has been associated with the severe declines and extinctions of amphibians in Costa Rica that primarily occurred during the 1980s and 1990s. However, the current impact of *Bd* infection on amphibian species in Costa Rica is unknown. We aimed to update the list of amphibian species in Costa Rica and evaluate the prevalence and infection intensity of *Bd* infection across the country to aid in the development of effective conservation strategies for amphibians. We reviewed taxonomic lists and included new species descriptions and records for a total of 215 amphibian species in Costa Rica. We also sampled for *Bd* at nine localities from 2015–2018 and combined these data with additional *Bd* occurrence data from multiple studies conducted in amphibian communities across Costa Rica from 2005–2018. With this combined dataset, we found that *Bd* was common (overall infection rate of 23%) across regions and elevations, but infection intensity was below theoretical thresholds associated with mortality. *Bd* was also more prevalent in Caribbean lowlands and in terrestrial amphibians with an aquatic larval stage; meanwhile, infection load was the highest in direct-developing species (forest and stream-dwellers). Our findings can be used to prioritize regions and taxonomic groups for conservation strategies.

Keywords: amphibian; chytridiomycosis; conservation; disease; enzootics; epizootics; population declines

---

1. Introduction

Anthropogenic threats including habitat destruction, pollution, climate change, introduction of invasive species, and pathogens are causing a rapid and severe decline in global biodiversity [1]. Scientific consensus states that we are in the midst of a sixth mass extinction event [2,3]. Within vertebrates, amphibians are the most endangered taxonomic class with approximately 41% of described species classified as "globally threatened" [4,5]. The majority of the amphibian declines have occurred in the tropics of Australia, Central America, and South America [6,7], and have been observed even in seemingly pristine and protected environments [8,9]. However, information is still lacking regarding which species are suffering the greatest declines and which abiotic and biotic factors are contributing the most [10]. Identifying threatened species and factors contributing to global amphibian declines is vital for effective conservation and management efforts [11,12].

Costa Rica, with an area of only 51,100 km^2, is home to a great diversity of amphibians [13]. More than 200 of the approximately 8000 described amphibian species are present in Costa Rica [14], and new species continue to be described. The vast species richness confined to a relatively small area is due to complex biogeographic events and climatic conditions throughout the country, and a long history of work has been done by in-country taxonomic specialists [13,15]. Costa Rica is also an example of a country where numerous amphibian population declines have occurred in response to multiple environmental threats [16], highlighted by the enigmatic disappearance of the golden toad (*Incilius periglenes*) [17]. However, several species that catastrophically declined in the last thirty years, such as the harlequin frog, the Golfito Robber frog, and the Holdridge's toad, have been recently rediscovered in viable populations [18–20]. These findings suggest that highly susceptible species can recover from or at least persist when faced with deadly threats [21]. Thus, Costa Rica is an excellent location to study not only how amphibian communities are affected by environmental threats but also their resistance and resilience from declines [22].

One widespread cause of amphibian declines is the introduction of the pathogen *Batrachochytrium dendrobatidis* (*Bd*) [23]. This fungus causes chytridiomycosis [24], a potentially deadly skin disease that has contributed to the decline of at least 500 amphibian species globally [10]. In Central America, amphibian declines peaked during the 1980s and 1990s and have been linked to the introduction of *Bd*, which caused deadly outbreaks of chytridiomycosis (i.e., epizootics) [16,25,26]. It has been suggested that *Bd*-driven epizootic declines mostly affected species in highland lotic environments because moisture and temperature in these sites matches the optimal conditions for *Bd* growth in the lab [27,28]. However, it is also well known that some amphibian species suffered unexpected and unexplained declines in lowland environments (<700 m above sea level) during the 1980s and 1990s, likely due to chytridiomycosis [16,19,29–31]. After the declines, the evolution of resistance and tolerance mechanisms in amphibian communities [32], and/or the evolution of less-pathogenic strains of *Bd* [33], might have allowed susceptible amphibians to persist with endemic *Bd* infection (i.e., enzootics) [22,34–36]. However, susceptible species are still at a high risk of extinction under endemic infection if conditions shift in favor of the pathogen. For example, the introduction of an invasive species that is also a competent reservoir might amplify infection in the environment to epizootic levels [37–39]. Thus, examining the life history traits and conditions that may favor outbreaks of *Bd* is the key to understanding the underlying mechanisms behind why some infected species declined more severely than others and which species are most vulnerable to future outbreaks [40,41].

In this study, we present an updated list of all the amphibian species of Costa Rica, quantifying species diversity in each herpetological province and describing their conservation status. We also identified the effect of geography (herpetological province and altitudinal belt) and life-history traits associated with foraging and reproduction on current infection with *Bd*. For this, we sampled for *Bd* at nine tropical localities across Costa Rica from 2015–2018. In addition, we built a robust dataset by adding records from studies that detected *Bd* across Costa Rica from 2005–2018 in multi-species amphibian assemblages. We hypothesized that *Bd* is widespread across herpetological provinces and altitudinal belts in Costa Rica and would exhibit an infection intensity below theoretical thresholds associated with mass mortalities [42]. To compare across life-history traits, we developed an index that combines foraging habitat, reproductive habitat, and type of development. We hypothesized that *Bd* infection would vary across habitats, with the highest prevalence and infection intensity found in species with the greatest use of cool and humid environments [43,44]. The knowledge from this work will aid policy-makers in identifying the most threatened regions and taxonomic clades to develop better conservation strategies in Costa Rica [22,45,46].

2. Materials and Methods

2.1. Species Assessment

We updated the last official list of amphibian species in Costa Rica published in 2011 [47] by consulting the Herpetological Database ("Herp Database") of the Museo de Zoología at Universidad de Costa Rica (http://museo.biologia.ucr.ac.cr/) and taxonomists' lists [48,49]. In addition, we georeferenced the distribution of all amphibian species within the five Costa Rican herpetological provinces (see

Section 2.3.1). For this, we extracted all collection points for each species from the "Herp Database" (Datum WGS1984) and mapped them using a shapefile of the Costa Rican herpetological provinces and QGIS software 3.8.1 (QGIS Development Team, http://qgis.osgeo.org).

For every species, we showed their status in Costa Rica [50] according to the International Union of Conservation of Nature (IUCN) [51] as follows: NA = "not applicable," DD = "data deficient," LC = "least concerned," NT = "near threatened," VU = "vulnerable," EN = "endangered," CR = "critically endangered," and EX = "extinct in the wild" (for additional details see http://www.iucnredlist.org/). We also included environmental vulnerability scores (EVS) [52], a regional vulnerability index that classifies amphibians and reptiles into four levels of risk: "no immediate risk" (EVS < 3), "low vulnerability" (EVS of 3–9), "medium vulnerability" (EVS of 10–13), and "high vulnerability" (EVS of 14–17). A high EVS indicates species that are restricted in distribution, occur in a single life zone, and have a highly derived reproductive mode. The EVS for Costa Rican amphibians reported here were extracted from Sasa et al. [49]. Finally, we compiled a list of all the species that have been screened for *Bd* and the methods used for detection: histology or polymerase chain reaction (PCR).

2.2. Field Dataset

To add to existing datasets of amphibian distribution and *Bd* infection, we surveyed nine amphibian assemblages across Costa Rica in both versants (Caribbean and Pacific) and at elevations ranging from sea level to 1385 m (Figure S1). All surveys were conducted during the months of June and July between 2016–2018, except in the locality of Alto Lari, which was sampled in March 2015. At each site, we conducted visual and acoustic encounter surveys searching for amphibians in streams, ponds/puddles, and forest (leaflitter and canopy), and then caught individuals to screen them for *Bd* (see below). In total, we screened for *Bd* from 267 amphibians from 33 species (see Tables S1 and S2, Figure S2). Four of those species were classified in threatened categories: *Oophaga granulifera* (VU), *Ptychohyla legleri* (EN), *Craugastor ranoides* (CR), and *C. taurus* (CR).

All observed amphibians were collected using nitrile gloves and temporally placed individually in clean, unused plastic bags. Each individual was inspected for visible signs of chytridiomycosis, such as hyperplasia, hyperkeratosis, abnormal shedding, depigmentation, and lethargic behavior [24,53]. We swabbed (using MW-113 swabs) each individual's skin to detect *Bd* as follows: five strokes on one hand, five strokes on the ventral patch, five strokes on one foot, and five strokes along the inner thigh. The swabs were placed into 1.5 mL screw-cap tubes and stored dry at −20 °C until fungal DNA extraction. Once swabbed, all animals were released back to the site of their collection. During this study we followed field protocols [54,55] approved by the National System of Conservation Areas of Costa Rica (SINAC), the Comisión Nacional para la Gestión de la Biodiversidad (CONAGEBIO), and animal care protocols from the Purdue Animal Care and Use Committee (PACUC 1604001392), ensuring that all animals were being cared for in accordance with standard protocols and treated in an ethical manner (research permits 001-2012-SINAC, R-019-2016-OT-CONAGEBIO, R-023-2016-OT-CONAGEBIO, R-057-2016-OT-CONAGEBIO, R-060-2016-OT-CONAGEBIO).

We conducted diagnostic quantitative polymerase chain reaction (qPCR) on each swab to quantify *Bd* infection load following standard protocols [56], with the following modifications: (1) the fungal DNA was extracted using 60 µL of PrepMan Ultra, and (2) an internal positive control (IPC) was used to detect inhibitors [57]. Fungal DNA was diluted 1:10 in 0.25× TE buffer and run in singlicate [58] using a Step One Plus (Applied Biosystems, Woburn, MA, USA). Negative controls (DNase/RNase-free water) were run in triplicate on every 96-well qPCR plate. We classified samples as positive when both dyes (*Bd* probe and IPC) amplified in each well. Samples absent of IPC amplification were considered inhibited. In order to eliminate inhibitors, we diluted 5 µL of a new dilution in 0.25×TE buffer in a proportion of 1:100. Ten samples were classified as inhibited and then determined to be negative after dilution. Quantification curves for genomic equivalents were constructed using 1000, 100, 10, and 1 zoospore quantification standards derived from a gBlock® Gene fragment (Integrated DNA Technologies, Coralville, IA, USA). In order to calculate the zoospore genomic equivalents in the original sample, we multiplied the qPCR score by the dilution factor of 120 (dilution factor = 60 × 20 × 1/10). We estimated prevalence with 95% binomial confidence intervals (CIs) by locality.

2.3. Combined Dataset

We generated a dataset from multiple studies that screened for *Bd* in multiple amphibian assemblages in Costa Rica after the year 2000 using conventional PCR and qPCR methods [59–65] (Figure 1, Table 1) (including the 267 individuals from the 33 species we tested in the "field dataset" (see Section 2.2 and supporting data). In total, this "combined dataset" consisted of 1750 individual records from 79 species and 20 localities at elevations ranging from sea level to 2000 m. We identified the year 2000 as the starting of post-decline because most epizootic outbreaks of *Bd* occurred during the 1980s and early 1990s [26,30,66]. We also assumed that *Bd* expanded its range across Costa Rica by 2000 due to the rapid rate of spreading that this pathogen exhibits in tropical locations [67,68]. Although *Bd* was detected in 405 swabs in this dataset, quantification through qPCR was conducted only in 351 *Bd*-positive swabs (from the "field dataset" and three of the seven reviewed studies [61–63]). We did not consider studies that used histology as a method of detection because most of these studies evaluated samples that were taken before 2000. We also excluded records of individuals that were identified only at the genus level, e.g., *Craugastor* spp. [64] and *Agalychnis* spp. [61], and cases where only one species was screened for *Bd* (e.g., *Atelopus varius* in the locality of Uvita [65]). Finally, we classified all sampled amphibians according to herpetological province, altitudinal belt, and life history traits (foraging habitat, reproductive habitat, and type of development).

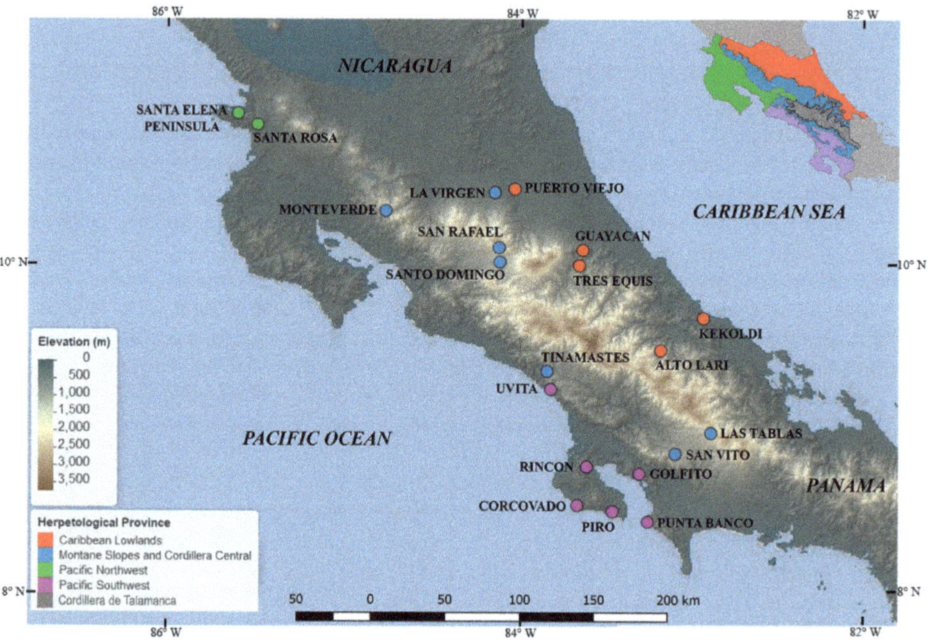

Figure 1. Map of 20 survey sites across Costa Rica. Sites are color-coded by herpetological province.

Table 1. Summary of studies where *Batrachochytrium dendrobatidis* (*Bd*) was detected in multi-species amphibian assemblages using conventional PCR and quantitative PCR (qPCR) in Costa Rica between 2005–2018. The table shows surveyed localities, herpetological province, sampling period, percentage of infection, and Holdridge's altitudinal belt. Symbology: CL—Caribbean Lowlands, MSCC—Montane Slopes and Cordillera Central, PN—Pacific Northwest, and PS—Pacific Southwest.

Study Site (Elevation m) and Herpetological Provinces	Sampling Period	% of Infection (n Sampled)	Altitudinal Belt	Reference
Monteverde (1400–2000), MSCC	July 2005	12.2 (41)	Lower montane	[59]
San Vito de Coto Brus (1120–1385), MSCC		9.3 (43)	Premontane	
Rincón de Osa (0–100), PS	May–June 2006	0.1 (91)	Lowland	[60]
Piro (0–100), PS		0.0 (62)	Lowland	
Corcovado (0–100), PS		0.1 (25)	Lowland	
Kekoldi (0–100), CL	January 2008	7.9 (126)	Lowland	[64]
La Virgen de Sarapiquí (0–200), CL	January–March 2011	21.3 (253)	Lowland	[61]
Santa Elena Peninsula (0–200), PN	January–March 2007–2008	0.0 (310)	Lowland	[62]
Santa Rosa (0–200), PN		9.0 (100)	Lowland	
Punta Banco–Burica (0–100), PS	November–December 2011	68.6 (35)	Lowland	[63]
Rincón de Osa (0–100), PS		0.0 (25)	Lowland	
Puerto Viejo de Sarapiquí (0–200), CL		67.4 (144)	Lowland	
Guayacán de Siquirres (400–600), CL		47.9 (144)	Lowland	
San Vito de Coto Brus (1120–1385), MSCC	Unknown/not indicated	10.5 (19)	Lowland	[65]
Punta Banco–Burica (0–100), PS		0.0 (20)	Lowland	
Guayacán de Siquirres (400–600), CL		5.3 (19)	Lowland	
San Rafael de Heredia (1800), MSCC		66.7 (15)	Lower montane	
Santo Domingo de Heredia (1000–1200), MSCC		45.5 (11)	Premontane	
Las Tablas (1350), MSCC		28.6 (14)	Lower montane	

2.3.1. Herpetological Provinces

We classified all surveyed assemblages within the five herpetological provinces proposed by Savage [13] and modified by Sasa and colleagues [49].

Caribbean Lowlands: This faunal area represents 30% of Costa Rica and includes the lowlands of the Caribbean versant and the northernmost region of the country, predominantly consisting of lowland wet forest. Sampling for *Bd* through PCR has been conducted in the localities of La Virgen de Sarapiquí, Puerto Viejo de Sarapiquí, Tres Equis de Turrialba, Guayacán de Siquirres, Kekoldi, and the remote Alto Lari. Alto Lari was surveyed as part of a recent expedition following an enigmatic path that connects the Caribbean Lowlands with the highlands of Cordillera de Talamanca and is known as "the Gabb's route" [69].

Pacific Northwest: This herpetological province includes the lowlands of the Pacific Northwest and extends into the western side of the Central Valley, in the Meseta Central Occidental (Central Valley) up to the base of Cerros de Ochomogo. The Pacific Northwest consists of predominantly Lowland Dry Forest vegetation and constitutes 24% of Costa Rica's area. This province contains a distinctive dry season that lasts five to six months. Within the Pacific Northwest, sampling for *Bd* has been conducted in the tropical dry forest at Guanacaste National Park (Santa Rosa and Santa Elena Peninsula stations).

Pacific Southwest: Encompassing the lowlands of the Pacific central and south, the herpetological provinces consist primarily of lowland wet forest and lowland moist forest and accounts for 15% of the country's area. This herpetological province is biogeographically related to the Caribbean Lowlands and species have more recently been differentiated between these herpetological provinces due to isolation caused by the uplifting of the Cordillera de Talamanca. Within the Pacific Southwest, sampling for *Bd* through PCR has been conducted in the localities of Punta Banco–Burica, Golfito, Rincón de Osa, Piro, Corcovado, and Uvita.

Montane Slopes and Cordillera Central: This area represents 23% of Costa Rica and occurs along all of Costa Rica's mountain ranges from 500–2100 m elevation in Cordillera de Guanacaste, 500–3400 in Cordillera Central, and 500–1600 in Cordillera de Talamanca (Lower Talamanca). The Montane Slopes and Cordillera Central province includes regions that receive the highest annual precipitation in the country. Sampling for *Bd* through PCR has been conducted in the localities of Monteverde, San Vito de Coto Brus, Las Tablas, Tinamastes de Pérez Zeledón, San Rafael de Heredia, and Santo Domingo de Heredia.

Cordillera de Talamanca: Found at elevations above 1600 m (Upper Talamanca). This is the smallest faunal area (8% of Costa Rica) and consists primarily of montane rainforest and subalpine pluvial paramo. This faunal area is the least explored herpetological province of Costa Rica and there is no published data for *Bd* detection through PCR or qPCR in this faunal province.

2.3.2. Altitudinal Belt

We classified species according to the Holdridge's life zone system [70], which divides Costa Rica into five altitudinal belts: lowland, premontane, lower montane, montane, and subalpine. Due to the elevational limits for altitudinal belts being slightly different among regions in Costa Rica, we established the limits of each belt as follows: lowland (0–700 m), premontane (700–1500 m), lower montane (1500–2700 m), montane (2700–3500 m), and subalpine (>3500 m).

2.3.3. Foraging-Reproduction Habitat Index

To compare *Bd* infection dynamics across taxonomic groups, we developed a foraging–reproduction habitat index (FRHI). The FRHI was created to classify species with a system of three letters that represented life history traits associated with foraging and reproduction (Table 2). First, we classified species according to their development into indirect- (I) or direct-developing amphibians (D). Second, we classified species according to their foraging habitat into terrestrial (T), arboreal (A), pond/puddle-dwellers (P), stream-dwellers (R), and phytotelma (F). Finally, species were classified according to their reproductive habitat into terrestrial (T), arboreal (A), pond/puddle-breeders (P), stream-breeders (R), and phytotelma (F).

Table 2. Categories and taxonomic examples of the foraging–reproduction habitat index (FRHI) that we developed for this study to analyze current prevalence of *Batrachochytrium dendrobatidis* across taxonomic groups. Symbology: First letter represents development: (I) indirect- or (D) direct-developing amphibians. Second letter represents foraging habitat: terrestrial (T), arboreal (A) pond/puddle-dwellers (P), stream-breeders (R), and phytotelma (F). Third letter represents reproductive habitat: terrestrial (T), arboreal (A), pond/puddle-dwellers (P), stream-breeders (R), and phytotelma (F).

FRHI	Species	Taxonomic Group (Examples)
DAA	2	*Diasporus* spp. (dink frogs, e.g., *Diasporus diastema*)
DAT	5	*Pristimantis* spp. (rain frogs, e.g., *Pristimantis cerasinus*)
DRT	3	*Craugastor punctariolus* clade (robber frogs, e.g., *Craugastor taurus*) *C. fitzingeri* (Pacific side)
DTT	13	*Craugastor* spp. (leaf-litter frogs. e.g., *Craugastor bransfordi*) *C. fitzingeri* (Caribbean side) Plethodontidae (e.g., *Oedipina gracilis*)
IAP	17	Hylidae (pond-breeding treefrogs, e.g., *Boana rufitela*)
IAR	15	Centrolenidae (glass frogs, e.g., *Teratohyla pulverata*) Hylidae (stream-breeding treefrogs, e.g., *Duellmanohyla rufioculis*)
ITF	4	Dendrobatidae (Poison-dart frogs, e.g., *Oophaga pumilio*)
ITP	12	Leptodactylidae (Leptodactylid frogs, e.g., *Leptodactylus melanonotus*) Microhylidae (sheep frogs, e.g., *Hypopachus variolosus*) Ranidae (Ranid frogs, e.g., *Lithobates forreri*)
ITR	7	Bufonidae (toads, e.g., *Incilius coccifer*) Bufonidae (river toads, e.g., *Rhaebo haematiticus*)

2.4. Statistical Analysis

We reduced our "combined dataset" to 1741 samples from 74 species and 20 localities because there was insufficient information to accurately classify nine records of the species *Diasporus tigrillo, D. vocator, Hyloscirtus palmeri, Triprion spinosus,* and *Cruziohyla calcarifer* in the FRHI. For our analyses, we pooled all species together instead of using species as a predictor because the samples sizes per species were highly variable (from 1–177), which could produce significant models that may be an artifact of opportunistic sampling instead of a real pattern. Instead, we used the FRHI, which is highly correlated with taxonomic group. We were unable to include time of sampling as a predictor in our analyses because these data were missing in several of the amphibian assemblages sampled. All our analyses were performed with the R package "stats" [71].

We analyzed *Bd* prevalence with fixed-effects generalized linear models (GLMs) using infected status as a binomial response variable (uninfected or infected) and herpetological province, altitudinal belt, and the FRHI as predictors. Ranking of the candidate GLMs followed the Akaike's information criterion (AIC) where the model with the lowest AIC was considered the most robust [72]. To analyze *Bd* infection intensity (estimated as the number of genomic equivalents), we analyzed the 351 *Bd*-positive swabs where *Bd* infection intensity was quantified through qPCR (see Section 2.3). We used linear models (LMs) to compare *Bd* infection intensity (response variable) across herpetological provinces, altitudinal belts, and FRHI (predictor variables). We log-transformed the *Bd* infection intensity to reduce skewness. Statistical significance of models was tested with ANOVA. For both, GLMs and LMs, we conducted post hoc, pairwise comparisons (Tukey's honestly significant difference; HSD-test) to confirm where the differences occurred between significant predictors. We were unable to run mix-effects models or fixed-effects interaction models because some combinations of predictors presented missing or low values, causing convergence difficulties.

3. Results

3.1. Species Assessment

The previous list of Costa Rican amphibians from 2011 included 196 species; however, we excluded the species *Pristimantis educatoris* [73] due to taxonomic uncertainty [74] for a total of 195 species. Our new list of amphibians in Costa Rica included a total of 215 species grouped in three orders, 16 families, and 48 genera (Table S3). This represented and addition of 20 species (10 anurans, 9 salamanders, and 1 caecilian; Table 3). The order Anura (frogs and toads) is the most diverse in Costa Rica, being 72% of the total species (13 families and 41 genera). Salamanders (order Caudata) are represented by only one family (Plethodontidae), three genera, and 53 species. Finally, caecilians (order Gymnophiona) are represent by two families, four genera, and eight species. A total of 63 species are endemic to Costa Rica (36 salamanders, 24 anurans, and 3 caecilians). We also included five anuran species that are not native to Costa Rica (*Eleutherodactylus coqui, E. johnstoni, E. planirostris, Lithobates catesbeianus,* and *Osteopilus septentrionalis*).

Table 3. List of new additions to the updated list of amphibians of Costa Rica.

Order	Family	Species	Source
	Centrolenidae	*Hyalinobatrachium dianae*	[75]
	Craugastoridae	*Craugastor aenigmaticus*	[76]
		Craugastor gabbi	[77]
		Craugastor zunigai	[78]
Anura	Eleutherodactylidae	*Diasporus amirae*	[79]
		Eleutherodactylus planirostris *	[80]
	Hylidae	*Ecnomiohyla bailarina*	[81]
		Ecnomiohyla veraguensis	Unpublished
		Smilisca manisorum	[82]
	Phyllomedusidae	*Cruziohyla sylviae*	[83]

Table 3. *Cont.*

Order	Family	Species	Source
		Bolitoglossa aurae	[84]
		Bolitoglossa aureogularis	[85]
		Bolitoglossa kamuk	[85]
		Bolitoglossa pygmaea	Unpublished
Caudata	Plethodontidae	*Bolitoglossa splendida*	[85]
		Nototriton costaricense	[86]
		Nototriton matama	[85]
		Oedipina berlini	[87]
		Oedipina nimaso	[85]
Gymnophiona	Caeciliidae	*Caecilia volcani*	[88]

* *Eleutherodactylus planirostris* is an invasive species that have been found in the Caribbean Lowlands of Costa Rica.

Regionally (Figure 2a, Table S3), the Cordillera de Talamanca is the most diverse province in terms of species per area (88 species, 2.2 species/100 km^2). It contains 23 species of amphibians that only occurred within this herpetological province (e.g., *Diasporus ventrimaculatus*, *Nototriton costaricense*). The Montane Slopes and Cordillera Central is the most diverse herpetological province (188 species, 1.3 species/100 km^2), with 27 species that are exclusively found within this herpetological province (e.g., *Cochranella euknemos*, *Nototriton picadoi*). The Caribbean Lowlands (101 species, 0.7 species/100 km^2) includes 20 species that are only found within this herpetological province (e.g., *Cruziohyla calcarifer*, *Caecilia volcani*). The Pacific Southwest (71 species, 0.9 species/100 km^2), has five species that only occur within this province (e.g., *Craugastor taurus*, *Oophaga granulifera*). Finally, the Pacific Northwest (66 species, 0.5 species/100 km^2) includes only two species that are found exclusively within this province (*Rhinophrynus dorsalis* and *Eleutherodactylus johnstonei*). A total of 20 species occur in all five herpetological provinces (e.g., *Craugastor fitzingeri*, *Diasporus diastema*, *Dendropsophus ebraccatus*, *Hyalinobatrachium fleischmanni*, *Lithobates warszewitschii*, *Rhinella horribilis*, *Smilisca sordida*).

Overall, 200 species have been classified into IUCN categories at the country level and 15 species need future assessment. A total of 155 species do not fulfill the criteria to be considered in the threatened categories, including 136 species listed as LC, 18 as DD, and one as NA (a category for taxa that occur in the region but have been excluded from the regional Red List for a specific reason). Within threatened categories, two species are listed as EX, 24 as CR, ten as EN, seven as VU, and two as NT. Regionally, lowlands exhibited the lowest percentage (0–2%) of DD species (Figure 2b). Similarly, about 75–80% of species occurring in lowlands are listed as LC (Figure 2c). In highlands, 6–10% of species are categorized as DD (Figure 2b) and 26% of species in Cordillera de Talamanca are classified within threatened categories (Figure 2d). According to EVS, a total of 81 species were classified as "no immediate risk," four species at "low vulnerability," 50 species at "medium vulnerability," and 48 species at "high vulnerability" (Table S3).

In our review, we found a total of 105 amphibian species (49%) that have been screened for *Bd* in Costa Rica (103 anurans and only 2 species of salamanders) (Table S4). In the field, the most common method used to detect *Bd* was qPCR, especially after 2005. Conventional PCR was used only in one study in the Caribbean Lowlands [64]. Histology and qPCR have also been used in retrospective studies on preserved specimens from declined and extinct species.

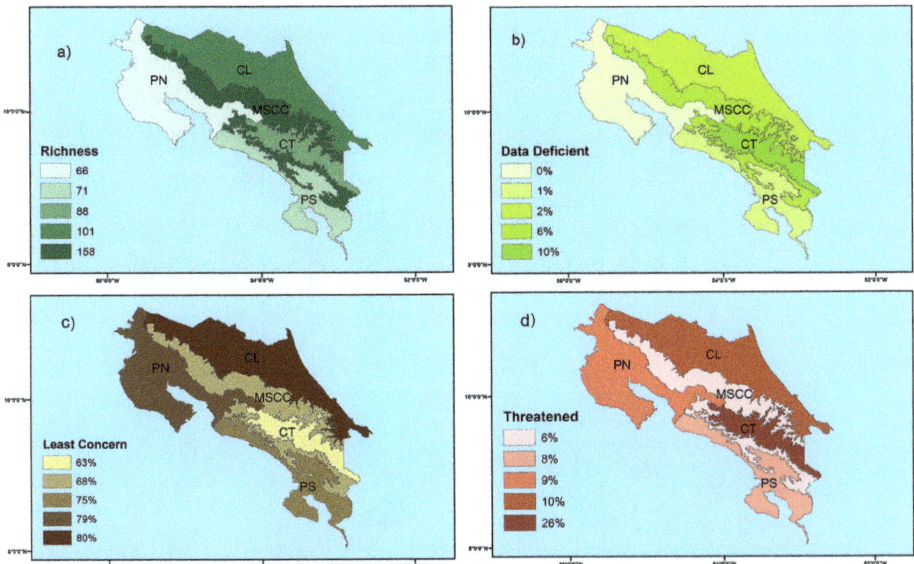

Figure 2. Map of Costa Rica showing (**a**) amphibian species richness for each herpetological province and percentage of amphibian species classified as (**b**) data deficient, (**c**) least concerned, and (**d**) threatened categories (near threatened, vulnerable, endangered, critically endangered, and extinct in the wild) for each herpetological province according to the Red List of Threatened Species from the International Union of Conservation of Nature (IUCN). Symbology: CL—Caribbean Lowlands, MSCC—Mountain Slopes and Cordillera Central, PN—Pacific Northwest, and PS—Pacific Southwest.

3.2. Endemic Dynamics

Overall, *Bd* prevalence in Costa Rica was estimated to be 0.23 (60% of species tested positive for *Bd*) (Table S5). The most robust GLM found both herpetological province and the FRHI as significant predictors of *Bd* prevalence (AIC = 1700, $p < 0.001$, Table S6). Among herpetological provinces (Figure 3), the highest percentage of infected individuals was found in the Caribbean Lowlands (34%) and the lowest in the Pacific Northwest (4%). The Mountain Slopes, Cordillera Central, and Pacific Southwest had a similar percentage of infected individuals ($\approx$23%). Furthermore, *Bd* was proportionally more prevalent in amphibians with terrestrial foraging and larval stage in phytotelma (ITF), pond-breeding treefrogs (IAP), and direct-developing species that breed in the forest (leaf-litter frogs DTT, rain frogs, DAT) (Figure 4a).

The species *Craugastor taurus* (the Golfito robber frog) was the species that had the highest average infection load (average *Bd* load of 11632.4 versus 571.6 genomic equivalents or 2.51 versus 1.18 after log transformation) (Table 4). We found an effect of the FRHI on infection load ($F_{8342} = 7.91$, $p < 0.01$, Table S6). Direct-developing frogs with terrestrial reproduction (robber frogs and leaf-litter frogs; DTR and DTT respectively) had the highest *Bd* loads (Figure 4b, Table 4).

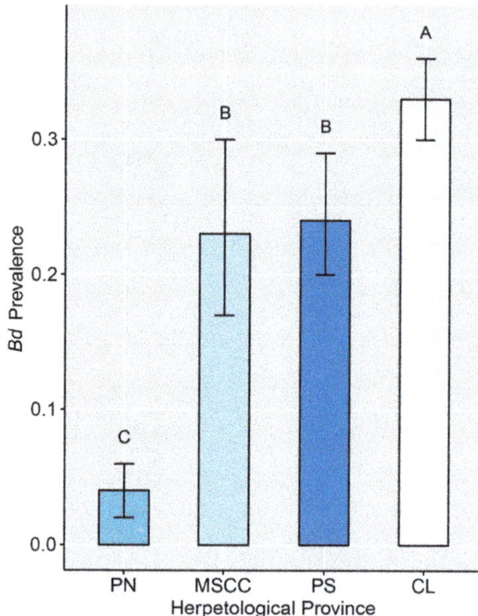

Figure 3. Mean prevalence of infection with *Batrachochytrium dendrobatidis* (*Bd*) in amphibian assemblages at four herpetological provinces in Costa Rica (with 95% binomial CI). Means followed by a common letter are not significantly different according to the Tukey's honestly significant difference (HSD) test at the 5% level of significance. The plot does not display results for Cordillera de Talamanca because no sampling has been conducted for *Bd* in that province. Symbology: CL—Caribbean Lowlands, MSCC—Mountain Slopes and Cordillera Central, PN—Pacific Northwest, and PS—Pacific Southwest.

Table 4. Infection intensity in the 351 individuals where *Batrachochytrium dendrobatidis* (*Bd*) was quantified using qPCR in Costa Rica between 2000–2018. For every species, the table shows the foraging–reproduction habitat index (FRHI), the number of *Bd* positive swabs, the average (SE), and $\log_{10}$(SE) of genomic equivalents of *Bd* zoospores quantified per species estimated from the *Bd* + swabs. Symbology: First letter represents development: (I) indirect or (D) direct-developing amphibians. Second letter represents foraging habitat: terrestrial (T), arboreal (A), pond/puddle-dwellers (P), stream-breeders (R), and phytotelma (F). Third letter represents reproductive habitat: terrestrial (T), arboreal (A), pond/puddle-dwellers (P), stream-breeders (R), and phytotelma (F).

Species (FRHI)	*Bd* + Swabs	*Bd* Load Average (SE)	$\log_{10}$ *Bd* Load Average (SE)
Agalychnis callidryas (IAP)	4	8.19 (3.81)	0.77 (0.21)
Agalychnis spurrelli (IAP)	5	39.83 (32.47)	1.10 (0.30)
Boana rufitela (IAP)	8	8.41 (4.12)	0.53 (0.23)
Bolitoglossa colonnea (DTT)	1	1.83 (0.00)	0.26 (0.00)
Cochranella granulosa (IAR)	1	3.95 (0.00)	0.60 (0.00)
Craugastor bransfordi (DTT)	23	1007.06 (483.50)	1.78 (0.25)
Craugastor crassidigitus (DTT)	5	1636.74 (1583.43)	1.69 (0.64)
Craugastor fitzingeri (DTT, DRT)	44	951.48 (310.27)	1.97 (0.17)
Craugastor megacephalus (DTT)	1	0.62 (0.00)	−0.21 (0.00)
Craugastor mimus (DTT)	9	125.48 (74.69)	1.01 (0.46)
Craugastor ranoides (DRT)	3	187.40 (174.13)	1.65 (0.55)
Craugastor stejnegerianus (DTT)	2	2.18 (0.95)	0.29 (0.20)
Craugastor taurus (DRT)	12	11,632.50 (6564.67)	2.51 (0.41)
Dendropsophus ebraccatus (IAP)	34	315.85 (194.09)	1.00 (0.18)
Diasporus diastema (DAA)	2	14.44 (4.12)	1.14 (0.13)
Duellmanohyla rufioculis (IAR)	1	3.65 (0.00)	0.56 (0.00)
Engystomops pustulosus (ITP)	11	34.83 (13.26)	1.11 (0.3)

Table 4. *Cont.*

Species (FRHI)	*Bd* + Swabs	*Bd* Load Average (SE)	Log₁₀ *Bd* Load Average (SE)
Espadarana prosoblepon (IAR)	3	3691.59 (3684.73)	1.06 (1.75)
Hyalinobatrachium colymbiphyllum (IAR)	1	0.01 (0.00)	−2.00 (0.00)
Hyalinobatrachium valerioi (IAR)	2	8.38 (2.04)	0.91 (0.11)
Incilius melanochlorus (ITR)	2	23.27 (19.93)	1.08 (0.56)
Leptodactylus melanonotus (ITP)	4	11.86 (0.66)	1.07 (0.02)
Leptodactylus poecilochilus (ITP)	1	1073.45 (0.00)	3.03 (0.00)
Leptodactylus savagei (ITP)	1	33.49 (0.00)	1.52 (0.00)
Lithobates forreri (ITP)	2	569.24 (241.10)	2.71 (0.20)
Lithobates warszewitschii (ITR)	14	978.92 (801.60)	1.47 (0.31)
Oophaga granulifera (ITF)	9	23.92 (11.31)	1.20 (0.11)
Oophaga pumilio (ITF)	34	1765.81 (778.67)	1.71 (0.25)
Pristimantis cerasinus (DAT)	9	14.82 (10.97)	0.47 (0.32)
Pristimantis ridens (DAT)	7	48.37 (32.34)	0.69 (0.50)
Rhaebo haematiticus (ITR)	22	239.20 (178.56)	0.70 (0.26)
Scinax boulengeri (IAP)	1	195.20 (0.00)	2.29 (0.00)
Scinax elaeochroa (IAP)	5	1384.15 (1350.51)	1.78 (0.58)
Smilisca phaeota (IAP)	4	37.25 (19.15)	1.44 (0.18)
Smilisca sordida (IAP)	46	14.96 (9.27)	0.24 (0.16)
Teratohyla pulverata (IAR)	2	34.53 (22.92)	1.41 (0.35)
Teratohyla spinosa (IAR)	5	937.99 (825.54)	1.90 (0.57)
Tlalocohyla loquax (IAP)	11	144.66 (107.28)	1.22 (0.30)

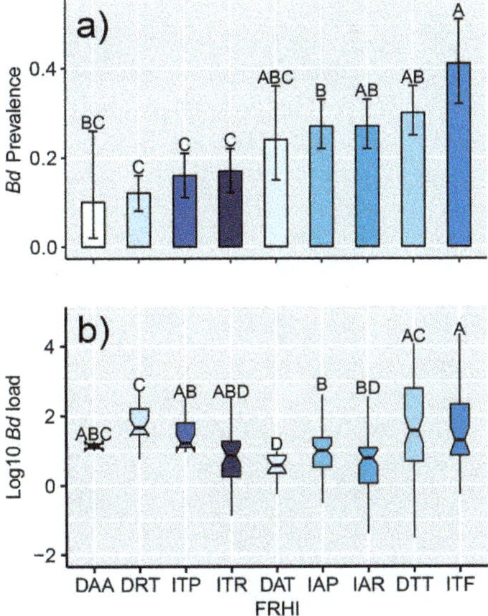

Figure 4. (**a**) Mean prevalence of infection with *Batrachochytrium dendrobatidis (Bd)* in amphibian assemblages (with 95% binomial CI) according to the foraging–reproduction habitat index (FRHI); (**b**) box plots with whiskers and notches that describe infection intensity for the 351 individuals where *Bd* was quantified using qPCR in Costa Rica between 2000–2018 according to the foraging–reproduction habitat index (FRHI). The box displays the inter-quantile range (25th–75th percentiles) with a center line representing the median (50th percentile). Notches show the median confidence region, and whiskers display the highest and lowest points. Means followed by a common letter are not significantly different according to the Tukey's honestly significant difference (HSD) test at the 5% level of significance. Symbology: First letter represents development: (I) indirect or (D) direct-developing amphibians. Second letter represents foraging habitat: terrestrial (T), arboreal (A), pond/puddle-dwellers (P), stream-breeders (R), and phytotelma (F). Third letter represents reproductive habitat: terrestrial (T), arboreal (A), pond/puddle-dwellers (P), stream-breeders (R), and phytotelma (F).

4. Discussion

4.1. Species Assessment

We presented the first updated list of Costa Rican amphibians since 2011 [47]. Compared to the last list, we added 10 anurans, 9 salamanders, and 1 caecilian (Table 3) for a total of 215 species (Table S3). As is common throughout the world, anurans exhibited the highest amphibian species richness in Costa Rica, with 72% of listed species. However, the richness of salamander species is also high (25%). In Costa Rica, the diversity and endemism of amphibians (especially salamanders) increase with elevation and complex mountain topography [13]. Proportionally, in terms of number of species per unit area (km^2), the richest herpetological province was the Cordillera de Talamanca. In this herpetological province, the number of species continued to increase and most of the newly described species in our report came from this remote and almost inaccessible province [69]. The Montane Slopes and Cordillera Central presented the highest number of species (158 species). Within this province, numerous mountain ranges provide multiple microhabitats for niche differentiation and further speciation [13,49]. In lowlands, the highest number of amphibian species occurred in the Caribbean Lowlands with 101 species. However, the Pacific Southwest presented more species per unit area. The Pacific Northwest only had 66 species, which was also the lowest number of species per unit area. This pattern may be attributed to the warm and dry conditions that occur in most part of this herpetological province [13,49].

According IUCN, the species *Craugastor escoces* and *Incilius periglenes* are classified as EX in Costa Rica. However, *C. escoces* was recently rediscovered [89]. Similarly, several species that remained undetected after the 1980s and 1990s such as *Incilius holdridgei* [20], *Craugastor taurus* [19], and *Atelopus varius* [18] have been rediscovered in peripheral populations during the last few years. However, the number of extinct species could be higher because multiple threatened species still remain undetected in the field (e.g., *Craugastor andi, Incilius fastidiosus, Atelopus senex*). We recommend expedition surveys to find populations of declined and data-deficient species [90] and captive-breeding for species where ex situ reproduction has been successful (e.g., harlequin frogs.) [91]. Although we acknowledge that there are limited funds available for these types of conservation efforts, knowledge from these sites is essential to be able to identify conditions that favor the persistence of threatened species and identify species that should be targeted for future conservation efforts [92].

Lowlands of Costa Rica exhibited the lowest proportion of DD species (0–2%; Figure 2b) and the highest proportion of LC species (75–80%; Figure 2c). On the other hand, highlands exhibited the highest percentage of DD species (6–10%; Figure 2b). Similarly, Cordillera de Talamanca had the highest percentage of threatened species (26%; Figure 2d). Based on these findings, we strongly recommend increasing the sampling effort in the montane and subalpine altitudinal belts (>2800 m) that exclusively occur in Cordillera de Talamanca and Montane Slopes and Cordillera Central. These herpetological provinces present the highest rate of endemism (especially for salamanders) and contain several of the recently described species [85,86]. Conducting expeditions and long-term studies in highlands will aid in monitoring threatened species and reducing information gaps, allowing for more accurate assessments of amphibian species.

To better evaluate the vulnerability of amphibian species, we utilized EVS (Table S3). This index relies on ecological information for categorizing threat levels, which makes application easy for most species in a specific region [52]. Unlike the IUCN Red List of Threatened Species, previous evaluations of threats are not considered by this index [49]. For that reason, species that are classified as LC by IUCN can be classified as highly vulnerable in this index (e.g., *Duellmanohyla rufioculis*). We categorized 48 species (24 salamanders, 21 anurans, and 3 caecilians) in "high vulnerability" (e.g., *Atelopus senex, Craugastor andi, Bolitoglossa pesrubra, Nototriton guanacaste*, and *Oscaecilia osae*). These species exhibited the highest EVS values because their habitats are restricted and because they exhibit complex reproductive modes. Quantifying environmental threats and combining information from both indexes will help policy-makers to prioritize conservation actions for threatened species.

In Central America, habitat destruction is the most important threat impacting amphibian populations [49,52]. Although approximately 30% of Costa Rica remains forested and protected, rapid urbanization, extensive agriculture, excessive pesticide use, illegal traffic, and inappropriate waste management negatively affect numerous amphibian populations. However, even in seemingly pristine locations, amphibian declines have occurred [93]. Additionally, climate change has been associated with the decline of several amphibian species in Costa Rica by affecting their reproduction and likely increasing susceptibility to pathogens [28]. Although it has not been found in Central America, we recommend screening for the recently emerged fungus *Batrachochytrium salamandrivorans* [94], which causes chytridiomycosis in salamanders. Conveniently, swabbing methods and qPCR allow for accurate detection of both fungi species in the same assay, which may facilitate rapid population assessments. For a fully detailed review of the environmental threats for amphibian communities in Costa Rica, we recommend the work of Sasa et al. [49].

4.2. Post-Decline Dynamics

In this study, we found strong evidence that *Bd* is widespread in Costa Rica. Our results also suggest that post-decline *Bd* exhibited enzootic dynamics, characterized by high prevalence of infection across regions and pathogenic loads below thresholds associated with mass mortalities [39,42]. We found *Bd* in all the herpetological provinces and altitudinal belts surveyed (Figures 3 and 4, Tables S5 and S6), for a total infection rate of 23%. We also found that *Bd* was more prevalent in terrestrial amphibians with an aquatic larval stage and direct-developing frogs exhibited the highest pathogenic loads.

We found the lowest infection rate (<5%) in the Pacific Northwest and the highest (33%) in the Caribbean Lowlands (Figure 3). However, these values may be an effect of sampling periods. A study conducted at La Selva Biological Station in the Caribbean Lowlands [95] reported infection rates varying from <5% during the warmest months (May to early November) to 35% in the coolest months (mid-November to January) in three common amphibian species. In addition, the gradual population declines observed at La Selva over several decades [31] and opportunistic observations of small-scale mortality events during cold periods [96] suggest that *Bd* may be causing mortality in amphibians long after its initial invasion. Similar mortality events in response to seasonality could be occurring in amphibian communities in "refuges from decline" in the Pacific Northwest [62,97] and Pacific Southwest [63] of Costa Rica; however, they have not yet been documented. There, seasonal changes in precipitation and temperature caused *Bd* prevalence to vary from >5% in the peak of the dry season (March and April) to 80% in the coldest months (November–December). Therefore, we recommend follow-up studies at these sites to identify whether seasonal disease dynamics are causing mortality events in regions that have been considered unsuitable for *Bd* [98].

Bd was found across all altitudinal belts across Costa Rica. Similar results of high *Bd* prevalence across all elevations has been found in Panama [99–101]. These findings suggest that current environmental conditions are suitable for *Bd* at most elevations in Central America [63]. It is also plausible that *Bd*-driven declines during the 1980s and 1990s were not exclusively restricted to highlands [30] but were relatively undetected at lower elevations. Another hypothesis is that species with high susceptibility historically occupied high elevations sites, but severely declined or went extinct after *Bd* was introduced, leaving only species with mid-to-low susceptibility across elevations [102]. On the other hand, the absence of samples from montane and subalpine belts and uncontrolled variables (e.g., changes in species composition, climatic disturbances) could have reduced the statistical power to determine changes in *Bd* prevalence across altitudinal belts. We suggest that future studies increase sampling at high elevations (>2700 m) to better understand the local spatial dynamics of *Bd* across elevations.

Our results showed that infection with *Bd* was common in amphibians across all life-history traits evaluated in the FRHI (Table 2). However, *Bd* was significantly more prevalent in terrestrial amphibians with a larval stage (Figure 4a), especially those that complete metamorphosis in phytotelma (ITF). All the species within the ITF category belonged to the family Dendrobatidae, which have previously been shown to easily acquire *Bd* infection [62,95]. The high susceptibly of the Dendrobatidae family is likely due to their preferred habitat (e.g., water-filled bromeliads for many species), as it offers suitable conditions for *Bd* infection [103]. In addition, dendrobatid adults forage in the tropical forest floor and stream-associated low vegetation, which are environments that can sustain *Bd* [104]. Regarding infection intensity (Figure 4b), the FRHI showed similar results to studies that have used the aquatic index [43]. We found that direct-developing species with terrestrial reproduction had significantly higher infection load than other species with different life-history traits (Figure 4b, Table 4). This life-history trait is exhibited by leaf-litter frogs and all the species within the *Craugastor punctariolus* clade (robber frogs), which is one of the clades most affected by chytridiomycosis in Central America [29,105]. Robber frogs spend a majority of their life cycle along fast-flowing streams [106], an aquatic environment that seems highly suitable for *Bd* in Central America [107]. In addition, these frogs appear to be highly susceptible to *Bd*-driven mass mortalities outside warm and dry ecosystems [19,30,108]. Our results suggest the FRHI is particularly useful for identifying taxonomic units that are more susceptible to *Bd* [92].

5. Conclusions

Our results demonstrated that the number of identified amphibian species in Costa Rica is still growing, and there may be potential future additions (e.g., *Bolitoglossa anthracina* [109] and *B. indio* [110]). A continuous assessment of species and regions is needed to identify continuing threats to amphibian biodiversity. We found that *Bd* was widespread across species, herpetological provinces, and altitudinal belts in samples collected since 2000. Conducting more studies in remote regions, such as Cordillera de Talamanca, may help to better describe spatial dynamics of both amphibian hosts and *Bd*. In addition, future studies should test whether seasonal disease dynamics are causing mortality events in regions that are considered unsuitable for *Bd*. Under potential scenarios of climate change, environmental conditions may shift to ideal ranges for *Bd* infection [28] and seasonal regions that sustain critically endangered species (e.g., tropical dry forest) may experience future outbreaks of chytridiomycosis [111]. We also recommend continuous surveillance of invasive species, which might amplify *Bd* in the environment, causing future epizootics [37]. This vital information will aid in the development of more effective conservation strategies for amphibians across a broader range of habitats [46,112–114].

Supplementary Materials: The following are available online at http://www.mdpi.com/1424-2818/11/8/129/s1. Figure S1: Map of Costa Rica showing elevational gradient and nine study sites surveyed for *Batrachochytrium dendrobatidis*, Figure S2: Prevalence and intensity of infection of *Batrachochytrium dendrobatidis* in amphibian assemblages from the nine surveyed sites in Costa Rica, Table S1: List of species where *Batrachochytrium dendrobatidis* was surveyed in Costa Rica for our "field dataset", Table S2: Prevalence and infection intensity of *Batrachochytrium dendrobatidis* at nine sites in Costa Rica, Table S3: Distribution of amphibians in Costa Rica according to herpetological province and elevation, Table S4: List of species that have been screened for *Batrachochytrium dendrobatidis* in Costa Rica, Table S5: List of species where *Batrachochytrium dendrobatidis* was surveyed in Costa Rica in our combined dataset, Table S6: Candidacy generalized linear models (GLMs) and linear models (LMs) used to determine the best predictors of prevalence and infection intensity of *Batrachochytrium dendrobatidis* in amphibian assemblages from Costa Rica.

Author Contributions: Conceptualization, H.Z.; Data curation, H.Z., K.N., A.G., G.C., E.A., F.B., S.W., and C.S.; Formal analysis, H.Z., K.N., A.G., and G.C.; Methodology, H.Z., K.N., A.G., G.C., E.A., F.B., S.W., and C.S.; Project administration, C.S.; Supervision, H.Z. and C.S.; Writing—Original draft, H.Z. and K.N.; Writing—Review and editing, H.Z., K.N., A.G., G.C., E.A., F.B., S.W., and C.S.

Funding: H.Z. was funded by (1) the Department of Biological Sciences at Purdue University through the Lindsay Fellowship and the Andrews Environmental Travel Grant, (2) the Osa Conservation through the Álvaro Ugalde Scholarship, and (3) IdeaWild. K.N. was funded by the Department of Biological Sciences at Purdue University through the McAtee Scholarship and a Cable-Silkman award. A.G. is currently supported by a postdoctoral fellowship from Dirección General de Asuntos del Personal Académico (DGAPA) at Instituto de Biología, Universidad Nacional Autónoma de México. The field trip following the Gabb's route was funded by the National Geographic Society (grant number W-346-14).

Acknowledgments: We thank B. Araya, F. Ureña, and, P. Piza for field assistance, and D. Vasquez for statistical counseling. We also acknowledge the Costa Rican Ministry of Environment and Energy (MINAE)) for providing the corresponding scientific collecting permits for this work (research permits 001-2012-SINAC, R-019-2016-OT-CONAGEBIO, R-023-2016-OT-CONAGEBIO, R-057-2016-OT-CONAGEBIO, R-060-2016-OT-CONAGEBIO).

Conflicts of Interest: The authors declare no conflict of interest. The funders had no role in the design of the study; in the collection, analyses, or interpretation of data; in the writing of the manuscript; or in the decision to publish the results.

References

1. Novacek, M.J.; Cleland, E.E. The current biodiversity extinction event: Scenarios for mitigation and recovery. *Proc. Natl. Acad. Sci. USA* **2001**, *98*, 5466–5470. [CrossRef] [PubMed]
2. Barnosky, A.D.; Matzke, N.; Tomiya, S.; Wogan, G.O.U.; Swartz, B.; Quental, T.B.; Marshall, C.; McGuire, J.L.; Lindsey, E.L.; Maguire, K.C.; et al. Has the Earth's sixth mass extinction already arrived? *Nature* **2011**, *471*, 51–57. [CrossRef] [PubMed]
3. Wake, D.B.; Vredenburg, V.T. Are we in the midst of the sixth mass extinction? A view from the world of amphibians. *Proc. Natl. Acad. Sci. USA* **2008**, *105*, 11466–11473. [CrossRef] [PubMed]
4. Monastersky, R. Life—A status report. *Nature* **2014**, *516*, 158–161. [CrossRef] [PubMed]
5. Stuart, S.N.; Chanson, J.; Cox, N.A.; Young, B.E.; Rodrigues, A.S.L.; Fischman, D.; Waller, R. Status and trends of amphibian declines and extinctions worldwide. *Science* **2004**, *306*, 1783–1786. [CrossRef]
6. Catenazzi, A. State of the World's Amphibians. *Annu. Rev. Environ. Resour.* **2015**, *40*, 91–119. [CrossRef]
7. Daszak, P.; Berger, L.; Cunningham, A.A.; Hyatt, A.D.; Green, D.E.; Speare, R. Emerging infectious diseases and amphibian population declines. *Emerg. Infect. Dis.* **1999**, *5*, 735–748. [CrossRef] [PubMed]
8. Collins, J.P. Amphibian decline and extinction: What we know and what we need to learn. *Dis. Aquat. Organ.* **2010**, *92*, 93–99. [CrossRef] [PubMed]
9. La Marca, E.; Lips, K.R.; Lotters, S.; Puschendorf, R.; Ibanez, R.; Rueda-Almonacid, J.V.; Schulte, R.; Marty, C.; Castro, F.; Manzanilla-Puppo, J.; et al. Catastrophic population declines and extinctions in Neotropical harlequin frogs (Bufonidae: *Atelopus*). *Biotropica* **2005**, *37*, 190–201. [CrossRef]
10. Scheele, B.C.; Pasmans, F.; Skerratt, L.F.; Berger, L.; Martel, A.; Beukema, W.; Acevedo, A.A.; Burrowes, P.A.; Carvalho, T.; Catenazzi, A.; et al. Amphibian fungal panzootic causes catastrophic and ongoing loss of biodiversity. *Science* **2019**, *363*, 1459–1463. [CrossRef]
11. Gerber, B.D.; Converse, S.J.; Muths, E.; Crockett, H.J.; Mosher, B.A.; Bailey, L.L. Identifying species conservation strategies to reduce disease-associated declines. *Conserv. Lett.* **2018**, *11*, e12393. [CrossRef]
12. Meredith, H.; Van Buren, C.; Antwis, R.E. Making amphibian conservation more effective. *Conserv. Evid.* **2016**, *13*, 1–6.
13. Savage, J.M. *The Amphibians and Reptiles of Costa Rica: A Herpetofauna between Two Continents, between Two Deas*; University of Chicago Press: Chicago, IL, USA, 2002.
14. Frost, D.R. Amphibian Species of the World: An Online Reference. Version 6.0. Available online: http://research.amnh.org/vz/herpetology/amphibia/ (accessed on 27 June 2019).
15. Bagley, J.C.; Johnson, J.B. Phylogeography and biogeography of the lower Central American Neotropics: Diversification between two continents and between two seas. *Biol. Rev.* **2014**, *89*, 767–790. [CrossRef] [PubMed]
16. Bolaños, F. Situación de los anfibios de Costa Rica. *Biocenosis* **2009**, *22*, 95–108.
17. Pounds, J.A.; Crump, M.L. Amphibian declines and climate disturbance: The case of the golden toad and the harlequin frog. *Conserv. Biol.* **1994**, *8*, 72–85. [CrossRef]
18. González-Maya, J.F.; Belant, J.L.; Wyatt, S.A.; Schipper, J.; Cardenal, J.; Corrales, D.; Cruz-Lizano, I.; Hoepker, A.; Escobedo-Galván, A.H.; Castañeda, F.; et al. Renewing hope: The rediscovery of *Atelopus varius* in Costa Rica. *Amphib. Reptil.* **2013**, *34*, 573–578. [CrossRef]
19. Chaves, G.; Zumbado-Ulate, H.; García-Rodríguez, A.; Gómez, E.; Vredenburg, V.T.; Ryan, M.J. Rediscovery of the critically endangered streamside frog, *Craugastor taurus* (Craugastoridae), in Costa Rica. *Trop. Conserv. Sci.* **2014**, *7*, 628–638. [CrossRef]
20. Abarca, J.; Chaves, G.; García-Rodríguez, A.; Vargas, R. Reconsidering extinction: Rediscovery of *Incilius holdridgei* (Anura: Bufonidae) in Costa Rica after 25 years. *Herpetol. Rev.* **2010**, *41*, 150.

21. Hero, J.-M.; Williams, S.E.; Magnusson, W.E. Ecological traits of declining amphibians in upland areas of eastern Australia. *J. Zool.* **2005**, *267*, 221–232. [CrossRef]

22. Mendelson, J.R.; Whitfield, S.M.; Sredl, M.J. A recovery engine strategy for amphibian conservation in the context of disease. *Biol. Conserv.* **2019**, *236*, 188–191. [CrossRef]

23. Longcore, J.E.; Pessier, A.P.; Nichols, D.K. *Batrachochytrium dendrobatidis* gen. et sp. nov., a chytrid pathogenic to amphibians. *Mycologia* **1999**, *91*, 219–227. [CrossRef]

24. Berger, L.; Speare, R.; Daszak, P.; Green, D.E.; Cunningham, A.A.; Goggin, C.L.; Slocombe, R.; Ragan, M.A.; Hyatt, A.D.; McDonald, K.R.; et al. Chytridiomycosis causes amphibian mortality associated with population declines in the rain forests of Australia and Central America. *Proc. Natl. Acad. Sci. USA* **1998**, *95*, 9031–9036. [CrossRef] [PubMed]

25. Lips, K.R.; Green, D.E.; Papendick, R. Chytridiomycosis in wild frogs from southern Costa Rica. *J. Herpetol.* **2003**, *37*, 215–218. [CrossRef]

26. Puschendorf, R.; Bolaños, F.; Chaves, G. The amphibian chytrid fungus along an altitudinal transect before the first reported declines in Costa Rica. *Biol. Conserv.* **2006**, *132*, 136–142. [CrossRef]

27. Piotrowski, J.S.; Annis, S.L.; Longcore, J.E. Physiology of *Batrachochytrium dendrobatidis*, a chytrid pathogen of amphibians. *Mycologia* **2004**, *96*, 9–15. [CrossRef] [PubMed]

28. Pounds, A.J.; Bustamante, M.R.; Coloma, L.A.; Consuegra, J.A.; Fogden, M.P.L.; Foster, P.N.; La Marca, E.; Masters, K.L.; Merino-Viteri, A.; Puschendorf, R.; et al. Widespread amphibian extinctions from epidemic disease driven by global warming. *Nature* **2006**, *439*, 161–167. [CrossRef] [PubMed]

29. Zumbado-Ulate, H.; Bolaños, F.; Willink, B.; Soley-Guardia, F. Population status and natural history notes on the critically endangered stream dwelling frog *Craugastor ranoides* (Craugastoridae) in a Costa Rican tropical dry forest. *Herpetol. Conserv. Biol.* **2011**, *6*, 455–464.

30. Puschendorf, R.; Carnaval, A.C.; VanDerWal, J.; Zumbado-Ulate, H.; Chaves, G.; Bolaños, F.; Alford, R.A. Distribution models for the amphibian chytrid *Batrachochytrium dendrobatidis* in Costa Rica: Proposing climatic refuges as a conservation tool. *Divers. Distrib.* **2009**, *15*, 401–408. [CrossRef]

31. Whitfield, S.M.; Bell, K.E.; Philippi, T.; Sasa, M.; Bolaños, F.; Chaves, G.; Savage, J.M.; Donnelly, M.A. Amphibian and reptile declines over 35 years at La Selva, Costa Rica. *Proc. Natl. Acad. Sci. USA* **2007**, *104*, 8352–8356. [CrossRef]

32. Christie, M.R.; Searle, C.L. Evolutionary rescue in a host–pathogen system results in coexistence not clearance. *Evol. Appl.* **2018**, *11*, 681–693. [CrossRef]

33. Retallick, R.W.R.; Miera, V. Strain differences in the amphibian chytrid *Batrachochytrium dendrobatidis* and non-permanent, sub-lethal effects of infection. *Dis. Aquat. Organ.* **2007**, *75*, 201–207. [CrossRef] [PubMed]

34. Briggs, C.J.; Vredenburg, V.T.; Knapp, R.A.; Rachowicz, L.J. Investigating the population-level effects of chytridiomycosis: An emerging infectious disease of amphibians. *Ecology* **2005**, *86*, 3149–3159. [CrossRef]

35. Rachowicz, L.J.; Knapp, R.A.; Morgan, J.A.T.; Stice, M.J.; Vredenburg, V.T.; Parker, J.M.; Briggs, C.J. Emerging infectious disease as a proximate cause of amphibian mass mortality. *Ecology* **2006**, *87*, 1671–1683. [CrossRef]

36. Retallick, R.W.R.; McCallum, H.; Speare, R. Endemic infection of the amphibian chytrid fungus in a frog community post-decline. *PLoS Biol.* **2004**, *2*, e351. [CrossRef] [PubMed]

37. Searle, C.L.; Biga, L.M.; Spatafora, J.W.; Blaustein, A.R. A dilution effect in the emerging amphibian pathogen *Batrachochytrium dendrobatidis*. *Proc. Natl. Acad. Sci. USA* **2011**, *108*, 16322–16326. [CrossRef] [PubMed]

38. O'Brien, V.A.; Moore, A.T.; Young, G.R.; Komar, N.; Reisen, W.K.; Brown, C.R. An enzootic vector-borne virus is amplified at epizootic levels by an invasive avian host. *Proc. R. Soc. B Biol. Sci.* **2011**, *278*, 239–246. [CrossRef] [PubMed]

39. Briggs, C.J.; Knapp, R.A.; Vredenburg, V.T. Enzootic and epizootic dynamics of the chytrid fungal pathogen of amphibians. *Proc. Natl. Acad. Sci. USA* **2010**, *107*, 9695–9700. [CrossRef]

40. Miller, C.A.; Tasse Taboue, G.C.; Ekane, M.M.P.; Robak, M.; Sesink Clee, P.R.; Richards-Zawacki, C.; Fokam, E.B.; Fuashi, N.A.; Anthony, N.M. Distribution modeling and lineage diversity of the chytrid fungus *Batrachochytrium dendrobatidis* (Bd) in a central African amphibian hotspot. *PLoS ONE* **2018**, *13*, e0199288. [CrossRef]

41. Hitchman, S.M.; Mather, M.E.; Smith, J.M.; Fencl, J.S. Identifying keystone habitats with a mosaic approach can improve biodiversity conservation in disturbed ecosystems. *Glob. Chang. Biol.* **2018**, *24*, 308–321. [CrossRef]

42. Vredenburg, V.T.; Knapp, R.A.; Tunstall, T.S.; Briggs, C.J. Dynamics of an emerging disease drive large-scale amphibian population extinctions. *Proc. Natl. Acad. Sci. USA* **2010**, *107*, 9689–9694. [CrossRef]

43. Brem, F.; Lips, K. *Batrachochytrium dendrobatidis* infection patterns among Panamanian amphibian species, habitats and elevations during epizootic and enzootic stages. *Dis. Aquat. Organ.* **2008**, *81*, 189–202. [CrossRef] [PubMed]

44. Kriger, K.M.; Hero, J.-M. The chytrid fungus *Batrachochytrium dendrobatidis* is non-randomly distributed across amphibian breeding habitats. *Divers. Distrib.* **2007**, *13*, 781–788. [CrossRef]

45. Heard, G.W.; Scroggie, M.P.; Ramsey, D.S.L.; Clemann, N.; Hodgson, J.A.; Thomas, C.D. Can habitat management mitigate disease impacts on threatened amphibians? *Conserv. Lett.* **2018**, *11*, e12375. [CrossRef]

46. Scheele, B.C.; Hunter, D.A.; Grogan, L.F.; Berger, L.; Kolby, J.E.; Mcfadden, M.S.; Marantelli, G.; Skerratt, L.F.; Driscoll, D.A. Interventions for reducing extinction risk in chytridiomycosis-threatened amphibians. *Conserv. Biol.* **2014**, *28*, 1195–1205. [CrossRef] [PubMed]

47. Bolanos, F.; Savage, J.M.; Chaves, G. Anfibios y Reptiles de Costa Rica. Listas Zoológicas Actualizadas UCR. Available online: http://museo.biologia.ucr.ac.cr/Listas/Anteriores/HerpCREsp.htm (accessed on 27 June 2019).

48. Savage, J.M.; Bolaños, F. A checklist of the amphibians and reptiles of Costa Rica: Additions and nomenclatural revisions. *Zootaxa* **2009**, *2005*, 1–23. [CrossRef]

49. Sasa, M.; Chaves, G.; Porras, L.W. The Costa Rican herpetofauna: Conservation status and future perspectives. In *Conservation of Mesoamerican Amphibians and Reptiles*; Townsend, J.H., Johnson, J.D., Eds.; Eagle Mountain Press: Salt Lake City, UT, USA, 2010; pp. 510–603.

50. Chaves, G.; Bolaños, F.; Rodríguez, J.E.; Matamoros, Y. *Actualización de las Listas Rojas Nacionales de Costa Rica. Anfibios y Reptiles*; Conservation Breeding Specialist Group (SSC/IUCN)/CBSG Mesoamerica): San José, Costa Rica, 2014.

51. IUCN The IUCN Red List of Threatened Species. Version 2019-1. Available online: https://www.iucnredlist.org/en (accessed on 27 June 2019).

52. Wilson, L.D.; McCranie, J.R. The conservation status of the herpetofauna of Honduras. *Amphib. Reptile Conserv.* **2004**, *3*, 6–33. [PubMed]

53. Voyles, J.; Young, S.; Berger, L.; Campbell, C.; Voyles, W.F.; Dinudom, A.; Cook, D.; Webb, R.; Alford, R.A.; Skerratt, L.F.; et al. Pathogenesis of chytridiomycosis, a cause of catastrophic amphibian declines. *Science* **2009**, *326*, 582–585. [CrossRef]

54. Kriger, K.M.; Hines, H.B.; Hyatt, A.D.; Boyle, D.G.; Hero, J.-M. Techniques for detecting chytridiomycosis in wild frogs: Comparing histology with real-time Taqman PCR. *Dis. Aquat. Organ.* **2006**, *71*, 141–148. [CrossRef]

55. Skerratt, L.; Berger, L.; Hines, H.; McDonald, K.; Mendez, D.; Speare, R. Survey protocol for detecting chytridiomycosis in all Australian frog populations. *Dis. Aquat. Organ.* **2008**, *80*, 85–94. [CrossRef]

56. Boyle, D.G.; Boyle, D.B.; Olsen, V.; Morgan, J.A.T.; Hyatt, A.D. Rapid quantitative detection of chytridiomycosis (*Batrachochytrium dendrobatidis*) in amphibian samples using real-time Taqman PCR assay. *Dis. Aquat. Organ.* **2004**, *60*, 141–148. [CrossRef]

57. Hyatt, A.D.; Boyle, D.G.; Olsen, V.; Boyle, D.B.; Berger, L.; Obendorf, D.; Dalton, A.; Kriger, K.; Hero, M.; Hines, H.; et al. Diagnostic assays and sampling protocols for the detection of *Batrachochytrium dendrobatidis*. *Dis. Aquat. Organ.* **2007**, *73*, 175–192. [CrossRef] [PubMed]

58. Kriger, K.M.; Hero, J.-M.; Ashton, K.J. Cost efficiency in the detection of chytridiomycosis using PCR assay. *Dis. Aquat. Organ.* **2006**, *71*, 149–154. [CrossRef] [PubMed]

59. Picco, A.M.; Collins, J.P. Fungal and viral pathogen occurrence in Costa Rican amphibians. *J. Herpetol.* **2007**, *41*, 746–749. [CrossRef]

60. Goldberg, C.S.; Hawley, T.J.; Waits, L.P. Local and regional patterns of amphibian chytrid prevalence on the Osa Peninsula, Costa Rica. *Herpetol. Rev.* **2009**, *40*, 309–311.

61. Whitfield, S.M.; Geerdes, E.; Chacon, I.; Ballestero Rodriguez, E.; Jimenez, R.; Donnelly, M.; Kerby, J. Infection and co-infection by the amphibian chytrid fungus and ranavirus in wild Costa Rican frogs. *Dis. Aquat. Organ.* **2013**, *104*, 173–178. [CrossRef] [PubMed]

62. Zumbado-Ulate, H.; Bolaños, F.; Gutiérrez-Espeleta, G.; Puschendorf, R. Extremely low prevalence of *Batrachochytrium dendrobatidis* in frog populations from Neotropical dry forest of Costa Rica supports the existence of a climatic refuge from disease. *EcoHealth* **2014**, *11*, 593–602. [CrossRef] [PubMed]

63. Zumbado-Ulate, H.; García-Rodríguez, A.; Vredenburg, V.T.; Searle, C.L. Infection with *Batrachochytrium dendrobatidis* is common in tropical lowland habitats: Implications for amphibian conservation. *Ecol. Evol.* **2019**, *9*, 4917–4930. [CrossRef] [PubMed]

64. Saenz, D.; Adams, C.K.; Pierce, J.B.; Laurencio, D. Occurrence of *Batrachochytrium dendrobatidis* in an anuran community in the southeastern Talamanca region of Costa Rica. *Herpetol. Rev.* **2009**, *40*, 311–313.

65. Abarca, J.G. Quitridiomicosis en Costa Rica: Aislamiento y Descripción de Cepas Circulantes del Patógeno y Análisis de la Microbiota del Hospedero como Posible Factor en la Incidencia de la Enfermedad. Master's Thesis, Universidad de Costa Rica, Heredia, Costa Rica, 2018.

66. Lips, K.R.; Diffendorfer, J.; Mendelson, J.R.; Sears, M.W. Riding the wave: Reconciling the roles of disease and climate change in amphibian declines. *PLoS Biol.* **2008**, *6*, e72. [CrossRef]

67. Lips, K.R.; Burrowes, P.A.; Mendelson, J.R.; Parra-Olea, G. Amphibian declines in Latin America: Widespread population declines, extinctions, and impacts. *Biotropica* **2005**, *37*, 163–165. [CrossRef]

68. Whitfield, S.M.; Lips, K.R.; Donnelly, M.A. Amphibian decline and conservation in Central America. *Copeia* **2016**, *104*, 351–379. [CrossRef]

69. Arias, E.; Chaves, G. 140 years after William M. Gabb's climb to Cerro Pico Blanco. *Mesoam. Herpetol.* **2014**, *1*, 176–180.

70. Holdridge, L.R. *Life Zone Ecology*; Tropical Science Center: San Jose, Costa Rica, 1967.

71. R Core Team. *R: A Language and Environment for Statistical Computing*; R Foundation for Statistical Computing: Vienna, Austria, 2019; Available online: https://www.r-project.org/ (accessed on 22 February 2019).

72. Burnham, K.P.; Anderson, D.R. Multimodel inference: Understanding AIC and BIC in model selection. *Sociol. Methods Res.* **2004**, *33*, 261–304. [CrossRef]

73. Ryan, M.; Lips, K.; Giermakowski, J.T. New species of *Pristimantis* (Anura: Terrarana: Strabomantinae) from lower Central America. *J. Herpetol.* **2010**, *44*, 193–200. [CrossRef]

74. Batista, A.; Hertz, A.; Köhler, G.; Mebert, K.; Vesely, M. Morphological variation and phylogeography of frogs related to *Pristimantis caryophyllaceus* (Anura: Terrarana: Craugastoridae) in Panama. *Salamandra* **2014**, *50*, 155–171.

75. Kubicki, B.; Salazar, S.; Puschendorf, R. A new species of glassfrog, genus *Hyalinobatrachium* (Anura: Centrolenidae), from the Caribbean foothills of Costa Rica. *Zootaxa* **2015**, *3920*, 069–084. [CrossRef]

76. Arias, E.; Chaves, G.; Parra-Olea, G. A new species of *Craugastor* (Anura: Craugastoridae) from the montane rainforest of the Cordillera de Talamanca, Costa Rica. *Phyllomedusa J. Herpetol.* **2018**, *17*, 211–232. [CrossRef]

77. Arias, E.; Chaves, G.; Crawford, A.J.; Parra-Olea, G. A new species of the *Craugastor podiciferus* species group (Anura: Craugastoridae) from the premontane forest of southwestern Costa Rica. *Zootaxa* **2016**, *4132*, 347–363. [CrossRef]

78. Arias, E.; Hertz, A.; Parra-Olea, G. Taxonomic assessment of *Craugastor podiciferus* (Anura: Craugastoridae) in lower Central America with the description of two new species. *Amphib. Reptile Conserv.* **2019**, *13*, 173–197.

79. Arias, E.; Chaves, G.; Salazar, S.; Salazar-Zúñiga, J.A.; García-Rodríguez, A. A new species of dink frog, genus *Diasporus* (Anura: Eleutherodactylidae), from the Caribbean foothills of the Cordillera de Talamanca, Costa Rica. *Zootaxa* **2019**, *4609*, 269–288. [CrossRef]

80. Barquero, M.D.; Araya, M.F. First record of the Greenhouse frog, *Eleutherodactylus planirostris* (Anura: Eleutherodactylidae), in Costa Rica. *Herpetol. Notes* **2016**, *9*, 145–147.

81. Kubicki, B.; Salazar, S. Discovery of the golden-eyed fringe-limbed treefrog, *Ecnomiohyla bailarina* (Anura: Hylidae), in the Caribbean foothills of southeastern Costa Rica. *Mesoam. Herpetol.* **2015**, *2*, 76–86.

82. McCranie, J.R. Morphological and systematic comments on the Caribbean lowland population of *Smilisca baudinii* (Anura: Hylidae: Hylinae) in northeastern Honduras, with the resurrection of *Hyla manisorum*. *Mesoam. Herpetol.* **2017**, *4*, 513–526.

83. Gray, A.R. Review of the genus *Cruziohyla* (Anura: Phyllomedusidae), with description of a new species. *Zootaxa* **2018**, *4450*, 401–426. [CrossRef] [PubMed]

84. Kubicki, B.; Arias, E. A beautiful new yellow salamander, genus *Bolitoglossa* (Caudata: Plethodontidae), from the northeastern slopes of the Cordillera de Talamanca, Costa Rica. *Zootaxa* **2016**, *4184*, 329–346. [CrossRef] [PubMed]

85. Boza-Oviedo, E.; Rovito, S.M.; Chaves, G.; Garcia-Rodriguez, A.; Artavia, L.G.; Bolaños, F.; Wake, D.B. Salamanders from the eastern Cordillera de Talamanca, Costa Rica, with descriptions of five new species (Plethodontidae: *Bolitoglossa, Nototriton,* and *Oedipina*) and natural history notes from recent expeditions. *Zootaxa* **2012**, *3309*, 36–61. [CrossRef]

86. Arias, E.; Kubicki, B. A new moss salamander, genus *Nototriton* (Caudata: Plethodontidae), from the Cordillera de Talamanca, in the Costa Rica-Panama border region. *Zootaxa* **2018**, *4369*, 487–500. [CrossRef] [PubMed]

87. Kubicki, B. A new species of salamander (Caudata: Plethodontidae: *Oedipina*) from the central Caribbean foothills of Costa Rica. *Mesoam. Herpetol.* **2016**, *3*, 819–840.

88. Kubicki, B.; Arias, E. Vulcan's Slender Caecilian, *Caecilia volcani*, in Costa Rica. *Mesoam. Herpetol.* **2017**, *4*, 488–492.

89. Jiménez, R.; Alvarado, G. *Craugastor escoces* (Anura: Craugastoridae) reappears after 30 years: Rediscovery of an "extinct" Neotropical frog. *Amphib. Reptil.* **2017**, *38*, 257–259. [CrossRef]

90. García-Rodríguez, A.; Chaves, G.; Benavides-Varela, C.; Puschendorf, R. Where are the survivors? Tracking relictual populations of endangered frogs in Costa Rica. *Divers. Distrib.* **2012**, *18*, 204–212. [CrossRef]

91. Lewis, C.H.R.; Richards-Zawacki, C.L.; Ibáñez, R.; Luedtke, J.; Voyles, J.; Houser, P.; Gratwicke, B. Conserving Panamanian harlequin frogs by integrating captive-breeding and research programs. *Biol. Conserv.* **2019**, *236*, 180–187. [CrossRef]

92. Searle, C.L.; Gervasi, S.S.; Hua, J.; Hammond, J.I.; Relyea, R.A.; Olson, D.H.; Blaustein, A.R. Differential host susceptibility to *Batrachochytrium dendrobatidis*, an emerging amphibian pathogen. *Conserv. Biol.* **2011**, *25*, 965–974. [CrossRef] [PubMed]

93. Young, B.E.; Lips, K.R.; Reaser, J.K.; Ibáñez, R.; Salas, A.W.; Cedeño, J.R.; Coloma, L.A.; Ron, S.; La Marca, E.; Meyer, J.R.; et al. Population declines and priorities for amphibian conservation in Latin America. *Conserv. Biol.* **2001**, *15*, 1213–1223. [CrossRef]

94. Martel, A.; Spitzen-van der Sluijs, A.; Blooi, M.; Bert, W.; Ducatelle, R.; Fisher, M.C.; Woeltjes, A.; Bosman, W.; Chiers, K.; Bossuyt, F.; et al. *Batrachochytrium salamandrivorans* sp. nov. causes lethal chytridiomycosis in amphibians. *Proc. Natl. Acad. Sci. USA* **2013**, *110*, 15325–15329. [CrossRef] [PubMed]

95. Whitfield, S.M.; Kerby, J.; Gentry, L.R.; Donnelly, M.A. Temporal variation in infection prevalence by the amphibian chytrid fungus in three species of frogs at La Selva, Costa Rica. *Biotropica* **2012**, *44*, 779–784. [CrossRef]

96. Whitfield, S.M.; Zoo Miami, FL, USA. Personal Communication, 2019.

97. Whitfield, S.M.; Alvarado, G.; Abarca, J.; Zumbado-Ulate, H.; Zuñiga, I.; Wainwright, M.; Kerby, J. Differential patterns of *Batrachochytrium dendrobatidis* infection in relict amphibian populations following severe disease-associated declines. *Dis. Aquat. Organ.* **2017**, *126*, 33–41. [CrossRef]

98. Puschendorf, R.; Hoskin, C.J.; Cashins, S.D.; McDonald, K.; Skerratt, L.F.; Vanderwal, J.; Alford, R.A. Environmental refuge from disease-driven amphibian extinction. *Conserv. Biol.* **2011**, *25*, 956–964. [CrossRef]

99. Perez, R.; Richards-Zawacki, C.L.; Krohn, A.R.; Robak, M.; Griffith, E.J.; Ross, H.; Gratwicke, B.; Ibanez, R.; Voyles, J. Field surveys in Western Panama indicate populations of *Atelopus varius* frogs are persisting in regions where *Batrachochytrium dendrobatidis* is now enzootic. *Amphib. Reptile Conserv.* **2014**, *8*, 30–35.

100. Woodhams, D.C.; Kilburn, V.L.; Reinert, L.K.; Voyles, J.; Medina, D.; Ibáñez, R.; Hyatt, A.D.; Boyle, D.G.; Pask, J.D.; Green, D.M.; et al. Chytridiomycosis and amphibian population declines continue to spread eastward in Panama. *EcoHealth* **2008**, *5*, 268–274. [CrossRef]

101. Kilburn, V.L.; Ibáñez, R.; Sanjur, O.; Bermingham, E.; Suraci, J.P.; Green, D.M. Ubiquity of the pathogenic chytrid fungus, *Batrachochytrium dendrobatidis*, in anuran communities in Panamá. *EcoHealth* **2010**, *7*, 537–548. [CrossRef] [PubMed]

102. Acosta-Chaves, V.J.; Madrigal-Elizondo, V.; Chaves, G.; Morera-Chacón, B.; Garcia-Rodriguez, A.; Bolaños, F. View of shifts in the diversity of an amphibian community from a premontane forest of San Ramón, Costa Rica. *Rev. Biol. Trop.* **2019**, *67*, 259–273.

103. Garner, T.W.J.; Walker, S.; Bosch, J.; Leech, S.; Marcus Rowcliffe, J.; Cunningham, A.A.; Fisher, M.C. Life history tradeoffs influence mortality associated with the amphibian pathogen *Batrachochytrium dendrobatidis*. *Oikos* **2009**, *118*, 783–791. [CrossRef]

104. Kolby, J.E.; Ramirez, S.D.; Berger, L.; Richards-Hrdlicka, K.L.; Jocque, M.; Skerratt, L.F. Terrestrial dispersal and potential environmental transmission of the amphibian chytrid fungus (*Batrachochytrium dendrobatidis*). *PLoS ONE* **2015**, *10*, e0125386. [CrossRef] [PubMed]

105. Ryan, M.J.; Lips, K.R.; Eichholz, M.W. Decline and extirpation of an endangered Panamanian stream frog population (*Craugastor punctariolus*) due to an outbreak of chytridiomycosis. *Biol. Conserv.* **2008**, *141*, 1636–1647. [CrossRef]

106. Campbell, J.A.; Savage, J.M. Taxonomic reconsideration of Middle American frogs of the *Eleutherodactylus rugulosus* group (Anura: Leptodactylidae): A reconnaissance of subtle nuances among frogs. *Herpetol. Monogr.* **2000**, *14*, 186–292. [CrossRef]

107. Lips, K.R.; Reeve, J.D.; Witters, L.R. Ecological traits predicting amphibian population declines in Central America. *Conserv. Biol.* **2003**, *17*, 1078–1088. [CrossRef]

108. Köhler, G.; Batista, A.; Carrizo, A.; Hertz, A. Field notes on *Craugastor azueroensis* (Savage, 1975) (Amphibia: Anura: Craugastoridae). *Herpetol Notes* **2012**, *5*, 157.

109. Hanken, J.; Wake, D.B.; Savage, J.M. A solution to the large black salamander problem (genus *Bolitoglossa*) in Costa Rica and Panamá. *Copeia* **2005**, *2005*, 227–245. [CrossRef]

110. Sunyer, J.; Wake, D.B.; Obando, L. Distributional data for *Bolitoglossa* (Amphibia, Caudata, Plethodontidae) from Nicaragua and Costa Rica. *Herpetol. Rev.* **2012**, *43*, 564–568.

111. AlMutairi, B.S.; Grossmann, I.; Small, M.J. Climate model projections for future seasonal rainfall cycle statistics in Northwest Costa Rica. *Int. J. Climatol.* **2019**, *39*, 2933–2946. [CrossRef]

112. Grenyer, R.; Orme, C.D.L.; Jackson, S.F.; Thomas, G.H.; Davies, R.G.; Davies, T.J.; Jones, K.E.; Olson, V.A.; Ridgely, R.S.; Rasmussen, P.C.; et al. Global distribution and conservation of rare and threatened vertebrates. *Nature* **2006**, *444*, 93–96. [CrossRef] [PubMed]

113. Woodhams, D.C.; Bosch, J.; Briggs, C.J.; Cashins, S.; Davis, L.R.; Lauer, A.; Muths, E.; Puschendorf, R.; Schmidt, B.R.; Sheafor, B.; et al. Mitigating amphibian disease: Strategies to maintain wild populations and control chytridiomycosis. *Front. Zool.* **2011**, *8*, 8. [CrossRef] [PubMed]

114. Garner, T.W.J.; Schmidt, B.R.; Martel, A.; Pasmans, F.; Muths, E.; Cunningham, A.A.; Weldon, C.; Fisher, M.C.; Bosch, J. Mitigating amphibian chytridiomycosis in nature. *Philos. Trans. R. Soc. B Biol. Sci.* **2016**, *371*, 20160207. [CrossRef] [PubMed]

Article

Conservation Status of *Brachycephalus* Toadlets (Anura: Brachycephalidae) from the Brazilian Atlantic Rainforest

Marcos R. Bornschein [1,2,*], Marcio R. Pie [2,3] and Larissa Teixeira [1]

[1] Instituto de Biociências, Universidade Estadual Paulista (UNESP), Praça Infante Dom Henrique s/no, São Vicente, São Paulo, CEP 11330-900, Brazil

[2] Mater Natura—Instituto de Estudos Ambientais, Rua Lamenha Lins 1080, Curitiba, Paraná, CEP 80250-020, Brazil

[3] Departamento de Zoologia, Universidade Federal do Paraná, Curitiba, Paraná, CEP 81531-980, Brazil

* Correspondence: bornschein.marcao@gmail.com or marcos.bornschein@unesp.br; Tel.: +55 13-98156-9582

Received: 1 July 2019; Accepted: 23 August 2019; Published: 27 August 2019

Abstract: The number of described anurans has increased continuously, with many newly described species determined to be at risk. Most of these new species inhabit hotspots and are under threat of habitat loss, such as *Brachycephalus*, a genus of small toadlets that inhabits the litter of the Brazilian Atlantic Rainforest. Of 36 known species, 22 were described in the last decade, but only 11 have been assessed according to the IUCN Red List categories, with just one currently listed as Critically Endangered. All available data on occurrence, distribution, density, and threats to *Brachycephalus* were reviewed. The species extent of occurrence was estimated using the Minimum Convex Polygon method for species with three or more records and by delimiting continuous areas within the altitudinal range of species with up to two records. These data were integrated to assess the conservation status according to the IUCN criteria. Six species have been evaluated as Critically Endangered, five as Endangered, 10 as Vulnerable, five as Least Concern, and 10 as Data Deficient. Deforestation was the most common threat to imperiled *Brachycephalus* species. The official recognition of these categories might be more readily adopted if the microendemic nature of their geographical distribution is taken into account.

Keywords: deforestation; timber harvest; fire; invasion of exotic plants; conservation; public policy; protected areas; critically endangered; data deficient

---

1. Introduction

Frogs and toads (Anura) comprise more than 7000 species worldwide [1]. Special attention has been given to this group due to the large number of new species described each year as well as due to the increasing number of endangered species [2,3]. According to the IUCN Red List criteria [4,5], there are 1825 species of anurans at risk of extinction (25% of all species), making Anura the vertebrate order with the highest proportion of endangered species [5]. Since 1980, there have been records of a rapid population decline of nearly 450 anuran species [6–8]. The decline of these species can be mainly attributed to habitat loss and pathogens, such as chytrid fungi and Ranavirus [6,7,9–11]. Recently, Ranavirus has been reported in natural populations of frogs in South America, but the effects in wild anuran populations are still unknown [11]. Unlike Ranavirus, chytrid fungi (*Batrachochytrium dendrobatidis*) has been commonly reported as a cause of population decline in high altitude locations in Costa Rica and Panama [9]. Due to the rapid rate of the description of a new species, the proportion of endangered species, and sensitivity, Anura is the priority order for a conservation assessment, particularly in countries with a high level of deforestation, such as in Brazil [3].

The Atlantic Rainforest, a biodiversity hotspot [12], is the largest in area after the Amazon forest, with its original extent covering more than 1.3 million km^2 [13,14]. It is located on the eastern coast of South America, stretching from northeastern to southern Brazil, with inland extensions to the east of Paraguay, northeast of Argentina, and central Brazil. This biome has been experiencing massive habitat loss due to agricultural expansion, urbanization, and historic loss of natural habitats [15]. Currently, only 28% of the original extent remains if secondary forests and forests affected by the edge effects are included [15]. The Atlantic Rainforest houses nearly 2500 species of vertebrates, including 550 anurans, of which 323 are endemic (63%) and 15 are currently considered to be threatened by extinction [1,5,16].

The genus *Brachycephalus* (Fitzinger, 1826) is endemic to the Atlantic Rainforest and includes small (less than 2.5 cm in snout-vent length) diurnal toadlets with a reduced number of digits, bright colors, neurotoxins in the skin, and direct development, and they live in leaf litter, specifically that of montane forests [17–23]. There are currently 36 recognized species of *Brachycephalus* [1], of which 22 have been described in the last decade [1]. Most have extremely restricted geographical distributions of less than 100 ha [12,24,25]. *Brachycephalus* is divided into three phenetic groups [26], two of which (*B. ephippiumsi* and *B. pernix* groups) are montane with few records at lower altitudes, whereas the remaining group (*B. didactylus* group) includes more ecologically plastic species that occur from the sea level up to high altitudes [23,27]. The dependence on a colder climate and isolation in the mountains as sky islands have been hypothesized as the reason that montane groups have diverged into so many species (19 of *B. pernix* and 12 of *B. ephippiumsi* groups), whereas the *B. didactylus* group includes only four species [23,28,29]. Another species (*B. atelopoide*) cannot be compared to any of the groups due to the unavailability of the holotype [23,30].

Species descriptions of *Brachycephalus* have not been accompanied by corresponding assessments of the conservation status. Only 11 species have been assessed for the IUCN Red List to date [31–41]: eight as Data Deficient (DD) and three as Least Concern (LC). The Ministério do Meio Ambiente (MMA, the Ministry of the Environment of the Brazilian government) evaluated only four species and categorized one as Critically Endangered (CR), two as DD, and one as Near Threatened (NT) [42–45]. The absence of conservation status assessments of most species and the evaluation of some of them as DD highlight the need for a comprehensive effort to assess the risk of extinction of the *Brachycephalus* species, most notably the microendemic taxa found in the *B. pernix* and *B. ephippiumsi* species groups (*sensu* [26]). Species evaluated as DD should be prioritized to generate enough data to properly classify them into a conservation category [46,47].

One way to direct effective initiatives for conservation species is through a prior assessment of their conservation status [3]. There is a widely adopted IUCN methodology for proposing a conservation status [3], which serves an important role in allowing for comparisons and for classifying conservation actions as well the proposition of public policies. The objectives of the study were (1) to review data on occurrence, altitudinal distribution, density, and threats to the *Brachycephalus* species, (2) to compile new data from the literature and unpublished observations, (3) to generate systematized data on geographic distribution, population sizes, and threats to place them into IUCN conservation categories, and (4) to discuss conservation priorities and future management actions.

2. Material and Methods

All available occurrence records of *Brachycephalus* spp. were compiled from the literature up to the time of compilation (June 2019). The data encompassed toponymy, species identification, geographical coordinates of the occurrence record, and altitude of the corresponding site. Data on altitudinal range were also considered when available. The process began with the latest compilation of locality and altitude data for *Brachycephalus* provided by Bornschein et al. [23], and the same selection criteria were adopted for subsequent records. For example, those associated with precise localities were retained, and records that included only municipality names as occurrence information were discarded. Finally, the authors' previously unpublished data were included.

Occurrence records were plotted using Google Earth Pro v. 7.1.4.1529 and connected to form a polygon using the Minimum Convex Polygon approach (MCP; [48]) with modifications suggested by Reinert et al. [49] and adopted by Bornschein et al. [23]. These modifications allow for the exclusion of inappropriate habitats, such as bodies of water, pastures, silvicultures, urban areas, rock areas, and/or forest areas, beyond the altitudinal range of occurrence of the species.

Polygon delimitation required three or more occurrence records. For species with one or two records, polygons encompassing the altitudinal range of the species were created [23]. A continuous topography inside the polygon was considered a location (*sensu* IUCN and as IUCN [48]) that could potentially contain one or more records of a given species. The topography was considered discontinuous if it was isolated by altitudes beyond the altitudinal range of the respective species.

The MCP and altitudinal polygons were measured using GEPath v. 1.4.5 to obtain the extent of occurrence (EO; IUCN [48]; see also [23,25,50]) of each species. Because some species have such reduced EO, they could potentially also be ranked by area of occupancy (AO), although AO was not measured in this study; however, species with less than 1000 ha of EO could also be categorized based on the criterion of an AO of less than 1,000 ha (criteria B2, for CR [48]) as well as species with an AO less than 50,000 ha (criteria B2, for EN [48]) because AO is always smaller than EO and is located within the EO polygon [48].

Population size was inferred for each species based on the estimates of area in m^2 inhabited by one individual compiled by Bornschein et al. [24]. Based on estimates of the number of calling males [24], a sex ratio of one female per male [24] was assumed. In cases with distinct estimates of densities per species [24], the mean density was used. The mean area in which one individual per species can be found and its respective EO was then used to calculate the population size.

Data on EO, number of locations, population size, and threats of the species were integrated to evaluate and to categorize its conservation status according to the IUCN Red List and Criteria [48]. For the recognition of threats, data from the literature, personal field experience of the authors collected in the EO of 29 species, and information on land use, forest quality, and trends of deforestation over the previous 10 years were considered. For temporal trends in land use, the time series of satellite images of Google Earth Pro v. 7.1.4.1529 was analyzed.

In the treatment of the data in relation to the IUCN criteria, the flow chart presented in Figure 1 was used. Six pathways were developed beginning with the evaluation of the number of localities (one to two; three or more). If the species had up to two recorded localities, its altitudinal range was calculated. If an altitudinal range was not associated with the record, this prevented creating a polygon and estimating the EO. The species was then considered DD (pathway 3 of Figure 1). If the records were associated with altitudinal range, an EO was created based on the lower and upper altitudinal limits. It was not always possible to infer the EO without encompassing inland areas far west of the record and outside the assumed natural range, sometimes nearly reaching Argentina, which is clearly unrealistic. In these situations, the species were considered DD (pathway 2 of Figure 1). When there were up to two records associated with an altitudinal range that encompassed a realistic polygon for EO (as indicated), the status of the species was evaluated (pathway 1 of Figure 1). Further pathways related to the procedure can be observed in Figure 1.

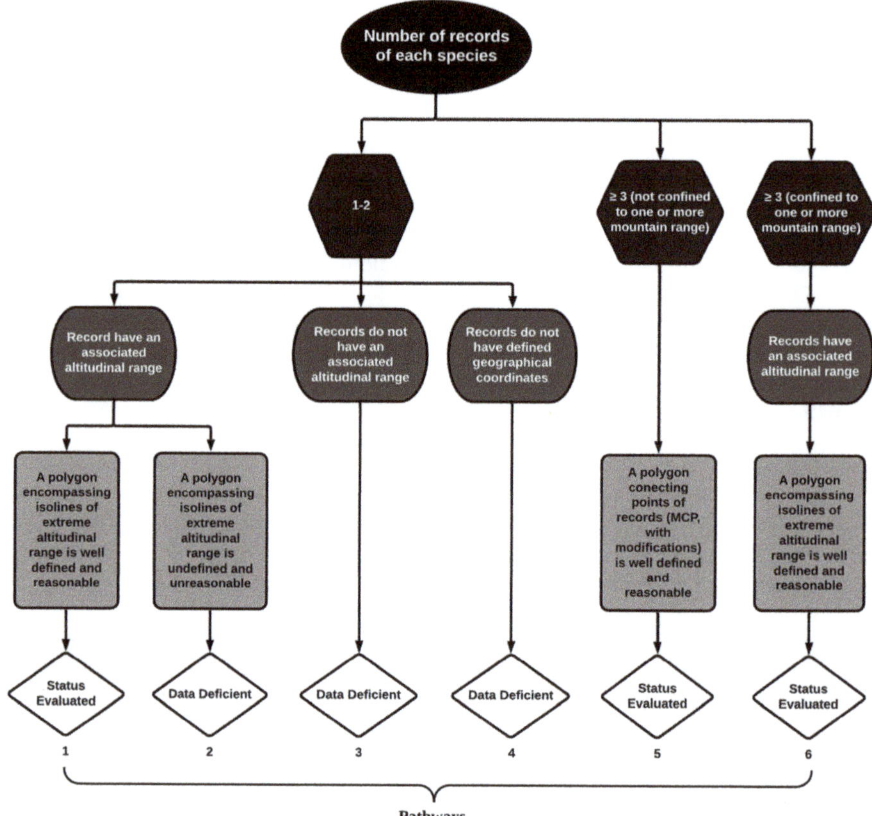

Figure 1. Flow chart indicating the approach to creating polygons of the extent of occurrence to compare the results with IUCN's species extinction risk classification criteria [48].

3. Results

A total of 185 locality records representing all 36 currently recognized *Brachycephalus* species in addition to 32 *Brachycephalus* sp. were generated (Table 1). An unidentified *Brachycephalus* species represented one between two described species that could not be adequately identified (i.e., old museum material collected before certain species were described) as well as new species awaiting formal description. Hereafter, only the described species are analyzed, leaving any evaluations to their own descriptors. The EO for 26 species (Table 1) was estimated, comprising several highly restricted EOs as well as larger ones: 23.8 ha for *B. fuscolineatus*, 37.4 ha for *B. coloratus*, 38.8 ha for *B. boticario*, 41.4 ha for *B. tridactylus*, 56.8 ha for *B. mirissimus* (all from the *B. pernix* group), 143,325.0 ha for *B. hermogenesi*, 702,983.4 ha for *B. didactylus*, 3,021,786.1 ha for *B. sulfuratus* (*B. didactylus* group), and 1,792,535.1 ha for *B. ephippiumsi* (*B. ephippiumsi* group). The population sizes of eight species (Table 1) was also estimated. All were highly abundant with population sizes ranging from 78,344 individuals for *B. mirissimus* and 302,178,610 individuals for *B. sulfuratus*.

Table 1. Locality records of *Brachycephalus*.

Species	Group	Locality and State	Altitude [1]	Source
B. didactylus	*didactylus*	Monumento Natural Serra das Torres (21°00′04″ S, 41°13′17″ W), municipality of Atílio Vivácqua, Espírito Santo	600–900?	[51] as *B. didactylus*; [52] as *B. didactylus*
B. didactylus	*didactylus*	Fazenda Santa Bárbara (22°25′17″ S, 42°35′01″ W), Parque Estadual dos Três Picos, municipality of Cachoeiras de Macacu, Rio de Janeiro	500–800	[53] as *B. didactylus*; [54] as *B. didactylus*
B. didactylus	*didactylus*	Reserva Ecológica de Guapiaçu (22°24′00″ S, 42°44′00″ W), municipality of Cachoeiras de Macacu, Rio de Janeiro	300–520	[55] as *B. didactylus*]
B. didactylus	*didactylus*	Reserva Ecológica Rio das Pedras (22°59′00″ S, 44°06′45″ W), municipality of Mangaratiba, Rio de Janeiro	200–1110	[23] as *B. didactylus*; [54] as *B. didactylus*; [56] as *B. didactylus*; [57] as *B. didactylus*
B. didactylus	*didactylus*	Sacra Família do Tinguá (22°29′11″ S, 43°36′18″ W), municipality of Engenheiro Paulo de Frontin, Rio de Janeiro	600	[17] as *B. didactylus*; [27] as *B. didactylus*; [58] as *B. didactylus*; [59] as *B. didactylus*; [60] as *B. didactylus*; [61] as *B. didactylus*; [62] as *B. didactylus*; [63] as *B. didactylus*; [64] as *B. didactylus*; [65] as *B. didactylus*; [66] as *B. didactylus*
B. didactylus	*didactylus*	Theodoro de Oliveira (first position: 22°22′11″ S, 42°33′25″ W), Parque Estadual dos Três Picos, municipality of Nova Friburgo, Rio de Janeiro	1100–1400?	[23] as *B. didactylus*; [67] as *B. didactylus*
B. didactylus	*didactylus*	Tinguá (22°35′51″ S, 43°24′54″ W), municipality of Nova Iguaçu, Rio de Janeiro	35	[17] as *B. didactylus*
B. didactylus	*didactylus*	Vila Dois Rios (23°11′01″ S, 44°12′23″ W), Ilha Grande, municipality of Angra dos Reis, Rio de Janeiro	220–240	[23] as *B. didactylus*; [68] as *B. didactylus*; [69] as *B. didactylus*; [70] as *B. didactylus*
B. hermogenesi	*didactylus*	Corcovado (23°28′20″ S, 45°11′41″ W), municipality of Ubatuba, São Paulo	30–250	This study, [18] as *B. hermogenesi*; [23] as *B. hermogenesi*; [25] as *B. hermogenesi* collected at Picinguaba; [27] as *B. hermogenesi*; [63] as *B. hermogenesi*
B. hermogenesi	*didactylus*	Estação Biológica de Boracéia (23°39′10″ S, 45°53′05″ W), municipality of Salesópolis, São Paulo	825–900	[23] as *B. hermogenesi*; [27] as *B. hermogenesi*; [63] as *B. hermogenesi*; [71] as *B. hermogenesi*; [72] as *B. hermogenesi*
B. hermogenesi	*didactylus*	Fazenda Capricórnio (23°23′27″ S, 45°04′26″ W), municipality of Ubatuba, São Paulo	60	[18] as *B. hermogenesi*; [23] as *B. hermogenesi*; [27] as *B. hermogenesi*; [63] as *B. hermogenesi*; [72] as *B. hermogenesi*
B. hermogenesi	*didactylus*	Morro Cuscuzeiro (23°17′50″S, 44°47′21″ W), on the border of municipalities of Paraty, Rio de Janeiro, and Ubatuba, São Paulo	730–1090	This study
B. hermogenesi	*didactylus*	Morro do Corcovado (23°27′06″ S, 45°12′03″ W), Parque Estadual da Serra do Mar, municipality of Ubatuba, São Paulo	250–1060	This study
B. hermogenesi	*didactylus*	Municipality of Paraibuna (c. 23°23′34″ S, 45°39′42″ W), São Paulo	?	[72] as *B. hermogenesi*
B. hermogenesi	*didactylus*	Núcleo Cunha (23°15′48″S, 45°02′39″W), Parque Estadual da Serra do Mar, municipality of Cunha, São Paulo	1045–1140	This study
B. hermogenesi	*didactylus*	Núcleo Picinguaba (23°22′21″S, 44°49′53″W), Parque Estadual da Serra do Mar, municipality of Ubatuba, São Paulo	0–700	[18] as *B. hermogenesi*; [23] as *B. hermogenesi*]; [27] as *B. hermogenesi*; [29] as *B. hermogenesi*; [63] as *B. hermogenesi*; [64] as *B. hermogenesi*; [71] as *B. hermogenesi*; [72] as *B. hermogenesi*

Table 1. *Cont.*

Species	Group	Locality and State	Altitude [1]	Source
B. hermogenesi	*didactylus*	Sertão da Cutia (not located), municipality of Ubatuba, São Paulo	?	[72] as *B. hermogenesi*
B. hermogenesi	*didactylus*	Trilha do Corisco (23°16′38″ S, 44°46′39″ W), municipality of Paraty, Rio de Janeiro	350–725	This study
B. hermogenesi	*didactylus*	Trilha do Ipiranga 50 m from the Rio Ipiranga (23°20′41″ S, 45°08′21″ W), Núcleo Santa Virgínia, Parque Estadual da Serra do Mar, municipality of São Luiz do Paraitinga, São Paulo	920–940	This study
B. pulex	*didactylus*	Serra Bonita (15°23′28″ S, 39°33′59″ W), municipality of Camacan, Bahia	800–930	[20] as *B. pulex*
B. sulfuratus	*didactylus*	Base of the Serra Água Limpa (24°28′52″ S, 48°47′12″ W), municipality of Apiaí, São Paulo	920	[23] as *Brachycephalus* sp. 1; [25] as *B. sulfuratus*; [28] without species identification; [50] as *B. sulfuratus*; [73] without species identification; [74] as *B. sulfuratus*
B. sulfuratus	*didactylus*	Biquinha (24°17′43″ S, 47°36′26″ W), municipality of Juquiá, São Paulo	40	This study
B. sulfuratus	*didactylus*	Braço do Norte (26°07′29″ S, 48°43′48″ W), municipality of Itapoá, Santa Catarina	240	[75] as *B. sulfuratus*
B. sulfuratus	*didactylus*	Caratuval, near the Parque Estadual das Lauráceas (24°51′17″ S, 48°43′43″ W), municipality of Adrianópolis, Paraná	900	[23] as *Brachycephalus* sp. 1; [25] as *B. sulfuratus*; [27] as *Brachycephalus* sp. nov. 1; [28] without species identification; [50] as *B. sulfuratus*; [73] without species identification; [74] as *B. sulfuratus*
B. sulfuratus	*didactylus*	Caratuval, Parque Estadual das Lauráceas (24°51′14″ S, 48°42′01″ W), municipality of Adrianópolis, Paraná	890	[23] as *Brachycephalus* sp. 1; [27] as *Brachycephalus* sp. nov. 1
B. sulfuratus	*didactylus*	Castelo dos Bugres (26°13′47″ S, 49°03′20″ W), municipality of Joinville, Paraná	790–860	[23] as *Brachycephalus* sp. 1; [27] as *Brachycephalus* sp. nov. 1; [72] as *B. sulfuratus*; [75] as *B. sulfuratus*
B. sulfuratus	*didactylus*	Centro de Estudos e Pesquisas Ambientais da Univille (26°13′39″ S, 48°41′31″ W), Vila da Glória, Distrito do Saí, municipality of São Francisco do Sul, Santa Catarina	125	[72] as *B. sulfuratus*
B. sulfuratus	*didactylus*	Corvo (25°20′17″ S, 48°54′56″ W), municipality of Quatro Barras, Paraná	930	[23] as *Brachycephalus* sp. 1; [25] as *B. sulfuratus*; [27] as *Brachycephalus* sp. nov. 1; [28] without species identification; [29] as *B. sulfuratus*; [50] as *B. sulfuratus*; [73] without species identification; [74] as *B. sulfuratus*
B. sulfuratus	*didactylus*	Estância Hidroclimática Recreio da Serra (25°27′14″ S, 49°00′28″ W), Serra da Baitaca, municipality of Piraquara, Paraná	1150–1205	This study
B. sulfuratus	*didactylus*	Fazenda Thalia (25°30′58″ S, 49°40′12″ W), municipality of Balsa Nova, Paraná	1025	[23] as *Brachycephalus* sp. 1; [25] *as B. sulfuratus*; [27] as *Brachycephalus* sp. nov. 1; [28] without species identification; [50] as *B. sulfuratus*; [73] without species identification; [74] as *B. sulfuratus*
B. sulfuratus	*didactylus*	near the Jurupará dam (23°56′30″ S, 47°23′45″ W), municipality of Piedade, São Paulo	690	[25] as *B. sulfuratus*

Table 1. *Cont.*

Species	Group	Locality and State	Altitude [1]	Source
B. sulfuratus	*didactylus*	Mananciais da Serra (25°29′32″ S, 48°59′33″ W), municipality of Piraquara, Paraná	970–1050	[23] as *Brachycephalus* sp. 1; [25] as *B. sulfuratus*; [27] as *Brachycephalus* sp. nov. 1; [50] as *B. sulfuratus*; [74] as *B. sulfuratus*
B. sulfuratus	*didactylus*	Morro Anhangava (25°22′51″ S, 49°01′26″ W), municipality of Quatro Barras, Paraná	915	[72] as *B. sulfuratus*; [75] as *B. sulfuratus*
B. sulfuratus	*didactylus*	Morro do Cantagalo (26°10′31″ S, 48°42′44″ W), Vila da Glória, Distrito do Saí, municipality of São Francisco do Sul, Santa Catarina	160	[72] as *B. sulfuratus*
B. sulfuratus	*didactylus*	Morro do Garrafão (26°28′23″ S, 49°15′57″ W), municipality of Corupá, Santa Catarina	500–530	[25] as *B. sulfuratus*; [76] as *B. sulfuratus*
B. sulfuratus	*didactylus*	Morro Garuva (26°02′29″ S, 48°53′14″ W), municipality of Garuva, Santa Catarina	215–495	This study
B. sulfuratus	*didactylus*	Núcleo Itutinga-Pilões (23°54′17″ S, 46°29′22″ W), Parque Estadual da Serra do Mar, municipality of Cubatão, São Paulo	55	This study
B. sulfuratus	*didactylus*	Parque Estadual da Ilha do Cardoso (25°06′53″ S, 47°55′40″ W), municipality of Cananéia, São Paulo	385	[63] as possibly *B. hermogenesi*; [72] as *B. sulfuratus*
B. sulfuratus	*didactylus*	Parque Estadual Intervales (24°16′33″ S, 48°25′04″ W), municipality of Iporanga, São Paulo	820	This study
B. sulfuratus	*didactylus*	Recanto das Hortências (25°33′24″ S, 48°59′38″ W), municipality of São José dos Pinhais, Paraná	975	[23] as *Brachycephalus* sp. 1; [25] as *B. sulfuratus*; [50] as *B. sulfuratus*; [74] as *B. sulfuratus*
B. sulfuratus	*didactylus*	Reserva Particular do Patrimônio Natural Salto Morato (25°09′14″ S, 48°18′06″ W), municipality of Guaraqueçaba, Paraná	40–880	[23] as *Brachycephalus* sp. 1; [77] as *B. hermogenesi*; [78] as *B. hermogenesi*; [79] as *B. hermogenesi*
B. sulfuratus	*didactylus*	Salto do Inferno (25°00′02″ S, 48°37′07″ W), Rio Capivari, municipality of Bocaiúva do Sul, Paraná	610	[25] as *B. sulfuratus*; [50] as *B. sulfuratus*; [74] as *B. sulfuratus*
B. sulfuratus	*didactylus*	Serra do Guaraú (24°47′12″ S, 48°07′11″ W), on the border of the municipalities of Cajati and Jacupiranga, São Paulo	680–835	This study
B. sulfuratus	*didactylus*	Serra do Pico (26°08′31″ S, 48°57′19″ W), municipality of Joinville, Santa Catarina	340–720	This study
B. sulfuratus	*didactylus*	Torre Embratel (24°52′46″ S, 48°15′27″ W), municipality of Cajati, São Paulo	960–990	This study
B. sulfuratus	*didactylus*	Truticultura (26°01′33″ S, 48°52′02″ W), municipality of Garuva, Paraná	90	[23] as *Brachycephalus* sp. 1; [27] as *Brachycephalus* sp. nov. 1
B. alipioi	*ephippium*	Fazenda Aoki or Fazenda dos Japoneses (20°28′24″ S, 41°00′36″ W), boundary of the municipalities of Vargem Alta and Domingos Martins, Espírito Santo	1070–1100	[27] as *B. alipioi*; [64] as *B. alipioi*; [66] as *B. alipioi*; [80] as *B. alipioi*; [81] as *B. alipioi*; [82] as *B. alipioi*
B. alipioi	*ephippium*	Forno Grande (20°31′41″ S, 41°06′51″ W), Parque Estadual de Forno Grande, municipality of Castelo, Espírito Santo	1430?	[27] as *B. alipioi*

Table 1. *Cont.*

Species	Group	Locality and State	Altitude [1]	Source
B. alipioi	*ephippium*	Alto Castelinho (20°30′34″ S, 41°00′33″ W), municipality of Vargem Alta, Espírito Santo	1100	This study, [25] as *B. alipioi*
B. bufonoides	*ephippium*	Serra de Macaé (22°18′02″ S, 42°18′20″ W), municipality of Nova Friburgo, Rio de Janeiro	1100?	[30] as *B. bufonoides*; [66] as *B. bufonoides*; [83] as *B. ephippiumsi bufonoides*
B. crispus	*ephippium*	Bacia B, Núcleo Cunha, Parque Estadual da Serra do Mar (23°15′15″ S, 45°01′58″ W), municipality of Cunha, São Paulo	800–1190	This study, [84] as *B. crispus*
B. darkside	*ephippium*	Mata do Pai Inácio (20°46′44″ S, 42°29′10″ W), Parque Estadual da Serra do Brigadeiro, municipality Miradouro, Minas Gerais	1340	[66] as *B. ephippiumsi*; [85] as *B. ephippiumsi*; [86] as *B. darkside*
B. darkside	*ephippium*	Trilha do Cruzeiro (20°52′41″ S, 42°31′15″ W), Parque Estadual da Serra do Brigadeiro, boundary of the municipalities of Ervália and Muriaé, Minas Gerais	1265–1500	[86] as *B. darkside*
B. ephippiumsi	*ephippium*	Condomínio Ermida (23°14′13″ S, 46°58′52″ W), Serra do Japi, municipality of Jundiaí, São Paulo	1225	[27] as *B. ephippiumsi*
B. ephippiumsi	*ephippium*	Hotel Fazenda Pé da Serra (22°51′56″ S, 45°31′40″ W), municipality of Pindamonhangaba, São Paulo	700	[27] as *B. ephippiumsi*
B. ephippiumsi	*ephippium*	Lago Azul (22°27′23″ S, 44°36′34″ W), Parque Nacional do Itatiaia, municipality of Itatiaia, Rio de Janeiro	750	[27] as *B. ephippiumsi*
B. ephippiumsi	*ephippium*	Maromba (22°25′43″ S, 44°37′11″ W), Parque Nacional do Itatiaia, municipality of Itatiaia, Rio de Janeiro	1125	[27] as *B. ephippiumsi*
B. ephippiumsi	*ephippium*	Monteiro Lobato (22°57′07″ S, 45°50′20″ W), municipality of Monteiro Lobato, São Paulo	700	[66] as *B. ephippiumsi*
B. ephippiumsi	*ephippium*	Observatório de Capricórnio (22°53′54″ S, 46°49′01″ W), Serra das Cabras, Joaquim Egídio District, boundary of the municipalities of Campinas and Morungaba, São Paulo	1085	[19] as *B. ephippiumsi*]; [27] as *B. ephippiumsi*; [66] as *B. ephippiumsi*; [87] as *B. ephippiumsi*
B. ephippiumsi	*ephippium*	Parque Municipal de Itapetinga (Grota Funda) (23°11′07″ S, 46°31′47″ W), municipality of Atibaia, São Paulo	900–1250	[27] as *B. ephippiumsi*; [64] as *B. ephippiumsi*; [81] as *B. ephippiumsi*; [88] as *B. ephippiumsi*; [89] as *B. ephippiumsi*
B. ephippiumsi	*ephippium*	Reserva Biológica da Serra do Japi (23°17′07″ S, 47°00′05″ W), Serra do Japi, boundary of the municipalities of Jundiaí and Cabreúva, São Paulo	1000	[27] as *B. ephippiumsi*; [64] as *B. ephippiumsi*; [66] as *B. ephippiumsi*; [90] as *B. ephippiumsi*
B. ephippiumsi	*ephippium*	Reserva Ecológica do Trabiju (22°48′01″ S, 45°32′03″ W), Trabiju, municipality of Pindamonhangaba, São Paulo	1000?	[66] as *B. ephippiumsi*
B. ephippiumsi	*ephippium*	Reserva Pedra Branca (22°56′22″ S, 45°41′04″ W), municipality of Tremembé, São Paulo	890?	[66] as *B. ephippiumsi*
B. ephippiumsi	*ephippium*	Santo Antônio do Pinhal (22°49′28″ S, 45°40′20″ W), municipality of Santo Antônio do Pinhal, São Paulo	1080	[66] as *B. ephippiumsi*

Table 1. *Cont.*

Species	Group	Locality and State	Altitude [1]	Source
B. ephippiumsi	*ephippium*	São Francisco Xavier (22°53′44″ S, 45°58′04″ W), municipality of São José dos Campos, São Paulo	1000	[27] as *B. ephippiumsi*; [66] as *B. ephippiumsi*; [91] as *B. ephippiumsi*; [92] as *B. ephippiumsi*
B. ephippiumsi	*ephippium*	Serra Negra (21°57′28″ S, 43°47′20″ W), municipality of Santa Bárbara do Monte Verde, Minas Gerais	?	[23] as *B. ephippiumsi*; [65] as BMV MG2
B. ephippiumsi	*ephippium*	Serra da Concórdia (22°20′30″ S, 43°44′04″ W), Parque Estadual Serra da Concórdia, Barão de Juparanã, municipality of Valença, Rio de Janeiro	900?	[66] as *B. ephippiumsi*
B. ephippiumsi	*ephippium*	Alto do Soberbo (22°27′15″ S, 42°59′21″ W), municipality of Teresópolis, Rio de Janeiro	1250	[66] as *B. ephippiumsi*
B. ephippiumsi	*ephippium*	Comary (22°27′22″ S, 42°58′24″ W), municipality of Teresópolis, Rio de Janeiro	990	[66] as *B. ephippiumsi*
B. ephippiumsi	*ephippium*	Floresta dos Macacos (22°58′15″ S, 43°15′24″ W), municipality of Rio de Janeiro, Rio de Janeiro	450?	[66] as *B. ephippiumsi*
B. ephippiumsi	*ephippium*	Garrafão (22°28′04″ S, 43°01′52″ W), municipality of Guapimirim, Rio de Janeiro	1785?	[66] as *B. ephippiumsi*
B. ephippiumsi	*ephippium*	Pedra Branca (22°55′55″ S, 43°28′23″ W), Serra da Pedra Branca, municipality of Rio de Janeiro, Rio de Janeiro	1000	[58] as *B. ephippiumsi*; [66] as *B. ephippiumsi*
B. ephippiumsi	*ephippium*	Represa do Rio Grande (22°55′58″ S, 43°26′36″ W), Parque Estadual da Pedra Branca, municipality of Rio de Janeiro, Rio de Janeiro	150?	[27] as *B. ephippiumsi*; [66] as *B. ephippiumsi*
B. ephippiumsi	*ephippium*	Reserva Ecológica Rio das Pedras (22°59′00″ S, 44°06′45″ W), municipality of Mangaratiba, Rio de Janeiro	200–1110	[56] as *B. ephippiumsi*
B. ephippiumsi	*ephippium*	Riacho Beija-flor (22°27′04″ S, 43°00′04″ W), Parque Nacional da Serra dos Órgãos, municipality of Teresópolis, Rio de Janeiro	1195	[27] as *B. ephippiumsi*
B. ephippiumsi	*ephippium*	Rocio District (22°28′23″ S, 43°14′38″ W), municipality of Petrópolis, Rio de Janeiro	950	[27] as *B. ephippiumsi*
B. ephippiumsi	*ephippium*	Serra do Tinguá (22°35′31″ S, 43°28′16″ W), municipality of Nova Iguaçu, Rio de Janeiro	950?	[66] as *B. ephippiumsi*
B. ephippiumsi	*ephippium*	Vale da Revolta (22°26′17″ S, 42°56′19″ W), municipality of Teresópolis, Rio de Janeiro	1035	[66] as *B. ephippiumsi*
B. ephippiumsi	*ephippium*	Varginha (22°24′34″ S, 42°52′11″ W), municipality of Teresópolis, Rio de Janeiro	825?	[66] as *B. ephippiumsi*
B. ephippiumsi	*ephippium*	Bonito (22°42′51″ S, 44°34′39″ W), Serra da Bocaina, municipality of São José do Barreiro, São Paulo	1660?	[66] as *B. ephippiumsi*
B. ephippiumsi	*ephippium*	Estação Ecológica de Bananal (22°48′05″ S, 44°22′12″ W), Serra da Bocaina, municipality of Bananal, São Paulo	1200?	[93] as *B. ephippiumsi*

Table 1. *Cont.*

Species	Group	Locality and State	Altitude [1]	Source
B. ephippiumsi	*ephippium*	Lídice District (22°50'01" S, 44°11'32" W), municipality of Rio Claro, Rio de Janeiro	650?	[58] as *B. ephippiumsi*; [66] as *B. ephippiumsi*
B. ephippiumsi	*ephippium*	Pedra Branca (23°10'38" S, 44°47'19" W), Serra da Bocaina, municipality of Parati, Rio de Janeiro	630?	[58] as *B. ephippiumsi*; [66] as *B. ephippiumsi*
B. ephippiumsi	*ephippium*	Reserva Florestal de Morro Grande (23°42'08" S, 46°58'22" W), municipality of Cotia, São Paulo	990?	[94] as *B. ephippiumsi*
B. garbeanus	*ephippium*	Alto Caledônia (22°20'10" S, 42°33'20" W), municipality of Nova Friburgo, Rio de Janeiro	1070?	[66] as *B. garbeanus*
B. garbeanus	*ephippium*	Baixo Caledônia (22°21'33" S, 42°34'12" W), municipality of Nova Friburgo, Rio de Janeiro	1600–1900	[66] as *B. garbeanus*; [67] as *B. garbeanus*; [95] as *B. garbeanus*; [96] as *B. garbeanus*
B. garbeanus	*ephippium*	Macaé de Cima (22°21'37" S, 42°17'50" W), municipality of Nova Friburgo, Rio de Janeiro	1130	[27] as *B. garbeanus*; [64] as *B. ephippiumsi*; [66] as *B. garbeanus*; [81] as *B. garbeanus*; [91] as *B. ephippiumsi*; [92] as *B. ephippiumsi*
B. garbeanus	*ephippium*	Morro São João (22°22'47" S, 42°30'34" W), municipality of Nova Friburgo, Rio de Janeiro	1550?	[66] as *B. garbeanus*
B. garbeanus	*ephippium*	Serra de Macaé (22°18'02" S, 42°18'20" W), municipality of Nova Friburgo, Rio de Janeiro	1100?	[30] as *B. garbeanus*; [66] *as B. garbeanus*; [83] as *B. ephippiumsi garbeanus*
B. garbeanus	*ephippium*	Serra Nevada (22°21'46" S, 42°32'48" W), municipality of Nova Friburgo, Rio de Janeiro	1190	[66] as *B. garbeanus*
B. garbeanus	*ephippium*	Theodoro de Oliveira (second position: 22°21'48" S, 42°33'13" W), Parque Estadual dos Três Picos, municipality of Nova Friburgo, Rio de Janeiro	1400	[66] as *B. garbeanus*; [67] as *B. garbeanus*; [95] as *B. garbeanus*
B. guarani	*ephippium*	Morro Prumirim (23°20'50" S, 45°01'37" W), municipality of Ubatuba, São Paulo	500–900	[82] as *B. guarani*; [84] as *B. guarani*; [88] as *Brachycephalus* sp.
B. margaritatus	*ephippium*	Castelo Country Club (22°32'21" S, 43°13'08" W), municipality of Petrópolis, Rio de Janeiro	980	[66] as *B. margaritatus*
B. margaritatus	*ephippium*	Castelo Montebello (22°24'24" S, 42°58'06" W), municipality of Teresópolis, Rio de Janeiro	920	[66] as *B. margaritatus*
B. margaritatus	*ephippium*	Independência (22°32'58" S, 43°12'27" W), municipality of Petrópolis, Rio de Janeiro	860	[66] as *B. margaritatus*
B. margaritatus	*ephippium*	Morro Azul (22°28'34" S, 43°34'40" W), municipality of Engenheiro Paulo de Frontin, Rio de Janeiro	620	[65] as BPF RJ2; [66] as *B. margaritatus*
B. margaritatus	*ephippium*	Quitandinha (22°31'47" S, 43°12'26" W), municipality of Petrópolis, Rio de Janeiro	925	[66] as *B. margaritatus*
B. margaritatus	*ephippium*	Sacra Família do Tinguá (22°29'11" S, 43°36'18" W), municipality of Engenheiro Paulo de Frontin, Rio de Janeiro	600	[17] as *B. ephippiumsi*; [27] as *B. ephippiumsi*; [58] as *Brachycephalus* cf. *ephippium*; [66] as *B. margaritatus*

Table 1. *Cont.*

Species	Group	Locality and State	Altitude [1]	Source
B. nodoterga	ephippium	Estação Biológica de Boracéia (second position: 23°38′00″S, 45°52′00″W), municipality of Salesópolis, São Paulo	945	[27] as B. nodoterga; [30] as B. nodoterga; [58] as B. nodoterga; [59] as B. nodoterga; [60] as B. nodoterga; [61] as B. nodoterga; [66] as B. nodoterga; [97] as B. nodoterga; [98] as B. nodoterga; [99] as B. nodoterga
B. nodoterga	ephippium	Fazenda Paiva Ramos (23°28′21″ S, 46°47′25″ W), municipality of Osasco, São Paulo	820	[99] as B. nodoterga
B. nodoterga	ephippium	Pico do Ramalho (23°51′42″ S, 45°21′28″ W), Ilha de São Sebastião, municipality of Ilhabela, São Paulo	700–900	[27] as B. nodoterga; [66] as B. nodoterga; [99] as B. nodoterga; [100] as Brachycephalus sp. aff. nodoterga
B. nodoterga	ephippium	Santana de Parnaíba (23°26′19″ S, 46°56′06″ W), municipality of Santana de Parnaíba, São Paulo	730	[99] as B. nodoterga
B. nodoterga	ephippium	Serra da Cantareira (23°27′13″ S, 46°38′11″ W), Parque Estadual da Cantareira, municipality of São Paulo, São Paulo	850?	[30] as B. nodoterga; [59] as B. nodoterga; [60] as B. nodoterga; [61] as B. nodoterga; [64] as B. nodoterga; [66] as B. nodoterga; [81] as B. nodoterga; [82] as B. nodoterga, [83] as B. ephippiumsi nodoterga; [84] as B. nodoterga, [98] as B. nodoterga; [99] as B. nodoterga
B. pitanga	ephippium	Fazenda Capricórnio (23°22′36″ S, 45°04′07″ W), municipality of Ubatuba, São Paulo	450?	[27] as B. pitanga; [61] as B. pitanga; [65] as B. pitanga; [101] as Brachycephalus sp. 2
B. pitanga	ephippium	Núcleo Santa Virgínia (23°19′23″ S, 45°05′19″ W), Parque Estadual da Serra do Mar, municipality of São Luis do Paraitinga, São Paulo	980–1140	[102] as B. pitanga; [103] as B. pitanga
B. pitanga	ephippium	SP 125—municipality of São Luís do Paraitinga (23°22′57″ S, 45°09′59″ W), São Paulo	935–950	[23] as B. pitanga
B. pitanga	ephippium	Trilha do Ipiranga 50 m from the Rio Ipiranga (23°20′39″ S, 45°08′16″ W), Núcleo Santa Virgínia, Parque Estadual da Serra do Mar, municipality of São Luis do Paraitinga, São Paulo	900–960	[27] as B. pitanga; [61] as B. pitanga; [64] as B. pitanga; [81] as B. pitanga; [102] as B. pitanga; [104] as B. pitanga
B. toby	ephippium	Morro do Corcovado (23°27′22″ S, 45°11′53″ W), Parque Estadual da Serra do Mar, municipality of Ubatuba, São Paulo	750–1060	This study, [27] as B. toby; [81] as B. toby; [82] as B. toby; [84] as B. toby; [98] as B. toby
B. vertebralis	ephippium	Morro Cuscuzeiro (23°17′51″ S, 44°47′20″ W), on the border of municipalities of Paraty, Rio de Janeiro, and Ubatuba, Sao Paulo	760–1110	This study, [27] as B. vertebralis; [81] as B. vertebralis; [84] as B. vertebralis
B. vertebralis	ephippium	Pedra Branca (23°10′38″ S, 44°47′19″ W), Serra da Bocaina, municipality of Parati, Rio de Janeiro	630?	[27] as B. vertebralis; [30] as B. vertebralis; [58] as. B. vertebralis; [64] as B. vertebralis
B. actaeus	pernix	Braço do Norte (26°07′29″ S, 48°43′48″ W), municipality of Itapoá, Santa Catarina	240	[75] as B. actaeus
B. actaeus	pernix	Centro de Estudos e Pesquisas Ambientais da Univille (CEPA) (26°13′39″ S; 48°41′31″ W), Vila da Glória, Distrito do Saí, municipality of São Francisco do Sul, Santa Catarina	120	[75] as B. actaeus
B. actaeus	pernix	Estrada do Saí (26°12′06″ S; 48°41′37″ W), Distrito do Saí, municipality of São Francisco do Sul, Santa Catarina	100	[75] as B. actaeus

Table 1. *Cont.*

Species	Group	Locality and State	Altitude [1]	Source
B. actaeus	*pernix*	Fazenda Morro Grande (26°17′47″ S; 48°37′10″ W), Morro Grande, Ilha de São Francisco, municipality of São Francisco do Sul, Santa Catarina	60	[75] *as B. actaeus*
B. actaeus	*pernix*	Fazenda Palmito Juriti (26°08′09″ S; 48°43′54″ W), municipality of Itapoá, Santa Catarina	100–170	[75] *as B. actaeus*
B. actaeus	*pernix*	Serra da Palha (26°17′50″ S; 48°40′28″ W), Laranjeiras, Ilha de São Francisco, municipality of São Francisco do Sul, Santa Catarina	20–90	[75] *as B. actaeus*
B. actaeus	*pernix*	Serra da Tiririca (26°07′42″ S, 48°44′32″ W), municipality of Itapoá, Santa Catarina	170–530	This study
B. albolineatus	*pernix*	Morro Azul (26°45′52″ S, 49°12′20″ W), on the border between the municipalities of Pomerode and Rio dos Cedros, Santa Catarina	725–740	This study
B. albolineatus	*pernix*	Morro Boa Vista (26°30′58″ S, 49°03′14″ W), on the border between the municipalities of Jaraguá do Sul and Massaranduba, Santa Catarina	790–835	[74] as *B. albolineatus*; [105] as *B. albolineatus*
B. albolineatus	*pernix*	Morro do Garrafão (26°30′58″ S, 49°03′14″ W), municipality of Corupá, Santa Catarina	500–530	[76] as *B. albolineatus*
B. albolineatus	*pernix*	Morro do Schmidt (26°39′55″ S, 49°12′55″ W), municipality of Pomerode, Santa Catarina	810–870	This study
B. auroguttatus	*pernix*	Pedra da Tartaruga (26°00′21″S, 48°55′25″W), municipality of Garuva, Santa Catarina	1070–1100	[23] as *B. auroguttatus*; [26] as *B. auroguttatus*; [28] as *B. auroguttatus*; [29] *as B. auroguttatus*; [73] without species identification
B. boticario	*pernix*	Morro do Cachorro (26°46′42″ S, 49°01′57″ W), boundary of the municipalities of Blumenau, Gaspar, and Luiz Alves, Santa Catarina	685–795	[23] as *B. boticario*; [26] as *B. boticario*; [28] as *B. boticario*; [29] as *B. boticario*; [73] without species identification
B. brunneus	*pernix*	Abrigo 1 (25°13′29″ S, 48°51′17″ W), municipality of Campina Grande do Sul, Paraná	1440–1640	This study, [28] as not identified; [29] as *B. brunneus*
B. brunneus	*pernix*	Camapuã (25°15′59″ S, 48°50′16″ W), Serra dos Órgãos, boundary of the municipalities of Campina Grande do Sul and Antonina, Paraná	1595	[27] as *B. brunneus*; [28] as *B. brunneus*]; [29] as *B. brunneus*; [73] without species identification; [106] as *B. brunneus*
B. brunneus	*pernix*	Caranguejeira (25°20′27″ S, 48°54′31″ W), Serra da Graciosa, municipality of Quatro Barras, Paraná	1095–1110	[23] as *B. brunneus*; [73] without species identification
B. brunneus	*pernix*	Caratuva (25°14′33″ S, 48°50′04″ W), Serra dos Órgãos, municipality of Campina Grande do Sul, Paraná	1300–1770	[27] as *B. brunneus*; [28] as *B. brunneus*; [29] as *B. brunneus*, [59] as *B. brunneus*; [64] as *B. brunneus*; [65] as *B. brunneus*; [66] as *B. brunneus*; including Pico Paraná; [73] without species identification; including Pico Paraná; [81] as *B. brunneus*; [106] as *B. brunneus*; [107] as *B. brunneus*
B. brunneus	*pernix*	Getúlio (25°14′18″ S, 48°50′13″ W), Serra dos Órgãos, municipality of Campina Grande do Sul, Paraná	1310–1490	[23] as *B. brunneus*; [27] as *B. brunneus*

Table 1. *Cont.*

Species	Group	Locality and State	Altitude [1]	Source
B. brunneus	*pernix*	Mãe Catira (25°20′51″ S, 48°54′25″ W), Serra da Graciosa, municipality of Quatro Barras, Paraná	1135–1405	This study, [27] as *Brachycephalus* sp. nov. 2; [28] as not identified; [73] without species identification
B. coloratus	*pernix*	Estância Hidroclimática Recreio da Serra (25°27′14″ S, 49°00′27″ W), Serra da Baitaca, municipality of Piraquara, Paraná	1145–1230	[50] as *B. coloratus*
B. curupira	*pernix*	Morro do Canal (25°30′55″ S, 48°58′56″ W), municipality of Piraquara, Paraná	1320	This study, [23] as *Brachycephalus* sp. 4; [28] as not identified]; [73] without species identification
B. curupira	*pernix*	Morro do Vigia (25°30′33″ S, 48°58′58″ W), municipality of Piraquara, Paraná	1250	[23] as *Brachycephalus* sp. 4; [27] as *Brachycephalus* sp. nov. 3; [28] as not identified; [29] as *B. curupira*; [73] without species identification
B. curupira	*pernix*	Serra do Salto (25°42′07″ S, 49°03′44″ W), Malhada District, municipality of São José dos Pinhais, Paraná	1095–1160	[23] as *Brachycephalus* sp. 6; [27] as *Brachycephalus* sp. 2; [28] as not identified; [29] as *B. curupira*; [50] as *B. curupira*, [73] without species identification
B. ferruginus	*pernix*	Olimpo (25°27′03″ S, 48°54′59″ W), Serra do Marumbi, municipality of Morretes, Paraná	965–1470	[27] as *B. ferruginus*; [28] as *B. ferruginus*; [29] as *B. ferruginus*, [60] as *B. ferruginus*; [64] as *B. ferruginus*; [66] as *B. ferruginus*; [73] without species identification; [81] as *B. ferruginus*
B. fuscolineatus	*pernix*	Morro Braço da Onça (26°44′58″ S, 48°55′41″ W), municipality of Luiz Alves, Santa Catarina	525–530	[24] as *B. fuscolineatus*
B. fuscolineatus	*pernix*	Morro do Baú (26°47′58″ S, 48°55′47″ W), municipality of Ilhota, Santa Catarina	640–790	[26] as *B. fuscolineatus*; [27] as *Brachycephalus* sp. nov. 9; [28] as *B. fuscolineatus*; [29] as *B. fuscolineatus*; [73] without species identification
B. izecksohni	*pernix*	Torre da Prata, Serra da Prata (25°37′25″ S, 48°41′31″ W), boundary of the municipalities of Morretes, Paranaguá, and Guaratuba, Paraná	980–1340	[27] as *B. izecksohni*; [28] as *B. izecksohni*; [29] as *B. izecksohni*; [59] as *B. izecksohni*; [64] as *B. izecksohni*; [66] as *B. izecksohni*; [73] without species identification; [81] as *B. izecksohni*
B. leopardus	*pernix*	Morro dos Perdidos (25°53′22″ S, 48°57′22″ W), municipality of Guaratuba, Paraná	1340–1420	[26] as *B. leopardus*; [27] as *Brachycephalus* sp. nov. 4; [28] as *B. leopardus*; [73] without species identification
B. leopardus	*pernix*	Serra do Araçatuba (25°54′07″ S, 48°59′47″ W), municipality of Tijucas do Sul, Paraná	1640–1645	[26] as *B. leopardus*; [27] as *Brachycephalus* sp. nov. 4; [28] as *B. leopardus*; [73] without species identification
B. mariaeterezae	*pernix*	Reserva Particular do Patrimônio Natural Caetezal, top of the Serra Queimada (26°06′51″ S, 49°03′45″ W), municipality of Joinville, Santa Catarina	1265–1270	[26] as *B. mariaeterezae*; [27] as *Brachycephalus* sp. nov. 6; [28] as *B. mariaeterezae*; [29] as *B. mariaeterezae*; [73] without species identification
B. mirissimus	*pernix*	Morro Santo Anjo (26°37′41″ S, 48°55′50″ W), municipality of Massaranduba, Santa Catarina	470–540	[25] as *B. mirissimus*
B. olivaceus	*pernix*	Base of the Serra Queimada (26°04′57″ S, 49°03′59″ W), municipality of Joinville, Santa Catarina	985	[17] as *Brachycephalus* sp. nov. 7; [26] as *B. olivaceus*
B. olivaceus	*pernix*	Castelo dos Bugres (second position: 26°13′59″S, 49°03′13″W), municipality of Joinville, Santa Catarina	800–835	[26] as *B. olivaceus*; [27] as *Brachycephalus* sp. nov. 7; [28] as *B. olivaceus*; [73] without species identification; [108] as *B. olivaceus*
B. olivaceus	*pernix*	Morro do Boi (26°24′42″ S, 49°12′59″ W), municipality of Corupá, Santa Catarina	650–920	[23] as *B. olivaceus*; [27] as *Brachycephalus* sp. 3; [29] as *B. olivaceus*

Table 1. *Cont.*

Species	Group	Locality and State	Altitude [1]	Source
B. olivaceus	*pernix*	Pico Jurapê (26°16′27″ S, 49°00′13″ W), municipality of Joinville, Santa Catarina	650–780	This study
B. pernix	*pernix*	Anhangava (25°23′19″ S, 49°00′15″ W), Serra da Baitaca, municipality of Quatro Barras, Paraná	1135–1405	[27] as *B. pernix*; [28] as *B. pernix*; [29] as *B. pernix*; [62] as *B. pernix*; [64] as *B. pernix*; [65] as *B. pernix*; [66] as *B. pernix*; [73] without species identification; [81] as *B. pernix*; [97] as *B. pernix*; [101] as *B. pernix*; [109] as *B. pernix*; [110] as *B. pernix*; [111] as *B. pernix*
B. pombali	*pernix*	Morro dos Padres (25°36′40″ S, 48°51′22″ W), Serra da Igreja, municipality of Morretes, Paraná	1060–1300	[27] as *B. pombali*; [28] as *B. pombali*; [29] as *B. pombali*; [60] as *B. pombali*; [64] as *B. pombali*; [73] without species identification; [81] as *B. pombali*
B. pombali	*pernix*	trail to Morro dos Padres (25°35′58″ S, 48°51′57″ W), municipality of Morretes, Paraná	845–1060	[27] as *B. pombali*
B. quiririensis	*pernix*	Serra do Quiriri (26°01′17″ S, 48°59′47″ W), municipality of Campo Alegre, Santa Catarina	1240–1270	[23] as *B. quiririensis*; [27] as *Brachycephalus* sp. nov. 5; [28] as *B. quiririensis*; [29] as *B. quiririensis*; [73] without species identification; [112] as *B. quiririensis*
B. quiririensis	*pernix*	Serra do Quiriri (first position: 26°01′42″ S, 48°57′11″ W; second position: 26°01′32″ S, 48°58′24″ W), municipality of Garuva, Santa Catarina	1320–1380	[27] as *Brachycephalus* sp. nov. 5; [108] as *B. quiririensis*
B. tridactylus	*pernix*	Serra do Morato (25°08′09″ S, 48°17′59″ W), Reserva Natural Salto Morato, municipality of Guaraqueçaba, Paraná	805–910	[23] as *B. tridactylus*; [28] as *B. tridactylus*; [113] as *B. tridactylus*; [114] as *B. tridactylus*
B. verrucosus	*pernix*	Morro da Tromba (26°12′44″ S, 48°57′29″ W), municipality of Joinville, Santa Catarina	455–945	[23] as *B. verrucosus*; [26] as *B. verrucosus*; [27], as *Brachycephalus* sp. nov. 8; [28] as *B. verrucosus*; [29] as *B. verrucosus*; [73] without species identification
B. atelopoide	?	Piquete, São Paulo	?	[30] as *B. atelopoide*; [83] as *B. ephippiumsi atelopoide*]
Brachycephalus sp. (cf. *B. sulfuratus*)	*didactylus*	Alto Quiriri (26°05′34″ S, 48°59′41″ W), municipality of Garuva, Santa Catarina	240	[23] as *Brachycephalus* sp. 1; [27] as *Brachycephalus* sp. nov. 1
Brachycephalus sp. (cf. *B. sulfuratus*)	*didactylus*	Colônia Castelhanos (25°47′58″ S, 48°54′40″ W), municipality of Guaratuba, Paraná	290	[23] as *Brachycephalus* sp. 1; [27] as *Brachycephalus* sp. nov. 1; [72] as *B. sulfuratus*; [115] as *Brachycephalus* aff. *hermogenesi*]; [116] as *B. hermogenesi*
Brachycephalus sp. (cf. *B. sulfuratus*)	*didactylus*	Dona Francisca (26°09′52″ S, 48°59′23″ W), municipality of Joinville, Santa Catarina	150	[23] as *Brachycephalus* sp. 1; [27] as *Brachycephalus* sp. nov. 1
Brachycephalus sp. (*B. sulfuratus* or *B. hermogenesi*)	*didactylus*	Estação Ecológica Juréia-Itatins (c. 24°27′ S, 47°24′ W), municipality of Iguape, São Paulo	?	[63] as *B. hermogenesi*
Brachycephalus sp. (cf. *B. sulfuratus*)	*didactylus*	Estrada do Rio do Júlio (26°17′02″ S, 49°06′08″ W), municipality of Joinville, Santa Catarina	650	[23] as *Brachycephalus* sp. 1; [117] as *Brachycephalus* sp.
Brachycephalus sp. (cf. *B. sulfuratus*)	*didactylus*	Fazenda Pico Paraná (25°13′29″ S, 48°51′17″ W), municipality of Campina Grande do Sul, Paraná	1050–1085	[23] as *Brachycephalus* sp. 1; [27] as *Brachycephalus* sp. nov. 1

Table 1. *Cont.*

Species	Group	Locality and State	Altitude [1]	Source
Brachycephalus sp. (cf. *B. sulfuratus*)	*didactylus*	Fazenda Primavera (24°53′08″ S, 48°45′51″ W), municipality of Tunas do Paraná, Paraná	1060	[23] as *Brachycephalus* sp. 1; [27] as *Brachycephalus* sp. nov. 1
Brachycephalus sp. (*B. sulfuratus* or *B. hermogenesi*)	*didactylus*	Municipality of Ibiúna (c. 23°39′ S, 47°13′ W), São Paulo	?	[72] as *B. hermogenesi*
Brachycephalus sp. (*B. sulfuratus* or *B. hermogenesi*)	*didactylus*	Municipality of Juquitiba (c. 23°56′ S, 47°04′ W), São Paulo	?	[63] as *B. hermogenesi*; [72] as *B. hermogenesi*
Brachycephalus sp. (cf. *B. hermogenesi*)	*didactylus*	Municipality of Paraty (c. 23°13′07″ S, 44°43′15″ W), Rio de Janeiro	?	[18] as *B. hermogenesi*
Brachycephalus sp. (*B. sulfuratus* or *B. hermogenesi*)	*didactylus*	Municipality of Peruíbe (24°18′ S, 46°59′ W), São Paulo	?	[72] as *B. hermogenesi*
Brachycephalus sp. (*B. sulfuratus* or *B. hermogenesi*)	*didactylus*	Municipality of Piedade (c. 23°54′S, 47°25′ W), São Paulo	?	[81] as *B. hermogenesi*; [118] as *B. hermogenesi*
Brachycephalus sp. (*B. sulfuratus* or *B. hermogenesi*)	*didactylus*	Municipality of Registro (c. 24°30′ S, 47°51′ W), São Paulo	?	[72] as *B. hermogenesi*
Brachycephalus sp. (*B. sulfuratus* or *B. hermogenesi*)	*didactylus*	Municipality of Ribeirão Grande (c. 24°06′ S, 48°22′ W), São Paulo	?	[63] as *B. hermogenesi*
Brachycephalus sp. (*B. sulfuratus* or *B. hermogenesi*)	*didactylus*	Municipality of Tapiraí (c. 23°57′55″ S, 47°30′19″ W), São Paulo	870	[63] as *B. hermogenesi*; [118] as *B. hermogenesi*
Brachycephalus sp. (*B. sulfuratus* or *B. hermogenesi*)	*didactylus*	Parque Estadual de Jacupiranga (c. 24°38′ S, 48°24′ W), municipality of Eldorado, São Paulo	?	[72] as *B. hermogenesi*
Brachycephalus sp. (*B. hermogenesi* or *B. sulfuratus*)	*didactylus*	Parque Natural Municipal Nascentes de Paranapiacaba (23°46′10″ S, 46°17′36″ W), municipality of Santo André, São Paulo	840	[119] as *B. hermogenesi*
Brachycephalus sp. (cf. *B. sulfuratus*)	*didactylus*	Pico Agudinho (25°36′24″ S, 48°43′33″ W), Serra da Prata, municipality of Morretes, Paraná	385	[23] as *Brachycephalus* sp. 1; [27] as *Brachycephalus* sp. nov. 1

Table 1. *Cont.*

Species	Group	Locality and State	Altitude [1]	Source
Brachycephalus sp. (*B. sulfuratus* or *B. hermogenesi*)	*didactylus*	Reserva Betary (24°33′08″ S, 48°40′49″ W), municipality of Iporanga, São Paulo	190	This study
Brachycephalus sp. (*B. hermogenesi* or *B. sulfuratus*)	*didactylus*	Reserva Biológica do Alto da Serra de Paranapiacaba (23°46′40″ S, 46°18′45″ W), municipality of Santo André, São Paulo	800–850	[23] as *B. hermogenesi*; [63] as *B. hermogenesi*; [119] as *B. hermogenesi*
Brachycephalus sp. (*B. sulfuratus* or *B. hermogenesi*)	*didactylus*	Reserva Florestal de Morro Grande (23°42′08″ S, 46°58′22″ W), municipality of Cotia, São Paulo	990?	[23] as *B. hermogenesi*, [63] as *B. hermogenesi*; [72] as *B. hermogenesi*; [94] as *B. hermogenesi*
Brachycephalus sp. (cf. *B. sulfuratus*)	*didactylus*	Sítio Ananias (25°47′08″ S, 48°43′03″ W), municipality of Guaratuba, Paraná	25	[23] as *Brachycephalus* sp. 1; [27] as *Brachycephalus* sp. nov. 1
Brachycephalus sp.	*ephippium*	Paranapiacaba (23°46′30″ S, 46°17′57″ W), municipality of Santo André, São Paulo	825	[27] as *Brachycephalus* sp. 1; [66] as *B. ephippiumsi*
Brachycephalus sp.	*ephippium*	Parque Natural Municipal Nascentes de Paranapiacaba (23°46′10″ S, 46°17′36″ W), municipality of Santo André, São Paulo	800–1164?	[120] as *Brachycephalus* sp.
Brachycephalus sp.	*ephippium*	Penísula do Bororé (23°47′11″ S, 46°38′45″ W), Represa Billings, Grajaú District, municipality of São Paulo, São Paulo	780	[27] as *Brachycephalus nodoterga*; [99] as another species than *B. nodoterga* of [27]
Brachycephalus sp.	*ephippium*	Reserva Biológica do Alto da Serra de Paranapiacaba (23°46′40″ S, 46°18′45″ W), municipality of Santo André, São Paulo	800	[27] as *Brachycephalus* sp. 1
Brachycephalus sp.	*ephippium*	Theodoro de Oliveira (first position: 22°22′11″ S, 42°33′25″ W), Parque Estadual dos Três Picos, municipality of Nova Friburgo, Rio de Janeiro	1100–1200	[67] as *Brachycephalus* sp.; [95] as *Brachycephalus* sp. nov.
Brachycephalus sp.	*pernix*	Pedra Branca do Araraquara (25°56′00″ S, 48°52′50″ W), Serra do Araraquara, municipality of Guaratuba, Paraná	1000	[23] as *Brachycephalus* sp. 5
Brachycephalus sp.	*pernix*	Pico Paraná (25°15′10″ S, 48°48′32″ W), Serra dos Órgãos, municipality of Antonina, Paraná	1880	This study
Brachycephalus sp.	*pernix*	Serra Canasvieiras (25°36′58″ S, 48°46′59″ W), boundary of the municipalities of Guaratuba and Morretes, Paraná	1080	[23] as *Brachycephalus* sp. 5; [25] as *B.* sp. Canasvieiras; [28] as not identified; [73] without species identification
Brachycephalus sp.	*pernix*	Tupipiá (25°15′13″ S, 48°48′20″ W), Serra dos Órgãos, municipality of Antonina, Paraná	1560	This study, [27] as *B. brunneus*; [28] as *B. brunneus*; [29] as *B.* sp. Tupipiá, [73] without species identification
Brachycephalus sp. (cf. *B. darkside* juvenile)	?	Parque Estadual da Serra do Brigadeiro (cf. 20°43′16″ S, 42°29′05″ W), municipality of Araponga, Minas Gerais	1330?	[85] as *Brachycephalus* cf. *didactylus*

[1] Data with "?" were not available in literature.

The main threat to the species of *Brachycephalus* is deforestation, affecting not only their EO but also other aspects of their biology, such as population size and individual health. Indeed, deforestation affects 20 species. Other species are under threat due to their small EO. Forests within EOs were converted into agricultural areas (e.g., for coffee and palm plantations - *Archontophoenix alexandrae* H. Wendl. & Drude), tree monocultures (*Pinus* spp. and *Eucalyptus* spp.), urban areas and, more frequently, pastures. Some species also have part of their EO flooded by dams (e.g., *B. nodoterga*) or affected by landslides (e.g., *B. izecksohni*). Fire, edge effects, timber harvest, grazing, and the invasion of exotic plants are impacts that reduce the quality of EO. For instance, deforestation and loss of habitat quality are important threats to *B. mariaeterezae*, whose type locality suffers from fire, grazing, and timber harvests. Fire and grazing substantially affect the quality of forests, even the cloud forests of *B. quiririensis*. The estimated population sizes were above those used in the IUCN criteria and therefore were not useful to rank the studied species regarding their conservation status.

The conservation status of all described species (Figure 2, Table 2) were determined. Twenty-one species (58.3% of all species) were classified as at risk of extinction: six as CR (28.6%), five as Endangered (EN; 23.8%), and 10 as Vulnerable (VU; 47.6%). Five species (13.9% of all species) were classified as non-threatened (= LC) and the remaining ten species (27.8%) as DD. The reduced EO (criteria B1) contributes to the ranking status of the conservation of 16 species associated with the number of locations (criteria B1a; 16 species), threats that reduce the area of EO (criteria B1b(i); 16 species), and quality of the area of EO (B1b(iii); 16 species; Table 2). B2 criteria were adopted for eight species (Table 2). For the B2 criteria, the number of locations (criteria B2a; eight species) and the threats that reduced the AO area (criteria B2b(ii); eight species) and quality (criteria B2b(iii); eight species) were also considered. Only one additional criterion (D2) for five species with less than 2000 ha of AO was used, and no threat could be assessed for this AO.

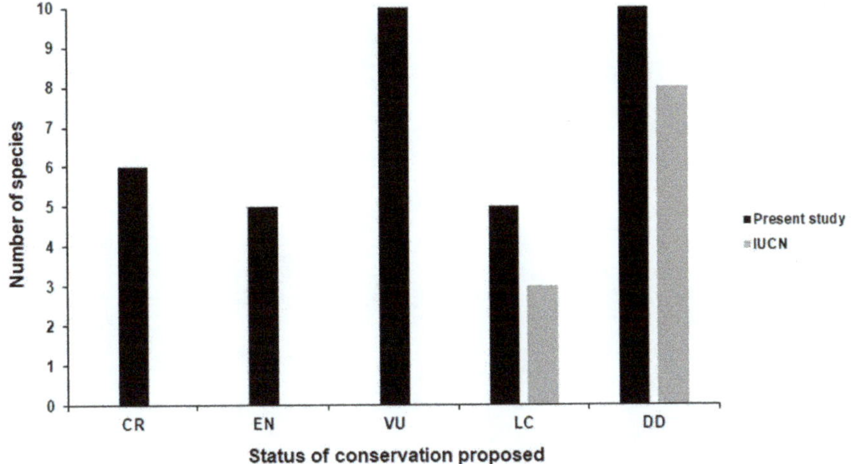

Figure 2. Status of conservation for the 36 species of *Brachycephalus* proposed in this study according to the IUCN [48] criteria and categories proposed by IUCN [31–41]. Abbreviation: CR = Critically Endangered; EN = Endangered; VU = Vulnerable; LC = Least Concern; DD = Data Deficient.

Table 2. Status of conservation of identified species of *Brachycephalus* according to IUCN [48]. Abbreviations: EO = extent of occurrence (see text for details); MMA = Ministério do Meio Ambiente (Brazil).

Species [1]	Localities [1]	Altitudinal Range (m a.s.l.) [1,2]	Evaluation of EO (ha)		Flow Chart Pathway [3]	Population		Conservation Status—Criteria			
			Previous [2]	This Study [2]		Locations [2]	Individuals [2,3]	IUCN	MMA	Others	This Study
B. didactylus group											
B. didactylus	8	35–1110	—	702,983.4	5	4	79,655,049	LC [41]	—	—	VU - B1ab(i,iii)
B. hermogenesi	11	0–1090	567,589.9 [23]	143,325.0	5	1	?	LC [33]	—	—	VU - B1ab(i,iii)
B. pulex	1	800–930	488.2 [23]	482.3	1	1	?	—	—	—	VU - D2
B. sulfuratus	26	40–1205	778,458.4 [23]	3,021,786.1	5	1	302,178,610	—	—	—	LC
B. ephippiumsi group											
B. alipioi	3	1070–1100	38,950.0 [47], 27,930.0 [43]	1,706.1	3	1	?	DD [35]	NT [43]	—	CR - B1ab(i,iii)
B. bufonoides	1	?	—	?	4	?	?	—	—	—	DD
B. crispus	1	800–1190	?	?	2	1	?	—	—	—	DD
B. darkside	2	1265–1500	–	5,700.8	1	1	?	—	—	—	CR - B1ab(i,iii)
B. ephippiumsi	31	200–1250	?	1,792,535.1	5	6	13,336,461	LC [38]	—	—	VU - B1ab(i,iii)
B. garbeanus	7	1130–1900	12,268.0 [23]	6,426.5	5	2	?	—	—	—	EN - B1ab(i,iii)+2ab(ii,iii)
B. guarani	1	500–900	?	?	2	1	?	—	—	—	DD
B. margaritatus	6	600–980	18,272.9 [23]	10,710.5	5	2	?	—	—	—	EN - B1ab(i,iii)
B. nodoterga	5	700–900	9,690.0 [47], 108,280.0 [42]	28.458.1	5	3	?	DD [31]	DD [42]	—	VU - B1ab(i,iii)
B. pitanga	4	900–1140	2,377.1 [23]	2,245.1	5	1	29,157,136	—	—	—	LC
B. toby	1	750–1060	?	?	2	1	?	—	—	—	DD
B. vertebralis	2	760–1110	161,990.0 [47], 18,580.0 [44]	?	2	2	?	DD [37]	DD [44]	—	DD
B. pernix group											
B. actaeus	7	20–530	—	15,841.6	6	2	?	—	—	—	EN - B1ab(i,iii)+2ab(ii,iii)
B. albolineatus	4	500–835	34.4 [74]	2,784.4	5	12	1,076,087	—	—	DD [74]	VU - B1ab(i,iii)
B. auroguttatus	1	1070–1100	—	?	3	1	?	—	—	—	DD
B. boticario	1	685–795	11.1 [23]	38.8	1	1	?	—	—	—	CR - B1ab(i,iii)+2ab(ii,iii)
B. brunneus	6	1095–1770	1,100.0 [47], 5,687.1 [23]	5,385.6	6	2	?	DD [39]	—	—	LC
B. coloratus	1	1145–1230	—	37.4	1	1		—	—	DD [50]	VU - D2
B. curupira	3	1095–1320	2,211.54 [23]	4,751.4	6	2	21,117,312	—	—	DD [50]	LC
B. ferruginus	1	965–1,470	38,950.0 [47], 5,475.5 [23]	5,994.3	1	1	?	DD [34]	—	—	LC
B. fuscolineatus	2	525–790	23.63 [23], 23.8 [24]	23.8	1	2	119,000	—	—	—	CR - B1ab(i,iii)+2ab(ii,iii)
B. izecksohni	1	980–1340	1,100.0 [47], 350.4 [23]	378.3	1	1	?	DD [40]	—	—	VU - D2
B. leopardus	2	1340–1645	176.7 [23]	363.1	1	3	?	—	—	—	EN - B1ab(i,iii)+2ab(ii,iii)

Table 2. *Cont.*

| Species [1] | Localities [1] | Altitudinal Range (m a.s.l.) [1,2] | Evaluation of EO (ha) | | Flow Chart Pathway [3] | Population | | Conservation Status—Criteria | | | |
			Previous [2]	This Study [2]		Locations [2]	Individuals [2,3]	IUCN	MMA	Others	This Study
B. mariaeterezae	1	1265–1270	—	?	3	1	?	—	—	—	DD
B. mirissimus	1	470–540	56.8 [25]	56.8	1	1	78,344	—	—	CR [25]	CR - B1ab(i,iii)+2ab(ii,iii)
B. olivaceus	4	650–985	12,531.6 [23]	18,850.1	5	2	?	—	—	—	EN - B1ab(i,iii)+2ab(ii,iii)
B. pernix	1	1135–1405	1,950.0 [47], 432.1 [23], 400 [45]	389.4	1	1	?	DD [32]	CR - B1ab(iii)+2ab(iii) [45]	—	VU - D2
B. pombali	2	845–1300	31,300.0 [47]	?	2	1	?	DD [36]	—	—	DD
B.quiririensis	2	1240–1380	1339.0 [23]	629.0	1	1	?	—	—	—	CR - B1ab(i,iii)+2ab(ii,iii)
B. tridactylus	1	805–910	41.4 [23]	41.4	1	1	?	—	—	—	VU - D2
B. verrucosus	1	455–945	—	?	2	1	?	—	—	—	DD
Incertae sedis											
B. atelopoide	1	?	—	?	4	?	?	—	—	—	DD

[1] According Table 1. [2] Data with "?" could not be estimate because of lack of information in literature. [3] According Figure 1.

4. Discussion

Based on the assessments, the number of endangered species of *Brachycephalus* should increase from one to 21 (Table 2). This is a significant shift and poses the question regarding why only one species had been formally recognized as threatened until now (Table 2). There are two possible reasons: (1) a delay due to the fact that many species have only been recently described and (2) a resistance based on current policies of the MMA to change a species conservation status in a short period of time (see below). Twenty-two species have been described in the last 10 years, 14 of which are only known from their type locality, and there is a natural tendency to expect them to be more widely distributed; however, studies in recent years have been gradually revealing new species rather than new records of known species, and new records of already described species have not substantially altered their geographical distributions (e.g., [24,29,76,99]). For example, a new locality record for *B. fuscolineatus* published after its description increased its EO by just 0.19 ha [24], and this species still has the smallest estimated EO for any *Brachycephalus* species (Table 2). Two new records of *B. curupira* (Table 1, [29]) double its EO, which remains small (= 4751.4 ha; Table 2). A new record of *B. albolineatus* published after its description [76] and two new localities included in Table 1 substantially extend its EO, but as in the case of *B. curupira*, this remains small (= 2784.4 ha; Table 2). A new record for *B. nodoterga* [99] did not change the EO of the species because it is located within its EO polygon. *Brachycephalus* was not found in 13 localities from southern São Paulo to northeastern Santa Catarina with an altitude comparable to other localities where *Brachycephalus* populations were present. Overall, the reduced geographical distributions of *Brachycephalus* is the rule for the montane species of the genus, i.e., the *B. ephippiumsi* and *B. pernix* groups [23]. *Brachycephalus ephippium* is the only exception of a montane species group with a large EO, but it is expected that some, if not all, populations may be identified as distinct species in future studies [23].

With respect to the resistance to incorporate drastic changes into the official number of endangered *Brachycephalus* species, this proposal is warranted despite the current policy of the MMA indicated. The MMA joined several international agreements that set targets for the conservation of the country's threatened biodiversity, and these efforts have been implemented in the successful execution of National Action Plans for the Conservation of Brazilian Endangered Species (Planos de Ação Nacional - PANs). The national scientific community and the MMA have been working together to list threats and conservation actions to all threatened species of the country and to review and to monitor these actions annually. This is possibly the reason that the MMA tended to prefer moving forward with conservation strategies of species that are already listed as threatened rather than revising the list. The effort to prioritize conservation initiatives prior to substantial updates to the list of endangered species is recognized, but the need for MMA to revise the list and to recognize the species listed in this article as threatened is also acknowledged given that they are not yet legally protected.

The most prevalent threat to *Brachycephalus* is deforestation, much of which is no longer done with heavy machinery and chainsaws. Recently, deforestation in the Atlantic Rainforest has become more subtle and involves the selective removal of trees and shrubs, particularly through inconspicuous strategies, such as bark girdling, which leads to the opening of the canopy and an increased tendency for wind to knock down additional trees. These actions are deliberately conducted a few meters into the forest edge to avoid detection by environmental inspectors. This type of deforestation has been carried out at an alarming rate in Paraná and in the northeast of Santa Catarina for at least 25 years to cultivate bananas, and more recently, to cultivate palm trees (*Archontophoenix alexandrae*). Deforestation for agricultural activities could also result in soil contamination, affecting species that depend on specific microhabitats and that have permeable skin [121]. Finally, deforestation could exacerbate edge effects, altering microhabitats and microclimatic conditions, which changes sunlight exposure, soil moisture, and plant species composition in the edges [15]. Indeed, *B. fuscolineatus* was not encountered in forest edges but only in more nuclear vegetation [24].

Deforestation in lowlands can lead to a decrease in the altitudinal distribution of cloud forests [122], potentially shifting the distributions of montane species of *Brachycephalus* to higher altitudes. This possibility of altitudinal species displacement could also be driven by climate change [23]. In tropical forests, temperatures can vary from 0.4 °C to 0.7 °C per 100 m altitude variation [123]. The thermal variation in the altitudinal gradient in a site with the occurrence of *Brachycephalus* (*B. pernix*) was determined as 0.56 °C of the reduction every 100 m of altitude [124]. A difference in the precipitation levels at this site was also evaluated, with an increase of 40 mm in mean annual precipitation every 100 m at elevation [124]. Studies on litter anurans of the Atlantic Rainforest, including *Brachycephalus*, have shown that population densities are particularly affected by air humidity, air temperature, and altitude [52,55,78,90]. This climatic dependence and its relationship with the altitude gradient raises concerns for the long-term conservation of *Brachycephalus* species that occur in mountains with a restricted altitudinal amplitude.

Lowering the category of threat for *B. pernix* from CR [45] to VU (Table 2) is proposed. The effects of trampling and timber harvests by tourists in the type locality of the species are likely to be minor, which is entirely distributed within a protected area (Parque Estadual Serra da Baitaca; Table S1). There is a threat of fire in part of the EO of the species, but the vegetation cover is regenerating well in this area after many years of management by volunteer mountaineers, reducing fire susceptibility.

It is recognized that there is some level of subjectivity to apply EO and the number of locations of threatened species. This is because each parameter shows some overlap between EN and VU categories. *Brachycephalus hermogenesi* and *B. nodoterga* fit the EN category, but both are recommended to be considered for the VU category because part of their EO is in protected areas (Table S1).

The presence of threatened *Brachycephalus* in protected areas is a useful tool to rank the species for which conservation actions are more urgent. In Table S1, 10 species without records in protected areas are recognized with three classified as CR (*B. boticario*, *B. mirissimus*, and *B. quiririensis*), two as EN (*B. actaeus* and *B. leopardus*), one as VU (*B. albolineatus*), and four as DD (*B. atelopoide*, *B. auroguttatus*, *B. bufonoides*, *B. leopardus*, and *B. verrucosus*). There are no known living populations of two species (*B. atelopoide* and *B. bufonoides*). The remaining eight species belonging to the *B. pernix* group occur in southern Paraná (*B. leopardus*) and Santa Catarina (*B. actaeus*, *B. albolineatus*, *B. auroguttatus*, *B. boticario*, *B. mirissimus*, *B. quiririensis*, and *B. verrucosus*). Also, it is argued that DD species need special attention to direct further studies to complete adequate assessments of their conservation status as soon as possible.

Santa Catarina stands out as the state in which emergency conservation actions should converge. Creating protected areas is an important way to protect species, however, the conservation of the top three priority species would require the creation of three new protected areas. A protected area for the CR *B. quiririensis* could already house other species of *Brachycephalus* that are not in any reserve, namely *B. leopardus* (EN) and *B. auroguttatus* (DD). Nonetheless, to be effective, a protected area would first require the expropriation of the land in addition to management actions aimed at recovering forest quality. Given that there are dozens of protected areas waiting for expropriation, this path to conservation does not seem likely at the moment. Private protected areas are an alternative (e.g., Reserva Particular do Patrimônio Natural—RPPN), and some of them already protect two species of *Brachycephalus* (*B. mariaeterezae* and *B. tridactylus*; Table 1 and Table S1). This is the most stable category of protected areas in Brazil and cannot be undone; however, one aspect that does not stimulate the creation of more private protected areas is the lack of government incentives to private owners, except for exemption from territorial taxes. There is an impediment to transferring public financial resources to private persons, even if they are addressing conservation measures.

The conservation of *Brachycephalus* should also include alternatives to the creation of protected areas. One approach would be to lease land with the occurrence of threatened *Brachycephalus* at a percentage of the regional value of production per hectare of mountainous lands, which would be an incentive for landowners to leave their land intact. This must be governed by a renewable contract. For this strategy to be put into practice, it is vital to attract international resources. It would also be interesting to attract additional resources of the lease value to promote environmental recovery. The management of invasive alien species, both plants and animals, is unfortunately incipient in Brazil due to the high involved costs, thus discouraging the proposition of new management projects. The state of Santa Catarina has emphasized its concern with the conservation of microendemic anurans in its region, creating a specific program for this purpose (Portaria Instituto Estadual do Meio Ambiente - IMA N° 283/2018 - 19/12/2018). This is an interesting effort that can put actions discussed into practice and can also result in other effective and innovative actions for the conservation of *Brachycephalus* in Santa Catarina. In the long term, successful practices can be replicated in other regions of Brazil.

5. Conclusions

Advances in knowledge regarding the geographical distribution of the *Brachycephalus* species have confirmed that they are in fact restricted, and this restriction is the reason for classifying 58% of the species of the genus as threatened according to IUCN criteria. Restricted geographical distributions should be considered an attribute of the species of the *Brachycephalus* montane groups. This coincides with the tendency of species with small ranges to be geographically concentrated and disproportionately under the threat of extinction [125] as well as with the tendency of newly described species to be more threatened than those described earlier [3]. With an increased understanding of the nature of most *Brachycephalus* species as microendemic species, international (IUCN) and national (MMA) agencies might be more likely to update their conservation status based on this proposal. Furthermore, Brazil has the highest amphibian richness in the world and the highest description of new species in recent years, but it is one of the countries with the lowest update rates of conservation status [3].

Deforestation and loss of habitat quality impact almost all species of *Brachycephalus* (22 species). Species of the genus are locally highly abundant, but they respond in density and geographical distribution to temperature and humidity [23,24,27], which vary along the altitudinal gradient [122,124]. Climate change can influence climatic conditions along the altitudinal gradient, confining the distribution of species even further to higher altitudes in the future.

The common action to protect endangered species in Brazil is to create protected areas. The creation of a new protected area in southern Paraná (Serra do Araçatuba) and adjacent to Santa Catarina (Serra do Quiriri) is proposed, but only because it would protect three species (*B. quiririensis*—CR, *B. leopardus*—EN, and *B. auroguttatus*—DD). In the marshes and grasslands associated with the forest of occurrence of these three *Brachycephalus* species is another endangered frog, *Melanophryniscus biancae*, which is a candidate for EN [126,127]. One reserve including the distribution of these four species would have about 11,000 ha—6,000 ha of forests, and 5,000 ha of grasslands [126]—and would also protect the springs of important rivers, such as the Negro, Cubatão, and Pirabeiraba. The creation of several other protected areas to safeguard the remaining threatened species without occurrence in reserves is impractical in the current Brazilian economic scenario. A program to lease strategic private land for owners to keep them intact with the support of international resources is a possibility for the conservation of the species in the short and medium term.

Supplementary Materials: The following are available online at http://www.mdpi.com/1424-2818/11/9/150/s1, Table S1: Records and distribution of the extent of occurrence (EO) of the species of Brachycephalus in relation to protected areas (PA).

Author Contributions: Conceptualization, M.R.B.; methodology, M.R.B.; formal analysis, M.R.B.; investigation, M.R.B.; data curation, M.R.B.; writing—original draft preparation, M.R.B. and L.T.; writing—review and editing, M.R.B., L.T. and M.R.P.

Funding: Field researches was partially funded by Fundação Grupo Boticário de Proteção à Natureza (through grants 0895_20111 and A0010_2014, conducted by Mater Natura—Instituto de Estudos Ambientais) and National Geographic Society (through the grant EC–50722R-18 to L.T.) L.T. was supported through a grant from CAPES / Reitoria and M.R.P. through a grant from CNPq/MCT (571334/2008–3).

Acknowledgments: Luiz F. Ribeiro provided valuable assistance during the field work. We thank two anonymous reviewers for comments on the manuscript.

Conflicts of Interest: The authors declare no conflict of interest. The funders had no role in the design of the study, the collection, analyses, or interpretation of data, the writing of the manuscript, or the decision to publish the results.

References

1. Frost, D.R. Amphibian Species of the World: An Online Reference, Version 6.0. 2019. Available online: http://research.amnh.org/herpetology/amphibia/index.html (accessed on 22 June 2019).
2. Köhler, K.; Vieites, D.R.; Bonett, R.M.; Garcia, F.H.; Glaw, F.; Steinke, D.; Vences, M. New Amphibians and Global Conservations: A boost in species discoveries in a highly endangered vertebrate group. *BioScience* **2005**, *55*, 693–696. [CrossRef]
3. Tapley, B.; Michaels, C.J.; Gumbs, R.; Böhm, M.; Luedtke, J.; Pearce-Kelly, P.; Rowley, J.J.L. The disparity between species description and conservation assessment: A case study in taxa with high rates of species discovery. *Biol. Conserv.* **2018**, *220*, 209–214. [CrossRef]
4. Wake, D.B.; Vredenburg, V.T. Are we in the midst of the sixth mass extinction? A view from the world of amphibians. *Proc. Natl. Acad. Sci. USA* **2008**, *105*, 11466–11473. [CrossRef] [PubMed]
5. IUCN. The IUCN Red List of Threatened Species. Available online: https://www.iucnredlist.org (accessed on 28 June 2019).
6. Stuart, S.; Chanson, J.; Cox, N. Status and trends of amphibian declines and extinctions worldwide. *Science* **2004**, *306*, 1783–1786. [CrossRef] [PubMed]
7. Skerrat, L.F.; Berger, L.; Speare, R.; Cashins, S.; McDonald, K.R. Spread of Chytridiomycosis Has Caused the Rapid Global Decline and Extinction of Frogs. *EcoHealth* **2007**, *4*, 126. [CrossRef]
8. Stuart, S.N.; Hoffmann, M.; Chanson, J.S.; Cox, N.A.; Berridge, R.J.; Ramani, P.; Young, B.E. *Threatened Amphibians of the World*; Lynx Editions: Barcelona, Spain, 2008; pp. 1–151.
9. Lips, K.P.R.; Burrowes, P.A.; Mendelson, J.R., III; Parra-Olea, G. Amphibian declines in Latins America: A synthesis. *Biotropica* **2005**, *37*, 222–226. [CrossRef]
10. Ruland, F.; Jeschke, J.M. Threat-dependent traits of endangered frogs. *Biol. Conserv.* **2017**, *206*, 310–313. [CrossRef]
11. Ruggeri, J.; Ribeiro, L.P.; Pontes, M.R.; Toffolo, C.; Candido, M.; Carriero, M.M.; Zanella, N.; Sousa, R.L.M.; Toledo, L.F. First case of wild amphibians infected with *Ranavirus* in Brazil. *J. Wildlife Dis.* **2019**, *55*. Preprint.
12. Myers, N.; Mittermeier, R.A.; Mittermeier, C.G.; Fonseca, G.A.B.; Kent, J. Biodiversity hotspots for conservation priorities. *Nature* **2000**, *403*, 853–858. [CrossRef] [PubMed]
13. Dov-Por, F. *Sooretema, the Atlantic Rain Forest of Brazil*; SPB Academic Publishing: The Hague, The Netherlands, 1992; Volume 128.
14. MMA. *Mapa de Vegetação Nativa na Área de Aplicação da Lei no. 11.428/2006—Lei da Mata Atlântica (ano base 2009)*; Ministério do Meio Ambiente: Brasília, Brazil, 2015; pp. 1–85.
15. Rezende, C.L.; Scarano, F.R.; Assad, E.D.; Joly, C.A.; Metzger, J.P.; Strassburg, B.B.N.; Tabarelli, M.; Fonseca, G.A.; Mittermeier, R.A. From hotspots to hopespot: An opportunity for the Brazilian Atlantic Forest. *Perspect. Ecol. Conserv.* **2018**, *16*, 204–214. [CrossRef]
16. Ministério do Meio Ambiente. *Livro Vermelho da Fauna Brasileira Ameaçada de Extinção—Sumário Executivo*; Instituto Chico Mendes de Conservação da Biodiversidade—ICMBio: Brasília, Brazil, 2016; pp. 1–76.
17. Izecksohn, E. Novo gênero e nova espécie de Brachycephalidae do Estado do Rio de Janeiro, Brasil. *Boletim do Museu Nacional Zoologia* **1971**, *280*, 1–12.
18. Giaretta, A.A.; Sawaya, R.J. Second species of *Psyllophryne* (Anura: Brachycephalidae). *Copeia* **1998**, *1998*, 985–987. [CrossRef]
19. Pombal, J.P., Jr. Oviposição e desenvolvimento de *Brachycephalus ephippium* (Spix) (Anura, Brachycephalidae). *Rev. Brasil. Zool.* **1999**, *16*, 967–976. [CrossRef]
20. Napoli, M.F.; Caramaschi, U.; Cruz, C.A.G.; Dias, I.R. A new species of flea-toad, genus *Brachycephalus* Fitzinger (Amphibia: Anura: Brachycephalidae), from the Atlantic Rainforest of southern Bahia, Brazil. *Zootaxa* **2011**, *2739*, 33–40. [CrossRef]
21. Yeh, J. The effect of miniaturized body size on skeletal morphology in frogs. *Evolution* **2002**, *56*, 628–641. [CrossRef] [PubMed]
22. Schwartz, C.A.; de Souza Castro, M.; Pires Júnior, O.R.; Maciel, N.M.; Schwartz, E.N.F.; Sebben, A. Princípios bioativos da pele de anfíbios: Panorama atual e perspectivas. In *Herpetologia no Brasil II*; Nascimento, L.B., Oliveira, M.E., Eds.; Sociedade Brasileira de Herpetologia: Belo Horizonte, Brasil, 2007; pp. 146–168.

23. Bornschein, M.R.; Firkowski, C.R.; Belmonte-Lopes, R.; Corrêa, L.; Ribeiro, L.F.; Morato, S.A.A.; Antoniazzi, R.L., Jr.; Reinert, B.L.; Meyer, A.L.S.; Cini, F.A.; et al. Geographical and altitudinal distribution of *Brachycephalus* (Anura: Brachycephalidae) endemic to the Brazilian Atlantic Rainforest. *PeerJ* **2016**, *4*, e2490. [CrossRef] [PubMed]

24. Bornschein, M.R.; Teixeira, L.; Ribeiro, L.F. New record of *Brachycephalus fuscolineatus* Pie, Bornschein, Firkowski, Belmonte-Lopes & Ribeiro, 2015 (Anura, Brachycephalidae) from Santa Catarina state, Brazil. *Check List* **2019**, *15*, 379–385.

25. Pie, M.R.; Ribeiro, L.F.; Confetti, A.E.; Nadaline, M.J.; Bornschein, M.R. A new species of *Brachycephalus* (Anura: Brachycephalidae) from southern Brazil. *PeerJ* **2018**, *6*, e5683. [CrossRef] [PubMed]

26. Ribeiro, L.F.; Bornschein, M.R.; Belmonte-Lopes, R.; Firkowski, C.R.; Morato, S.A.A.; Pie, M.R. Seven new microendemic species of *Brachycephalus* (Anura: Brachycephalidae) from southern Brazil. *PeerJ* **2015**, *3*, e1011. [CrossRef] [PubMed]

27. Pie, M.R.; Meyer, A.L.S.; Firkowski, C.R.; Ribeiro, L.F.; Bornschein, M.R. Understanding the mechanisms underlying the distribution of microendemic montane frogs (*Brachycephalus* spp., Terrarana: Brachycephalidae) in the Brazilian Atlantic Rainforest. *Ecol. Model.* **2013**, *250*, 165–176. [CrossRef]

28. Firkowski, C.R.; Bornschein, M.R.; Ribeiro, L.F.; Pie, M.R. Species delimitation, phylogeny and evolutionary demography of co-distributed, montane frogs in the southern Brazilian Atlantic Forest. *Mol. Phylogenet. Evol.* **2016**, *100*, 345–360. [CrossRef] [PubMed]

29. Pie, M.R.; Faircloth, B.C.; Bornschein, M.R.; McComarck, J.E. Phylogenomics of montane frogs of the Brazilian Atlantic Forest is consistent with isolation in sky islands followed by climatic stability. *Biol. J. Linn. Soc.* **2018**, *125*, 72–82. [CrossRef]

30. Pombal, J.P., Jr. A posição taxonômica das "variedades" de *Brachycephalus ephippium* (Spix, 1824) descritas por Miranda-Ribeiro, 1920 (Amphibia, Anura, Brachycephalidae). *Bol. Mus. Nac. Zool.* **2010**, *526*, 1–12.

31. Silvano, D.; Heyer, R.; Caramaschi, U. *Brachycephalus nodoterga*. The IUCN Red List of Threatened Species 2004: e.T54454A11149387. Available online: https://www.iucnredlist.org/species/54454/11149387 (accessed on 13 July 2019).

32. Silvano, D.; Garcia, P.; Segalla, M.V. *Brachycephalus pernix*. The IUCN Red List of Threatened Species 2004, e.T54455A11149530. Available online: https://www.iucnredlist.org/species/54455/11149530 (accessed on 13 July 2019).

33. Silvano, D.; Caramaschi, U. *Brachycephalus hermogenesi*. The IUCN Red List of Threatened Species 2010, e.T29487A9501270. Available online: https://www.iucnredlist.org/species/29487/9501270 (accessed on 13 July 2019).

34. Angulo, A. *Brachycephalus ferruginus*. The IUCN Red List of Threatened Species 2006, e.T135912A4229152. Available online: https://www.iucnredlist.org/species/135912/4220152 (accessed on 13 July 2019).

35. Angulo, A. *Brachycephalus alipioi*. The IUCN Red List of Threatened Species 2008, e.T135774A4199662. Available online: https://www.iucnredlist.org/species/135774/4199662 (accessed on 13 July 2019).

36. Angulo, A. *Brachycephalus pombali*. The IUCN Red List of Threatened Species 2008, e.T135830A4208137. Available online: https://www.iucnredlist.org/species/135830/4208137 (accessed on 13 July 2019).

37. Caramaschi, U.; Carvalho-e-Silva, S.P. *Brachycephalus vertebralis*. The IUCN Red List of Threatened Species 2004, e.T54456A11134718. Available online: https://www.iucnredlist.org/species/54456/11134718 (accessed on 13 July 2019).

38. Sluys, M.V.; Rocha, C.F. *Brachycephalus ephippium*. The IUCN Red List of Threatened Species 2010: e.T54453A11149233. Available online: https://www.iucnredlist.org/species/54453/11149233 (accessed on 13 July 2019).

39. Stuart, S. *Brachycephalus brunneus*. The IUCN Red List of Threatened Species 2006, e.T61745A12553521. Available online: https://www.iucnredlist.org/species/61746/12553521 (accessed on 13 July 2019).

40. Stuart, S. *Brachycephalus izecksohni*. The IUCN Red List of Threatened Species 2006, e.T61747A12553638. Available online: https://www.iucnredlist.org/species/61747/12553638 (accessed on 13 July 2019).

41. Telles, A.M.; Carvalho-e-Silva, S.P. *Brachycephalus didactylus*. The IUCN Red List of Threatened Species 2004, e.T54452A11148997. Available online: https://www.iucnredlist.org/species/54452/11148997 (accessed on 13 July 2019).

42. Haddad, C.F.B.; Machado, I.F.; Giovanelli, J.G.R.; Bataus, Y.S.L.; Ublig, V.M.; Batista, F.R.Q.; Cruz, C.A.G.; Conte, C.E.; Zank, C.; Strüsmann, C.; et al. Avaliação do Risco de Extinção de *Brachycephalus nodoterga* Miranda-Ribeiro, 1920. In *Processo de Avaliação do Risco de Extinção da Fauna Brasileira*; Instituto Chico Mentes de conservação da Biodiversidade—ICMbio: Brasília, Brazil, 2016.

43. Haddad, C.F.B.; Machado, I.F.; Giovanelli, J.G.R.; Bataus, Y.S.L.; Uhlig, V.M.; Batista, F.R.Q.; Cruz, C.A.G.; Conte, C.E.; Zank, C.; Strüsmann, C.; et al. Avaliação do Risco de Extinção de *Brachycephalus alipioi* Pombal & Gasparini, 2006. In *Processo de Avaliação do Risco de Extinção da Fauna Brasileira*; Instituto Chico Mentes de conservação da Biodiversidade—ICMbio: Brasília, Brazil, 2016.

44. Haddad, C.F.B.; Machado, I.F.; Giovanelli, J.G.R.; Bataus, Y.S.L.; Uhlig, V.M.; Batista, F.R.Q.; Cruz, C.A.G.; Conte, C.E.; Zank, C.; Strüsmann, C.; et al. Avaliação do Risco de Extinção de *Brachycephalus vertebralis* Pombal, 2001. In *Processo de Avaliação do Risco de Extinção da Fauna Brasileira*; Instituto Chico Mentes de conservação da Biodiversidade—ICMbio: Brasília, Brazil, 2016.

45. Haddad, C.F.B.; Segalla, M.V.; Bataus, Y.S.L.; Uhlig, V.M.; Batista, F.R.Q.; Garda, A.; Hudson, A.A.; Cruz, C.A.G.; Strüsmann, C.; Brasileiro, C.A.; et al. Avaliação do Risco de Extinção de *Brachycephalus pernix* Pombal, Wistuba & Bornschein, 1998. In *Processo de Avaliação do Risco de Extinção da Fauna Brasileira*; Instituto Chico Mentes de conservação da Biodiversidade—ICMbio: Brasília, Brazil, 2016.

46. Bland, L.M.; Colle, B.; Orme, C.D.L.; Bielby, J. Data uncertainty and the selectivity of extinction risk in freshwater invertebrates. *Divers. Distrib.* **2012**, *18*, 1211–1220. [CrossRef]

47. Morais, A.R.; Siqueira, M.N.; Lemes, P.; Maciel, N.M.; De Marco, P.; Brito, D. Unraveling the conservations status of Data Deficient species. *Biol. Conserv.* **2013**, *166*, 98–102. [CrossRef]

48. IUCN. *IUCN Red List Categories and Criteria: Version 3.1.*, 2nd ed.; International Union for Conservation of Nature—IUCN: Gland, Switzerland; Cambridge, UK, 2012.

49. Reinert, B.L.; Bornschein, M.R.; Firkowski, C. Distribuição, tamanho populacional, hábitat e conservação do bicudinho-do-brejo *Stymphalornis acutirostris* Bornschein, Reinert e Teixeira, 1995 (Thamnophilidae). *Rev. Bras. Ornitol.* **2007**, *15*, 493–519.

50. Ribeiro, L.F.; Blackburn, D.C.; Stanley, E.L.; Pie, M.R.; Bornschein, M.R. Two new species of the *Brachycephalus pernix* group (Anura: Brachycephalidae) from the state of Paraná, southern Brazil. *PeerJ* **2017**, *5*, e3603. [CrossRef] [PubMed]

51. De Oliveira, J.C.F.; Coco, L.; Pagotto, R.V.; Pralon, E.; Vrcibradic, D.; Pombal, J.P., Jr.; Rocha, C.F.D. Amphibia, Anura, *Brachycephalus didactylus* (Izecksohn, 1971) and *Zachaenus parvulus* (Girard, 1853): Distribution extension. *Check List* **2012**, *8*, 242–244. [CrossRef]

52. Oliveira, J.C.F.; Pralon, E.; Coco, L.; Pagotto, R.V.; Rocha, C.F.D. Environmental humidity and leaf-litter depth affecting ecological parameters of a leaf-litter frog community in an Atlantic Rainforest area. *J. Nat. Hist.* **2013**, *47*, 2115–2124. [CrossRef]

53. Siqueira, C.C.; Vrcibradic, D.; Almeida-Gomes, M.; Borges, V.N.T., Jr.; Almeida-Santos, P.; Almeida-Santos, M.; Ariani, C.V.; Guedes, D.M.; Goyannes-Araújo, P.; Dorigo, T.A.; et al. Density and richness of the leaf litter frogs of an Atlantic Rainforest area in Serra dos Órgãos, Rio de Janeiro State, Brazil. *Zoologia* **2009**, *26*, 97–102. [CrossRef]

54. Almeida-Santos, M.; Siqueira, C.C.; Van Sluys, M.; Rocha, C.D.F. Ecology of the Brazilian flea frog *Brachycephalus didactylus* (Terrana: Brachycephalidae). *J. Hepertol.* **2011**, *45*, 251–255.

55. Siqueira, C.C.; Vrcibradic, D.; Nogueira-Costa, P.; Martins, A.R.; Dantas, L.; Gomes, V.L.R.; Bergallo, H.G.; Rocha, C.F.D. Environmental parameters affecting the structure of leaf-litter frog (Amphibia: Anura) communities in tropical forests: A case study from an Atlantic Rainforest area in southeastern Brazil. *Zoologia* **2014**, *31*, 147–152. [CrossRef]

56. Carvalho-e-Silva, A.M.T.; Silva, G.R.; Carvalho-e-Silva, S.P. Anuros da Reserva Rio das Pedras, Mangaratiba, RJ, Brasil. *Biota Neotrop.* **2008**, *8*, 199–209. [CrossRef]

57. Rocha, C.F.D.; Vrcibradic, D.; Kiefer, M.C.; Almeida-Gomes, M.; Borges-Junior, V.N.T.; Menezes, V.A.; Ariani, C.V.; Pontes, J.A.L.; Goyannes-Araújo, P.; Marra, R.V.; et al. The leaf-litter frog community from Reserva Rio das Pedras, Mangaratiba, Rio de Janeiro State, Southeastern Brazil: Species richness, composition and densities. *North West. J. Zool.* **2013**, *9*, 151–156.

58. Pombal, J.P., Jr. A new species of *Brachycephalus* (Anura: Brachycephalidae) from Atlantic Forest of southeastern Brazil. *Amphib. Reptil.* **2001**, *22*, 179–185. [CrossRef]

59. Ribeiro, L.F.; Alves, A.C.R.; Haddad, C.F.B. Two new species of *Brachycephalus* Günther, 1858 from the state of Paraná, southern Brazil (Amphibia, Anura, Brachycephalidae). *Bol. Mus. Nac. Zool.* **2005**, *519*, 1–18.

60. Alves, A.C.R.; Ribeiro, L.F.; Haddad, C.F.B.; Reis, S.F. Two new species of *Brachycephalus* (Anura: Brachycephalidae) from the Atlantic Forest in Paraná State, southern Brasil. *Hepertologica* **2006**, *62*, 221–233. [CrossRef]

61. Alves, A.C.R.; Sawaya, R.J.; Reis, S.F.; Haddad, C.F.B. New species of *Brachycephalus* (Anura: Brachycephalidae) from the Atlantic Rain Forest in São Paulo State, Southeastern Brazil. *J. Hepertol.* **2009**, *43*, 212–219. [CrossRef]

62. Da Silva, H.R.; Campos, L.A.; Sebben, A. The auditory region of *Brachycephalus* and its bearing on the monophyly of the genus (Anura: Brachycephalidae). *Zootaxa* **2007**, *1422*, 59–68. [CrossRef]

63. Verdade, V.K.; Rodrigues, M.T.; Cassimiro, J.; Pavan, D.; Liou, N.; Lange, M. Advertisement call, vocal activity, and geographic distribution of *Brachycephalus hermogenesi* (Giaretta and Sawaya, 1998) (Anura, Brachycephalidae). *J. Herpetol.* **2008**, *42*, 542–549. [CrossRef]

64. Clemente-Carvalho, R.G.B.; Antoniazzi, M.M.; Jared, C.; Haddad, C.F.B.; Alvez, A.C.R.; Rocha, H.S.; Pereira, G.R.; Oliveira, D.F.; Lopes, R.T.; Reis, S.F. Hyperossification in miniaturized toadlets of the genus *Brachycephalus* (Amphibia: Anura: Brachycephalidae): Microscopic structure and macroscopic patterns of variation. *J. Morphol.* **2009**, *270*, 1285–1295. [CrossRef] [PubMed]

65. Campos, L.A. Sistemática Filogenética do Gênero *Brachycephalus* Ftzinger, 1826 (Anura Brachycephalidae) Com Base Em Dados Morfológicos. Ph.D. Thesis, Universidade de Brasília, Brasília, Brazil, 2011.

66. Pombal, J.P., Jr.; Izecksohn, E. Uma nova espécie de *Brachycephalus* (Anura, Brachycephalidae) do Estado do Rio de Janeiro. *Pap. Av. Zool.* **2011**, *51*, 443–451. [CrossRef]

67. Siqueira, C.C.; Vrcibradic, D.; Dorigo, T.A.; Rocha, C.F.D. Anurans from two high-elevation areas of Atlantic Forest in the state of Rio de Janeiro, Brazil. *Zoologia* **2011**, *28*, 457–464. [CrossRef]

68. Rocha, C.F.D.; Van Sluys, M.; Alves, M.A.S.; Bergallo, H.G.; Vrcibradic, D. Activity of leaf-litter frogs: When should frogs be sampled? *J. Herpetol.* **2000**, *34*, 285–287. [CrossRef]

69. Rocha, C.F.D.; Van Sluys, M.; Alves, M.A.S.; Bergallo, H.G.; Vrcibradic, D. Estimates of forest floor litter frog communities: A comparison of two methods. *Austral. Ecol.* **2001**, *26*, 14–21. [CrossRef]

70. Van Sluys, M.; Vrcibradic, D.; Alves, M.A.S.; Bergallo, H.G.; Rocha, C.F.D. Ecological parameters of the leaf-litter frog community of an Atlantic Rainforest area at Ilha Grande, Rio de Janeiro state, Brazil. *Austral. Ecol.* **2007**, *32*, 254–260. [CrossRef]

71. Pimenta, B.V.S.; Bérnils, R.S.; Pombal, J.P., Jr. Amphibia, Anura, Brachycephalidae, *Brachycephalus hermogenesi*: Filling gap and geographic distribution map. *Check List* **2007**, *3*, 277–279. [CrossRef]

72. Condez, T.H.; Monteiro, J.P.C.; Comitti, E.J.; Garcia, P.C.A.; Amaral, I.B.; Haddad, C.F.B. A new species of flea-toad (Anura: Brachycephalidae) from southern Atlantic Forest, Brazil. *Zootaxa* **2016**, *4083*, 40–56. [CrossRef]

73. Firkowski, C.R. Diversification and microendemism in montane refugia from the Brazilian Atlantic Forest. Master's Thesis, Universidade Federal do Paraná, Curitiba, Brazil, 2013.

74. Bornschein, M.R.; Ribeiro, L.F.; Blackburn, D.C.; Stanley, E.L.; Pie, M.R. A new species of *Brachycephalus* (Anura: Brachycephalidae) from Santa Catarina, southern Brazil. *PeerJ* **2016**, *4*, e2629. [CrossRef]

75. Monteiro, J.P.C.; Condez, T.H.; Garcia, P.C.A.; Comitti, E.J.; Amaral, I.B.; Haddad, C.F.B. A new species of *Brachycephalus* (Anura, Brachycephalidae) from the coast of Santa Catarina State, southern Atlantic Forest, Brazil. *Zootaxa* **2018**, *4407*, 483–505. [CrossRef]

76. Teixeira, L.; Ribeiro, L.F.; Côrrea, L.; Confetti, A.E.; Pie, M.R.; Bornschein, M.R. A second record of the recently described *Brachycephalus albolineatus* Bornschein, Ribeiro, Blackburn, Stanley & Pie, 2016 (Anura, Brachycephalidae). *Check List* **2018**, *14*, 1013–1016.

77. Pereira, M.S.; Candaten, A.; Milani, D.; Oliveira, F.B.; Gardelin, J.; Rocha, C.F.D.; Vrcibradic, D. Geographic distribution: *Brachycephalus hermogenesi*. *Herpetol. Rev.* **2010**, *41*, 506.

78. Santos-Pereira, M.; Candaten, A.; Milani, D.; Oliveira, F.B.; Gardelin, J.; Rocha, C.F.D. Seasonal variation in the leaf-litter frog community (Amphibia: Anura) from an Atlantic Forest area in the Salto Morato Natural Reserve, southern Brazil. *Zoologia* **2011**, *28*, 755–761. [CrossRef]

79. Santos-Pereira, M.; Milani, D.; Barata-Bittencourt, L.F.; Iapp, T.M.; Rocha, C.F.D. Anuran species of the Salto Morato Nature Reserve in Paraná, southern Brazil: Review of the species list. *Check List* **2016**, *12*, 1907. [CrossRef]

80. Pombal, J.P., Jr.; Gasparini, J.L. A new *Brachycephalus* (Anura: Brachycephalidae) from the Atlantic Rainforest of Espírito Santo, southeastern Brazil. *S. Am. J. Herpet.* **2006**, *1*, 87–93. [CrossRef]

81. Clemente-Carvalho, R.G.B.; Klaczko, J.; Perez, S.R.; Alves, A.C.R.; Haddad, C.F.B.; Reis, S.F. Molecular phylogenetic relationships and phenotypic diversity in miniaturized toadlets, genus *Brachycephalus* (Amphibia: Anura: Brahycephalidae). *Mol. Phylogenet. Evol.* **2011**, *61*, 79–89. [CrossRef]

82. Clemente-Carvalho, R.B.G.; Giaretta, A.A.; Condez, T.H.; Haddad, C.F.B.; Reis, S.F. A new species of miniaturized toadlet, genus *Brachycephalus* (Anura: Brachycephalidae), from the Atlantic Forest of southeastern Brazil. *Herpetologica* **2012**, *68*, 365–374. [CrossRef]

83. Miranda-Ribeiro, A. Os Brachycephalideos do Museu Paulista (com tres especies novas). *Rev. Mus. Paulista* **1920**, *12*, 306–318.

84. Condez, T.H.; Clemente-Carvalho, R.B.G.; Haddad, C.F.B.; Reis, S.F. A new species of *Brachycephalus* (Anura: Brachycephalidae) from the highlands of the Atlantic Forest, southeastern Brazil. *Herpetologica* **2014**, *70*, 89–99. [CrossRef]

85. Moura, M.R.; Motta, A.P.; Fernandes, V.D.; Feio, R.N. Herpetofauna from Serra do Brigadeiro, an Atlantic Forest remain in the state of Minas Gerais, southeastern Brazil. *Biota Neotrop.* **2012**, *12*, 209–235. [CrossRef]

86. Guimarães, C.S.; Luz, S.; Rocha, P.C.; Feio, R.N. The dark side of pumpkin toadlet: A new species of *Brachycephalus* (Anura: Brachycephalidae) from Serra do Brigadeiro, southeastern Brazil. *Zootaxa* **2017**, *4258*, 327–344. [CrossRef]

87. Pombal, J.P., Jr.; Sazima, I.; Haddad, C.F.B. Breeding behavior of the pumpkin toadlet, *Brachycephalus ephippium* (Brachycephalidae). *J. Herpetol.* **1994**, *28*, 516–519. [CrossRef]

88. Giaretta, A.A. Diversidade e Densidade de Anuros de Serapilheira Num Gradiente Altitudinal na Mata Atlântica Costeira. Ph.D. Thesis, Universidade Estadual de Campinas, Campinas, Brazil, 1999.

89. Giaretta, A.A.; Facure, K.G.; Sawaya, R.J.; Meyer, J.H.M.; Chemin, N. Diversity and abundance of litter frogs in a montane forest of Southeastern Brazil: Seasonal and altitudinal changes. *Biotropica* **1999**, *31*, 669–674. [CrossRef]

90. Giaretta, A.A.; Sawaya, R.J.; Machado, G.; Araújo, M.S.; Facure, K.G.; Medeiros, H.F.; Nunes, R. Diversity and abundance of litter frogs at altitudinal sites at Serra do Japi, Southeastern Brazil. *Rev. Brasil. Zool.* **1997**, *14*, 341–346. [CrossRef]

91. Clemente-Carvalho, R.B.G.; Monteiro, L.R.; Bonato, V.; Rocha, H.S.; Pereira, G.R.; Oliveira, D.F.; Lopes, R.T.; Haddad, C.F.B.; Martins, E.G.; Reis, S.F. Geographic variation in cranial shape in the Pumpkin Toadlet (*Brachycephalus ephippium*): A geometric analysis. *J. Herpetol.* **2008**, *42*, 176–185. [CrossRef]

92. Clemente-Carvalho, R.G.B.; Alves, A.C.R.; Perez, S.I.; Haddad, C.F.B.; Reis, S.F. Morphological and molecular variation in the Pumpkin Toadlet *Branchycephalus ephippium* (Anura: Brachycephalidae). *J. Herpetol.* **2011**, *45*, 94–99. [CrossRef]

93. Zaher, H.; Aguiar, E.; Pombal, J.P. *Paratelmatobius gaigeae* (Cochran, 1938) rediscovered (Amphibia, Anura, Leptodactylidae). *Arquiv. Mus. Nac.* **2005**, *63*, 321–328.

94. Dixo, M.; Verdade, V.K. Herpetofauna de serrapilheira da Reserva Florestal de Morro Grande, Cotia (SP). *Biota Neotrop.* **2006**, *6*, 1–20. [CrossRef]

95. Siqueira, C.C.; Vrcibradic, D.; Rocha, C.F.D. Altitudinal records of data-deficient and threatened frog species from the Atlantic Rainforest of the Serra dos Órgãos mountains, in southeastern Brazil. *Braz. J. Biol.* **2013**, *73*, 229–230. [CrossRef]

96. Dorigo, T.A.; Siqueira, C.C.; Vrcibradic, D.; Maia-Carneiro, T.; Almeida-Santos, M.; Rocha, C.F.D. Ecological aspects of the pumpkin toadlet, *Brachycephalus garbeanus* Miranda-Ribeiro, 1920 (Anura: Neobatrachia: Brachycephalidae), in a highland forest of southeastern Brazil. *J. Nat. Hist.* **2012**, *46*, 2497–2507. [CrossRef]

97. Pombal, J.P., Jr.; Wistuba, E.M.; Bornschein, M.R. A new species of brachycephalid (Anura) from the Atlantic Rain Forest of Brazil. *J. Herpetol.* **1998**, *32*, 70–74. [CrossRef]

98. Haddad, C.F.B.; Alves, A.C.R.; Clemente-Carvalho, R.B.G.; Reis, S.F. A new species of *Brachycephalus* from the Atlantic Rain Forest in São Paulo state, southeastern Brazil (Amphibia: Anura: Brachycephalidae). *Copeia* **2010**, 410–420. [CrossRef]

99. Abegg, A.D.; Ortiz, F.R.; Rocha, B.; Condes, T.H. A new record for *Brachycephalus nodoterga* (Amphibia, Anura Brachycephalidae) in the state of São Paulo, Brazil. *Check List* **2015**, *11*, 1769. [CrossRef]

100. Ribeiro, R.S. Ecologia Alimentar das Quatro Espécies Dominantes da Anurofauna de Serapilheira Em Um Gradiente Altitudinal na Ilha de São Sebastião, SP. Master Thesis, Universidade Estadual Paulista "Júlio de Mesquita Filho", Rio Claro, São Paulo, Brazil, 2006.

101. Campos, L.A.; Silva, H.R.; Sebben, A. Morphology and development of additional bony elements in the genus *Brachycephalus* (Anura: Brachycephalidae). *Biol. J. Linn. Soc.* **2010**, *99*, 752–767. [CrossRef]

102. Oliveira, E.G. História Natural de *Brachycephalus pitanga* no Núcleo Santa Virgínia, Parque Estadual da Serra do Mar, Estado de São Paulo. Master Thesis, Universidade Estadual Paulista "Júlio de Mesquita Filho", Rio Claro, Brazil, 2013.

103. Tandel, M.C.F.F.; Loibel, S.; Oliveira, E.G.; Haddad, C.F.B. Diferenciação de 3 tipos de vocalizações (cantos) na espécie *Brachycephalus pitanga*. *Revista da Estatística da Universidade Federal de Ouro Preto* **2014**, *3*, 374–386.

104. Araújo, C.B.; Guerra, T.J.; Amatuzzi, M.C.O.; Campos, L.A. Advertisement and territorial calls of *Brachycephalus pitanga* (Anura: Brachycephalidae). *Zootaxa* **2012**, *3302*, 66–67. [CrossRef]

105. Bornschein, M.R.; Ribeiro, L.F.; Rollo, M.M., Jr.; Confetti, A.E.; Pie, M.R. Advertisement call of *Brachycephalus albolineatus* (Anura: Brachycephalidae). *PeerJ* **2018**, *6*, e5273. [CrossRef]

106. Fontoura, P.L.; Ribeiro, L.F.; Pie, M.R. Diet of *Brachycephalus brunneus* (Anura: Brachycephalidae) in the Atlantic Rainforest of Paraná, southern Brazil. *Zoologia* **2011**, *28*, 687–689. [CrossRef]

107. Pie, M.R.; Ströher, P.R.; Bornschein, M.R.; Ribeiro, L.F.; Faircloth, B.C.; Mccormack, J.E. The mitochondrial genome of *Brachycephalus brunneus* (Anura: Brachycephalidae), with comments on the phylogenetic position of Brachycephalidae. *Biochem. Syst. Ecol.* **2017**, *71*, 26–31. [CrossRef]

108. Monteiro, J.P.C.; Condez, T.H.; Garcia, P.C.A.; Haddad, C.F.B. The advertisement calls of two species of *Brachycephalus* (Anura: Brachycephalidae) from southern Atlantic Forest, Brazil. *Zootaxa* **2018**, *4415*, 183–188. [CrossRef]

109. Wistuba, E.M. História Natural de *Brachycephalus pernix* Pombal, Wistuba e Bornschein, 1998 (Anura) no Morro Anhangava, Município de Quatro Barras, Estado do Paraná. Ph.D. Thesis, Universidade Federal do Paraná, Curitiba, Brazil, 1998.

110. Pires, O.R., Jr.; Sebben, A.; Schwartz, E.F.; Morales, R.A.V.; Bloch, C., Jr.; Schwartz, C.A. Further report of the occurrence of tetrodotoxin and new analogues in the Anuran family Brachycephalidae. *Toxicon* **2005**, *45*, 73–79. [CrossRef] [PubMed]

111. Ribeiro, L.F.; Ströher, P.R.; Firkowski, C.R.; Cini, F.A.; Bornschein, M.R.; Pie, M.R. *Brachycephalus pernix* (Anura: Brachycephalidae), a new host of *Ophiotaenia* (Eucestoda: Proteocephalidea). *Herpetol. Notes* **2014**, *7*, 291–294.

112. Pie, M.R.; Ribeiro, L.F. A new species of *Brachycephalus* (Anura: Brachycephalidae) from the Quiriri mountain range of southern Brazil. *PeerJ* **2015**, *3*, e1179. [CrossRef] [PubMed]

113. Garey, M.V.; Lima, A.M.X.; Hartmann, M.T.; Haddad, C.F.B. A new species of miniaturized toadlet, genus *Brachycephalus* (Anura: Brachycephalidae), from southern Brazil. *Herpetologica* **2012**, *68*, 266–271. [CrossRef]

114. Bornschein, M.R.; Rollo, M.M., Jr.; Pie, M.R.; Confetti, A.E.; Ribeiro, L.F. Redescription of the advertisement call of *Brachycephalus tridactylus* (Anura: Brachycephalidae). *Phyllomedusa* **2019**, *18*, 3–12. [CrossRef]

115. Cunha, A.K.; Oliveira, I.S.; Hartmann, M.T. Anurofauna da Colônia Castelhanos, na Área de Proteção Ambiental de Guaratuba, Serra do Mar paranaense, Brasil. *Biotemas* **2010**, *23*, 123–134.

116. De Oliveira, A.K.C.; Oliveira, I.S.; Hartmann, M.T.; Silva, N.R.; Toledo, L.F. Amphibia, Anura, Brachycephalidae, *Brachycephalus hermogenesi* (Giaretta and Sawaya, 1998): New species record in the state of Paraná, southern Brazil and geographic distribution map. *Check List* **2011**, *7*, 17–18. [CrossRef]

117. Mariotto, L.R. Anfíbios de Um Gradiente Altitudinal Em Mata Atlântica. Master Thesis, Universidade Federal do Paraná, Curitiba, Brazil, 2014.

118. Condez, T.H.; Sawaya, R.J.; Dixo, M. Herpetofauna dos remanescentes de Mata Atlântica da região de Tapiraí e Piedade, SP, sudeste do Brasil. *Biota Neotrop.* **2009**, *9*, 157–185. [CrossRef]

119. Verdade, V.K.; Rodrigues, M.T.; Pavan, D. *Anfíbios Anuros da Região da Estação Biológica do Alto da Serra de Paranapiacaba. Patrimônio da Reserva Biológica do Alto da Serra de Paranapiacaba: A antiga Estação Biológica do Alto da Serra*; Governo do Estado de São Paulo, Secretaria do Meio Ambiente: São Paulo, Brazil, 2009; pp. 579–603.

120. Trevine, V.; Forlani, M.C.; Haddad, C.F.B.; Zaher, H. Herpetofauna of Paranapiacaba: Expanding our knowledge on a historical region in the Atlantic forest of southeastern Brazil. *Zoologia* **2014**, *31*, 126–146. [CrossRef]

121. Collins, J.P. Amphibian decline and extinction: What we know and what we need to learn. *Dis. Aquat. Organ.* **2010**, *92*, 93–99. [CrossRef]

122. Walter, H. *Zonas de Vegetación y Clima: Breve Exposición Desde el Punto de Vista Causal y Global*; Omega: Barcelona, Espanha, 1977; pp. 1–245.

123. Walsh, R.P.D. Climate. In *The Tropical Rain Forest*; Richards, P.W., Ed.; Cambridge University Press: Cambridge, UK, 1979; pp. 159–205.

124. Roderjan, C.V. O Gradiente da Floresta Ombrófila Densa no Morro Anhangava, Quatro Barras, PR—Aspectos Climáticos, Pedológicos e Fitossociológicos. Ph.D. Thesis, Universidade Federal do Paraná, Curitiba, Brazil, 1994.

125. Pimm, S.L.; Jenkins, C.N.; Abell, R.; Brooks, T.M.; Gittleman, J.L.; Joppa, L.N.; Raven, P.H.; Roberts, C.M.; Sexton, J.O. The biodiversity of species and their rates of extinction, distribution, and protection. *Science* **2014**, *344*, 1246752. [CrossRef] [PubMed]

126. Bornschein, M.R.; Firkowski, C.R.; Baldo, D.; Ribeiro, L.F.; Belmonte-Lopes, R.; Corrêa, L.; Morato, S.A.A.; Pie, M.R. Three new species of phytotelm-breeding *Melanophryniscus* from the Atlantic Rainforest of southern Brazil (Anura: Bufonidae). *PLoS ONE* **2015**, *10*, e0142791. [CrossRef] [PubMed]

127. Nadaline, M.J.; Ribeiro, L.F.; Teixeira, L.; Vannuchi, F.S.; Bornschein, M.R. New record of *Melanophryniscus biancae* Bornschein, Baldo, Pie, Firkowski, Ribeiro & Corrêa, 2015 (Anura: Bufonidae) from Paraná, Brazil, with comments on its phytotelm-breeding ecology. *Check List*, in press.

 *diversity*

Article

Impact of Habitat Loss and Mining on the Distribution of Endemic Species of Amphibians and Reptiles in Mexico

Fernando Mayani-Parás [1,*]**, Francisco Botello** [1]**, Saúl Castañeda** [2] **and Víctor Sánchez-Cordero** [1]

[1] Departamento de Zoología, Instituto de Biología UNAM, Circuito Exterior s/n, Ciudad Universitaria, CDMX 04510, Mexico; francisco.botello@ib.unam.mx (F.B.); victor@ib.unam.mx (V.S.-C.)
[2] Departamento de monitoreo biológico y planeación de conservación, Conservación Biológica y Desarrollo Social, A. C., CDMX 04870, Mexico; saulcastaneda@conbiodes.com
* Correspondence: fermayani@gmail.com

Received: 20 August 2019; Accepted: 12 October 2019; Published: 5 November 2019

Abstract: Mexico holds an exceptional diversity and endemicity of amphibian and reptile species, but several factors pose a threat to their conservation. Here, we produced ecological niche models for 179 Mexican endemic amphibian and reptile species and examined the impact of habitat loss and mining activities on their projected potential distributions, resulting in their extant distributions. We compared extant species distributions to the area required to conserve a minimum proportion of the species distribution. The combined impact of habitat loss and mining on extant species distribution was significantly higher than the impact of habitat loss alone. Only 40 species lost <30% of their distribution, while 83 species lost between 30–50%, 54 species lost between 50–80%, and 2 species lost more than 80% of their distribution. Furthermore, the size and configuration of the area required to conserve 20% of the extant species distributions changed considerably by increasing the number of fragments, with a potential increase in local population extirpations. Our study is the first to address the combined impact of habitat loss and mining on a highly vulnerable rich endemic species group, leading to a decrease in their potential distribution and a potential increase in the extinction risk of species.

Keywords: ecological niche modeling; potential species distribution; extant species distribution; conservation areas

1. Introduction

Mexico holds an exceptional species richness and endemicity of amphibians and reptiles, ranking in the top three countries worldwide [1]. It has over 376 species of amphibians and 864 species of reptiles, and approximately 65% and 57% of these species, respectively, are endemic [2,3]. However, it has been argued that both amphibians and reptiles are the two most vulnerable groups of terrestrial vertebrates, being at significantly higher risk than mammals and birds [4–6] for threats such as habitat loss and fragmentation. One third of the species of amphibians worldwide are threatened with extinction according to IUCN [7], and only 35% of Neotropical and Nearctic species are in the IUCN Least Concern Category [5,6]. Species of reptiles have been less studied, but they are also highly vulnerable to these threats [8], and it is likely that reptiles are at high risk of extinction as well [4]. In general, both groups are highly dependent on the environmental conditions and are very vulnerable to pathogens, invasive species, ultraviolet-B exposure, and pollution [9–12]. Several factors, such as habitat loss and fragmentation, water pollution, climate change, and mining have been identified to negatively affect their breeding activities, reproduction, and survival performance [13,14], leading to the reduction of species range size and local population extirpation.

Mexico has lost over 13.5 million ha of natural vegetation in the last 50 years [1] due to annual deforestation rates greater than 1% nationwide [15]. This habitat loss and fragmentation affects most ecosystems and species of flora and fauna, increasing the vulnerability and extinction risk of species [16]. Since species of amphibians and reptiles have dispersal limitations [17], their movements between habitat fragments are limited and going from unfavorable to favorable habitats is unlikely [13,14]. As a result, the Global Amphibian Assessment suggests that habitat loss impacts 89% of the threatened amphibian species in the Americas, which is three times the impact of any other threat [18]. Recent studies have proposed that human-induced habitat loss is important, affecting the diversity and abundance of amphibian and reptile species [19] in Neotropical habitats [20], xeric habitats [21], and dry plains [22]. Furthermore, small-range species are more likely to show population declines [23]. For example, 70% of Mexican species of amphibians and 80% of species of reptiles have restricted distributional ranges and high environmental specialization, increasing their vulnerability [24]. In fact, the distributions of endemic amphibian and reptile species have declined 80% and 70%, respectively [25].

In addition to habitat loss, mining activities are suspected to impose a high risk to species of amphibians and reptiles, but this has been poorly studied. Mexico ranks second in silver production worldwide and is one of the countries with the largest production of gold, zinc, copper, and other minerals [26]. There are currently 1531 mining projects (884 more than in 2010), of which 1113 projects are in the exploration stage (where perforations are made to determine the available minerals), 63 are under the construction of the mine, 274 are in the production stage where the minerals are being extracted, and 81 have been postponed [27]. After exploitation, environmental regulations recommend closing and restoring the affected areas. However, the companies are not forced to elaborate an integrated plan of mining closure and restoration [27]. Most mining projects are open pit mining, which is conducted at large scale, generating pollution of rivers and aquifers with heavy metals, large quantities of polluting debris, acid drainage, continuous emissions of gases and dust into the atmosphere, and the local removal of all plant and animal species [28,29]. Mining activities are considered of public utility. They are prioritized over any other use or activity in the territory and can be conducted regardless of the regime of land tenure, such as territories of indigenous people, urban areas, and private and social property [29]. There are currently more than 24,000 terrestrial active mining concessions covering over 20 million ha, and 14 marine concessions covering approximately 740 marine ha [30–32]. A total of 85% of mining activities are located in areas with vegetation cover holding ecological integrity [33,34]. Furthermore, the legislation does not restrict the possibility of establishing mining activities in most categories of protected areas [29], which has resulted in 73 mining projects covering more than 2 million ha inside protected areas and Ramsar sites; while 60,000 ha are located inside the core zones of protected areas [31,35,36].

Academic and NGO organizations assessing species extinction risk at a global level (IUCN) and at a national level (Mexican governmental ecological regulations) consider habitat loss and habitat fragmentation as the main variables to assign species extinction risk; the higher the proportion of habitat loss in their distributions, the higher the category assigned for species extinction risk [37]. However, other potential factors affecting the conservation status of species, such as mining activities, are largely underestimated. In this study, we (1) analyzed the combined impact of habitat loss and mining activities on potential species distributions of Mexican endemic amphibian and reptile species, and (2) determined the modifications and area needed to conserve a minimum proportion (20%) of the extant distribution of these species.

2. Materials and Methods

2.1. Point Occurrence Data

The study included Mexican endemic species of amphibians and reptiles. We compiled point occurrence data for 275 species of amphibians and 474 species of reptiles from the Global Biodiversity Information Facility website (GBIF; https://www.gbif.org/; accessed on 25 January 2018). We excluded

(1) all point occurrence data prior to 1970; (2) points that had a resolution lower than 2 decimals of degree or no geographic coordinates (decimal latitud = 0, empty, 99, −99); (3) fossil records; (4) alive specimens (from zoos); (5) data obtained from iNaturalist (www.iNaturalist.com.mx), since those records do not have collected and verifiable specimens, and (6) records that were found within the same pixel of the bioclimatic variables from the WorldClim, used for constructing the models (see below; 1km^2). In ArcMap, we eliminated all points that did not coincide with the currently recognized distribution of the species. Once the databases were refined, we included only the species with a minimum of 10 records. The minimum number of 10 records per species was defined based on published information for an adequate species distribution modeling approach in Maxent [38].

2.2. Potential Species Distributions

For each species, the polygons of the Mexican terrestrial ecoregions, including occurrence points, were selected, leaving a 50 km buffer zone surrounding them to be used as the modeling area (M region) [39,40]. The environmental variables used to construct potential species distributions were nineteen bioclimatic variables (~1 km^2) from the WorldClim database (https://www.worldclim.org/; accessed on 31 January 2018) [41]. Variables with a correlation r > 0.7 were considered redundant and only one was included [42].

We generated the ecological niche models following the methodology described by Sánchez-Cordero [43]. Using the ENMeval library [44] in the R software [45], 10,000 background points were selected within the modeling area to parameterize the model, the block method was used to divide the presence data into training and testing groups [46], and 5 regularization multipliers and 13 feature classes were established to adjust the models. From a total of 65 models per species, the best model was selected based on the omission rate and the area under the curve (AUC), and projected into a discrete presence/absence map through a maximum sensitivity plus specificity threshold [47]. The area of each potential species distribution was obtained using the Consnet software package [16,48,49] by obtaining the number of cells occupied by each species and multiplying this number by 0.78, the size in km^2 of the used rack cells.

2.3. Extant Species Distributions

From the potential species distribution models, we obtained two scenarios, as follows: (1) Extant species distributions due to habitat loss, and (2) extant species distributions due to habitat loss and mining activities. Habitat loss was estimated based on the land use and vegetation coverage map [33], which contains information on habitat transformation since 1968, and includes transformed areas into agricultural, rural, or urban settlements. For mining activities, we used the official mining concession map, which has information of all mining concessions since 1942 that are still active today [50]. Furthermore, we used the software package ConsNet on potential species distribution, extant species distribution due to habitat loss, and extant species distribution due to habitat loss and mining activities to analyze the area of occupancy of each species under each scenario.

Potential species distributions were compared with extant species distributions under each scenario and the percentage of the reduction in their distributions was obtained. We divided species in four groups according to percentage ranges of their distribution loss, as follows: (1) Species that lost <30% of their distribution; (2) species that lost between 30–50% of their distribution; (3) species that lost between 50–80% of their distribution, and (4) species that lost >80% of their distribution.

2.4. Selection of Priority Areas for Conservation

For each scenario of potential species distribution, extant species distributions due to habitat loss, and extant species distribution due to both habitat loss and mining activities, we obtained the selection of priority areas for conservation using the ConsNet software package, which allows the identification of conservation solutions by defining multiple previously set criteria [16,48,49]. ConsNet allows searching for the best solutions for different objectives according to the required conservation plan. For example, it can search for the minimum selected area and the best surrogate representation without considering any other restrictions, such as shape or connectivity. Similarly, a conservation solution can be searched for by considering, for example, the best representation, minimum area, connectivity, and/or shape configuration. The conservation target for all species was set to 20% of their distribution under the three scenarios, considering that all species are endemic and have limited distributions. We searched for the best representation with the minimum area and shape using the RF4 adjacency algorithm with a basic neighbor selection, and running 200,000 iterations to find the best solution. We obtained the total area (km^2) of conservation, perimeter, number of clusters, shape, and total representation of the species on the conservation area network.

2.5. Statistical Analysis

Using the statistical package StatSoft STATISTICA [51], we analyzed the differences in the reduction of extant species distributions due to habitat loss and extant species distribution due to habitat loss and mining activities, respectively (Student's *t*-test). We also performed two different one-way ANOVAs to determine differences between species of amphibians and reptiles under each scenario.

3. Results

Of a total of 749 endemic species of amphibians and reptiles, we produced robust potential species distributions for 62 species of amphibians and 117 species of reptiles. There were a total of 10,079 records, ranging from 10 as in *Sceloporus zosteromus*, *Phyllodactylus unctus*, *Lithobates sierramadrensis*, *Craugastor pozo*, and *Incilius cavifrons*, to 514 as in *Sceloporus torquatus*.

3.1. Extant Species Distributions

A total of 49 species showed a reduction of less than 30% of their potential distribution due to habitat loss. The remaining 130 species lost more than 30% of habitat, as follows: A total of 79 species lost between 30–50% of their distribution, 49 species lost between 50–80%, and 2 species lost more than 80% (Figure 1; Figure 2; Supplementary Material: Table S1). There was no significant difference in the impact of habitat loss on species of amphibians and reptiles ($F = 0.203$, $df = 1$, $p = 0.65$).

The combined impact of habitat loss and mining activities in extant species distribution was significantly higher (extant species distribution due to habitat loss: 40.56 ± 16.15; extant species distribution due to habitat loss and mining: 42.16 ± 15.67; $t = -20.10$, $df = 178$, $p < 0.001$), but it did not differ between species of amphibians and reptiles ($F = 0.341$, $df = 1$, $p = 0.56$). Of the 179 endemic species of amphibians and reptiles, only 40 species lost less than 30% of their distribution, while 83 species lost between 30–50%, 54 species lost between 50–80%, and two species lost more than 80% of their distribution (Figure 1; Table S1). The contribution of mining activities increased the percentage of distribution loss of all species, and resulted in nine species increasing from a loss <30% to a loss between 30–50%, and five species increased from a loss between 30–50% to a loss between 50–80% of their distribution (Figure 1; Table S1).

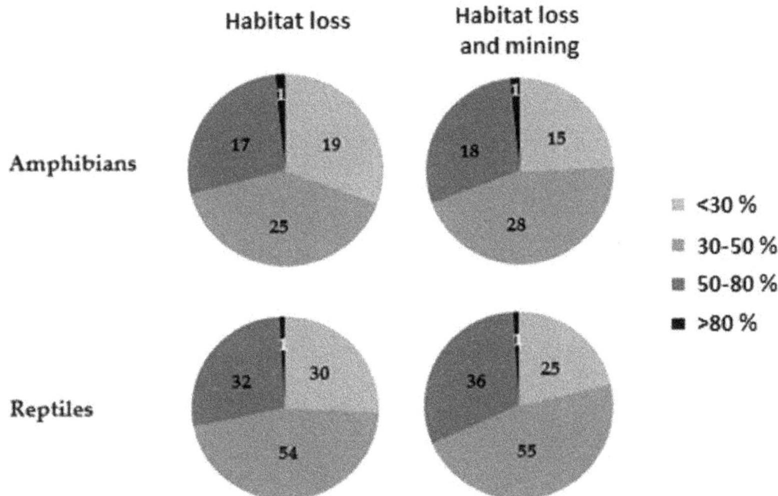

Figure 1. Number of Mexican endemic species of amphibians and reptiles in each group of percentage of distribution lost under each scenario.

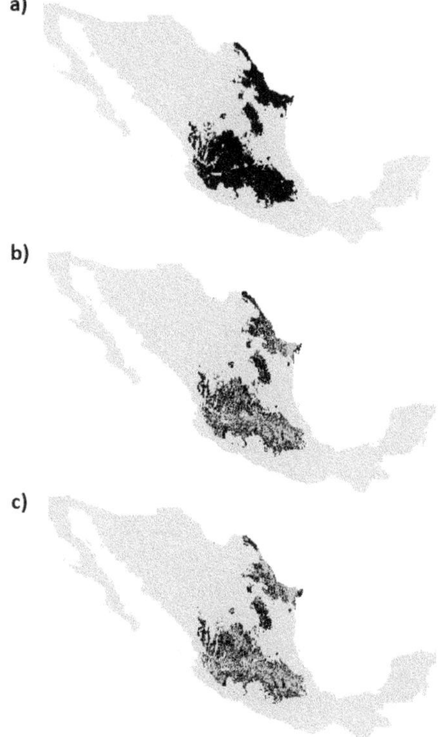

Figure 2. Distribution of *Anaxyrus compactilis* under three scenarios: (**a**) Potential species distribution; (**b**) extant species distribution due to habitat loss; and (**c**) extant species distribution due to habitat loss and mining activities.

3.2. Conservation Area Network under Three Scenarios

Under the scenario of potential species distribution, the conservation area network representing 20% of species distributions, resulted in a total of 237,195 km^2 contained in 2624 clusters, with a perimeter of 222,715.21 km, and a shape of 0.50 (perimeter/area). Under the scenario of extant species distribution due to habitat loss in a total of 250,563 km^2 contained in 8010 clusters, a perimeter of 222,715.21 km and a shape of 0.88 resulted. Under the scenario of extant species distribution due to habitat loss and mining activities in a total of 251,678.80 km^2, a perimeter of 237,148.20 km, 8706 clusters and a shape of 0.94 resulted (Figure 3).

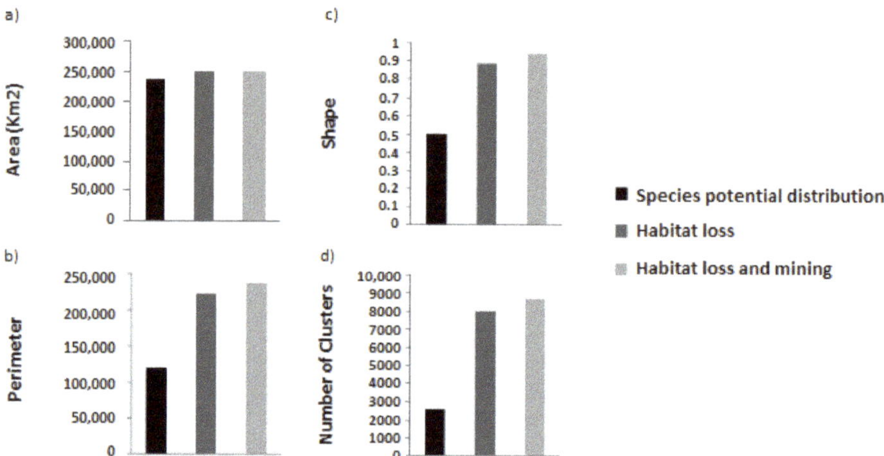

Figure 3. Area and configuration of the conservation area network (area, perimeter, shape, and number of clusters or fragments) including 20% of the 179 species potential distribution, species extant distribution due to habitat loss, and species extant distributions due to habitat loss and mining activities, respectively.

4. Discussion

Habitat loss has been identified as a major factor negatively affecting biodiversity and threatening species conservation worldwide [13–15], but there is a need to study the impact of other large-scale factors such as mining. Here, we based our analyses assuming that habitat loss and habitat fragmentation affects the conservation status of the endemic species of amphibians and reptiles included in this study. We argue that it is more convenient for conservation purposes to impose a higher risk for a species due to habitat loss (even if this does not harm a species), than to leave the species under the same risk category under the assumption that habitat loss does not harm that species [19,52,53]. This argument is particularly important for endemic species showing small ranges of distribution.

Our results show that the combined negative impact of habitat loss and mining activities increased the loss of distribution of all species of amphibians and reptiles. For example, when considering only habitat loss, 49 species out of 179 species of amphibians and reptiles retained enough remnant natural habitat in their distribution, but 130 species lost more than 30% of their distributions, which could increase their vulnerability. When we added the combined impact of habitat loss and mining, all species distributions were reduced. Only 40 species showed less than 30% of habitat loss in their distributions, while 83 lost more than 30% of their distribution, 54 lost more than half of their potential distribution, and two species are in a critical situation where they only have less than 20% of their distribution remaining (Figure 1; Table S1).

Mining activities have become a relevant threat and could be causing species to increase their extinction risk. Of the 179 species included in our study, only 50 species (35 amphibians and 15 reptiles) are currently included in the IUCN red list of endangered species [4]. However, we believe that if other factors such as mining were systematically included into the assessment of species extinction risk, more species of amphibians and reptiles should be included in an extinction risk category. Furthermore, of the 179 species included in our study, 10 species have not been assessed by the IUCN or are under the data deficient (DD) category. When all of the species occurring in Mexico are considered, the number increases to over 38 amphibians (10% of all species in Mexico) in the DD category [19] and 307 reptiles (36.2% of all species in Mexico) in the DD or not evaluated (NE) categories [54]. Moreover, IUCN does not consider species of amphibians extinct in the wild, but some reports suggest that at least 35 species are possibly extinct, and IUCN categorizes them as DD, EN, or CR [55,56]. Clearly, more efforts are needed to improve the assessment of species of amphibians and reptiles given that many species have not been assigned with a proper risk category.

Our study only analyzed the impact of mining over species distributions, but other factors caused by mining activities, such as pollution of rivers and aquifers, polluting debris, acid drainage, gas and dust emissions, and local removal of all vegetation [28], could increase the impact of mining over species and further increase species extinction risks. It has been recently reported that 84 of the 632 highly contaminated sites in Mexico are caused by mining activities, of which 11 are found inside protected areas [57]. Therefore, further studies should focus on obtaining an integrated scenario of the impact of mining activities on biodiversity conservation. In Mexico, more than 20 million ha have concessions to carry out mining activities, and this area could double by the end of 2019 [29]. There is an increasing concern that current law regulations consider mining activities as a priority, to the point that they can even be established inside protected areas. In order to adequately conserve biodiversity and meet the international conservation commitments to conserve 17% of the territory and not to allow mining activities inside protected areas, the Mexican government must urgently change sections of the environmental legislation.

Besides reducing species range size, habitat fragmentation has other implications on species conservation status. The configuration of the conservation area network solution ranked worst when including the combined impact of habitat loss and mining activities by increasing its area, perimeter, number of habitat fragments, and shape, with expected negative consequences for endemic and micro endemic species with dispersal limitations [58]. This increases risks of local population extirpation, and a decreasing genetic diversity of species [59]. Other factors, such as climate change, have also been considered to have a negative impact on species of amphibians and reptiles. It has been proposed that climate change will significantly increase the extinction risks in the short term. For example, it has been reported that an average reduction of about 64% in the current geographical range of endemic amphibians could be expected by the year 2080, with 50% of the species losing more than 60% of their distributions [60].

Protected areas are keystone initiatives to conserve biodiversity worldwide [61,62]. Thus, their adequate management, to ensure their long-term viability and to support the strategic development of conservation area networks, is essential [63,64]. Worldwide, these areas have helped to protect more than 2000 million ha [65] and in Mexico they protect around 25 million ha, representing more than 12% of the Mexican territory [66]. However, protected areas do not appear to adequately represent most biodiversity in Mexico, as shown by some studies using well-studied faunistic groups [16,63,67], and some biodiversity hotspots remain unprotected [67]. The population decline of species of amphibians and reptiles has been well documented, but these groups are not often taken into account when establishing conservation objectives [68,69]. Only 31% of the amphibians (29% of endemics) and 76% of the reptiles (46% of endemics) living in Mexico occur inside protected areas [70]. Furthermore, biodiversity representation in protected areas will be inadequate under current and climate change scenarios [71–73]. Our study provides baseline information suggesting that, as a result of the combined impact of habitat loss and mining activities, species of amphibians and reptiles are in greater danger of extinction than previously known and these factors should be included in more integrated criteria for adequately assigning species conservation status.

Supplementary Materials: The following are available online at http://www.mdpi.com/1424-2818/11/11/210/s1, Table S1: Percentage of distribution lost caused by the impact of habitat loss and habitat loss and mining activities for the Mexican endemic species of amphibians and reptiles.

Author Contributions: Conceptualization, F.M.-P. and F.B. methodology, S.C. and F.M.-P.; data analyses, S.C. and F.M.-P.; discussion and writing, F.M.-P., F.B., and V.S.-C.

Funding: F. Mayani-Parás was supported by a scholarship (Posgrado en Ciencias Biológicas of Universidad Nacional Autónoma de México and CONACyT (CVU 853134). This research was funded by the Instituto de Biología, Universidad Nacional Autónoma de México.

Acknowledgments: We thank the suggestions made by three reviewers that improved the quality of this manuscript.

Conflicts of Interest: The authors declare no conflict of interest.

References

1. SEMARNAT. *Informe de la Situación del Medio Ambiente en México. Compendio de Estadísticas Ambientales. Indicadores Clave, de Desempeño Ambiental y de Crecimiento Verde*; Edición 2015; Semarnat: Mexico City, México, 2016.

2. Flores-Villela, O.; García-Vázquez, U.O. Biodiversidad de Reptiles en México. *Rev. Mex. Biodivers.* **2014**, *85*, 31. [CrossRef]

3. Olea, G.P.; Flores-Villela, O.; Almeralla, C.M. Biodiversidad de Anfibios en México. *Rev. Mex. Biodivers.* **2014**, *85*, 33.

4. IUCN. *IUCN Red List of Threatened Species 2006*; International Union for the Conservation of Nature (IUCN): Gland, Switzerland, 2006.

5. Stuart, S.N.; Chanson, J.S.; Cox, N.A.; Young, B.E.; Rodrigues, A.S.L.; Fischman, D.L.; Waller, R.W. Status and Trends of Amphibian Declines and Extinctions Worldwide. *Science* **2004**, *306*, 1783–1786. [CrossRef]

6. Beebee, T.J.C.; Griffiths, R.A. The Amphibian Decline Crisis: A Watershed for Conservation Biology? *Biol. Conserv.* **2005**, *125*, 271–285. [CrossRef]

7. Baillie, J.E.M.; Hilton-Taylor, C.; Stuart, S.N. *2004 IUCN Red List of Threatened Species. A Global Species Assessment*; IUCN: Gland, Switzerland; Cambridge, UK, 2004.

8. Gibbons, J.W.; Scott, D.E.; Ryan, T.J.; Buhlmann, K.A.; Tuberville, T.D.; Metts, B.S.; Greene, J.L.; Mills, T.; Leiden, Y.; Poppy, S.; et al. The Global Decline of Reptiles, Déjà Vu Amphibians: Reptile Species are Declining on a Global Scale. Six Significant Threats to Reptile Populations are Habitat Loss and Degradation, Introduced Invasive Species, Environmental Pollution, Disease, Unsustainable Use, and Global Climate Change. *Bioscience* **2000**, *50*, 653–667. [CrossRef]

9. Broomhall, S.D.; Osborne, W.S.; Cunningham, R.B. Comparative Effects of Ambient Ultraviolet-B Radiation on Two Sympatric Species of Australian Frogs. *Conserv. Biol.* **2000**, *14*, 420–427. [CrossRef]

10. Kiesecker, J.M.; Blaustein, A.R.; Belden, L.K. Complex Causes of Amphibian Population Declines. *Nature* **2001**, *410*, 681–684. [CrossRef]

11. Hecnar, S.J. Acute and Chronic Toxicity of Ammonium Nitrate Fertilizer to Amphibians from Southern Ontario. *Environ. Toxicol. Chem. Int. J.* **1995**, *14*, 2131–2137. [CrossRef]

12. Davidson, C.; Shaffer, H.B.; Jennings, M.R. Declines of the California Red-Legged Frog: Climate, UV-B, Habitat, and Pesticides Hypotheses. *Ecol. Appl.* **2001**, *11*, 464–479. [CrossRef]

13. Gardner, T.A.; Barlow, J.; Peres, C.A. Paradox, Presumption and Pitfalls in Conservation Biology: The Importance of Habitat Change for Amphibians and Reptiles. *Biol. Conserv.* **2007**, *138*, 166–179. [CrossRef]

14. Becker, C.G.; Fonseca, C.R.; Baptista-Haddad, C.F.; Fernándes-Batista, R.; Prado, P.I. Habitat Split and the Global Decline of Amphibians. *Nature* **2007**, *318*, 1775–1777. [CrossRef]

15. Sarukhán, J.; Koleff, P.; Carabias, J.; Soberón, J.; Dirzo, R.; Llorente-Bousquets, J.; Halffter, G.; González, R.; March, I.; Mohar, A.; et al. *Capital Natural de México. Síntesis: Conocimiento Actual, Evaluación y Perspectivas de Sustentabilidad*; Comisión Nacional Para el Conocimiento y Uso de la Biodiversidad: Mexico City, México, 2009.

16. Botello, F.; Sánchez-Cordero, V.; Ortega-Huerta, M.A. Disponibilidad de Hábitats Adecuados Para Especies de Mamíferos a Escalas Regional (Estado de Guerrero) y Nacional (México). *Rev. Mex. Biodivers.* **2015**, *86*, 226–237. [CrossRef]

17. Ochoa-Ochoa, L.M.; Rodríguez, P.; Mora, F.; Flores-Villela, O.; Whittaker, R.J. Climate Change and Amphibian Diversity Patterns in Mexico. *Biol. Conserv.* **2012**, *150*, 94–102. [CrossRef]

18. Young, B.E.; Stuart, S.N.; Chanson, J.S.; Cox, N.A.; Boucher, T.M. Disappearing Jewels: The Status of New World Amphibians. *Appl. Herpetol.* **2005**, *2*, 429–435.

19. Frías-Alvarez, P.; Zúniga-Vega, J.J.; Flores-Villela, O. A General Assessment of the Conservation Status and Decline Trends of Mexican Amphibians. *Biodivers. Conserv.* **2010**, *19*, 3699–3742. [CrossRef]

20. Rovito, S.M.; Parra-Olea, G.; Vásquez-Almazán, C.R.; Papenfuss, T.J.; Wake, D.B. Dramatic Declines in Neotropical Salamander Populations are an Important Part of the Global Amphibians Crisis. *Proc. Natl. Acad. Sci. USA* **2009**, *106*, 3231–3236. [CrossRef]

21. Lovich, R.E.; Grismer, L.L.; Danemann, G. Conservation Status of the Herpetofauna of Baja California, México and Associated Islands in the Sea of Cortez and Pacific Ocean. *Herpetol. Conserv. Biol.* **2009**, *4*, 358–378.

22. Sigala-Rodríguez, J.J.; Greene, H.W. Landscape Change and Conservation Priorities: Mexican Herpetofaunal Perspectives at Local and Regional Scales. *Rev. Mex. Biodivers.* **2009**, *80*, 231–240.

23. Botts, E.A.; Erasmus, B.F.; Alexander, G.J. Small Range Size and Narrow Niche Breadth Predict Range Contractions in South African Frogs. *Glob. Ecol. Biogeogr.* **2013**, *22*, 567–576. [CrossRef]

24. Flores-Villela, O. Análisis de la Distribución de la Herpetofauna de México. Ph.D. Thesis, Facultad de Ciencias, UNAM, Mexico City, México, 1991.

25. Ochoa-Ochoa, L.M.; Flores-Villela, O. *Áreas de Diversidad y Endemismo de la Herpetofauna Mexicana.* Report; UNAM CONABIO: Mexico City, México, 2006.

26. INEGI. *La Minería en México 2014*; INEGI: Mexico City, Mexico, 2014.

27. Fundar, Centro de Análisis e Investigación, A.C. *Anuario 2018. Las Actividades Extractivas en México: Desafíos Para la 4T*; Fundar, Centro de Análisis e Investigación, A.C.: Mexico City, Mexico, 2018.

28. Armendáriz-Villegas, E.J. *Áreas Naturales Protegidas y Minería en México: Perspectivas y Recomendaciones*; CIBNOR: Mexico City, Mexico, 2016.

29. Fundar, Centro de Análisis e Investigación, A.C. *Anuario 2017. Las Actividades en México: Minería e Hidrocarburos Hacia el fin del Sexenio*; Fundar, Centro de Análisis e Investigación, A.C.: Mexico City, Mexico, 2018.

30. Secretaría de Economía. *Cartografía de Concesiones Mineras en el Territorio Nacional*; Secretaría de Economía: Mexico City, Mexico, 2016.

31. Secretaría de Economía. *Cartografía de Concesiones Mineras en el Territorio Nacional*; Secretaría de Economía: Mexico City, Mexico, 2017.

32. INEGI. *Marco Geoestadístico Nacional 2016*; INEGI: Mexico City, Mexico, 2016.

33. INEGI. *2014 Carta de Uso del Suelo y Vegetación Serie VI, Escala 1:250,000*; INEGI: Mexico City, México, 2017.

34. Equihua, M. *Integridad Ecosistémica 2013*; Conabio; Conafor: Mexico City, México, 2016.

35. Comisión Nacional de Áreas Naturales Protegidas (Conanp). *Zonificación Primaria*; Conanp: Mexico City, México, 2017.

36. Bezaury-Creel, J.E.; Torres-Origel, J.F.; Ochoa-Ochoa, L.M.; Castro-Campos, M.; Moreno-Díaz, N.; Llano, M.; Flores, C. *Áreas Naturales Protegidas Estatales, del Distrito Federal, Municipales y Áreas de Valor Ambiental en México [Base de Datos Geográfica]*; The Nature Conservancy/Comisión Nacional para el Conocimiento y Uso de la Biodiversidad: Mexico City, México, 2017.

37. IUCN. *IUCN Red List Categories and Criteria: Version 3.1*, 2nd ed.; IUCN: Gland, Switzerland; Cambridge, UK, 2012.

38. Wisz, M.S.; Hijmans, R.J.; Li, J.; Peterson, A.T.; Graham, C.H.; Guisan, A. NCEAS Predicting Species Distributions Working Group. Effects of Sample Size on the Performance of Species Distribution Models. *Divers. Distrib.* **2008**, *14*, 763–773. [CrossRef]

39. Soberón, J.; Peterson, A.T. Interpretation of Models of Fundamental Ecological Niches and Species' Distributional Areas. *Biodivers. Inform.* **2005**, *2*, 1–10. [CrossRef]

40. Barve, N.; Barve, V.; Jiménez-Valverde, A.; Lira-Noriega, A.; Maher, S.P.; Peterson, A.T.; Soberón, J.; Villalobos, F. The Crucial Role of the Accessible Area in Ecological Niche Modeling and Species Distribution Modeling. *Ecol. Model.* **2011**, *222*, 1810–1819. [CrossRef]

41. Hijmans, R.J.; Cameron, S.E.; Parra, J.L.; Jones, P.G.; Jarvis, A. Very High Resolution Interpolated Climate Surfaces for Global Land Areas. *Int. J. Climatol.* **2005**, *25*, 1965–1978. [CrossRef]

42. Venette, R.C. Climate Analyses to Assess Risks from Invasive Forest Insects: Simple Matching to Advanced Models. *Curr. For. Rep.* **2017**, *3*, 255–268. [CrossRef]

43. Sánchez-Cordero, V. *Propuesta Metodológica Para Evaluar la Vulnerabilidad Actual y Futura Ante el Cambio Climático de la Biodiversidad en México: El Caso de las Especies Endémicas, Prioritarias y en Riesgo de Extinción;* Instituto Nacional de Ecología y Cambio Climático: Mexico City, México, 2017.

44. Muscarella, R.; Galante, P.J.; Soley-Guardia, M.; Boria, R.A.; Kass, J.M.; Uriarte, M.; Anderson, R.P. ENMeval: An R Package for Conducting Spatially Independent Evaluations and Estimating Optimal Model Complexity for Maxent Ecological Niche Models. *Methods Ecol. Evol.* **2014**, *5*, 1198–1205. [CrossRef]

45. R Core Team. *R: A Language and Environment for Statistical Computing;* R Foundation for Statistical Computing: Vienna, Austria, 2014.

46. Hijmans, R.J. Cross-Validation of Species Distribution Models: Removing Spatial Sorting Bias and Calibration with a Null Model. *Ecology* **2012**, *93*, 679–688. [CrossRef]

47. Liu, C.; Berry, P.M.; Dawson, T.P.; Pearson, R.G. Selecting Thresholds of Occurrence in the Prediction of Species Distributions. *Ecography* **2005**, *28*, 385–393. [CrossRef]

48. Ciarleglio, M.; Barnes, J.W.; Sarkar, S. ConsNet: New Software for the Selection of Conservation Area Networks with Spatial and Multi-Criteria Analyses. *Ecography* **2009**, *32*, 205–209. [CrossRef]

49. Ciarleglio, M.; Barnes, J.W.; Sarkar, S. ConsNet-A Tabu Search Approach to the Spatially Coherent Conservation Area Network Design Problem. *J. Heuristics* **2010**, *16*, 537–557. [CrossRef]

50. Secretaría de Economía, Gobierno de México. Available online: https://datos.gob.mx/busca/dataset/cartografia-minera-de-se (accessed on 1 June 2019).

51. StatSoft STATISTICA. *Data Analysis Software System*, version 8.0; StatSoft, Inc.: Tulsa, OK, USA, 2007; Available online: http://www.statsoft.com (accessed on 20 June 2019).

52. Sánchez-Cordero, V.; Illoldi-Rangel, P.; Linaje, M.; Sahotra, S.; Peterson, A.T. Deforestation and Extant Distributions of Mexican Endemic Mammals. *Biol. Conserv.* **2005**, *126*, 465–473. [CrossRef]

53. Sánchez-Cordero, V.; Illoldi-Rangel, P.; Escalante, T.; Figueroa, F.; Rodríguez, G.; Linaje, M.; Fuller, T.; Sarkar, S. Deforestation and Biodiversity Conservation in Mexico. In *Endangered Species: New Research;* Columbus, A., Kuznetsov, L., Eds.; Nova Science Publishers: New Haven, CT, USA, 2009; pp. 279–297. [CrossRef]

54. Wilson, L.D.; Mata-Silva, V.; Johnson, J.D. A Conservation Reassessment of the Reptiles of Mexico Based on the EVS Measure. *Amphib. Reptile Conserv.* **2013**, *7*, 1–47.

55. Baena, M.L.; Halffter, G.; Lira-Noriega, A.; Soberón, J. Extinción de Especies. Conocimiento Actual de la Biodiversidad. In *Capital Natural de México. Vol. 1;* Soberón, J., Halffter, G., Llorente-Bousquets, J., Eds.; Comisión Nacional para el Conocimiento y Uso de la Biodiversidad: Distrito Federal, Mexico, 2008.

56. Ochoa-Ochoa, L.; Urbina-Cardona, J.N.; Vázquez, L.B.; Flores-Villela, O.; Bezaury-Creel, J. The Effects of Governmental Protected Areas and Social Initiatives for Land Protection on the Conservation of Mexican Amphibians. *PLoS ONE* **2009**, *4*, e6878. [CrossRef] [PubMed]

57. Secretaría de Medio Ambiente y Recursos Naturales (Semarnat). *Sitios Contaminados;* Semarnat: Mexico City, México, 2012.

58. deMaynadier, P.G.; Hunter, M.L., Jr. Road Effects on Amphibian Movements in a Forested Landscape. *Nat. Areas J.* **2000**, *20*, 56–65.

59. Reh, W.; Seitz, A. The Influence of Land Use on the Genetic Structure of Populations of the Common Frog (Rana Temporaria). *Biol. Conserv.* **1990**, *54*, 239–249. [CrossRef]

60. García, A.; Ortega-Huerta, M.A.; Martinez-Meyer, E. Potential Distributional Changes and Conservation Priorities of Endemic Amphibians in Western Mexico as a Result of Climate Change. *Environ. Conserv.* **2014**, *41*, 1–12. [CrossRef]

61. Chape, S.; Spalding, M.; Taylor, M.; Putney, A.; Ishwaran, N.; Thorsell, J.; Blasco, D.; Robertson, J.; Bridgewater, P.; Harrison, J.; et al. History, Definitions, Value and Global Perspective. In *The World's Protected Areas: Status, Values and Prospect in the 21st Century;* Chape, S., Spalding, M., Jenkins, M., Eds.; University of California Press: Los Angeles, CA, USA, 2008.

62. Gray, C.L.; Hill, S.L.; Newbold, T.; Hudson, L.N.; Börger, L.; Contu, S.; Scharlemann, J.P. Local Biodiversity is Higher Inside than Outside Terrestrial Protected Areas Worldwide. *Nat. Commun.* **2016**, *7*, 12306. [CrossRef]

63. Margules, C.R.; Sarkar, S. *Systematic Conservation Planning*; Cambridge University Press: Cambridge, UK, 2007.

64. Sarkar, S.; Dyer, J.S.; Margules, C.; Ciarleglio, M.; Kemp, N.; Wong, G.; Juhn, D.; Supriatna, J. Developing an Objectives Hierarchy for Multicriteria Decisions on Land Use Options, with a Case Study of Biodiversity Conservation and Forestry Production from Papua, Indonesia. *Environ. Plan. B Urban Anal. City Sci.* **2017**, *44*, 464–485. [CrossRef]

65. UNEP; WCMC. *State of the Wolrd´s Protected Areas 2007: Of Global Conservation Progress*; UNEP-WCMC: Cambridge, UK, 2008.

66. OCDE. *OECD Environmental Data. Compendium 2008*; Organisation for Economic Co-operation and Development: Paris, France, 2008.

67. Comisión Nacional para el Conocimiento y Uso de la Biodiversidad (CONABIO). *Capital Natural de México, vol II: Estado de Conservación y Tendencias de Cambio*; Comisión Nacional para el Conocimiento y Uso de la Biodiversidad: Mexico City, Mexico, 2009.

68. Pawar, S.; Koo, M.S.; Kelley, C.; Ahmed, M.F.; Chaudhuri, S.; Sarkar, S. Conservation Assessment and Prioritization of Areas in Northeast India: Priorities for Amphibians and Reptiles. *Biol. Conserv.* **2007**, *136*, 346–361. [CrossRef]

69. Urbina-Cardona, J.N. Conservation of Neotropical Herpetofauna: Research Trends and Challenges. *Trop. Conserv. Sci.* **2008**, *1*, 359–375. [CrossRef]

70. Santos-Barrera, G.; Pacheco, C.; Ceballos, G. La Conservación de los Anfibios y Reptiles de Mexico. *Biodiversitas* **2004**, *57*, 1–6.

71. Ackerly, D.D.; Loarie, S.R.; Cornwell, W.K.; Weiss, S.B.; Hamilton, H.; Branciforte, R.; Kraft, N.J.B. The Geography of Climate Change: Implications for Conservation Biogeography. *Divers. Distrib.* **2010**, *16*, 476–487. [CrossRef]

72. Alagador, D.; Cerdeira, J.O. Meeting Species Persistence Targets under Climate Change: A Spatially Explicit Conservation Planning Model. *Divers. Distrib.* **2017**, *23*, 703–713. [CrossRef]

73. Monzón, J.; Moyer-Horner, L.; Baron-Palmar, M. Climate Change and Species Range Dynamics in Protected Areas. *BioScience* **2011**, *11*, 752–761. [CrossRef]

MDPI

St. Alban-Anlage 66

4052 Basel

Switzerland

Tel. +41 61 683 77 34

Fax +41 61 302 89 18

www.mdpi.com

Diversity Editorial Office

E-mail: diversity@mdpi.com

www.mdpi.com/journal/diversity